Dissection Experience

Customizable images enliven presentations and quizzes for lecture or lab.

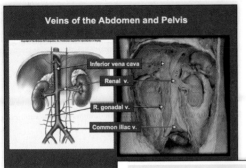

Veins of the Abdomen and Pelvis

Inferior vena cava
Renal v.
R. gonadal v.
Common iliac v.

6. Which muscle's origin is highlighted in red?

a. Infraspinatus
b. *Subscapularis*
c. Teres minor
d. Teres major
e. Supraspinatus

Bonus: Which of the above muscles is not part of the rotator cuff?

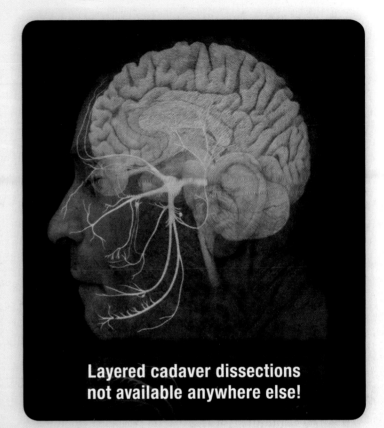

Layered cadaver dissections not available anywhere else!

www.aprevealed.com

Portable cadavers can replace or enhance the laboratory experience.

Imaging
Correlate dissected anatomy with radiologic images like X-ray, MRI, and CT scans.

Self-Test
Gauge proficiency with quizzes and simulated practical exams.

Search
Type in a term to quickly locate any structure in the program.

McGraw Hill

Mader's

UNDERSTANDING

HUMAN
ANATOMY&
PHYSIOLOGY

SEVENTH EDITION

Susannah Nelson Longenbaker

Columbus State Community College, Columbus, OH

Mc
Graw
Hill

Connect
Learn
Succeed™

MADER'S UNDERSTANDING HUMAN ANATOMY & PHYSIOLOGY, SEVENTH EDITION

Published by McGraw-Hill, a business unit of The McGraw-Hill Companies, Inc., 1221 Avenue of the Americas, New York, NY 10020. Copyright © 2011 by The McGraw-Hill Companies, Inc. All rights reserved. Previous editions © 2008, 2005, and 2001. No part of this publication may be reproduced or distributed in any form or by any means, or stored in a database or retrieval system, without the prior written consent of The McGraw-Hill Companies, Inc., including, but not limited to, in any network or other electronic storage or transmission, or broadcast for distance learning.

Some ancillaries, including electronic and print components, may not be available to customers outside the United States.

This book is printed on acid-free paper.

1 2 3 4 5 6 7 8 9 0 DOW/DOW 1 0 9 8 7 6 5 4 3 2 1 0

ISBN 978–0–07–122201–3
MHID 0–07–122201–4

All credits appearing on page or at the end of the book are considered to be an extension of the copyright page.

CONTENTS

PART 3

Integration and Coordination

PART 4

Maintenance of the Body

PART 5
Reproduction and Development

About the Authors

Susannah Nelson Longenbaker

After acquiring a B.S. in Biology from St. Mary's College (Notre Dame, IN), and a M.S. in Physiology from Ohio State University, Sue Longenbaker began her educational career at Columbus State Community College, teaching biology and human anatomy and physiology. During her tenure there, she accepted two promotions, and co-authored and received several grants. She was the recipient of both the Columbus State Community College Distinguished Teaching Award and Ohio Magazine's Excellence in Education Award, all while authoring four different lab manuals, and serving as a contributing author on the fifth edition of *Understanding Human Anatomy and Physiology*. She is now the primary author for this market-leading textbook. Sue continues to teach at the college and loves every minute of it. She feels privileged to pass along the gift of learning that she received from many excellent teachers.

Sue is honored to be named the author to continue Dr. Mader's work on this textbook and looks forward to hearing suggestions or comments from both instructors and students alike. Feel free to contact her at the following address:

Sue Longenbaker
Department of Biological/Physical Science
Columbus State Community College
Columbus, OH 43215
(614) 287-2430

Dr. Sylvia Mader

In her 20-year career with McGraw-Hill, Dr. Sylvia Mader has written an impressive collection of textbooks. In addition to five editions of *Understanding Human Anatomy and Physiology*, she is the author of *Biology*, ninth edition; *Human Biology*, tenth edition, and *Inquiry into Life*, twelfth edition, through which she has successfully helped innumerable students learn biology as well as human anatomy and physiology.

Dr. Mader's interest in anatomy and physiology began when she took courses at the Medical School of St. Andrews University, in Scotland, during her junior year abroad. As a fledgling faculty member, she was called upon to teach a variety of courses, among them human anatomy and physiology. As a textbook writer she discovered that the teaching and learning techniques she so successfully used in the classroom were also appropriate for her biology texts and then later for her anatomy and physiology text. Dr. Mader's direct writing style and carefully constructed pedagogy combine to provide students with unmatched opportunities to master the basics of biology and anatomy and physiology.

Preface

With this seventh edition, I'm once again working on a project that I love, and one that seeks to continue to develop and expand upon a text by Dr. Sylvia Mader. I have been privileged and honored to be entrusted with *Mader's Understanding Human Anatomy and Physiology*. Sylvia and I share the same vision for the book: one that is creative, informative, accurate, and most important, *relevant* to today's students. This book is designed to appeal to a wide audience, from students in pre-nursing and allied health fields, to nonscience majors who want a clear and concise explanation about how their bodies work. It's my goal to capture the reader's interest right away, and keep that reader's attention as he or she learns something new about how we humans work.

Mader's Understanding Human Anatomy and Physiology remains the perfect text for a one-semester course. Each chapter beings with a short introduction that seeks to immediately seize attention, and then draw the reader into a more complete exploration. For example, the introductory topic for chapter 8 on the nervous system describes the cause for "brain-freeze," the piercing headache that everyone has likely experienced while eating something very cold. Historical anecdotes, such as the story of Henrietta Lacks that begins chapter 4, will also intrigue the reader. Next, chapter objectives are carefully constructed to be achievable to students who have no prior training in anatomy and physiology. Each chapter is enriched with many real-world clinical applications to engage curiosity. Throughout the text, "Content Check-Ups" allow the reader to test comprehension before continuing on to a new section. "Begin Thinking Clinically" is a new feature that fosters critical thinking skills. Each "Begin Thinking Clinically" question asks the reader to conduct further investigation into the chapter's subject matter. A great deal of thought and attention has gone into the conclusion of each chapter. Basic and applied medical terminology exercises help to build a working vocabulary, thus facilitating comprehension and increasing student confidence. Chapter summaries and study questions can be used as a checklist to ensure that important concepts are well understood. Finally, the student can "take the test," using the Study Questions to craft a short essay, followed by Learning Objective Questions that replicate short answer quiz questions (matching, true-false, multiple choice, and the like).

My students love to talk about anatomy, physiology, and pathophysiology examples they've seen in the media or come across on the job. The new feature, "In Case of Emergency," is tailored for all students, but particularly those with an interest in emergency response or paramedic careers. I've added to the "Focus on Forensics" readings on forensic dentistry and the Innocence Project, which seeks to use DNA evidence to acquit individuals who have been wrongfully imprisoned. The "Medical Focus" and "What's New" features have been updated to reflect advancements in technology and clinical practice. But, perhaps the most important improvement to the seventh edition is a new illustration package. The new line drawings are realistic, detailed, and colorful. You'll find some of the best artwork in the industry in this edition of *Mader's Understanding Human Anatomy and Physiology*.

I have enjoyed the best job in the world—being a classroom biological sciences teacher—for over 25 years. Though it may sound like a cliché, it's still true that students teach me something new every day. Being in the classroom helps me to develop new and better ways to describe a concept using real-world examples and analogies. I know my explanation works when a student's puzzled expression changes to "Oh...*now* I get it." It's my goal to share with teachers and students alike some of the ideas that work for me. I know that this text will help you, the instructor, to engage and excite your students about the fascinating study of the human body.

Acknowledgements

When you have an amazing support team like the one I have at McGraw-Hill, you can have a lot of fun with a project such as this one. Each of these individuals deserves recognition for his or her hard work: Sponsoring Editor, Jim Connely; Marketing Manager, Lynn Breithaupt; Project Manager, Sandy Wille; Photo Research Coordinator, Lori Hancock; Design Coordinator, Brenda Rolwes; Production Supervisor, Sandy Ludovissy; and Media Project Manager, Jodi Banowetz. Testbank author Debbie McCool read every word in this edition and offered valuable feedback as she completed each chapter.

And once again, I owe a tremendous debt of gratitude to my developmental editor, Kathy Loewenberg. Kathy, thanks for keeping me focused, for encouraging me, and for your never-failing good humor. You're the best.

The seventh edition of *Mader's Understanding Human Anatomy and Physiology* would not have been possible without the valuable input from both students and teaching professionals. I would like to give a special thanks to Benjamin Gateno and Claudia Cazenave, two students who took the time to contact me with feedback about the book. Many colleagues have provided comments and suggestions in their reviews, and I would like to acknowledge the following individuals and thank all of them for contributing their time and talents. The observations and detailed recommendations they shared with me have served as a guideline to update chapter content. I deeply appreciate their help. As an author, it's very comforting to know that you have skilled and talented peer educators to review your content and help to improve it.

— *Sue Longenbaker*

Reviewers

Marilyn Banta
University of Northern Colorado

Patty Bostwick Taylor
Florence Darlington Technical College

Robert J. Caron
Bristol Community College

Lu Ann Clark
Lansing Community College

Cynthia A. Conaway Mavroidis
Northwest State Community College

Sonya Conway
Northern Illinois University

Janie Corbitt
Central Georgia Technical College

Elizabeth D. Crandall
University of Arkanasas Community College at Batesville

Tammy Filliater
University of Montana

Maria Florez
Cy-Fair College

Ruby Fogg
New Hampshire Community Tech College

Andrew Goliszek
North Carolina A & T State University

Sandra Grauer
Limestone College

Eric S. Green
Salt Lake Community College

Shari Harden
MCC Blue River Community College

Michael J. Harman
North Harris College

Krista L. Hoekstra
Hennepin Technical College

Gary W. Hunt
Tulsa Community College

Edward W. Johnson
Central Oregon Community College

Thomas Jordan
Pima Community College

Mary Jane Keith
Wichita State University

Diane M. Kelly
Broome Community College

Kathleen P. Lauber
College of Southern Maryland

William A. Lefrancois
Mount Wachusett Community College

Marty Lowe
Bergen Community College

Janet McCausey
Heritage College of Health Careers

Michael W. Orick
Schoolcraft College

Ruth Peterson
Hennepin Technical College

Donna R. Potacco
William Paterson University

David Quadagno
Florida State University

David Rice
Career College of Northern Nevada

Donna Ritch
University of Wisconsin–Green Bay

Susan Rohde
Triton College

Janet Anne Sherman
Pennsylvania College of Technology

Phillip Snider Jr.
Gadsden State Community College

Roxianne Snodgrass
Pittsburgh Technical Institute

Marty D. Spranger
Wayne State University

Michael Squires
Columbus State Community College

Focus Group Participants

Pegge Alciatore
University of Louisiana–Lafayette

Sara Brenizer
Shelton State Community College

Kathy Bruce
Parkland College

Jorge Cortese
Durham Technical Community College

Juville Dario-Becker
Central Virginia Community College

Smruti Desai
Cy-Fair College

Don Steve Dutton
Amarillo College

Glen Early
Jefferson Community and Technical College

Clair Eckersell
Brigham Young University–Idaho

Jim Ezell
J. Sargeant Reynolds Community College

Deanna Ferguson
Gloucester County College

Pamela Fouche
Walters State Community College

Mary Fox
University of Cincinnati

Bagie George
Georgia Perimeter College–Lawrenceville Campus

Peter Germroth
Hillsborough Community College–Dale Mabry Campus

Richard Griner
Augusta State University

Mark Hubley
Prince George's Community College

Sister Carol Makravitz
Luzerne County Community College

Ronald Markle
Northern Arizona University

Carl McAllister
Georgia Perimeter College

Alfredo Munoz
University of Texas at Brownsville

Necia Nicholas
Calhoun Community College

Robyn O'Kane
LaGuardia Community College/CUNY

Karen Payne
Chattanooga State Technical Community College

Joseph Schiller
Austin Peay State University

Colleen Sinclair
Towson University

Mary Elizabeth Torrano
American River College

Cinnamon VanPutte
Southwestern Illinois College

Chuck Venglarik
Jefferson State Community College

Mark Wygoda
McNeese State University

About This Book

Understanding is the key . . .

- Students develop an understanding of basic concepts, laying the foundation for a working knowledge of structure and function, via an *accessible writing style*.

- Self-confidence increases as students master important medical terminology with *simply designed art* and *built-in study aids*.

- *Clinical applications* build on core principles and introduce students to anatomy and physiology put into practice.

More important than any other component of a textbook, the writing must be appropriate for the level of the reader. *Mader's Understanding Human Anatomy and Physiology* features the **perfect writing style for the one-semester course**. It has always been written and designed for the one-semester course, not adapted from a two-semester textbook. Paragraph introductions, explanations, comparisons, and relevant, every-day examples are used with these students in mind. The flow of the text is logical and accessible without being overly "chatty," and consistently makes use of relevant examples and analogies.

Easy-to-Understand Art covers what's important but leaves out unnecessary, confusing detail. Good examples of this are the homeostasis illustrations—instead of lots of various colored arrows and boxes with explanations, these simple visual pieces get the message across beautifully. Another example is stepped-out art, which shows key stages of an illustration identified by numbered circles. This type of explanation builds comprehension sequentially. And, in this seventh edition—over half of the art is NEW!

"This chapter [Ch 19] is a revelation. It is the chapter I have been looking for to supplement my genetics lectures. This topic is so much fun for me but it is a nightmare to non-majors. The chapter has excellent explanations for topics that interest students with strong but not overwhelming factual information. The illustrations and captions only strengthen the ability to comprehend the facts dealing with Human Genetics.
—Marty Lowe, *Bergen Community College*

"One of the reasons I have decided to adopt this text is because of the writing style. The two chapters I reviewed were excellent, concise, and understandable. The greatest strengths of these chapters were the writing and the illustration program, and how the two were so well coordinated. The material was presented in such a coherent and interesting manner that reading the chapters was a breeze."
—Andrew Gloiszek, *North Carolina A & T State University*

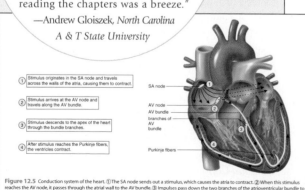

① Stimulus originates in the SA node and travels across the walls of the atria, causing them to contract.

② Stimulus arrives at the AV node and travels along the AV bundle.

③ Stimulus descends to the apex of the heart through the bundle branches.

④ After stimulus reaches the Purkinje fibers, the ventricles contract.

Figure 12.5 Conduction system of the heart. ①The SA node sends out a stimulus, which causes the atria to contract. ②When this stimulus reaches the AV node, it passes through the atrial wall to the AV bundle. ③Impulses pass down the two branches of the atrioventricular bundle to the Purkinje fibers, and ④thereafter, the ventricles contract.

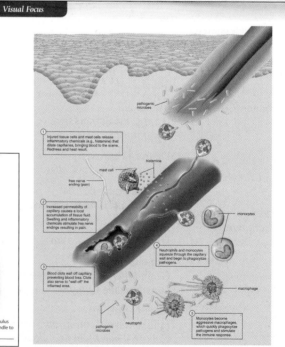

Visual Focus

Figure 13.3 **Steps of the inflammatory reaction.** (1) When there is an injury, tissue cells and mast cells release inflammatory chemicals (such as histamine) that dilate blood vessels and cause them to leak. More blood flows to the area, and the skin appears red and feels warm. (2) Because capillaries are more permeable, edema forms. Swelling and inflammatory chemicals stimulate pain receptors (free nerve endings). (3) Blood clotting seals off the capillary, preventing blood loss and walling off the inflamed area. (4) Monocytes (which become macrophages) and neutrophils squeeze through the capillary wall, and phagocytosis begins. (5) Monocytes transform to macrophages, which stimulate specific immune responses in addition to destroying pathogens.

Built-in Study Aids such as the "Content Check-Up" and the new "Thinking Clinically" features allow students to test themselves over major sections of text before continuing. Add these review questions all together and students have a practice quiz over the entire chapter. Key terms are divided into basic and clinical terms, and include page references. Two levels of additional questions, along with exercises that reinforce medical terminology, are also included with every chapter.

The authors of *Understanding Human Anatomy and Physiology* recognize that students need to see applications of the important concepts and terminology presented in lectures.

Unsurpassed Clinical Coverage

is evident all through this text. Years of teaching experience, combined with countless hours of research, have given Susannah Longenbaker a rich supply of examples, articles, and data to demonstrate the hands-on side of anatomy and physiology. Everyday scenarios, cutting-edge forensic methods, just-released discoveries, and new medical procedures all combine to present clinical connections that are unrivaled among one-semester A & P textbooks.

"What's New," "Thinking Clinically," "Medical Focus," "I.C.E. Emergency" boxes, and "Focus on Forensics" readings or study aids are written to relate the very latest research and developments in applied aspects of anatomy and physiology to important concepts in the text. Examples include "Improvements in Transfusion Technology," "Necrotizing Fasciitis" and "Swine Flu—The Global Pandemic." The "Focus on Forensics" and I.C.E. Emergency readings engage students in real-life scenarios that challenge them to use, and expand upon, their recently acquired knowledge.

FOCUS on FORENSICS

The Stories That Teeth Can Tell

In the aftermath of the tragic events of September 11, 2001, officials in New York City faced the tremendous task of determining the identity of the 2,819 deceased victims of the terrorist attacks. Only 289 bodies were found relatively intact and were therefore easily identifiable; most of the bodies were charred beyond recognition. Further complicating the situation was the fact that over 19,000 separate body parts—mostly bony structures and teeth—were recovered from the rubble. In order to return these remains to the victims' families, forensic dentists from all over the world were called in to help identify the remains. The forensic dentists relied on the unique properties of human teeth—the hardest substance of the human body—to help them with their monumental identification task (so monumental that it continues to the present day).

Dentition—the makeup of a person's set of teeth and jaws—is even more individualized than one's DNA structure (identical twins have identical DNA but unique dentition). Thus, it is an invaluable tool to use in the identification of unknown bodies. Many characteristics combine to make each person's dentition unique. There are normally 32 teeth in a human adult, and each tooth has five different surfaces, giving a total of 160 distinct surfaces in each person's teeth. Also, the structure of tooth roots and the bone surrounding each tooth differs from one person to the next. (Additional information about a deceased person can be

gained if tooth roots happen to con[...] features personalize the bones of t[...] the person's teeth may be missing [...] canal. In the case of extreme heat, s[...] attacks, man-made dental structures [...] teeth begin to burn at 400°F, and a[...] vidual metal dental work (such as [...] remains. A forensic dentist carefully [...] of teeth, jaws, and metal dental w[...] deceased's identity.

In forensic science, dentiti[...] fication of the deceased. It ca[...] suspect through analysis of th[...] or her victim. Forensic dentis[...] ing details such as the distan[...] the shape of each tooth's ma[...] missing or out of alignment, [...] features can then be compared [...] tion, saliva left in the bite may co[...] viding further information.) One [...] was largely convicted by a bite m[...] 1970's serial murderer, was finally [...] bite marks found on her body.

What's New in This Edition?

Mader's Understanding Human Anatomy and Physiology has undergone a thorough revision, particularly with the art program. Every chapter has been improved to include new material, pedagogy, examples, and data based on the latest research available and valued reviewer feedback.

A **revised art program** includes new renderings for over half of the art in the textbook! Images now look more realistic and modern, enabling students to have a better visual picture of all of the major structures and organs. New photos also join the illustrations, including most of the chapter openers.

Figure 10.2 Action of a peptide hormone. The binding of a peptide hormone leads to cAMP and then to activation of an enzyme cascade. In this example, the hormone causes glycogen to be broken down to glucose.

Figure 10.3 Action of a steroid hormone. A steroid hormone results in a hormone-receptor complex that activates DNA and protein synthesis.

New **"I.C.E. Emergency"** readings are scenarios that connect topics in the chapter to emergency situations students may very well face in health-related careers. For EMTs, this is a direct application to their futures!

I.C.E. — IN CASE OF EMERGENCY

Lung Collapse

Imagine that you're a military medic who's called upon to respond when troops have been injured due to a bomb blast. As you arrive at the scene, two fallen soldiers need your attention. One has an open chest wound, caused by shrapnel cutting his chest. The second was nearby when the blast occurred, but has no obvious wounds. Yet, both have the same symptoms: sharp pain when they inhale, difficulty speaking, and a feeling of breathlessness. Both soldiers' blood pressure is low and pulse is rapid, indicating that they might slip into shock. You take a quick history from both victim

Right away, you suspect technical term for a collapsed are held up against the chest surface tension. If air enters fail, and the lungs will collaps is allowing air from the atm section of the second soldier high pressure from the bomb rax from the hole in his lung. filling the chest compresses th ing with blood. This is terme the thorax. You'll need to act f shock.

With the help of the first s airtight pressure bandage over prevent additional air from er bleeding. Next, you'll start his tening to the second soldier's c be able to tell where the lung h hollow. When you trained as a *cocentesis*, and you'll rapidly i dier's ribs to let the trapped a your patients are ready for their for more advanced care.

I.C.E. — IN CASE OF EMERGENCY

Cardiopulmonary Resuscitation and Automated External Defibrillators

How many people's lives have been saved during cardiac arrest by cardiopulmonary resuscitation (CPR)? The exact number would probably be impossible to track, according to the American Heart Association (AHA). Regrettably, it's easier to track the tragedies that occur without it: Fewer than 30% of people who suffer cardiac arrest ever receive bystander help. Onlookers simply don't know what to do, or perhaps are afraid that they might do something wrong. For this reason, the AHA has published the *Chain of Survival*, a new set of four guidelines that make it easier to help a person in cardiac arrest:

- Recognize that you're in an emergency, and immediately call for help. Call 9-1-1, or use the EMS service in your area.
- Start CPR. The traditional method alternates chest compressions with mouth-to-mouth breathing, also known as rescue breathing. If you're not sure how to do rescue breathing, simply pump fast (100 or more times a minute) and hard on the center of the person's chest. Recent research from the AHA has shown that simple manual chest compression can be as effective as traditional CPR (chest compression accompanied by rescue breathing).

- If you have access to an Automatic External Defibrillator (AED), use it. This computerized device is available in many public places, such as airports and shopping malls, and is prominently labeled. An AED will explain, step by step, how to check for the person's breathing and a pulse first. Next, it will describe how to apply pads to the victim's chest so that the computer can analyze the heart activity. If a shock is needed to restart the victim's heart, the AED delivers a burst of intense electrical current to the chest. The rescuer simply moves back and pushes a button when prompted. Voice instructions will also detail how to do CPR until paramedics arrive.
- Transfer to advanced care. Clinicians refer to th after admission to an emergency room as th It's not surprising that research has show receive the best care during this critical likely to survive.

The AHA recommends that everyone lea use an AED. The Red Cross and many introductory and refresher classes in and online. With training, you just mig person's life!

New **"Begin Thinking Clinically"** questions encourage students to push their recently acquired knowledge beyond the text and into real-life applications similar to those they may face in their future careers. These questions also encourage "aha" moments as students practice critical thinking skills and make connections in comprehension by applying information to clinical situations.

? Begin Thinking Clinically

Questions will be scattered within each chapter. These are single questions and their design.

? Begin Thinking Clinically

During an actual delivery, a pregnant woman's uterine contractions will sometimes begin, seemingly strengthen, and then stop abruptly. What might cause this to occur? What should be done to correct this?

Answers and discussion in Appendix C, page 466.

The Endocrine System

10

CHAPTER

John F. Kennedy, the youngest man to be elected president, appeared healthy, vigorous, and active throughout his entire political career. Photos of the president showed a handsome, tanned sailor at the family estate in Massachusetts; others showed the Kennedy family playing football. However, throughout Kennedy's entire political life, he and his staff took great pains to hide his many ailments—in particular, the fact that Kennedy suffered from Addison's disease. This rare illness is caused by a deficiency of adrenal cortex hormones, and you can read about it on page 217. (Ironically, one symptom of Addison's disease is skin that looks like JFK's signature suntan.) Although he was hospitalized many times, Kennedy repeatedly denied having the disease or any other serious health problems. Yet, medical records tell a different story—the president was in almost constant pain, in part because his vertebrae had been destroyed by the medication for Addison's disease. At autopsy, his adrenal glands were shrunken and nonfunctional. Still, there is no evidence that Kennedy's illness affected his performance as president, and he left a legacy that includes the space program and the Peace Corps.

The seventh edition unveils new **chapter introductions,** with interesting stories to begin each chapter. Fresh bright interior colors and pleasing table designs give this edition an engaging, contemporary feel.

Chapter Outline & Learning Objectives
After you have studied this chapter, you should be able to:

206

Changes to This Edition

Chapter-by-Chapter Changes

Chapter 1
- *NEW* chapter introduction
- *NEW* "Begin Thinking Clinically" questions
- *NEW* description of serous membranes

Chapter 2
- *NEW* chapter introduction
- *NEW* "Begin Thinking Clinically" questions
- shortened discussion of nutrition

Chapter 3
- *NEW* chapter introduction
- *NEW* "Begin Thinking Clinically" questions

Chapter 4
- *NEW* chapter introduction
- *NEW* "Begin Thinking Clinically" questions
- expanded discussion on types of connective tissue, adding greater detail
- *NEW* Medical Focus: "Necrotizing Fasciitis"
- updated What's New: "Blocking Blood Supply to Treat Cancer"

Chapter 5
- *NEW* chapter introduction
- *NEW* "Begin Thinking Clinically" questions
- *NEW* I.C.E.: "In Case of Emergency: Burns"
- expanded discussion of epidermis of skin

Chapter 6
- *NEW* chapter introduction
- *NEW* "Begin Thinking Clinically" questions
- additional illustrations of bone, with accompanying photos
- *NEW* I.C.E.: "In Case of Emergency: Broken Bones"
- *NEW* What's New: "Oh, My Aching Back: Options for Back Pain"
- updated information on osteoporosis
- updated table on the surface features of bones
- revised content on knee repairs

Chapter 7
- *NEW* chapter introduction
- *NEW* "Begin Thinking Clinically" questions
- revised Figure 7.2
- revised Figure 7.3
- expanded Table 7.2, Muscles of the Head and Neck

Chapter 8
- *NEW* chapter introduction
- *NEW* "Begin Thinking Clinically" questions
- *NEW* What's New: "Magnetic Therapy for Clinical Depression"
- *NEW* I.C.E.: "In Case of Emergency: Traumatic Brain Injury"
- heavy revision to sections on nerve signal conduction, resting potential, and action potential
- updated Medical Focus: "Alzheimer Disease"
- shortened and updated "Spinal Cord Injuries—Cause for Hope"
- completely revised Figures 8.3 and 8.4

Chapter 9
- *NEW* chapter introduction
- *NEW* "Begin Thinking Clinically" questions
- *NEW* Medical Focus: "Eye Diseases and Disorders"
- *NEW* What's New: "Detecting Alzheimer Disease with Laser Eye Exams"

Chapter 10
- *NEW* chapter introduction
- *NEW* "Begin Thinking Clinically" questions
- *NEW* I.C.E.: "In Case of Emergency: Insulin Shock and Diabetic Ketoacidosis"
- revised What's New: "Options for Type I Diabetics: Insulin Pills and More"
- rewritten section on hormone control
- updated Medical Focus: "Side Effects of Anabolic Steroids"

Chapter 11
- *NEW* chapter introduction
- *NEW* "Begin Thinking Clinically" questions
- per reviewers' request, moved the capillary exchange section to Chapter 12
- *NEW* What's New: "Improvements in Transfusion Technology"
- *NEW* I.C.E.: "In Case of Emergency: Hemorrhage"
- rewrote section on sickle cell trait/sickle cell anemia

Chapter 12
- *NEW* chapter introduction
- *NEW* "Begin Thinking Clinically" questions
- updated Medical Focus: "Arteriosclerosis, Atherosclerosis, and Coronary Artery Disease"
- *NEW* I.C.E.: "In Case of Emergency: Cardiopulmonary Resuscitation and Automated External Defibrillators"
- *NEW* Figure 12.1
- updated Medical Focus: "Stroke: Beware of 'Brain Attack'"

Chapter 13
- *NEW* chapter introduction
- *NEW* "Begin Thinking Clinically" questions
- revised section on inflammatory reaction
- rewrote the table on 3 lines of defense
- updated Medical Focus: "AIDS Epidemic"
- updated Medical Focus: "Immunization: The Great Protector"
- *NEW* What's New: "Swine Flu—The Global Pandemic"

Chapter 14
- *NEW* chapter introduction
- *NEW* "Begin Thinking Clinically" questions
- *NEW* I.C.E.: "In Case of Emergency: Lung Collapse"
- Deleted "Medical Focus: Emphysema: Medical and Surgical Treatment," but put pertinent information into the Effects of Aging section
- Per reviewer request, rewrote the discussion on pH and respiratory control
- Update to "Medical Focus: The Most-Often-Asked Questions About Tobacco and Health"

Chapter 15
- *NEW* chapter introduction
- *NEW* "Begin Thinking Clinically" questions
- revised Medical Focus: "Swallowing a Camera?"
- *NEW* Focus on Forensics: "Stories That Teeth Can Tell"
- rewrote discussion on regulation of contraction and secretion in the digestive tract

Chapter 16
- *NEW* chapter introduction
- *NEW* "Begin Thinking Clinically" questions
- rewrote sections on nephrons, loop of Henle, and collecting duct
- *NEW* descriptions of vasa recta and aquaporins
- shortened the homeostasis section to eliminate redundancy

Chapter 17
- *NEW* chapter introduction
- *NEW* "Begin Thinking Clinically" questions
- rewrote the description of meiosis
- updated Table 17.3
- rewrote Medical Focus: "Endocrine-Disrupting Contaminants"

Chapter 18
- *NEW* chapter introduction
- *NEW* "Begin Thinking Clinically" questions
- rewrote description of how polyspermy is prevented

Chapter 19
- *NEW* chapter introduction
- *NEW* "Begin Thinking Clinically" questions
- *NEW* Focus on Forensics: "The Innocence Project"
- *NEW* section on using a Punnett Square
- updated Medical Focus: "Preimplantation Genetic Studies"

Teaching and Learning Supplements

McGraw-Hill offers various tools and technology products to support the seventh edition of *Understanding Anatomy and Physiology*. Students can order supplemental study materials by contacting their campus bookstore. Instructors can obtain teaching aids by calling the McGraw-Hill Customer Service Department at 1-800-338-3987, visiting our A&P catalog at *www.mhhe.com/ap*, or contacting their local McGraw-Hill sales representative.

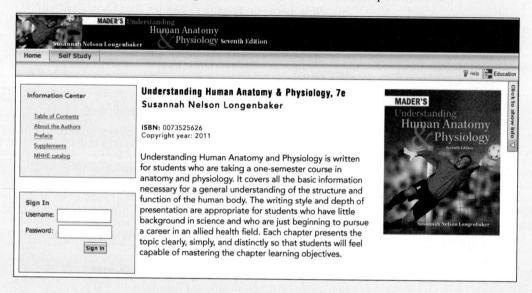

Helpful Online Content for Instructors and Students

Instructors—Build instructional materials wherever, whenever, and however you want!

Part of this edition's website, Presentation Center contains assets such as artwork, animations, PowerPoints, and other media resources which can be used to create customized lectures, visually enhance tests and quizzes, and design compelling course websites. Nothing could be easier!

The dynamic search engine allows instructors to explore by discipline, course, textbook chapter, asset type, or keyword. Simply browse, select, and download the files needed to build engaging course presentations.

- **Art** Full-color digital files of all illustrations in the book can be readily incorporated into lecture presentations, exams, or custom-made classroom materials. In addition, all files are pre-inserted into blank PowerPoint slides for ease of lecture preparation.

- **Tables** Every table that appears in the text has been saved in electronic form for use in classroom presentations and/or quizzes.

- **Animations** Numerous full-color animations illustrating physiological processes are also provided. Harness the visual impact of processes in motion by importing these files into classroom presentations or online course materials.

- **Lecture Outlines** Specially prepared custom outlines for each chapter are offered in easy-to-use PowerPoint slides. New with this edition—now there is a version of the Lecture PowerPoints with embedded animations!

- **Instructor's Testbank** A flexible and easy-to-use electronic testing program, this testbank provides a wealth of resources for the instructor. Instructors can create tests from book specific items, and choose from a wide range of question types, including the option to add their own questions. Multiple versions of the test can be created and any test can be exported for use with course management systems such as WebCT, BlackBoard, Angel, etc. The **Instructor's Manual** is also included.

Students—Increase your grade and decrease your study time!

The website that accompanies this textbook includes tutorials, animations, practice quizzing, case studies, study games, helpful Internet links and more—a whole semester's worth of help for your anatomy and physiology course.

Check out *www.mhhe.com/longenbaker7* and start benefiting today!

McGraw-Hill's Anatomy & Physiology | REVEALED®

McGraw-Hill's Anatomy & Physiology | REVEALED® is a unique multimedia study aid designed to help students learn and review human anatomy using digital cadaver specimens. Dissections, animations, imaging, and self-tests all work together as an exceptional tool for the study of structure and function. This text has the opportunity to be packaged with the most unique and popular student tutorial in the anatomy & physiology market today—now available online! Visit *www.mhhe.com/aprevealed* for more information.

"I would rate it [this textbook] superior to our current text as a stand alone, but with the addition of AP Revealed there is no comparison–APR far excels above the other ancillary electronic dissection DVD's available."

—Ruth Peterson, *Hennepin Technical College*

Physiology Interactive Lab Simulations (Ph.I.L.S.) 3.0

This unique student study tool is the perfect way to reinforce key physiology concepts with powerful lab experiments. Created by Dr. Phil Stephens of Villanova University, the program offers 37 laboratory simulations that may be used to supplement or substitute for wet labs. Students can adjust variables, view outcomes, make predictions, draw conclusions, and print lab reports. The easy-to-use software offers the flexibility to change the parameters of the lab experiment—there is no limit to the amount of times an experiment can be repeated.

MediaPhys Tutorial

This physiology study aid offers detailed explanations, high-quality illustrations, and amazing animations to provide a thorough introduction to the world of physiology. MediaPhys is filled with interactive activities and quizzes to help reinforce physiology concepts that are often difficult for students to understand.

Virtual Anatomy Dissection Review

This multimedia program contains high-quality cat dissection photographs that are correlated to illustrations and photos of human structures. The format makes it easy to identify and review cat anatomy, and also relate the cat specimen to the corresponding human structures. Available online at *www.mhhe.com/watersdemo*.

Organization of the Body

PART I

CHAPTER 1

Recognize anything familiar? The little face, chest, and hands shown here belong to a baby in the womb, detailed using ultrasound, one of several imaging techniques described on page 16. All modern imaging methods have evolved from simple two-dimensional X rays, invented in 1895. The three-dimensional ultrasound procedure used to create this photo reveals structures with great detail and clarity. As you can see, this small person is enjoying a quiet moment, contentedly sucking his thumb. Ultrasound technology can monitor the infant's development and help to ensure a safe delivery for baby and mother.

Chapter Outline & Learning Objectives

After you have studied this chapter, you should be able to:

1.1 The Human Body (p. 2)
- Define anatomy and physiology, and explain how they are related.
- Describe each level of organization of the body with reference to an example.

1.2 Anatomical Terms (p. 2)
- Use anatomical terms to describe the relative positions of the body parts, the regions of the body, and the planes by which the body can be sectioned.

1.3 Body Cavities and Membranes (p. 5)
- List the cavities of the body, and state their locations.
- Name the organs located in each of the body cavities.
- Name the membranes that line each body cavity and adhere to the organs.

1.4 Organ Systems (p. 8)
- List the organ systems of the body, and state the major organs associated with each.
- Describe in general the functions of each organ system.

1.5 Homeostasis (p. 10)
- Describe how a feedback system maintains homeostasis.
- Describe the role of each body system in the maintenance of homeostasis.

Medical Focus
Meningitis and Serositis (p. 9)
Imaging the Body (p. 16)

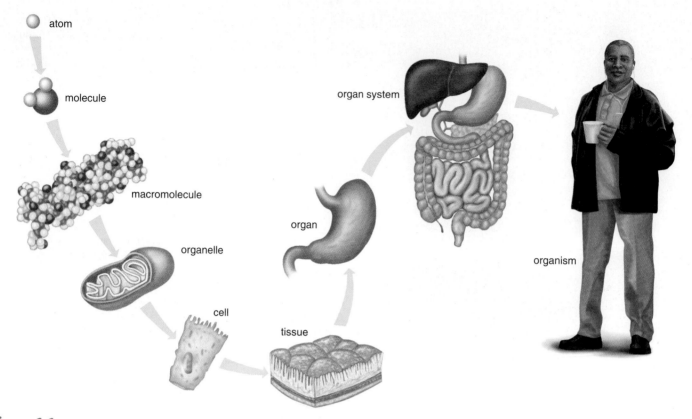

Figure 1.1 **Levels of organization of the human body.** Each level is more complex than the previous level.

atom

molecule

macromolecule

organelle

cell

tissue

organ

organ system

organism

1.1 The Human Body

Anatomy and physiology both involve the study of the human body. **Anatomy** is concerned with the structure of a part. For example, the stomach is a J-shaped, pouchlike organ (Fig. 1.1). The stomach wall has thick folds, which disappear as the stomach expands to increase its capacity. **Physiology** is concerned with the function of a part. For example, the stomach temporarily stores food, secretes digestive juices, and passes on partially digested food to the small intestine.

Anatomy and physiology are closely connected in that the structure of an organ suits its function. For example, the stomach's pouchlike shape and ability to expand are well-suited to its function of storing food. In addition, the microscopic structure of the stomach wall is suitable to its secretion of digestive juices, as we shall see in Chapter 15.

Organization of Body Parts

The structure of the body can be studied at different *levels of organization* (Fig. 1.1). First, all substances, including body parts, are composed of chemicals made up of submicroscopic particles called **atoms**. Atoms join to form **molecules**, which can in turn join to form **macromolecules.** For example, molecules called amino acids join to form macromolecules called proteins. Different proteins make up the bulk of our muscles.

Macromolecules compose the cellular **organelles,** which are found within all cells. Organelles are tiny structures that perform cellular functions. For example, the organelle called the nucleus is especially concerned with cell reproduction. Another organelle, called the mitochondrion, supplies the cell with energy. **Cells** are the basic units of living things.

Tissues are the next level of organization. A **tissue** is composed of similar types of cells and performs a specific function. An **organ** is composed of several types of tissues and performs a particular function within an **organ system.** For example, the stomach is an organ that is a part of the digestive system. It has a specific role in this system, whose overall function is to supply the body with the nutrients needed for growth and repair. The other systems of the body (see pages 14 and 15) also have specific functions.

All of the body systems together make up the **organism—** for example, a human being. Human beings are complex animals, but this complexity can be broken down and studied at ever simpler levels. Each simpler level is organized and constructed in a particular way.

1.2 Anatomical Terms

Certain terms are used to describe the location of body parts, regions of the body, and imaginary planes by which the body can be sectioned. You should become familiar with these

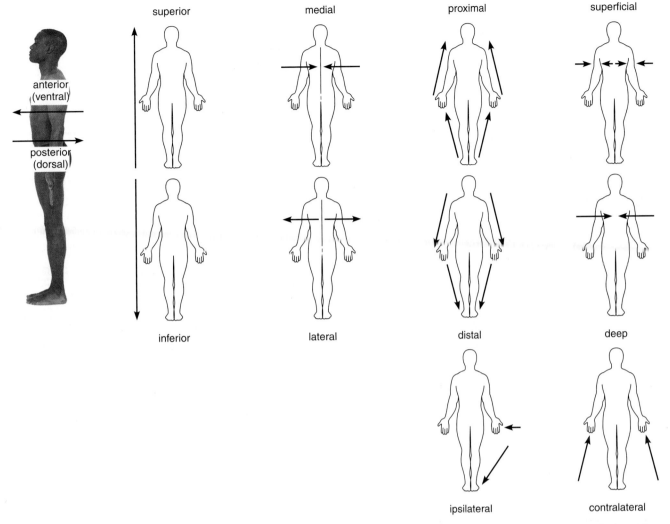

superior · medial · proximal · superficial

anterior (ventral) · posterior (dorsal)

inferior · lateral · distal · deep

ipsilateral · contralateral

Figure 1.2 Directional terms. Directional terms tell us where body parts are located with reference to the body in anatomical position.

terms before your study of anatomy and physiology begins. Anatomical terms are useful only if everyone has in mind the same position of the body and is using the same reference points. Therefore, we will assume that the body is in the **anatomical position**: standing erect, with face forward, arms at the sides, and palms and toes directed forward, as illustrated in Figure 1.3a.

Directional Terms

Directional terms are used to describe the location of one body part in relation to another (Fig. 1.2):

Anterior (ventral)—a body part is located toward the front. The windpipe (trachea) is *anterior* to the esophagus.

Posterior (dorsal)—a body part is located toward the back. The heart is *posterior* to the sternum (breastbone).

Superior—a body part is located above another part, or toward the head. The face is *superior* to the neck.

Inferior—a body part is below another part, or toward the feet. The navel is *inferior* to the chin.

Medial—a body part is nearer than another part to an imaginary midline of the body. The bridge of the nose is *medial* to the eyes.

Lateral—a body part is farther away from the midline. The eyes are *lateral* to the nose.

Proximal—a body part is closer to the point of attachment or closer to the trunk. The elbow is *proximal* to the hand.

Distal—a body part is farther from the point of attachment or farther from the trunk or torso. The hand is *distal* to the elbow.

Superficial (external)—a body part is located near the surface. The skin is *superficial* to the muscles.

Deep (internal)—a body part is located away from the surface. The intestines are *deep* to the spine.

Central—a body part is situated at the center of the body or an organ. The *central* nervous system is *centrally* located along the main axis of the body.

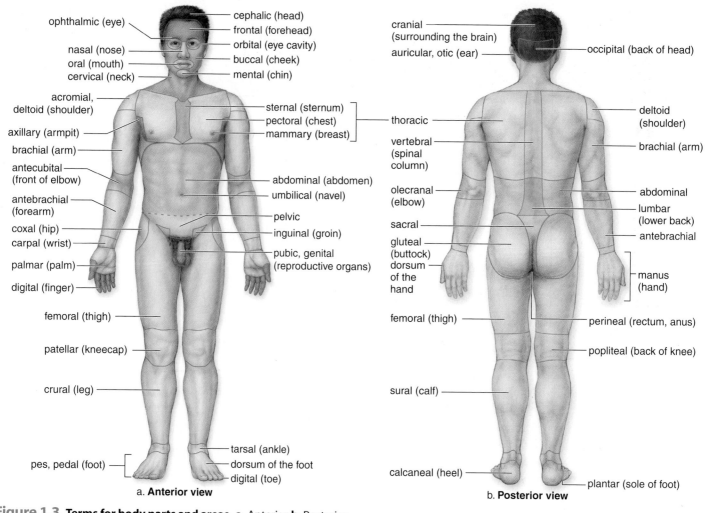

ophthalmic (eye)
nasal (nose)
oral (mouth)
cervical (neck)
acromial, deltoid (shoulder)
axillary (armpit)
brachial (arm)
antecubital (front of elbow)
antebrachial (forearm)
coxal (hip)
carpal (wrist)
palmar (palm)
digital (finger)
femoral (thigh)
patellar (kneecap)
crural (leg)
pes, pedal (foot)

cephalic (head)
frontal (forehead)
orbital (eye cavity)
buccal (cheek)
mental (chin)
sternal (sternum)
pectoral (chest)
mammary (breast)
abdominal (abdomen)
umbilical (navel)
pelvic
inguinal (groin)
pubic, genital (reproductive organs)
tarsal (ankle)
dorsum of the foot
digital (toe)

a. **Anterior view**

cranial (surrounding the brain)
auricular, otic (ear)
thoracic
vertebral (spinal column)
olecranal (elbow)
sacral
gluteal (buttock)
dorsum of the hand
femoral (thigh)
sural (calf)
calcaneal (heel)

occipital (back of head)
deltoid (shoulder)
brachial (arm)
abdominal
lumbar (lower back)
antebrachial
manus (hand)
perineal (rectum, anus)
popliteal (back of knee)
plantar (sole of foot)

b. **Posterior view**

Figure 1.3 **Terms for body parts and areas. a.** Anterior. **b.** Posterior.

Peripheral—a body part is situated away from the center of the body or an organ. The *peripheral* nervous system is located outside the central nervous system.

Ipsilateral—a body part is on the same side of the body as another body part. The right hand is *ipsilateral* to the right foot.

Contralateral—a body part is on the opposite side of the body from another body part. The right hand is *contralateral* to the left hand.

Regions of the Body

The human body can be divided into axial and appendicular portions. The **axial portion** includes the head, neck, spinal column, and ribs. The trunk can be divided into the thorax, abdomen, and pelvis. The pelvis is that part of the trunk associated with the hips. The **appendicular portion** of the human body includes the limbs—that is, the upper limbs and the lower limbs.

The human body is further divided as shown in Figure 1.3. The labels in Figure 1.3 do not include the word "region." It is understood that you will supply the word region in each case. The scientific name for each region is followed by the common name for that region. For example, the cephalic region is commonly called the head.

Notice that the upper arm includes among other parts the brachial region (arm) and the antebrachial region (forearm), and the lower limb includes among other parts the femoral region (thigh) and the crural region (leg). In other words, contrary to common usage, the terms arm and leg refer to only a part of the upper limb and lower limb, respectively.

Most likely, it will take practice to learn the terms in Figure 1.3. One way to practice might be to point to various regions of your own body and see if you can give the scientific name for that region. Check your answer against the figure.

Planes and Sections of the Body

To observe the structure of an internal body part, it is customary to section (cut) the body along a plane. A plane is an imaginary flat surface passing through the body.

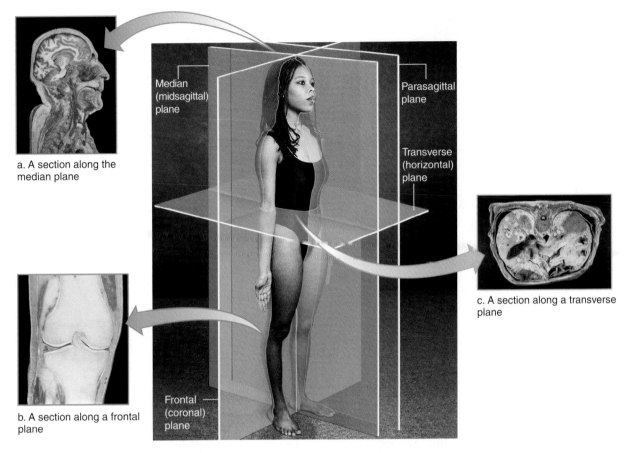

Median (midsagittal) plane

Parasagittal plane

Transverse (horizontal) plane

Frontal (coronal) plane

a. A section along the median plane

b. A section along a frontal plane

c. A section along a transverse plane

Figure 1.4 **Body planes and sections.** Observation of internal parts requires sectioning the body along various planes.

The body is customarily sectioned along the following planes (Fig. 1.4):

A **sagittal** (median) **plane** extends lengthwise and divides the body into right and left portions. A midsagittal plane passes exactly through the midline of the body. The head and neck are shown in a midsagittal section (Fig. 1.4*a*). Sagittal cuts that are not along the midline are called parasagittal (paramedian) sections.

A **frontal** (coronal) **plane** also extends lengthwise, but it is perpendicular to a sagittal plane and divides the body or an organ into anterior and posterior portions. Here, the knee joint is shown in frontal section (Fig. 1.4*b*).

A **transverse** (horizontal) **plane** is perpendicular to the body's long axis and therefore divides the body horizontally to produce a cross section. A transverse cut divides the body or an organ into superior and inferior portions. Figure 1.4*c* is a transverse section of abdomen at the level of the umbilicus (navel).

The terms *longitudinal section* and *cross section* are often applied to body parts that have been removed and cut either lengthwise or straight across, respectively.

1.3 Body Cavities and Membranes

During embryonic development, the body is first divided into two internal cavities: the posterior (dorsal) body cavity and the anterior (ventral) body cavity. Each of these major cavities is then subdivided into smaller cavities. The cavities, as well as the organs in the cavities (called the **viscera**), are lined by membranes.

Posterior (Dorsal) Body Cavity

The posterior body cavity is subdivided into two parts: (1) The **cranial cavity,** enclosed by the bony cranium, contains

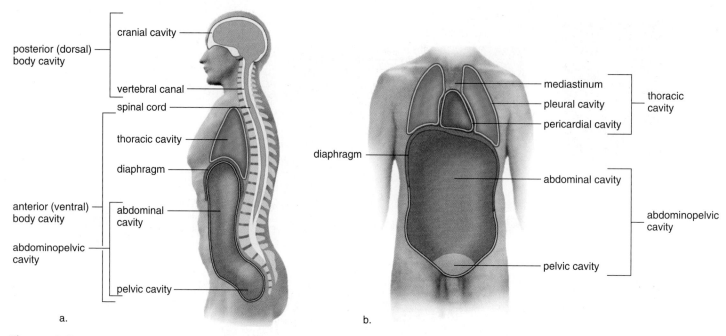

a.   b.

Figure 1.5 **The two major body cavities and their subdivisions. a.** Left lateral view **b.** Frontal view.

the brain. (2) The **vertebral canal,** enclosed by vertebrae, contains the spinal cord (Fig. 1.5a).

The posterior body cavity is lined by three membranous layers collectively called the **meninges** (sing., *meńinx*). The innermost, or deepest, of the meninges is tightly bound to the surface of the brain and the spinal cord. The space between this layer and the next layer is filled with cerebrospinal fluid. Cerebrospinal fluid supports and nourishes the brain and the spinal cord, and enables their cells to transmit electrical signals. The most superficial of the three meninges lies directly under the bone of the skull and vertebral column.

Anterior (Ventral) Body Cavity

The large anterior body cavity is subdivided into the superior **thoracic cavity** and the inferior **abdominopelvic cavity** (Fig. 1.5a). A muscular partition called the **diaphragm** separates the two cavities. Membranes that line these cavities are called **serous membranes** because they secrete a fluid that is similar to blood **serum.** Serum is the fluid that remains if all of the clotting proteins are removed from the blood. **Serous fluid** between the smooth serous membranes reduces friction when the viscera rub against each other or against the body wall.

To understand the relationship among serous membranes, the outer body wall, and an organ, consider the following example. Imagine a soft, pliable balloon (the serous membrane) filled with a small amount of fluid (serous fluid). The balloon sits inside a container (the inner body wall), tightly pressed to all sides of the container. An organ (the lung, for example) is placed on top of this balloon and is then covered by the balloon (Fig. 1.6). As a result, two layers of

serous membrane are created, separated from each other by the serous fluid. The balloon's outermost layer (covering the inner body wall) is termed the **parietal serous membrane.** The inner layer covering the organ is the **visceral serous membrane.** Inflammation of the serous membrane, or infection of the serous fluid in the body cavities causes serious and potentially fatal illness (see Medical Focus, p. 9).

Thoracic Cavity

The thoracic cavity is enclosed by the rib cage and has three portions: the left, right, and medial portions. The medial portion, called the **mediastinum,** contains the heart, thymus gland, trachea, esophagus, and other structures (Fig. 1.5b).

The right and left portions of the thoracic cavity contain the lungs. The lung tissue is covered by a serous membrane—

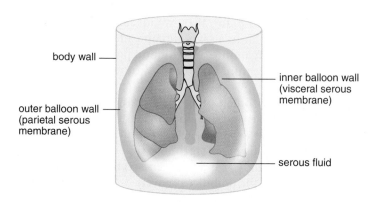

Figure 1.6 **Relationship between the body wall, serous membranes and organs.**

TABLE 1.1 Body Cavities and Membranes

Name of Cavity	Contents of Cavity	Membranes	
POSTERIOR BODY CAVITY			
Cranial cavity	Brain	Meninges	
Vertebral canal	Spinal cord	Meninges	
ANTERIOR BODY CAVITY			
Thoracic Cavity		*Parietal Membrane*	*Visceral Membrane*
Pleural cavity	Lungs	Parietal pleura	Visceral pleura
Pericardial cavity	Heart	Parietal pericardium	Visceral pericardium (epicardium)
Abdominopelvic Cavity			
Abdominal cavity	Digestive organs, liver, kidneys	Parietal peritoneum	Visceral peritoneum
Pelvic cavity	Reproductive organs, urinary bladder, rectum	Parietal peritoneum	Visceral peritoneum

the **visceral pleura.** The **parietal pleura** lines the thoracic cavity. In between these two **pleurae** is the pleural cavity, which contains a small amount of pleural fluid. Similarly, in the medial thoracic cavity, the heart is covered by the **visceral pericardium.** The visceral pericardium contributes to the outermost connective tissue layer of the heart. Forming a tough connective tissue sac around the heart is the **parietal pericardium,** which creates the **pericardial cavity.** The heart, inside of its visceral pericardial sac, is separated from the outer parietal pericardium by a small amount of pericardial fluid.

Abdominopelvic Cavity

The abdominopelvic cavity has two portions: the superior **abdominal cavity** and the inferior **pelvic cavity.** The stomach, liver, spleen, gallbladder, and most of the small and large intestines are in the abdominal cavity. The pelvic cavity contains the rectum, the urinary bladder, the internal reproductive organs, and the rest of the large intestine. Males have an external extension of the abdominal wall, called the **scrotum,** where the testes are located.

Many of the organs of the abdominopelvic cavity are covered by the **visceral peritoneum,** whereas the wall of the abdominal cavity is lined with the **parietal peritoneum.** Peritoneal fluid fills the cavity between the visceral and parietal peritoneum. Table 1.1 summarizes our discussion of body cavities and membranes.

It is important that clinicians use the same terminology to reference various regions of the abdominopelvic cavity. Either of two systems can be used. The first uses nine regions (imagine a "tic-tac-toe" grid, with the umbilicus (navel) in the center square). The upper regions are right hypochondriac, epigastric, and left hypochondriac. The center regions are right lumbar, umbilical, and left lumbar. The lower regions are right inguinal (iliac), pubic, and left inguinal (iliac). Note that the terms used are those for each body area, as illustrated in Figure 1.3.

Alternatively, the abdominopelvic cavity can be divided into four quadrants by running a horizontal plane across the median plane at the point of the navel (Fig. 1.7a).

Begin Thinking Clinically

?

Imagine that you are caring for a small child, who tells you that his stomach hurts. However, he points to his umbilical region, about an inch below his navel. What organ is more likely the source of his pain?

Answer and discussion in Appendix C, page 461.

Physicians commonly use these quadrants to identify the locations of patient's symptoms. The four quadrants are: (1) right upper quadrant, (2) left upper quadrant, (3) right lower quadrant, and (4) left lower quadrant.

Figure 1.7b compares the two methods of referencing the abdominopelvic region and shows the organs within each region.

Content CHECK-UP!

4. Match each of the serous membranes to its function.

   parietal pleura _____   a. covers the heart

   visceral pericardium _____   b. covers walls of right and left portions of thoracic cavity

   visceral peritoneum _____   c. covers the abdominal organs

5. Pleurisy refers to infection or inflammation of the pleurae. The most logical symptom of pleurisy is_____.

Answers in Appendix C, page 461.

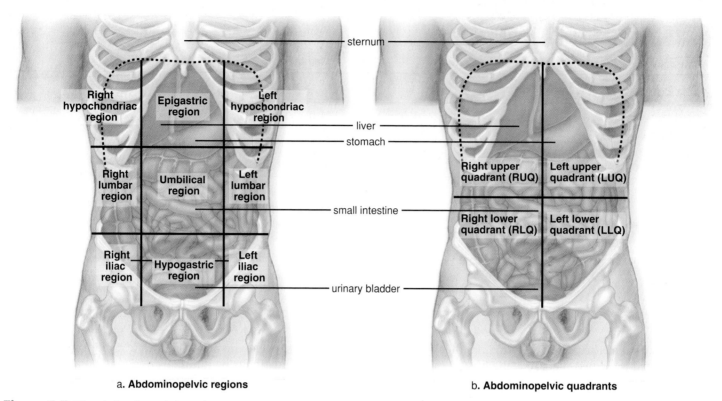

| a. Abdominopelvic regions | b. Abdominopelvic quadrants |

Figure 1.7 The abdominopelvic cavity can be subdivided into a. nine regions or **b.** four quadrants.

1.4 Organ Systems

The organs of the body work together in systems. The reference figures found on pages 14 and 15 summarize the functions of each of these systems. Corresponding figures that complete each organ system chapter show how that particular system interacts with all the other systems. In this text, the organ systems of the body have been divided into four categories, as discussed next.

Support, Movement, and Protection

The **integumentary system,** discussed in Chapter 5, includes the skin and accessory organs, such as the hair, nails, sweat glands, and sebaceous glands. The skin protects underlying tissues, prevents infection and water loss, helps regulate body temperature, contains sense organs, and even synthesizes certain chemicals that affect the rest of the body.

The **skeletal system** and the **muscular system** give the body support and the ability to move. The skeletal system, discussed in Chapter 6, consists of the bones of the skeleton and associated cartilage, as well as the ligaments that bind these structures together. The skeleton protects body parts. For example, the skull forms a protective encasement for the brain, as does the rib cage for the heart and lungs. Some bones produce blood cells, and all bones are a storage area for calcium and phosphorus salts. The skeleton as a whole serves as a place of attachment for the muscles.

Contraction of *skeletal muscles,* discussed in Chapter 7, accounts for our ability to move voluntarily and to respond to

outside stimuli. These muscles also maintain posture and are responsible for the production of body heat. *Cardiac muscle* and *smooth muscle* are called involuntary muscles because they contract automatically. Cardiac muscle makes up the heart, and smooth muscle is found within the walls of internal organs and blood vessels.

Integration and Coordination

The **nervous system,** discussed in Chapter 8, consists of the brain, spinal cord, and associated nerves. The nerves conduct sensory nerve signals to the brain and spinal cord. They also conduct nerve signals from the brain and spinal cord to the muscles, glands, and organs.

The sense organs (discussed in Chapter 9) provide us with information about our internal and external environments. The brain and spinal cord then process this information, and the individual responds to environmental stimuli using organs, glands, and/or muscles.

The **endocrine system,** discussed in Chapter 10, consists of the hormonal glands, which secrete chemicals that serve as messengers between body parts. Both the nervous and endocrine systems help maintain a relatively constant internal environment by coordinating and regulating the functions of the body's other systems. The nervous system acts quickly but has a short-lived effect; the endocrine system acts more slowly but has a more sustained effect on body parts. The endocrine system also helps maintain the proper functioning of the male and female reproductive organs.

Meningitis and Serositis

The anterior and posterior body cavities are enclosed areas that are protected by bone, muscle, connective tissues, and skin. Inflammation of the membranes lining these cavities is a fairly rare, but serious, illness. If body defenses are overcome by bacteria, viruses, or other microbes, the result is a serious, potentially fatal infection and inflammation of the meninges (meningitis) or the serous membranes (**serositis**). Pleurisy, pericarditis, and peritonitis are all forms of serositis.

Meningitis is the term for inflammation of the meninges—linings of the posterior body cavity that cover the brain and spinal cord. The most dangerous form is caused by bacteria that commonly inhabit the nose. In the bacterial meningitis patient, a previous viral infection (which may be a simple common cold) allows these bacteria to enter the bloodstream and infect the meninges. Symptoms of bacterial meningitis include a severe headache and stiff neck, sensitivity to light, fever, weakness, and fatigue. Even with aggressive antibiotic treatment, bacterial meningitis is fatal in 25% of adults. The best treatment is prevention by immunization—especially important for young college students living in the close quarters of a college dorm.

Pleurisy is an inflammation of the pleurae—linings of the thoracic cavity that also cover the lungs. It is often caused by a cold virus, although it can signal the presence of more serious infections or even lung cancer. Its symptoms include chest pain that worsens with deep breathing, and *pleural friction rub*—a rough, grating sound in the chest that can be heard with a stethoscope placed over the painful area. Treatment for pleurisy depends on its cause; most often, pleurisy that results from a common cold requires only pain medication such as aspirin or ibuprofen. Treatment for bacterial infection requires antibiotics.

Pericarditis affects the linings surrounding the heart. Like meningitis, it often results from previous infections and can be extremely dangerous. It is a common complication in drug abusers who use dirty needles for injections. Symptoms include severe chest pain (which may be mistaken for a heart attack), fever, and weakness. Physicians can hear *pericardial friction rub* by placing a stethoscope over the patient's heart. Fluid accumulation inside the pericardial sac surrounding the heart may interfere with blood flow to and from the heart. Bacterial pericarditis is treated with antibiotics, pain medications, and drugs that reduce swelling.

Peritonitis affects the lining of the abdominopelvic cavity. It usually results from bacterial infection; a common cause of infection is a ruptured appendix from appendicitis. Severe pain, fever, elevated white blood cell counts, and tenderness are common symptoms. Aggressive treatment with antibiotics is necessary to prevent bacteria from invading the blood.

a. **Midsagittal view**

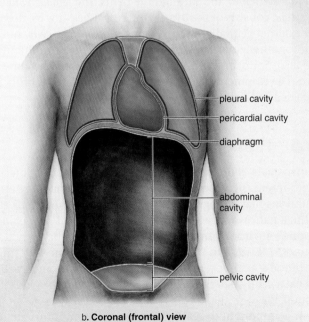

b. **Coronal (frontal) view**

Figure 1A **Meningitis and serositis. a.** Meningitis is infection or inflammation of the linings of the cranial cavity and vertebral canal. **b.** Serositis is infection or inflammation of the ventral body cavities. Pleurisy affects the pleural cavities, pericarditis affects the pericardial cavity, and peritonitis affects the abdominopelvic cavities.

Maintenance of the Body

The internal environment of the body is the blood within the blood vessels and the tissue fluid that surrounds the cells. Five systems add substances to and/or remove substances from the blood and tissue fluid: the cardiovascular, lymphatic, respiratory, digestive, and urinary systems.

The **cardiovascular system,** discussed in Chapter 12, consists of the heart and the blood vessels that carry blood through the body. Blood transports nutrients and oxygen to the cells, and removes waste molecules to be excreted from the body. Blood also contains cells produced by the **lymphatic system,** discussed in Chapter 13. The lymphatic system protects the body from disease.

The **respiratory system,** discussed in Chapter 14, consists of the lungs and the tubes that take air to and from the lungs. The respiratory system brings oxygen into the lungs and takes carbon dioxide out of the lungs.

The **digestive system** (see Fig. 1.1), discussed in Chapter 15, consists of the mouth, esophagus, stomach, small intestine, and large intestine (colon), along with the accessory organs: teeth, tongue, salivary glands, liver, gallbladder, and pancreas. This system receives food and digests it into nutrient molecules, which can enter the cells of the body.

The **urinary system,** discussed in Chapter 16, contains the kidneys and the urinary bladder. This system rids the body of nitrogenous (nitrogen containing) wastes and helps regulate the fluid level and chemical content of the blood.

Reproduction and Development

The male and female **reproductive systems,** discussed in Chapter 17, contain different organs. The *male reproductive system* consists of the testes, other glands, and various ducts that conduct semen to and through the penis. The *female reproductive system* consists of the ovaries, uterine tubes, uterus, vagina, and external genitalia. Both systems produce sex cells, but in addition, the female system receives the sex cells of the male and also nourishes and protects the fetus until the time of birth. Development before birth and the process of birth are discussed in Chapter 18.

1.5 Homeostasis

Homeostasis is the relative constancy of the body's internal environment. Because of homeostasis, even though external conditions may change dramatically, internal conditions stay within a narrow range. For example, regardless of how cold or hot one's environment gets, the temperature of the body stays around 37°C (97° to 99°F). No matter how acidic your meal, the blood's acidity remains relatively constant, and even if you eat a candy bar, the amount of sugar in your blood is just about 0.1%.

It is important to realize that internal conditions are not absolutely constant; they tend to fluctuate above and below a particular value. Therefore, the internal state of the body is often

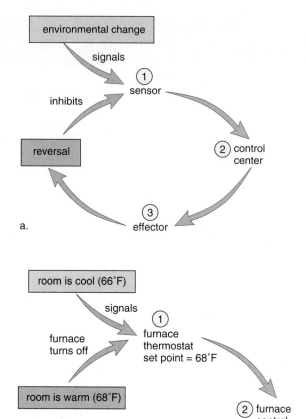

Figure 1.8 Negative feedback. In each example, a sensor (1) detects external or internal change, and signals a control center (2). The control center activates an effector (3), which reverses the change. **a.** The general pattern. **b.** A mechanical example. *(continued)*

described as one of *dynamic* equilibrium. If internal conditions change to any great degree, illness results. This makes the study of homeostatic mechanisms medically important.

Negative Feedback

Negative feedback is the primary homeostatic mechanism that keeps a variable close to a particular value, or set point. A homeostatic mechanism has three components: a sensor, a control center, and an effector (Fig. 1.8a). As you study processes controlled by negative feedback, note that all follow the same basic pattern of events: (1) The sensor detects a change in the external or internal environment and alerts the control center; (2) the control center activates the effector; and (3) the effector reverses the initial change and brings conditions back to normal again. Now, the sensor is no longer activated. These steps are illustrated in Figures 1.8 and 1.9

Mechanical Example

A home heating system illustrates how a negative feedback mechanism works (Fig. 1.8b). You set the thermostat at, say,

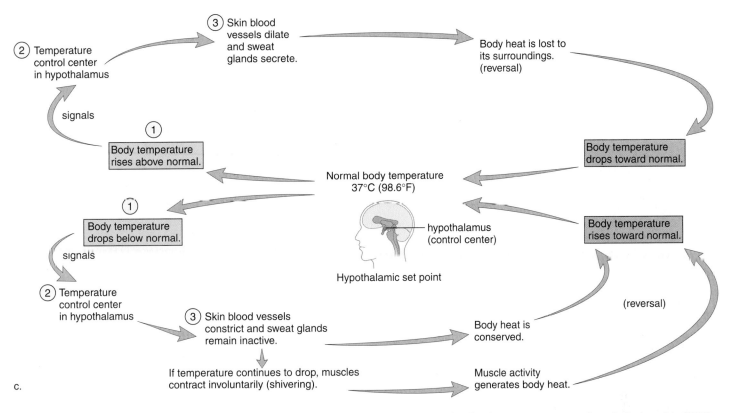

Figure 1.8 *(continued)* **c.** Body temperature regulation, a human example of negative feedback. Temperature remains relatively stable (37°C), and is returned to normal when it fluctuates above or below the set point.

68°F. This is the set point. The thermostat contains a thermometer, a sensor that detects when the room temperature falls below the set point. The thermostat is also the control center; it turns the furnace on. The furnace plays the role of the effector. The heat given off by the furnace raises the temperature of the room to 70°F. Now, the furnace turns off because the sensor is no longer activated.

Notice that a negative feedback mechanism prevents change in the same direction; the room does not get warmer and warmer because warmth inactivates the system.

Human Example: Regulation of Body Temperature

The thermostat for body temperature is located in a part of the brain called the hypothalamus. When the body temperature falls below normal, the control center directs (via nerve signals) the blood vessels of the skin to constrict (Fig. 1.8). This conserves heat. If body temperature falls even lower, the control center sends nerve signals to the skeletal muscles, and shivering occurs. Shivering generates heat, and gradually body temperature rises to 37°C. When the temperature rises to normal, the regulatory center is inactivated.

When the body temperature is higher than normal, the control center directs the blood vessels of the skin to dilate. This allows more blood to flow near the surface of the body, where heat can be lost to the environment. In addition, the nervous system activates the sweat glands, and the evapora-

tion of sweat helps lower body temperature. Gradually, body temperature decreases to 37°C.

Human Example: Regulation of Blood Pressure

Numerous other examples of negative feedback control help to maintain homeostasis. For example, when blood pressure falls, sensory receptors signal a control center in the brain (Fig. 1.9). This center sends out nerve signals to the arterial

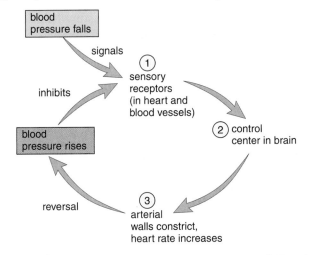

Figure 1.9 **Homeostasis and blood pressure control.** Negative feedback mechanisms control blood pressure so that it remains relatively constant.

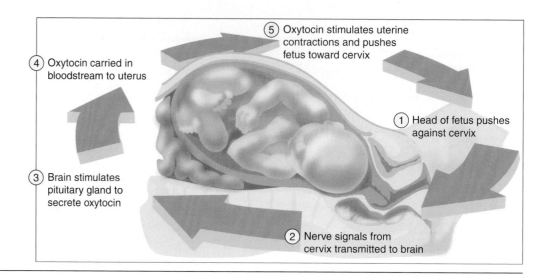

Figure 1.10 Positive feedback. Positive feedback loops assist the body to complete a process with a definite cut-off point, like labor and delivery.

⑤ Oxytocin stimulates uterine contractions and pushes fetus toward cervix

④ Oxytocin carried in bloodstream to uterus

① Head of fetus pushes against cervix

③ Brain stimulates pituitary gland to secrete oxytocin

② Nerve signals from cervix transmitted to brain

walls so that they constrict. Once the blood pressure rises, the system is inactivated.

Positive Feedback

Like negative feedback, in a **positive feedback** mechanism a sensor will trigger the control center. The control center in turn activates the effector. However, the effector in this type of feedback mechanism produces a response that continues to stimulate the sensor. Thus, positive feedback brings about an even greater change in the same direction. A positive feedback mechanism can be harmful, as when a fever causes metabolic changes that push the fever still higher. Death occurs at a body temperature of 45°C because cellular proteins are destroyed at this temperature and metabolism stops.

Still, positive feedback loops such as those involved in blood clotting, the stomach's digestion of protein, and childbirth assist the body in completing a process that has a definite cutoff point.

Consider that when a woman is giving birth, the head of the baby begins to press against the cervix, stimulating sensory receptors there. When nerve signals reach the brain (control center), the brain causes the pituitary gland (effector) to secrete the hormone oxytocin. Oxytocin travels in the blood and causes the uterus to contract. As labor continues, the cervix is ever more stimulated, and uterine contractions become ever stronger until birth occurs (Fig. 1.10). Once the birth process is complete, the sensory receptors of the cervix are no longer stimulated. However, excessive bleeding in the mother might occur immediately after birth if uterine contraction stops completely. Thus, continued secretion of oxytocin is necessary to safeguard the mother. Stimulating the nipples by allowing the baby to suckle signals the brain, and continues oxytocin secretion. This process is also controlled by positive feedback.

Homeostasis and Body Systems

The internal environment of the body consists of blood and tissue fluid. Tissue fluid bathes all the cells of the body.

Oxygen and nutrients move from blood to tissue fluid, and wastes move from tissue fluid into the blood (Fig. 1.11). Tissue fluid remains constant only as long as blood composition remains constant. All systems of the body contribute toward maintaining homeostasis and, therefore, a relatively constant internal environment.

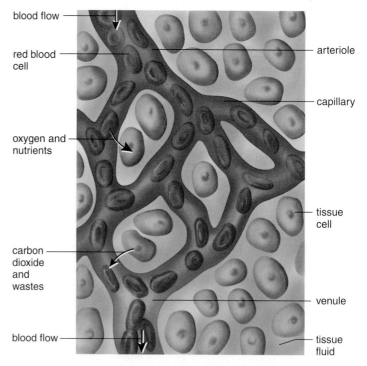

blood flow

red blood cell

oxygen and nutrients

carbon dioxide and wastes

blood flow

arteriole

capillary

tissue cell

venule

tissue fluid

Figure 1.11 Regulation of tissue fluid composition. Cells are surrounded by tissue fluid (blue), which is continually refreshed because oxygen and nutrient molecules constantly exit the bloodstream, and carbon dioxide and waste molecules continually enter the bloodstream.

As you use the text, you will be introduced to each of the organ systems. Remember the job of body systems in maintaining homeostasis as you study each:

Integumentary System (skin)—*Support and protection* of delicate internal structures.

Skeletal System—*Movement.* In addition, the skeleton stores minerals and produces the blood cells.

Muscular System—*Movement.* Skeletal muscles and bones work together for movement. Smooth muscle inside internal structures moves substances inside a tube, as when a meal is moved through the digestive tract. Muscle activity also generates heat, thus helping to maintain body temperature.

Nervous and Sensory Systems—*Control.* We have already seen that in negative and positive feedback mechanisms, sensory receptors send nerve signals to control centers in the brain, which then activate effectors: muscles, glands, or organs. The nervous system can cause rapid change, if needed, to maintain homeostasis.

Endocrine System—*Control.* Endocrine glands secrete hormones into the blood. Endocrine hormones bring about a slower, more lasting change that keeps the internal environment relatively stable.

Blood and Cardiovascular System—*Transportation and defense.* Red blood cells and blood plasma (the liquid fraction of blood) transport oxygen, carbon dioxide, nutrients, and wastes. Platelets in blood participate in the clotting process, preventing excess blood loss. White blood cells defend against infection. The cardiovascular system conducts blood to and away from capillaries, where exchange occurs. The heart pumps the blood and thereby keeps it moving toward the capillaries.

Lymphatic System—*Transportation and defense.* The lymphatic system assists the cardiovascular system. Lymphatic capillaries collect excess tissue fluid, which is returned via lymphatic vessels to the cardiovascular veins. Lymph nodes help to purify lymph and keep it free of pathogens. This action is assisted by the white blood cells that are housed within lymph nodes.

Respiratory System—*Gas exchange.* The respiratory system adds oxygen to and removes carbon dioxide from the blood. It also plays a role in regulating acid-base balance in blood and tissue fluid. Removal of CO_2 helps to eliminate acid and prevent acidosis.

Digestive System—*Nourishment and waste removal.* The digestive system takes in and digests food, providing nutrient molecules to replace the nutrients that are constantly being used by the body cells. Substances that cannot be digested are eliminated. The liver, an accessory digestive organ, also manufactures urea, a waste product of protein digestion. The liver also removes toxic chemicals such as alcohol and other drugs. Additionally, the liver regulates blood glucose (sugar). As glucose enters the blood after a meal, any excess is removed by the liver and stored as glycogen. Later, the glycogen can be broken down to replace the glucose used by the body cells. In this way, the glucose composition of blood remains constant.

Urinary System—*Waste removal.* Urea and other metabolic waste molecules are excreted by the kidneys, which are a part of the urinary system. Urine formation by the kidneys is extremely critical to the body, not only because it rids the body of unwanted substances but also because urine formation offers an opportunity to carefully regulate blood volume, salt balance, and acid-base balance.

Reproductive System—*Survival of the species.* Although individuals can survive and thrive without reproducing, the human species cannot continue without this vital system.

The contributions of each of the body's systems are summarized in the Human Systems Work Together illustration on page 14.

Disease

Disease occurs when homeostasis fails and the body (or part of the body) no longer functions properly. The effects may be limited or widespread. A **local disease** is more or less restricted to a specific part of the body. On the other hand, a **systemic disease** affects the entire body or involves several organ systems. For example, streptococcal tonsillitis, or "strep throat," is a local disease. If not effectively treated with antibiotics, strep throat can progress to become a dangerous systemic disease—rheumatic fever. Diseases may also be classified on the basis of their severity and duration. **Acute diseases** occur suddenly and generally last a short time. **Chronic diseases** tend to be less severe, develop slowly, and are long term. Acute bronchitis is a bacterial infection of lung airways, and a frequent complication of the common cold; chronic bronchitis is typically caused by years of smoking.

The medical profession has many ways of diagnosing disease including imaging internal body parts (see Medical Focus, page 16).

Content **CHECK-UP!**

6. The two organ systems responsible for controlling other body systems are the _____ and _____ systems.

7. The two organ systems that defend the body against infection are:

   a. cardiovascular and urinary

   b. cardiovascular and respiratory

   c. cardiovascular and skeletal

   d. cardiovascular and lymphatic

8. A person with osteoarthritis, a degenerative disorder that causes stiff, painful joints, developed this condition at age 60 after decades of "wear and tear." Osteoarthritis can best be described as:

   a. acute

   b. chronic

Answers in Appendix C, page 461.

Human Systems Work Together

Inegumentary System

External support and protection of body; helps maintain body temperature.

Skeletel System

Internal support and protection; body movement; production of blood cells.

Muscular System

Body movement; production of heat that maintains body temperature.

Nervous System

Regulatory centers for control of all body systems; learning and memory.

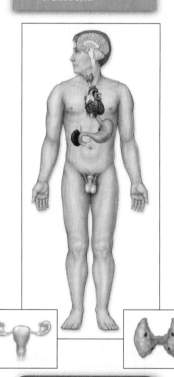

Endocrine System

Secretion of hormones for chemical regulation of all body systems.

Blood/Cardiovascular System

Transport of nutrients to body cells and transport of wastes away from cells

Lymphatic System/Immunity

Drainage of tissue fluid; purifies tissue fluid and keeps it free of pathogens.

Respiratory System

Rids the blood of carbon dioxide and supplies the blood with oxygen; helps maintain the pH of the blood.

Digestive System

Breakdown of food and absorption of nutrients into blood.

Urinary System

Maintenance of volume and chemical composition of blood.

Reproductive System

Production of sperm and egg; transfer of sperm to female system where development occurs.

Imaging the Body

Imaging the body for diagnosis of disease has certainly changed since Wilhelm Roentgen's accidental invention of the X ray in 1895. Many new techniques allow clinicians to visualize internal structures with great accuracy. The least expensive and most widely used imaging technique remains the X ray. X rays are produced when high-speed electrons strike a heavy metal. Dense structures (bone, teeth, cancer tumors) absorb X rays well and show up as light areas on photographic film. Soft tissues absorb X rays to a lesser extent and show up as dark areas. Injecting opaque dye into blood vessels allows X-ray imaging of blood vessels, as in a coronary angiogram (an X ray of the arteries supplying the heart). X-ray imaging of the digestive tract is possible on patients who have first consumed opaque dye solutions.

A bone density scan is another specialized type of X ray. The patient is first injected with a harmless radioactive tracer, which is rapidly taken up by bone. Subsequent X rays can reveal areas of increased bone metabolism (as in a cancerous tumor) or decreased metabolism (as in osteoporosis). A similar technique uses radioactive iodine to study the thyroid gland. The thyroid is the only tissue to use the element iodine (Fig. 1B, 1C).

During computed tomography, or CT scan (previously called computerized axial tomography), X rays are sent through the body at various angles, and a computer uses the X-ray information to form a series of cross sections (Fig. 1D). CT scanning has reduced the need for exploratory surgery and can guide the surgeon in visualizing complex body structures during surgical procedures.

PET (positron emission tomography) is a variation on CT scanning (Fig. 1D). Radioactively labeled sugar is injected into the body. Metabollically active tissues tend to take up this substance and then emit gamma rays. A computer uses the gamma ray information to again generate cross-sectional images of the body, but this time the image indicates metabolic activity, not structure. PET scanning is used to diagnose brain disorders, such as a brain tumor, Alzheimer disease, epilepsy, or stroke (Fig. 1D).

During MRI (magnetic resonance imaging), the patient lies in a massive, hollow, cylindrical magnet and is exposed to short bursts of a powerful magnetic field. This causes the protons in the nuclei of hydrogen atoms to align. When the protons are exposed to strong radio waves, they move out of alignment and produce signals. A computer changes these signals into an image. Tissues with many hydrogen atoms (such as fat) show up as bright areas, while tissues with few hydrogen atoms (such as bone) appear black. This is the opposite of an X ray, which is why MRI is more useful than an X ray for imaging soft tissues. However, many people cannot undergo MRI because the magnetic field can actually pull a metal object (like a tooth filling or an artificial hip) out of the body!

The least-expensive method of creating tissue images is sonography, or ultrasound. High-frequency soundwaves are transmitted into tissues. Organs and tissues reflect the soundwaves; a receiver creates an image using the reflected waves. Use of sonography avoids any radiation exposure to the patient. The rare adverse reaction to dyes or radioactive compounds can also

Figure 1B **CT (computed tomography) of the brain.** This transverse section is at eye level.

thyroid gland

trachea (windpipe)

a. **Drawing of thyroid**   b. **Scan of thyroid**

Figure 1C **Use of radiation to aid a diagnosis.** After the administration of radioactive iodine, a scan of the thyroid reveals pathology. The missing portion of the gland is cancerous and therefore failed to take up the iodine.

a. **Patient entering PET scanner**   b. **PET scan**

Figure 1D **Use of radiation to study the brain.** After the administration of radioactively labeled glucose, a PET scan reveals which portions of the brain are most active.

be avoided. Ultrasound is safe for imaging the fetus in a pregnant woman and can show an amazing amount of detail, as you can see in the chapter introduction.

SELECTED NEW TERMS

Basic Key Terms

abdominal cavity (ab-dom'ĭ-nal kav'ĭ-te), p. 7

abdominopelvic cavity (ab-dom"ĭ-no-pel'vik kav'ĭ-te), p. 6

anatomical position (an'uh-tom"ĭ-kăl pō-zish'un), p. 3

anatomy (uh-nat'o-me), p. 2

anterior (ăn-tîr'ē-ər), p. 3

appendicular portion (ăp'ən-dĭk'yə-lər pôr'shən), p. 4

atoms (ăt'əms), p. 2

axial portion (ăk'sē-əl pôr'shən), p. 4

cells (sĕls), p. 2

central (cen'trəl), p. 3

contralateral (kŏn'trə-lăt'ər-əl), p. 4

cranial cavity (kra'ne-al kav'ĭ-te), p. 5

deep (dēp), p. 3

diaphragm (dī'ə-frăm'), p. 6

distal (dis'tal), p. 3

frontal plane (frŭn'tl plăn), p. 5

homeostasis (ho"me-o-sta'sis), p. 10

inferior (ĭn-fîr'ē-ər), p. 3

ipsilateral (ĭp'sə-lăt'ər-əl), p. 4

lateral (lat'er-al), p. 3

macromolecules (măk'rŏ-mŏl'ĭ-kyoōl's), p. 2

medial (me'de-al), p. 3

mediastinum (me"de-uh-sti'num), p. 6

meninges (muh-nĭn' jēz), p. 6

molecules (mŏl'ĭ-kyoōl's), p. 2

negative feedback (neg'uh-tiv fēd'bak), p. 10

organ (ôr'gən), p. 2

organelles (ôr'gə-nĕls), p. 2

organism (ôr'gə-nĭz'əm), p. 2

organ system (ôr'gən sĭs'təm), p. 2

parietal pericardium (per"ĭ-kar'de-um), p. 7

parietal peritoneum (pe-rī'ĭ-təl pĕr'ĭ-tə-nē'əm), p. 7

parietal pleurae (pə-rī'ĭ-təl ploŏr'ə), p. 7

parietal serous membrane (pə-rī'ĭ-təl sîr'əs mĕm'brān'), p. 6

pelvic cavity (pel'vik kav'ĭ te), p. 7

pericardial cavity (pĕr'ĭ-kăr'dē-əl kăv'ĭ-tē), p. 7

peripheral (pə-rĭf'ər-əl), p. 4

peritoneum (per"ĭ-to-ne'um), p. 7

physiology (fiz"e-ol'o-je), p. 2

pleurae (ploo'rā), p. 7

positive feedback (poz'ĭ-tiv fēd'bak), p. 12

posterior (pŏ-stîr'ē-ər), p. 3

proximal (prok'sĭ-mal), p. 3

sagittal plane (saj'ĭ-tal plăn), p. 5

scrotum (skrō'tŭm), p. 7

serous fluid (sîr'əs floō'ĭd), p. 6

serous membrane (sēr'us mem'brān), p. 6

serum (sîr'əm), p. 6

superficial (soō'par-fish'əl), p. 3

superior (soō-pîr'ē-ər), p. 3

thoracic cavity (tho-ras'ik kav'ĭ-te), p. 6

tissue (tĭsh'oō), p. 2

transverse plane (trans-vers' plăn), p. 5

vertebral canal (vŭr'tə-brəl kə-năl), p. 6

viscera, visceral pleura (vis'-er-uh, vis'-er-ŭl), p. 7

visceral pericardium (vĭs'ər-əl pĕr'ĭ-kär'dē-əm), p. 7

visceral peritoneum (vĭs'ər-əl pĕr'ĭ-tə-nē'əm), p. 7

visceral serous membrane (vĭs'ər-əl sîr'əs), p. 6

Clinical Key Terms

acute (uh-kyoot') disease, p. 13

chronic (kron'-ik) disease, p. 13

disease (di-zez'), p. 13

local disease (lō'kəl dĭ-zĕz'), p. 13

meningitis (men"in-ji'tis), p. 9

pericarditis (per"i-kar-di'tis), p. 9

peritonitis (per"i-to-ni'tis), p. 9

pleurisy (ploo'ri-see), p. 9

serositis (se"ro-si'tis), p. 9

systemic disease (sis-tem'ik di-zez'), p. 13

SUMMARY

1.1 **The Human Body**
  A. Anatomy is the study of the structure of body parts, and physiology is the study of the function of these parts. Structure is suited to the function of a part.
  B. The body has levels of organization that progress from atoms to molecules, macromolecules, organelles, cells, tissues, organs, organ systems, and finally, the organism.

1.2 **Anatomical Terms**
  Various terms are used to describe the location of body organs when the body is in the **anatomical position** (standing erect, with face forward, arms at the sides, and palms and toes directed forward).
  A. The terms *anterior/posterior, superior/inferior, medial/lateral, proximal/distal, superficial/deep, central/peripheral,* and

contralateral/ipsilateral describe the relative positions of body parts.
  B. The body can be divided into axial and appendicular portions, each of which can be further subdivided into specific regions. For example, *brachial* refers to the arm, and *pedal* refers to the foot.
  C. The body or its parts may be sectioned (cut) along certain planes. A median or midsagittal (vertical) cut divides the body into equal right and left portions. A parasagittal or paramedian section is a sagittal section parallel to the midline. A frontal (coronal) cut divides the body into anterior and posterior parts. A transverse (horizontal) cut is a cross section.

1.3 **Body Cavities and Membranes**
  The human body has two major cavities: the posterior (dorsal) body

cavity and the anterior (ventral) body cavity. Each is subdivided into smaller cavities, within which specific viscera are located. Specific serous membranes line body cavities and adhere to the organs within these cavities.

1.4 **Organ Systems**
  The body has a number of organ systems. These systems have been characterized as follows:
  A. Support, movement, and protection. The integumentary system, which includes the skin, not only protects the body, but also has other functions. The skeletal system contains the bones, and the muscular system contains the three types of muscles. The primary function of the skeletal and muscular systems is support and movement, but they have other functions as well.

B. Integration and coordination. The nervous system contains the brain, spinal cord, and nerves. Because the nervous system communicates with both the sense organs and the muscles, it allows us to respond to outside stimuli. The endocrine system consists of the hormonal glands. The nervous and endocrine systems coordinate and regulate the activities of the body's other systems.

C. Maintenance of the body. The cardiovascular system (heart and vessels), lymphatic system (lymphatic vessels and nodes, spleen, and thymus), respiratory system (lungs and conducting tubes), digestive system (mouth, esophagus, stomach, small and large intestines, and associated organs), and urinary system (kidneys and bladder) all perform specific processing and transporting functions to maintain the normal conditions of the body.

D. Reproduction and development. The reproductive system in males (testes, other glands, ducts, and penis) and in females (ovaries, uterine tubes, uterus, vagina, and external genitalia) carries out those functions that give humans the ability to reproduce.

1.5 Homeostasis
Homeostasis is the relative constancy of the body's internal environment, which is composed of blood and the tissue fluid that bathes the cells.

A. Negative feedback mechanisms help maintain homeostasis. Positive feedback occurs in processes with a definite cut-off point.

B. All of the body's organ systems contribute to homeostasis. Some, including the respiratory, digestive, and urinary systems, remove and/or add substances to blood.

C. The nervous and endocrine systems regulate the activities of other systems. Negative feedback is a self-regulatory mechanism by which systems and conditions of the body are controlled.

STUDY QUESTIONS

1. Distinguish between the study of anatomy and the study of physiology. (p. 2)
2. Give an example that shows the relationship between the structure and function of body parts. (p. 2)
3. List the levels of organization within the human body in reference to a specific organ. (p. 2)
4. What purpose is served by directional terms as long as the body is in anatomical position? (pp. 2, 3)
5. Distinguish between the axial and appendicular portions of the body.

State at least two anatomical terms that pertain to the head, thorax, abdomen, and limbs. (p. 4)
6. Distinguish between a midsagittal section, a parasagittal section, a transverse section, and a coronal section. (p. 5)
7. Distinguish between the posterior and anterior body cavities, and name two smaller cavities that occur within each. (pp. 5–7)
8. Name the four quadrants and the nine regions of the abdominopelvic cavity. (pp. 7–8)

9. Name the major organ systems, and describe the general functions of each. (pp. 8, 10)
10. List the major organs found within each organ system. (pp. 8, 10)
11. Define homeostasis, and give examples of negative feedback and positive feedback mechanisms. (pp. 10–12)
12. Discuss the contribution of each body system to homeostasis. (pp. 13, 14–15)

LEARNING OBJECTIVE QUESTIONS

I. **Match the terms in the key to the relationships listed in questions 1–5.**

Key:
  a. anterior
  b. posterior
  c. superior
  d. inferior
  e. medial
  f. lateral
  g. proximal
  h. distal

1. the esophagus in relation to the stomach
2. the ears in relation to the nose
3. the shoulder in relation to the hand
4. the intestines in relation to the vertebrae

5. the rectum in relation to the mouth

II. **Match the terms in the key to the body regions listed in questions 6–12.**

Key:
  a. oral
  b. occipital
  c. gluteal
  d. carpal
  e. palmar
  f. cervical
  g. axillary

6. buttocks
7. palm
8. back of head
9. mouth
10. wrist
11. armpit
12. neck

III. **Match the terms in the key to the organs listed in questions 13–18.**

Key:
  a. cranial cavity
  b. vertebral canal
  c. thoracic cavity
  d. abdominal cavity
  e. pelvic cavity

13. stomach
14. heart
15. urinary bladder
16. brain
17. liver
18. spinal cord

IV. Match the organ systems in the key to the organs listed in questions 19–25.

Key:

    a. digestive system
    b. urinary system
    c. respiratory system
    d. cardiovascular system
    e. reproductive system
    f. nervous system
    g. endocrine system

19. thyroid gland
20. lungs
21. heart
22. ovaries
23. brain
24. stomach
25. kidneys

V. Fill in the blanks.
26. A(n) _____ is composed of several types of tissues and performs a particular function.

27. The imaginary plane that passes through the midline of the body is called the _____ plane.
28. All the organ systems of the body together function to maintain _____, a relative constancy of the internal environment.

MEDICAL TERMINOLOGY EXERCISE

Consult Appendix B for help in pronouncing, analyzing, and filling in the blanks to give a brief meaning to the terms that follow.

1. Suprapubic (su″pruh-pyū′bik) means _____ the pubis.
2. Infraorbital (in″fruh-or′bǐ-tal) means _____ the eye orbit.
3. Gastrectomy (gas-trek′to-me) means excision of the _____.
4. Celiotomy (se″le-ot′o-me) means incision (cut into) of the _____.
5. Macrocephalus (mak″ro-sef′uh-lus) means large _____.
6. Transthoracic (trans″tho-ras′ik) means across the _____.

7. Bilateral (bi-lat′er-al) means two or both _____.
8. Ophthalmoscope (of-thal′mo-skōp) is an instrument to view inside the _____.
9. Dorsalgia (dor-sal′je-uh) means pain in the _____.
10. Endocrinology (en″do-krǐ-nol′o-je) is the _____ of the endocrine system.
11. The pectoralis (pek-to-ral′is) muscle can be found on the _____.
    a. chest b. head c. buttocks
    d. thigh

12. The sacral (sa′krul) nerves are located in the _____.
    a. lower back b. neck c. upper back
    d. head
13. Hematuria (he-muh-tu′re-uh) means _____ in the urine.
14. Nephritis (nef-ri′tis) is inflammation of the _____.
    a. lungs b. heart c. liver d. kidneys
15. Tachypnea (tak-ip-ne′uh) is a breathing rate that is _____.
    a. faster than normal b. slower than normal

WEBSITE LINK

Visit this text's website at http://www.mhhe.com/longenbaker7 for additional quizzes, interactive learning exercises, and other study tools.

Chemistry of Life

Why is a poison poisonous? Like the sweat of the poison arrow frog, most poisons are *neurotoxins*—attacking the nervous system and interfering with communication between a nerve and another nerve, muscle, or organ. An additional shared trait of most poisons is that they are *alkaloids*—salts with a basic pH, made by reacting an acid with a base. (You can read more about acids, bases, and salts on page 25). The chemical structure of alkaloid poisons allows them to dissolve in water, and in blood as well (because blood is mostly water). Once dissolved in a person's blood, the poison can have its fatal effects. However, if the dosage is lowered enough, alkaloid poisons can have medicinal or recreational value. You'll recognize the names of some: nicotine, caffeine, morphine, codeine, and cocaine.

Chapter Outline & Learning Objectives

After you have studied this chapter, you should be able to:

2.1 Basic Chemistry (p. 21)
- Describe how an atom is organized, and tell why atoms interact.
- Define radioactive isotope, and describe how they can be used in the diagnosis and treatment of disease.
- Distinguish between an ionic bond and a covalent bond.

2.2 Water, Acids, and Bases (p. 24)
- Describe the characteristics of water and three functions of water in the human body.
- Explain the difference between an acid and a base using examples.
- Use and understand the pH scale.

2.3 Molecules of Life (p. 27)
- List the four classes of macromolecules in cells, and distinguish between a dehydration reaction and a hydrolysis reaction.
- Name the individual subunits that comprise carbohydrates, lipids, proteins, and nucleic acids.

2.4 Carbohydrates (p. 27)
- Give some examples of different types of carbohydrates and their specific functions in cells.

2.5 Lipids (p. 30)
- Describe the composition of a neutral fat, and give examples of how lipids function in the body.

2.6 Proteins (p. 32)
- State the major functions of proteins, and tell how globular proteins are organized.

2.7 Nucleic Acids (p. 35)
- Describe the structure and function of DNA and RNA in cells.
- Explain the importance of ATP in the body.

Medical Focus
Nutrition Labels (p. 28)
Prions: Malicious Proteins? (p. 34)

2.1 Basic Chemistry

Matter is anything that takes up space and has mass. It can be a solid, a liquid, or a gas. Therefore, not only are we humans matter, but so are the water we drink and the air we breathe.

Elements and Atoms

All matter is composed of basic substances called **elements.** It's quite remarkable that there are only 92 naturally occurring elements. It is even more surprising that over 90% of the human body is composed of just four elements: carbon, nitrogen, oxygen, and hydrogen.

Every element has a name and a symbol consisting of one or two letters. When the symbol consists of two letters, only the first letter is capitalized. For example, carbon has been assigned the atomic symbol C (Fig. 2.1a). Some of the symbols we use for elements are derived from Latin. For example, the symbol for sodium is Na because the Latin word *natrium* means sodium.

Elements are composed of tiny particles called atoms. The same name is given to both an element and its atoms.

Atoms

An **atom** is the smallest unit of an element that still retains the chemical and physical properties of the element. Although it is possible to split an atom by physical means, an atom is the smallest unit to enter into chemical reactions. For our purposes, it is satisfactory to think of each atom as having a central nucleus and pathways about the nucleus called *shells* (sometimes called *energy levels*). The subatomic particles called **protons** and **neutrons** are located in the nucleus, and **electrons** orbit about the nucleus in the shells (Fig. 2.1b). Most of an atom is empty space. If we could draw an atom the size of a football stadium, the nucleus would be like a gumball in the center of the field, and the electrons would be tiny specks whirling about in the upper stands.

The **atomic number** of an atom tells you how many protons an atom has. All atoms of any particular element have the same number of protons. Protons carry a positive (+) charge, and electrons have a negative (−) charge. Neutrons have no charge. When the atom is electrically neutral, the numbers of protons and electrons are equal. For example, the atomic number of carbon is six. Therefore, when carbon is neutral it has six protons and six electrons. How many electrons are in each shell of an atom? The inner shell is the lowest energy level and can hold only two electrons; after that, each shell, for the atoms noted in Figure 2.1a, can hold up to eight electrons. Using this information, we can calculate that carbon has two shells and that the outer shell has four electrons.

The number of electrons in the outer shell determines the chemical properties of an atom, including how readily it enters into chemical reactions. As we shall see, an atom is most stable when the outer shell has eight electrons. (Hydrogen and helium, with only one shell, are exceptions to this statement. Atoms with only one shell are stable when this shell contains two electrons.)

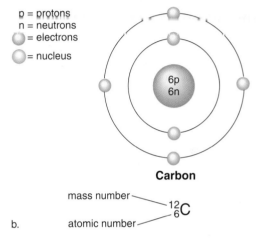

Common Elements in Living Things				
Element	Atomic Symbol	Atomic Number	Atomic Mass	Comment
hydrogen	H	1	1	These elements make up most biological molecules.
carbon	C	6	12	
nitrogen	N	7	14	
oxygen	O	8	16	
phosphorus	P	15	31	
sulfur	S	16	32	
sodium	Na	11	23	These elements occur mainly as dissolved salts.
magnesium	Mg	12	24	
chlorine	Cl	17	35	
potassium	K	19	39	
calcium	Ca	20	40	

a.

p = protons
n = neutrons
= electrons
= nucleus

Carbon

mass number ——— $^{12}_{\ 6}C$
atomic number ———

b.

Figure 2.1 Elements and atoms. a. The atomic symbol, number, and mass number are given for common elements in the body. **b.** The structure of carbon shows that an atom contains the subatomic particles called protons (p) and neutrons (n) in the nucleus (colored pink) and electrons (colored blue) in shells about the nucleus.

The subatomic particles are so light that their weight is indicated by special designations called *atomic mass units.* Protons and neutrons each have a weight of one atomic mass unit, and electrons have almost no mass. Therefore, the mass number of an atom generally tells you the number of protons plus the number of neutrons. How could you calculate that carbon (C) has six neutrons? Carbon's mass number is 12, and you know from its atomic number that it has six protons. Therefore, carbon has six neutrons (Fig. 2.1b).

Also, as shown in Figure 2.1b, the atomic number of an atom is often written as a subscript to the lower left of the atomic symbol. The mass number is often written as a superscript to the upper left of the atomic symbol.

In everyday life, length is measured in units such as centimeters, meters, inches, and feet. Mass, or weight, is measured in grams or pounds. Chemists have their own terminology for measuring the number of atoms of an element or molecules of a compound. The unit is a **mole,** and one mole is a specific, very large number: 6.02×10^{23}. (This number, termed *Avogadro's number,* is the number of carbon atoms in exactly 12 grams of carbon 12). Thus, when a chemist refers to a mole of helium, he or she is referring to 6.02×10^{23} helium atoms. Likewise, a mole of oxygen is 6.02×10^{23} molecules of oxygen.

Isotopes

Isotopes of the same element differ in the number of neutrons and therefore in their weight. For example, the element carbon has three common isotopes:

$$^{12}_{6}C \qquad\qquad ^{13}_{6}C \qquad\qquad ^{14}_{6}C*$$
*radioactive

Carbon 12 has six neutrons, carbon 13 has seven neutrons, and carbon 14 has eight neutrons. (Note that the mass number minus the atomic number equals the number of neutrons.) Unlike the other two isotopes of carbon, carbon 14 is unstable and breaks down over time. As carbon 14 decays, it releases various types of energy in the form of rays and subatomic particles, and therefore it is a **radioactive isotope.** The radiation given off by radioactive isotopes can be detected in various ways. You may be familiar with the use of a Geiger counter to detect radiation.

Low Levels of Radiation

The importance of chemistry to biology and medicine is nowhere more evident than in the many uses of radioactive isotopes. A radioactive isotope behaves the same as do the stable isotopes of an element. This means that you can put a small amount of radioactive isotope in a sample, and it becomes a **tracer** by which to detect molecular changes (see the Medical Focus on page 16 for a description of the uses of radioactive tracers).

High Levels of Radiation

Radioactive substances in the environment can harm cells, damage DNA, and cause cancer. The release of radioactive particles following a nuclear power plant accident can have far-reaching and long-lasting effects on human health. The harmful effects of radiation can also be put to good use, however. Radiation from radioactive isotopes has been used for many years to sterilize medical and dental products. Radiation has even been used to sterilize the U.S. mail to free it of possible pathogens, such as anthrax spores.

The ability of radiation to kill cells is often applied to cancer cells. Radioisotopes can be introduced into the body in a way that allows radiation to destroy only the cancerous cells, with little risk to the rest of the body.

Molecules and Compounds

Atoms often bond with each other to form a chemical unit called a **molecule.** A molecule can contain atoms of the same kind, as when an oxygen atom joins with another oxygen atom to form oxygen gas. Or the atoms can be different, as when an oxygen atom joins with two hydrogen atoms to form water. When the atoms are different, a **compound** results.

Two types of bonds join atoms: the **ionic bond** and the **covalent bond.** Ionic bonds are created by electrical attraction between **ions**—electrically charged atoms. Covalent bonds are created by the sharing of electrons between atoms.

Ionic Bonds

Recall that atoms with more than one shell are most stable when the outer shell contains eight electrons. Sometimes during a reaction, atoms give up or take on electron(s) in order to achieve a stable outer shell.

Figure 2.2 depicts a reaction between a sodium (Na) atom and a chlorine (Cl) atom. Sodium, with one electron in the outer shell, reacts with a single chlorine atom. Why? Because once the reaction is finished and sodium loses one electron to chlorine, its outer shell will have eight electrons. Similarly, a chlorine atom, which has seven electrons already, needs only to acquire one more electron to have a stable outer shell.

Ions are particles that carry either a positive ($+$) or a negative ($-$) charge. When the reaction between sodium and chlorine is finished, the sodium ion carries a positive charge because it now has 11 positively charged protons and only 10 negatively charged electrons (11^+, 10^-; net charge: 1^+). Meanwhile, the negatively charged chloride ion has 17 protons and 18 electrons (17^+, 18^-; net charge: 1^-). The attraction between oppositely charged sodium ions and chloride ions forms an **ionic bond.** The resulting compound, sodium chloride, is table **salt,** which we use to improve the taste of foods. Salts characteristically form an **ionic lattice** that *dissociates* (separates into ions) after dissolving in water (Fig. 2.2b). When dissolved in water, salts are referred to as *electrolytes.*

In contrast to sodium, why would calcium, with two electrons in the outer shell, react with two chlorine atoms? Because whereas calcium needs to lose two electrons, each chlorine, with seven electrons already, requires only one more electron to have a stable outer shell. The resulting salt ($CaCl_2$) is called calcium chloride.

The balance of various ions in the body is important to our health. Too much sodium in the blood can contribute to **hypertension** (high blood pressure); not enough calcium leads to **rickets** (a bowing of the legs) in children; too much or too little potassium results in **arrhythmia** (heartbeat irregularities). Bicarbonate, hydrogen, and hydroxide ions are all involved in maintaining the acid-base balance of the body (see pages 25–26).

Covalent Bonds

In a covalent bond, atoms share electrons instead of losing or gaining them. The overlapping outermost shells in Figure 2.3 indicate that the atoms are sharing electrons. Just as two hands participate in a handshake, each atom contributes one electron to the pair that is shared. These electrons spend part of their time in the outer shell of each atom; therefore, they are counted as belonging to both bonded atoms.

Covalent bonds can be represented in a number of ways. In contrast to the diagrams in Figure 2.3, structural formulas use straight lines to show the covalent bonds between the atoms. Each line represents a pair of shared electrons. A molecular formula describes the molecule's composition, but

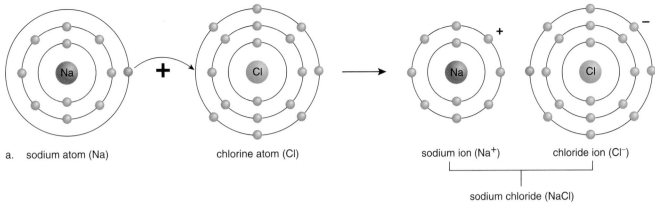

a. sodium atom (Na)　　　chlorine atom (Cl)　　　sodium ion (Na⁺)　　　chloride ion (Cl⁻)

sodium chloride (NaCl)

Figure 2.2 Ionic reaction. a. During the formation of sodium chloride, an electron is transferred from the sodium atom to the chlorine atom. At the completion of the reaction, each atom has eight electrons in the outer shell, but each also carries a charge as shown. **b.** In a sodium chloride crystal, bonding between ions creates a three-dimensional lattice in which each Na^+ ion is surrounded by six Cl^- ions, and each Cl^- is surrounded by six Na^+.

b. 1 mm

Na^+

Cl^-

molecular formulas indicate only the number of each type of atom making up a molecule. A comparison follows:

Structural formula: Cl—Cl
Molecular formula: Cl_2

Double and Triple Bonds In addition to a single bond, in which atoms share only a pair of electrons, a double or a triple bond can form. In a double bond, atoms share two pairs of electrons, and in a triple bond, atoms share three pairs of electrons between them. For example, in Figure 2.3, each nitrogen atom (N) requires three electrons to achieve a total of eight electrons in the outer shell. Notice that six electrons are placed in the outer overlapping shells in the diagram and that three straight lines are in the structural formula for nitrogen gas (N_2).

What would be the structural and molecular formulas for carbon dioxide? Carbon, with four electrons in the outer shell, requires four more electrons to complete its outer shell. Each oxygen, with six electrons in the outer shell, needs only two electrons to complete its outer shell. Therefore, carbon shares two pairs of electrons with each oxygen atom, and the formulas are as follows:

Structural formula: O=C=O
Molecular formula: CO_2

Polar and Non-Polar Covalent Bonds

In atoms that form a covalent molecule, electrons are shared in single, double, or triple pairs. Custody of the shared electrons

oxygen　　2 hydrogen　　　　water

nitrogen

Figure 2.3 Covalent reactions. After a covalent reaction, each atom will have filled its outer shell by sharing electrons. To determine this, it is necessary to count the shared electrons as belonging to both bonded atoms. Oxygen and nitrogen are most stable with eight electrons in the outer shell. Hydrogen is most stable with two electrons in the outer shell.

is not always equal. One of the elements often keeps the electrons for a longer period of time. Such elements are called **electronegative** (think "electron-grabbers"). When these electronegative elements such as oxygen are teamed with smaller, weaker elements like hydrogen, the resulting compound (water, H_2O) forms a **polar covalent bond.** Because the electrons spend most of their time with the electronegative element (oxygen), this element acquires a partial negative charge. The weaker hydrogen atoms gain a partial positive charge because they surrender their electrons for a longer period of time.

What if the atoms in the molecule share their electrons equally, so that each atom keeps the electrons for the same amount of time? This arrangement results in a **non-polar covalent bond.** Molecules linked by this type of bond do not have an electrical charge. As you'll see, the bonding properties of covalent molecules determine that molecule's characteristic role in living organisms.

Content **CHECK-UP!**

1. One particular isotope of the element potassium has the following symbol: ^{39}K. The atomic number (number of protons) for potassium is 19. How many neutrons are contained in this isotope?

2. Fluorine gas contains two atoms of fluorine that share two electrons. Draw the structural and molecular formulas for fluorine gas.

Answers in Appendix C, page 461.

Begin Thinking Clinically

? **The world's worst nuclear accident, at Chernobyl, Russia, in 1986, released radioactive strontium into the air. Strontium is chemically similar to calcium. What body tissue might be damaged by radioactive strontium?**

Answer and discussion in Appendix C, page 461.

2.2 Water, Acids, and Bases

Water is the most abundant molecule in living organisms, usually making up about 60–70% of the total body weight. Even so, water is an **inorganic molecule** because it does not contain carbon atoms. Carbon atoms are common to **organic molecules.**

As mentioned previously, in water the electrons spend more time circling the larger oxygen (O) atom than the smaller hydrogen (H) atoms. This imparts a slight negative charge (symbolized as δ^-) to the oxygen and a slight positive charge (symbolized as δ^+) to the hydrogen atoms. Therefore, water is a **polar molecule** with negative and positive ends:

The diagram on the left shows the structural formula of water, and the one on the right is called a space-filling model.

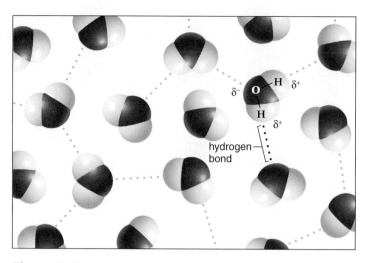

Figure 2.4 **Hydrogen bonding between water molecules.** The polarity of the water molecules causes hydrogen bonds (dotted lines) to form between the molecules.

Hydrogen Bonds

A **hydrogen bond** occurs whenever a covalently bonded hydrogen has a slight positive charge, and is attracted to a slight negatively charged atom nearby. A hydrogen bond is represented by a dotted line because it is relatively weak and can be broken rather easily. In Figure 2.4, you can see that each hydrogen atom, being slightly positive, bonds to the slightly negative oxygen atom of another water molecule nearby.

Properties of Water

Polarity and hydrogen bonding cause water to have many properties beneficial to life, including the three to be mentioned here.

1. Water is a **solvent** for polar (charged) molecules and thereby facilitates chemical reactions both outside and within our bodies.

When ions and molecules disperse in water, they move about and collide, allowing reactions to occur. Therefore, water is a solvent that facilitates chemical reactions. For example, when a salt such as sodium chloride (NaCl) is put into water, the negative ends of the water molecules are attracted to the sodium ions, and the positive ends of the water molecules are attracted to the chloride ions. This causes the sodium ions and the chloride ions to dissolve and to separate in water:

The salt NaCl dissolves in water.

Ions and molecules that are water-soluble are said to be **hydrophilic** ("water loving"). Nonionized and nonpolar molecules that do not interact with water are said to be **hydrophobic** ("water fearing").

2. Water molecules are both cohesive and adhesive.

Water molecules cling together because of hydrogen bonding, a property called **cohesion.** Water can also cling to other surfaces by hydrogen bonding, and this behavior is **adhesion.** Thus, water flows freely and water-based solutions fill vessels, such as blood vessels. These characteristics of water make it an excellent transport medium. Within our bodies, the blood that fills our arteries and veins is 92% water. Blood transports oxygen and nutrients to the cells and removes wastes such as carbon dioxide from the cells.

3. Water has a high **specific heat capacity** and a high **heat of vaporization.**

Specific heat capacity is the amount of heat energy needed to change an object's temperature by exactly 1°C. Water has the highest specific heat of almost any substance in nature. Thus, water can absorb a tremendous amount of heat energy without changing its temperature. This is due to the hydrogen bonds that link water molecules together. Maintaining a fairly constant body temperature is possible, in part, because the body is roughly 60% water. High water content helps to insulate the body from extreme heat and cold in the environment. Hydrogen bonding also accounts for water's high heat of vaporization—the amount of energy needed to turn water into steam. As water evaporates during sweating, the body can be cooled rapidly and effectively.

Acids and Bases

When water molecules dissociate (break up), they release an equal number of hydrogen ions (H^+) and hydroxide ions (OH^-):

$$H\text{-}O\text{-}H \rightleftharpoons H^+ + OH^-$$

water   hydrogen ion   hydroxide ion

Only a few water molecules at a time dissociate, and the actual number of H^+ and OH^- is very small (1×10^{-7} moles/liter).

Acids are substances that dissociate in water, releasing hydrogen ions (H^+). For example, an important inorganic acid is hydrochloric acid (HCl), which dissociates in this manner:

$$HCl \longrightarrow H^+ + Cl^-$$

Dissociation is almost complete; therefore, HCl is called a *strong acid.* If hydrochloric acid is added to a beaker of water, the number of hydrogen ions (H^+) increases greatly. Lemon juice, vinegar, tomatoes, and coffee are all acidic solutions.

Bases are substances that either take up hydrogen ions (H^+) or release hydroxide ions (OH^-). For example, an important inorganic base is sodium hydroxide (NaOH), which dissociates in this manner:

$$NaOH \longrightarrow Na^+ + OH^-$$

Dissociation is almost complete; therefore, sodium hydroxide is called a *strong base.* If sodium hydroxide is added to a beaker of water, the number of hydroxide ions increases. Milk of magnesia and ammonia are common basic solutions.

When an acid and a base are combined, the reaction produces a salt and water. For example, if hydrochloric acid and sodium hydroxide are combined:

$$HCl + NaOH \longrightarrow NaCl + H_2O$$

In the chapter introduction, alkaloid poisons are mentioned. These compounds are also salts formed by reacting an acid with a base, and are water soluble.

pH Scale

In chemistry, pH is a chemist's "shorthand." For practical purposes, it's useful to think of pH as a measure of hydrogen ion concentration. Technically, pH is defined as the negative of the base 10 logarithm of free hydrogen ion concentration. Suppose a chemist stirs up a solution containing 0.00001 moles per liter of free hydrogen ions. The solution's concentration can be written (using scientific notation) as 1×10^{-5} moles per liter. The base 10 logarithm is the exponent, -5. Using the pH system, concentration is expressed as negative times (-5), or pH 5. (Recall that a negative number multiplied by a negative number equals a positive number.)

It's important to remember that as pH *decreases*, H^+ concentration *increases*. For example, a solution of pH 2 (0.01 moles/liter) has 100 times *more* H^+ than a solution of pH 4 (0.0001 moles/liter).

The **pH scale,** which ranges from 0 to 14, is used to indicate the acidity and basicity (alkalinity) of a solution. pH 7, which is the pH of water, is neutral pH because water releases an equal number of hydrogen ions (H^+) and hydroxide ions (OH^-). Notice in Figure 2.5 that any pH above 7 is a base, with more hydroxide ions than hydrogen ions. Any pH below 7 is an acid, with more hydrogen ions than hydroxide ions. As we move toward a higher pH, each unit has 10 times the basicity of the previous unit, and as we move toward a lower pH, each unit has 10 times the acidity of the previous unit. This means that even a small change in pH represents a large change in the proportional number of hydrogen and hydroxide ions in the body.

The pH of body fluids needs to be maintained within a narrow range, or else health suffers. The pH of our blood when we are healthy is always about 7.4—that is, just slightly basic (alkaline). If the pH value drops below 7.35, the person is said to have **acidosis.** If it rises above 7.45, the condition is called **alkalosis.** The pH stability is normally possible because the body has built-in mechanisms to prevent pH changes. The respiratory and urinary systems cooperate to maintain normal pH (see Chapters 14 and 16). Additionally, **buffers** in blood and body fluids help to keep pH within its normal range.

When There's a Poison Involved

The world can be a pretty dangerous place for a child. Toddlers will examine a new object visually at first, and then it usually goes right into their mouths, often regardless of how awful the taste or smell might be. Small children also mimic the older people in their lives, watching as adults consume drinks or medication. Sometimes, curiosity leads to indiscriminate tasting of substances such as lipstick, crayons, hand lotion, shaving cream, modeling clay, and chalk (all are not toxic in small doses, but ingesting any of them still warrants a call to a poison control center). Unfortunately, accidental poisoning in children happens all too frequently when toxins are accessible.

Accidental poisoning occurs in adults as well, especially to seniors who may be confused about prescription medications. Obviously, poisoning is intentional when an older child, teen, or adult attempts suicide. When date rape or murder is the motive, poisoning is also deliberate (see **Focus on Forensics: Rape** in Chapter 17 for further discussion of date rape).

First responders must quickly take action with the **A-B-Cs** of emergency care: A → establish an Airway and make sure the person is Breathing, and maintain Circulation. (The techniques of cardiopulmonary resuscitation are described in Chapter 12). Other effects of the poison demand a quick reaction as well. Alkaloid poisons like nicotine (see the chapter introduction) often increase pulse rate and blood pressure to dangerous levels, while other toxins will cause the opposite effect. Further, many poisons cause delirium, seizures, and coma. In these cases, rapid transport to the emergency room can mean the difference between life and death.

Whenever possible, the patient's history should be recorded to give clues about the ingested substance. The victim's caregivers, relatives, friends, and co-workers can help. Partially empty containers, spilled material, and evidence of recreational drug use such as needles and syringes provide valuable information about the poison's possible identity. Once the poison is identified, local and national poison control centers can help, suggesting antidotes and therapies.*

EMS personnel shouldn't attempt to cause vomiting (unless a physician orders an *emetic,* a drug that causes vomiting). In the emergency room, clinicians often have the patient drink a solution of ground-up charcoal, called *activated charcoal,* which combines with many different types of chemicals. Stomach pumping—repeated filling and emptying of the stomach to wash out the poison, with or without charcoal solution—may be also be used. However, if the poison is a chemical like gasoline or kerosene, or a burning chemical like drain cleaner, these procedures can't be used because they might further damage the nose, throat, and esophagus. In very serious poisonings, the patient may need dialysis, a technique for cleansing blood artificially.

*The U.S. National Poison Control Center hotline telephone number is 1-800-222-1222.

Buffers are chemicals, or pairs of chemicals, that take up excess hydrogen ions (H^+) or hydroxide ions (OH^-). One of the most important buffer pairs in the body is the combination of carbonic acid (H_2CO_3) and bicarbonate ions (HCO_3^-), which helps to keep the pH of the blood within normal range. Carbonic acid is a weak acid—so called because not all H^+ separates from the molecules. Carbonic acid can separate and re-form in the following manner:

H_2CO_3	dissociates ⇌ re-forms	H^+	+	HCO_3^-
carbonic acid		hydrogen ion		bicarbonate ion

Note that the arrows shown in this chemical equation go forward and backwards. This indicates that the reaction is reversible. When hydrogen ions are added to blood (let's say, because you drink a can of pop) the following reaction occurs:

$$H^+ + HCO_3^- \longrightarrow H_2CO_3$$

When hydroxide ions (OH^-) are added to blood, this reaction occurs:

$$OH^- + H_2CO_3 \longrightarrow HCO_3^- + H_2O$$

In this way, the carbonic acid–bicarbonate buffer pair can temporarily prevent significant changes in the pH of blood.

Electrolytes

As we have seen, salts, acids, and bases are molecules that dissociate; that is, they ionize in water. For example, when a salt such as sodium chloride is put in water, the Na^+ ion separates from the Cl^- ion.

Substances that release ions when put into water are called **electrolytes** because the ions can conduct an electrical current. The electrolyte balance in the blood and body tissues is important for good health because it affects the functioning of vital organs such as the heart and the brain.

Figure 2.5 **The pH scale.** The proportionate amount of hydrogen ions to hydroxide ions is indicated by the diagonal line. Any solution with a pH above 7 is basic, while any solution with a pH below 7 is acidic.

2.3 Molecules of Life

Four categories of molecules called carbohydrates, lipids, proteins, and nucleic acids are unique to cells. They are called *macromolecules*, or *polymers*, because each is composed of many smaller subunits, or *monomers*.

Category	Polymer	Monomer(s) (subunits)
Carbohydrates	Polysaccharide	Monosaccharide
Lipids	Fat	Glycerol and fatty acids
Proteins	Polypeptide	Amino acid
Nucleic acid	DNA, RNA	Nucleotide

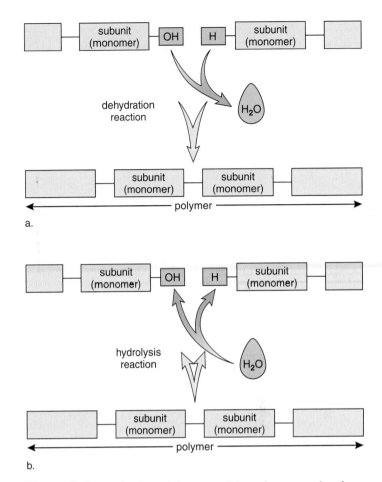

Figure 2.6 **Synthesis and decomposition of macromolecules. a.** In cells, synthesis often occurs when subunits bond following a dehydration reaction (removal of H_2O). **b.** Decomposition occurs when the subunits in a macromolecule separate after a hydrolysis reaction (addition of H_2O).

During synthesis of macromolecules, the cell uses a **dehydration reaction,** so called because an —OH (hydroxyl group) and an —H (hydrogen atom)—the equivalent of a water molecule—are removed as the molecule forms (Fig. 2.6a). The result is reminiscent of a train whose length is determined by how many boxcars are hitched together. To break up macromolecules, the cell reverses dehydration, using a **hydrolysis reaction.** In hydrolysis ("splitting with water"), the components of water are added (Fig. 2.6b).

2.4 Carbohydrates

Carbohydrates are organic molecules that always contain carbon (C), hydrogen (H), and oxygen (O) atoms. Carbohydrate molecules are characterized by the presence of the atomic grouping H—C—OH, in which the ratio of hydrogen atoms (H) to oxygen atoms (O) is approximately 2:1. Because this ratio is the same as the ratio in water, the name

MEDICAL FOCUS

Nutrition Labels

No, it's not quite as interesting as reading your morning newspaper, but tucked into the corner of your box of *Aqua Puffs* (and all other packaged food) is a piece of important information. Reading nutrition labels, like the one shown in Figure 2A, should become habit for anyone interested in health and wellness. You'll find the nutritional breakdown for a single serving (in this example, a serving is 1¼ cup, or 57 grams) on these labels. Labels also tell us what share of daily nutrients and Calories is found in that serving, based on a diet of 2,000 Calories. Calories are a measurement of food energy. One serving of the cereal provides 220 Calories.

Carbohydrates

Carbohydrates are the most readily available source of energy, and the best source for brain and nerve tissue. Sugar molecules are *simple* carbohydrates. *Complex* carbohydrates found in breads, cereals, vegetables, and fruits have the highest content of dietary fiber. Fiber is found in two forms, and each has distinct health benefits. *Soluble fiber* combines with cholesterol, preventing it from being absorbed into the body. *Insoluble fiber* has a laxative effect. Current recommendations include 25–30 grams of fiber daily. *Aqua Puffs* has a fairly high fiber content at 6 grams, but it's awfully sweet: 11 grams of sugar—empty calories!

Fats

Dietary fat makes food more appealing and tasty, but it's also essential for health. Fat is needed as an energy source, for vitamin absorption, and for manufacturing cell lipids and steroids.

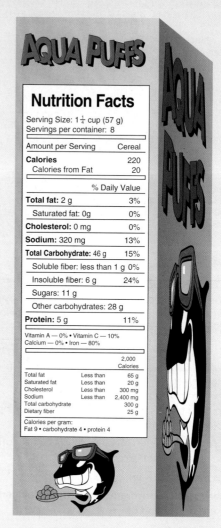

Figure 2A Nutrition label on the side panel of a cereal box.

"hydrates of carbon" seems appropriate. **Carbohydrates** first and foremost function for quick, short-term cellular energy in all organisms, including humans. The role of carbohydrates and other nutrients in maintaining good health is discussed in Medical Focus, above.

Simple Carbohydrates

If the number of carbon atoms in a carbohydrate is low (between three and seven), it is called a simple sugar, or **monosaccharide**. The designation **pentose** means a 5-carbon sugar, and the designation **hexose** means a 6-carbon sugar.

Glucose, the hexose our bodies use as an immediate source of energy, can be written in any one of these ways:

However, any excess nutrient energy is stored as fat, which could ultimately lead to obesity. Further, nutrition research continues to prove the danger of excess saturated fat, such as butter and animal fat. High levels have been implicated in cancer of the colon, pancreas, ovary, prostate, and breast. Saturated fat and cholesterol contribute to atherosclerosis, the major cause of hypertension (high blood pressure), heart attack, and stroke in the Western world (atherosclerosis is described on page 254). Better choices for fat intake are monosaturated fats like canola and olive oils. Regardless of the type of fat, a 2,000 Calorie-per-day diet should have no more than 65 grams (585 Calories) of fat. *Aqua Puffs* is a healthy option, with only 2 fat grams and zero cholesterol and saturated fat.

Proteins

A serving of *Aqua Puffs* provides 5 grams of protein. In a normal, healthy adult, a daily serving of protein is 45 grams—about the same size as a deck of playing cards. In adults, protein is typically needed only for tissue growth and repair. Proteins from animal sources, though a staple of the Western diet, may be high in saturated fat. It's better to rely on lower-fat protein from plant origins (e.g., whole-grain cereals like *Aqua Puffs*, dark breads, rice, and legumes such as beans). Rapid growth requires a diet with a higher percentage of protein, so children and women who are pregnant or nursing should increase protein intake.

Vitamins and Minerals

Vitamins are organic molecules required in small amounts for good health. Antioxidant vitamins (A, C, E and beta-carotene)

can neutralize free radicals, which are unstable molecules generated during cellular metabolism. Free radicals damage cell components, and are linked to cancer and atherosclerosis. Although *Aqua Puffs* lacks vitamins A and E, it does contain some vitamin C. Calcium, iron, and sodium are important minerals to consider. Calcium is required for bone manufacture, blood clotting, and proper nerve function. Iron is needed for hemoglobin, the oxygen transport protein found in red blood cells. However, sodium consumption should be limited—excess sodium can worsen hypertension. Recent research has shown that it can also cause kidney damage.

Bottom line: How nutritious is that bowl of *Aqua Puffs*? It's a low-fat, high-fiber cereal with no cholesterol or saturated fat. It provides 80% of daily dietary iron and 10% of vitamin C. Add a cup of low-fat milk to supply calcium and additional protein, and the cereal will be a decent, quick breakfast. But that sugar! It's bound to be tasty for someone with a sweet tooth, but simple sugars add calories. Further, you'd probably still need a vitamin supplement. A breakfast of fresh fruit, along with a dairy product for protein, would be a better way to match the cereal's nutrients without the sugar. Moreover, if you eat the recommended 2 cups of fruit and 3 cups of vegetables daily, you'll be getting those same vitamins, along with other valuable compounds you can't get from a supplement.

Current nutritional guidelines from the United States Department of Agriculture can be found in Chapter 15.

Other common hexoses are fructose, found in fruits, and galactose, a constituent of milk. A **disaccharide** (*di*, two; *saccharide*, sugar) is made by joining only two monosaccharides together by a dehydration reaction (see Fig. 2.6*a*). **Maltose** is a disaccharide that contains two glucose molecules:

maltose

When glucose and fructose join, the disaccharide **sucrose** forms. Sucrose—ordinary table sugar—is derived from sugar cane or sugar beets. **Lactose,** or milk sugar, is formed from glucose and galactose. (Individuals who are *lactose intolerant* cannot digest this sugar. If they consume dairy products, they experience intestinal cramping and diarrhea.)

Complex Carbohydrates (Polysaccharides)

Macromolecules such as starch, glycogen, and cellulose are **polysaccharides** that contain many glucose units. Although polysaccharides can contain other sugars, we will study the ones that use glucose.

Figure 2.7 Starch structure and function. Starch has straight chains of glucose molecules. Some chains are also branched, as indicated. The electron micrograph shows starch granules in potato cells. Starch is the storage form of glucose in plants.

Figure 2.8 Glycogen structure and function. Glycogen is more branched than starch. The electron micrograph shows glycogen granules in liver cells. Glycogen is the storage form of glucose in humans.

Starch and Glycogen

Starch and **glycogen** are ready storage forms of glucose in plants and animals, respectively. Some of the macromolecules in starch are long chains of up to 4,000 glucose units. Starch has fewer side branches, or chains of glucose that branch off from the main chain, than does glycogen, as shown in Figures 2.7 and 2.8. Flour, usually acquired by grinding wheat and used for baking, is high in starch, and so are potatoes.

After we eat starchy foods such as potatoes, bread, and pasta, glucose enters the bloodstream and the liver stores excess glucose as glycogen. In between meals, the liver releases glucose so that the blood glucose concentration is always about 0.1%. Maintaining a fairly constant blood glucose concentration is important because the brain and nerve tissues work best when they use glucose as their energy source.

Cellulose

The polysaccharide **cellulose** is found in plant cell walls. In cellulose, the glucose units are joined by a slightly different type of linkage from that in starch or glycogen. Although this might seem to be a technicality, actually it is important because humans are unable to digest foods containing this type of linkage. Cellulose largely passes through our digestive tract as fiber, or roughage. It is believed that fiber in the diet is necessary to good health, and some researchers have suggested it may even help prevent colon cancer.

2.5 Lipids

Lipids contain more energy per gram than other biological molecules, and some function as long-term energy storage molecules in organisms. Other lipids form cell membranes that enclose individual cells, as well as membranes that surround the organelles found inside cells. **Steroids** are a large class of lipids that includes, among other molecules, the sex hormones.

Lipids are diverse in structure and function, but they have a common characteristic: They do not dissolve in water. Their low solubility in water is due to an absence of polar groups. They contain little oxygen and consist mostly of carbon and hydrogen atoms.

Fats and Oils

The most familiar lipids are those found in fats and oils. **Fats,** which are usually of animal origin (e.g., lard and butter), are solid at room temperature. **Oils,** which are usually of plant origin (e.g., corn oil and soybean oil), are liquid at room temperature. Fat has several functions in the body: It is used for long-term energy storage, it insulates against heat loss, and it forms a protective cushion around major organs.

Fats and oils form when one **glycerol** molecule reacts with three fatty acid molecules (Fig. 2.9). A fat is sometimes called

Figure 2.9 Synthesis and degradation of a fat molecule. Fatty acids can be saturated (no double bonds between carbon atoms) or unsaturated (have double bonds, colored yellow, between carbon atoms). When a fat molecule forms, three fatty acids combine with glycerol, and three water molecules are produced.

a **triglyceride** because of its three-part structure, or a **neutral fat** because the molecule is nonpolar and carries no charge.

Emulsification

Emulsifiers can cause fats to mix with water. They contain molecules with a nonpolar end and a polar end. The molecules position themselves about an oil droplet so that their nonpolar ends project. Now the droplet disperses in water, which means that **emulsification** has occurred.

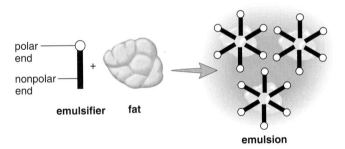

Emulsification takes place when dirty clothes are washed with soaps or detergents. Also, prior to the digestion of fatty foods, fats are emulsified by bile. The liver manufactures bile, and the gallbladder stores it for emulsifying fats prior to the digestive process.

Saturated and Unsaturated Fatty Acids

A **fatty acid** is a carbon–hydrogen chain that ends with the acidic group —COOH (Fig. 2.9). Most of the fatty acids in cells contain 16 or 18 carbon atoms per molecule, although smaller ones with fewer carbons are also known.

Fatty acids are either saturated or unsaturated. **Saturated fatty acids** have only single covalent bonds because the carbon chain is saturated, so to speak, with all the hydrogens it can hold. Saturated fatty acids account for the solid nature at room

temperature of fats such as lard and butter. **Unsaturated fatty acids** have double bonds between carbon atoms wherever fewer than two hydrogens are bonded to a carbon atom. Unsaturated fatty acids account for the liquid nature of vegetable oils at room temperature. Hydrogenation of vegetable oils can convert them to margarine and products such as Crisco.

Phospholipids

Phospholipids, as their name implies, contain a phosphate group (Fig. 2.10). Essentially, they are constructed like fats, except that in place of the third fatty acid, there is a phosphate group or a grouping that contains both phosphate and nitrogen. Phospholipid molecules are not electrically neutral, as are fats, because the phosphate and nitrogen-containing groups are ionized. They form the so-called hydrophilic head of the molecule, while the rest of the molecule becomes the hydrophobic tails. Phospholipids are the backbone of cellular membranes; they spontaneously form a bilayer in which the hydrophilic heads face outward toward watery solutions and the tails form the hydrophobic interior.

Steroids

Steroids are lipids that have an entirely different structure from those of fats. Steroid molecules have a backbone of four fused carbon rings. Each one differs primarily by the side-chain molecules, called **functional groups,** attached to the rings.

Cholesterol is a component of an animal cell's outer membrane and is the precursor of several other steroids, such as the sex hormones estrogen and testosterone. The male sex hormone, testosterone, is formed primarily in the testes, and the female sex hormone, estrogen, is formed primarily in the ovaries. Testosterone and estrogen differ only by the functional groups attached to the same carbon backbone, yet they have a profound effect on the body and on

a. Phospholipid structure

Figure 2.10 Phospholipid structure and function.
a. Phospholipids are structured like fats, but one fatty acid is replaced by a polar phosphate group. **b.** Therefore, the head is polar while the tails are nonpolar. **c.** This causes the molecule to arrange itself as shown when exposed to water.

Figure 2.11 Steroids. All steroids have four rings, but they differ by attached groups. The effects of **(a)** testosterone and **(b)** estrogen on the body largely depend on the difference in the attached groups shown in red.

our sexuality (Fig. 2.11a,b). Testosterone is a steroid that causes males to have greater muscle strength than females. Taking synthetic testosterone for this purpose, however, is dangerous to your health, as will be discussed in Chapter 10.

We know that a diet high in saturated fats and cholesterol can cause fatty material to accumulate inside the lining of blood vessels, thereby reducing blood flow. As discussed in the Medical Focus on page 28, nutrition labels are now required to list the calories from fat per serving and the percent daily value from saturated fat and cholesterol.

2.6 Proteins

Proteins perform wide range of functions, including the following:

- Collagen (found in connective tissue and skin) and keratin (found in skin, hair, and nails) are fibrous structural proteins.
- Many hormones, which are messengers that influence cellular metabolism, are proteins.

- The proteins actin and myosin account for the movement of cells and the ability of our muscles to contract.
- Some proteins transport molecules in the blood; for example, hemoglobin is a complex protein in our blood that transports oxygen.
- Antibodies in blood and other body fluids are proteins that combine with pathogens or their toxins.
- Enzymes are globular (round) proteins that speed chemical reactions.

Structure of Proteins

Proteins are macromolecules composed of amino acid subunits. An **amino acid** has a central carbon atom bonded to a hydrogen atom and three groups. The name of the molecule is appropriate because one of these groups is an amino group and another is an acidic group. The third group is called an *R* group because it is the *remainder* of the molecule (Fig. 2.12a).

Amino acids differ from one another by the R group, which varies in structure from a single hydrogen atom to a complicated ring. When two amino acids join, a **dipeptide** results. **Polypeptides** contain three or more amino acids (Fig. 2.12b) Proteins are large polypeptides. Twenty different amino acids compose all human polypeptides. The sequence of amino acids in a polypeptide is called its **primary structure** (Fig. 2.12b).

The bond between amino acids is termed a **peptide bond.** The atoms of a peptide bond share electrons unevenly, making hydrogen bonding possible in a polypeptide. Thus, polypeptide molecules are polar. Due to hydrogen bonding, the polypeptide often twists to form a coil or folds into a fan shape.

amino group   acid group

$$H_2N - \underset{\underset{R}{|}}{\overset{\overset{H}{|}}{C}} - COOH$$

a. Amino acid

peptide bond   amino acids

b. Primary structure

hydrogen bond

c. Secondary structure

iron

heme group

helical shape of the polypeptide molecule

d. Tertiary structure

e. Quaternary structure

Figure 2.12 **Levels of polypeptide structure. a.** Amino acids are the subunits of polypeptides. Note that an amino acid contains nitrogen. **b.** Polypeptides differ by the sequence of their amino acids, which are joined by peptide bonds. **c.** A polypeptide often twists to become a coil due to hydrogen bonding between members of the peptide bonds. **d.** The tertiary structure of polypeptide structure is due to various types of bonding between the *R* groups of the amino acids. **e.** The grouping of two or more small proteins creates the quaternary structure of a large protein molecule.

Coiling or folding creates the **secondary structure** of the protein (Fig. 2.12c) Finally, the coil bends and twists into a particular shape because of bonding between R groups. Hydrogen, ionic, and covalent bonding all occur in polypeptides, creating the **tertiary structure** (Fig. 2.12d) Hydrophobic portions of a polypeptide are found inside the molecule, while the hydrophilic portions are outside where they can contact water.

Some proteins have just one polypeptide, while others have more than one polypeptide, each with its own primary, secondary, and tertiary structures. If a protein has more than one polypeptide, arrangement of individual polypeptides gives a protein its **quaternary structure.**

The final three-dimensional shape of a protein is very important to its function. When proteins are exposed to extremes in heat and pH, they undergo an irreversible change in shape called **denaturation.** For example, addition of acid to milk (as when milk goes "sour") causes curdling. Heating causes egg white protein, called *albumin*, to coagulate (form clumps). Denaturation occurs because the normal bonding between the R groups has been disturbed. Once a protein loses its normal shape, it is no longer able to function normally. Researchers hypothesize that an alteration in protein organization may be the cause of Alzheimer disease (see the Medical Focus in Chapter 8) and variant Creutzfeldt-Jacob disease (see the Medical Focus on page 34).

Enzymatic Reactions

Metabolism is the sum of all of the chemical reactions that occur in a cell. Most cellular reactions cannot happen unless an **enzyme** is present. Enzymes are protein *catalysts*—molecules that enable a particular metabolic reaction to occur at the

Prions: Malicious Proteins?

Infectious diseases are known to be caused by minute organisms that successfully dodge the body's defense mechanisms. These invaders include parasites such as liver flukes and parasitic worms; smaller, one-celled parasites like disease-causing strains of *Amoeba*; and the smallest unicellular organisms, the bacteria. Viruses are disease-causing particles that are even smaller than bacteria. All of these disease agents share a common trait: All contain the nucleic acids DNA or RNA, or both. Nucleic acids are large molecules that comprise the genetic material of parasites, bacteria, and viruses. Until recently, scientists believed that infectious organisms had to contain genetic material.

However, disease outbreaks in England, Canada, and the United States have raised concerns that even smaller agents, consisting only of a protein molecule, are capable of causing disease. These proteinaceous infectious particles, called **prions** (pronounced pree-ahns), were probably responsible for an outbreak of variant Creutzfeld-Jakob (croits-feld yay-kob) disease, or "mad-cow" disease. Twelve British people contracted the illness between 1994 and 1996 after eating meat from infected cattle, and all died. Upon autopsy, scientists noted that holes pockmarked the brains of all human and animal victims, giving the tissue a spongy, Swiss-cheese appearance. Prions are now generally accepted to be the cause for a set of diseases called transmissible spongiform encephalopathies (TSEs), diseases that cause progressive death of brain tissue. *Scrapie* (a disease of sheep and goats) and *kuru* in humans are also TSE diseases.

How can something as small as a protein molecule cause disease? Scientists theorize that prions cause normal brain cell proteins to change into prion proteins. Accumulating prions kill brain cells and spread throughout brain tissue. Symptoms indicate brain destruction: "Mad" cows wobble, stagger, refuse food, and ultimately starve. Sheep with scrapie will frantically rub wool off their bodies. Infected humans lose memory, muscle control, and finally stop breathing. At this time, there is no treatment—all TSEs are fatal.

Preventing the spread of prions requires improved methods of raising beef cattle. Researchers believe that prions were spread from sheep to cattle, and later among infected cattle, when the animals ate feed that contained waste tissue from slaughterhouses. The waste tissue, which consists of brains and spinal cords, had been ground up and mixed with the animal feed. (It is interesting to note that kuru in humans was spread by cannibalism among the Fore' tribe of New Guinea, who honored dead relatives by eating their brains!) The addition of waste tissue to animal feed has been outlawed in the United States, Canada, and Europe, and infected animals have been destroyed. However, many puzzling questions about prions remain, and much additional research is needed.

body's normal temperature. Atoms and molecules frequently do not react unless they are first activated in some way. *Energy of activation* is the energy needed to start a reaction. In the lab, heat is often used to supply the energy needed for a reaction. For example, sugar can be broken down to form carbon dioxide and water, but doing so in a laboratory requires tremendous heat. In the body, enzymes lower the energy of activation by forming a complex with particular molecules. Just as a mutual friend can cause particular people to interact at a crowded party, enzymes in cells bring together atoms or molecules and cause reactions. Our bodies use enzymes to convert sugar into CO_2 and water (just like in a lab setting) but the reaction can occur at body temperature.

Enzyme-Substrate Complex

In any reaction, the molecules that interact are called *reactants*, while the substances that form as a result of the reaction are the *products*. The reactants in an enzymatic reaction are its **substrate(s)**. Enzymes are often named for their substrate(s); for example, maltase is the enzyme that digests maltose. Enzymes have a specific region, called an *active site*, where the reaction occurs. An enzyme's specificity is caused by the shape and chemical composition of its active site. There, the enzyme and

its substrate(s) fit together, much like pieces of a jigsaw puzzle (Fig. 2.13). After a reaction is complete and the products are released, the enzyme is ready to catalyze its reaction again:

$$E + S \rightarrow ES \rightarrow E + P$$

(where E = enzyme, S = substrate, ES = enzyme-substrate complex, and P = product).

Many enzymes require **cofactors.** Cofactors assist an enzyme and may even accept or contribute atoms to the reaction. Some cofactors are inorganic metals, such as copper, zinc, and iron. Other cofactors are organic, nonprotein molecules called coenzymes. Vitamins are often components of coenzymes.

Types of Reactions

Certain types of chemical reactions are common to metabolism.

Synthesis Reactions During **synthesis reactions,** two or more reactants combine to form a larger and more complex product (Fig. 2.13a). The dehydration synthesis reaction we have already studied (i.e., the joining of subunits to form a macromolecule) is an example of a synthesis reaction (see Fig. 2.8). When glucose molecules join in the liver, forming glycogen, dehydration synthesis has occurred. Notice that synthesis reactions always involve bond formation and require energy.

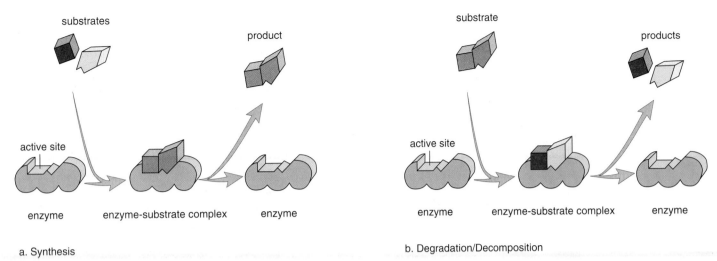

a. Synthesis

b. Degradation/Decomposition

Figure 2.13 Enzymatic action. An enzyme has an active site where the substrates come together and react. The products are released, and the enzyme is free to act again. **a.** In synthesis, the substrates join to produce a larger product. **b.** In degradation/decomposition, the substrate breaks down to smaller products.

Degradation (Decomposition) Reactions During degradation reactions (also called decomposition reactions), a larger and more complex molecule breaks down into smaller, simpler products (Fig. 2.13b). The hydrolysis reactions that break down macromolecules into their subunits are examples of degradation reactions. When protein is digested to amino acids in the stomach, a degradation reaction has occurred.

Replacement (Exchange) Reactions Replacement reactions involve both degradation decomposition and synthesis. Reactions between acids and bases are replacement reactions. For example, when hydrochloric acid (HCl) is reacted with sodium hydroxide (NaOH), the chloride ion (Cl^-) trades places with the hydroxide ion (OH^-). Table salt (NaCl) and water (H_2O) are formed as a result.

Content CHECK-UP!

5. The sequence of amino acids found in a protein is that protein's _____ structure.

   a. primary   c. tertiary

   b. secondary   d. quaternary

6. Why is it possible for humans to digest starch, but not cellulose?

   a. Starch is made of glucose molecules but cellulose isn't.

   b. Starch comes from plants, but cellulose doesn't.

   c. The glucose units found in cellulose are linked differently than those in starch.

7. What type of reaction is the digestion of glucose to form carbon dioxide and water?

   a. synthesis   c. replacement

   b. degradation/decomposition

Answers in Appendix C, page 461.

2.7 Nucleic Acids

Nucleic acids are huge macromolecules composed of nucleotides. Every **nucleotide** is a molecular complex of three types of subunit molecules—a phosphate (phosphoric acid), a pentose sugar, and a nitrogen-containing base:

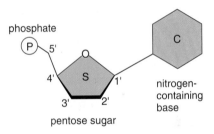

Nucleic acids contain hereditary information that determines which proteins a cell will have. Two classes of nucleic acids are in cells: **DNA (deoxyribonucleic acid)** and **RNA (ribonucleic acid)**. DNA makes up the hereditary units called **genes**. Genes pass on from generation to generation the instructions for replicating DNA, making RNA, and joining amino acids to form the proteins of a cell. RNA is an intermediary in the process of protein synthesis, conveying information from DNA regarding the amino acid sequence in proteins. As an analogy, think of DNA as a "blueprint"— a set of instructions for assembling an entire building. Using this example, RNA would be the directions, made from the blueprint, describing how to put beams together to make a wall.

The nucleotides in DNA contain the 5-carbon sugar deoxyribose; the nucleotides in RNA contain the sugar ribose. This difference accounts for their respective names. As indicated in Figure 2.14, there are four different types of bases in DNA: A = adenine, T = thymine, G = guanine, and C = cytosine. The base can have two rings (like adenine or guanine) or

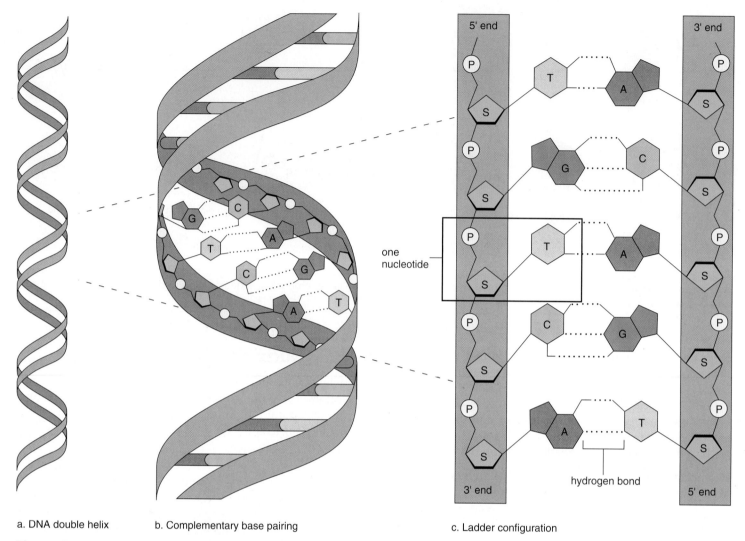

a. DNA double helix

b. Complementary base pairing

c. Ladder configuration

Figure 2.14 Overview of DNA structure. a. Double helix. **b.** Complementary base pairing between strands. **c.** Ladder configuration. Notice that the uprights are composed of phosphate and sugar molecules and that the rungs are complementary paired bases.

one ring (like thymine or cytosine). In RNA, the base uracil replaces the base thymine.

The bases in DNA and RNA are nitrogen-containing bases—that is, a nitrogen atom is a part of the ring. Like other bases, the presence of the nitrogen-containing base in DNA and RNA raises the pH of a solution.

The nucleotides in DNA and RNA form a linear molecule called a *strand*. A strand has a backbone made up of phosphate-sugar-phosphate-sugar, with the bases projecting to one side of the backbone. Because the nucleotides occur in a definite order, so do the bases. Any particular DNA or RNA has a definite sequence of bases, although the sequence can vary between molecules. RNA is usually single-stranded, while DNA is usually double-stranded, with the two strands twisted about each other in the form of a double helix (like a spiral staircase). The molecular differences between DNA and RNA are listed in Table 2.1.

TABLE 2.1	DNA Structure Compared to RNA Structure	
	DNA	**RNA**
Sugar	Deoxyribose	Ribose
Bases	Adenine, guanine, thymine, cytosine	Adenine, guanine, uracil, cytosine
Strands	Double-stranded	Single-stranded
Helix	Yes	No

In DNA, the two strands are held together by hydrogen bonds between the bases (see Fig. 2.14). When unwound, DNA resembles a stepladder. The sides of the ladder are made entirely of phosphate and sugar molecules, and the rungs of the ladder are made only of complementary paired bases.

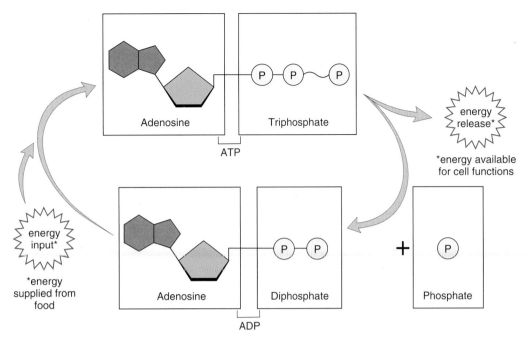

Figure 2.15 **Breakdown and formation of ATP.**

Thymine (T) always pairs with adenine (A), and guanine (G) always pairs with cytosine (C) (see Fig. 2.14). This is called **complementary base pairing.**

Complementary bases pair because they have shapes that fit together. We shall see that complementary base pairing allows DNA to replicate in a way that ensures the sequence of bases will remain the same. When RNA is produced, complementary base pairing occurs between DNA and RNA, but uracil takes the place of thymine. Then, the sequence of the bases in RNA determines the sequence of amino acids in a protein because every three bases code for a particular amino acid (see Chapter 3, pp. 56–57). The code is nearly universal and is the same in other organisms as it is in humans.

ATP (Adenosine Triphosphate)

Individual nucleotides can have metabolic functions in cells. Some nucleotides are important in energy transfer. When adenosine (adenine plus ribose) is modified by the addition of three phosphate groups, it becomes **ATP (adenosine triphosphate),** the primary energy carrier in cells.

Cells require a constant supply of ATP. To obtain it, they continually break down glucose (and other food molecules as well) and convert the energy that is released into ATP molecules. As an analogy, the energy in glucose is like a $100 bill, and the energy in ATP is like a $10 bill. The larger bill (glucose energy), when broken down, will yield ten smaller bills (ATP energy). Cells "spend" ATP energy when cellular reactions require energy. Therefore, ATP is called the *energy currency* of cells.

However, converting food energy into ATP energy is not 100% complete. To continue the comparison, imagine that a store won't accept your $100 bill because it's too large. You go to a bank and exchange your $100 bill for $10 bills. Naturally, you receive ten $10 bills from the bank—but you lose six $10 bills on the way home.

A similar situation occurs for body cells. Energy in foods must be released, but during the conversion about 60% of food energy is not converted to ATP. In the body, this energy is given off as heat. Thus, energy released from the breakdown of food fuels cellular reactions and warms the body at the same time.

Cells use ATP energy when macromolecules such as carbohydrates and proteins are synthesized. In muscle cells, ATP is used for muscle contraction. In nerve cells, it is used for the conduction of nerve signals. Likewise, all other body cells use ATP energy to fuel their various functions.

ATP is a high-energy molecule because the last two phosphate bonds are unstable and easily broken. In cells, the terminal phosphate bond is usually hydrolyzed, releasing energy for cell reactions. The terminal bond is sometimes called a high-energy bond, symbolized by a wavy line. But this terminology is misleading—the breakdown of ATP *releases* energy because the products of hydrolysis are *more* stable than ATP. Breakdown of ATP leaves one molecule of ADP (adenosine diphosphate) and a molecule of inorganic phosphate, Ⓟ (Fig. 2.15).

After ATP breaks down and the energy is used for a cellular purpose, ATP is rebuilt by the addition of Ⓟ to ADP again (Fig. 2.15). There is enough energy in one glucose molecule to build roughly 36 to 40 ATP molecules in this way. Homeostasis is only possible because cells continually produce and use ATP molecules. The use of ATP as the energy currency of cells also occurs in other organisms, ranging from bacteria to humans.

SELECTED NEW TERMS

Basic Key Terms

acid (as'id), p. 25

adhesion (ăd-hē'-zhŭn), p. 25

amino acid (uh-me'no as'id), p. 32

atomic number, p. 21

ATP (adenosine triphosphate) (uh-den'o-sēn tri-fos'fāt), p. 37

base (bās), p. 25

buffer (buf'er), p. 25

carbohydrate (kar"bo-hi'drāt), p. 28

cellulose (sĕl'yə-lōs'), p. 30

cofactors (kō'făk'tərz), p. 34

cohesion (kō-hē'-zhŭn), p. 25

complementary base pairing (kŏm'plə-mĕn'tə-rĕ bās pâr-ing), p. 37

compound (kŏm-pound'), p. 22

covalent bond (ko-va'lent bond), p. 22

degradation decomposition reactions (dĕg'rə-dā'shən dē-kŏm'pə-zish'ən rē-ăk'shə-nz), p. 35

dehydration reaction (de'hī-drā'shən rē-ăk'shənz), p. 27

denaturation (dē-nā'chər-ā'shən), p. 33

dipeptide, p. 32

disaccharide (di-sak'uh-rīd), p. 29

DNA (deoxyribonucleic acid) (de-oks'e-ri"-bo-nu-kla"ikas'id), p. 35

electrolyte (e-lek'tro-līt), p. 26

electronegative (ĭ-lĕk'trō-nĕg'-tĭv), p. 23

electrons (ĭ-lĕk'tron'z), p. 21

elements (ĕl'ə-məntz), p. 21

emulsification (ĭ-mŭl'sə-fə-cā'-shən), p. 31

enzyme (en'zĭm), p. 33

fats (fătz), p. 30

fatty acid (fat'e as'id), p. 31

functional groups (fŭngk'shə-nəl groopz), p. 31

gene (jēn), p. 35

glucose (gloo' kōs), p. 28

glycerol (glis'er-ol), p. 30

glycogen (gli'ko-jen), p. 30

heat of vaporization (hēt ŭv vā'pə-rī-zā'shən), p. 25

hexose (hĕk'sōs'), p. 28

hydrogen bond (hi'dro-jen bond), p. 24

hydrolysis reaction (hi-drol'ĭ-sis re-ak'shun), p. 27

hydrophilic (hĭ'drə-fĭl'ĭk), p. 25

hydrophobic (hĭ'drə-fō'bĭk), p. 25

inorganic molecule (in-or-gan'ik mol'e-kyūl), p. 24

ion (i'on), p. 22

ionic bond (i-on'ik bond), p. 22

ionic lattice (īŏn'-ĭk lă'-tis), p. 22

isotope (i'so-tōp), p. 22

lactose (lăk'tōs), p. 29

lipid (lip'id), p. 30

maltose (măl'tōs), p. 29

metabolism (mə-tăb'ō-lĭzm), p. 33

mole (mōl), p. 21

monosaccharide (mon"o-sak'ah-rĭd), p. 28

neutral fat (nōō'trəl făt), p. 31

neutrons (nōō'trŏn'z), p. 21

non-polar covalent bond (nŏn pō'lĕr kō'vā'lent bŏnd), p. 24

nucleic acid (nu-kla'ik as'id), p. 35

nucleotide (nōō'klēətid), p. 35

oils (oil'z), p. 30

organic molecule (or-gan'ik mol'ĕ-kyūl), p. 24

pentose (pĕn'tōs), p. 28

peptide bond (pep'tĭd bond), p. 32

pH scale (péăch' skāl), p. 25

phospholipids (fŏs'fō-lĭp'ĭdz), p. 31

polar covalent bond (pō'lĕr kō'vā'lent bŏnd), p. 24

polar molecule (pō'lər mŏl'ĭ-kyōōl), p. 24

polypeptide (pŏl-ē–pəp'tĭd), p. 32

polysaccharide (pol"e-sak'uh-rīd), p. 29

primary structure (prī'măr-ē strŭk'chŭr), p. 32

protein (pro'tēn), p. 32

protons (prō'tŏn'z), p. 21

quaternary structure (kwă'tĕr-năr-ē strŭk'chŭr), p. 33

radioactive isotope (ra"de-o-ak'tiv i'so-tōp), p. 22

replacement reactions (rĭ-plās'mənt rē-ăk'shənz), p. 35

RNA (ribonucleic acid) (ri"bo-nu-kla'ik as'id), p. 35

salt (sawlt), p. 22

saturated fatty acids (sătch'ū-rāt-əd făt'ēăs'idz), p. 31

secondary structure (sək'ŏn-dār-ēstrŭk'chŭr), p. 33

solvent (sŏl'vənt), p. 24

specific heat capacity (spĭ-sĭf'ĭk hēt kə-păs'ĭ-tē), p. 25

starch (stärch), p. 30

steroids (stĕr'oid'z), p. 30

substrate (sŭb'strāt'), p. 34

synthesis reactions (sĭn'thĭ-sĭs rē-ăk'shənz), p. 34

tertiary structure (tĕr'shē-ă-rē strŭk'char), p. 33

triglyceride (trĭ-glĭs'-ə-rīd), p. 31

unsaturated fatty acids (ŭn'sătch"ŭ-rāt-əd făt'ē ăs'ĭdz), p. 31

Clinical Key Terms

acidosis (as"ĭ-do'sis), p. 25

alkalosis (al"kuh-lo'sis), p. 25

arrhythmia (uh-rith'me-uh), p. 22

hypertension (hi"per-ten'shun), p. 22

prion (prî'-ŏn), p. 34

rickets (rik'ets), p. 22

tracer (trā'sĕr), p. 22

SUMMARY

2.1 Basic Chemistry

A. All matter is composed of elements, each made up of just one type of atom. An atom has an atomic symbol, atomic number (number of protons and, therefore, electrons when neutral), and atomic mass (number of protons and neutrons). The isotopes of some atoms are radioactive and have biological and medical applications.

B. Atoms react with one another to form molecules. Following an ionic reaction, charged ions are attracted to one another. Following a covalent reaction, atoms share electrons. Polar covalent bonding results from unequal sharing of electrons. Non-polar covalent bonds occur when electrons are shared equally.

2.2 Water, Acids, and Bases

A. In water, the electrons are shared unequally, and the result is a polar molecule. Hydrogen bonding can occur between polar molecules.

B. Water is a polar molecule and acts as a solvent; it dissolves various chemical substances and facilitates chemical reactions. Because of hydrogen bonding, water molecules are cohesive and adhesive. Water heats up and cools down slowly. This helps keep body temperature within normal limits.

C. Substances such as salts, acids, and bases that dissociate in water are called electrolytes. The electrolyte balance in the blood and body tissues is important for good health.

D. Acids have a pH less than 7, and bases have a pH greater than 7. The actions of the lungs and kidneys, and the presence of buffers, help to keep the pH of body fluids around pH 7.

2.3 Molecules of Life
A. Carbohydrates, lipids, proteins, and nucleic acids are the molecules of life.

B. A monosaccharide, such as glucose, is a subunit (monomer) for larger carbohydrates. Glycerol and fatty acids are monomers for fat. Amino acids are monomers for proteins, and nucleotides are monomers for nucleic acids.

2.4 Carbohydrates
Glucose is an immediate source of energy in cells. Glycogen stores energy in the body, starch is a dietary source of energy, and cellulose is fiber in the diet.

2.5 Lipids
Lipids include neutral fat (a long-term, energy-storage molecule that forms from glycerol and three fatty acids) and the related phospholipids, which have a charged group. Fatty acids can be saturated or unsaturated. Steroids have an entirely different structure from that of fats.

2.6 Proteins
A. Proteins, which are composed of one or more long polypeptides, have both structural and physiological functions. Polypeptides have several levels of structure. Their three-dimensional shape is necessary to their function.

B. Enzymes are proteins necessary to metabolism. The reaction occurs at the active site of an enzyme.

2.7 Nucleic Acids
A. Both DNA and RNA are polymers of nucleotides; DNA is usually double-stranded. DNA makes up the genes, and along with RNA, specifies protein synthesis.

B. ATP is the energy "currency" of cells because its breakdown supplies energy for many cellular processes.

STUDY QUESTIONS

1. Name the subatomic particles of an atom. Using the carbon atom as an example, describe their charge, atomic mass unit, and location. (p. 21)
2. What is an isotope? A radioactive isotope? Discuss the clinical uses of radioactive isotopes. (p. 22)
3. Give an example of an ionic reaction, and explain it. (p. 22)
4. Give an example of a covalent reaction, and explain it. (pp. 22–23)
5. Relate three characteristics of water to its polarity and hydrogen bonding between water molecules. (pp. 24–25)
6. What is an acid? A base? (pp. 25–26)
7. On the pH scale, which numbers indicate a basic solution? An acidic solution? Why? (p. 25)

8. What are buffers, and how do they function? (p. 26)
9. Name the four categories of macromolecules (polymers) in cells; give an example for each category, and name the subunits (monomers) of each. (p. 27)
10. Tell how macromolecules are built up and broken down. (p. 27)
11. Name some monosaccharides, disaccharides, and polysaccharides, and give the functions for each. (pp. 28–30)
12. What is a lipid? A saturated fatty acid? An unsaturated fatty acid? What is the function of fats? (p. 30)
13. Relate the structure of a phospholipid to that of a neutral fat. What is the function of a phospholipid? (p. 31)

14. Name two steroids that function as sex hormones in humans. (p. 31)
15. What are some functions of proteins? Why do proteins stop functioning if exposed to the wrong pH or high temperature? (pp. 32–33)
16. Discuss the levels of protein structure. (pp. 32–33)
17. How do enzymes function? Name three types of metabolic reactions. (pp. 33–35)
18. Discuss the structure and function of the nucleic acids DNA and RNA. (pp. 35–37)

LEARNING OBJECTIVE QUESTIONS

Fill in the blanks.

1. _____ are the smallest units of matter nondivisible by chemical means.
2. Isotopes differ by the number of _____ in the nucleus.
3. The two primary types of reactions and bonds are _____ and _____.
4. A type of weak bond, called a _____ bond, exists between water molecules.

5. Acidic solutions contain more _____ ions than basic solutions, but they have a _____ pH.
6. Glycogen is a polymer of _____, molecules that serve to give the body immediate _____.
7. A fat hydrolyzes to give one _____ molecule and three _____ molecules.

8. A polypeptide has levels of structure. The first level is the sequence of _____; the second level is very often a _____; the third level is its final _____.
9. _____ speed chemical reactions in cells.
10. Genes are composed of _____, a nucleic acid made up of _____ joined together.

MEDICAL TERMINOLOGY EXERCISE

Consult Appendix B for help in pronouncing and analyzing the meaning of the terms that follow.

1. anisotonic (an-i″so-ton′ik)
2. dehydration (de″hi-dra′shun)
3. hypokalemia (hi″po-kă-le′me-uh)
4. hypovolemia (hi″po-vo-le′me-uh)
5. nonelectrolyte (non″ē-lek′tro-lit)
6. lipometabolism (lip″o-mě-tab′o-lizm)
7. hyperlipoproteinemia (hi″per-lip″o-pro″te-in-e′me-uh)
8. hyperglycemia (hi″per-gli-se′me-uh)
9. hypoxemia (hi″pok-se′me-uh)
10. hydrostatic pressure (hi″dro-stat′-ik-presh′ur)
11. galactosemia (guh-lak-to-se′me-uh)
12. hypercalcemia (hi″per-kal-se′me-uh)
13. hyponatremia (hi″po-nuh-tre′me-uh)
14. gluconeogenesis (glu″ko-ne-o-jen′uh-sis)
15. edema (uh-de′muh)

WEBSITE LINK

Visit this text's website at http://www.mhhe.com/longenbaker7 for additional quizzes, interactive learning exercises, and other study tools.

Cell Structure and Function

CHAPTER 3

Need another reason to quit using tobacco? The fine, hairlike cilia you see on these cells from the trachea, or windpipe, are exquisitely tailored organelles with an important protective function. Sticky mucus covering the tracheal walls traps harmful pollutants like dust and mold spores before they can reach the lungs. Cilia push the mucus upward toward the throat, and you can either spit it out or swallow it. In either case, the mucus and trapped pollutants are usually harmless. Now consider this: nicotine (remember, it's an alkaloid poison!) temporarily poisons delicate cilia. A deep, hacking *smoker's cough* becomes the only way to clear mucus from the airways. You'll find tips to help you stop smoking in Chapter 14.

Chapter Outline & Learning Objectives
After you have studied this chapter, you should be able to:

3.1 Cellular Organization

Every human cell has a plasma membrane, a single nucleus, and cytoplasm. (Some exceptions to this rule exist. A mature erythrocyte, or red blood cell, ejects its nucleus as it develops. Thus, erythrocytes are *anucleate*. Cells of skeletal muscle, liver, and other tissues may have up to 50 nuclei and are *multinucleate*.) The **plasma membrane,** which surrounds the cell and keeps it intact, regulates what enters and exits a cell. The plasma membrane is a phospholipid bilayer that is said to be *semipermeable* (or *selectively* permeable) because it allows certain molecules but not others to enter the cell. Proteins present in the plasma membrane play important roles in allowing substances to enter the cell.

The **nucleus** is a large, centrally located structure that can often be seen with a light microscope. The nucleus contains the chromosomes and is the control center of the cell. It controls the metabolic functioning and structural characteristics of the cell. The **nucleolus** is a region inside the nucleus.

The **cytoplasm** is the portion of the cell between the nucleus and the plasma membrane. The matrix of cytoplasm is a gelatinous, semifluid medium that contains water and various types of molecules suspended or dissolved in the medium. The presence of proteins accounts for the semifluid nature of cytoplasm.

The cytoplasm contains various **organelles** (Table 3.1 and Fig. 3.1). Organelles are small, usually membranous structures that are best seen with an electron microscope.[1]

Each type of organelle has a specific function. For example, one type of organelle transports substances, and another type produces ATP for the cell. *Organelles* compartmentalize the cell, keeping the various cellular activities separated from one another. Just as the rooms in your house have particular pieces of furniture that serve a particular purpose, organelles have a structure that suits their function.

Cells also have a **cytoskeleton,** a network of interconnected filaments and microtubules in the cytoplasm. The name *cytoskeleton* is convenient in that it allows us to compare the cytoskeleton to our bones and muscles. Bones and muscles give us structure and produce movement. Similarly, the elements of the cytoskeleton maintain cell shape and allow the cell and its contents to move. Some cells move by using cilia and flagella, which are made up of microtubules.

The Plasma Membrane

Our cells are surrounded by an outer plasma membrane. The plasma membrane separates the inside of the cell, termed the *cytoplasm*, from the outside. Plasma membrane integrity is necessary to the life of the cell.

---

[1]Electron microscopes are high-powered instruments that are used to generate detailed photographs of cellular contents. The photographs are called electron micrographs. Scanning electron micrographs have depth while transmission electron micrographs are flat (see Fig. 3.3). Light microscopes are used to generate photomicrographs that are often simply called micrographs.

TABLE 3.1 Structures in Human Cells

Name	Composition	Function
MEMBRANOUS STRUCTURES		
Plasma membrane	Phospholipid bilayer with embedded proteins	Cell border; selective passage of molecules into and out of cell; location of cell markers, cell receptors
Nucleus	Nuclear membrane (envelope) surrounding nucleoplasm, chromatin, and nucleolus	Storage of genetic information; control center of cell; cell replication
Nucleolus	Concentrated area of chromatin, RNA, and proteins	Ribosomal formation
Ribosome	Two subunits composed of protein and RNA	Protein synthesis
Endoplasmic reticulum	Complex system of tubules, vesicles, and sacs	Synthesis and/or modification of proteins and other substances; transport by vesicle formation
Rough endoplasmic reticulum	Endoplasmic reticulum studded with ribosomes	Protein synthesis for export
Smooth endoplasmic reticulum	Endoplasmic reticulum without ribosomes	Varies: lipid and/or steroid synthesis; calcium storage
Golgi apparatus	Stacked, concentrically folded membranes	Processing, packaging, and distribution of molecules
Vacuole	Small membranous sac	Isolates substances inside cell
Vesicle	Small membranous sac	Storage and transport of substances in/out of cell
Lysosome	Vesicle containing digesting enzymes	Intracellular digestion; self-destruction of the cell
Peroxisome	Vesicle containing oxidative enzymes	Detoxifies drugs, alcohol, etc.; breaks down fatty acids
Mitochondrion	Inner membrane within outer membrane	Cellular respiration
Cytoskeleton	Microtubules, actin filaments	Shape of cell and movement of its parts
Cilia and flagella	9 + 2 pattern of microtubules	Movement by cell; movement of substances inside a tube
Centriole	9 + 0 pattern of microtubules	Formation of basal bodies for cilia and flagella; formation of spindle in cell division

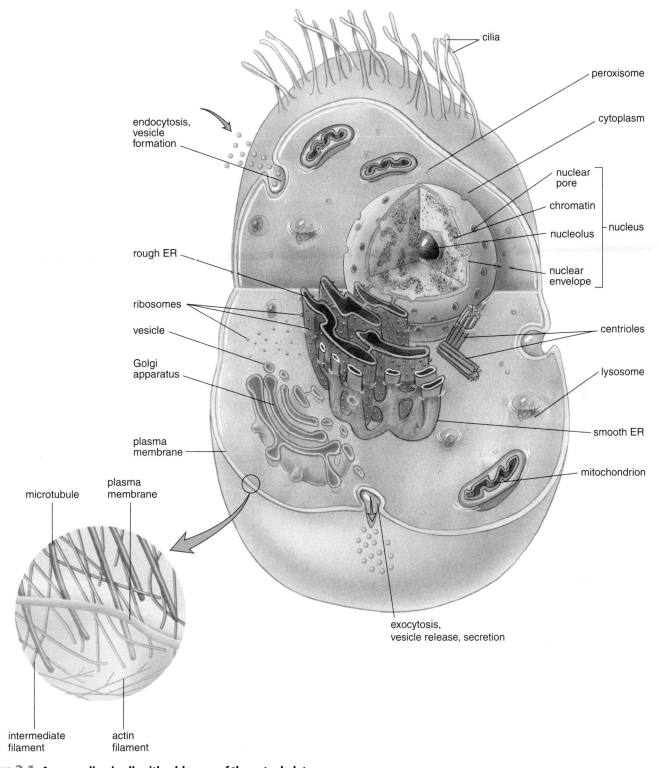

Figure 3.1 A generalized cell, with a blowup of the cytoskeleton.

The plasma membrane is a phospholipid bilayer with attached (also called *peripheral*) or embedded (also called *integral*) proteins. The phospholipid molecule has a polar head and nonpolar tails (Fig. 3.2*a*). Because the polar heads are charged, they are *hydrophilic* (water-loving) and face outward, where they are likely to encounter a watery environment.

The nonpolar tails are *hydrophobic* (water-fearing) and face inward, where there is no water. When phospholipids are placed in water, they naturally form a spherical bilayer because of the chemical properties of the heads and the tails.

At body temperature, the phospholipid bilayer is a liquid; it has the consistency of olive oil, and the proteins are able to

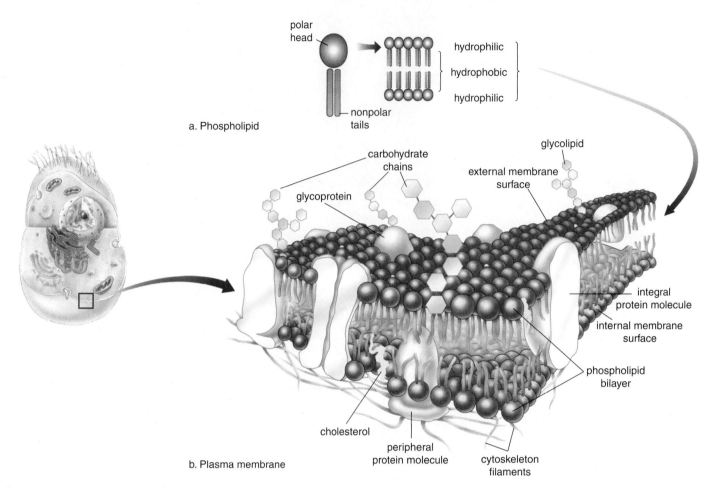

polar
head

nonpolar
tails

hydrophilic

hydrophobic

hydrophilic

a. Phospholipid

carbohydrate
chains

glycoprotein

glycolipid

external membrane
surface

integral
protein molecule

internal membrane
surface

phospholipid
bilayer

cholesterol

peripheral
protein molecule

cytoskeleton
filaments

b. Plasma membrane

Figure 3.2 **Fluid-mosaic model of the plasma membrane. a.** In the phospholipid bilayer, the polar (hydrophilic) heads project outward and the nonpolar (hydrophobic) tails project inward. **b.** Proteins are embedded in the membrane. Glycoproteins have attached carbohydrate chains, as do glycolipids.

change their positions by moving laterally. The *fluid-mosaic model,* a working description of membrane structure, suggests that the protein molecules have a changing pattern (form a mosaic) within the fluid phospholipid bilayer (Fig. 3.2*b*). Our plasma membranes also contain a substantial number of *cholesterol* molecules. These molecules stabilize the phospholipid bilayer and prevent a drastic decrease in fluidity at low temperatures.

Short chains of sugars are attached to the outer surfaces of some proteins and lipids within the plasma membrane. This creates molecules termed *glycoproteins* and *glycolipids,* respectively. Unique glycoprotein molecules, specific to each cell, mark the cell as belonging to a particular individual. Such cell markers account for such characteristics as blood type or why a patient's system sometimes rejects an organ transplant. Some glycoproteins have a special configuration that allows them to act as a receptor for a chemical messenger such as a hormone. Some integral plasma membrane proteins form channels through which certain substances can enter cells, while others are carriers involved in the passage of molecules through the membrane.

The Nucleus

The nucleus is a prominent structure in human cells. The nucleus is of primary importance because it stores the genetic information that determines the characteristics of the body's cells and their metabolic functioning. The unique chemical composition of each person's DNA forms the basis for DNA fingerprinting (see Focus on Forensics, p. 59). Every cell contains a copy of genetic information, but each cell type has certain genes turned on and others turned off. Activated DNA, with messenger RNA (mRNA) acting as an intermediary, controls protein synthesis (see page 56). The proteins of a cell determine its structure and the functions it can perform.

When you look at the nucleus, even in an electron micrograph, you cannot see DNA molecules, but you can see chromatin (Fig. 3.3). Chemical analysis shows that **chromatin** contains DNA and much protein, as well as some RNA. Chromatin undergoes coiling into rodlike structures called **chromosomes** just before the cell divides. Chromatin is immersed in a semifluid medium called nucleoplasm.

Most likely, too, when you look at an electron micrograph of a nucleus (Fig. 3.3), you'll see one or more regions that look

Figure 3.3 The nucleus. The nuclear envelope with pores (arrows) surrounds the chromatin. The nucleus has a special region called the nucleolus where rRNA is produced and ribosomal subunits are assembled.

darker than the rest of the chromatin. These are **nucleoli (sing., nucleolus)** where another type of RNA, called ribosomal RNA (rRNA), is produced and where rRNA joins with proteins to form the subunits of ribosomes. (Ribosomes are small bodies in the cytoplasm that contain rRNA and proteins.)

The nucleus is separated from the cytoplasm by a double membrane known as the **nuclear envelope,** which is continuous with the endoplasmic reticulum (Fig. 3.4) discussed on this page. The nuclear envelope has **nuclear pores** of sufficient size to permit the passage of proteins into the nucleus and ribosomal subunits out of the nucleus. Once again, remember the most important function of the nuclear envelope: to surround and contain DNA.

Ribosomes

Ribosomes are composed of two subunits, one large and one small. Each subunit has its own mix of proteins and rRNA. Protein synthesis occurs at the ribosomes.

Ribosomes are found free within the cytoplasm either singly or in groups called **polyribosomes** (polysomes). Ribosomes are often attached to the endoplasmic reticulum, a membranous system of saccules and channels discussed next (Fig. 3.4).

Proteins synthesized by cytoplasmic ribosomes are used inside the cell for various purposes. Those produced by ribosomes attached to endoplasmic reticulum may eventually be secreted from the cell.

Endomembrane System

The **endomembrane system** consists of the nuclear envelope, the endoplasmic reticulum, the Golgi apparatus, lysosomes,

Figure 3.4 Endoplasmic reticulum. Rough endoplasmic reticulum is studded with ribosomes where protein synthesis occurs. Smooth endoplasmic reticulum, which has no attached ribosomes, produces lipids and often has other functions as well in particular cells.

and **vesicles** (tiny membranous sacs) (Fig. 3.5). These components of the cell work together to produce and secrete a product.

The Endoplasmic Reticulum

The **endoplasmic reticulum (ER),** a complicated system of membranous channels and saccules (flattened vesicles), is physically continuous with the outer membrane of the nuclear envelope. Rough ER is studded with ribosomes on the side of the membrane that faces the cytoplasm. Here proteins are synthesized and enter the ER interior where processing and modification begin. Some of these proteins are incorporated into the membrane, and some are for export. Smooth ER, which is

Figure 3.5 **The endomembrane system.** Transport vesicles from the ER bring proteins and lipids to the Golgi apparatus where they are modified and repackaged into secretory vesicles. Secretion occurs when vesicles fuse with the plasma membrane. Lysosomes made at the Golgi apparatus digest macromolecules after fusing with incoming vesicles.

continuous with rough ER, does not have attached ribosomes. Smooth ER synthesizes the phospholipids that occur in membranes and has various other functions, depending on the particular cell. In the testes, it produces testosterone, and in the liver it helps detoxify drugs.

Regardless of any specialized function, ER also forms **vesicles** in which large molecules are transported to other parts of the cell. Often these vesicles are on their way to the plasma membrane or the Golgi apparatus.

The Golgi Apparatus

The **Golgi apparatus** is named for Camillo Golgi, who discovered its presence in cells in 1898. The Golgi apparatus consists of a stack of 3 to 20 slightly curved saccules whose appearance can be compared to a stack of pancakes (Fig. 3.5). In animal cells, one side of the stack (the inner face) is directed toward the ER, and the other side of the stack (the outer face) is directed toward the plasma membrane. Vesicles can frequently be seen at the edges of the saccules.

The Golgi apparatus receives protein and/or lipid-filled vesicles that bud from the ER. Some biologists believe that these fuse to form a saccule at the inner face and that this saccule remains a part of the Golgi apparatus until the molecules are repackaged in new vesicles at the outer face. Others believe that the vesicles from the ER proceed directly to the outer face of the Golgi apparatus, where processing and packaging occur within its saccules. The Golgi apparatus contains enzymes that modify proteins and lipids. For example, it can add a chain of sugars to proteins and lipids, thereby making them glycoproteins and glycolipids, which are molecules found in the plasma membrane.

The vesicles that leave the Golgi apparatus move to other parts of the cell. Some vesicles proceed to the plasma membrane where they discharge their contents. Because this is secretion, note that the Golgi apparatus is involved in processing, packaging, and secretion. Other vesicles that leave the Golgi apparatus are lysosomes.

Lysosomes

Lysosomes, membranous sacs produced by the Golgi apparatus, contain hydrolytic digestive enzymes. Sometimes macromolecules are brought into a cell by vesicle formation

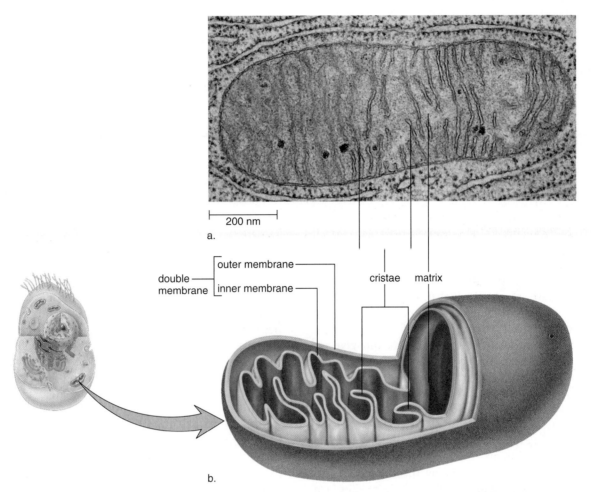

Figure 3.6 **Mitochondrion structure.** Mitochondria are involved in cellular respiration. **a.** Electron micrograph of a mitochondrion. **b.** Generalized drawing in which the outer membrane and portions of the inner membrane have been cut away to reveal the cristae.

at the plasma membrane (Fig. 3.5). When a lysosome fuses with such a vesicle, the lysosome's contents are digested by lysosomal enzymes into simpler subunits that then enter the cytoplasm. Even parts of a cell are digested by its own lysosomes (called autodigestion). Normal cell rejuvenation most likely takes place in this manner, but autodigestion is also important during development. For example, when a tadpole becomes a frog, lysosomes digest away the cells of the tail. The fingers of a human embryo are at first webbed, but they are freed from one another as a result of lysosomal action.

Occasionally, a child is born with **Tay-Sachs disease,** a metabolic disorder involving a missing or inactive lysosomal enzyme in nerve cells. In these cases, the lysosomes fill to capacity with macromolecules that cannot be broken down. The nerve cells become so crowded with these nonfunctional lysosomes that they can no longer function. Nerve signals in the brain and spinal cord are no longer transmitted. The brain stops developing and the child dies. Someday soon, it may be possible to provide the missing enzyme for these children.

Peroxisomes and Vacuoles

Although **peroxisomes** are not part of the endomembranous system, they are similar to lysosomes in structure. Like lysosomes, they are vesicles that contain enzymes. Peroxisome enzymes function to detoxify drugs, alcohol, and other potential toxins. The liver and kidneys contain large numbers of peroxisomes because these organs help to cleanse the blood. Peroxisomes also break down fatty acids from fats so that fats can be metabolized by the cell.

Vacuoles are occasionally found within human cells, where they isolate substances captured inside the cell. For example, vacuoles may contain parasites that are awaiting digestion by lysosomes.

Mitochondria

Although the size and shape of mitochondria (sing., **mitochondrion**) can vary, all are bounded by a double membrane. The inner membrane is folded to form little shelves called *cristae*, which project into the *matrix*, an inner space filled with a gel-like fluid (Fig. 3.6).

Mitochondria are the site of ATP (adenosine triphosphate) production involving complex metabolic pathways. As described in section 2.7, ATP molecules are the common carrier of energy in cells. A shorthand way to indicate the chemical transformation that involves mitochondria is as follows:

Read as follows: As carbohydrate is broken down to carbon dioxide and water, ATP molecules are built up.

Mitochondria are often called the powerhouses of the cell: Just as a powerhouse burns fuel to produce electricity, the mitochondria convert the chemical energy of carbohydrate molecules into the chemical energy of ATP molecules. In the process, mitochondria use up oxygen and give off carbon dioxide and water. The oxygen you breathe in enters cells and then mitochondria; the carbon dioxide you breathe out is released by mitochondria. Because oxygen is used up and carbon dioxide is released, we say that mitochondria carry on **cellular respiration.**

Fragments of digested carbohydrate, protein, and lipid enter the mitochondrial matrix from the cytoplasm. The matrix contains enzymes for metabolizing these fragments to carbon dioxide and water. Energy released from metabolism is used for ATP production, which occurs at the cristae. The protein complexes that aid in the conversion of energy are located in an assembly-line fashion on these membranous shelves.

Every cell uses a certain amount of ATP energy to synthesize molecules, but many cells use ATP to carry out their specialized functions. For example, muscle cells use ATP for muscle contraction, which produces movement, and nerve cells use it for the conduction of nerve impulses, which make us aware of our environment.

The Cytoskeleton

Several types of filamentous protein structures form a cytoskeleton that helps maintain the cell's shape and either anchors the organelles or assists their movement as appropriate. The cytoskeleton includes microtubules, intermediate filaments, and actin filaments (see Fig. 3.1).

Microtubules are hollow cylinders whose wall is made up of 13 longitudinal rows of the globular protein tubulin. Remarkably, microtubules can assemble and disassemble. Microtubule assembly is regulated by the centrosome, which lies near the nucleus. The centrosome is the region of the cell that contains the centrioles. Microtubules radiate from the centrosome, helping to maintain the shape of the cell and acting as tracks along which organelles move. It is well known that during cell division, microtubules form spindle fibers, which assist the movement of chromosomes.

Intermediate filaments differ in structure and function. Because they are tough and resist stress, intermediate filaments often form cell-to-cell junctions. Intermediate filaments join skin cells in the outermost skin layer, the epidermis. **Actin filaments** are long, extremely thin fibers that usually occur in bundles or other groupings. Actin filaments have been isolated from various types of cells, especially those in which movement occurs. Microvilli, which project from certain cells and can shorten and extend, contain actin filaments. Actin filaments, like microtubules, can assemble and disassemble.

Centrioles

Centrioles are short cylinders with a 9 + 0 pattern of microtubules, meaning that there are nine outer microtubule triplets and no center microtubules (see Fig. 3.8b). Each cell has a pair of centrioles in the centrosome, a region near the nucleus. The members of each pair of centrioles are at right angles to one another. Before a cell divides, the centrioles duplicate, and the members of the new pair are also at right angles to one another. During cell division, the pairs of centrioles separate so that each daughter cell gets one centrosome.

Centrioles are directed by the centrosome in the formation of the spindle apparatus, which functions during cell division. Their role in forming the spindle is described later in the Mitosis/Meiosis section of this chapter on pages 57–58, 60. A single centriole forms the anchor point, or **basal body**, for each individual cilium or flagellum. Basal bodies direct the formation of cilia and flagella as well.

Cilia and Flagella

Cilia and flagella (sing., **cilium, flagellum**) are projections of cells that can move either in an undulating fashion, like a whip, or stiffly, like an oar. Cilia are shorter than flagella (Fig. 3.7). Cells that have these organelles are capable of self-movement or moving material along the surface of the cell. For example, sperm cells, carrying genetic material to the egg, move by means of flagella. The cells that line our respiratory tract are ciliated. These cilia sweep debris trapped within mucus back up the throat, and this action helps keep the lungs clean. Within a woman's uterine tubes, ciliated cells move the ovum toward the uterus (womb), where a fertilized ovum grows and develops.

Recall that each cilium and flagellum is anchored by its basal body, which lies in the cytoplasm of the cell. Basal bodies, like centrioles, have a 9 + 0 pattern of microtubule triplets (Fig. 3.8a). They are believed to organize the structure of cilia and flagella even though cilia and flagella have a 9 + 2 pattern of microtubules. In cilia and flagella, nine microtubule doublets surround two central microtubules. This arrangement is believed to be necessary to their ability to move (Fig. 3.8b).

a.

b.

Figure 3.7 Cilia and flagella. a. Cilia are common on the surfaces of certain tissues, such as the one that forms the inner lining of the respiratory tract. **b.** Flagella form the tails of human sperm cells.

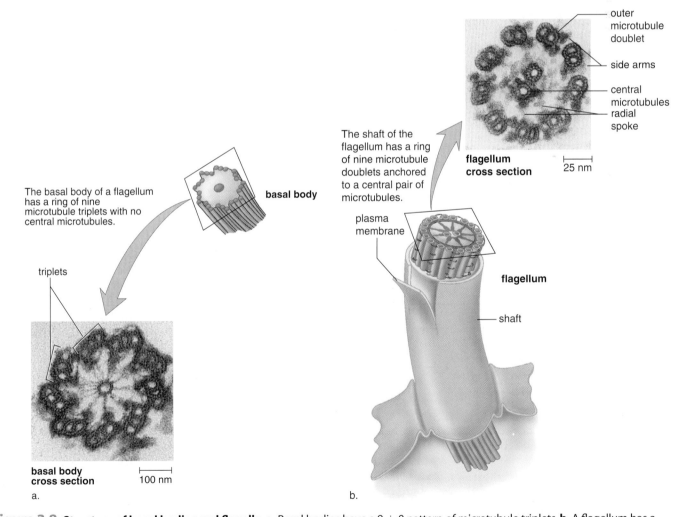

The basal body of a flagellum has a ring of nine microtubule triplets with no central microtubules.

basal body

triplets

basal body
cross section

100 nm

a.

The shaft of the flagellum has a ring of nine microtubule doublets anchored to a central pair of microtubules.

outer
microtubule
doublet

side arms

central
microtubules
radial
spoke

flagellum
cross section

25 nm

plasma
membrane

flagellum

shaft

b.

Figure 3.8 Structure of basal bodies and flagella. a. Basal bodies have a 9 + 0 pattern of microtubule triplets. **b.** A flagellum has a 9 + 2 pattern of microtubules.

Content CHECK-UP!

1. Match each part of the plasma membrane to its function:

   a. phospholipid molecules _____

   b. glycoproteins _____

   c. cholesterol _____

   1. stabilize the phospholipid bilayer
   2. form a bilayer
   3. account for a person's blood type

2. The endomembrane system consists of which structures?

   a. peroxisomes

   b. Golgi apparatus

   e. b and c

   c. rough endoplasmic reticulum

   d. a and c

3. The 9 + 0 microtubule pattern is found in:

   a. basal bodies and centrioles.

   b. basal bodies and flagella.

   c. cilia and flagella.

Answers in Appendix C, page 461.

3.2 Crossing the Plasma Membrane

The plasma membrane keeps a cell intact. It allows only certain molecules and ions to enter and exit the cytoplasm freely; therefore, the plasma membrane is said to be **selectively permeable** (semipermeable). Both passive and active methods are used to cross the plasma membrane (Table 3.2). A passive transport mechanism is one that doesn't require cellular energy. By contrast, active transport must be fueled by cellular energy (usually in the form of ATP).

Simple Diffusion

Simple diffusion is the random movement of simple atoms or molecules from an area of higher concentration to an area of lower concentration until they are equally distributed. To illustrate diffusion, imagine putting a tablet of dye into water. The water eventually takes on the color of the dye as the dye molecules diffuse (Fig. 3.9).

The chemical and physical properties of the plasma membrane allow only a few types of molecules to enter and exit a cell by simple diffusion. Lipid-soluble molecules such as alcohols can diffuse through the membrane because lipids are the membrane's main structural components. Gases can also diffuse through the lipid bilayer; this is the mechanism by which oxygen enters cells and carbon dioxide exits cells. For example, consider the movement of oxygen from the lungs to the bloodstream. When you inhale, oxygen fills the tiny air sacs, or *alveoli*, within your lungs. Neighboring lung capillaries contain red blood cells with a very low oxygen concentration. Oxygen diffuses from the area of highest concentration to the area of lowest concentration: first through alveolar cells, then lung capillary cells, and finally into the red blood cells.

When atoms or molecules diffuse from areas of higher to lower concentration across plasma membranes, no cellular energy is involved. Instead, kinetic or thermal energy of matter is the energy source for diffusion.

Osmosis

Osmosis is the diffusion of water across a plasma membrane. Osmosis occurs whenever an unequal concentration of water exists on either side of a selectively permeable membrane. (Recall that a selectively permeable membrane allows water to pass freely, but not most dissolved substances.) In a solution, water is *more* concentrated when it contains fewer dissolved substances, or **solutes** (and thus is closest to pure water). Water is *less* concentrated as solute concentration increases. **Osmotic pressure** is the force exerted on a selectively permeable membrane because water has moved from the area of

TABLE 3.2 Crossing the Plasma Membrane			
Name	**Direction**	**Requirement**	**Examples**
PASSIVE METHODS			
Simple diffusion	High to low concentration	Concentration gradient	Lipid-soluble molecules, gases
Osmosis	High to low concentration	Semipermeable membrane, water concentration gradient	Absorption of water from digestive tract to bloodstream
Facilitated diffusion	High to low concentration	Carrier molecule plus concentration gradient	Sugars and amino acids
ACTIVE METHODS			
Active transport	Low to high concentration	Carrier molecule plus cell energy	Ions, sugars, amino acids
Endocytosis			
Phagocytosis	Into the cell ("cell eating")	Vesicle formation	Bacterial cells, viruses, cell debris
Pinocytosis	Into the cell ("cell drinking")	Vesicle formation	Breast milk absorption in infants
Exocytosis	Out of the cell	Vesicle fuses with plasma membrane	Hormones, messenger chemicals, other macromolecules

water
molecules
(solvent)

dye
molecules
(solute)

a. Crystal of
dye placed
in water

b. Diffusion of
water and dye
molecules

c. Equal distribution
of molecules
results

Figure 3.9 Diffusion. As dye molecules move from areas of higher to lower concentration, the water takes on the color of the dye.

plasma
membrane

Animal cells

a. In an isotonic solution, there is no net movement of water. A normal red blood cell is shown.

b. In a hypotonic solution, water enters the cell, which may burst (lysis). Note the cell is swollen.

c. In a hypertonic solution, water leaves the cell, which shrivels (crenation).

Figure 3.10 Tonicity. The arrows indicate the movement of water.

higher water concentration to the area of lower water concentration (higher concentration of solute).

Tonicity is the degree to which a solution's concentration of solute-versus-water causes water to move into or out of cells. Normally, body fluids are **isotonic** to cells (Fig. 3.10a)—that is, there is an equal concentration of solutes (dissolved substances) and solvent (water) on both sides of the plasma membrane, and cells maintain their usual size and shape. Medically administered intravenous solutions usually have this tonicity. Body fluids which are *not* isotonic to body cells are the result of dehydration or water intoxication (see Medical Focus, p. 53).

Solutions (solute plus solvent) that cause cells to swell or even to burst due to an intake of water are said to be **hypotonic**

solutions. If red blood cells are placed in a hypotonic solution, which has a higher concentration of water (lower concentration of solute) than do the cells, water enters the cells and they swell to bursting (Fig. 3.10b). The term **lysis** refers to breaking or disintegration of cells. **Hemolysis** is the disintegration of red blood cells.

Solutions that cause cells to shrink or to shrivel due to a loss of water are said to be **hypertonic** solutions. If red blood cells are placed in a hypertonic solution, which has a lower concentration of water (higher concentration of solute) than do the cells, water leaves the cells and they shrink (Fig. 3.10c). The term **crenation** refers to red blood cells in this condition.

Begin Thinking Clinically

An intravenous hypertonic solution may help relieve swelling in the brain in patients with head trauma. How does this work?

Answer and discussion in Appendix C, page 461.

Filtration

Capillaries, described in detail in Chapter 12, are the smallest blood vessels. Because capillary walls are only one cell thick, small molecules (e.g., water or small solutes) tend to passively diffuse across these walls, from areas of higher concentration to those of lower concentration. However, blood pressure aids matters by pushing water and dissolved solutes out of the capillary, through tiny pores between capillary cells. This process, called **filtration,** is the movement of liquid from high pressure to low pressure.

You can observe filtration in a drip coffeemaker. Water moves from the area of high pressure (the water reservoir) to the area of low pressure (the coffee pot). Large substances (the coffee grounds) remain behind in the coffee filter, but small molecules (caffeine, flavor) and water pass through.

Transport by Carriers

Most solutes do not simply diffuse across a plasma membrane; rather, they are transported by means of protein carriers within the membrane. During **facilitated diffusion** (facilitated transport), a molecule (e.g., an amino acid or glucose) is transported across the plasma membrane from the side of higher concentration to the side of lower concentration. The cell does not need to expend energy for this type of transport because the molecules are moving down their concentration gradient.

During **active transport,** a molecule is moving contrary to the normal direction—that is, from lower to higher concentration (Fig. 3.11). For example, iodine collects in the cells of the thyroid gland; sugar is completely absorbed from the gut by cells that line the digestive tract; and sodium (Na^+) is sometimes almost completely withdrawn from urine by cells lining kidney tubules. Active transport requires a protein carrier and the use of cellular energy obtained from the breakdown of ATP. When ATP is broken down, energy is released, and in this case the energy is used by a carrier to carry out active transport. Therefore, it is not surprising that cells involved in active transport have a large number of mitochondria near the plasma membrane at which active transport is occurring.

Proteins involved in active transport often are called pumps because just as a water pump uses energy to move water against the force of gravity, proteins use energy to move substances against their concentration gradients. One type of pump that is active in all cells but is especially associated

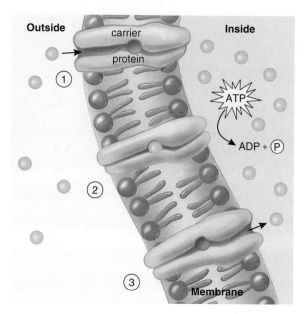

Figure 3.11 **Active transport through a plasma membrane.** Active transport allows a molecule to cross the membrane from lower concentration to higher concentration. ① Molecule enters carrier. ② Breakdown of ATP induces a change in shape that ③ drives the molecule across the membrane.

with nerve and muscle cells moves sodium ions (Na^+) to the outside of the cell and potassium ions (K^+) to the inside of the cell.

The passage of salt (NaCl) across a plasma membrane is of primary importance in cells. First, sodium ions are pumped across a membrane; then, chloride ions simply diffuse through channels that allow their passage. Chloride ion channels malfunction in persons with **cystic fibrosis,** and this leads to the symptoms of this inherited (genetic) disorder.

Endocytosis and Exocytosis

During **endocytosis,** a portion of the plasma membrane forms an inner pocket to envelop a substance, and then the membrane pinches off to form an intracellular vesicle (see Fig. 3.1, *top*). Two forms of endocytosis exist: **phagocytosis,** or "cell eating," is a mechanism whereby the cell can ingest solid particles. White blood cells consume bacterial cells by phagocytosis. Once inside the cell, the bacterial cell can be destroyed. **Pinocytosis,** or "cell drinking," allows the cell to consume solutions. An infant's intestinal lining ingests breast milk by pinocytosis, allowing the mother's protective antibodies to enter the baby's bloodstream.

During **exocytosis,** a vesicle fuses with the plasma membrane as secretion occurs (see Fig. 3.1, *bottom*). This is the way insulin leaves insulin-secreting cells, for instance. Table 3.2 summarizes the various ways molecules cross the plasma membrane.

Dehydration and Water Intoxication

Dehydration is due to a loss of water. The solute concentration in extracellular fluid increases—that is, tissue fluid becomes hypertonic to cells, and water leaves the cells. A common cause of dehydration is excessive sweating, perhaps during exercise, without any replacement of the water lost. Dehydration can also be a side effect of any illness that causes prolonged vomiting or diarrhea.

The signs of moderate dehydration are a dry mouth, sunken eyes, and skin that will not bounce back after light pinching. If dehydration becomes severe, the pulse and breathing rate are rapid, the hands and feet are cold, and the lips are blue. Although dehydration leads to weight loss, it is never a good idea to deliberately dehydrate to lose weight.

Figure 3A **Dehydration versus water intoxication. a.** If extracellular fluid loses too much water, cells lose water by osmosis and become dehydrated. **b.** If extracellular fluid gains too much water, cells gain water by osmosis and water intoxication occurs.

Water intoxication is due to excessive consumption of pure water. The tissue fluid becomes hypotonic to the cells, and water enters the cells. Water intoxication can lead to pulmonary edema (excess tissue fluid in the lungs) and swelling in the brain. In extreme cases, it is fatal. Water intoxication is not nearly as common in adults as is dehydration. It can result from a mental disorder termed *psychogenic polydipsia*. Another cause can be the intake of too much pure water during vigorous exercise: for example, a marathon race. Marathoners who collapse and have nausea and vomiting after a race may be suffering from water intoxication. The cure, an intravenous solution containing high amounts of sodium, is the opposite of that for dehydration. Therefore, it is important that physicians be able to diagnose water intoxication in athletes who have had an opportunity to drink fluids over a period of a few hours. To prevent both dehydration and water intoxication, athletes should replace lost fluids continuously. Pure water is a good choice if the exercise period is short. Low-sodium solutions, such as sports drinks, are a good choice for longer-duration events like marathons.

4. Which process requires cellular ATP energy?

    a. osmosis

    b. facilitated diffusion (facilitated transport)

    c. active transport

    d. simple diffusion

5. A researcher studying the white blood cells of a patient infected with tuberculosis (TB) bacteria notices the bacteria are in *vesicles* in the cytoplasm. How did the bacteria come to be inside the cell?

    a. pinocytosis   c. exocytosis

    b. phagocytosis

6. The cell organelle that is needed to destroy the TB bacterium discussed in question 5 is a:

    a. ribosome.   c. centrosome.

    b. lysosome.

Answers in Appendix C, page 461.

3.3 The Cell Cycle

The **cell cycle** is an orderly set of stages that take place between the time a cell divides and the time the resulting daughter cells also divide. The cell cycle is controlled by internal and external signals. A signal is a molecule that stimulates or inhibits a metabolic event. For example, *growth factors* are external signals received at the plasma membrane that cause a resting cell to undergo the cell cycle. When blood platelets release a growth factor, skin fibroblasts in the vicinity finish the cell cycle, thereby repairing an injury. Other signals ensure that the stages follow one another in the normal sequence and that each stage is properly completed before the next stage begins.

The cell cycle has a number of checkpoints, places where the cell cycle stops if all is not well. Any cell that did not successfully complete mitosis and is abnormal undergoes apoptosis at the *restriction checkpoint.* **Apoptosis** is often defined as *programmed cell death* because the cell progresses through a series of events that bring about its destruction. The cell rounds up and loses contact with its neighbors. The nucleus fragments and the plasma membrane develops blisters. Finally, the cell fragments and its bits and pieces are engulfed by white blood cells and/or neighboring cells. The enzymes that bring about apoptosis are ordinarily held in check by inhibitors, but are unleashed by either internal or external signals.

Following a certain number of cell cycle revolutions, cells are apt to become specialized and no longer go through the cell cycle. For example, once they have matured, muscle and nerve cells rarely, if ever, go through the cell cycle. At the other extreme, some cells in the body, called stem cells, are always immature and go through the cell cycle repeatedly. There is a great deal of interest in stem cells today because it may be possible to control their future development into particular tissues and organs.

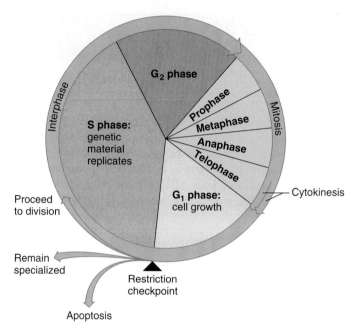

Figure 3.12 The cell cycle. The cell cycle consists of interphase, during which cellular components duplicate, and a mitotic stage, during which the cell divides. Interphase consists of two so-called "growth" phases (G_1 and G_2) and a DNA synthesis (S) phase. The mitotic stage consists of the phases noted plus cytokinesis.

Cell Cycle Stages

The cell cycle has two major portions: **interphase** and the **mitotic stage** (Fig. 3.12).

Interphase

The cell in Figure 3.1 is in interphase because it is not dividing. During interphase, the cell carries on its regular activities, and it also gets ready to divide if it is going to complete the cell cycle. For these cells, interphase has three stages, called G_1 phase, S phase, and G_2 phase.

G_1 Phase Early microscopists named the phase before DNA replication G_1, and they named the phase after DNA replication G_2. G stood for "gap." Now that we know how metabolically active the cell is, it is better to think of G as standing for "growth." During the G_1 phase, cells return to their normal, pre-mitotic condition. Each cell enlarges and doubles its number of organelles, such as mitochondria and ribosomes. Proteins are synthesized and other materials are accumulated in preparation for DNA synthesis. Most important, it is during the G_1 phase that the cell performs all its "tasks"—endocrine cells make and secrete hormones, white blood cells defend against invaders, and so on. As you read about the functions of cells in a particular tissue, remember that these are cells in the G_1 phase.

S Phase Following G_1, the cell enters the S (for "synthesis") phase. During the S phase, DNA replication occurs. At the beginning of the S phase, each chromosome is composed of one

Parental DNA molecule contains so-called old strands hydrogen-bonded by complementary base pairing.

Region of replication. Parental DNA is unwound and unzipped. New nucleotides are pairing with those in old strands.

new strand   old strand

Replication is complete. Each double helix is composed of an old (parental) strand and a new (daughter) strand.

Figure 3.13 Overview of DNA replication. During replication, an old strand serves as a template for a new strand. The new double helix is composed of an old (parental) strand and a new (daughter) strand.

Figure 3.14 Ladder configuration and DNA replication. Use of the ladder configuration better illustrates how complementary nucleotides available in the cell pair with those of each old strand before they are joined together to form a daughter strand.

DNA double helix, which is equal to a chromatid. At the end of this phase, each chromosome has two identical DNA double helix molecules, and therefore is composed of two sister chromatids. Another way of expressing these events is to say that DNA replication has resulted in duplicated chromosomes.

G₂ Phase During this phase, the cell synthesizes proteins that will assist cell division, such as the protein found in microtubules. The cell also completes replication of the centrioles. The role of centrioles and microtubules in cell division is described later in this section.

Events During Interphase

Two significant events during interphase are replication of DNA and protein synthesis.

Replication of DNA

During **replication**, an exact copy of a DNA helix is produced. (DNA and RNA structure are described on pp. 35–37.) The double-stranded structure of DNA aids replication because each strand serves as a template for the formation of a complementary

strand. A **template** is most often a mold used to produce a shape opposite to itself. In this case, each old (parental) strand is a template for each new (daughter) strand.

Figures 3.13 and 3.14 show how replication is carried out. Figure 3.14 uses the ladder configuration of DNA for easy viewing.

1. Before replication begins, the two strands that make up parental DNA are hydrogen-bonded to one another.
2. During replication, the old (parental) DNA strands unwind and "unzip" (i.e., the weak hydrogen bonds between the two strands break).
3. New complementary nucleotides, which are always present in the nucleus, pair with the nucleotides in the old strands. A pairs with T and C pairs with G. The enzyme DNA polymerase joins the new nucleotides forming new (daughter) complementary strands.
4. When replication is complete, the two resulting double helix DNA molecules are identical.

Each strand of a double helix is equal to a **chromatid**, which means that at the completion of replication each chromosome is composed of two sister chromatids. They are

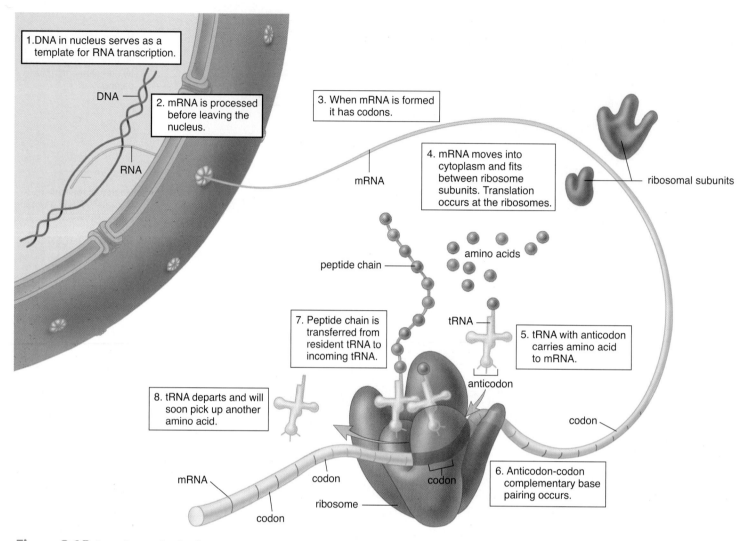

1. DNA in nucleus serves as a template for RNA transcription.

DNA

RNA

2. mRNA is processed before leaving the nucleus.

3. When mRNA is formed it has codons.

mRNA

4. mRNA moves into cytoplasm and fits between ribosome subunits. Translation occurs at the ribosomes.

ribosomal subunits

amino acids

peptide chain

tRNA

5. tRNA with anticodon carries amino acid to mRNA.

7. Peptide chain is transferred from resident tRNA to incoming tRNA.

anticodon

8. tRNA departs and will soon pick up another amino acid.

codon

mRNA

codon

codon

ribosome

codon

6. Anticodon-codon complementary base pairing occurs.

Figure 3.15 Protein synthesis. The two steps required for protein synthesis are transcription, which occurs in the nucleus, and translation, which occurs in the cytoplasm at the ribosomes.

called **sister chromatids** because they are identical. The chromosome is called a **duplicated chromosome** (see Fig. 3.16).

Protein Synthesis

DNA not only serves as a template for its own replication, but is also a template for RNA formation and for the construction of proteins by the cell. Protein synthesis requires two steps: transcription and translation. During **transcription,** a messenger RNA (mRNA) molecule is produced, using DNA as the template. During **translation,** this mRNA specifies the order of amino acids in a particular polypeptide (Fig. 3.15). A gene (i.e., DNA) contains coded information for the sequence of amino acids in a particular polypeptide. The code is a **triplet code:** Every three bases in DNA (and therefore in mRNA) stand for a particular amino acid.

Transcription and Translation

During transcription, DNA unwinds and unzips, as though it is preparing for replication. Complementary RNA nucleotides

from an RNA nucleotide pool in the nucleus pair with the DNA nucleotides of one strand. The RNA nucleotides are joined by an enzyme called RNA polymerase, and an RNA molecule results (Fig. 3.15, *step 1*). There are three forms of RNA: messenger, transfer, and ribosomal; however, only messenger RNA determines amino acid sequence. When mRNA forms, it has a sequence of bases complementary to those of DNA. A sequence of three bases that is complementary to the DNA triplet code is a **codon** (Fig. 3.15, *step 3*).

Translation requires several enzymes and all three types of RNA. Messenger RNA, containing the polypeptide's code, is sandwiched between the two ribosome subunits (Fig. 3.15, *step 4*). **Transfer RNA (tRNA)** molecules deliver amino acids to the ribosomes, which are composed of **ribosomal RNA (rRNA)** and protein (Fig. 3.15, *step 5*). There is at least one tRNA molecule for each of the 20 amino acids found in proteins. The amino acid binds to one end of the molecule, and the entire complex is designated as tRNA-amino acid.

Recall that the primary sequence of a protein is determined by the ordering of its amino acids (see Chapter 2,

Stages of Mitosis

Early Prophase
Chromatin is condensing into chromosomes.

chromatin
centrioles have replicated
nucleolus
nuclear pore
nuclear envelope

Prophase
Duplicated chromosomes are scattered.

sister chromatids
spindle
nucleolus
centromere
nuclear envelope fragments

Metaphase
Chromosomes are aligned at the equator of the spindle.

spindle
aster

Anaphase
Daughter chromosomes are moving to the poles.

chromosome

Telophase
Daughter nuclei are forming and spindle is disappearing.

cleavage furrow
centromere

Cytokinesis complete
Chromosomes are decondensing.

daughter cells

Figure 3.16 **The mitotic stage of the cell cycle.** Humans have 46 chromosomes; four are shown here. The blue chromosomes were originally inherited from the father, and the red were originally inherited from the mother.

page 32). Transfer RNA must deliver the correct amino acid so that the primary sequence is also correct. At the other end of each tRNA molecule is a specific **anticodon,** a group of three bases that is complementary to an mRNA codon. A tRNA molecule comes to the ribosome, where its anticodon pairs with an mRNA codon (Fig. 3.15, *step 6*). For example, if the codon is ACC, the anticodon is UGG and the amino acid is threonine. (The codes for each of the 20 amino acids are known.) Notice that the order of the codons of the mRNA determines the order that tRNA-amino acids come to a ribosome, and therefore the final sequence of amino acids in a polypeptide.

Mitotic Stage

Following interphase, the cell enters the M (for mitotic) stage. This cell division stage includes **mitosis** (division of

the nucleus) and **cytokinesis** (division of the cytoplasm). The **parental cell** is the cell that divides, and the **daughter cells** are the cells that result. During mitosis, chromosomes are distributed to two separate nuclei. When cytokinesis is complete, two daughter cells are present.

As mitosis begins (Fig. 3.16, *upper left*), the centrioles have doubled in preparation for mitosis. Each chromosome is duplicated—it is composed of two chromatids held together at a central region, called the centromere. Mitosis is divided into four phases: prophase, metaphase, anaphase, and telophase (Fig. 3.16).

Prophase

Several events occur during **prophase** that visibly indicate the cell is about to divide. The two pairs of centrioles outside the nucleus begin moving away from each other toward opposite ends of the nucleus. Spindle fibers appear between

Figure 3.17 Micrographs of mitosis occurring in a whitefish embryo.

the separating centriole pairs, the nuclear envelope begins to fragment, and the nucleolus begins to disappear.

As mentioned previously, the disorganized chromatin DNA found in the nucleus now coils into chromosomes. The chromosomes are now fully visible. Although humans have 46 chromosomes, only four are shown in Figure 3.16 for ease in following the phases of mitosis. Spindle fibers attach to the centromeres as the chromosomes continue to shorten and thicken. During prophase, chromosomes are randomly placed in the nucleus. The nuclear membrane begins to fragment, as shown in Figure 3.16.

Structure of the Spindle At the end of prophase, a cell has a fully formed spindle. A **spindle** has poles, asters, and fibers. The **asters** are arrays of short microtubules that radiate from the **poles** and the **fibers** are bundles of microtubules that stretch between the poles. (A spindle resembles a lopsided bicycle wheel; the asters are the "spokes.") Centrioles are located in centrosomes, at opposite poles of the cell. Centrosomes are believed to organize the spindle.

Metaphase

During **metaphase,** the nuclear envelope is completely fragmented, and the spindle occupies the region formerly occupied by the nucleus. The paired chromosomes are now at the equator (center) of the spindle. Metaphase is characterized by a fully formed spindle, and the chromosomes, each with two sister chromatids, are aligned at the equator (Fig. 3.17).

Anaphase

At the start of **anaphase,** the sister chromatids separate. *Once separated, the chromatids are called chromosomes.* Separation of the sister chromatids ensures that each cell receives a copy of each type of chromosome and thereby has a full complement of genes. During anaphase, the daughter chromosomes move to the poles of the spindle. Anaphase is characterized by the movement of chromosomes toward each pole and thus, to opposite sides of the cell.

Function of the Spindle The spindle brings about chromosome movement. Two types of spindle fibers are involved in the movement of chromosomes during anaphase. One type extends from the poles to the equator of the spindle; there, they overlap. As mitosis proceeds, these fibers increase in length, and this helps push the chromosomes apart. The chromosomes themselves are attached to other spindle fibers that simply extend from their centromeres to the poles. These fibers get shorter and shorter as the chromosomes move toward the poles. Therefore, they pull the chromosomes apart.

Spindle fibers, as stated earlier, are composed of microtubules. Microtubules can assemble and disassemble by the addition or subtraction of tubulin (protein) subunits. This is what enables spindle fibers to lengthen and shorten, and it ultimately causes the movement of the chromosomes.

Telophase and Cytokinesis

Telophase begins when the chromosomes arrive at the poles. During telophase, the chromosomes become indistinct

DNA Fingerprinting

It's almost a cliché of modern life: Look no further than the evening news if you want to view evidence of human cruelty. Tragic scenarios are played out daily in emergency rooms and morgues across North America, as increasing numbers of innocent people become victims of violent crimes—beatings, shootings, stabbings, sexual assault. However, recent advancements in biotechnology make the likelihood of catching and convicting criminals greater than ever before. DNA identification technology, also called DNA fingerprinting, identifies the DNA samples recovered from a crime scene. Like actual fingerprints, the composition of DNA is unique to each individual. Only identical twins share the same DNA.

The process of identifying human DNA begins with sample collection at the crime scene. Blood, hair, bone, tissue, nail clippings, saliva, even vomit—all contain human cells. An individual's DNA is contained within the nucleus and the mitochondrion of the cell, although investigators use nuclear DNA whenever possible because of its larger sample size. Exceedingly small samples of only a nanogram (one-billionth of a gram, approximately 160 human cells) are sufficient, making it possible to identify human DNA from saliva on a cigarette butt or a single human hair.

DNA is first extracted from the sample; if necessary, it can be reproduced from a very small sample to create adequate amounts for testing. The DNA is digested into fragments of differing sizes using restriction enzymes, which act as a molecular "scissors." Each fragment is polar—possessing either a positive or negative charge. A technique called *gel electrophoresis* separates the fragments by size and electrical charge, spreading the

fragments throughout a gel medium. Identifying the exact nucleotide sequence for each fragment is never completed—to do so would be extremely expensive and time consuming. Instead, the best DNA analysis technique identifies STRs, or "short tandem repeats." STRs are segments of repeating nucleotides (for example, A-A-A-A or T-T-T-T). They can be found in all human DNA, but their exact location in the molecule is unique to each individual. Each STR is tagged with radioactive molecules. Finally, the radioactively labeled gel is exposed to X-ray film. The result is a pattern similar to a grocery store "bar code," with alternating light and dark bands.

If a suspect's DNA matches the crime scene DNA, the probability of finding another person on earth with the same DNA pattern is 1 in 300 billion (a number roughly equal to the number of human beings found on earth!). Thus, DNA fingerprinting is recognized as a very powerful tool in the courtroom for convicting a guilty individual. However, the same technique can also be used to clear the innocent. If there is no match between the crime scene DNA and that of the suspect, the suspect can be declared innocent of the crime.

Crime labs at the state, national, and international levels now maintain computerized databases of DNA fingerprints. These databases allow rapid electronic comparison of DNA samples from crime scenes to the DNA fingerprints of known and suspected felons. Laws currently exist in all 50 states that require felons convicted of murder, assault, sexual offenses, etc., to submit DNA samples on demand to the Combined DNA Index System (CODIS), which is maintained by the FBI.

9	Marker
8	PST Control
7	Suspect 2
6	Suspect 1
5	Marker
4	Evidence 2
3	Evidence 1
2	Victim
1	Marker

FORENSIC TEST

Figure 3B **Actual DNA fingerprint, from a rape trial.** Lanes 1, 2, 5, 8, and 9 are controls. Lane 2 is the victim's DNA, taken from a blood sample. Lanes 3 and 4 were created using semen samples taken from the victim's body after the rape was committed. Study lanes 6 and 7 carefully—who committed the crime—suspect 1 or suspect 2?

chromatin again. The spindle disappears as nucleoli appear, and nuclear envelope components reassemble in each cell. Telophase is characterized by the presence of two daughter nuclei.

Cytokinesis is division of the cytoplasm and organelles. In human cells, a slight indentation called a **cleavage fur-**

row passes around the circumference of the cell. Actin filaments form a contractile ring, and as the ring gets smaller and smaller, the cleavage furrow pinches the cell in half. As a result, each cell becomes enclosed by its own plasma membrane.

Importance of Mitosis

Because of mitosis, each cell in our body is genetically identical, meaning that it has the same number and kinds of chromosomes. Mitosis is important to the growth and repair of multicellular organisms. When a baby develops in the mother's womb, mitosis occurs as a component of growth. As a wound heals, mitosis occurs, and the damage is repaired.

Meiosis: Reduction-Division

Mitosis is the process for growing new body cells, whereas meiosis is the process for producing a person's *gametes*, or sex cells—sperm cells in males and ova in females—are produced by meiosis. In meiosis, the stages of mitosis—prophase, metaphase, anaphase, and telophase—are repeated twice. When meiosis is complete, the sperm or ova that result have half the normal chromosome number, or 23 (instead of 46 chromosomes in a body cell). When the ovum is joined by a sperm at conception, 23 ovum chromosomes join with 23 sperm chromosomes to form the genetic makeup of the new zygote. (See Chapter 17 for a complete discussion of meiosis.)

Content **CHECK-UP!**

7. How many chromosomes are found in a human cell after the S phase of the cell cycle is complete?

   a. 46

   b. 23

   c. 92 (46 × 2)

8. The nuclear envelope fragments and the nucleolus begins to disappear during:

   a. anaphase.

   b. prophase.

   c. telophase.

   d. metaphase.

9. Formation of a cleavage furrow signals the beginning of which stage?

   a. anaphase

   b. metaphase

   c. telophase

   d. cytokinesis

Answers in Appendix C, page 461.

SELECTED NEW TERMS

Basic Key Terms

actin filaments (ăk′tĭn fĭl′ə-məntz), p. 48

active transport (ak′tiv trans′port), p. 52

anaphase (ăn′ə-fāz), p. 58

anticodon (ăn′tī′-kō′dŏn′), p. 57

apoptosis (ap″o-to′-sis), p. 54

asters (ăs′tərz), p. 58

basal body (bā′sel bod′ē) p. 48

cell cycle (sel sī′-kl), p. 54

cellular respiration (sĕl′yə-lər rĕs′pə-rā′shən), p. 48

centriole (sen′tre-ōl), p. 48

chromatid (krō′mə-tĭd), p. 55

chromatin (kro′muh-tin), p. 44

chromosome (kro′mo-sōm), p. 44

cilium (sĭl′ē-əm), p. 48

cleavage furrow (klĕv′ij fur′o), p. 59

codon (kō′dŏn′), p. 56

crenation (krē-nā′-shun), p. 51

cytokinesis (si′to-kĭ-ne′sis), p. 57

cytoplasm (si′to-plazm), p. 42

cytoskeleton (si′to-skel″ĕ-tun), p. 42

dehydration (dē′hī-drā′shən), p. 53

endocytosis (ĕn′dō-sī′tō-sĭs), p. 52

endomembrane system (en″do-mem′brān sis′tem), p. 45

endoplasmic reticulum (en-do-plaz′mic rĕ-tik′yū-lum), p. 45

exocytosis (eks′ō-sī-tō′sis) p. 52

facilitated diffusion (fuh-sil′-ĭ-tāt′id di-fyŭ-zhun), p. 52

fibers (fī′ bĕrz), p. 58

filtration (fil-tra′shun), p. 52

flagellum (flə-jĕl′əm), p. 48

Golgi apparatus (gol′je ap″uh-rā′tus), p. 46

hemolysis (hē-mŏl′ĭ-sĭs), p. 51

hypertonic (hī′ per-tŏn-ĭk), p. 51

hypotonic (hī′ pō-tŏn-ĭk), p. 51

intermediate filaments (ĭn′tər-mē′dē-ĭt fĭl′ə-məntz), p. 48

interphase (ĭn′tər-fāz′), p. 54

isotonic (ī′ sō-tŏn-ĭk), p. 51

lysis (lī′sĭs), p. 51

lysosome (li′so-sōm), p. 46

metaphase (mĕt′ə-fāz), p. 58

microtubule (mi″kro-tu′byūl), p. 48

mitochondrion (mi″to-kon′dre-on), p. 47

mitosis (mi-to′sis), p. 57

mitotic stage (mī-tŏt′ĭc stāj), p. 54

nuclear envelope (nu′kle-er en′vĕ-lōp), p. 45

nuclear pore (nu′kle-er por), p. 45

nucleolus (nu-kle′o-lus), p. 42

nucleus (nu′kle-us), p. 42

organelle (or′guh-nel), p. 42

osmosis (oz-mo′sis), p. 50

osmotic pressure (ŏz-mŏt′ĭk prĕsh′ər), p. 50

peroxisome (per-ok′-si-sōm), p. 47

phagocytosis (făg′ ə-sī-tō′sĭs), p. 52

pinocytosis (pĭn′ ə-sī-tō′sĭs), p. 52

plasma membrane (plaz′muh mem′brān), p. 42

poles (pōlz), p. 58

polyribosomes (pōl-ē-rī′bō-sōmz), p. 45

prophase (prō′fāz′), p. 57

replication (rĕp′lĭ-kā′shən), p. 55

ribosomal RNA (ri′bo-sōm′al RNA), p. 56

ribosome (ri′bo-sōm), p. 45

selectively permeable (se-lĕk′tiv-le per′me-uh-bl), p. 50

simple diffusion (sĭm′pāl dĭ-fyoō′zhən), p. 50

solute (sol′ūt), p. 50

spindle (spin′dl), p. 58

telophase (tĕl′ ə-fāz), p. 58

template (tĕm′plĭt), p. 55

tonicity (tō-nĭs′ĭ-tē), p. 51

transcription (trans-krip′shun), p. 56

transfer RNA (trans′fer RNA), p. 56

translation (trans-la′shun), p. 56

triplet code (trip′let cōd), p. 56

vesicle (ves′ĭ-kl), p. 46

Clinical Key Terms

cystic fibrosis (sis′tik fī-brō′-sis), p. 52

Tay-Sachs (tā saks), p. 47

water intoxication (wŏh′-tar ĭn-tŏk′sĭ-kāt′shən), p. 53

SUMMARY

Cells differ in shape and function, but even so, a generalized cell can be described.

3.1 Cellular Organization

All human cells, despite varied shapes and sizes, have a plasma membrane. Most have a single nucleus, although some are multinucleate. Mature red blood cells are anucleate. Cell cytoplasm contains organelles and a cytoskeleton.

A. The plasma membrane, composed of phospholipid and protein molecules, regulates the entrance and exit of other molecules into and out of the cell.

B. The nucleus contains chromatin, which condenses into chromosomes just prior to cell division. Genes, composed of DNA, are on the chromosomes, and they code for the production of proteins in the cytoplasm. The nucleolus is involved in ribosome formation.

C. Ribosomes are small organelles where protein synthesis occurs. Ribosomes occur in the cytoplasm, both singly and in groups. Numerous ribosomes are attached to the endoplasmic reticulum.

D. The endomembrane system consists of the endoplasmic reticulum (ER), the Golgi apparatus, and the lysosomes and various transport vesicles.

E. The ER is involved in protein synthesis (rough ER) and various other processes such as lipid synthesis (smooth ER). Molecules produced or modified in the ER are eventually enclosed in vesicles that take them to the Golgi apparatus.

F. The Golgi apparatus processes and packages molecules, distributes them within the cell, and transports them out of the cell. It is also involved in secretion.

G. Lysosomes are produced by the Golgi apparatus, and their hydrolytic enzymes digest macromolecules from various sources. Peroxisomes digest toxins and fatty acids.

H. Mitochondria are involved in cellular respiration, a metabolic pathway that provides ATP molecules to cells.

I. Notable among the contents of the cytoskeleton are microtubules, intermediate filaments, and actin filaments. The cytoskeleton maintains the shape of the cell and also directs the movement of cell parts.

J. Centrioles lie near the nucleus and are involved in the production of the spindle during cell division and in the formation of cilia and flagella.

3.2 Crossing the Plasma Membrane

When substances enter and exit cells by simple diffusion, osmosis, or filtration, no carrier is required. Facilitated diffusion (facilitated transport) and active transport do require a carrier.

A. Some substances can simply diffuse across a plasma membrane. The diffusion of water is called osmosis. In an isotonic solution, cells neither gain nor lose water. In a hypotonic solution, cells swell. In a hypertonic solution, cells shrink.

B. During filtration, movement of water and small molecules out of a blood vessel is aided by blood pressure.

C. During facilitated diffusion (facilitated transport), a carrier is required, but energy is not because the substance is moving from higher to lower concentration. Active transport, which requires a carrier and ATP energy, moves substances from lower to higher concentration.

D. Endocytosis involves the uptake of substances by a cell through vesicle formation. Phagocytosis allows solids to be taken up, whereas pinocytosis is ingestion of a liquid. Exocytosis involves the release of substances from a cell as vesicles within the cell cytoplasm fuse with the plasma membrane.

3.3 The Cell Cycle

The cell cycle consists of interphase (G_1 phase, S phase, G_2 phase) and the mitotic stage, which includes mitosis and cytokinesis.

A. During interphase, DNA replication and protein synthesis take place. DNA serves as a template for its own replication: The DNA parental molecule unwinds and unzips, and new (daughter) strands form by complementary base pairing. Protein synthesis consists of transcription and translation. During transcription, DNA serves as a template for the formation of RNA. During translation, mRNA, rRNA, and tRNA are involved in polypeptide synthesis.

B. Mitosis consists of a number of phases, during which each newly formed cell receives a copy of each kind of chromosome. Later, the cytoplasm divides by furrowing. Mitosis occurs during tissue growth and repair.

STUDY QUESTIONS

1. What are the three main parts to any human cell? (p. 42)
2. Describe the fluid-mosaic model of membrane structure. (pp. 43–44)
3. Describe the nucleus and its contents, and include the terms *DNA* and *RNA* in your description. (pp. 44–45)
4. Describe the structure and function of ribosomes. (p. 45)
5. What is the endomembrane system? What organelles belong to this system? (pp. 45–47)
6. Describe the structure and function of endoplasmic reticulum (ER). Include the terms *smooth ER, rough ER,* and *ribosomes* in your description. (pp. 45–46)
7. Describe the structure and function of the Golgi apparatus. Mention vesicles and lysosomes in your description. (p. 46)

8. Describe the structure and function of mitochondria. Mention the energy molecule ATP in your description. (pp. 47–48)
9. What is the cytoskeleton? What role does the cytoskeleton play in cells? (p. 48)
10. Describe the structure and function of centrioles. Mention the mitotic spindle in your description. (pp. 48–49)
11. Contrast passive transport (diffusion, osmosis, filtration) with active transport of molecules across the plasma membrane. (p. 50)
12. Define osmosis, and discuss the effects of placing red blood cells in isotonic, hypotonic, and hypertonic solutions. (pp. 50–51)
13. What is the cell cycle? What stages occur during interphase? What happens during the mitotic stage? (pp. 54–58, 60)
14. Describe the structure of DNA. How does this structure contribute to the process of DNA replication? (pp. 55–56)
15. Briefly describe the events of protein synthesis. (pp. 56–57)
16. List the phases of mitosis. What happens during each phase? (pp. 57–60)
17. Discuss the importance of mitosis in humans. (p. 60)

LEARNING OBJECTIVE QUESTIONS

I. Match the organelles in the key to the functions listed in questions 1–5.

Key:
- a. mitochondria
- b. nucleus
- c. Golgi apparatus
- d. rough ER
- e. centrioles

1. packaging and secretion
2. cell division
3. powerhouse of the cell
4. protein synthesis
5. control center for the cell

II. Fill in the blanks.

6. The fluid-mosaic model of membrane structure says that _____ molecules drift about within a double layer of _____ molecules.
7. Rough ER has _____, but smooth ER does not.

8. Basal bodies that organize the microtubules within cilia and flagella are believed to be derived from _____.
9. Water will enter a cell when it is placed in a _____ solution.
10. Active transport requires a protein _____ and _____ for energy.
11. Vesicle formation occurs when a cell takes in material by _____.
12. At the conclusion of mitosis, each newly formed cell in humans contains _____ chromosomes.
13. The _____, which is the substance outside the nucleus of a cell, contains bodies called _____, each with a specific structure and function.

III. Match the organelles in the key to the functions listed in questions 14–17.

Key:
- a. DNA
- b. mRNA
- c. tRNA
- d. rRNA

14. Joins with proteins to form subunits of a ribosome.
15. Contains codons that determine the sequence of amino acids in a polypeptide.
16. Contains a code and serves as a template for the production of RNA.
17. Brings amino acids to the ribosomes during the process of transcription.

MEDICAL TERMINOLOGY EXERCISE

Consult Appendix B for help in pronouncing and analyzing the meaning of the terms that follow.

1. hemolysis (he″mol′ĭ-sis)
2. cytology (si-tol′o-je)
3. cytometer (si-tom′ĕ-ter)
4. nucleoplasm (nu′kle-o-plazm)
5. pancytopenia (pan″si-to-pe′ne-uh)
6. cytogenic (si-to-jen′ik)
7. erythrocyte (ĕ-rith′ro-sit)
8. atrophy (at′ro-fe)
9. hypertrophy (hi-per′tro-fe)
10. oncotic pressure, colloid osmotic pressure (ong-kot′ik presh′er)(kol′oyd oz-mah′-tik presh′er)
11. hyperplasia (hi-per-pla′zhe-uh)

WEBSITE LINK

Visit this text's website at <u>http://www.mhhe.com/longenbaker7</u> for additional quizzes, interactive learning exercises, and other study tools.

Body Tissues and Membranes

Henrietta Lacks (1920–1951).

C H A P T E R 4

How long can human cells survive? In the case of cancer, some types of cells might, in fact, be "immortal." Consider the cells of Henrietta Lacks, a young and very poor African American woman who died of cervical cancer in 1951. Cells taken from her original tumor, designated *HeLa* (from *Henrietta Lacks*) became the first human cells to grow easily in a laboratory. In their 50-plus years of survival, thousands of research projects have used these sturdy cells. Developing vaccines, studying drug effects, investigating virus behavior, developing tests for genetic disorders, and of course, research into cancer—these are only a few of their uses. They can be found in tissue culture laboratories all over the world, including those on the international space station. If you're interested in cell research, *HeLa* cells can even be purchased from catalogs. *HeLa* cells—from the cancer of a poor woman, buried in an unmarked grave—continue to serve humanity. You can read more about cancer in Medical Focus: *Cancer—The Traitor Inside* on page 70.

Chapter Outline & Learning Objectives
After you have studied this chapter, you should be able to:

A tissue is composed of specialized cells of similar structure that perform a common function in the body. There are four major types of tissues: (1) Epithelial tissue, also called epithelium, covers body surfaces and organs, and lines body cavities; (2) connective tissue binds and supports body parts; (3) muscular tissue contracts; and (4) nervous tissue responds to stimuli and transmits impulses from one body part to another. You can use the "4Cs" mnemonic to remember these functions: *Tissues Cover-Connect-Contract-Communicate.* Body organs are typically composed of all four tissue types. For example, the heart is covered by *epithelium*; its valves are supported by *connective tissue*; cardiac *muscle* pumps blood, and *nerves* regulate how rapidly the heart beats.

4.1 Epithelial Tissue

Epithelial tissues share a set of common characteristics. In all epithelial tissues, the cells are tightly packed, with little space between them. Externally, this tissue protects the body from drying out, injury, and bacterial invasion. On internal surfaces, epithelial tissue protects, but it also may have an additional function. For example, in the respiratory tract, epithelial tissue sweeps up impurities by means of cilia. Along the digestive tract, it secretes mucus, which protects the lining from digestive enzymes. In kidney tubules, its absorptive function is enhanced by the presence of fine, cellular extensions called microvilli.

Epithelial cells readily divide to produce new cells that replace lost or damaged ones. Skin cells as well as those that line the stomach and intestines are continually being replaced. To support its very high rate of reproduction, epithelial tissue must get its nutrients from underlying connective tissues. Epithelial tissues are *avascular*—they lack blood vessels. Thus, they can be shed safely without the risk of bleeding.

Because epithelial tissue covers surfaces and lines cavities, it always has a *free surface*. The other surface is attached to underlying tissue by a layer of carbohydrates and proteins called the *basement membrane*.

Epithelial tissues are classified according to the shape of the cells and the number of cell layers (Table 4.1). **Simple** epithelial tissue is composed of a single layer, and **stratified** epithelial tissue is composed of two or more layers. **Squamous** epithelium has flattened cells; **cuboidal** epithelium has cube-shaped cells; and **columnar** epithelium has elongated cells. In addition, there are two specialized epithelial tissues: transitional epithelium and pseudostratified columnar epithelium.

Squamous Epithelium

Simple squamous epithelium is composed of a single layer of flattened cells, and therefore its protective function is not as significant as that of other epithelial tissues (Fig. 4.1). It is found in areas where simple diffusion occurs. (Recall that simple diffusion is movement of a substance from areas of high to low concentration.) For example, simple squamous epithelium forms the tiny air sacs (called alveoli) of the lungs, where oxygen and carbon dioxide are exchanged. Capillaries (the smallest blood vessels) are tubes of simple squamous epithelium. Nutrients and wastes are exchanged between capillaries and neighboring cells by diffusion.

Stratified squamous epithelium has many cell layers and does play a protective role (Fig. 4.2). While the deeper cells may be cuboidal or columnar, the outer layer is composed of squamous-shaped cells (Note that stratified epithelial tissues are named according to the cells on their outer layer). The outer portion of skin is stratified squamous epithelium. New cells produced in a basal layer become reinforced by keratin, a protein that waterproofs and provides strength, as they move toward the skin's surface. Aside from skin, stratified squamous epithelium is found lining the various orifices of the body.

TABLE 4.1 Epithelial Tissues		
Type of Tissue	**Description**	**Location**
Simple squamous	One layer of flattened cells	Blood capillaries; air sacs (alveoli) of lungs
Stratified squamous	Many layers; cells flattened at free surface	Skin and body orifices
Simple cuboidal	One layer of cube-shaped cells	Secreting glands, ovaries, linings of kidney tubules
Stratified cuboidal	Two or more layers of cube-shaped cells	Linings of salivary gland and mammary gland ducts
Simple columnar	One layer of elongated cells	Lining of digestive organs; lining of uterine tubes
Stratified columnar	Two or more layers of elongated cells	Pharynx (back of throat); male urethra (tube that carries urine out of body)
Pseudostratified columnar	One layer of elongated, tapered cells; appear stratified	Air passages of the respiratory system
Transitional	Many layers; when tissue stretches, layers become fewer	Urinary bladder (stores urine); ureters and urethra (tubes carrying urine)

free surface

plasma membrane

nucleus

basement membrane

Simple Squamous Epithelium

Location:
Lines air sacs of lungs; blood vessels; heart

Function:
Filtration; diffusion; osmosis

Figure 4.1 **Simple squamous epithelium.**
The thin and flat cells are tightly joined.
The nuclei tend to be broad and thin.

keratinized stratified squamous epithelial cells

living stratified squamous epithelial cells

basement membrane

connective tissue

keratinized stratified squamous epithelial cells

living stratified squamous epithelial cells

basement membrane

connective tissue

LM 100x

Stratified Squamous Epithelium

Location:
Lining of esophagus, mouth, vagina; epidermis

Function:
Protection from abrasion & dessication in keratinized form

Figure 4.2 **Stratified squamous epithelial tissue.** This type of tissue is found in the skin.

Cuboidal Epithelium

Simple cuboidal epithelium (Fig. 4.3) consists of a single layer of cube-shaped cells attached to a basement membrane. This type of epithelium is frequently found in glands, such as salivary glands, the thyroid gland, and the pancreas, where its function is secretion. Simple cuboidal epithelium also covers the ovaries and lines most of the kidney tubules, the portion of the kidney where urine is formed. In one part of the kidney tubule, it absorbs substances from the tubule, and in another part it secretes substances into the tubule. Tubular absorption and secretion are both forms of active transport. Thus, the cuboidal epithelial cells contain many mitochondria, which supply the ATP needed for active transport. Additionally, where the cells function in reabsorption, microvilli (tiny, fanlike folds in the plasma membrane) increase the surface area of the cells.

Stratified cuboidal epithelium is mostly found lining the larger ducts of certain glands, such as the mammary glands and the salivary glands. Often this tissue has only two layers.

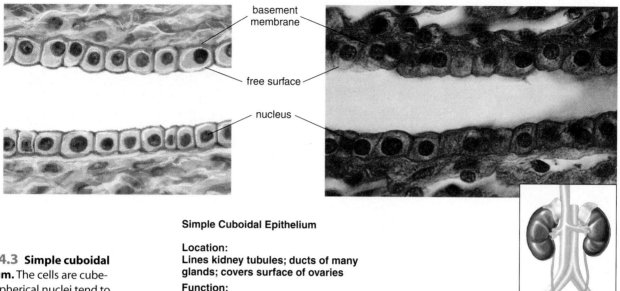

basement membrane

free surface

nucleus

Simple Cuboidal Epithelium

Location:
Lines kidney tubules; ducts of many glands; covers surface of ovaries

Function:
Secretion; absorption

Figure 4.3 Simple cuboidal epithelium. The cells are cube-shaped. Spherical nuclei tend to be centrally located.

free surface

mucus

goblet cell

nucleus

basement membrane

Simple Columnar Epithelium

Location:
Lines gastrointestinal tract; the ducts of many glands

Function:
Protection; secretion; absorption

Figure 4.4 Simple columnar epithelium. The cells are longer than they are wide. The nuclei are in the lower half of the cells.

Columnar Epithelium

Simple columnar epithelium (Fig. 4.4) has cells that are longer than they are wide. They are modified to perform particular functions. Some of these cells are goblet cells that secrete mucus onto the free surface of the epithelium.

This tissue is well known for lining digestive organs, including the small intestine, where microvilli expand the surface area and aid in absorbing the products of digestion. Simple columnar epithelium also lines the uterine tubes. Here, many cilia project from the cells and propel the egg toward the uterus, or womb.

Stratified columnar epithelium is not very common but does exist in parts of the pharynx (back of the throat) and the male urethra.

Pseudostratified Columnar Epithelium

Pseudostratified columnar epithelium is so named because it appears to be layered (*pseudo,* false; *stratified,* layers). However, true layers do not exist because each cell touches the basement membrane. Each cell is tapered and narrow at one end; the opposite end contains the nucleus. The irregular

cilia
mucus
goblet cell
nucleus
basement membrane

Pseudostratified Ciliated Columnar Epithelium

Figure 4.5 Pseudostratified ciliated columnar epithelium. The cells have cilia and appear to be stratified, but each actually touches the basement membrane.

Location:
Lines respiratory tract; parts of the reproductive tracts
Function:
Protection; secretion; movement of mucus and sex cells

placement of the nuclei creates the appearance of several layers where only one exists.

Pseudostratified ciliated columnar epithelium (Fig. 4.5) lines parts of the reproductive tract as well as the air passages of the respiratory system, including the nasal cavities and the trachea (windpipe) and its branches. Mucous-secreting goblet cells are scattered among the ciliated epithelial cells. A surface covering of mucus traps foreign particles, and upward ciliary motion carries the mucus to the back of the throat, where it may be either swallowed or expectorated.

Begin Thinking Clinically

?

In long-term smokers, the pseudostratified ciliated columnar epithelium in the bronchi is gradually converted to stratified squamous epithelium. Why do you think this change occurs?

Answer and discussion in Appendix C, page 462.

Transitional Epithelium

The term **transitional epithelium** implies changeability, and this tissue changes in response to tension. It forms the lining of the urinary bladder, the ureters (tubes that carry urine from the kidneys to the bladder), and part of the urethra (the single tube that carries urine to the outside). All are organs that may need to stretch. When the walls of the bladder are relaxed, the transitional epithelium consists of several layers of cuboidal cells. When the bladder is distended with urine, the epithelium stretches, and the outer cells take on a squamous appearance. It's interesting to observe that the cells in transitional epithelium of the bladder are physically able to slide in relation to one another while at the same time forming a barrier that prevents any part of urine from diffusing into the internal environment.

Content **CHECK-UP!**

1. Which of the following are characteristics of epithelial tissues?
   a. possess a blood supply.
   b. rapid rate of mitosis.
   c. very tightly packed.
   d. b and c are correct.
   e. all are correct.

2. A single layer of tall, skinny epithelial cells would be best described as:
   a. stratified columnar epithelium.
   b. simple cuboidal epithelium.
   c. simple squamous epithelium.
   d. simple columnar epithelium.

3. A student studying slides of epithelial tissues under a microscope observes the presence of microvilli. She guesses that the function of the tissue is probably:
   a. secreting mucus.
   b. moving substances through a tube.
   c. absorbing something inside a tube.

Answers in Appendix C, page 461.

4.2 Connective Tissue

Connective tissue binds structures together, provides support and protection, fills spaces, produces blood cells, and stores fat. The body uses this stored fat for energy, insulation, and organ protection. As a rule, connective tissue cells are widely separated by a nonliving, extracellular **matrix** composed of an *organic ground substance* that contains *fibers* and varies in consistency from solid to semifluid to fluid. Whereas the functional and physical properties of epithelial tissues are derived from

TABLE 4.2 Classification of Connective Tissue

Type	Structure of Matrix	Example of Location
FIBROUS CONNECTIVE TISSUE		
Loose connective tissue		
Areolar	Collagen, elastin, and reticular fibers	Between tissues and organs
Adipose	Fibroblasts enlarge and store fat; very little matrix	Beneath skin; around organs
Dense connective tissue		
Regular	Bundles of parallel collagen fibers	Tendons; ligaments; aponeuroses
Irregular	Bundles of nonparallel collagen fibers	Dermis of skin; joint capsules
Reticular connective tissue	Reticular fibers	Lymphatic organs and liver
CARTILAGE		
Hyaline cartilage	Fine collagen fibers	Ends of long bones; rib cartilages; nose
Elastic cartilage	Many elastin fibers	External ear
Fibrocartilage	Strong collagen fibers	Between vertebrae of spine
BONE		
Compact	Collagen plus calcium salts; arranged in *osteons*	Skeleton
Spongy	Collagen plus calcium salts; arranged in *trabeculae*	Ends of long bones
BLOOD	Plasma plus red and white blood cells, platelets	Inside blood vessels

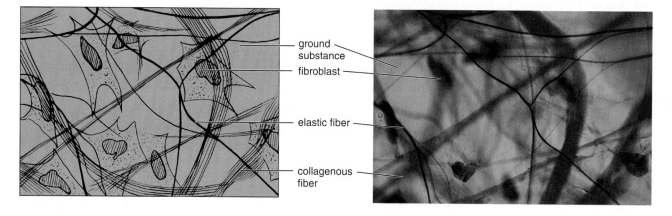

ground substance

fibroblast

elastic fiber

collagenous fiber

Loose (Areolar) Connective Tissue

Location:
Between muscles; beneath the skin;
beneath most epithelial layers

Function:
Binds organs together

Figure 4.6 Loose (areolar) connective tissue. This tissue has a loose network of fibers.

their cells, connective tissue properties are largely derived from the characteristics of the matrix (Table 4.2).

The fibers within the matrix are of three types. **Collagen fibers** contain the fibrous protein *collagen*, a substance that gives the fibers flexibility and tremendous strength. **Elastic fibers** contain the protein *elastin*, which is not as strong as collagen but is more elastic. **Reticular fibers** are very thin, highly branched, collagenous fibers that form delicate supporting networks.

Fibrous Connective Tissue

Fibrous connective tissue includes loose connective tissue and dense connective tissue. The body's membranes are composed of an epithelium and fibrous connective tissue (see page 6).

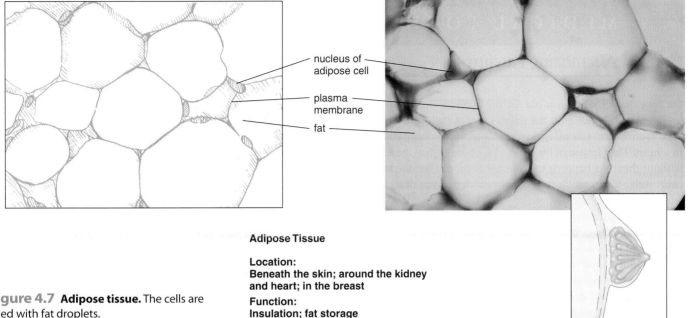

nucleus of adipose cell

plasma membrane

fat

Adipose Tissue

Location:
Beneath the skin; around the kidney and heart; in the breast

Function:
Insulation; fat storage

Figure 4.7 Adipose tissue. The cells are filled with fat droplets.

fibroblasts

collagenous fibers

Dense Connective Tissue

Location: Tendons; ligaments

Function: Binds organs together

Figure 4.8 Dense regular connective tissue. Parallel bundles of collagenous fibers are closely packed.

Loose (areolar) connective tissue commonly lies between other tissues or between organs, binding them together. Areolar tissue has a fine, spider-web appearance. The cells of this tissue are mainly **fibroblasts**—large, star-shaped cells that produce extracellular fibers (Fig. 4.6). The cells are located some distance from one another because they are separated by a matrix with a jellylike ground substance that contains many collagen and elastin fibers. The collagen fibers occur in bundles and are strong and flexible. The elastin fibers form a highly elastic network that returns to its original length after stretching. *Adipose tissue* (Fig. 4.7) is a type of loose connective tissue in which the fibroblasts enlarge and store fat, and there is limited extracellular matrix. Adipose protects and

cushions many organs, including the eye and kidney. It functions to store energy as well.

Dense connective tissue (Fig. 4.8) has a matrix produced by fibroblasts that contains thick bundles of collagen fibers. In *dense regular connective tissue,* the bundles are parallel as in **tendons** (which connect muscles to bones), **ligaments** (which connect bones to other bones at joints), and **aponeuroses** (sing., *aponeurosis*; join muscle to muscle). In *dense irregular connective tissue,* the bundles run in different directions. This type of tissue is found in the inner portion of the skin (called the *dermis*) and in joint capsules (Fig. 4.9).

The fibroblasts of **reticular connective tissue** are called reticular cells, and the matrix contains only reticular fibers.

Compact Bone

Location: Bones of skeleton

Function: Support; protection; mineral storage

Figure 4.11 **Compact bone.** Cells are arranged in a cylindrical manner about a central canal.

friction between joints. Fibrocartilage is found in structures that withstand tension and pressure, such as the disks between the vertebrae in the backbone and the pads in the knee joint.

Bone

Bone is the most rigid of the connective tissues. It has an extremely hard matrix of mineral salts, notably calcium salts, deposited around collagen fibers. The minerals give bone rigidity. The collagen fibers provide elasticity and strength, much like steel rods reinforce concrete.

The outer portion of a long bone contains compact bone. **Compact bone** consists of many cylindrical-shaped units called *osteons,* or *Haversian systems* (Fig. 4.11). In an osteon, matrix is deposited in thin layers called *lamellae* that form a concentric pattern around tiny tubes called *central canals.* The canals contain nerve fibers and blood vessels. The blood vessels bring nutrients to bone cells (called *osteocytes*) that are located in small hollows called *lacunae* between the lamellae. The nutrients can reach all of the cells because minute canals (*canaliculi*) containing thin extensions of the osteocytes connect the osteocytes with one another and with the central canals.

The ends of a long bone contain spongy bone, which has an entirely different structure. **Spongy bone** contains numerous bony bars and plates called **trabeculae** separated by irregular spaces. Although lighter than compact bone, spongy bone is still designed for strength. Like braces used for support in buildings, the solid plates of spongy bone follow lines of stress. Blood cells are formed within red marrow found in spongy bone at the ends of certain long bones.

Begin Thinking Clinically

?

Two young, otherwise healthy basketball players are in the emergency room. One has an ankle sprain (tearing of a ligament). The second has an ankle bone fracture. Who do you think might heal more quickly?
Answer and discussion in Appendix C, page 462.

Blood

Blood (Fig. 4.12) is a connective tissue composed of **formed elements** suspended in a liquid matrix called **plasma.** There are three types of formed elements: **red blood cells (erythrocytes),** which carry oxygen; **white blood cells (leukocytes),** which aid in fighting infection; and **platelets (thrombocytes),** which are important to the initiation of blood clotting. Platelets are not complete cells; rather, they are fragments of giant cells called **megakaryocytes,** which are found in the bone marrow.

In red bone marrow, stem cells continually divide to produce new cells that mature into the different types of blood cells. The rate of cell division is high because blood cells have a relatively short life span and must be replaced constantly.

Blood is unlike other types of connective tissue in that the extracellular matrix (*plasma*) is not made by the cells of the tissue. Plasma is a mixture of different types of molecules that enter blood at various organs.

plasma

buffy coat

formed elements

platelet

erythrocytes
(red blood cells)

leukocytes
(white blood
cells)

LM 720x

Location:
In the blood vessels

Function:
Supplies cells with nutrients and oxygen
and takes away their wastes; fights infection

Figure 4.12 **Blood.** When a blood sample is centrifuged, the formed elements settle out below the plasma. White blood cells form a layer termed the "buffy coat" on top of red blood cells. Plasma is the liquid portion of the blood. Red blood cells, white blood cells, and platelets are called the formed elements.

TABLE 4.3 Classification of Muscular Tissue			
Type	**Fiber Appearance**	**Location**	**Control**
Skeletal	Striated	Attached to skeleton	Voluntary
Smooth	Spindle-shaped	Wall of hollow organs (e.g., intestine, urinary bladder, uterus, and blood vessels)	Involuntary
Cardiac	Striated and branched	Heart	Involuntary

4.3 Muscular Tissue

Muscular (contractile) tissue is composed of cells called muscle fibers (Table 4.3). Muscle fibers contain actin and myosin, which are protein filaments whose interaction accounts for movement. The three types of vertebrate muscles are skeletal, smooth, and cardiac.

Skeletal Muscle

Skeletal muscle, also called *voluntary muscle* (Fig. 4.13), is attached by tendons to the bones of the skeleton. When skeletal muscle contracts, the muscle shortens, and body parts such as arms and legs move. Contraction of skeletal muscle is generally under one's conscious control. Skeletal muscle fibers are cylindrical and quite long—sometimes they run the length of the muscle. They arise during development when several cells fuse, resulting in one fiber with multiple nuclei. The nuclei are located at the periphery of the cell, just inside the plasma membrane. The fibers have

striation

nucleus

Skeletal Muscle

Fiber appearance: Striated

Location:
Usually attached to skeleton

Control: Voluntary

Figure 4.13 **Skeletal muscle.** The cells are long, cylindrical, and multinucleated.

Cancer—The Traitor Inside

The life of almost every person has been touched, either directly or indirectly, by the specter of cancer. **Cancer** is not one disease, but perhaps several hundred diseases, all sharing a common characteristic: rapid, uncontrolled, and disorganized growth of tissue cells. Thus, any cell in any of the body's tissues can be the starting point for cancer. Cancers are classified according to the type of tissue from which they arise. **Carcinomas,** the most common type, are cancers of epithelial tissues (skin and linings); **sarcomas** are cancers that arise in connective tissue (muscle, bone, and cartilage); **leukemias** are cancers of the blood; and **lymphomas** are cancers of reticular connective tissue. The chance of cancer occurring in a particular tissue is related to the rate of cell division. Because epithelial cells reproduce at a high rate, carcinomas account for 90% of all human cancers.

In the body, a cancer cell divides to form a **malignant neoplasm** ("new tissue"), or a **malignant tumor,** that invades and destroys neighboring tissue. Cancer cells can also detach and spread to other sites by invading the blood vessels or the lymphatic vessels. Through this process, called **metastasis,** cancer tumors colonize healthy tissue elsewhere in the body. By contrast, noncancerous, or **benign tumors** are encapsulated and stay in one place. To support their growth, cancer cells release a growth factor that causes neighboring blood vessels to branch into the cancerous tissue. This phenomenon is called **vascularization,** and some modes of cancer treatment are aimed at preventing vascularization (see What's New on page 79).

Cancer development seems to occur by a two-step process involving (1) initiation and (2) promotion. Initiation of cancer is caused by a change, or a **mutation,** in the DNA (genes) of a cell, which results in runaway cell growth. Agents that are known to cause DNA mutations are called **carcinogens.** Known carcinogens include viruses, excessive radiation, and certain chemicals. For example, cigarette smoke contains chemical carcinogens that may initiate cancers of the lung, throat, mouth, and urinary bladder. A cancer **promoter** is any influence that causes a mutated cell to start growing in an uncontrolled manner. A promoter might cause a second mutation or provide the environment for cells to form a tumor. Some evidence suggests that a diet rich in saturated fats and cholesterol promotes colon cancer. Considerable time may elapse between initiation and promotion, and this is one reason why cancer is seen more often in older people.

Cancer can be detected by physical examination, assisted by various means of viewing the internal organs. Mammograms can detect early breast cancer using low-level X ray, and thyroid cancer is diagnosed using radioactive iodine (see the Medical Focus on page 16). Specific blood tests exist for tumors that secrete a particular chemical in the blood. For example, the level of prostate-specific antigen (PSA) appears to increase in the blood according to the size of a prostate tumor. Tissue samples can also detect early malignancy. During a **Pap smear** (named for George Papanicolaou, the Greek doctor who first described the test), epithelial tissue lining the cervix at the opening of the uterus is obtained using a cotton swab. A **biopsy** is the removal of sample tissue using a plunger-like device. A **pathologist** is skilled at recognizing the abnormal characteristics that allow for cancer **diagnosis.** If cancer is found and treated before metastasis occurs, chances for a complete cure are greatly increased.

---

are microglia, astrocytes, oligodendrocytes, and ependymal cells. *Microglia,* in addition to supporting neurons, engulf bacterial and cellular debris. *Astrocytes* provide nutrients to neurons and produce a hormone known as glia-derived growth factor, which someday might be used as a cure for Parkinson disease and other diseases caused by neuron degeneration. *Oligodendrocytes* form myelin, a protective layer of fatty insulation. *Ependymal cells* line the hollow cavities, or ventricles, of the brain.

Schwann cells are the type of neuroglia that enclose all long nerve fibers located outside the brain or spinal cord. Each Schwann cell covers only a small section of a nerve fiber. The gaps between Schwann cells are called **nodes of Ranvier.** Collectively, the Schwann cells provide nerve fibers with a myelin sheath separated by the nodes. The **myelin sheath** speeds conduction because the nerve impulse jumps from node to node. Because the myelin sheath is white, all myelinated nerve fibers appear white.

Content **CHECK-UP!**

4. Bundles of nonparallel collagen fibers can be found in:
   a. dense regular connective tissue.
   b. dense irregular connective tissue.
   c. reticular connective tissue.
   d. adipose tissue.

5. Gap junctions and intercalated disks allow this type of muscle to contract as a unit:
   a. skeletal muscle.　　　　　c. cardiac muscle.
   b. smooth muscle.

6. Complete this statement: In a nerve cell, a(n) _____ receives signals while a(n) _____ conducts impulses.
   a. axon, dendrite　　　　　c. dendrite, cell body
   b. dendrite, axon

Answers in Appendix C, page 461.

Tumors can often be removed surgically, but there is always the danger that they have metastasized. For this reason, surgery is often preceded or followed by radiation therapy and/or chemotherapy. The goal of both is to destroy rapidly dividing cancer cells. Radiation destroys the more rapidly dividing cancer cells, but may also damage the more slowly dividing normal cells. The use of radioactive protons is preferred over X ray because proton beams can be aimed directly at the tumor, like an automatic rifle hitting the bull's-eye of a target.

Chemotherapy is the use of drugs to kill actively growing cancer cells. Sometimes, cancer cells become resistant to chemotherapy (even when several drugs are used in combination). The plasma membrane in resistant cells contains a carrier that pumps toxic chemicals out of the cell. Researchers are testing drugs known to poison the pump in an effort to restore sensitivity to chemotherapy. Unfortunately, both chemotherapy and radiation therapy kill normal cells in addition to the cancer. The patient will suffer the negative side effects of therapy: nausea, vomiting, hair loss, weight loss, anemia, etc. Thus, the use of chemotherapy and radiation must be balanced carefully: It must be strong enough to kill cancer, but not so strong as to cause the person's death.

Immunotherapy and gene therapy are new, experimental ways of treating cancer. Immunotherapy is the use of an immune system component to treat a disease. For example, cancer patients are sometimes given *cytotoxins*—chemicals released by lymphocytes, a type of white blood cell. Gene therapy is the substitution of "good genes" for defective or missing genes to treat a disease. The hope is that, one day, cancer can be cured by replacing a normal gene for the mutated gene in cancerous cells.

Individuals should be aware of the seven danger signals for cancer (Table 4A) and inform their doctor when any one of these are observed. Further, the evidence is clear that the risk of certain types of cancer can be reduced by adopting certain behaviors. For example, avoiding excessive sunlight reduces the risk of skin cancer, and abstaining from cigarettes, cigars, and chewing tobacco reduces the risk of oral, throat, and lung cancers, as well as other types of cancer. Exercise and a healthy diet are also believed to be important. Recommendations include:

1. Lowering total fat intake
2. Eating more high-fiber foods
3. Increasing consumption of foods rich in vitamins A and C
4. Reducing consumption of salt-cured and smoked foods
5. Including vegetables of the cabbage family in the diet
6. Consuming only moderate amounts of alcohol

TABLE 4.A Danger Signals for Cancer

C hange in bowel or bladder habits
A sore that does not heal
U nusual bleeding or discharge
T hickening or lump in breast or elsewhere
I ndigestion or difficulty in swallowing
O bvious change in wart or mole
N agging cough or hoarseness

4.5 Extracellular Junctions, Glands, and Membranes

Extracellular Junctions

The cells of a tissue can function in a coordinated manner when the plasma membranes of adjoining cells interact. The junctions that occur between cells help cells function as a tissue.

A **tight junction** forms an impermeable barrier because adjacent plasma membrane proteins actually join, producing a zipperlike fastening (Fig. 4.17*a*). In the stomach, digestive secretions are contained, and in the kidneys, the urine stays within kidney tubules because epithelial cells are joined by tight junctions. A **gap junction** forms when two adjacent plasma membrane channels join (Fig. 4.17*b*). This lends strength, but it also allows ions, sugars, and small molecules to pass between the two cells. Gap junctions in heart and smooth muscle ensure synchronized contraction. In an **adhesion junction** (desmosome), the adjacent plasma membranes do not touch but are held together by extracellular filaments firmly attached to cytoplasmic plaques, composed of dense protein material (Fig. 4.17*c*). Desmosomes that join heart muscles cells prevent the cells from tearing apart during contraction. Similarly, desmosomes in the *cervix*, the opening to the uterus (womb), prevent the cervix from ripping when a woman gives birth.

Glands

A **gland** consists of one or more cells that produce and secrete a product. Most glands are composed primarily of epithelium in which the cells secrete their product by exocytosis. During secretion, the contents of a vesicle are released when the vesicle fuses with the plasma membrane.

The mucous-secreting goblet cells within the columnar epithelium lining the digestive tract are single cells (see Fig. 4.4).

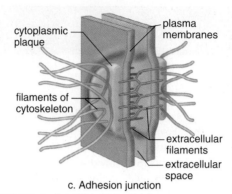

plasma membranes

tight-junction proteins

extracellular space

a. Tight junction

plasma membranes

membrane channels

extracellular space

b. Gap junction

cytoplasmic plaque

filaments of cytoskeleton

plasma membranes

extracellular filaments

extracellular space

c. Adhesion junction

Figure 4.17 Extracellular junctions. Tissues are held together by **a.** tight junctions that are impermeable; **b.** gap junctions that allow materials to pass from cell to cell; and **c.** adhesion junctions that allow tissues to stretch.

Simple | Example: Sweat gland | Compound | Example: Mammary gland | Compound | Example: Pancreas

Duct Secretory portion

Figure 4.18 Multicellular exocrine glands. Exocrine glands have ducts that can be simple or compound. Compound glands vary according to the placement of secretory portions.

Glands with ducts that secrete their product onto the outer surface (e.g., sweat glands and mammary glands) or into a cavity (e.g., pancreas) are called *exocrine glands.* Ducts can be simple or compound, as illustrated in Figure 4.18.

Glands that no longer have a duct are appropriately known as the ductless glands, or endocrine glands. *Endocrine glands* (e.g., pituitary gland and thyroid) secrete their products internally so they are transported by the bloodstream. Endocrine glands produce hormones that help promote homeostasis. Each type of hormone influences the metabolism of a particular target organ or cells.

Glands are composed of epithelial tissue, but they are supported by connective tissue, as are other epithelial tissues.

Membranes

Membranes line the internal spaces of organs and tubes that open to the outside, and they also line the body cavities discussed on pages 6 and 7.

Mucous Membranes

Mucous membranes line the interior walls of the organs and tubes that open to the outside of the body, such as those of the digestive, respiratory, urinary, and reproductive systems. These membranes consist of an epithelium overlying a layer of loose connective tissue. The epithelium contains goblet cells that secrete mucus.

The mucus secreted by mucous membranes ordinarily protects interior walls from invasion by bacteria and viruses;

for example, more mucus is secreted when a person has a cold, resulting in a "runny nose." In addition, mucus usually protects the walls of the stomach and small intestine from digestive juices, but this protection breaks down when a person develops an *ulcer.*

Serous Membranes

As also discussed on pages 6 and 7, **serous membranes** line cavities, including the thoracic and abdominopelvic cavities, and cover internal organs such as the intestines. The term **parietal** refers to the wall of the body cavity, while the term visceral pertains to the internal organs. Therefore, parietal membranes line the interior of the thoracic and abdominopelvic cavities, and visceral membranes cover the organs.

Serous membranes consist of a layer of simple squamous epithelium overlying a layer of loose connective tissue. They secrete serous fluid, which keeps the membranes lubricated. Serous membranes support the internal organs and tend to compartmentalize the large thoracic and abdominopelvic cavities. This helps hinder the spread of any infection.

In the thorax, the **pleurae** are serous membranes that form a double layer around the lungs. The parietal pleura lines the inside of the thoracic wall, while the visceral pleura adheres to the surface of the lungs. Similarly a double-layered serous membrane is a part of the **pericardium,** a covering for the heart (see Fig. 12.2).

The **peritoneum** is the serous membrane within the abdomen. The parietal peritoneum lines the abdominopelvic wall, and the visceral peritoneum covers the organs. In between

Blocking Blood Supply to Treat Cancer

All tissue cells share some basic requirements for survival: a constant supply of oxygen and nutrients, and a mechanism to eliminate wastes. Epithelial tissues meet these needs by simple diffusion from underlying tissues, but the rest of the body's tissues must have a continuous blood supply. The basic needs of cancer cells are further increased by their furious pace of growth and metabolism. Cancers guarantee their own continuous blood supply by stimulating nearby blood vessels to sprout new capillaries, in a process called *angiogenesis* (*angio*, blood vessel; *genesis*, to create).

Angiogenesis occurs in normal tissues as well as cancerous growths. Blood vessels proliferate and grow during embryonic and fetal development, as a child grows, whenever a wound heals, and as a woman's uterus goes through the early stage of her menstrual cycle. Blood vessel growth requires that cells produce specific chemicals called growth factors. Cancer cells have been shown to produce large quantities of a specific growth factor, termed vascular endothelial growth factor, or VEGF. This chemical will induce the blood vessels that neighbor a tumor to branch toward the tumor, creating a vascular network. The completed network not only supplies the tumor's metabolic needs—it also serves as a "super-highway" that allows tumor cells to metastasize to other parts of the body.

The U.S. Food and Drug Administration (FDA) has recently approved a new medication for the treatment of colorectal cancer as well as certain types of lung cancer. The drug, called bevacizumab (Avastatin®), has successfully blocked the action of VEGF and other angiogenesis growth factors in these types of cancers. Similar medications for treating breast, pancreatic, liver, and other cancers are in clinical trials. These treatments consist of *antibodies*—small immune system proteins—that combine with VEGF. Thus, VEGF is prevented from signaling the growth of blood vessels to supply a tumor—essentially starving the tumor. Researchers studying angiogenesis inhibitors have also found that using the drugs boosts the effectiveness of traditional cancer-fighting methods like chemotherapy and radiation.

Blocking a tumor's blood supply by using immune system proteins is one aspect of *immunotherapy*. Research in immunotherapy is promising because this type of treatment allows cancer cells to be specifically targeted, without causing collateral damage to healthy tissue.

the organs, the visceral peritoneum comes together to form a double-layered **mesentery** that supports these organs (see Fig. 15.5).

Synovial Membranes

Synovial membranes line freely movable joint cavities and are composed of connective tissues. They secrete synovial fluid into the joint cavity. This fluid lubricates the ends of the bones so that they can move freely (see Fig. 6.21). In *rheumatoid arthritis*, the synovial membrane becomes inflamed and grows thicker. Fibrous tissue then invades the joint and may eventually become bony so that the bones of the joint are no longer capable of moving.

Meninges

The **meninges** are membranes found within the posterior cavity (see Fig. 1.5). They are composed only of connective tissue and serve as a protective covering for the brain and spinal cord. *Meningitis* is a life-threatening infection of the meninges (see the Medical Focus on page 9).

Cutaneous Membrane

The **cutaneous membrane,** or skin, forms the outer covering of the body. It consists of an outer portion of keratinized stratified squamous epithelium attached to a thick underlying layer of dense irregular connective tissue. The skin is discussed in detail in Chapter 5.

Content **CHECK-UP!**

7. Match each type of extracellular junction to its function:

a. desmosome _____	1. allows ions and small molecules to pass between cells
b. tight junction _____	2. prevents adjacent cells from tearing apart
c. gap junction _____	3. forms an impermeable barrier

8. Match each type of membrane to its function:

a. synovial membrane _____	1. forms outer covering of the body
b. cutaneous membrane _____	2. lines walls of organs that open to the outside of the body
c. mucous membrane _____	3. lines the interior of a joint capsule

Answers in Appendix C, page 461.

SELECTED NEW TERMS

Basic Key Terms

adhesion junction (ăd-hē′zhŭn jŭngk′shən), p. 77

aponeuroses (ap′ ō noo-rō-sis), p. 69

blood (blŭd), p. 72

bone (bōn), p. 72

cartilage (kar′tĭ-lij), p. 70

collagen fibers (kŏl′ə-jən fī′bərz), p. 68

columnar (kə-lŭm′nər), p. 64

connective tissue (kŏ-nek′tiv tish′u), p. 67

cuboidal (kyoō-boid′l), p. 64

cutaneous membrane (kyū-ta′ne-us mem′brān), p. 79

elastic cartilage (ī-lăs′tĭk kär′tl-ĭj), p. 70

elastic fibers (ī-lăs′tĭk fī′bərz), p. 68

epithelial tissue (epi″ī-the′le-al tish′u), p. 68

fibroblasts (fī′brə-blăst′), p. 69

fibrocartilage (fī′brō-kär′tl-ĭj), p. 70

gap junction (găp jŭngk′shən), p. 77

hyaline cartilage (hī′ə-lĭn kär′tl-ĭj), p. 70

intercalated disk (in-tĕr-kĕ-lāt-ĕd disk), p. 75

lacunae (luh-ku′nā), p. 70

ligaments (lĭg′a-məntz), p. 69

lymphatic tissue (lĭm-făt′ĭk tĭsh′oō), p. 70

matrix (ma′triks), p. 67

meninges (mĕ-nin′jēz), p. 79

mesentery (mes′en-tĕr″e), p. 79

mucous membrane (myŭ′kus mem′brān), p. 78

muscular tissue (mus′kyŭ-ler tish′u), p. 73

myelin sheath (mi′ĕ-lin shēth), p. 76

nervous tissue (ner′vus tish′u), p. 75

neuroglia (nu-rog′le-uh), p. 75

neuron (nu′ron), p. 75

pericardium (pēr′ī-kär′dē-əm), p. 78

pseudostratified columnar (su″do-strat′ī-fīd kə-lŭm′n ər), p. 66

reticular connective tissue (re-tik′ŭ-lār ko-nek′tiv tish′ū), p. 69

reticular fibers (rĭ-tīk′yə-lər fī′b ərz), p. 68

serous membrane (sĕr′us mem′brān), p. 78

simple (sĭm′pəl), p. 64

squamous (skwā′məs), p. 64

stratified (strat′ī-fīd), p. 64

stratified cuboidal epithelium (strat′ī-fīd kyoō-boyd′l ep-ī″thē′ lē-yŭm), p. 65

synovial membrane (sĭ-no′ve-al mem′brān), p. 79

tendons (tĕn′dənz), p. 69

tight junction (tīt jŭnāk′shən), p. 77

transitional epithelium (trăn-zīsh′ ən-əl ĕp′ ə-thē′lē-əm), p. 67

Clinical Key Terms

benign (bē-nīn′) tumor, p. 76

biopsy (bi′op-se), p. 76

cancer (kăn′sər), p. 76

carcinogen (kar-sin′ō-jen), p. 76

carcinoma (kar-sĭ-no′muh), p. 76

diagnosis (di-ahg-no′sis), p. 76

leukemia (lu-ke′me-uh), p. 76

lymphoma (lim-fo′muh), p. 76

malignant (mă-lig′-nant) tumor, p. 76

metastasis (mĕ-tas′tă-sis), p. 76

mutation (myoō-tā′shən), p. 76

neoplasm (nē′ ə-plăz′ əm), p. 76

Pap smear (pap smēr), p. 76

pathologist (puh-thol′uh-jist), p. 76

promoter (prə-mō′tər), p. 76

sarcoma (sar-ko′muh), p. 76

vascularization (văs′kyə-lər-ĭ-zā′shən), p. 76

SUMMARY

4.1 Epithelial Tissue
A. Body tissues are categorized into four types: epithelial, connective, muscular, and nervous.
B. Epithelial is classified according to cell shape and number of layers. The cell shape can be squamous, cuboidal, or columnar. Simple tissues have one layer of cells, and stratified tissues have several layers.

4.2 Connective Tissue
A. In connective tissue, cells are separated by a matrix (organic ground substance plus fibers).
B. Fibrous connective tissue can be loose connective tissue, in which fibroblasts are separated by a jellylike ground substance, or dense connective tissue, which contains bundles of collagenous fibers. Adipose tissue is a type of loose connective tissue in which the fibroblasts enlarge and store fat.
C. Cartilage and bone are support tissues. Cartilage is more flexible

than bone because the matrix is rich in protein, rather than the mineral salts found in bone.
D. Blood is a connective tissue in which the matrix is plasma.

4.3 Muscular Tissue
Muscular tissue contains actin and myosin protein filaments. These form a striated pattern in skeletal and cardiac muscle, but not in smooth muscle. Cardiac and smooth muscle are under involuntary control. Skeletal muscle is generally under voluntary control.

4.4 Nervous Tissue
Nervous tissue contains conducting cells called neurons. Neurons have processes called axons and dendrites. In the brain and spinal cord, axons are organized into tracts. Outside the brain and spinal cord, axons (fibers) are found in nerves.

4.5 Extracellular Junctions, Glands, and Membranes
A. In a tissue, cells can be joined by tight junctions, gap junctions, or adhesion junctions (desmosomes).

B. Glands are composed of epithelial tissue that produces and secretes a product, usually by exocytosis. Glands can be unicellular or multicellular. Multicellular exocrine glands have ducts and secrete onto surfaces; endocrine glands are ductless and secrete into the bloodstream.
C. Mucous membranes line the interior of organs and tubes that open to the outside. Serous membranes line the thoracic and abdominopelvic cavities, and cover the organs within these cavities. Synovial membranes line certain joint cavities. Meninges are membranes that cover the brain and spinal cord. The skin forms a cutaneous membrane.

STUDY QUESTIONS

1. What is a tissue? (p. 64)
2. Name the four major types of tissues. (p. 64)
3. What are the functions of epithelial tissue? Name the different kinds of epithelial tissue, and give a location for each. (pp. 64–67)
4. What are the functions of connective tissue? Name the different kinds of connective tissue, and give a location for each type. (pp. 67–72)
5. Contrast the structure of cartilage with that of bone, using the words *lacunae* and *central canal* in your description. (pp. 70–72)
6. Describe the composition of blood, and give a function for each type of blood cell. (p. 72)
7. What are the functions of muscular tissue? Name the different kinds of muscular tissue, and give a location for each. (pp. 73–75)
8. What types of cells does nervous tissue contain? Which organs in the body are made up of nervous tissue? (pp. 75–76)
9. Name three types of junctions, and state the function of each with examples. (p. 77)
10. Describe the structure of a gland. What is the difference between an exocrine gland and an endocrine gland? (pp. 77–78)
11. Name the different types of body membranes, and associate each type with a particular location in the body. (pp. 78–79)

LEARNING OBJECTIVE QUESTIONS

I. Fill in the blanks.

1. Most organs contain several different types of _____.
2. Pseudostratified ciliated columnar epithelium contains cells that appear to be _____, have projections called _____, and are _____ in shape.
3. Connective tissue cells are widely separated by a _____ that usually contains _____.
4. Both cartilage and blood are classified as _____ tissue.
5. A mucous membrane contains _____ tissue overlying _____ tissue.

II. Match the organs in the key to the epithelial tissues listed in questions 6–9.

Key:
   a. kidney tubules
   b. small intestine
   c. air sacs of lungs
   d. trachea (windpipe)

6. simple squamous
7. simple cuboidal
8. simple columnar
9. pseudostratified ciliated columnar

III. Match the muscle tissues in the key to the descriptions listed in questions 10–12.

Key:
   a. skeletal muscle
   b. smooth muscle
   c. cardiac muscle

10. striated and branched, involuntary
11. striated and voluntary
12. visceral and involuntary

MEDICAL TERMINOLOGY EXERCISE

Consult Appendix B for help in pronouncing and analyzing the meaning of the terms that follow.

1. epithelioma (ep″ĭ the″le-o′muh)
2. fibrodysplasia (fi″bro-dis-pla′se-uh)
3. meningoencephalopathy (mĕ-ning″go-en-sef″ul-lop′uh-the)
4. pericardiocentesis (per″i-kar″de-o-sen-te′sis)
5. peritonitis (per″ĭ-to-ni′tis)
6. intrapleural (in″tra-plur′al)
7. neurofibromatosis (nu″ro-fi″bro″muh-to′sis)
8. submucosa (sub″myū-ko′suh)
9. polyarthritis (pol″e-ar-thri′tis)
10. cardiomyopathy (kar′de-o-mi-ah′puh-the)
11. encephalitis (en-sef′uh-li-tis)
12. glioma (gle-o′muh)
13. pleurisy (plūr′ĭ-se)
14. chondroblast (kon′dro-blast)
15. osteology (os′te-ol′o-je)

WEBSITE LINK

Visit this text's website at http://www.mhhe.com/longenbaker7 for additional quizzes, interactive learning exercises, and other study tools.

The Integumentary System

PART II

CHAPTER 5

Normally, about 100 hairs a day fall out from hair follicles (described on pages 85–87), like the one pictured here in this photomicrograph. But if you're a man between the ages of 20 and 50, and more than 100 hairs are trapped in your hairbrush every day, you're most likely experiencing male-pattern baldness (**androgenetic alopecia**). This form of hair loss is caused by an excess of the male hormone dihydrotestosterone, and it's inherited. Lotions containing minoxidil (Rogaine©) may increase blood flow to the hair follicles and promote hair growth—but the beneficial effect takes 6–9 months of constant daily use, and only 30–40% of men experience any effect. However, there's good news for bald guys—in today's society, bald is a definite fashion statement!

Chapter Outline & Learning Objectives

After you have studied this chapter, you should be able to:

5.1 Structure of the Skin (p. 83)
- Describe the regions of the skin and the hypodermis.
- Name two main epidermal layers, and describe their structure and function.

5.2 Accessory Structures of the Skin (p. 85)
- Describe the structure and growth of hair and nails.
- Name three glands of the skin, and describe their structure and function.

5.3 Disorders of the Skin (p. 88)
- Name the three types of skin cancer, and state their risk factor.

- Name and describe four types of burns with regard to depth.
- Describe how the "rule of nines" may be used to estimate the extent of a burn.
- Describe the steps by which a skin wound heals.

5.4 Effects of Aging (p. 90)
- Describe the anatomical and physiological changes that occur in the integumentary system as we age.

5.5 Homeostasis (p. 91)
- List and discuss four functions of the skin that contribute to homeostasis.

Medical Focus
Decubitus Ulcers (p. 85)

I.C.E—In Case of Emergency
Burns (p. 92)

Human Systems Work Together
Integumentary System (p. 94)

Medical Focus
Body Art: Buyer Beware! (p. 93)

5.1 Structure of the Skin

The skin, sometimes called the **cutaneous membrane** or the **integument,** covers the entire surface of the human body. In an adult, the skin has a surface area of about 1.8 square meters (20.83 square feet). Thus, it is the largest organ in the human body. It is important to note that the skin is an organ, comprised of all four tissue types: epithelial, connective, muscle, and nervous tissue. Because the skin has several accessory organs, it is also technically an organ system sometimes referred to as the **integumentary system.**

The skin (Fig. 5.1) has two regions: the epidermis and the dermis. These layers are joined together by a basement membrane. The **hypodermis,** a subcutaneous tissue, is found between the skin and any underlying structures, such as muscle. Usually, the hypodermis is only loosely attached to underlying muscle tissue, but where no muscles are present, the hypodermis attaches directly to bone. For example, there are *flexion creases* where the skin attaches directly to the joints of the fingers.

Epidermis

The **epidermis** is the outer and thinner region of the skin. It is made up of stratified squamous epithelium divided into five separate layers, or *strata* (sing., *stratum*). From deepest to most superficial, the layers are *stratum basale, stratum spinosum, stratum granulosum, stratum lucidum*, and *stratum corneum*. Only the stratum basale, stratum lucidum, and stratum corneum will be discussed here. Like all epithelial tissues, epidermis lacks blood vessels and has tightly packed cells.

Stratum Basale

The basal cells of the **stratum basale** lie just superficial to the dermis and are constantly dividing and producing new cells that are pushed to the surface of the epidermis in two to four weeks. As the cells move away from the dermis, they get progressively farther away from the blood vessels in the dermis. Because these cells are not being supplied with nutrients and oxygen (since epidermis lacks blood vessels), they eventually die and are sloughed off.

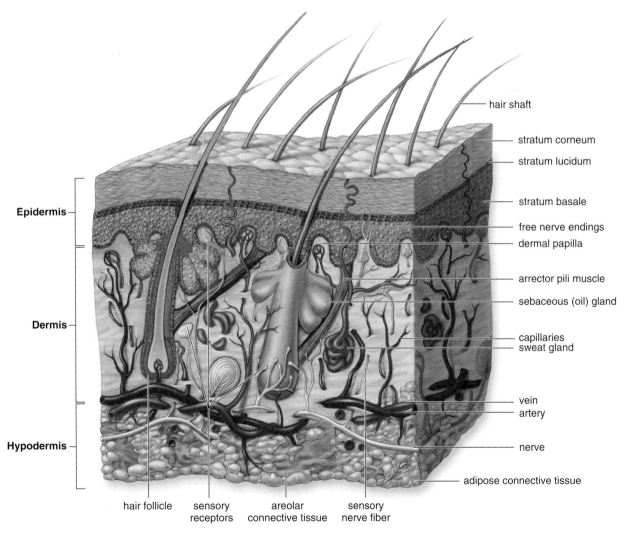

Figure 5.1 Skin anatomy. Skin is composed of two regions: the epidermis and the dermis. The hypodermis, or subcutaneous layer, is located beneath the skin.

Langerhans cells are macrophages found deep in the epidermis. Macrophages are a type of white blood cell (leukocyte). These cells phagocytize microbes and then travel to lymphatic organs, where they stimulate the immune system to react.

Melanocytes are another type of specialized cell located in the deeper epidermis. Melanocytes produce **melanin**, the pigment primarily responsible for skin color. Because the number of melanocytes is about the same in all individuals, variation in skin color is due to the amount of melanin produced and its distribution. When skin is exposed to the sun, melanocytes produce more melanin to protect the skin from the damaging effects of the ultraviolet (UV) radiation in sunlight. The melanin is passed to other epidermal cells, and the result is tanning, or in some people, the formation of patches of melanin called freckles. A hereditary trait characterized by the lack of ability to produce melanin is known as **albinism**. Individuals with this disorder lack pigment not only in the skin, but also in the hair and eyes. Another pigment, called carotene, is present in epidermal cells and in the dermis and gives the skin of certain Asians its yellowish hue. The pinkish color of fair-skinned people is due to the pigment hemoglobin in the red blood cells in the capillaries of the dermis.

Sensory nerves also supply the stratum basale. **Free nerve endings** supply pain and temperature sensations to the brain (Fig. 5.1). **Tactile cells** (also called **Merkel cells**) signal the brain that an object has touched the skin.

Stratum Spinosum and Stratum Granulosum

Immediately superior to the stratum basale are two small layers of cells—the stratum spinosum and stratum granulosum. Like the stratum basale, stratum spinosum cells can reproduce by mitosis. Their name is derived from their spiny appearance, which is created by keratin filaments. Stratum granulosum cells are flattened cells containing still more keratin. They are tightly sealed together, which allows these cells to serve as a barrier.

Stratum Lucidum

As you examine your skin, you can likely identify areas where constant abrasion has created calluses. These are areas where the epidermis has formed **stratum lucidum**, just deep to the stratum corneum. This additional layer is found only in thick skin: the palms of the hands, soles of the feet, elbows, etc. The stratum lucidum provides protection from constant friction.

Stratum Corneum

As cells are pushed toward the surface of the skin, they become flat and hard, forming the tough, uppermost layer of the epidermis, the **stratum corneum**. Hardening is caused by keratinization, the cellular production of a fibrous, waterproof protein called **keratin**. Over much of the body, keratinization is minimal. However, in areas containing an underlying stratum lucidum, a particularly thick layer of dead, keratinized cells affords extra protection.

The waterproof nature of keratin protects the body from water loss and water gain. The stratum corneum allows us to live in a desert or a tropical rain forest without damaging our inner cells. The stratum corneum also serves as a mechanical barrier against microbe invasion. This protective function of skin is assisted by the secretions of *sebaceous glands* (discussed in section 5.2).

> ### Begin Thinking Clinically
>
> **?** Psoriasis is a disorder of the immune system that causes immune cells to target the epidermis. The cells of the epidermis reproduce much more rapidly than normal. What do you think will happen to the skin's appearance?
>
> *Answer and discussion in Appendix C, page 462.*

Dermis

The **dermis**, a deeper and thicker region than the epidermis, is composed of dense irregular connective tissue. The upper layer of the dermis has fingerlike projections called *dermal papillae*. Dermal papillae project into and anchor the epidermis. In the overlying epidermis, dermal papillae cause ridges, resulting in spiral and concentric patterns commonly known as "fingerprints." The function of the epidermal ridges is to increase friction and thus provide a better gripping surface. Because they are unique to each person, fingerprints and footprints can be used for identification purposes.

The dermis contains collagen and elastic fibers. The *collagen fibers* are flexible but offer great resistance to overstretching; they prevent the skin from being torn. The *elastic fibers* stretch to allow movement of underlying muscles and joints, but they maintain normal skin tension. The dermis also contains blood vessels that supply oxygen and nutrients to its cells, and those of the epidermis as well. Blood rushes into these vessels when a person blushes; *pallor* (pale skin) develops when blood flow to dermal vessels is reduced. If blood is not adequately supplied with oxygen (perhaps because of lung disease), the person turns *cyanotic*, or "blue."

Extended periods of diminished blood flow to the dermis can cause the formation of decubitus ulcers, or bedsores (see Medical Focus, p. 85).

There are also numerous sensory nerve fibers in the dermis that take nerve impulses to and from the accessory structures of the skin, which are discussed in section 5.2.

Hypodermis

As mentioned previously, the hypodermis lies below the dermis. From the names for this layer, we get the terms **subcutaneous injection**, performed with a **hypodermic needle**.

The hypodermis is composed of loose connective tissue, including adipose (fat) tissue. Fat is an energy storage form that can be called upon when necessary to supply the body with molecules for cellular respiration. Adipose tissue also helps insulate the body. A well-developed hypodermis gives the body a rounded appearance and provides protective padding against external assaults. Excessive development of adipose tissue in the hypodermis layer results in obesity.

Decubitus Ulcers

Decubitus ulcers, or *bedsores*, are a critical problem that should concern *anyone* who is a care provider for patients requiring long-term or extended care. Bedsores develop when there is constant, unrelieved pressure on a single area of the skin. Blood supply to the dermis is blocked by the continuous pressure. Because the epidermis relies on diffusion of oxygen from underlying dermis, epidermal cells will begin to die. Areas of the body where bone is close to the skin are common areas for development of bedsores. These include the sacral and coxal areas (see Fig. 1.3), as well as the ankle, heel, shoulder, and elbow.

Populations at risk for bedsores include elderly patients, quadriplegics and paraplegics, and brain-injured individuals. Additional risk factors include anemia (decreased red blood cell count), urinary/fecal incontinence, and malnutrition. Signs of injury to the skin begin with a sunburn-like redness; as injury worsens, blisters develop. Progressive injury will cause skin loss, first involving the epidermis and subsequently spreading to the dermis and hypodermis (Fig. 5A). Full-thickness bedsores may penetrate to underlying muscle tissue or even to the bone itself. Patients will have a fever and increased white blood cell count. If the patient is able to feel pain, the sore is very painful; however, many quadriplegics and paraplegics have no pain sensation. Infection of the bedsore can slow the healing process, and in severe cases will be fatal.

Once formed, decubitus ulcers are very difficult to cure. Keeping pressure off the affected area is critical. Moist sterile dressings are applied to the wound to protect growing skin. Therapy using high-pressure oxygen is sometimes helpful. Severe cases may require a skin graft to replace lost tissue.

Figure 5A Decubitus ulcer (bedsore). The most frequent sites for bedsores are in the skin overlying a bony projection, such as on the hip, ankle, heel, shoulder, or elbow.

The best treatment for bedsores is prevention. Bedridden patients must be turned and repositioned hourly. Wet, soiled clothing must be changed immediately to prevent further irritation to the skin. Bony areas at risk for bedsore development should be well padded. Massaging the skin gently to stimulate blood flow is helpful. Proper nutrition is important to provide the nutrients needed by rapidly growing skin cells.

Content **CHECK-UP!**

1. Blood vessels can be found in:
   a. all three layers of the skin.
   b. the dermis only.
   c. the hypodermis and the dermis.
2. Which of the following white blood cells found in the epidermis phagocytize microbes and stimulate the immune system?
   a. Merkel cells.   c. melanocytes.
   b. Langerhans cells.
3. The deepest layer of the epidermis is the:
   a. stratum corneum.   c. stratum basale.
   b. stratum lucidum.

Answers in Appendix C, page 462.

5.2   Accessory Structures of the Skin

Hair, nails, and glands are structures of epidermal origin, even though some parts of hair and glands are largely in the dermis.

Hair and Nails

Hair is found on all body parts except the palms, soles, lips, nipples, and portions of the external reproductive organs. Most of this hair is fine and downy, but the hair on the head includes stronger types as well. After puberty, when sex hormones are made in quantity, there is noticeable hair in the axillary and pubic regions of both sexes. In the male, a beard develops, and other parts of the body may also become quite hairy. When women produce more male sex hormone than usual, they can develop **hirsutism,** a condition characterized by excessive body and facial hair.

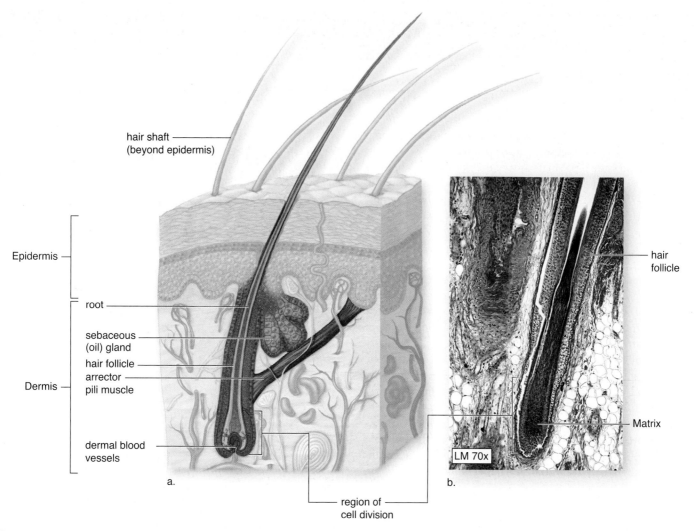

hair shaft
(beyond epidermis)

Epidermis

root

sebaceous
(oil) gland
hair follicle
arrector
pili muscle

Dermis

dermal blood
vessels

a.

hair
follicle

Matrix

LM 70x

b.

region of
cell division

Figure 5.2 **Hair follicle and hair shaft. a.** A hair grows from the base of a hair follicle where epidermal cells produce new cells as older cells move outward and become keratinized. **b.** A hair shaft magnified 70x.

Hormonal injections and procedures to kill hair roots are possible treatments.

Hairs project from complex structures called **hair follicles.** These hair follicles are formed from epidermal cells but are located in the dermis of the skin (Fig. 5.2). Hair follicle cells at the base of the hair continually divide, producing new cells that form a hair. At first, the cells are nourished by dermal blood vessels, but as the hair grows up and out of the follicle, they are pushed farther away from this source of nutrients, become keratinized, and die. The portion of a hair within the follicle is called the *root,* and the portion that extends beyond the skin is called the *shaft.*

The life span of any particular hair is usually three to four months for an eyelash and three to four years for a scalp hair; then it is shed and regrows. In males, baldness occurs when the hair on the head fails to regrow. **Alopecia,** meaning hair loss, can have many causes. As you read in the chapter introduction, male pattern baldness, or **androgenetic**

alopecia, is an inherited condition. *Alopecia areata* is characterized by the sudden onset of patchy hair loss. It is most common among children and young adults, and can affect either sex.

Each hair has one or more oil, or sebaceous, glands, whose ducts empty into the follicle. A smooth muscle, the **arrector pili,** attaches to the follicle in such a way that contraction of the muscle causes the hair to stand on end. If a person has had a scare or is cold, "goose bumps" develop due to contraction of these muscles.

Nails grow from special epithelial cells at the base of the nail in the region called the *nail root* (Fig 5.3). These cells become keratinized as they grow out over the nail bed. The visible portion of the nail is called the *nail body.* The cuticle is a fold of skin that hides the nail root. Ordinarily, nails grow only about 1 millimeter per week.

The pink color of nails is due to the vascularized dermal tissue beneath the nail. The whitish color of the half-moon-shaped

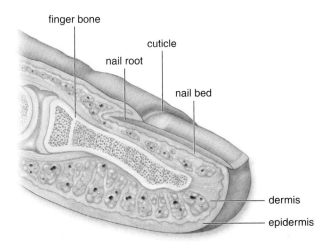

Figure 5.3 **Sagittal section of a nail.** Cells produced by the nail root become keratinized, forming the nail body.

base, or **lunula,** results from the thicker layer of rapidly reproducing cells in this area.

Glands

The *glands* in the skin are groups of cells specialized to produce and secrete a substance into ducts.

Sweat Glands

Sweat glands, or sudoriferous glands, are present in all regions of the skin. There can be as many as 90 glands per square centimeter on the leg, 400 glands per square centimeter on the palms and soles, and an even greater number on the fingertips. A sweat gland is tubular. The tubule is coiled, particularly at its origin within the dermis. These glands become active when a person is under stress.

Two types of sweat glands are shown in Figure 5.4. Both types secrete their products by exocytosis (see Chapter 3).

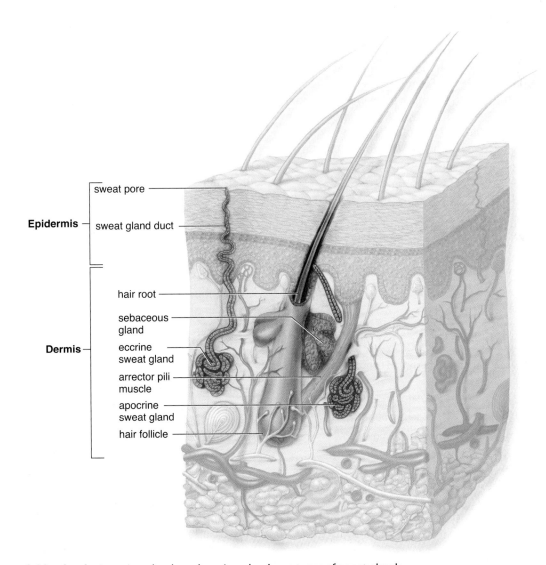

Figure 5.4 **Types of skin glands.** Apocrine glands and eccrine glands are types of sweat glands.

Apocrine glands open into hair follicles in the anal region, groin, and armpits. These glands begin to secrete at puberty, and a component of their secretion may act as a sex attractant. *Eccrine glands* open onto the surface of the skin. They become active when a person is hot, helping to lower body temperature as sweat evaporates. The sweat (perspiration) produced by these glands is mostly water, but it also contains salts and some urea, a waste substance. Therefore, sweat is a form of excretion. The opening into the ear, called the external ear canal, contains modified sweat glands, called ceruminous glands, which produce cerumen, or earwax.

Sebaceous Glands

Most **sebaceous glands** are associated with a hair follicle. These glands secrete an oily substance called **sebum** that flows into the follicle and then out onto the skin surface. This secretion lubricates the hair and skin, and helps waterproof them. Sebum also weakens or kills bacteria on the skin surface.

Particularly on the face and back, the sebaceous glands may fail to discharge sebum, and the secretions collect, forming whiteheads or blackheads. If pus-inducing bacteria are also present, a boil or pimple may result.

Acne vulgaris, the most common form of acne, is an inflammation of the sebaceous glands that most often occurs during adolescence. Hormonal changes during puberty cause the sebaceous glands to become more active at this time.

Mammary Glands

Mammary glands are modified apocrine sweat glands that produce milk only after childbirth. The anatomy and physiology of mammary glands are discussed in Chapter 17.

Content CHECK-UP!

4. Which glands begin secretion at puberty and contain a sex attractant?

    a. eccrine glands   c. ceruminous glands

    b. apocrine glands   d. sebaceous glands

5. Which part of the hair contains actively growing cells?

    a. base of hair follicle   c. hair shaft

    b. hair root

6. Which of the following pathologies causes women to develop excessive body and facial hair?

    a. alopecia   c. acne vulgaris

    b. hirsutism

Answers in Appendix C, page 462.

5.3   Disorders of the Skin

The skin is subject to many disorders, some of which are more annoying than life-threatening. For example, **athlete's foot** is caused by a fungal infection that usually involves the skin of the toes and soles. **Impetigo** is a highly contagious disease occurring most often in young children. It is caused by a bacterial infection that results in pustules that crust over. **Psoriasis** is a chronic condition, possibly hereditary, in which the skin develops pink or reddish patches covered by silvery scales due to overactive cell division. **Eczema,** an inflammation of the skin, is caused by sensitivity to various chemicals (e.g., soaps or detergents), to certain fabrics, or even to heat or dryness. **Dandruff** is a skin disorder not caused by a dry scalp, as is commonly thought, but by an accelerated rate of keratinization in certain areas of the scalp, producing flaking and itching. **Urticaria,** or hives, is an allergic reaction characterized by the appearance of reddish, elevated patches and often by itching.

Individuals who choose to pierce or tattoo their skin must take care to avoid skin damage or life threatening infection (see Medical Focus, p. 93).

Skin Cancer

Skin cancer is categorized as either melanoma or nonmelanoma. Like all cancers, it begins with mutation of the skin cell DNA.

Nonmelanoma cancers, which include basal cell carcinoma and squamous cell carcinoma, are much less likely than melanoma cancer to metastasize. **Basal cell carcinoma** (Fig. 5.5*a*), the most common type of skin cancer, begins when ultraviolet (UV) radiation causes epidermal basal cells to form a tumor, while at the same time suppressing the immune system's ability to detect the tumor. The signs of a tumor are varied. They include an open sore that will not heal; a recurring reddish patch; a smooth, circular growth with a raised edge; a shiny bump; or a pale mark. About 95% of patients are easily cured by surgical removal of the tumor, but recurrence is common.

Squamous cell carcinoma (Fig. 5.5*b*) begins in the epidermis proper. While five times less common than basal cell carcinoma, it is more likely to spread to nearby organs, and death occurs in about 1% of cases. It, too, is triggered by excessive UV exposure. The signs of squamous cell carcinoma are the same as those for basal cell carcinoma, except that it may also show itself as a wart or scaly growth that bleeds and scabs.

Melanoma (Fig. 5.5*c*), the type that is more likely to be malignant, starts in the melanocytes and has the appearance of an unusual **mole** (Table 5.1). Unlike a normal mole, which is dark, circular, and confined, a melanoma mole looks like a spilled ink spot, and a single melanoma mole may display a variety of shades. A melanoma mole can also itch, hurt, or feel numb. The skin around the mole turns gray, white, or red. Melanoma is most common in fair-skinned persons, particularly if they have suffered occasional severe

a. Basal cell carcinoma

b. Squamous cell carcinoma

c. Melanoma

Figure 5.5 **Skin cancer.** In each of the three types shown, the skin clearly has an abnormal appearance.

burns as children. Melanoma risk increases with the number of moles a person has. Most moles appear before the age of 14, and their appearance is linked to sun exposure. Melanoma rates have doubled in the last decade. Each year, an estimated 62,000 new cases of melanoma are diagnosed, and melanoma is responsible for approximately 8,000 fatalities annually.

Protecting the skin from the harmful effects of ultraviolet radiation is the best strategy to prevent all forms of skin cancer:

- Use a broad-spectrum sunscreen with a sun-protective factor (SPF) of at least 15, and make sure it protects against both UV-A and UV-B radiation.
- Wear protective clothing: long-sleeved shirts made from fabrics with a tight weave, for example. Choose a wide-brimmed hat over a baseball cap, which doesn't protect the ears from sunburn. At high altitude or in areas of the Earth where the ozone layer is thin, extra care must be taken. For example, in Australia (where ozone is thin due to the Earth's rotation), children should wear long-sleeved shirts and wide-brimmed hats when they play outside.
- Wear sunglasses that reflect both forms of UV radiation. Children's sunglasses should also protect, so avoid sunglasses that are cute but not protective.
- Stay out of the sun between 10 A.M. and 3 P.M. Doing so will reduce annual exposure to the sun's rays by as much as 60%.
- Avoid tanning machines, unless prescribed by a physician for Seasonal Affective Disorder (SAD).

Kaposi's sarcoma is a form of skin cancer that is most commonly seen in patients with AIDS, and in others whose immune system defenses are weakened or non-functional. The tumors of Kaposi's sarcoma appear as red, blue, or black spots on the skin. The tumors respond to treatment with a combined set of drugs commonly referred to as the "AIDS cocktail."

Raised growths on the skin, such as moles and **warts**, usually are not cancerous. Moles are due to an overgrowth of melanocytes, and warts are due to a viral infection.

TABLE 5.1 Warning Signs for Melanoma: The ABCDE Rule
WHEN EXAMINING A MOLE ON THE BODY, NOTE THE FOLLOWING SIGNS OF MELANOMA:
A **Asymmetrical**—instead of being perfectly round, the mole has an oval or irregular shape.
B **Borders** of the mole are irregular and have notches or indentation in them.
C **Color** is uneven and several colors may be present: black, brown, tan, red, or blue.
D **Diameter** of the mole is greater than 6 mm (larger than a pencil eraser).
E **Elevation**—the mole is elevated above the surface of the skin.

Wound Healing

Injury to the skin will cause an inflammatory response, characterized by redness, swelling, heat, and pain (inflammation is discussed fully in Chapter 13). A wound that punctures a blood vessel will fill with blood. Chemicals released by damaged tissue cells will cause the blood to clot. The clot prevents pathogens and toxins from spreading to other tissues (Fig. 5.6a). The part of the clot exposed to air will dry and harden, gradually becoming a scab. White blood cells and fibroblasts move into the area. White blood cells help fight infection and fibroblasts are able to pull the margins of the wound together (Fig. 5.6b). Fibroblasts promote tissue regeneration: The basal layer of the epidermis begins to produce new cells at a faster than usual rate. The proliferating fibroblasts bring about scar formation; the scar may or may not be visible from the surface (Fig. 5.6c). A scar is a tissue composed of many collagen fibers arranged to provide maximum strength. A scar does not contain the accessory organs of the skin and is usually devoid of feeling. In any case, epidermis and dermis have now healed (Fig. 5.6d).

Figure 5.6 The process of wound healing. a. Tissue injury causes inflammation, with redness, swelling, heat, and pain. Deep wounds rupture blood vessels, and blood fills the wound. **b.** After a blood clot forms, a protective scab develops. Fibroblasts and white blood cells migrate to the wound site. **c.** New epidermis forms, and fibroblasts promote tissue regeneration. **d.** Freshly healed skin.

5.4 Effects of Aging

As aging occurs, the epidermis maintains its thickness, but the rate of cell mitosis decreases. The dermis becomes thinner, the dermal papillae flatten, and the epidermis is held less tightly to the dermis so that the skin is looser. Adipose tissue in the hypodermis of the face and hands also decreases, which means that older people are more likely to feel cold.

The fibers within the dermis change with age. The collagenous fibers become coarser, thicker, and farther apart; therefore, there is less collagen than before. Elastic fibers in the upper layer of the dermis are lost, and those in the lower dermis become thicker, less elastic, and disorganized. The skin wrinkles because (1) the epidermis is loose, (2) the fibers are fewer and those remaining are disorganized, and (3) the hypodermis has less padding.

With aging, homeostatic adjustment to heat is limited due to less vasculature (fewer blood vessels) and fewer sweat glands. The number of hair follicles decreases, causing the hair on the scalp and extremities to thin. Because of a reduced number of sebaceous glands, the skin tends to crack.

As a person ages, the number of melanocytes decreases. This causes the hair to turn gray and the skin to become paler. In contrast, some of the remaining pigment cells are larger, and pigmented blotches appear on the skin.

Many of the changes that occur in the skin as a person ages appear to be due to sun damage. Ultraviolet radiation causes rough skin, mottled pigmentation, fine lines and wrinkles, deep furrows, numerous benign skin growths, and the various types of skin cancer discussed in section 5.3.

a.   b.

Figure 5.7 **a. X ray of a normal child. b. X ray of a child with rickets.** Rickets develops from an improper diet and also from a lack of ultraviolet (UV) light (sunlight). Under these conditions, vitamin D does not form in the skin thereby reducing calcium uptake. Low calcium in bones weakens them and can cause bowing.

5.5 Homeostasis

The illustration on page 94, called Human Systems Work Together, tells how the functions of the skin assist the other systems of the body and how the other systems help the skin carry out these functions.

Functions of the Skin

Skin has a protective function. First and foremost, the skin forms a protective covering over the entire body, safeguarding underlying parts from physical trauma. The melanocytes in skin protect it from UV radiation. Further, sebaceous gland secretions, phagocytic Langerhans cells, and the outermost layer of dead cells all help to protect against invasion by pathogens.

Skin helps regulate water loss. Since outer skin cells are dead and keratinized, the skin is waterproof, thereby preventing water loss. The skin's waterproofing also prevents water from entering the body when the skin is immersed.

Skin assists the function of the urinary system. Sweat glands secrete water from the body through sensible and insensible perspiration. Sensible perspiration can be felt; insensible perspiration occurs without one's awareness as water evaporates from the body. This perspiration contains small amounts of salt, ammonia, urea, and other wastes. Thus, skin plays a minor role in waste elimination.

Skin produces vitamin D. When skin cells are exposed to sunlight, the ultraviolet (UV) rays assist them in producing vitamin D from a precursor molecule. Only a small amount of UV radiation is needed. Vitamin D leaves the skin and enters the liver and kidneys, where it is converted to a hormone called calcitriol. Calcitriol regulates calcium uptake by the digestive system and both calcium and phosphorus

metabolism in cells. Calcium and phosphorus are very important to bone development and mineralization. Most milk today is fortified with vitamin D, which helps prevent **rickets** characterized especially by soft and deformed bones (Fig. 5.7b).

Skin gathers sensory information. The sensory receptors in the epidermis and dermis specialized for touch, pressure, pain, hot, and cold supply the central nervous system with information about the external environment. The fingertips contain the greatest number of touch receptors, allowing the fingers to be used for delicate tasks. The sensory receptors also facilitate communication between people. For example, the touch receptors play a major role in sexual arousal, which assists the reproductive system.

Skin helps regulate body temperature. When muscles contract and ATP is broken down, heat is released. If body temperature starts to rise above normal, skin blood vessels (a part of the cardiovascular system) will dilate. The sweat glands will activate as blood moves to the skin's surface. Sweat absorbs body heat, and this heat is carried away as sweat evaporates. If the weather is humid, evaporation can be assisted by a cool breeze.

If the outer temperature is cool, the sweat glands remain inactive, and the blood vessels constrict so that less blood is brought to the skin's surface. Whenever the body's temperature falls below normal, the muscles start to contract, causing shivering, which produces heat. Further, arrector pili muscles attached to hair follicles cause goose bumps to form when a person is cold. The arrector pili response in animals (birds, cats, and dogs, for example) provides extra insulation when hair or feathers stand upright,

Burns

Burns are skin injuries that are usually caused by heat but can also be caused by radioactive, chemical, and electrical agents. Burns account for approximately 1 million emergency department visits yearly. Emergency caregivers must quickly assess burn severity and take appropriate supportive action.

Two criteria are used to estimate burn severity: depth and thickness. In *first-degree burns* (a moderate sunburn, for example), only the epidermis is affected. The person experiences redness and pain, but no blisters or swelling. The pain subsides within 48–72 hours, and the injury heals without further complications or scarring. Treatment involves pain management. The damaged skin peels off in about a week.

Second-degree burns extend through the entire epidermis and part of the dermis. The person experiences not only redness and pain, but also blistering. The deeper the burn, the more prevalent the blisters, which can enlarge after the injury. Unless they become infected, superficial second-degree burns heal in 10–14 days without complications and with little scarring. If the burn extends deep into the dermis, it heals more slowly over a period of 30–105 days. The healing epidermis is extremely fragile, and scarring is common. First- and second-degree burns are sometimes referred to as partial-thickness burns.

Third-degree burns, or full-thickness burns, destroy the entire thickness of the skin. The surface of the wound is leathery and may be brown or black. Pain receptors, blood vessels, sweat glands, sebaceous glands, and hair follicles are destroyed. *Fourth-degree burns* involve tissues down to the bone. Unless a very limited area of the body is affected, the patient will likely not survive.

The extent of the burn must also be estimated when judging a burn's severity. The "rule of nines" is used to divide the body into regions: the head and neck, 9% of the total body surface; each upper limb, 9%; each lower limb, 18%; the front and back portions of the trunk, 18% each; and the perineum, which includes the anal and urogenital regions, 1% (Fig. 5B). Physicians use the Lund-Browder chart to estimate the extent of burns in children. This system adjusts for the fact that a child's head is proportionally larger than an adult's.

Figure 5B The "rule of nines" for estimating the extent of burns.

A burn is a critical injury if: (1) Second-degree burns cover 25% or more of the patient's body; (2) third-degree burns cover 10% or more of the patient's body; (3) any portion of the body has a fourth-degree burn; or (4) third-degree burns occur on the face, hands, or feet. Facial burns may be accompanied by lung damage from smoke inhalation. Burns to the hands or feet can scar, resulting in loss of joint movement.

The first responder's reactions when caring for a burn patient (and for all other patients) can be summarized using the letters ABC:

A—establish the *airway*. A tube may need to be placed into the trachea (windpipe), especially if the patient might have suffered smoke inhalation. Smoke inhalation should be suspected whenever there are facial burns, or if the patient has trouble breathing.

B—make sure the person is *breathing*, using artificial respiration when necessary. If smoke inhalation is suspected, a paramedic will administer 100% oxygen.

C—*circulation* must be maintained. The patient's pulse is taken, and CPR is begun when needed. The patient must be carefully observed for signs of **shock**—a dangerously low drop in blood pressure (see Chapter 12 for discussion of shock). A paramedic may begin an IV (intravenous administration of a balanced salt solution) under a physician's direction.

First responders must also act quickly to stop the burn. All clothing is carefully removed from the body and any smoldering is extinguished. In the case of a chemical burn, the affected area is flushed continuously with water.

The major complications resulting from severe burns are fluid loss, heat loss, and bacterial infection. Fluid loss is counteracted by an IV. Heat loss is minimized by placing the burn patient in a warm environment. Application of antibacterial dressings and isolation help prevent infection. As soon as possible, the damaged tissue is removed and skin grafting is begun. The skin needed for grafting is usually taken from other parts of the patient's body. However, if the burned area is quite extensive, skin can be grown in the laboratory from only a few cells taken from the patient.

Body Art: Buyer Beware!

There's no doubt about it: Body art has gone mainstream. Piercing and tattooing the body are practices that have existed for thousands of years in many different human cultures. In the past decade, both practices have become commonplace in Western society as well. It is no longer surprising to see actors, sports figures, and other celebrities with visible tattoos, pierced navels, multiple earrings, or other forms of wearable art. As the trend continues, more and more individuals are tempted to get their own tattoo or piercing. Tattooing is the process of injecting inks into the dermal layer of the skin. The most common sites for piercing remain the earlobes and ear cartilage, but lips, eyebrows, navels, tongues, and genitals are also common sites.

Individuals must consider the decision to tattoo or pierce very carefully, and health-care providers must be prepared to give accurate information about both practices. With the increase in both tattooing and piercing, reports of complications have increased as well.

Tattooing and piercing can result in infections because both practices can potentially introduce bacteria into the skin, blood vessels, or lymphatic vessels. Complications from infection can range from minor skin irritations to life-threatening blood poisoning. Several cases on record have involved infection with *resistant* bacteria, which are extremely dangerous because they are not killed by common antibiotics. Fatalities have been reported from infections caused by piercing.

Further, the reaction of a person's skin to tattoo ink is unpredictable. Tattoo inks are not regulated by any form of federal agency, and many color additives have not been approved for contact with the skin. Some are inks used in paper printing or automobile paint. It is hardly surprising that many tattooed individuals develop allergies to ingredients in tattoo dyes. Tattooed skin may develop unattractive granulomas: hard connective tissue nodules under the skin. Keloids—scars that grow beyond normal boundaries—may also develop as the body reacts to chemicals in the ink.

Metals (especially nickel) used in body jewelry are another common cause for allergy. Once an allergic reaction develops, it typically persists for life. In such cases, the jewelry must be permanently removed if at all possible. Additionally, the jewelry may damage nerves or blood vessels as piercing occurs. Jewelry in and around the mouth may damage gums, break teeth or dentures, and interfere with chewing or swallowing. Mouth jewelry that becomes loose may be aspirated (sucked into the airways) where it can obstruct an airway and interfere with breathing.

Anyone considering body art must proceed with caution. If local, state, or provincial law regulates tattoo and piercing artists, use an approved artist or shop. Never try a "do-it-yourself" approach. Check the cleanliness of the shop and make sure that all equipment is steam-sterilized using a device called an *autoclave*. Insist that the artist or piercer use sterile gloves at all times. Make sure that jewelry is appropriate: not too heavy, and made of inert metals such as gold, titanium, or surgical stainless steel, which are less likely to cause allergic reactions. Keep the affected area scrupulously clean after piercing or tattooing and touch jewelry as little as possible. Allow adequate time for healing. Pierced navels, for example, may require up to a year to heal completely. Finally, if complications develop (excessive pain, redness, swelling, fever, etc.), seek medical attention promptly.

Most important, anyone considering body art must remember that the decision may very likely be *permanent*. Piercings may never reclose completely and they may leave a large scar. Although tattoos can be removed by laser therapy, removal is expensive and may leave large scars—and it often cannot completely remove the tattoo.

---

but the insulating effect is absent in humans because body hair is sparse. If the outside temperature is extremely cold and blood flow to the skin is severely restricted for an extended period, a portion of the skin will die, resulting in *frostbite*.

Hyperthermia and Hypothermia

Hyperthermia, a body temperature above normal, and **hypothermia,** a body temperature below normal, indicate that the body's regulatory mechanisms have been overcome. In *heat exhaustion*, blood pressure may be low, and salts may have been lost due to profuse sweating. Even so, body temperature remains high. *Heat stroke* is characterized by an elevated temperature, up to 43°C (110°F), with no sweating. *Fever* is a special case of hyperthermia that can be brought on by immune system response, and/or by a bacterial infection. When the fever "breaks," sweating occurs as the normal set point for body temperature returns.

At first, hypothermia is characterized by uncontrollable shivering, incoherent speech, and lack of coordination (body temperature 90°–95°F). Then the pulse rate slows, and hallucinations occur as unconsciousness develops (body temperature 80°–90°F). Breathing becomes shallow, and shivering diminishes as rigidity sets in. This degree of hypothermia is associated with a 50% mortality.

Human Systems Work Together

Skeletal System

Skin protects bones; helps provide vitamin D for Ca²⁺ absorption.

Bones provide support for skin.

Muscular System

Skin protects muscles; rids the body of or conserves heat produced by muscle contraction.

Muscle contraction provides heat to warm skin.

Nervous System

Skin protects nerves, helps regulate body temperature; skin receptors send sensory input to brain.

Brain controls nerves that regulate size of cutaneous blood vessels, activate sweat glands and arrector pili muscles.

Endocrine System

Skin helps protect endocrine glands.

Androgens activate sebaceous glands and help regulate hair growth.

Cardiovascular System

Skin prevents water loss; helps regulate body temperature; protects blood vessels.

Blood vessels deliver nutrients and oxygen to skin, carry away wastes; blood clots if skin is broken.

How the Integumentary System works with other body systems

hair

skin and associated glands

Lymphatic System/Immunity

Skin serves as a barrier to pathogen invasion; Langerhans cells phagocytize pathogens; protects lymphatic vessels.

Lymphatic vessels pick up excess tissue fluid; immune system protects against skin infections.

Respiratory System

Skin helps protect respiratory organs.

Gas exchange in lungs provides oxygen to skin and rids body of carbon dioxide from skin.

Digestive System

Skin helps to protect digestive organs; helps provide vitamin D for Ca²⁺ absorption.

Digestive tract provides nutrients needed by skin.

Urinary System

Skin helps regulate water loss; sweat glands carry on some excretion.

Kidneys compensate for water loss due to sweating; activate vitamin D precursor made by skin.

Reproductive System

Skin receptors respond to touch; mammary glands produce milk; skin stretches to accommodate growing fetus.

Androgens activate oil glands; sex hormones stimulate fat deposition, affect hair distribution in males and females.

SELECTED NEW TERMS

Basic Key Terms

arrector pili (ah-rek'tor pil'i), p. 86

dermis (der'mis), p. 84

epidermis (ep"ĭ-der'mis), p. 83

free nerve endings (frē nûrv ĕn'dĭngs), p. 84

hair follicle (hār fol'ĭ-kl), p. 86

hypodermis (hi"po-der'mis), p. 83

integument (in-teg'yū-ment), p. 83

integumentary system (in-teg"yū-men'tar-e sis'tem), p. 83

keratin (kĕr'uh-tin), p. 84

Langerhanse cells (lăhng'ĕrhăhnz sĕlz), p. 84

lunula (lu'nu-luh), p. 87

mammary glands (mă'mă rē glands), p. 88

melanin (mel'uh-nin), p. 84

melanocyte (mel'uh-no-sīt), p. 84

Merkel cells (mer'kelz sĕlz), p. 84

sebaceous gland (sĕ-ba'shus gland), p. 87

sebum (se'bum), p. 87

stratum basale (strā'təm bā'səlĕ'), p. 83

stratum corneum (strā'təm kor'nē-um), p. 84

stratum lucidum (strā'təm loo'sĭd-um), p. 84

subcutaneous tissue (sŭb'kyoo-tā'nē-ə s tĭsh'oo), p. 84

sweat glands (swet glands), p. 87

tactile cells (tăk'təl sĕlz), p. 84

Clinical Key Terms

acne vulgaris (ak'ne vul-ga'-ris), p. 87

albinism (al'bĭ-nizm), p. 84

alopecia (al-o-pe'she-uh), p. 87

androgenetic alopecia (an'-drŏ-juh-nĕ"-tik al-ō-pĕ'shē-uh) p. 82

athlete's foot (ath'lēts fút), p. 88

basal cell carcinoma (bās'al sel kar-sĭ-no'muh), p. 88

dandruff (dan'druf), p. 88

decubitus ulcer (de-kyŭ'bĭ-tus ul'ser), p. 85

eczema (ek'zĕ-muh), p. 88

hirsutism (her'suh-tizm), p. 85

hyperthermia (hi"per-ther'me-uh), p. 93

hypodermic needle (hi-po-der'mik ne'dl), p. 84

hypothermia (hi"po-ther'me-uh), p. 93

impetigo (im"pĕ-ti'go), p. 88

Kaposi's sarcoma (kä-pō'shĕz săr'kŏ-më), p. 89

melanoma (mel-uh-no'muh), p. 88

mole (mŏl), p. 88

psoriasis (so-ri'uh-sis), p. 88

shock (shăwk), p. 92

squamous cell carcinoma (skwa'mus sel kar-sĭ-no'muh), p. 88

subcutaneous injection (sub"kyū-ta'ne-us in-jek'shun), p. 84

urticaria (ur"tĭ-kār'e-uh), p. 88

SUMMARY

5.1 Structure of the Skin

The skin has two regions: the epidermis and the dermis. The hypodermis lies below the skin.

A. The epidermis, the outer region of the skin, is made up of stratified squamous epithelium. New cells continually produced in the stratum basale of the epidermis are pushed outward and become the keratinized cells of the stratum corneum. Stratum lucidum is found in thick skin of palms and soles.

B. The dermis, which is composed of dense irregular connective tissue, lies beneath the epidermis. It contains collagenous and elastic fibers, blood vessels, and nerve fibers.

C. The hypodermis is made up of loose connective tissue and adipose tissue, which insulates the body from heat and cold.

5.2 Accessory Structures of the Skin

Accessory structures of the skin include hair, nails, and glands.

A. Both hair and nails are produced by the division of epidermal cells and consist of keratinized cells.

B. Sweat glands are numerous and present in all regions of the skin.

Sweating helps lower the body temperature.

C. Sebaceous glands are associated with a hair follicle and secrete sebum, which lubricates the hair and skin.

D. Mammary glands located in the breasts produce milk after childbirth.

5.3 Disorders of the Skin

A. Skin cancer. Skin cancer, which is associated with ultraviolet radiation, occurs in three forms. Basal cell carcinoma and squamous cell carcinoma can usually be removed surgically. Melanoma is the most dangerous form of skin cancer.

B. Wound healing. The skin has regenerative powers and can grow back on its own if a wound is not too extensive.

C. Burns. The severity of a burn depends on its depth and extent. First-degree burns affect only the epidermis. Second-degree burns affect the entire epidermis and a portion of the dermis. Third-degree burns affect the entire epidermis and dermis. The "rule of nines" provides a means of estimating the extent of a burn injury.

5.4 Effects of Aging

Skin wrinkles with age because the epidermis is held less tightly, fibers in the dermis are fewer, and the hypodermis has less padding. The skin has fewer blood vessels, sweat glands, and hair follicles. Although pigment cells are fewer and the hair turns gray, pigmented blotches appear on the skin. Exposure to the sun results in many of the skin changes we associate with aging.

5.5 Homeostasis

A. Skin protects the body from physical trauma and bacterial invasion.

B. Skin helps regulate water loss and gain, which helps the urinary system. Also, sweat glands excrete some urea.

C. The skin produces vitamin D following exposure to UV radiation. A hormone derived from vitamin D helps regulate calcium and phosphorus metabolism involved in bone development.

D. The skin contains sensory receptors for touch, pressure, pain, hot, and cold, which help people to be aware of their surroundings. These receptors plates of the palatine bones. A cleft palate results when either (1) or (2) have failed to fuse.

Zygomatic Bones The two zygomatic bones form the sides of the orbits (Fig. 6.7a). They also contribute to the "cheekbones." Each zygomatic bone has a **temporal process.** A **zygomatic arch,** the most prominent feature of a cheekbone, consists of a temporal process connected to a zygomatic process (a portion of the temporal bone).

Lacrimal Bones The two small, thin lacrimal bones are located on the medial walls of the orbits (Fig. 6.6). A small opening between the orbit and the nasal cavity serves as a pathway for a duct that carries tears from the eyes to the nose.

Nasal Bones The two nasal bones are small, rectangular bones that form the bridge of the nose (Fig. 6.5). The ventral portion of the nose is cartilage, which explains why the nose is not seen on a skull.

Vomer Bone The vomer bone joins with the perpendicular plate of the ethmoid bone to form the inferior part of the nasal septum (Figs. 6.5 and 6.6a).

Inferior Nasal Conchae The two inferior nasal conchae are thin, curved bones that form a part of the inferior lateral wall of the nasal cavity (Fig. 6.6a). Like the superior and middle nasal conchae, they project into the nasal cavity and support the mucous membranes that line the nasal cavity.

Mandible The mandible, or lower jaw, is the only movable portion of the skull. The horseshoe-shaped front and horizontal sides of the mandible, referred to as the *body*, form the chin. The body has an **alveolar process** (Fig. 6.6a), which contains tooth sockets. Superior to the left and right angle of the mandible are upright projections called *rami*. Each ramus has the following:

mandibular condyle (Fig. 6.6b), which articulates with a temporal bone;
coronoid process (Fig. 6.6b), which serves as a place of attachment for the muscles used for chewing.

> **Begin Thinking Clinically**
>
> **?** You're treating an 11-year-old patient in the emergency room. His right eye was struck by a baseball bat, and he's rapidly developing a nasty black eye. What bones might have been broken by the injury?
>
> *Answer and discussion in Appendix C, page 462.*

Hyoid Bone

The U-shaped hyoid bone (Fig. 6.4) is located superior to the larynx (voice box) in the neck. It is the only bone in the body that does not articulate with another bone. Instead, it is suspended from the styloid processes of the temporal bones by the stylohyoid muscles and ligaments. It anchors the tongue and serves as the site for the attachment of several muscles associated with swallowing.

Vertebral Column (Spine)

The **vertebral column** extends from the skull to the pelvis. It consists of a series of separate bones, the **vertebrae,** separated by pads of fibrocartilage called the **intervertebral disks** (Fig. 6.8). The vertebral column is located in the middorsal region and forms the vertical axis. The skull rests on the superior end of the vertebral column, which also supports the rib cage and serves as a point of attachment for the pelvic girdle. The vertebral column

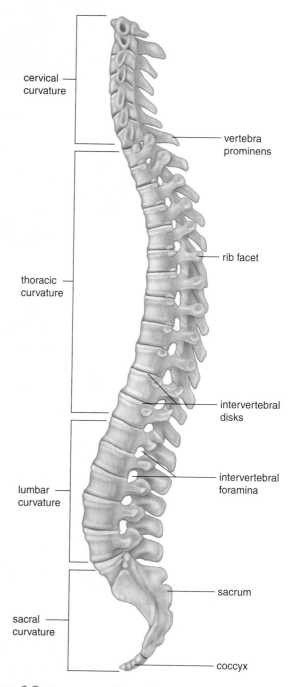

cervical curvature

vertebra prominens

rib facet

thoracic curvature

intervertebral disks

intervertebral foramina

lumbar curvature

sacral curvature

sacrum

coccyx

Figure 6.8 **Curvatures of the spine.** The vertebrae are named for their location in the body. Note the presence of the coccyx, also called the tailbone.

also protects the spinal cord, which passes through a vertebral canal formed by the vertebrae. The vertebrae are named according to their location: seven *cervical* (neck) *vertebrae,* twelve *thoracic* (chest) *vertebrae,* five *lumbar* (lower back) *vertebrae,* five *sacral vertebrae* fused to form the sacrum, and three to five *coccygeal vertebrae* fused into one coccyx.

When viewed from the side, the vertebral column has four normal curvatures, named for their location (Fig. 6.8). The cervical and lumbar curvatures are convex anteriorly, and the thoracic and sacral curvatures are concave anteriorly. In the fetus, the vertebral column has but one curve, and it is concave anteriorly. The cervical curve develops three to four months after birth, when the child begins to hold his or her head up. The lumbar curvature develops when a child begins to stand and walk, around one year of age. The curvatures of the vertebral column provide more support than a straight column would, and they also provide the balance needed to walk upright.

The curvatures of the vertebral column are subject to abnormalities (Fig. 6.9). An abnormally exaggerated lumbar curvature is called **lordosis,** or "swayback." People who are balancing a heavy midsection, such as pregnant women or men with "potbellies," may have swayback. An increased roundness of the thoracic curvature is **kyphosis,** or "hunchback." This abnormality sometimes develops in older people as the center sections of thoracic vertebrae become compressed. An abnormal lateral (side-to-side) curvature is called **scoliosis.** Occurring most often in the thoracic region, scoliosis is usually first seen during late childhood.

Intervertebral Disks

The fibrocartilaginous intervertebral disks located between the vertebrae act as a cushion. The disks are filled with gelatinous material, which prevents the vertebrae from grinding against one another and absorbs shock caused by such movements as

Lordosis ("swayback")
a.

Kyphosis ("hunchback")
b.

Scoliosis
c.

Figure 6.9 Abnormal curvatures of the vertebral column.
a. Lordosis, or "swayback." **b.** Kyphosis, or "hunchback." **c.** Scoliosis.

running, jumping, and even walking. The disks also allow motion between the vertebrae so that a person can bend forward, backward, and from side to side. Unfortunately, these disks become weakened with age, and can slip or even rupture (called a **herniated disk**). A damaged disk pressing against the spinal cord or the spinal nerves causes pain. Such a disk may need to be removed surgically. If a disk is removed, the vertebrae are fused together, limiting the body's flexibility.

Vertebrae

Figure 6.10*a* shows that a typical vertebra has an anteriorly placed *body* and a posteriorly placed vertebral *arch*. The vertebral arch forms the wall of a *vertebral foramen* (pl., *foramina*). The foramina become a canal through which the spinal cord passes.

The structure of a single vertebra can be likened to a house (the vertebral arch) sitting on a hill (the body of the vertebra). *Pedicles* are the upright walls of the house and the *laminae* form a slanting roof. The single *spinous process*, or *spine*, is like a flagpole, and the *transverse processes* are gutters projecting sideways at the corners where pedicles join laminae. Each of these bony projections on an actual vertebra serves as a site for muscle attachment. Additionally, articular processes (superior and inferior) function to join vertebrae.

The vertebrae have regional differences. For example, as the vertebral column descends, the bodies get bigger and are better able to carry more weight. In the cervical region, the spines are short and tend to have a split, or bifurcation. Cervical vertebrae all have an opening in the transverse process, called a transverse foramen. The thoracic spines are long and slender and project downward. Further, the transverse processes of thoracic vertebrae have articular facets for connecting to ribs. The lumbar spines are massive and square and project posteriorly.

Atlas and Axis The first two cervical vertebrae are not typical (Fig. 6.10*b*). The **atlas** supports and balances the head. It has two depressions that articulate with the occipital condyles, allowing movement of the head up and down (as though nodding "yes"). The **axis** has an *odontoid process* (also called the *dens*) that projects into the ring of the atlas. When the head moves from side to side, the atlas pivots around the odontoid process (as though shaking the head "no").

Sacrum and Coccyx The five sacral vertebrae are fused to form the **sacrum.** The sacrum articulates with the pelvic girdle and forms the posterior wall of the pelvic cavity (see Fig. 6.16). The **coccyx,** or tailbone, is the last part of the vertebral column. It is formed from a fusion of three to five vertebrae.

The Rib Cage

The **rib cage** (Fig. 6.11), sometimes called the thoracic cage, is composed of the thoracic vertebrae, ribs and associated cartilages, and sternum.

The rib cage demonstrates how the skeleton is protective but also flexible. The rib cage protects the heart and lungs; yet it swings outward and upward upon inspiration and then downward and inward upon expiration. The rib cage also

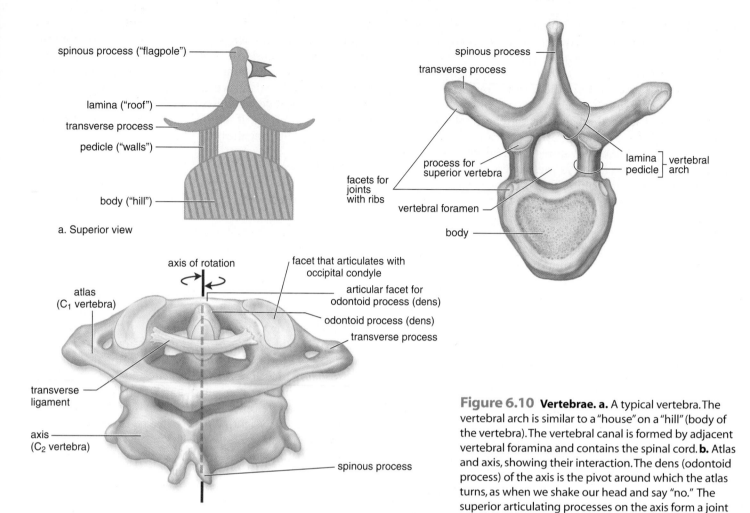

spinous process ("flagpole")

lamina ("roof")

transverse process

pedicle ("walls")

body ("hill")

a. Superior view

spinous process

transverse process

process for superior vertebra

facets for joints with ribs

vertebral foramen

body

lamina
pedicle } vertebral arch

axis of rotation

facet that articulates with occipital condyle

articular facet for odontoid process (dens)

odontoid process (dens)

transverse process

atlas (C₁ vertebra)

transverse ligament

axis (C₂ vertebra)

spinous process

b. Posterosuperior view

Figure 6.10 Vertebrae. a. A typical vertebra. The vertebral arch is similar to a "house" on a "hill" (body of the vertebra). The vertebral canal is formed by adjacent vertebral foramina and contains the spinal cord. **b.** Atlas and axis, showing their interaction. The dens (odontoid process) of the axis is the pivot around which the atlas turns, as when we shake our head and say "no." The superior articulating processes on the axis form a joint with the inferior articular processes on the atlas.

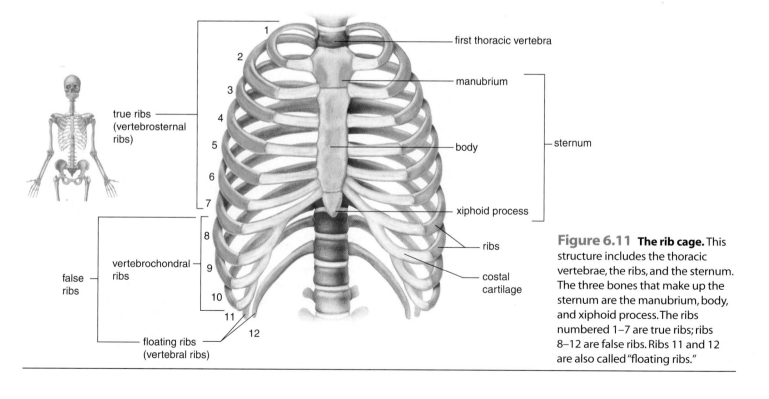

first thoracic vertebra

manubrium

body

sternum

xiphoid process

ribs

costal cartilage

true ribs (vertebrosternal ribs)

false ribs

vertebrochondral ribs

floating ribs (vertebral ribs)

Figure 6.11 The rib cage. This structure includes the thoracic vertebrae, the ribs, and the sternum. The three bones that make up the sternum are the manubrium, body, and xiphoid process. The ribs numbered 1–7 are true ribs; ribs 8–12 are false ribs. Ribs 11 and 12 are also called "floating ribs."

provides support for the bones of the pectoral girdle (see page 112).

The Ribs

There are 12 pairs of ribs. All 12 pairs connect directly to the thoracic vertebrae in the back. After connecting with thoracic vertebrae, each rib first curves outward and then forward and downward. The first pair of ribs attaches to the body of the first thoracic vertebra, or T1. It also attaches to the transverse process of T1, at the facet for the joint with the rib. The next eight pairs of ribs (ribs 2–9) attach to the vertebrae at three places (Fig. 6.10a). First, each attaches to the body of the same numbered vertebra (rib 4, for example, is attached to vertebral body 4). Next, each attaches to the body of the vertebra immediately superior (so, rib 4 also attaches to vertebral body 3). Finally, each rib attaches to the transverse process of the same numbered vertebra (rib 4 attaches to the transverse process of rib 4). Rib pairs 10 through 12 attach only to their respective vertebrae.

The upper seven pairs of ribs connect directly to the sternum by means of costal cartilages. These are called the "true ribs," or the *vertebrosternal* ribs. The next five pairs of ribs are called the "false ribs" because they attach indirectly to the sternum or are not attached at all. Ribs 8, 9, and 10 are called *vertebrochondral* ribs; each attaches its costal cartilage to the cartilage of the rib superior to it. All three ribs attach indirectly to the sternum using the costal cartilage of rib 7. Ribs 11 and 12 are *vertebral*, or "floating," ribs. These are short ribs with no attachment to the sternum (Fig. 6.11).

The Sternum

The **sternum,** or breastbone, is a flat bone that has the shape of a blade. The sternum, along with the ribs, helps protect the heart and lungs. During surgery the sternum may be split to allow access to the organs of the thoracic cavity. The sternum is also where a first responder will place his or her hands when performing CPR (see Chapter 14).

The sternum is composed of three bones that fuse during fetal development (Fig. 6.11). These bones are the manubrium, the body, and the xiphoid process. The *manubrium* is the superior portion of the sternum. The *body* is the middle and largest part of the sternum, and the *xiphoid process* is the inferior and smallest portion of the sternum. The manubrium joins with the body of the sternum at an angle. This joint is an important anatomical landmark because it occurs at the level of the second rib, and therefore allows the ribs to be counted. Counting the ribs is sometimes done to determine where the apex of the heart is located—usually between the fifth and sixth ribs.

The manubrium articulates with the costal cartilages of the first and second ribs; the body articulates with costal cartilages of the second through seventh ribs; and the xiphoid process doesn't articulate with any ribs.

The xiphoid process is the third part of the sternum. Composed of hyaline cartilage in the child, it becomes ossified in the adult. The variably shaped xiphoid process serves as an attachment site for the diaphragm, which separates the thoracic cavity from the abdominal cavity.

Content CHECK-UP!

4. Which of the following are cranial bones?

   a. frontal   c. maxilla

   b. sphenoid   d. a and b

   e. all of the above

5. The name for the first cervical vertebra is the:

   a. axis.   c. coccyx.

   b. atlas.   d. atlantis.

6. The most superior part of the sternum is the:

   a. xiphoid process.   c. manubrium.

   b. body.   d. costal cartilage.

Answers in Appendix C, page 462.

6.3 Appendicular Skeleton

The **appendicular skeleton** contains the bones of the pectoral girdle, upper limbs, pelvic girdle, and lower limbs.

Pectoral Girdle

The **pectoral girdle** (shoulder girdle) contains four bones: two clavicles and two scapulae (Fig. 6.12). It supports the arms and serves as a place of attachment for muscles that move the arms. The bones of this girdle are not held tightly together; rather, they are weakly attached and held in place by ligaments and muscles. This arrangement allows great flexibility but means that the arm is prone to dislocation at the shoulder joint.

Clavicles

The **clavicles** (collarbones) are slender and S-shaped (Fig. 6.12). Each clavicle articulates medially with the manubrium of the sternum. This is the only place where the pectoral girdle is attached to the axial skeleton.

Each clavicle also articulates with a scapula. The clavicle serves as a brace for the scapula and helps stabilize the

shoulder. It is structurally weak, however, and if undue force is applied to the shoulder, the clavicle will fracture.

Scapulae

The **scapulae** (sing., *scapula*), also called the shoulder blades, are broad bones that somewhat resemble triangles (Fig. 6.12b). One reason for the pectoral girdle's flexibility is that the scapulae are not joined to each other (see Fig. 6.4).

Each scapula has a spine. Note the following features as well:

acromion process, which articulates with a clavicle and provides a place of attachment for arm and chest muscles;

coracoid process, which serves as a place of attachment for arm and chest muscles;

glenoid cavity, which articulates with the head of the arm bone (humerus). The arm joint's flexibility is also a result of the glenoid cavity being smaller than the head of the humerus.

Upper Limb

The upper limb includes the bones of the arm (humerus), the forearm (radius and ulna), and the hand (carpals, metacarpals, and phalanges).[1]

Humerus

The **humerus** (Fig. 6.13) is the bone of the arm. It is a long bone with the following features at the proximal end:

head, which articulates with the glenoid cavity of the scapula;

greater and **lesser tubercles,** which provide attachments for muscles that move the arm and shoulder;

intertubercular groove, which holds the tendon from the biceps brachii, a muscle of the arm;

deltoid tuberosity, which provides an attachment for the deltoid, a muscle that covers the shoulder joint.

The humerus has the following features at the distal end:

capitulum, a lateral condyle that articulates with the head of the radius;

trochlea, a spool-shaped condyle that articulates with the ulna;

coronoid fossa, a depression for a process of the ulna when the elbow is flexed;

olecranon fossa, a depression for a process of the ulna when the elbow is extended.

Radius

The **radius** and **ulna** (see Figs. 6.12a and 6.14) are the bones of the forearm. The radius is on the lateral side of the forearm (the thumb side). When you turn your hand from the "palms

---

[1]The term *upper extremity* is used to include a clavicle and scapula (of the pectoral girdle), an arm, forearm, wrist, and hand.

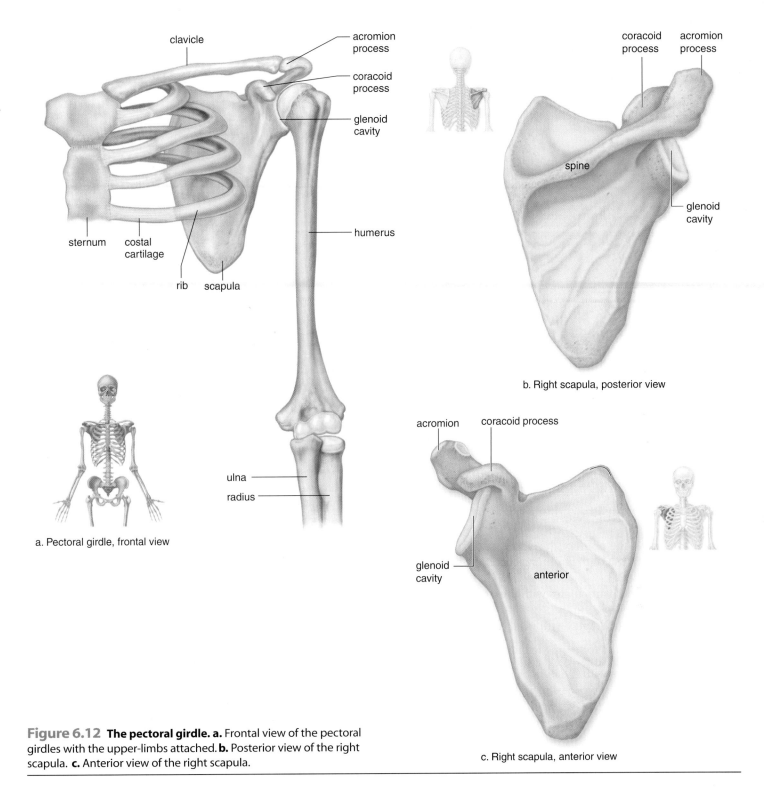

clavicle

acromion
process

coracoid
process

glenoid
cavity

sternum

costal
cartilage

humerus

rib   scapula

a. Pectoral girdle, frontal view

ulna

radius

coracoid
process

acromion
process

spine

glenoid
cavity

b. Right scapula, posterior view

acromion   coracoid process

glenoid
cavity

anterior

c. Right scapula, anterior view

Figure 6.12 **The pectoral girdle. a.** Frontal view of the pectoral girdles with the upper-limbs attached. **b.** Posterior view of the right scapula. **c.** Anterior view of the right scapula.

up" position to the "palms down" position, the radius crosses over the ulna, so the two bones are criss-crossed. Proximally, the radius has the following features:

head, which articulates with the capitulum of the humerus and fits into the radial notch of the ulna;
radial tuberosity, which serves as a place of attachment for a tendon from the biceps brachii;

Distally, the radius has the following features:

ulnar notch, which articulates with the head of the ulna;
styloid process, which serves as a place of attachment for ligaments that run to the wrist.

Ulna

The ulna is the longer bone of the forearm. Proximally, the ulna has the following features:

coronoid process, which articulates with the coronoid fossa of the humerus when the elbow is flexed;

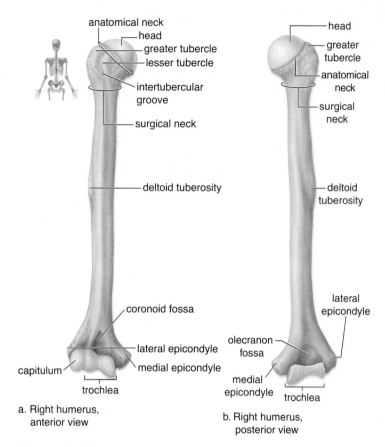

a. Right humerus,
anterior view

b. Right humerus,
posterior view

Figure 6.13 Right humerus. a. Anterior surface view. **b.** Posterior surface view.

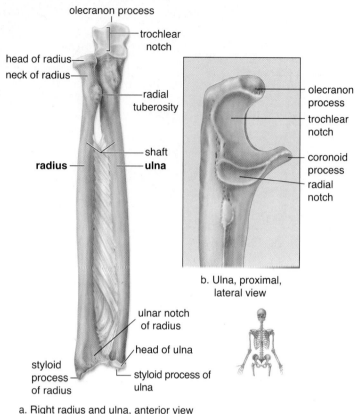

b. Ulna, proximal,
lateral view

a. Right radius and ulna, anterior view

Figure 6.14 Right radius and ulna. a. The head of the radius articulates with the radial notch of the ulna. The head of the ulna articulates with the ulnar notch of the radius. **b.** Lateral view of the proximal end of the ulna.

olecranon process, the point of the elbow, articulates with the olecranon fossa of the humerus when the elbow is extended;

trochlear notch, which articulates with the trochlea of the humerus at the elbow joint;

radial notch, which articulates with head of the radius.

Distally, the ulna has the following features:

head, which articulates with the ulnar notch of the radius;

styloid process, which serves as a place of attachment for ligaments that run to the wrist.

Hand

Each hand (Fig. 6.15) has a wrist, a palm, and five fingers, or digits.

The wrist, or carpus, contains eight small carpal bones, tightly bound by ligaments in two rows of four each. The distal forearm is where we wear a "wrist watch"—the true wrist is the proximal part of what we generally call the hand. Only two of the carpals (the scaphoid and lunate) articulate with the radius. Anteriorly, the concave region of the wrist is covered by a ligament called the flexor retinaculum, forming the so-called "carpal tunnel." Inflammation of the tendons running through this area is usually caused by abuse or overuse of the wrist area. The inflamed tendons compress a

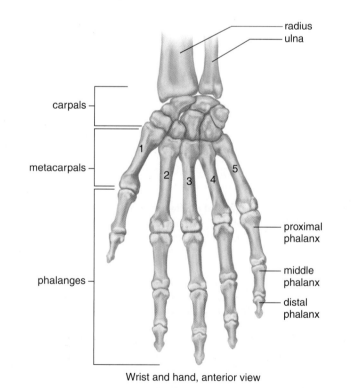

Wrist and hand, anterior view

Figure 6.15 Right wrist and hand.

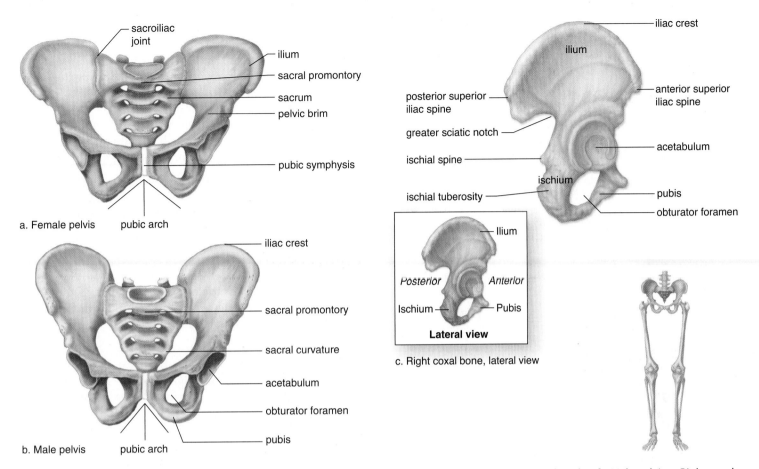

a. Female pelvis

b. Male pelvis

c. Right coxal bone, lateral view

Lateral view

Figure 6.16 **The pelvis.** The female pelvis is usually wider in all diameters than that of the male. **a.** Female pelvis. **b.** Male pelvis. **c.** Right coxal bone, lateral view.

nerve, and the resulting pain and numbness is *carpal tunnel syndrome.*

Five metacarpal bones, numbered 1 to 5 from the thumb side of the hand toward the little finger, fan out to form the palm. When the fist is clenched, the heads of the metacarpals, which articulate with the phalanges, become obvious. The first metacarpal is more anterior than the others, and this allows the thumb to touch each of the other fingers.

The fingers, including the thumb, contain bones called the *phalanges.* The thumb has only two phalanges (proximal and distal), but the other fingers have three each (proximal, middle, and distal).

Pelvic Girdle

The **pelvic girdle** contains two coxal bones (hipbones), as well as the sacrum and coccyx (Fig. 6.16a,b; see Fig. 6.8). The strong bones of the pelvic girdle are firmly attached to one another and bear the weight of the body. The pelvis also serves as the place of attachment for the lower limbs and protects the urinary bladder, the internal reproductive organs, and a portion of the large intestine.

Coxal Bones

Each **coxal bone** has the following three parts:

1. **ilium** (Fig. 6.16). The ilium, the largest part of a coxal bone, flares outward to give the hip prominence. The margin of the ilium is called the **iliac crest** (when you put your hands on your hips, you are resting them on the iliac crests). The end points of the iliac crest are called the **anterior superior iliac spine** and the **posterior superior iliac spine** (Fig. 6.16c). The **greater sciatic notch** is the site where blood vessels and the large sciatic nerve pass posteriorly into the leg. Each ilium connects posteriorly with the sacrum at a **sacroiliac joint.**

2. **ischium** (Fig. 6.16c). The ischium is the most inferior part of a coxal bone. Its posterior region, the **ischial tuberosity,** allows a person to sit. Near the junction of the ilium and ischium is the **ischial spine,** which projects into the pelvic cavity. The distance between the ischial spines is the size of the pelvic cavity.

3. **pubis** (Fig. 6.16c). The pubis is the anterior part of a coxal bone. The two pubic bones are joined by a fibrocartilage disk at the **pubic symphysis.** Posterior to where the pubis and the ischium join together is a large

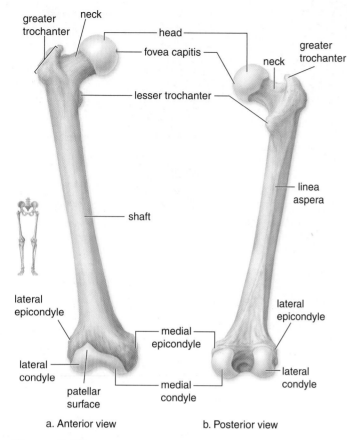

greater trochanter
neck
head
fovea capitis
neck
greater trochanter
lesser trochanter
linea aspera
shaft
lateral epicondyle
lateral epicondyle
medial epicondyle
lateral condyle
lateral condyle
patellar surface
medial condyle

a. Anterior view b. Posterior view

Figure 6.17 **Right femur. a.** Anterior view. **b.** Posterior view.

opening, the **obturator foramen,** through which blood vessels and nerves pass anteriorly into the leg.

Where the three parts of each coxal bone meet is a depression called the **acetabulum,** which receives the rounded head of the femur.

False and True Pelvises

The false pelvis is the portion of the trunk bounded laterally by the flared parts of the ilium. This space is much larger than that of the true pelvis. The true pelvis, which is inferior to the false pelvis, is the portion of the trunk bounded by the sacrum, lower ilium, ischium, and pubic bones. The true pelvis is said to have an upper inlet (also called the pelvic brim), and a lower outlet. The dimensions of these outlets are important for females because the outlets must be large enough to allow a baby to pass through during the birth process.

Gender Differences

Female and male pelvises (Fig. 6.16) usually differ in several ways, including the following:

1. Female iliac bones are more flared than those of the male; therefore, the female has broader hips.

2. The female pelvis is wider between the ischial spines and the ischial tuberosities.
3. The female inlet and outlet of the true pelvis are wider.
4. The female pelvic cavity is more shallow, while the male pelvic cavity is more funnel shaped.
5. Female bones are lighter and thinner.
6. The female pubic arch (angle at the pubic symphysis) is wider. In females, the pubic arch resembles an inverted letter U; in males, the pubic arch resembles an inverted V.

In addition to these differences in pelvic structure, male pelvic bones are larger and heavier, the articular ends are thicker, and the points of muscle attachment may be larger.

Lower Limb

The lower limb includes the bones of the thigh (femur), the kneecap (patella), the leg (tibia and fibula), and the foot (tarsals, metatarsals, and phalanges).[2]

Femur

The **femur** (Fig. 6.17), or thighbone, is the longest and strongest bone in the body. Proximally, note these structures on the femur:

head, which fits into the acetabulum of the coxal bone;
greater and **lesser trochanters,** which provide a place of attachment for the muscles of the thighs and buttocks;
linea aspera, a crest that serves as a place of attachment for several muscles.

Distally, the femur has the following features:

medial and **lateral epicondyles,** which serve as sites of attachment for muscles and ligaments;
lateral and **medial condyles,** which articulate with the tibia;
patellar surface, which is located between the condyles on the anterior surface, articulates with the **patella,** a small triangular bone that protects the knee joint.

Tibia

The **tibia** and **fibula** (Fig. 6.18) are the bones of the leg. The tibia, or shinbone, is medial to the fibula. It is thicker than the fibula and bears the weight from the femur, with which it articulates. Observe these structures on the tibia:

medial and **lateral condyles,** which articulate with the femur;
tibial tuberosity, where the patellar (kneecap) ligaments attach;
anterior crest, commonly called the shin;
medial malleolus, the bulge of the inner ankle, which articulates with the talus in the foot.

---

[2]The term *lower extremity* is used to include a coxal bone (of the pelvic girdle), the thigh, kneecap, leg, ankle, and foot.

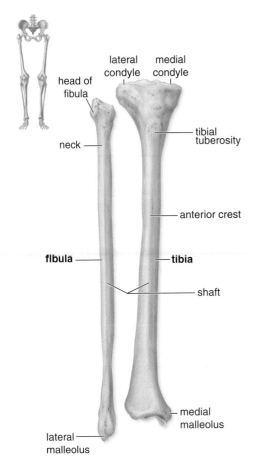

Figure 6.18 Bones of the right leg, viewed anteriorly.

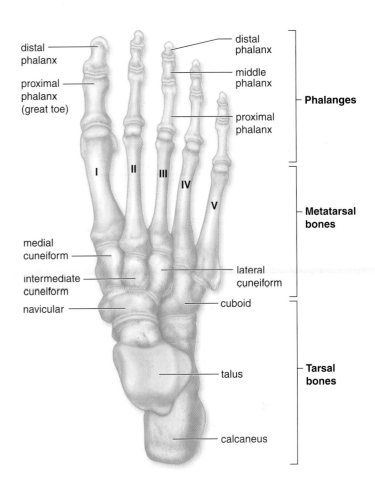

Figure 6.19 The right foot, viewed superiorly.

Fibula

The fibula is lateral to the tibia and is more slender. It has a head that articulates with the tibia just below the lateral condyle. Distally, the **lateral malleolus** articulates with the talus and forms the outer bulge of the ankle. Its role is to stabilize the ankle.

Foot

Each foot (Fig. 6.19) has an ankle, an instep, and five toes (also called digits).

The ankle has seven **tarsal bones;** together, they are called the tarsus. Only one of the seven bones, the **talus,** can move freely where it joins the tibia and fibula. The largest of the ankle bones is the **calcaneus,** or heel bone. Along with the talus, it supports the weight of the body.

The instep has five elongated **metatarsal bones.** The distal ends of the metatarsals form the ball of the foot. Along with the tarsals, these bones form the arches of the foot (longitudinal and transverse), which give spring to a person's step. If the ligaments and tendons holding these bones together weaken, fallen arches, or "flat feet," can result.

The toes contain the **phalanges.** The big toe has only two phalanges, but the other toes have three each.

6.4 Joints (Articulations)

Bones articulate at the joints. There are two systems for classifying joints: (1) according to the amount of movement they allow; and (2) according to their structure (the convention followed here). A joint called a **synarthrosis** is immovable; an **amphiarthrosis** allows slight movement; and a **diarthrosis** is freely movable.

Fibrous joints occur where fibrous connective tissue joins bone to bone. These joints are typically immovable (thus, synarthrosis joints) although exceptions exist.

Cartilaginous joints occur where fibrocartilage or hyaline cartilage joins bones. These are generally slightly movable (amphiarthrosis joints), with a few exceptions.

Synovial joints are formed when bone ends do not contact each other, but are enclosed in a capsule. These are usually freely movable (diarthrosis joints). Again, there are a few exceptions.

Fibrous Joints

Some bones, such as those that make up the adult cranium, are sutured together by a thin layer of fibrous connective tissue and are immovable. Review Figures 6.6 and 6.7, and note the following immovable *sutures*:

coronal suture, between the parietal bones and the frontal bone;

lambdoidal suture, between the parietal bones and the occipital bone;

squamosal suture, between each parietal bone and each temporal bone;

sagittal suture, between the parietal bones (not shown).

The joints formed by each tooth in its tooth socket are also fibrous joints.

Cartilaginous Joints

Where bones are joined by hyaline cartilage or fibrocartilage, the joint that forms is usually slightly movable. The ribs are joined to the sternum by costal cartilages, which are hyaline cartilage (see Fig. 6.11). The *bodies* of adjacent vertebrae are separated by fibrocartilage intervertebral disks, which increase vertebral flexibility. The pubic symphysis, the joint between the two pubic bones (see Fig. 6.16), consists largely of fibrocartilage. Due to hormonal changes, this joint becomes more flexible during late pregnancy, allowing the pelvis to expand during childbirth.

Synovial Joints

Synovial joints are generally freely movable because, unlike the joints discussed so far, the two bones are separated by a *joint cavity* (Figs. 6.20 and 6.21). The joint cavity is lined by a **synovial membrane,** which produces **synovial fluid,** a lubricant for the joint. The absence of tissue between the articulating bones allows them to be freely movable but means that the joint has to be stabilized in some way.

The joint is stabilized by the joint capsule, a sleevelike extension of the periosteum of each articulating bone. **Ligaments,** which are composed of dense regular connective tissue, bind the two bones to one another and add even more stability. Tendons, which are cords of dense regular connective tissue that connect muscle to bone, also help stabilize a synovial joint.

The articulating surfaces of the bones are protected in several ways. The bones are covered by a layer of articular (hyaline) cartilage. In addition, the joint, such as the knee, contains **menisci** (sing., *meniscus*), crescent-shaped pieces of cartilage, and fluid-filled sacs called **bursae,** which ease

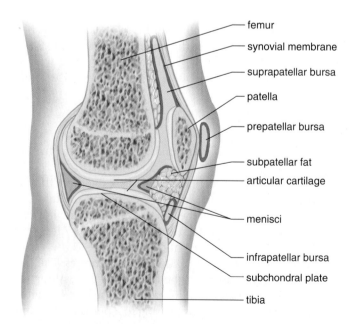

Knee joint, frontal view

Figure 6.20 Generalized anatomy of a synovial joint.

Knee joint, lateral view

Figure 6.21 The knee joint. Notice the menisci and bursae associated with the knee joint.

friction between all parts of the joint. Inflammation of the bursae is called **bursitis.** Tennis elbow is a form of bursitis. Articular cartilage can be worn away by years of constant use, such as when performing a sport.

Types of Synovial Joints

Different types of freely movable joints are listed here and depicted in Figure 6.22.

Saddle joint. Each bone is saddle-shaped and fits into the complementary regions of the other. A variety of movements are possible. *Example*: the joint between the carpal and metacarpal bones of the thumb.

Ball-and-socket joint. The ball-shaped head of one bone fits into the cup-shaped socket of another. Movement in all planes, as well as rotation, are possible. *Examples*: the shoulder and hip joints.

Pivot joint. A small, cylindrical projection of one bone pivots within the ring formed of bone and ligament of another bone. Only rotation is possible. *Examples*: the joint between the proximal ends of the radius and ulna, and the joint between the atlas and axis (see Fig. 6.10).

Hinge joint. The convex surface of one bone articulates with the concave surface of another. Up-and-down motion in one plane is possible. *Examples*: the elbow and knee joints.

Gliding joint. Flat or slightly curved surfaces of bones articulate. Sliding or twisting in various planes is possible. *Examples*: the joints between the bones of the wrist and between the bones of the ankle.

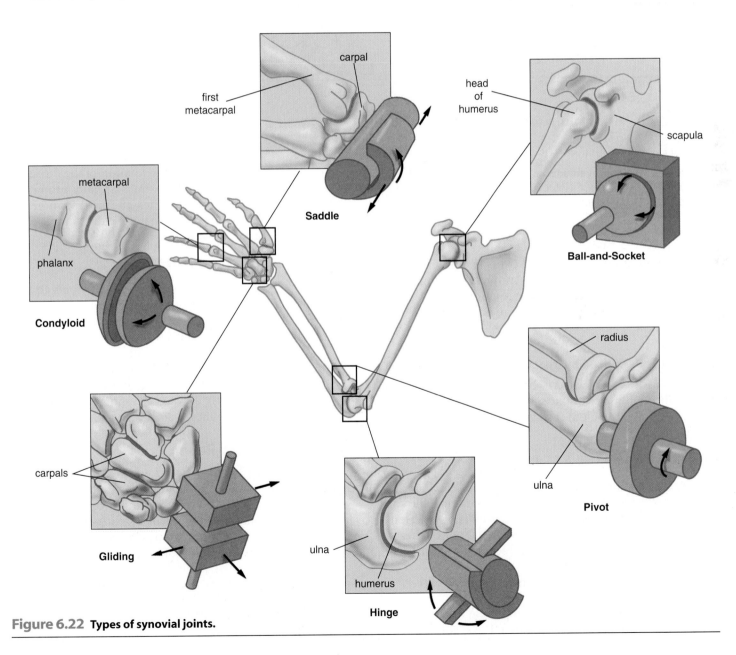

Figure 6.22 Types of synovial joints.

flexion of leg   extension of leg   adduction of thigh   abduction of thigh

a. Angular movements

rotation of arm   pronation of hand   inversion   eversion

supination

b. Circular movements   c. Special movements

Figure 6.23 **Joint movements. a.** Angular movements increase or decrease the angle between the bones of a joint. **b.** Circular movements describe a circle or part of a circle. **c.** Special movements are unique to certain joints.

Condyloid joint. The oval-shaped condyle of one bone fits into the elliptical cavity of another. Movement in different planes is possible, but rotation is not. *Examples*: the joints between the metacarpals and phalanges.

Movements Permitted by Synovial Joints

Skeletal muscles are attached to bones by tendons that cross joints. When a muscle contracts, one bone moves in relation to another bone. The more common types of movements are described here.

Angular Movements (Fig. 6.23a):

Flexion decreases the joint angle. Flexion of the elbow moves the forearm toward the arm; flexion of the knee moves the leg toward the thigh. *Dorsiflexion* is flexion of the foot upward, as when you stand on your heels; *plantar flexion* is flexion of the foot downward, as when you stand on your toes.

Extension increases the joint angle. Extension of the flexed elbow straightens the upper limb. *Hyperextension* occurs when a portion of the body part is extended beyond 180°. It is possible to hyperextend the head and the trunk of the body, and also the shoulder and wrist (arm and hand).

Adduction is the movement of a body part toward the midline. For example, adduction of the arms or legs moves them back to the sides, toward the body.

Abduction is the movement of a body part laterally, away from the midline. Abduction of the arms or legs moves them laterally, away from the body.

Circular Movements (Fig. 6.23b):

Circumduction is the movement of a body part in a wide circle, as when a person makes arm circles. Careful observation of the motion reveals that, because the proximal end of the arm is stationary, the shape outlined by the arm is actually a cone.

Rotation is the movement of a body part around its own axis, as when the head is turned to answer "no" or when the arm is twisted toward the trunk (medial rotation) and away from the trunk (lateral rotation).

Supination is the rotation of the forearm so that the palm is upward; **pronation** is the opposite—the movement of the forearm so that the palm is downward.

Special Movements (Fig. 6.23c):

Inversion and **eversion** apply only to the feet. Inversion is turning the foot so that the sole faces inward, and eversion is turning the foot so that the sole faces outward.

Elevation and **depression** refer to the lifting up and down, respectively, of a body part, as when you shrug your shoulders or move your jaw up and down.

Joint Damage and Repair

Joints can be damaged, and even destroyed, by overuse or chronic inflammation. Joint inflammation and destruction is termed **arthritis.** The most common form, **osteoarthritis,** is caused by deterioration of the articular cartilage of a joint. The damage might be caused by chronic overuse. Typical cases of osteoarthritis are seen in joints of older persons after decades of heavy use, or in knee joints of football players after constant misuse and/or abuse. **Rheumatoid arthritis** (RA) occurs when the synovial membrane becomes inflamed and grows thicker cartilage. RA is caused by an autoimmune reaction in which the body's immune system mistakenly attacks the

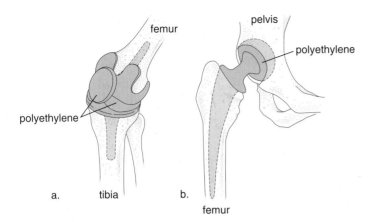

Figure 6.24 Artificial joints in which polyethylene replaces articular cartilage. a. Knee. **b.** Hip.

Content **CHECK-UP!**

10. A joint that is slightly movable is called:
   a. a synarthrosis.
   b. an amphiarthrosis.
   c. a diarthrosis.

11. The joint found at the elbow is what type of synovial joint?
   a. hinge
   b. ball-and-socket
   c. saddle
   d. condyloid

12. The joint found between the atlas and the axis is what type of synovial joint?
   a. hinge
   b. pivot
   c. ball-and-socket
   d. condyloid

Answers in Appendix C, page 462.

synovial membrane. RA is more common with age, but can occur in children and younger adults. **Gout,** or gouty arthritis, results from excessive buildup of uric acid (a metabolic waste) in the blood. Crystals of uric acid are deposited in the joints, causing inflammation.

Arthritis causes joint pain and stiffness. The afflicted joint is often swollen and may feel warm to the touch. Without the protective hyaline cartilage, exposed bone ends can grate against each other, and bone destruction occurs. On an X ray, the joint space is thinner and narrower than normal.

Treatment of arthritis should begin immediately to preserve function. Pain management is an important first step, followed by physical therapy and exercise. It's important that arthritis sufferers keep moving. Without regular exercise, the muscles around the joint will atrophy (shrink in size). Tendons and ligaments around a joint weaken, and the bones at a joint can be dislocated more easily.

Tissue culture (growing cells outside of the patient's body in a special medium) can help younger people and athletes with knee or ankle injuries to regenerate their own hyaline cartilage. In autologous chondrocyte implantation(ACI) surgery, a piece of healthy hyaline cartilage from the patient's joint is first removed surgically. The chondrocyte cells are grown outside the body in tissue culture medium, and then injected into the joint and left to grow. However, ACI isn't always successful and can't be used for elderly or overweight patients.

When other treatments for arthritis fail, surgical replacement of a joint can restore movement and relieve pain (see Fig. 6.24). Knees and hips are the most common joints to be replaced, but shoulders, elbows, ankles, and even finger joints can be replaced with surgical implants.

After ACI or joint replacement, the patient faces a lengthy rehabilitation. Physical therapy after ACI will stimulate cartilage growth without overstressing the area being repaired. In joint replacement patients, physical therapy will prevent muscle degeneration and promote bone growth around the new implant.

6.5 Effects of Aging

Both cartilage and bone tend to deteriorate as a person ages. The chemical nature of cartilage changes, and the bluish color typical of young cartilage changes to an opaque, yellowish color. The chondrocytes die, and reabsorption occurs as the cartilage undergoes calcification, becoming hard and brittle. Calcification interferes with the ready diffusion of nutrients and waste products through the matrix. The articular cartilage may no longer function properly and symptoms of arthritis can appear. As you know, decades of constant heavy use of a joint can speed development of arthritis. Osteoporosis, discussed in the Medical Focus on page 102, is present when weak and thin bones lead to fractures. Many of the degenerative changes seen in cartilage and bone can be slowed or stopped by regular weight-bearing exercise.

6.6 Homeostasis

The illustration in Human Systems Work Together on page 123 tells how the skeletal system assists other systems and how other systems assist the skeletal system. Let's review again the functions of the skeletal system, but this time as they relate to the other systems of the body.

Functions of the Skeletal System

The bones protect the internal organs. The rib cage protects the heart and lungs; the skull protects the brain; and the vertebrae protect the spinal cord. Certain endocrine organs such as the pituitary gland, pineal gland, and thymus, are also protected by bone.

The bones assist all phases of respiration. The rib cage assists the breathing process, enabling oxygen to enter the blood. Red bone marrow produces the blood cells, including the red blood cells that transport oxygen to the tissues. Without a supply of oxygen, the cells of the body could not efficiently produce ATP, the primary cellular energy source.

The bones store and release calcium. Hormones control the balance between the concentrations of calcium in the bones and in the blood. Calcium ions play a major role in muscle

"JOHN/JANE DOE, SKELETAL REMAINS, AGE UNKNOWN" is the initial identification given by law enforcement officials to the bones of an unidentified human being. The bones may have been found in the woods by a hiker or a hunter, or in a field after a farmer harvests his crops. Bones may be uncovered when a building is demolished, or if natural events such as floods or earthquakes disrupt the soil. Regardless of how human bones are found, questions must be answered. Who was this person? Was this a male or a female, and how old? What was the person's ethnicity? How did the person die, and how long ago? Was this person murdered, or did death come from natural causes?

It is the job of a *forensic anthropologist* to collect, analyze, and ultimately identify the remains. Forensic anthropologists typically have extensive training in the structure of the human skeleton and are able to examine the features of the recovered bones. These scientists rely on a national forensic analysis data bank that contains measurements and observations from thousands of skeletons. In addition, forensic anthropologists are routinely called upon to testify in criminal cases as to a victim's time and cause of death.

Clues about the identity and history of a deceased person can be found throughout the skeleton. Age is approximated by *dentition*—the structure of the teeth in the upper jaw (maxilla) and lower jaw (mandible). For example, infants aged 0–4 months have no teeth present; children aged approximately 6 through 10 have missing deciduous, or "baby" teeth; young adults acquire their last molars, or "wisdom teeth," around age 20. The age of older adults can be approximated by the number and location of missing or broken teeth.

In addition, ossification of bones—that is, replacement of a baby/child's incomplete cartilage skeleton with bone—continues in an orderly fashion until about the age of 20. Studying areas of bone ossification also gives clues to the age of the deceased at the time of death. In older adults, signs of joint breakdown provide additional information about age. Hyaline cartilage becomes worn, yellowed, and brittle with age, and the hyaline cartilages covering bone ends wear down over time. The amount of yellowed, brittle, or missing cartilage helps scientists to estimate the person's age.

If skeletal remains include the individual's pelvic bones, these provide the best method for determining an adult's gender (see pages 115–116). The long bones, particularly the humerus and femur, give information about gender as well. Long bones are thicker and denser in males, and points of muscle attachment are bigger and more prominent. The skull of a male has a square chin and more prominent ridges above the eye sockets or orbits (Figure 6B).

a. Female adult skull   b. Male adult skull

Figure 6B **Gender differences of the skull. a.** Note that the female skull is smaller, more delicate, and has a pointed chin. **b.** The male skull is large, bulky, and has a squared-off chin.

Determining ethnic origin of skeletal remains can be difficult because so many people have a mixed racial heritage. Forensic anatomists rely on observed racial characteristics of the skull. In general, individuals of African or African American descent have a greater distance between the eyes, eye sockets that are roughly rectangular, and a jaw that is large and prominent. Skulls of Native Americans typically have round eye sockets, prominent cheek (zygomatic) bones, and a rounded palate. Caucasian skulls usually have a U-shaped palate and a visible suture line between the frontal bones. Additionally, the external ear canals in Caucasians are long and straight, so that the auditory ossicles (tiny bones built inside the temporal bone and used for hearing) are visible.

Once the identity of the individual has been determined, the skeletal remains can be returned to the victim's family for proper burial. Although this can be a sorrowful event, return of physical remains provides closure and solace to many families. For this reason, special teams of forensic anthropologists employed by the U.S. military are currently researching the identities of bones from soldiers who fought in World War II, as well as the Korean, Vietnam, Gulf, and Iraq/Afghanistan wars. Ancestral remains from Native Americans are protected by the Native American Grave Protection and Repatriation Act, and must be returned to the leadership of the tribe.

Human Systems Work Together

Integumentary System

Bones provide support for skin.

Skin protects bones; helps provide vitamin D for Ca^{2+} absorption.

Muscular System

Bones provide attachment sites for muscles; store Ca^{2+} for muscle function.

Muscular contraction causes bones to move joints; muscles help protect bones.

Nervous System

Bones protect sense organs, brain, and spinal cord; store Ca^{2+} for nerve function.

Receptors send sensory input from bones to central nervous system.

Endocrine System

Bones provide protection for glands; store Ca^{2+} used as second messenger.

Growth hormone regulates bone development; parathyroid hormone and calcitonin regulate Ca^{2+} content.

Cardiovascular System

Rib cage protects heart; red bone marrow produces blood cells; bones store Ca^{2+} for blood clotting.

Blood vessels deliver nutrients and oxygen to bones, carry away wastes.

How the Skeletal System works with other body systems

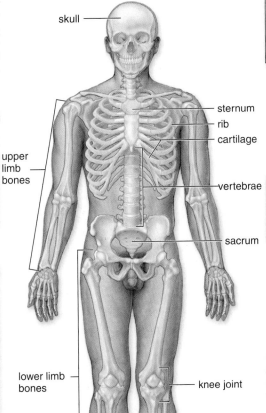

- skull
- sternum
- rib
- cartilage
- upper limb bones
- vertebrae
- sacrum
- lower limb bones
- knee joint

Lymphatic System/Immunity

Red bone marrow produces white blood cells involved in immunity.

Lymphatic vessels pick up excess tissue fluid; immune system protects against infections.

Respiratory System

Rib cage protects lungs and assists breathing; bones provide attachment sites for muscles involved in breathing.

Gas exchange in lungs provides oxygen and rids body of carbon dioxide.

Digestive System

Jaws contain teeth that chew food; hyoid bone assists swallowing.

Digestive tract provides Ca^{2+} and other nutrients for bone growth and repair.

Urinary System

Bones provide support and protection.

Kidneys provide active vitamin D for Ca^{2+} absorption and help maintain blood level of Ca^{2+} needed for bone growth and repair.

Reproductive System

Bones provide support and protection of reproductive organs.

Sex hormones influence bone growth and density in males and females.

contraction and nerve conduction. Calcium ions also help regulate cellular metabolism. Protein hormones, which cannot enter cells, are called the first messenger, and a second messenger such as calcium ions jump-starts cellular metabolism, directing it to proceed in a particular way.

The bones assist the lymphatic system and immunity. Red bone marrow produces the white cells, which congregate in the lymphatic organs. White blood cells defend the body against pathogens and cancerous cells. Without the ability to withstand foreign invasion, the body may quickly succumb to disease and die.

The bones assist digestion. The bones used for chewing break food into pieces small enough to be swallowed and chemically digested. Without digestion, needed nutrients would not enter the body.

The skeleton is necessary to locomotion. Our jointed skeleton forms a framework for muscle attachment, and muscle contraction causes joint movement. Thus, we can seek out and move to a more suitable external environment in order to maintain homeostasis.

Functions of Other Systems

How do the other systems of the body help the skeletal system function? The integumentary system and the muscles help the skeletal system protect internal organs. The digestive system absorbs the calcium from food and the plasma portion of blood transports calcium from the digestive system to the bones. The endocrine system regulates the storage of calcium in the bones, as well as the growth of bone and other tissues. The cardiovascular system transports oxygen and nutrients to bone, and wastes from bone. Movement of the bones would be impossible without contraction of the muscles. In these and other ways, the systems of the body help the skeletal system carry out its functions.

SELECTED NEW TERMS

Basic Key Terms

abduction (ab-duk′shun), p. 120

adduction (uh-duk′shun), p. 120

amphiarthrosis (am-fee-ar-thrō′sis), p. 117

appendicular skeleton (ap″en-dik′yū-ler skel′ĕ-ton), p. 112

appositional growth (ăp′ə-zĭsh′ən-əl grōth), p. 100

articular cartilage (ar-tik′yū-ler kar′tĭ-lij), p. 99

articulation (ar-tik″yū-la′shun), p. 98

atlas (ăt-ləs), p. 109

axial skeleton (ak′se-al skel′ĕ-ton), p. 103

axis (ăk′sĭs), p. 109

ball-and-socket joint (bôl and sŏk′ĭt joint), p. 119

bursae (bur′sā), p. 118

calcaneus (kăl-kā′nē-əas), p. 117

cartilaginious joints (kär′tl-ăj′ə-nas jointz), p. 118

circumduction (ser″kum-duk′shun), p. 120

clavicles (klăv′ĭ-kəls), p. 112

coccyx (kŏk′sĭks), p. 109

compact bone (kom′pakt bōn), p. 98

condyloid joint (kŏn′dĭl-oid joint), p. 120

coxal bone (kŏk-săl bōn), p. 118

depression (dĭ-prĕsh′ən), p. 121

diaphysis (di-af′ĭ-sis), p. 98

diarthrosis (dī-ar-thrō′-sis), p. 117

elevation (ĕl′ə-vā′shən), p. 121

endochondral ossification (ĕn′dō-kŏn′drəl ŏs′ə-fĭ-kā′shən), p. 100

epiphyseal plate (ep″ĭ-fĭz′e-al plăt), p. 100

epiphysis (ĕ-pif′ĭ-sis), p. 98

eversion (e-ver′zhun), p. 120

extension (ek-sten′shun), p. 120

femur (fē′mər), p. 116

fibrous joints (fī′brəs joint), p. 118

fibula (fĭb′yə-lə), p. 116

flexion (flek′shun), p. 120

fontanel (fon″tuh-nel′), p. 103

foramen magnum (for-a′-mən măg′nŭm), p. 104

gliding joint (glīd′-ing joint), p. 119

hematopoiesis (hem″ah-to-poi-e′sis), p. 98

hinge joint (hĭnj joint), p. 119

humerus (hyōō′mər-ās), p. 112

ilium (ĭl′ē-əm), p. 115

intervertebral disk (in″ter-ver′tĕ-bral disk), p. 108

intramembranous ossification (ĭn′trə-mĕm′brān-əs ŏs′ə-fĭ-kā′shən), p. 100

inversion (in-ver′zhun), p. 120

ischium (ĭs′kē-əm), p. 115

ligament (lig′uh-ment), p. 118

medullary cavity (med′u-lār″e kav′ĭ-te), p. 98

menisci (mĕ-nis′kī), p. 118

metatarsal bones (mĕt′ə-tär′səl bōn), p. 117

ossification (os′-ĭ-fĭ-ka′shun), p. 100

osteoblast (os′te-o-blast″), p. 99

osteoclast (os′te-o-klast″), p. 99

osteocyte (os′te-o-sīt), p. 99

osteon (ŏstē-ŏn), p. 98

osteoprogenitor cells (ŏs′tē-ə-prō-jĕn′ĭ-tər sĕlz), p. 99

patella (pə-tĕl′ə), p. 116

pectoral girdle (pek′tor-al ger′dl), p. 112

pelvic girdle (pel′vik ger′dl), p. 115

periosteum (per″e-os′te-um), p. 98

phalanges (fə-lăn′jēz), p. 117

pivot joint (pĭv′ət joint), p. 119

pronation (pro-na′shun), p. 120

pubic symphysis (pyōō′bĭk sĭm′fĭ-sĭs), p. 115

pubis (pyōō′bĭs), p. 115

radius (rā′dē-əs), p. 112

red bone marrow (red bōn măr′o), p. 98

rib cage (rĭb kāj), p. 109

rotation (ro-ta′shun), p. 120

sacrum (sā′krəm), p. 109

saddle joint (săd′l joint), p. 119

scapulae (skăp′yə-lə), p. 112

sinus (si′nus), p. 103

sinusitis (sī′nə-sī′tĭs), p. 103

spongy bone (spunj′e bōn), p. 98

sternum (stûr′nəm), p. 111

supination (su″pĭ-na′shun), p. 120

suture (su′cher), p. 103

synarthrosis (sĭn-ar-thrō′-sis), p. 117

synovial fluid (si-no′ve-al flu′id), p. 118

synovial joint (si-no′ve-al joint), p. 118

talus (tā′ləs), p. 121

tarsal bones (tär′səl bōns), p. 117

tibia (tĭb′ē-ə), p. 116

ulna (ŭl′nə), p. 112

vertebrae (vûr′tə-brē), p. 108

vertebral column (ver′tĕ-bral kah′lum), p. 108

Clinical Key Terms

arthritis (ărth-rī′tĭs), p. 121

bursitis (ber-si′tis), p. 119

fracture (frak′cher), p. 101

gout (gowt), p. 121

herniated disk (her′ne-a-ted disk), p. 109

kyphosis (ki-fo'sis), p. 109
lordosis (lor-do'sis), p. 109
mastoiditis (mas"toi-di'tis), p. 103

osteoarthritis (os"te-o-ar-thri'tis), p. 121
osteoporosis (os"te-o-po-ro'sis), p. 102
reduction (rĭ-duk'shən), p. 101

rheumatoid arthritis (RA) (ru'muh-toid
   ar-thri'tis), p. 121
scoliosis (sko"le-o'sis), p. 109

SUMMARY

6.1 Skeleton: Overview
  A. The skeleton supports and protects the body; produces red blood cells; serves as a storehouse for inorganic calcium and phosphate ions and fat; and permits flexible movement.
  B. A long bone has a shaft (diaphysis) and two ends (epiphyses), which are covered by articular cartilage. The diaphysis contains a medullary cavity with yellow marrow and is bounded by compact bone. The epiphyses contain spongy bone with red bone marrow that produces red blood cells.
  C. Bone is a living tissue. It develops, grows, remodels, and repairs itself. In all these processes, osteoclasts break down bone, and osteoblasts build bone.
  D. Fractures are of various types, but repair requires four steps: (1) hematoma, (2) fibrocartilaginous callus, (3) bony callus, and (4) remodeling.

6.2 Axial Skeleton
The axial skeleton lies in the midline of the body and consists of the skull, the hyoid bone, the vertebral column, and the thoracic cage.
  A. The skull is formed by the cranium and the facial bones. The cranium includes the frontal bone, two parietal bones, one occipital bone, two temporal bones, one sphenoid bone, and one ethmoid bone. The facial bones include two maxillae, two palatine bones, two zygomatic bones, two lacrimal bones, two nasal bones, the vomer bone, two inferior nasal conchae, and the mandible.
  B. The U-shaped hyoid bone is located in the neck. It anchors the tongue and does not articulate with any other bone.
  C. The typical vertebra has a body, a vertebral arch surrounding the vertebral foramen, and a spinous process. The first two vertebrae are the atlas and axis. The vertebral column has four curvatures and contains the cervical, thoracic, lumbar, sacral, and coccygeal vertebrae. Cervical, thoracic, and lumbar vertebrae are separated by intervertebral disks.
  D. The rib cage contains the thoracic vertebrae, ribs and associated cartilages, and the sternum.

6.3 Appendicular Skeleton
The appendicular skeleton consists of the bones of the pectoral girdle, upper limbs, pelvic girdle, and lower limbs.
  A. The pectoral (shoulder) girdle contains two clavicles and two scapulae.
  B. The upper limb contains the humerus, the radius, the ulna, and the bones of the hand (the carpals, metacarpals, and phalanges).
  C. The pelvic girdle contains two coxal bones, as well as the sacrum and coccyx. The female pelvis is generally wider and more shallow than the male pelvis.
  D. The lower limb contains the femur, the patella, the tibia, the fibula, and the bones of the foot (the tarsals, metatarsals, and phalanges).

6.4 Joints (Articulations)
  A. Joints are regions of articulation between bones. They are classified according to their structure and/or degree of movement. Some joints are immovable, some are slightly movable, and some are freely movable (synovial). The different kinds of synovial joints are ball-and-socket, hinge, condyloid, pivot, gliding, and saddle.
  B. Movements at joints are broadly classified as angular (flexion, extension, adduction, abduction); circular (circumduction, rotation, supination, and pronation); and special (inversion, eversion, elevation, and depression).

6.5 Effects of Aging
Two fairly common effects of aging on the skeletal system are arthritis and osteoporosis.

6.6 Homeostasis
  A. The bones protect the internal organs: The rib cage protects the heart and lungs; the skull protects the brain; and the vertebrae protect the spinal cord.
  B. The bones assist all phases of respiration. The rib cage assists the breathing process, and red bone marrow produces the red blood cells that transport oxygen.
  C. The bones store and release calcium. Calcium ions play a major role in muscle contraction and nerve conduction. Calcium ions also help regulate cellular metabolism.
  D. The bones assist the lymphatic system and immunity. Red bone marrow produces not only the red blood cells but also the white blood cells.
  E. The bones assist digestion. The bones used for chewing break food into pieces small enough for chemical digestion.
  F. The skeleton is necessary for locomotion. Humans have a jointed skeleton for the attachment of muscles that move the bones.

STUDY QUESTIONS

  1. What are five functions of the skeleton? (p. 98)
  2. What are five major categories of bones based on their shapes? (p. 98)
  3. What are the parts of a long bone? What are some differences between compact bone and spongy bone? (pp. 98–99)
  4. How does bone grow in children, and how is it remodeled in all age groups? (p. 99)

5. What are the various types of fractures? What four steps are required for fracture repair? (pp. 100–102)
6. List the bones of the axial and appendicular skeletons. (Fig. 6.4, p. 103)
7. What are the bones of the cranium and the face? What are the special features of the temporal bones, sphenoid bone, and ethmoid bone? (pp. 103–108)
8. What are the parts of the vertebral column, and what are its curvatures? Distinguish between the atlas, axis, sacrum, and coccyx. (pp. 108–110)
9. What are the bones of the rib cage? What are several functions of the rib cage? (pp. 111–112)
10. What are the bones of the pectoral girdle? Give examples to demonstrate the flexibility of the pectoral girdle. What are the special features of a scapula? (p. 112)
11. What are the bones of the upper limb? What are the special features of these bones? (pp. 112–115)
12. What are the bones of the pelvic girdle, and what are their functions? (pp. 115–116)
13. What are the false and true pelvises, and what are several differences between the male and female pelvises? (p. 116)
14. What are the bones of the lower limb? Describe the special features of these bones. (pp. 116–117)
15. How are joints classified? Give examples of each type of joint. (p. 117)
16. How can joint movements permitted by synovial joints be categorized? Give an example of each category. (pp. 117–121)
17. How does aging affect the skeletal system? (p. 121)
18. What functions of the skeletal system are particularly helpful in maintaining homeostasis? (pp. 121, 123)

LEARNING OBJECTIVE QUESTIONS

I. Match the items in the key to the bones listed in questions 1–6.
   Key:
   a. forehead
   b. chin
   c. cheekbone
   d. elbow
   e. shoulder blade
   f. hip
   g. leg
   1. temporal and zygomatic bones
   2. tibia and fibula
   3. frontal bone
   4. ulna
   5. coxal bone
   6. scapula

II. Match the items in the key to the bones listed in questions 7–13.

Key:
   a. external acoustic meatus
   b. cribriform plate
   c. xiphoid process
   d. glenoid cavity
   e. olecranon process
   f. acetabulum
   g. greater and lesser trochanters
   7. scapula
   8. sternum
   9. femur
   10. temporal bone
   11. coxal bone
   12. ethmoid bone
   13. ulna

III. Fill in the blanks.
   14. Long bones are _____ than they are wide.

15. The epiphysis of a long bone contains _____ bone, where red blood cells are produced.
16. The _____ are the air-filled spaces in the cranium.
17. The sacrum is a part of the _____, and the sternum is a part of the _____.
18. The pectoral girdle is specialized for _____, while the pelvic girdle is specialized for _____.
19. The term *phalanges* is used for the bones of both the _____ and the _____.
20. The knee is a freely movable (synovial) joint of the _____ type.

MEDICAL TERMINOLOGY EXERCISE

Consult Appendix B for help in pronouncing and analyzing the meaning of the terms that follow.

1. chondromalacia (kon"dro-muh-la′she-uh)
2. osteomyelitis (os"te-o-mi"e-li′tis)
3. craniosynostosis (kra"ne-o-sin"os-to′sis)
4. myelography (mi"ĕ-log′ruh-fe)
5. acrocyanosis (ak"ro-si"uh-no′sis)
6. syndactylism (sin-dak′tĭ-lizm)
7. orthopedist (or"tho-pe′dist)
8. prognathism (prog′nah-thizm)
9. micropodia (mi"kro-po′de-uh)
10. arthroscopic (ar"thro-skop′ik)
11. bursectomy (ber-sek′to-me)
12. synovitis (sin-o-vi′tis)
13. acephaly (a-sef′uh-le)
14. sphenoidostomy (sfe-noy-dos′to-me)
15. acetabuloplasty (as-ĕ-tab′yū-lo-plas-te)

WEBSITE LINK

Visit this text's website at http://www.mhhe.com/longenbaker7 for additional quizzes, interactive learning exercises, and other study tools.

The Muscular System

If you were asked to describe what skeletal muscles do, you'd probably talk about the conscious limb movements you need for everyday activities like walking, running, lifting, bending, twisting from side to side, performing sports activities, and so on. But here's another skeletal muscle function you might not have thought of: communication through facial expression. We use over 20 muscles just for facial expression (anatomists disagree on the exact number) and each performs a very precise movement. It's interesting to note that human infants mimic the expressions of their caregivers almost from birth. Expressing emotions—joy, fear, frustration, unhappiness, loneliness— without speaking is essential for the baby's survival, for without this ability the child couldn't communicate its needs. Further, facial expressions convey nonverbal messages between people of all races and cultures. Check out the expressions of the people pictured here—what do you think each is feeling? Then, study Figure 7.12 and see if you can determine which facial muscles are used to make each of these expressions.

Chapter Outline & Learning Objectives   *After you have studied this chapter, you should be able to:*

7.1 Functions and Types of Muscles

All muscles, regardless of the particular type, can contract—that is, shorten. When muscles contract, some part of the body or the entire body moves. Humans have three types of muscles: smooth, cardiac, and skeletal (Fig. 7.1). The contractile cells of these tissues are elongated and therefore are called **muscle fibers.**

Smooth Muscle

Smooth muscle is located in the walls of hollow internal organs and blood vessels. Its involuntary contraction moves materials through organs, and regulates blood flow in blood vessels. Smooth muscle fibers are narrow, tapered, rod-shaped cells, and each cell is uninucleate (has a single nucleus). The cells are usually arranged in parallel lines, forming sheets. Smooth muscle does not have the striations (bands of light and dark) seen in cardiac and skeletal muscle. Although smooth muscle is slower to contract than skeletal muscle, it can sustain prolonged contractions and does not fatigue easily.

Cardiac Muscle

Cardiac muscle forms the heart wall. Its fibers are uninucleated, striated, cylindrical, and branched, which allows the fibers to interlock at *intercalated disks.* Intercalated disks contain gap junctions, which permit contractions to spread quickly throughout the heart. Cardiac fibers relax completely between contractions, which prevents fatigue. Contraction of cardiac muscle fibers is rhythmical and occurs without requiring outside nervous stimulation. Thus, cardiac muscle contraction is involuntary. However, keep in mind the nerves that supply the heart do affect heart rate and strength of contraction.

Skeletal Muscle

Skeletal muscle fibers are cylindrical, multinucleated, and striated. They make up the skeletal muscles attached to the skeleton. Skeletal muscle fibers can run the length of a muscle and therefore can be quite long. Skeletal muscle is voluntary because its contraction is always stimulated and controlled by the nervous system. In this chapter, we will explore why skeletal muscle (and cardiac muscle) is striated.

Connective Tissue Coverings

Muscles are organs, and as such they contain other types of tissues, such as nervous tissue, blood vessels, and connective tissue. Connective tissue is essential to the organization of the fibers within a muscle (Fig. 7.2). First, each fiber is surrounded by a thin layer of areolar (loose) connective tissue called the *endomysium.* Blood capillaries and nerve fibers reach each muscle fiber by way of the endomysium. Second, the muscle fibers are grouped into bundles called *fascicles.* The fascicles have a sheath of connective tissue called the *perimysium.* Finally, the muscle itself is covered by a connective tissue layer called the *epimysium.* The epimysium becomes a part of the *fascia,* a layer of fibrous tissue that separates muscles from each other (deep fascia) and from the skin (superficial fascia).

Smooth muscle
- has narrow, tapered cylindrical-shaped cells.
- has nonstriated, uninucleated fibers.
- occurs in walls of internal organs and blood vessels.
- is involuntary.

Cardiac muscle
- has striated, cylindrical, branched, uninucleated fibers.
- occurs in walls of heart.
- is involuntary.

Skeletal muscle
- has striated, cylindrical, multinucleated fibers.
- is usually attached to skeleton.
- is voluntary.

Figure 7.1 Types of muscles. The three types of muscles in the body have the appearance and characteristics shown here.

Collagen fibers of the epimysium continue as a strong, fibrous **tendon** that attaches the muscle to a bone. The epimysium merges with the periosteum of the bone.

Begin Thinking Clinically

?

The deep fascia that surrounds a group of several muscles is called a *muscle compartment*. It functions as a tight sleeve containing the muscles, nerves, and blood vessels. In a **compartment syndrome**, swelling inside a compartment increases, choking off the blood supply to the muscle. This condition can cause complete muscle destruction and may be fatal. What type of injury could cause a compartment syndrome? How could it be corrected?

Answer and discussion in Appendix C, page 462.

Functions of Skeletal Muscles

This chapter concerns the skeletal muscles, and therefore it is fitting to consider their functions independent of the other types of muscles:

Skeletal muscles support the body. Skeletal muscle contraction opposes the force of gravity and allows us to remain upright. Some skeletal muscles are serving this purpose even when you think you are relaxed.

Skeletal muscles make bones and other body parts move. Muscle contraction accounts not only for the movement of limbs but also for eye movements, facial expressions, and breathing.

Skeletal muscles help maintain a constant body temperature. Skeletal muscle contraction causes ATP to break down, releasing heat that is distributed about the body.

Figure 7.2 Connective tissue of a skeletal muscle. a. Trace the connective tissue of a muscle from the endomysium to the perimysium to the epimysium, which becomes a part of the deep fascia and from which the tendon extends to attach a muscle to the periosteum of a bone. **b.** Cross section of the arm showing the arrangement of the muscles, which are separated from the skin by fascia. The superficial fascia contains adipose tissue. **c.** Photomicrograph of muscle fascicles from the tongue where the fascicles run in different directions. (c.s. = cross section; l.s. = longitudinal section.)

Skeletal muscle contraction assists movement in cardiovascular and lymphatic vessels. The pressure of skeletal muscle contraction keeps blood moving in cardiovascular veins and lymph moving in lymphatic vessels.

Skeletal muscles help protect bones and internal organs and stabilize joints. Muscles pad the bones, and the muscular wall in the abdominal region protects the internal organs. Muscle tendons help hold bones together at the joints.

Content CHECK-UP!

1. Choose the phrase you could use when describing skeletal muscle.
   a. uninucleate cells
   b. can be very long
   c. under voluntary control
   d. b. and c.
   e. all of the above

2. Choose the phrase you could use when describing smooth muscle.
   a. multinucleate
   b. cells arranged in sheets
   c. striated
   d. fatigues easily

3. Select the connective tissue covering of a muscle that is most superficial.
   a. perimysium
   b. endomysium
   c. epimysium

Answers in Appendix C, page 462.

7.2 Microscopic Anatomy and Contraction of Skeletal Muscle

We have already examined the structure of skeletal muscle as seen with the light microscope. As you know, skeletal muscle tissue has alternating light and dark bands, giving it a striated appearance. The electron microscope shows that these bands are due to the arrangement of protein filaments, called *myofilaments,* in a muscle fiber.

Muscle Fiber

A muscle fiber contains the usual cellular components, but special names have been assigned to some of these components (note that the terms used to describe muscle start with the prefixes *myo-* and *sarco-;* Table 7.1 and Fig. 7.3). Thus, the plasma membrane is called the *sarcolemma;* cytoplasm is the *sarcoplasm;* and the endoplasmic reticulum is the *sarcoplasmic reticulum.* A muscle fiber also has some unique anatomical characteristics. One feature is its T (for transverse) system; the sarcolemma forms **T (transverse) tubules** that penetrate, or dip down, into the cell so that they come into contact—but do not fuse—with expanded portions of the sarcoplasmic reticulum. The expanded portions of the sarcoplasmic reticulum are calcium storage sites. Calcium ions (Ca^{2+}), as we shall see, are essential for muscle contraction.

The sarcoplasmic reticulum encases hundreds and sometimes even thousands of **myofibrils,** which are bundles of myofilaments. Each myofibril is about 1 micrometer in diameter.

TABLE 7.1 Microscopic Anatomy of a Muscle

Name	Function
Sarcolemma	Plasma membrane of a muscle fiber that forms T tubules
Sarcoplasm	Cytoplasm of a muscle fiber that contains organelles, including myofibrils
Glycogen	A polysaccharide that stores energy for muscle contraction
Myoglobin	A red pigment that stores oxygen for muscle contraction
T tubule	Extension of the sarcolemma that extends into the muscle fiber and conveys nerve signals that cause Ca^{2+} to be released from the sarcoplasmic reticulum into the sarcoplasm
Sarcoplasmic reticulum	The smooth ER of a muscle fiber that stores Ca^{2+}
Myofibril	A bundle of myofilaments that contracts
Myofilament	Thick and thin filaments whose structure and functions account for muscle striations and contractions

Myofibrils are the contractile portions of the muscle fibers. Any other organelles, such as mitochondria, are located in the sarcoplasm between the myofibrils. The sarcoplasm also contains glycogen, which provides stored energy for muscle contraction, and the red pigment **myoglobin,** which binds oxygen until it is needed for muscle contraction.

Myofibrils and Sarcomeres

Myofibrils are cylindrical in shape and run the length of the muscle fiber. Each myofibril is composed of numerous **sarcomeres** (Fig. 7.3). As an analogy, imagine each myofibril as a single freight train. Each "boxcar" of the "train" is a sarcomere; thus, the myofibril "train" is composed of linked sarcomeres. Each sarcomere extends between two dark, vertical lines called **Z lines.** The horizontal stripes, or *striations,* of skeletal muscle fibers are formed by the placement of myofilaments within the sarcomeres. A sarcomere contains two types of protein myofilaments: The **thick filaments** are made up of a single protein called **myosin. Thin filaments** are made up of three proteins: a globular protein called **actin,** plus **tropomyosin** and **troponin.** In each sarcomere, the **I band** is light-colored because it contains only thin filaments attached to a Z line. Note that I bands overlap adjacent sarcomeres. The dark regions of the **A band,** found in the center of the sarcomere, contain overlapping thick and thin filaments. In the center of the A band is the lighter **H zone,** which contains only myosin filaments (Fig. 7.3*b* and *c*).

Myofilaments

The thick and thin filaments differ in the following ways:

Thick Filaments A thick filament is composed of several hundred molecules of the protein myosin. Each myosin molecule is

skeletal muscle

fascicle

muscle fibers

tendon

deep fascia

perimysium

endomysium

epimysium

T tubule

nucleus

sarcoplasmic reticulum

calcium storage sites

skeletal muscle fiber

one myofibril

sarcolemma

a. Muscle fiber has many myofibrils.

one sarcomere

Z line   Z line

b. Myofibril has many sarcomeres.

cross-bridge

thick filaments

thin filaments

c. Sarcomere is relaxed.

Z line

H zone

A band

I band

d. Sarcomere is contracted.

Figure 7.3 Anatomy of a muscle fiber. a. A muscle fiber contains many myofibrils with the components shown. **b.** A myofibril has many sarcomeres that contain myosin and actin filaments whose arrangement gives rise to the striations so characteristic of skeletal muscle. **c. d.** Muscle contraction occurs when sarcomeres contract and thin filaments slide past thick filaments.

composed of two protein strands, each shaped like a golf club. The straight portions of each strand coil around each other. Each myosin molecule ends in a double globular head, or *cross-bridge.* Cross-bridges are slanted away from the middle of a sarcomere, toward the thin filaments surrounding them. One portion of the cross-bridge will bind to actin in thin filaments during muscle contraction.

Thin Filaments A thin filament consists primarily of two strands of the globular protein actin, twisted around each other like intertwined bead necklaces. Double strands of tropomyosin coil over each actin strand. Troponin occurs at intervals on the tropomyosin strand (see Fig. 7.5a).

Sliding Filaments When muscles are activated by motor nerves, impulses travel down a T tubule, and calcium is released from the sarcoplasmic reticulum. Now the muscle fiber contracts as the sarcomeres within the myofibrils shorten. When a sarcomere shortens, the actin (thin) filaments slide past the myosin (thick) filaments and approach one another. This causes the I band to shorten and the H zone to almost or completely disappear (Fig. 7.3c and d). The movement of actin filaments in relation to myosin filaments is called the **sliding filament theory** of muscle contraction. During the sliding process, each sarcomere shortens even though the thick and thin filaments themselves remain the same length. Although the actin filaments slide past the myosin filaments, it is the myosin filaments that do the work. Myosin filaments break down ATP, which supplies the energy for muscle contraction. Myosin molecules have cross-bridges that pull the actin filaments toward the center of the sarcomere. Returning to the train analogy, when each sarcomere ("boxcar") shortens during a muscle contraction, the myofibril ("train") becomes shorter as well.

Skeletal Muscle Contraction

Muscle fibers are innervated—that is, they are stimulated to contract by motor neurons whose axons are found in nerves. Recall from Chapter 4 that an axon is the single, long extension of a nerve cell or neuron. The axon of one motor neuron has several branches and can stimulate from a few to several muscle fibers of a particular muscle. Each branch of the axon ends in an axon terminal that lies in close proximity to the sarcolemma of a muscle fiber. A small gap, called a *synaptic cleft,* separates the axon terminal from the sarcolemma. This entire region is called a **neuromuscular junction,** and it consists of the motor nerve axon and the muscle fibers it contacts (Fig. 7.4).

Axon terminals contain synaptic vesicles that are filled with the neurotransmitter acetylcholine (ACh). When nerve signals traveling down a motor neuron arrive at an axon terminal, the synaptic vesicles release neurotransmitter into the synaptic cleft, which quickly diffuses across the cleft and binds to receptors on the sarcolemma. Now the sarcolemma generates signals

Figure 7.4 Neuromuscular junction. The branch of a nerve axon ends in an axon terminal that meets but does not touch a muscle fiber. A synaptic cleft separates the axon terminal from the sarcolemma of the muscle fiber. Nerve impulses traveling down an axon cause synaptic vesicles to discharge acetylcholine, which diffuses across the synaptic cleft. When the neurotransmitter is received by the sarcolemma of a muscle fiber, impulses begin and lead to muscle fiber contractions. The motor control area of the brain signals muscles via motor nerves.

that spread over the sarcolemma and down T tubules to the sarcoplasmic reticulum, triggering calcium release. The release of calcium from the sarcoplasmic reticulum causes the filaments within the sarcomeres to slide past one another. Sarcomere contraction results in myofibril contraction, which in turn results in muscle fiber, and finally, muscle contraction.

The Role of Actin and Myosin

Figure 7.5a shows the placement of the three proteins that comprise a thin filament. Thin filaments are composed of a double row of twisted actin molecules. Threads of tropomyosin wind around each actin filament, and troponin occurs at intervals along the tropomyosin threads. A myosin binding site occurs on each actin molecule; when muscle is relaxed, these binding sites are covered by tropomyosin. Calcium ions (Ca^{2+}) that have been released from the sarcoplasmic reticulum combine with troponin. After binding occurs, the tropomyosin threads shift their position, and myosin binding sites are exposed (Fig. 7.5b).

Each of the cross-bridges of a myosin thick filament (recall that these are the paired globular heads of the filament) has two binding sites. One site binds to ATP and then functions as an ATPase enzyme, splitting ATP into ADP and Ⓟ (see Fig. 7.5c, *step 1*). This reaction activates the second binding site so that it will bind to actin. The ADP and Ⓟ remain on the myosin heads until the heads attach to actin (Fig. 7.5c, *step 2*). Now, ADP and Ⓟ are released, and this causes the cross-bridges to bend sharply (Fig. 7.5c, *step 3*). This is the **power stroke** that

a.
actin filament
troponin
tropomyosin

b.
myosin binding sites
Ca²⁺
Troponin-Ca²⁺ complex pulls tropomyosin away, exposing myosin binding sites.

actin filament
myosin filament
(P) ADP

① ATP is hydrolyzed when myosin head is unattached.

cross-bridge — myosin head

② ADP + (P) are bound to myosin as myosin head attaches to actin.

ATP

④ Binding of ATP causes myosin head to return to resting position.

③ ADP + (P) release causes head to change position and actin filament to move.

c.

Figure 7.5 The role of actin and myosin in muscle contraction. a. In relaxed muscle, tropomyosin covers myosin binding sites on actin. **b.** Upon release, calcium binds to troponin, exposing myosin binding sites. **c.** After breaking down ATP, myosin cross-bridges bind to actin filaments and, later, a power stroke causes the actin filament to move.

pulls the thin filaments toward the middle of the sarcomere. When another ATP molecule binds to a myosin head, the head detaches from actin (Fig. 7.5c, step 4). The cycle begins again; the thin filaments move nearer the center of the sarcomere each time the cycle is repeated. Note that ATP has two roles in this process: first to energize the myosin cross-bridge, and then to break the link between myosin and actin.

Contraction continues until nerve impulses cease and calcium ions are returned to their storage sites. The membranes of the sarcoplasmic reticulum contain active transport proteins that pump calcium ions back into the sarcoplasmic reticulum. Of course, this active transport process also requires ATP energy.

Contraction of Smooth Muscle

You will recall that smooth muscle cells are uninucleate, cylindrical cells with pointed ends, and that smooth muscle control is involuntary. Like skeletal muscle, smooth muscle contains thick and thin filaments. However, in smooth muscle these filaments are not arranged into myofibrils that create visible striations. Instead, thin filaments in smooth muscle are anchored directly to the sarcolemma or to protein molecules called *dense bodies*. Dense bodies are scattered through the sarcoplasm. When a smooth muscle cell contracts, its fibers shorten in all directions, causing the cylindrical cell to become more oval in shape. In turn, the entire smooth muscle shortens. Smooth muscle contraction occurs very slowly but can last for long periods of time without fatigue.

Energy for Muscle Contraction

ATP produced previous to strenuous exercise lasts a few seconds, and then muscles acquire new ATP in three different ways: creatine phosphate breakdown, cellular respiration, and fermentation (Fig. 7.6). Creatine phosphate breakdown and

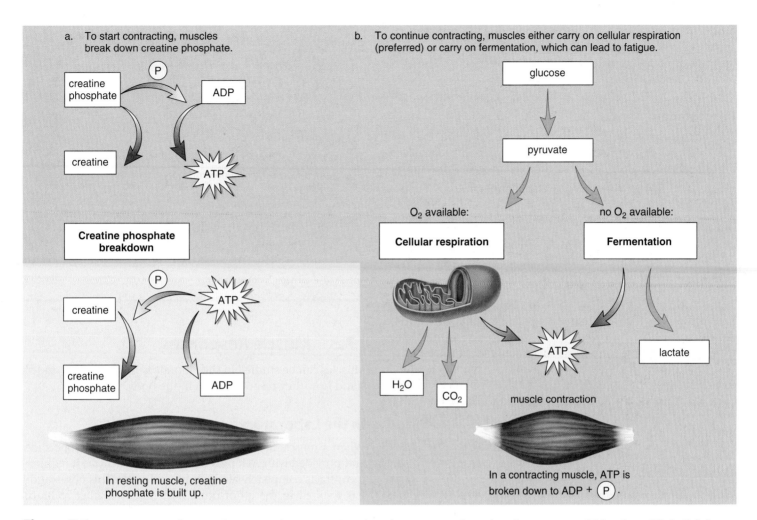

a. To start contracting, muscles break down creatine phosphate.

Creatine phosphate breakdown

In resting muscle, creatine phosphate is built up.

b. To continue contracting, muscles either carry on cellular respiration (preferred) or carry on fermentation, which can lead to fatigue.

glucose

pyruvate

O_2 available:

Cellular respiration

no O_2 available:

Fermentation

ATP

lactate

H_2O

CO_2

muscle contraction

In a contracting muscle, ATP is broken down to ADP + Ⓟ.

Figure 7.6 **Energy sources for muscle contraction. a.** Creatine phosphate can transfer a phosphate group to ADP to form ATP. **b.** Cellular respiration and fermentation also form ATP in muscle tissue.

fermentation are anaerobic, meaning that they do not require oxygen.

Creatine Phosphate Breakdown

Creatine phosphate is a high-energy compound built up when a muscle is resting. Creatine phosphate cannot participate directly in muscle contraction. Instead, it can regenerate ATP by transferring its phosphate to ADP, using the following reaction:

creatine phosphate   ⟶   creatine

This reaction occurs in the midst of sliding filaments, and therefore is the speediest way to make ATP available to muscles. Creatine phosphate provides enough energy for only about eight seconds of intense activity, and then it is spent. Creatine phosphate is rebuilt when a muscle is resting by transferring a phosphate group from ATP to creatine (see Fig. 7.6*a*).

Cellular Respiration

Cellular respiration completed in mitochondria usually provides most of a muscle's ATP. Glycogen and fat are stored in muscle cells. Therefore, a muscle cell can use glucose from glycogen and fatty acids from fat as fuel to produce ATP if oxygen is available:

glucose + oxygen   ⟶   carbon dioxide + water

Myoglobin, an oxygen carrier similar to hemoglobin in red blood cells, is synthesized in muscle cells. Its presence accounts for the reddish-brown color of skeletal muscle fibers. Myoglobin has a higher affinity for oxygen than does hemoglobin. Therefore, myoglobin can pull oxygen out of blood and make it available to muscle mitochondria that are carrying on cellular respiration. Then, too, the ability of myoglobin to temporarily store oxygen reduces a muscle's immediate need for oxygen when cellular respiration begins. The end products of cellular respiration (carbon dioxide and water) are usually no problem. Carbon dioxide leaves the body at the lungs, and water simply enters the extracellular space. The by-product—heat—keeps the entire body warm.

Fermentation

Fermentation, like creatine phosphate breakdown, supplies ATP without consuming oxygen. Thus, it is an *anaerobic* process. Fermentation produces the ATP necessary for short bursts of exercise—for example, a 50-yard dash or a run around the bases in a baseball game. During fermentation, glucose is broken down to lactate (lactic acid):

The accumulation of lactate in a muscle fiber makes the cytoplasm more acidic, and eventually enzymes cease to function well. If fermentation continues longer than two or three minutes, cramping and fatigue set in. Cramping seems to be due to lack of ATP. Recall that ATP is needed to pump calcium ions back into the sarcoplasmic reticulum and to break the linkages between the actin and myosin filaments so that muscle fibers can relax.

Oxygen Debt

When a muscle uses fermentation to supply its energy needs, it incurs an **oxygen debt.** Oxygen debt is obvious when a person continues to breathe heavily after exercising. The ability to run up an oxygen debt is one of muscle tissue's greatest assets. Brain tissue cannot last nearly as long without oxygen as muscles can.

Repaying an oxygen debt requires replenishing creatine phosphate supplies and disposing of lactic acid. Lactic acid can be changed back to a compound called pyruvic acid and metabolized completely in mitochondria, or it can be sent to the liver to reconstruct glycogen. A marathon runner who has just crossed the finish line is not exhausted due to oxygen debt. Instead, the runner has used up all the muscles', and probably the liver's, glycogen supply. It takes about two days to replace glycogen stores on a high-carbohydrate diet.

People who train rely more heavily on cellular respiration than do people who do not train. In people who train, the number of muscle mitochondria increases, and so fermentation is not needed to produce ATP. Their mitochondria can start consuming oxygen as soon as the ADP concentration starts rising during muscle contraction. Because mitochondria can break down fatty acid, instead of glucose, blood glucose is spared for the activity of the brain. (The brain, unlike other organs, ordinarily utilizes only glucose to produce ATP.) Because less lactate is produced in people who train, the pH of the blood remains steady, and there is less oxygen debt.

7.3 Muscle Responses

Muscles can be studied in the laboratory in an effort to understand how they respond when in the body.

In the Laboratory

When a muscle fiber is isolated, placed on a microscope slide, and provided with ATP plus the various electrolytes it requires, it contracts completely along its entire length. This observation has resulted in the **all-or-none law:** A muscle fiber contracts completely or not at all. In contrast, a whole muscle (made up of many muscle fibers) shows degrees of contraction. To study whole muscle contraction in the laboratory, an isolated muscle is stimulated electrically, and the mechanical force of contraction is recorded as a visual pattern called a *myogram.* When the strength of the stimulus is above a threshold level, the muscle contracts and then relaxes. This action—a single contraction that lasts only a fraction of a second—is called a **muscle twitch.** Figure 7.7 is a myogram of a muscle twitch, which is customarily divided into three stages: the latent period, or the period of time between stimulation and initiation of contraction; the contraction period, when the muscle shortens; and the relaxation period, when the muscle returns to its former length. It's interesting to use our knowledge of muscle fiber contraction to understand these events. From our study thus far, we know that a muscle fiber in an intact muscle contracts when calcium leaves storage sacs in the sarcoplasmic reticulum and relaxes when calcium returns to storage sacs.

But unlike the contraction of a muscle fiber, a muscle has degrees of contraction, and a twitch can vary in height (strength) depending on the degree of stimulation. Why should that be? Obviously, a stronger stimulation causes more individual fibers to contract than before.

If a whole muscle is given a rapid series of stimuli, it can respond to the next stimulus without relaxing completely. **Summation** is increased muscle contraction until maximal

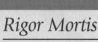

FOCUS *on* FORENSICS

Rigor Mortis

For everything there is a season, and a time for every matter under heaven: a time to be born, and a time to die . . .
—Ecclesiastes 3:1,2 KJV

When a person dies, the physiologic events that accompany death occur in an orderly progression. Respiration ceases, the heart ultimately stops beating, and tissue cells begin to die. The first tissues to die are those with the highest oxygen requirement. Brain and nervous tissues have an extremely high requirement for oxygen. Deprived of oxygen, these cells typically die after only six minutes, as the ATP energy reserve within the cell is used up. However, tissues that can produce ATP by fermentation (which does not require oxygen) can "live" for an hour or more before ATP is completely depleted. Muscle is capable of generating ATP by both fermentation and creatine phosphate breakdown; thus, muscle cells can survive for a time after clinical death occurs. Muscle death is signaled by a process termed *rigor mortis*—the "stiffness of death." Rigor mortis develops and reverses itself within a known time period of hours to days after death occurs. Forensic pathologists use this information to study the condition of a body to approximate the time of death.

Rigor mortis occurs as dying muscle cells deplete the last of their ATP energy reserve. Relaxing the muscle becomes impossible because ATP energy is needed to break the bond between an actin binding site and the myosin cross-bridge. In addition, ATP energy is required for muscle relaxation to return calcium ions from the sarcoplasm back to the sarcoplasmic reticulum. Thus, without ATP energy, the muscle remains fixed in the same state of contraction that preceded that person's death. If, for example, a murder victim dies while sitting at a desk, the body in rigor mortis will be frozen in the sitting position.

Forensic pathologists use body temperature and the presence or absence of rigor mortis to estimate time of death. For example, the body of someone who has been dead for 3 hours or less will still be warm (close to normal body temperature, 98.6°F or 37°C) and rigor mortis will be absent. After approximately 3 hours, the body will be significantly cooler than normal, and rigor mortis will begin to develop. The corpse of an individual who has been dead at least 8 hours will be in full rigor mortis, and the temperature of the body will be the same as the surroundings. Rigor mortis resolves approximately 24–36 hours after death. Muscles lose their stiffness because lysosomes inside the cell eventually rupture, releasing enzymes that break the myosin-actin bonds. Forensic pathologists know that a person has been dead for more than 24 hours if the body temperature is the same as the environment and there is no longer a trace of rigor mortis.

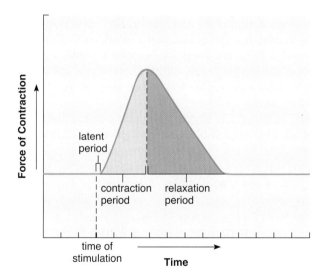

Figure 7.7 **A myogram showing a single muscle twitch.**

sustained contraction, called a **tetanic contraction,** is achieved (Fig. 7.8). The myogram no longer shows individual twitches; rather, the twitches are fused and blended completely into a straight line. Tetanus continues until the muscle *fatigues.*

Fatigue in muscles is apparent when the muscle relaxes even though stimulation continues. In the body, fatigue is a gradual weakening that occurs after repetitive use. There are several reasons why muscles become fatigued. First, ATP is depleted during constant use of a muscle; the muscle essentially "runs out of energy." Repetitive use causes production of lactic acid by fermentation, which lowers the pH of the sarcoplasm and inhibits muscle function. Additionally, the motor nerves that supply muscle can run out of their neurotransmitter, acetylcholine. Finally, the brain itself, through mechanisms not well understood, may signal a person to stop exercising even if the muscles are not truly fatigued. People who train can exercise for longer periods without experiencing fatigue.

In the Body

In the body, muscles are innervated to contract by nerves. As mentioned, each axon within a nerve stimulates a number of muscle fibers. A nerve fiber together with all of the muscle fibers it innervates is called a **motor unit.** A motor unit obeys the all-or-none law. Why? Because all the muscle fibers in a motor unit are stimulated at once, and they all either contract or do not contract. A variable of interest is the number of muscle fibers within a motor unit. For example, in the ocular

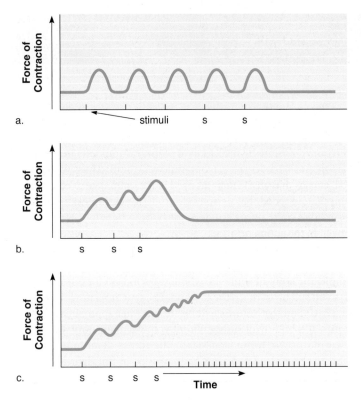

Figure 7.8 Myograms showing a. a series of twitches, **b.** summation, and **c.** a tetanic contraction. Note that an increased frequency of stimulations has resulted in these different responses.

muscles that move the eyes, the innervation ratio is one motor axon per 23 muscle fibers, while in the gastrocnemius muscle of the lower leg, the ratio is about one motor axon per 1,000 muscle fibers. No doubt, moving the eyes requires finer control than moving the legs.

Tetanic contractions ordinarily occur in the body because, as the intensity of nervous stimulation increases, more and more motor units are activated. This phenomenon, known as **recruitment,** results in stronger and stronger muscle contractions. But while some muscle fibers are contracting, others are relaxing. Because of this, intact muscles rarely fatigue completely. Even when muscles appear to be at rest, they exhibit **tone,** in which some of their fibers are always contracting. Muscle tone is particularly important in maintaining posture. If all the fibers within the muscles of the neck, trunk, and legs were to suddenly relax, the body would collapse.

Athletics and Muscle Contraction

Athletes who excel in a particular sport, and much of the general public as well, are interested in staying fit by exercising. The Medical Focus on page 150 gives suggestions for exercise programs according to age.

Exercise and Size of Muscles Muscles that are not used or that are used for only very weak contractions decrease in size, or atrophy. **Atrophy** can occur when a limb is placed in a cast

or when the nerve serving a muscle is damaged. If nerve stimulation is not restored, muscle fibers are gradually replaced by fat and fibrous tissue. Unfortunately, atrophy can cause muscle fibers to shorten progressively, leaving body parts contracted in contorted positions.

Forceful muscular activity over a prolonged period causes muscle to increase in size as the number of myofibrils within the muscle fibers increases. Increase in muscle size, called **hypertrophy,** occurs only if the muscle contracts to at least 75% of its maximum tension.

Some athletes take anabolic steroids, either testosterone or related chemicals, to promote muscle growth. This practice has many undesirable side effects, as discussed in the Medical Focus on page 151.

Slow-Twitch and Fast-Twitch Muscle Fibers We have seen that all skeletal muscle fibers metabolize both aerobically (using oxygen during cellular respiration) and anaerobically (without oxygen, using fermentation or creatine phosphate breakdown). Some muscle fibers, however, utilize one method more than the other to provide myofibrils with ATP. Slow-twitch and intermediate-twitch muscle fibers tend to be aerobic, and fast-twitch fibers tend to be anaerobic. Slow-twitch fibers are also referred to as type I fibers, intermediate-twitch fibers are type IIa, and fast-twitch fibers are called type IIb fibers.

Slow-twitch fibers have motor units with a smaller number of fibers. These muscle fibers are most helpful in endurance sports such as long-distance running or swimming in a triathlon. Because they produce most of their energy aerobically, they tire only when their fuel supply is gone. Slow-twitch fibers have many mitochondria that can maintain a steady, prolonged production of ATP when oxygen is available. An abundant supply of myoglobin, the respiratory pigment found in muscles, gives these fibers a dark color. They are also surrounded by dense capillary beds for a continuous supply of blood and oxygen. Slow-twitch fibers have a low maximum tension, which develops slowly, but these muscle fibers are highly resistant to fatigue.

Like slow-twitch fibers, *intermediate-twitch fibers* are well supplied with myoglobin and contain many mitochondria. Intermediate-twitch fibers have an extensive blood supply as well. However, they contract much more quickly and can be described as fast aerobic fibers. Activities that require moderate strength for shorter periods (such as walking, jogging, or biking) will employ these muscles.

Fast-twitch fibers have fewer capillaries to supply blood and oxygen, and tend to function anaerobically. These fibers are light in color because they have fewer mitochondria and little myoglobin. They provide explosions of energy and are most helpful in sports activities such as sprinting, weight lifting, swinging a golf club, or pitching a baseball. Fast-twitch fibers can develop maximum tension more rapidly and their maximum tension is greater. However, their dependence on anaerobic energy leaves them vulnerable to an accumulation of lactic acid that causes them to fatigue quickly.

Content **CHECK-UP!**

7. A motor unit is:
   a. a group of myofibrils.
   b. a group of skeletal muscles that perform a specific function.
   c. a group of motor nerves in the brain.
   d. a motor nerve and all of the muscle fibers it innervates.

8. A certain muscle produces most of its energy aerobically, contains many mitochondria, contracts quickly, and is resistant to fatigue. What kind is it?
   a. slow-twitch fiber
   b. intermediate-twitch fiber
   c. fast-twitch fiber

9. An increase in muscle size, caused by increasing the number of myofibrils in individual muscle cells, is called:
   a. recruitment.   c. atrophy.
   b. hypertrophy.   d. tetanic contraction.

Answers in Appendix C, page 462.

7.4   Skeletal Muscles of the Body

The human body has some 600 skeletal muscles, but this text will discuss only some of the most significant of these. First, let us consider certain basic principles of muscle contraction.

Basic Principles

When a muscle contracts at a joint, one bone remains fairly stationary, and the other one moves. The **origin** of a muscle is on the stationary bone, and the **insertion** of a muscle is on the bone that moves.

Frequently, a body part is moved by a group of muscles working together. Even so, one muscle does most of the work, and this muscle is called the **prime mover.** For example, in flexing the elbow, the prime mover is the biceps brachii (Fig. 7.9) The assisting muscles are called the **synergists.** The brachialis (see Fig. 7.15) is a synergist that helps the biceps brachii flex the elbow. A prime mover can have several synergists.

When muscles contract, they shorten. Therefore, muscles can only pull; they cannot push. However, muscles have **antagonists,** and antagonistic pairs work opposite one another to bring about movement in opposite directions. For example, the biceps brachii and the triceps brachii are antagonists; one flexes the forearm, and the other extends the forearm (Fig. 7.9). Later on in our discussion, we will encounter other antagonistic pairs.

Naming Muscles

You can more easily learn and remember the names of muscles by considering what the name means. The names of the various skeletal muscles are often combinations of

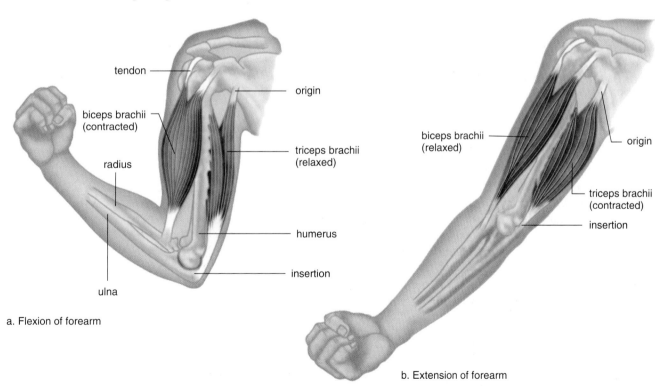

a. Flexion of forearm

b. Extension of forearm

Figure 7.9   Origin and insertion. The origin of a muscle is on a bone that remains stationary, and the insertion of a muscle is on a bone that moves when a muscle contracts. The two muscles shown here are antagonistic. **a.** When the biceps brachii contracts, the lower arm flexes. **b.** When the triceps brachii contracts, the lower arm extends.

the following terms used to characterize muscles:

1. **Size.** For example, the *gluteus maximus* is the largest muscle that makes up the buttocks. The *gluteus minimus* is the smallest of the gluteal muscles. Other terms used to indicate size are *vastus* (huge), *longus* (long), and *brevis* (short).

2. **Shape.** For example, the *deltoid* is shaped like a delta, or triangle, while the *trapezius* is shaped like a trapezoid. Other terms used to indicate shape are *latissimus* (wide) and *teres* (round).

3. **Direction of fibers.** For example, the *rectus abdominis* is a longitudinal muscle of the abdomen (*rectus* means straight). The *orbicularis oculi* is a circular muscle around the eye. Other terms used to indicate direction are *transverse* (across) and *oblique* (diagonal).

4. **Location.** For example, the *frontalis* overlies the frontal bone. The *external obliques* are located outside the internal obliques. Other terms used to indicate location are *pectoralis* (chest), *gluteus* (buttock), *brachii* (arm), and *sub* (beneath). You should also review these directional terms: *anterior, posterior, lateral, medial, proximal, distal, superficial,* and *deep*.

5. **Attachment.** For example, the *sternocleidomastoid* is attached to the sternum, clavicle, and mastoid process. The *brachioradialis* is attached to the brachium (arm) and the radius.

6. **Number of attachments.** For example, the *biceps brachii* has two attachments, or origins (and is located on the arm). The *quadriceps femoris* has four origins (and is located on the anterior femur).

7. **Action.** For example, the *extensor digitorum* extends the fingers or digits. The *adductor magnus* is a large muscle that adducts the thigh. Other terms used to indicate action are *flexor* (to flex), *masseter* (to chew), and *levator* (to lift).

Try to understand the name of every muscle you learn.

Skeletal Muscle Groups

In our discussion, the muscles of the body (Figs. 7.10 and 7.11) are grouped according to their location and their action. To better understand muscle groups, recall from Chapter 6 that the term *arm* refers to the humeral area, whereas the radius/ulna area is referred to as the *forearm*. Likewise, the femur is the *thigh* area, whereas the tibia/fibular area is referred to as the *leg*.

After you understand the meaning of a muscle's name, try to correlate its name with the muscle's location and the action it performs. Knowing the origin and insertion will also help you remember what the muscle does. Why? Because the insertion is on the bone that moves. You should review the various body movements listed and illustrated in Chapter 6 (see page 120). Only then will you be able to understand the actions of the muscles listed in Tables 7.2 to 7.5. Scientific terminology is necessary because it allows all persons to know the exact action being described for that muscle.

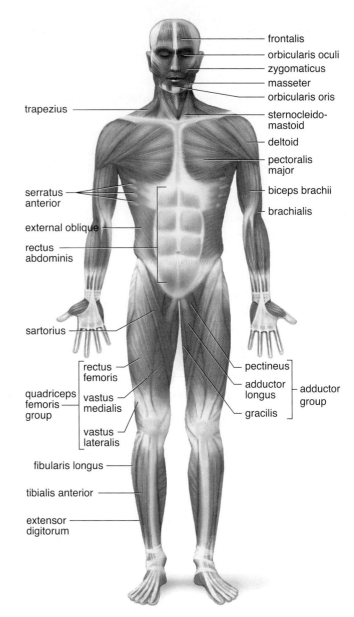

Figure 7.10 **Anterior view of the body's superficial skeletal muscles.**

Muscles of the Head

The muscles of the head and neck, the first group of muscles we will study, are illustrated in Figure 7.12 and listed in Table 7.2. The muscles of the head are responsible for facial expression and mastication (chewing). One muscle of the head and several muscles of the neck allow us to swallow. The muscles of the neck also move the head.

Muscles of Facial Expression

The muscles of facial expression are located on the scalp and face. These muscles are unusual in that they insert into and move the skin. Therefore, we expect them to move the skin and not a bone. As you know from the chapter introduction,

temporalis

occipitalis

sternocleidomastoid

trapezius

deltoid

teres major

triceps
brachii

biceps
femoris

hamstring
group

semi-
tendinosus

semi-
membranosus

gastrocnemius

calcaneal (Achilles)
tendon

infraspinatus

latissimus
dorsi

external
oblique

gluteus
medius

gluteus
maximus

adductor
magnus

gracilis

fibularis
longus

Figure 7.11 Posterior view of the body's superficial skeletal muscles.

these muscles communicate whether we are surprised, angry, fearful, happy, and so forth.

Frontalis lies over the frontal bone; it raises the eyebrows and wrinkles the brow. Frequent use results in furrowing of the forehead.

Orbicularis oculi is a ringlike band of muscle that encircles (forms an orbit about) the eye. It causes the eye to close or blink, and is responsible for "crow's feet" at the eye corners.

Orbicularis oris encircles the mouth and is used to pucker the lips, as in forming a kiss. Frequent use results in lines about the mouth.

Buccinator muscles are located in the cheek areas. When a buccinator contracts, the cheek is compressed, as when a person whistles or blows out air. Therefore, this muscle

is called the "trumpeter's muscle." Important to everyday life, the buccinator helps hold food in contact with the teeth during chewing. Babies use this muscle for suckling. It is also used in swallowing, as discussed next.

Zygomaticus extends from each zygomatic arch (cheekbone) to the corners of the mouth. It raises the corners of the mouth when a person smiles.

Levator anguli oris and **levator labii superioris** muscles lift the upper edge and corners of the lip. Simultaneously contracting these muscles on both sides of the mouth helps produce a smile. However, a person will sneer if he uses only the set on one side of his mouth.

Depressor anguli oris and **depressor labii oris** pull the lower edge and corners of the lip down, as when a person is frowning or a child is pouting.

Muscles of Mastication

We use the muscles of mastication when we chew food or bite something. Although there are four pairs of muscles for chewing, only two pairs are superficial and shown in Figure 7.12. As you might expect, both of these muscles insert on the mandible.

Each **masseter** has its origin on the zygomatic arch and its insertion on the mandible. The masseter is a muscle of mastication (chewing) because it is a prime mover for elevating the mandible.

Each **temporalis** is a fan-shaped muscle that overlies the temporal bone. It is also a prime mover for elevating the mandible. The masseter and temporalis are synergists.

Muscles of the Neck

Deep muscles of the neck (not illustrated) are responsible for swallowing. Superficial muscles of the neck move the head (see Table 7.2 and Figure 7.12).

Swallowing

Swallowing is an important activity that begins after we chew our food. First, the tongue (a muscle) and the buccinators squeeze the food back along the roof of the mouth toward the pharynx. An important bone that functions in swallowing is the hyoid (see page 104). The hyoid is the only bone in the body that does not articulate with another bone.

Muscles that lie superior to the hyoid, called the *suprahyoid muscles*, and muscles that lie inferior to the hyoid, called the *infrahyoid muscles*, move the hyoid. Because these muscles lie deep in the neck, they are not illustrated in Figure 7.12. The suprahyoid muscles pull the hyoid forward and upward toward the mandible. Because the hyoid is attached to the larynx, this pulls the larynx upward and forward. The epiglottis now lies over the glottis and closes the respiratory passages. Small palatini muscles (not illustrated) pull the soft palate backward, closing off the nasal passages. Pharyngeal constrictor muscles (not illustrated) push the bolus of food into the pharynx, which widens when the suprahyoid muscles move the hyoid. The hyoid bone and larynx are returned to their

Figure 7.18 **Muscles of the anterior right leg.**

Figure 7.19 **Muscles of the lateral right leg.**

Muscles That Move the Leg

The muscles that move the leg originate from the pelvic girdle or femur and insert on the tibia. They are listed in Table 7.5 and illustrated in Figures 7.16 and 7.17. Before studying these muscles, review the movement of the knee when the leg extends and when it flexes (see Chapter 6, page 120).

Quadriceps femoris group (rectus femoris, vastus lateralis, vastus medialis, vastus intermedius), also known as the "quads," is found on the anterior and medial thigh. The rectus femoris, which originates from the ilium, is external to the vastus intermedius, and therefore the vastus intermedius is not shown in Figure 7.16. These muscles are the primary extensors of the leg, as when you kick a ball by straightening your knee.

Sartorius is a long, straplike muscle that has its origin on the iliac spine and then goes across the anterior thigh to insert on the medial side of the knee (Fig. 7.16). Because this muscle crosses both the hip and knee joint, it acts on the thigh in addition to the leg. The

insertion of the sartorius is such that it flexes both the leg and the thigh. It also rotates the thigh laterally, enabling us to sit cross-legged, as tailors were accustomed to do in another era. Therefore, it is sometimes called the "tailor's muscle," and in fact, *sartor* means tailor in Latin.

Hamstring group (biceps femoris, semimembranosus, semitendinosus) is located on the posterior thigh (Fig. 7.17). Notice that these muscles also cross the hip and knee joint because they have origins on the ischium and insert on the tibia. They flex and rotate the leg medially, but they also extend the thigh. Their strong tendons can be felt behind the knee. These same tendons are present in hogs and were used by butchers as strings to hang up hams for smoking—hence, the name. Notice that the quadriceps femoris group and the hamstring group are antagonistic muscles in that the quads extend the leg and the hamstrings flex the leg.

Muscles That Move the Ankle and Foot

Muscles that move the ankle and foot are shown in Figures 7.18 and 7.19.

Gastrocnemius is a muscle of the posterior leg, where it forms a large part of the calf. It arises from the femur; distally, the muscle joins the strong calcaneal tendon, which attaches to the calcaneus bone (heel). The gastrocnemius is a powerful plantar flexor of the foot that aids in pushing the body forward during walking or running. It is sometimes called the "toe dancer's muscle" because it allows a person to stand on tiptoe.

Tibialis anterior is a long, spindle-shaped muscle of the anterior leg. It arises from the surface of the tibia and attaches to the bones of the ankle and foot. Contraction of this muscle causes dorsiflexion and inversion of the foot.

Fibularis muscles (fibularis longus, fibularis brevis) are found on the lateral side of the leg, connecting the fibula to the metatarsal bones of the foot. These muscles evert the foot and also help bring about plantar flexion.

Flexor (not shown) and **extensor digitorum longus** muscles are found on the lateral and posterior portion of the leg. They arise mostly from the tibia and insert on the toes. They flex and extend the toes, respectively, and assist in other movements of the feet.

Content **CHECK-UP!**

10. Which muscle is named for its shape?

 a. latissimus dorsi   c. gluteus maximus

 b. triceps brachii   d. frontalis

11. Which muscle of facial expression is also used for swallowing?

 a. frontalis   c. zygomaticus

 b. masseter   d. buccinator

12. Which arm muscle extends the forearm?

 a. biceps brachii   c. triceps brachii

 b. brachialis   d. biceps femoris

Answers in Appendix C, page 462.

7.5 Effects of Aging

Muscle mass and strength tend to decrease as people age. How much of this is due to lack of exercise and a poor diet has yet to be determined. Deteriorated muscle elements are replaced initially by connective tissue and, eventually, by fat. With age, degenerative changes take place in the mitochondria, and endurance decreases. Also, changes in the nervous and cardiovascular systems adversely affect the structure and function of muscles.

Muscle mass and strength can improve remarkably if elderly people undergo a training program. Exercise at any age appears to stimulate muscle buildup. As discussed in the Medical Focus, Benefits of Exercise on page 150, exercise has many other benefits as well. For example, exercise improves the cardiovascular system and reduces the risk of diabetes and glycation. During glycation, excess glucose molecules stick to body proteins so that the proteins no longer have their normal structure and cannot function properly. Exercise

burns glucose and, in this way, helps preve.. rioration.

7.6 Homeostasis

The illustration in Human Systems Work Together on page 152 tells how the muscular system works with other systems of the body to maintain homeostasis.

Cardiac muscle contraction accounts for the heartbeat, which creates blood pressure, the force that propels blood in the arteries and arterioles. The walls of the arteries and arterioles contain smooth muscle. Constriction of arteriole walls is regulated to help maintain blood pressure. Arterioles branch into the capillaries where exchange takes place that creates and cleanses tissue fluid. Blood and tissue fluid are the internal environment of the body, and without cardiac and smooth muscle contraction, blood would never reach the capillaries for exchange to take place. Blood is returned to the heart in cardiovascular veins, and excess tissue fluid is returned to the cardiovascular system within lymphatic vessels. Skeletal muscle contraction presses on the cardiovascular veins and lymphatic vessels, and this creates the pressure that moves fluids in both types of vessels. Without the return of blood to the heart, circulation would stop, and without the return of lymph to the blood vessels, normal blood pressure could not be maintained.

The contraction of sphincters composed of smooth muscle fibers temporarily prevents the flow of blood into a capillary. This is an important homeostatic mechanism because in times of emergency it is more important, for example, for blood to be directed to the skeletal muscles than to the tissues of the digestive tract. Smooth muscle contraction also accounts for peristalsis, the process that moves food along the digestive tract. Without this action, food would never reach all the organs of the digestive tract where digestion releases nutrients that enter the bloodstream. Smooth muscle contraction assists the voiding of urine, which is necessary for ridding the body of metabolic wastes and for regulating the blood volume, salt concentration, and pH of internal fluids.

Skeletal muscles protect internal organs, and their strength protects joints by stabilizing their movements. Skeletal muscle contraction raises and lowers the rib cage and diaphragm during the active phases of breathing. As we breathe, oxygen enters the blood and is delivered to the tissues, including the muscles, where ATP is produced in mitochondria with heat as a by-product. The heat produced by skeletal muscle contraction allows the body temperature to remain within the normal range for human beings.

Finally, skeletal muscle contraction moves bones and allows us to perform those daily activities necessary to our health and benefit. Although it may seem as if movement of our limbs does not affect homeostasis, it does so by allowing us to relocate our bodies to keep the external environment within favorable limits for our existence.

Benefits of Exercise

Exercise programs improve muscular strength, muscular endurance, and flexibility. Muscular strength is the force a muscle group (or muscle) can exert against a resistance in one maximal effort. Muscular endurance is judged by the ability of a muscle to contract repeatedly or to sustain a contraction for an extended period. Flexibility is tested by observing the range of motion about a joint.

As muscular strength improves, the overall size of the muscle, as well as the number of muscle fibers and myofibrils in the muscle, increases. The total amount of protein, the number of capillaries, and the amounts of connective tissue, including tissue found in tendons and ligaments, also increase. Physical training with weights can improve muscular strength and endurance in all adults, regardless of their age. Over time, increased muscle strength promotes strong bones.

A surprising finding, however, is that health benefits also accompany less strenuous programs, such as those described in Table 7A. A study of 12,000 men by Dr. Arthur Leon at the University of Minnesota showed that even moderate exercise lowered the risk of a heart attack by one-third. People with arthritis reported much less pain, swelling, fatigue, and depression after only four months of attending a twice-weekly, low-impact aerobics class. Increasing daily activity by walking to the corner store instead of driving and by taking the stairs instead of the elevator can improve a person's health.

The benefits of exercise are most apparent with regard to cardiovascular health. Brisk walking for 2.5 to 4 hours a week can raise the blood levels of high-density lipoprotein (HDL), a chemical that promotes healthy blood vessels (see Chapter 12). Exercise also helps prevent osteoporosis, a condition in which the bones are weak and tend to break. The stronger the bones are when a person is young, the less chance of osteoporosis as a person ages. Exercise promotes the activity of osteoblasts (as opposed to osteocytes) in young people, as well as older people. An increased activity level can also keep off unwanted pounds, which is a worthwhile goal because added body weight contributes to numerous conditions, such as type II diabetes (see page 221). Increased muscle activity is also helpful by causing glucose to be transported into muscle cells and making the body less dependent on the presence of insulin.

Cancer prevention and early detection involve eating properly, not smoking, avoiding cancer-causing chemicals and radiation, undergoing appropriate medical screening tests, and knowing the early warning signs of cancer. However, evidence indicates that because it helps prevent obesity, exercise also helps prevent certain kinds of cancer. Studies show that people who exercise are less likely to develop colon, breast, cervical, uterine, and ovarian cancer.

TABLE 7.A A Checklist for Staying Fit

Children, 7–12	Teenagers, 13–18	Adults, 19–65	Seniors, 65 and Up
Vigorous activity 1–2 hours daily	Vigorous activity 1 hour 3–5 days a week; otherwise, $\frac{1}{2}$ hour daily moderate activity	Vigorous activity 1 hour 3 days a week; otherwise, $\frac{1}{2}$ hour daily moderate activity	Moderate exercise 1 hour daily 3 days a week; otherwise, $\frac{1}{2}$ hour daily moderate activity
Free play	Build muscle with calisthenics	Exercise to prevent lower back pain: aerobics, stretching, yoga	Take a daily walk
Build motor skills through team sports, dance, swimming	Do aerobic exercise to control buildup of fat cells	Take active vacations: hike, bicycle, cross-country ski	Do daily stretching exercises
Encourage more exercise outside of physical education classes	Pursue tennis, swimming, horseback riding—sports that can be enjoyed for a lifetime	Find exercise partners: join a running club, bicycle club, outing group	Learn a new sport or activity: golf, fishing, ballroom dancing
Initiate family outings: bowling, boating, camping, hiking	Continue team sports, dancing, hiking, swimming		Try low-impact aerobics; before undertaking new exercises, consult your doctor

Muscular Disorders and Neuromuscular Disease

During the course of a lifetime, nearly everyone will suffer from some type of muscular disorder. Muscular disorders cover a wide spectrum in terms of severity. Minor muscle irritation, inflammation, or injury may resolve without any medical care. However, many diseases affecting the neuromuscular system are extremely serious and eventually prove to be fatal.

Spasms are sudden, involuntary muscular contractions, most often accompanied by pain. Spasms can occur in both smooth and skeletal muscles. A spasm of the intestinal tract is a "bellyache"; most such spasms are not serious. Multiple spasms of skeletal muscles are called a **convulsion. Cramps** are strong, painful spasms, especially of the leg and foot, usually due to strenuous athletic activity. Cramps typically occur after a strenuous workout, and may even occur when sleeping. Facial **tics,** such as periodic eye blinking or grimacing, are spasms that can be controlled voluntarily but only with great effort.

Muscles, joints, and their connective tissues are often subject to overuse injuries: strains, sprains, and tendinitis. A **strain** is caused by stretching or tearing of a muscle. A **sprain** is the twisting of a joint, leading to swelling and injury not only of muscles but also of ligaments, tendons, blood vessels, and nerves. The ankle and knee are two areas often subject to sprains. **Tendinitis** is inflammation of a tendon due to repeated athletic activity. The tendons most commonly affected are those associated with the shoulder, elbow, hip, and knee.

Overuse injuries are often minor and can be treated with pain medication and rest. However, an individual should seek medical attention if the injured area is extremely painful, hot, or swollen, or if accompanied by a fever.

Neuromuscular Diseases

Neuromuscular disease can result from pathologic changes to the muscle itself. It can also result from excessive motor nerve stimulation, or from damage or destruction of the motor neurons that supply the muscle.

Tetanus develops in persons who have not been properly immunized against the toxin of the tetanus bacterium. Tetanus toxin shuts down brain areas that normally inhibit unnecessary muscle contractions. As a result, excessive brain stimulation causes muscles to lock in a tetanic contraction (from which the disease gets its name). A rigidly locked jaw is one of the first signs

of bacterial infection and toxin production. Though antibiotics will kill the bacteria, once the toxin is circulating in the bloodstream, it cannot be removed or neutralized. Because muscles can't relax, the patient cannot breathe or swallow, and death may occur due to respiratory failure. Immunization and periodic booster shots will prevent the toxin's effects (see Immunization: The Great Protector, p. 296).

Fibromyalgia is a chronic condition whose symptoms include achy pain, tenderness, and stiffness of muscles. Its precise cause is not known, though 80–90% of sufferers are women. Substance P, a neurotransmitter (messenger chemical) of pain pathways in the brain, has been found in the bloodstream of affected individuals. Exercise seems to decrease blood levels of substance P. Therapeutic massage, over-the-counter pain medication, and muscle relaxants are also recommended.

Muscular dystrophy is a broad term applied to a group of disorders that causes progressive degeneration and weakening of muscles. As muscle fibers die, fat and connective tissue take their place. **Duchenne muscular dystrophy**, the most common type, is inherited through a flawed gene carried by the mother. It is now known that the lack of a protein called *dystrophin* causes the condition. When dystrophin is absent, calcium leaks into the cell and activates an enzyme that dissolves muscle fibers. Treatment includes muscle injections with immature muscle cells that do produce dystrophin.

Myasthenia gravis is an autoimmune disease characterized by weakness that especially affects the muscle of the eyelids, face, neck, and extremities. Muscle contraction is impaired because the immune system mistakenly produces antibodies that destroy acetylcholine receptors on the sarcolemma. (Recall that acetylcholine is the neurotransmitter released by motor neurons.) In many cases, the first signs of the disease are drooping eyelids and double vision. Treatment includes drugs that inhibit the enzyme that digests acetylcholine, thus allowing it to accumulate.

Amyotrophic lateral sclerosis (ALS) is often called Lou Gehrig's disease, after its most famous victim, the 1930s-era baseball player. ALS sufferers experience the gradual death of their motor neurons, thus losing the ability to walk, talk, chew, swallow, etc. Intellect and sensation are not affected, however. Drugs can slow the disease's progression, but ALS is always fatal.

Human Systems Work Together

Integumentary System

Muscle contraction provides heat to warm skin. Muscle moves skin of face.

Skin protects muscles; rids the body of heat produced by muscle contraction.

Skeletal System

Muscle contraction causes bones to move joints; muscles help protect bones.

Bones provide attachment sites for muscles; store Ca^{2+} for muscle function.

Nervous System

Muscle contraction moves eyes, permits speech, creates facial expressions.

Brain controls nerves that innervate muscles; receptors send sensory input from muscles to brain.

Endocrine System

Muscles help protect glands.

Androgens promote growth of skeletal muscle; epinephrine stimulates heart and constricts blood vessels.

Cardiovascular System

Muscle contraction keeps blood moving in heart and blood vessels.

Blood vessels deliver nutrients and oxygen to muscles, carry away wastes.

How the Muscular System works with other body systems

Lymphatic System/Immunity

Skeletal muscle contraction moves lymph; physical exercise enhances immunity.

Lymphatic vessels pick up excess tissue fluid; immune system protects against infections.

Respiratory System

Muscle contraction assists breathing; physical exercise increases respiratory capacity.

Lungs provide oxygen for, and rid the body of, carbon dioxide from contracting muscles.

Digestive System

Smooth muscle contraction accounts for peristalsis; skeletal muscles support and help protect abdominal organs.

Digestive tract provides glucose for muscle activity; liver metabolizes lactic acid following anaerobic muscle activity.

Urinary System

Smooth muscle contraction assists voiding of urine; skeletal muscles support and help protect urinary organs.

Kidneys maintain blood levels of Na^+, K^+, and Ca^{2+}, which are needed for muscle innervation, and eliminate creatinine.

Reproductive System

Muscle contraction occurs during orgasm and moves gametes; abdominal and uterine muscle contraction occurs during childbirth.

Androgens promote growth of skeletal muscle.

SELECTED NEW TERMS

Basic Key Terms

A band (ā bănd), p. 130

actin (ak′tin), p. 130

all-or-none law (ôl-ôr-nŭn lô), p. 136

antagonist (an-tag′o-nist), p. 139

cardiac muscle (kar′de-ak mus′el), p. 128

creatine phosphate (kre′uh-tin fos′făt), p. 135

H zone (H zōn), p. 130

I band (ī bănd), p. 130

insertion (in-ser′shun), p. 139

motor unit (mo′tor yū′nit), p. 137

muscle fiber (mus′el fi′ber), p. 128

muscle twitch (mus′el twich), p. 136

myofibril (mi″o-fi′bril), p. 130

myoglobin (mi″o-glo′bin), p. 130

myosin (mi′o-sin), p. 130

neuromuscular junction (nu″ro-mus′kyū-ler junk′shun), p. 132

origin (or′ĭ-jin), p. 139

oxygen debt (ok′sĭ-jen debt), p. 136

power stroke (pou′ər strōk), p. 133

prime mover (prĭm mu′ver), p. 139

recruitment (re-krūt′ment), p. 138

sarcomere (sar′ko-mĕr), p. 130

skeletal muscle (skel′ĕ-tal mus′el), p. 128

sliding filament theory (sli′ding fil′uh-ment the′o-re), p. 132

smooth muscle (smūth mus′el), p. 128

summation (sə-mā′shən), p. 136

synergist (sin′er-jist), p. 139

tendon (ten′don), p. 129

tetanic contraction (tĕ-tăn′ĭk kən-trăk′shən), p. 137

thick filaments (thĭk fil′ə mənts), p. 130

thin filaments (thĭn fil′ə-mənts), p. 130

tone (tōn), p. 138

T (transverse) tubules (tranz-vers′ tu′byūl), p. 130

tropomyosin (trŏ′pə-mī′ə-sĭn), p. 130

troponin (trŏ′pə-nĭn), p. 130

z line (z līne), p. 130

Clinical Key Terms

amyotrophic lateral sclerosis (ALS) (ă′mī-ə trŏ′fik lăt′ər-əl sklə-rō′sĭs), p. 151

atrophy (at′ro-fe), p. 138

compartment syndrome (kum-part′mant sin-drōm), p. 129

convulsion (kən-vŭl′shən), p. 151

cramps (krămpz), p. 151

fatigue (fə-tēg′), p. 137

fibromyalgia (fi′brō-mī-ăl′je-ə), p. 151

hypertrophy (hi-per′tro-fe), p. 138

muscular dystrophy (mus′kyū-ler dis′trĕ-fe), p. 151

myasthenia gravis (mi″as-the′ne-uh grah′vis), p. 151

spasm (spazm), p. 151

sprain (sprān), p. 151

strain (strān), p. 151

tendinitis (ten″dĕ-ni′tis), p. 151

tetanus (tet′uh-nus), p. 151

tics (tĭks), p. 151

SUMMARY

7.1 Functions and Types of Muscles

A. Muscular tissue is either smooth, cardiac, or skeletal. Skeletal muscles have tubular, multinucleated, and striated fibers that contract voluntarily.

B. Skeletal muscles support the body, make bones move, help maintain a constant body temperature, assist movement in cardiovascular and lymphatic vessels, and help protect internal organs and stabilize joints.

7.2 Microscopic Anatomy and Contraction of Skeletal Muscle

A. The sarcolemma, which extends into a muscle fiber, forms T tubules; the sarcoplasmic reticulum has calcium storage sites. The placement of actin and myosin in the contractile myofibrils accounts for the striations of skeletal muscle fibers.

B. Skeletal muscle innervation occurs at neuromuscular junctions. Impulses travel down the tubules of the T system and cause the release of calcium from calcium storage sites. The presence of calcium and ATP in muscle cells prompts actin myofilaments to slide past myosin myofilaments, shortening the length of the sarcomere.

C. ATP, required for muscle contraction, can be generated by way of creatine phosphate breakdown and fermentation. Lactic acid from fermentation represents an oxygen debt, because oxygen is required to metabolize this product. Cellular respiration, an aerobic process, is the best source of ATP.

7.3 Muscle Responses

A. In the laboratory, muscle fibers obey the all-or-none law, but whole muscles in the body do not. The occurrence of a muscle twitch, summation, or tetanic contraction depends on the frequency with which a muscle is stimulated.

B. In the body, muscle fibers belong to motor units that obey the all-or-none law. The strength of muscle contraction depends on the recruitment of motor units. A muscle has tone because some fibers are always contracting.

7.4 Skeletal Muscles of the Body

A. When muscles cooperate to achieve movement, some act as prime movers, others as synergists, and still others as antagonists.

B. The skeletal muscles of the body are divided into those that move: the head and neck (see Table 7.2); the trunk (see Table 7.3); the shoulder and arm (see Table 7.4); the forearm (see Table 7.4); the hand and fingers (see Table 7.4); the thigh (see Table 7.5); the leg (see Table 7.5); and the ankle and foot (see Table 7.5).

7.5 Effects of Aging

As we age, muscles become weaker, but exercise can help retain vigor.

7.6 Homeostasis

Smooth muscle contraction helps move the blood; cardiac muscle contraction pumps the blood. Skeletal muscle contraction produces heat and is needed for breathing.

8.1 Nervous System

The nervous system has three specific functions:

1. *Sensory input.* Sensory receptors in skin and organs respond to external and internal stimuli by generating nerve signals that travel to the brain and spinal cord. For example, temperature sensors in the skin may signal to the brain that the air surrounding the body is cold.
2. *Integration.* The brain and spinal cord interpret the data received from sensory receptors all over the body, and signal the appropriate nerve responses. To continue the example of body temperature, sensory information from temperature receptors is sent to the *hypothalamus,* the brain center that controls body temperature.
3. *Motor output.* The nerve impulses from the brain and spinal cord go to the effectors, which are muscles, glands, and organs. Muscle contractions, gland secretions, and changes in organ function are responses to stimuli received by sensory receptors. For example, when adjusting body temperature, the hypothalamus triggers shivering—skeletal muscles contract rhythmically, producing heat that warms the body.

It is important to stress that the nervous system maintains **homeostasis** by receiving sensory information, integrating that information, and making an appropriate response.

Divisions of the Nervous System

The nervous system has two major divisions: the central nervous system and the peripheral nervous system (Fig. 8.1). The **central nervous system (CNS)** includes the brain and spinal cord, which have a *central* location—they lie in the midline of the body. The **peripheral nervous system (PNS),** which is further divided into the *afferent* (sensory) and *efferent* (motor) divisions, includes all the cranial and spinal nerves. Nerves have a *peripheral* location in the body, meaning that they project out from the central nervous system. The division between the central nervous system and the peripheral nervous system is arbitrary; the two systems work together, as we shall see.

Nervous Tissue

Although exceedingly complex, nervous tissue is made up of just two principal types of cells: (1) **neurons,** also called nerve

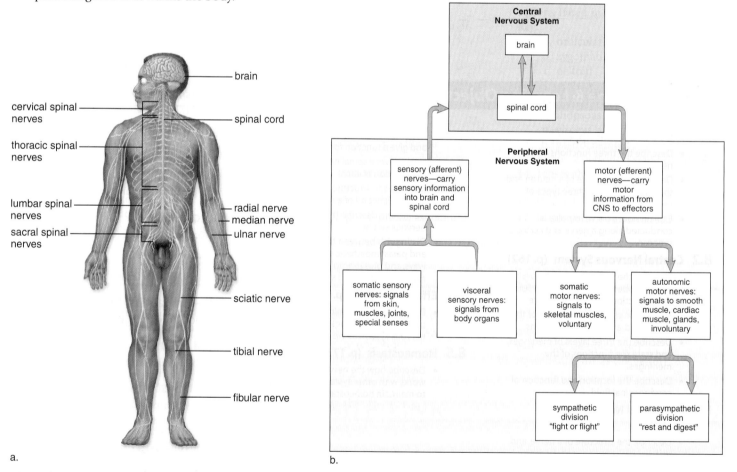

a.

b.

Figure 8.1 Organization of the nervous system in humans. a. This pictorial representation shows the central nervous system (CNS; composed of brain and spinal cord) and some of the nerves of the peripheral nervous system (PNS). **b.** The CNS and PNS communicate with each other. Somatic sensory and visceral sensory nerves carry information to the brain and spinal cord. Somatic motor nerves signal skeletal muscles and autonomic motor nerves signal smooth muscle, cardiac muscle, and glands.

cells, which transmit nerve impulses; and (2) **neuroglia**, which support and nourish neurons (see Chapter 4, page 75).

Neuron Structure

Neurons vary in appearance, but all of them have just three parts: a cell body, dendrite(s), and an axon. As shown in Figure 8.2, the

Figure 8.2 Neuron anatomy. a. Sensory neuron with dendritelike structures projecting from the peripheral end of the axon. **b.** Interneuron (from the cortex of the cerebellum) with highly branched dendrites. Note that the myelin sheath is absent. **c.** Motor neuron. Note the branched dendrites and the single, long axon, which branches only near its tip.

cell body of each type of neuron contains the nuc[...] as other organelles. **Dendrites** are shorter, branche[...] that receive signals from sensory receptors or other neu[...] the dendrites, signals can result in nerve impulses that are then conducted away from the cell body by an axon. The **axon** is the portion of a neuron that conducts nerve signals.

Axons can be grouped together in bundles. A bundle of parallel axons in the peripheral nervous system is called a **nerve**, whereas a similar axon bundle in the central nervous system is a **tract**. Axons found in nerves or in tracts may be covered by cells containing myelin, a lipid coating that insulates the axon. The myelin covering of axons in the PNS is formed by neuroglial cells called *Schwann cells* or *neurolemmocytes*. (Recall from Chapter 4 that neuroglial cells are support cells for the brain and nervous system.) *Oligodendrocytes*, another type of neuroglial cell, perform a similar function in the CNS. Because myelin is contained in cells, the cell's metabolism can influence the amount and composition of myelin. Gaps in the myelin sheath are called *nodes of Ranvier* (*neurofibril nodes*). These gaps greatly increase the speed of conduction for a nerve signal, as will be discussed shortly.

Types of Neurons

Neurons can be classified according to their function and structure. **Motor neurons** take nerve impulses from the CNS to muscles, organs, or glands. Motor neurons are said to be multipolar because they have many dendrites and a single axon (Fig. 8.2*a*). Motor neurons cause muscle fibers to contract, organs to modify their function, or glands to secrete, and therefore they are said to *innervate* these structures.

Sensory neurons take nerve impulses from sensory receptors to the CNS. The sensory receptor, which is the distal end of the long axon of a sensory neuron, may be as simple as a naked nerve ending (a pain receptor), or it may be a part of a highly complex organ, such as the eye or ear. Almost all sensory neurons have a structure that is termed unipolar (Fig. 8.2*b*). In unipolar neurons, the extension from the cell body divides into a branch that comes to the periphery and another that goes to the CNS. Because both branches are long and transmit nerve impulses, it is now generally accepted to refer to them collectively as an axon.

Interneurons, also known as association neurons, occur entirely within the CNS. Interneurons, which are typically multipolar (Fig. 8.2*c*), convey nerve impulses between various parts of the CNS. Some lie between sensory neurons and motor neurons, and some take messages from one side of the spinal cord to the other or from the brain to the cord, and vice versa. They also form complex pathways in the brain where processes accounting for thinking, memory, and language occur.

Nerve Signal Conduction

The function of neurons is to conduct nerve signals and activate the release of neurotransmitters, which are chemical messengers that allow neurons to communicate with other cells and transmit information. Although resting neurons

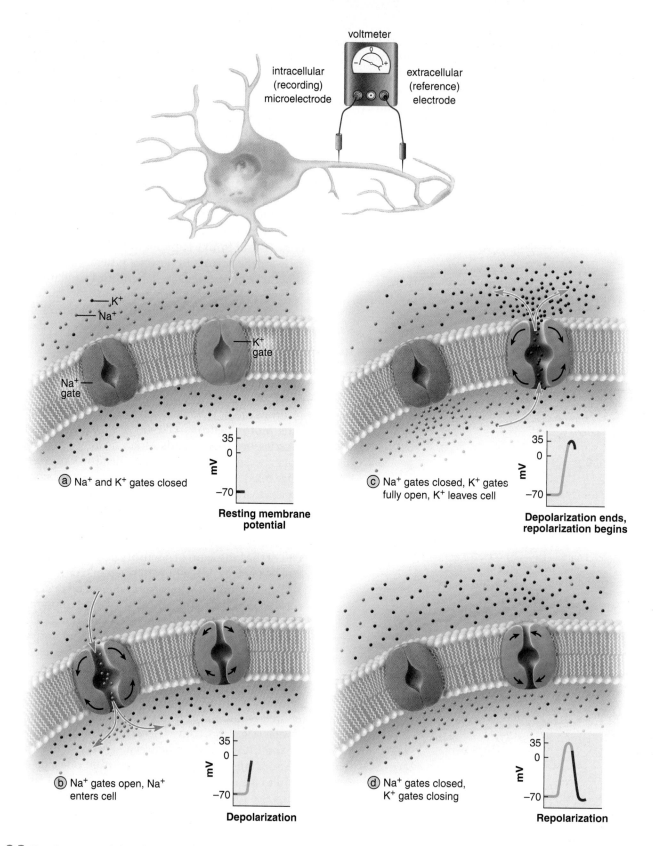

Figure 8.3 Resting potential and action potential in an unmyelinated axon. a. Resting potential. Sodium ions (red dots) are concentrated outside the cell; potassium (blue dots) and large anions are inside the cell. Potential is approximately 70 mV; the inside of the cell is negative compared to the outside. **b.** Action potential: depolarization. Na$^+$ gates open and sodium ions flow into cell; the inside of the cell becomes positive. **c.** Action potential: repolarization. K$^+$ gates open and potassium ions flow out. **d.** Repolarization: Both Na$^+$ and K$^+$ gates are closed. When the sodium potassium pump restores the ions to their places, repolarization is complete.

aren't sending nerve signals, they are constantly prepared to do so. A nerve signal, also called an *action potential*, is conducted by the axon of the neuron.

Resting Potential

Anyone who has ever used a battery has employed an energy source that is manufactured by separating positively charged ions across a membrane from negative ions. The battery's *potential energy* can be used to perform work: lighting a flashlight, for example. When a neuron is resting, it too possesses potential energy, much like a fully charged battery. This energy, called the **resting potential**, exists because the cell membrane is *polarized*: positively charged outside the cell, negatively charged inside. The cell's outside is positive because positively charged sodium (Na^+) ions cluster around the cell. Although there is a concentration gradient for sodium ions and the ions can diffuse into the cell, the cell membrane is relatively impermeable to sodium ion diffusion when the cell is at resting potential. In addition, positively charged potassium ions are concentrated inside the cell, and the cell membrane is permeable to potassium. Because of this concentration gradient, potassium ions are constantly leaking out of the cell. There, these ions contribute to the extracellular positive charge. The inside of the cell is negatively charged because of the presence of large, negatively charged proteins and other molecules that are stuck inside the cell because of their size.

Like a battery, the neuron's resting potential energy can be measured in volts. A D-size flashlight battery has 1.5 volts; a nerve cell typically has 0.07 volts (V), or 70 millivolts (mV), of stored energy (Fig. 8.3a). By convention, the cell's voltage measurement is assigned a negative value. This is because scientists compare the inside of the cell (where negatively charged proteins and other large molecules cluster) to the outside of the cell (where positively charged sodium ions and some potassium ions are gathered). Neurons, like rechargeable batteries, must maintain their resting potential to work effectively. To do so, neurons continually transport sodium ions out of the cell and return potassium to the cytoplasm. A protein carrier in the membrane, called the sodium-potassium pump, pumps sodium ions (Na^+) out of the neuron and potassium ions (K^+) into the neuron. However, the sodium-potassium pump doesn't perform an even exchange. Rather, for every three sodium ions pumped out of the cell, only two potassium ions are returned to the cell. Because potassium leaks out of the cell faster than sodium leaks in, and the pump returns fewer potassium ions to the cell, the pump allows the outside of the cell to keep its resting positive charge. Thus, the sodium-potassium pump constantly "recharges" the cell, keeping the inside of the cell negatively charged and the outside positively charged. Like a freshly charged battery, the cell is always ready to perform its work.

Action Potential

The resting potential energy of the neuron can be used to perform the work of the neuron: conduction of nerve signals. The process of conduction is termed an **action potential**, and it occurs in the axons of neurons. An action potential begins with a **stimulus,** which activates the neuron. (For example, poking the skin with a sharp pin would be a stimulus for pain neurons in the skin.) The stimulus causes protein channels for sodium located in the cell membrane to open, and sodium ions rush into the cell. Adding positively charged sodium ions causes the inside of the axon to become positive, compared to the outside (Fig. 8.3b). This change is called **depolarization.**

Immediately after depolarization, the channels for sodium close and a separate set of protein channels for potassium opens. Potassium rapidly leaves the cell (Fig. 8.3c). As positively charged potassium ions exit the cell, the inside of the cell becomes negative again (due to the presence of large negatively charged ions stuck inside the cell). This change in polarity is called **repolarization** (Fig. 8.3d). Once the action potential is complete, the sodium and potassium ions are rapidly restored to their proper place through the action of the sodium-potassium pump.

Conduction of Action Potentials

The conduction of an action potential is an all-or-none event—that is, either an axon conducts its action potential or it doesn't. The intensity of a message is determined by how many action potentials are generated within a given time span. A neuron can conduct a volley of action potentials very quickly because only a small number of ions is exchanged with each action potential.

If an axon is unmyelinated, an action potential at one location stimulates an adjacent part of the axon membrane to produce an action potential. Conduction along the entire axon in this fashion can be rather slow (approximately 1 meter/second in thin axons) because each section of the axon must be stimulated. In myelinated fibers, an action potential at one node of Ranvier causes an action potential at the next node, jumping over the entire myelin-coated portion of the axon (Fig. 8.4). This type of conduction, called **saltatory conduction,**

Figure 8.4 Conduction of an action potential in unmyelinated and myelinated axons. a. In unmyelinated axon, conduction is slow. **b.** In a myelinated axon, the action potential jumps from one node of Ranvier to the next, and the speed of conduction is much more rapid. Almost all axons are myelinated in humans.

Alzheimer Disease

Alzheimer disease (AD) is a disorder characterized by a gradual loss of reason that begins with memory lapses and ends with the inability to perform any type of daily activity. Personality changes signal the onset of AD. A normal 50- to 60-year-old might forget the name of a friend not seen for years. People with AD, however, forget the name of a neighbor who visits daily. People afflicted with AD become confused and tend to repeat the same question. Signs of mental disturbance eventually appear, and patients gradually become bedridden and die of a complication, such as pneumonia.

Researchers have discovered that in some families whose members have a 50% chance of AD, a genetic defect exists on chromosome 21. This is of extreme interest because Down syndrome results from the inheritance of three copies of chromosome 21, and people with Down syndrome tend to develop AD.

AD is characterized by the presence of abnormally structured neurons and a reduced amount of ACh. The AD neuron has two features: (1) Bundles of fibrous protein, called neurofibrillary tangles, surround the nucleus in the cells, and (2) protein-rich accumulations, called amyloid plaques, envelop the axon branches. These abnormal neurons are especially seen in the portions of the brain involved in reason and memory.

Treatment for Alzheimer Disease

Treatment for AD involves using one of two categories of drugs. Cholinesterase inhibitors (Aricept, Cognex, Exelon, Reminyl) work at neuron synapses in the brain, slowing the activity of the enzyme that breaks down acetylcholine (ACh). Allowing ACh to accumulate in synapses keeps memory pathways in the brain functional for a longer period of time. A second drug, memantine (Namenda), blocks *excitotoxicity:* the tendency of diseased neurons to self-destruct. This recently approved medication is used only in moderately to severely affected patients. Using the drug allows neurons involved in memory pathways to survive longer in affected patients. However, it is important to note that neither type of medication cures AD—both merely slow the progress of disease symptoms, allowing the patient to function independently for a longer period of time. Additional research is currently underway to test the effectiveness of anti-cholesterol *statin* drugs, as well as anti-inflammatory medications, in slowing the progress of the disease.

Much of current research on AD focuses on the prevention and cure of the disease. Scientists believe that curing AD will require an early diagnosis because it is thought that the disease may begin in the brain 15 to 20 years before symptoms ever develop. Currently, diagnosis can't be made with absolute certainty until the brain is examined at autopsy. Cerebrospinal fluid testing, currently being researched, may allow amyloid proteins to be detected before disease symptoms appear. The "What's New" reading in Chapter 9 describes an eye scan technique that might allow an earlier diagnosis. Researchers are also testing vaccines for AD that would target the patient's immune system to destroy amyloid.

Early findings have shown that risk factors for cardiovascular disease—heart attacks and stroke—also contribute to an increased incidence of AD. Risk factors for cardiovascular disease include elevated blood cholesterol and blood pressure, smoking, obesity, sedentary lifestyle, and diabetes mellitus. Thus, evidence suggests that a lifestyle tailored for good cardiovascular health may also prevent AD.

---

is much faster: in thick, myelinated fibers, the rate is more than 100 m/sec.

As soon as the action potential has passed by each successive portion of an axon, the sodium and potassium ions for that section are reversed and must be returned to their correct places by the sodium-potassium pump. The time needed to pump sodium outside the cell and potassium inside the cell is called the **refractory period.** During this short interval, the axon section is unable to conduct an action potential. This ensures the one-way direction of a signal, from the cell body down the length of the axon to the axon terminal.

It is interesting to observe that all functions of the nervous system, from our deepest emotions to our highest reasoning abilities, are dependent on our ability to start and conduct nerve signals and release neurotransmitters.

Transmission Across a Synapse

Every axon branches into many fine endings, each tipped by a small swelling called an **axon terminal.** Each swelling lies very close to another cell, often another neuron. This region of close proximity is called a **synapse** (Fig. 8.5). At a synapse between two neurons, the membrane of the first neuron is called the *pre*synaptic membrane, and the membrane of the next neuron is called the *post*synaptic membrane. The small gap between is the **synaptic cleft.** The neuromuscular junction, which you studied in Chapter 7, is a synapse between motor nerves and muscle cells.

Transmission across a synapse is carried out by molecules called **neurotransmitters,** which are stored in synaptic vesicles in the axon terminals. When nerve impulses traveling along an axon reach an axon terminal, channels

axon branches
of other neurons

cell body

axon terminals

dendrites

axon

presynaptic
neuron

path of action potential

direction of Ca²⁺
movement

synaptic
vesicles

axon terminal

dendrite

synaptic cleft

postsynaptic
neuron

After an action potential arrives at an
axon terminal and calcium ions enter
the cell, synaptic vesicles fuse with
the presynaptic membrane.

cell body of
postsynaptic cell

axon terminals of
presynaptic cell

Many axons synapse with each cell body.

neurotransmitter

synaptic vesicle

presynaptic
membrane

synaptic cleft

postsynaptic
membrane

receptor

Neurotransmitter molecules are released
and bind to receptors on the postsynaptic
membrane.

Na⁺

neurotransmitter

Figure 8.5 Synapse structure and function. Transmission across a
synapse from one neuron to another occurs when a neurotransmitter is
released at the presynaptic membrane, diffuses across a synaptic cleft, and
binds to a receptor in the postsynaptic membrane.

When a stimulatory neurotransmitter binds to a receptor,
Na⁺ diffuses into the postsynaptic neuron.

for calcium ions (Ca^{2+}) open, and calcium enters the terminal. This sudden rise in Ca^{2+} stimulates synaptic vesicles to merge with the presynaptic membrane, and neurotransmitter molecules are released into the synaptic cleft. This process is called *exocytosis* (see Chapter 3). Neurotransmitter molecules then diffuse across the cleft to the postsynaptic membrane, where they bind with specific receptor proteins.

Depending on the type of neurotransmitter and the type of receptor, the response of the postsynaptic neuron can be toward excitation or toward inhibition. After excitatory neurotransmitters combine with a receptor, a sodium ion channel opens, and Na^+ enters the neuron (Fig. 8.5). Other neurotransmitters have an inhibitory effect as described in the next section.

Graded Potentials and Synaptic Integration

A single neuron can have many dendrites plus the cell body, and both can synapse with many other neurons. Typically, a neuron is on the receiving end of many excitatory and inhibitory synapses (Fig. 8.5). Each of the small signals from a synapse is called a **graded potential.** An excitatory neurotransmitter opens the neuron's sodium channels, producing a graded potential that drives the polarity of a neuron closer to an action potential. An inhibitory neurotransmitter produces a graded potential that makes it harder for a neuron to have an action potential. Inhibitory signals can be created opening potassium channels and allowing potassium ions to leave the cell. Opening other channels that allow negatively charged chloride ions (Cl^-) to enter the cell will also create an inhibitory signal.

Neurons integrate these incoming signals. Integration is the summing up of excitatory and inhibitory signals. If a neuron receives many excitatory signals (either from different synapses, or at a rapid rate from one synapse), chances are the axon will transmit an action potential. On the other hand, if a neuron receives both inhibitory and excitatory signals, the summing up of these signals may prohibit the axon from firing.

Sensory receptors (for example, light receptors in the eye) have special graded potentials called **receptor potentials.** These will be discussed in Chapter 9.

Neurotransmitter Molecules

At least 75 different neurotransmitters have been identified, and more will likely be identified in the future. Two very well-known ones are **acetylcholine (ACh)** and **norepinephrine (NE).**

Once a neurotransmitter has been released into a synaptic cleft and has initiated a response, it is removed from the cleft. In some synapses, the postsynaptic membrane contains enzymes that rapidly inactivate the neurotransmitter. For example, the enzyme **acetylcholinesterase (AChE)** breaks down acetylcholine. In other synapses, the presynaptic membrane rapidly reabsorbs the neurotransmitters,

possibly for repackaging in synaptic vesicles or for molecular breakdown. The short existence of neurotransmitters at a synapse prevents continuous stimulation (or inhibition) of postsynaptic membranes.

The Medical Focus feature, Alzheimer Disease, discusses the disease, which is due in part to a lack of ACh in the brain. It is also of interest to note that many drugs are available that enhance or block the release of a neurotransmitter, mimic the action of a neurotransmitter or block the receptor, or interfere with the removal of a neurotransmitter from a synaptic cleft.

Content **CHECK-UP!**

1. Which of the following types of neurons are multipolar?
   a. motor neurons
   b. sensory neurons
   c. interneurons
   d. both motor neurons and interneurons
   e. motor neurons, sensory neurons, and interneurons

2. During an action potential, depolarization occurs because:
   a. channels for sodium ions in the cell membrane open, and sodium flows into the cell.
   b. channels for potassium ions in the cell membrane open, and potassium leaves the cell.
   c. the sodium-potassium pump returns sodium to the outside of the cell and potassium to the inside.

3. An excitatory neurotransmitter excites the postsynaptic cell because:
   a. it opens sodium channels so sodium can enter the cell.
   b. it opens potassium channels so potassium can leave the cell.
   c. it turns on the sodium-potassium pump.

Answers in Appendix C, page 463.

8.2 Central Nervous System

The CNS, consisting of the brain and spinal cord, is composed of gray matter and white matter. **Gray matter** is gray because it contains cell bodies and short, nonmyelinated fibers. **White matter** is white because it contains myelinated axons that run together in bundles called tracts. The myelin covering on these axons gives them a shiny, white appearance.

Meninges and Cerebrospinal Fluid

Both the spinal cord and the brain are wrapped in protective membranes known as **meninges** (sing., *meninx*). The outer meninx, the **dura mater,** is tough, white, fibrous connective tissue that lies next to the skull and vertebrae. The dura mater is constructed of two separate membrane layers that are fused. However, in several areas the layers separate to form the **dural venous sinuses.** The dural venous sinuses collect venous

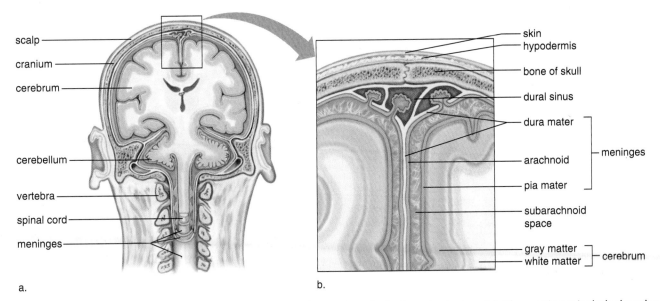

Figure 8.6 Meninges. a. Meninges are protective membranes that enclose the brain and spinal cord. **b.** The meninges include three layers: the dura mater, the arachnoid, and the pia mater.

blood and excess cerebrospinal fluid, returning both to the cardiovascular system. Deep to the dura mater is the **arachnoid mater** ("spider-like"), so-called because it consists of spider-web-like connective tissue. Thin strands of the arachnoid mater attach it to the **pia mater,** the deepest meninx. The pia mater is very thin and closely follows the contours of the brain and spinal cord (Fig. 8.6). Between the arachnoid mater and the pia mater is the subarachnoid space, which is filled with **cerebrospinal fluid.** This clear tissue fluid, similar to blood plasma, forms a protective cushion around and within the CNS.

Cerebrospinal fluid (CSF) is formed by the choroid plexus, which is a tissue created by folds of pia mater. These folds are lined with a dense layer of capillaries. Choroid plexus can be found in the hollow interconnecting cavities of the brain called **ventricles.** Cerebrospinal fluid fills brain ventricles as well as the **central canal** of the spinal cord. Normally, any excess CSF is drained into the dural venous sinuses to return to the cardiovascular system (Fig. 8.6b). However, blockages can occur. In an infant, the ventricles can enlarge due to cerebrospinal fluid accumulation. As a result, the skull enlarges and the brain tissue between the ventricles and skull will be compressed, sometimes resulting in brain damage. This condition, called **hydrocephalus** ("water on the brain"), can sometimes be corrected surgically.

The Spinal Cord

The **spinal cord** is a cylinder of nervous tissue that begins at the base of the brain and extends through a large opening in the skull called the **foramen magnum** (see Chapter 6, Fig. 6.7). The spinal cord is protected by the vertebral column, which is composed of individual vertebrae. When the vertebrae are stacked on top of each other, their vertebral foramina form the central canal. The spinal cord passes through the central canal, ending at approximately the first lumbar vertebra (see Fig. 6.7).

Structure of the Spinal Cord

Figure 8.7a shows how an individual vertebra protects the spinal cord. The spinal nerves extend from the cord between the vertebrae. Intervertebral disks, which are composed of tough fibrocartilage and filled with gelatinous material, separate each vertebra. If a disk is torn open, this **herniated disk** may compress spinal nerves. Pain and loss of function result.

A cross section of the spinal cord shows a central canal, gray matter, and white matter (Fig. 8.7b,c). The central canal contains cerebrospinal fluid, as do the meninges that protect the spinal cord. The gray matter is centrally located and shaped like the letter H. Portions of sensory neurons and motor neurons are found there, as are interneurons that communicate with these two types of neurons. The posterior (dorsal) root of a spinal nerve contains sensory fibers entering the gray matter, and the anterior (ventral) root of a spinal nerve contains motor fibers exiting the gray matter. The posterior and anterior roots join, forming a spinal nerve that leaves the vertebral canal. Spinal nerves are a part of the PNS.

The white matter of the spinal cord contains *ascending tracts*, which take sensory information to the spinal cord and brain, and *descending tracts*, which take motor information from the brain. Ascending tracts are generally located in the posterior white matter; descending tracts are found in the anterior white matter. Because the tracts typically cross just after

white matter

gray matter

central canal

a.

spinal cord

vertebra

gray matter

white matter

posterior (dorsal)
root

posterior root
ganglion

spinal
nerve

anterior
(ventral) root

vertebra

b.

central canal

posterior root

posterior
root
ganglion

posterior

gray matter

white matter

anterior

spinal
nerve

anterior root

meninges

c.

posterior (dorsal)
root ganglion

posterior (dorsal)
root branches

cut vertebrae

d. Posterior view of spinal cord and posterior roots of spinal nerves

Figure 8.7 **Spinal cord. a.** Photomicrograph of a cross section of the spinal cord. **b.** The spinal cord passes through the vertebral canal formed by the vertebrae. Spinal nerves branch off the spinal cord and project through openings between the vertebrae. **c.** The spinal cord has a central canal filled with cerebrospinal fluid, gray matter in an H-shaped configuration, and white matter elsewhere. The white matter contains tracts that take nerve impulses to and from the brain. **d.** Photo of posterior view of spinal cord and posterior roots of spinal nerve.

they enter and exit the brain, the left brain controls the right side of the body and the right brain controls the left side of the body.

Functions of the Spinal Cord

The spinal cord provides a means of communication between the brain and the peripheral nerves that leave the cord.

When someone touches your hand, sensory receptors generate action potentials that travel by way of sensory nerve axons to the spinal cord. One of several ascending tracts next carries the information to the sensory area of the brain. When you voluntarily move your limbs, action potentials originating in the motor control area of the brain pass down one of several descending tracts to the spinal cord and out to your muscles by way of motor nerve axons. The Medical Focus on page 166 discusses promising new therapies for patients whose spinal cord is injured.

We will see that the spinal cord is also the center for thousands of reflex arcs (see Fig. 8.13): A stimulus causes sensory receptors to generate action potentials that travel in sensory

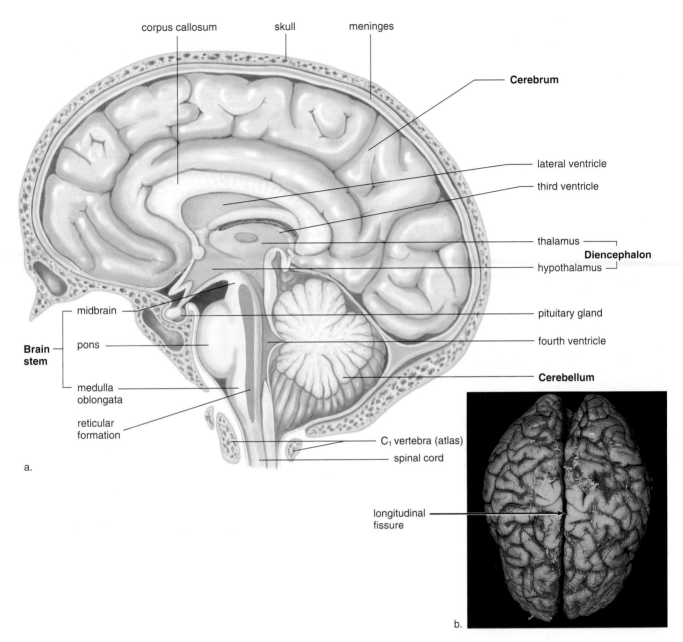

corpus callosum   skull   meninges

Cerebrum

lateral ventricle

third ventricle

thalamus ⎤
Diencephalon
hypothalamus ⎦

pituitary gland

fourth ventricle

Cerebellum

midbrain

pons

Brain stem

medulla oblongata

reticular formation

C₁ vertebra (atlas)

spinal cord

a.

longitudinal fissure

b.

Figure 8.8 **The human brain. a.** The cerebrum, seen here in sagittal section, is the largest part of the brain in humans. **b.** Superior view of the left and right cerebral hemispheres. They are connected by the corpus callosum.

neurons to the spinal cord. Interneurons integrate the incoming data and relay signals to motor neurons. A response to the stimulus occurs when motor axons cause skeletal muscles to contract. Each interneuron in the spinal cord has synapses with many other neurons, and therefore they send signals to several other interneurons (as well as to the brain) in addition to motor neurons.

The Brain

We will discuss the parts of the brain with reference to the cerebrum, the diencephalon, the cerebellum, and the brain stem. The brain's four ventricles are called, in turn, the two lateral ventricles, the third ventricle, and the fourth ventricle. It will be helpful for you to associate the cerebrum with the two lateral ventricles, the diencephalon with the third ventricle, and the brain stem and the cerebellum with the fourth ventricle (Fig. 8.8a).

The electrical activity of the brain can be recorded in the form of an **electroencephalogram (EEG)**. Electrodes are taped to different parts of the scalp, and an instrument records the so-called brain waves. The EEG is a diagnostic tool; for example, an irregular pattern can signify epilepsy or a brain tumor. Absence of electrical activity on an EEG signifies brain death.

Spinal Cord Injuries: Cause for Hope   *In Memoriam: Christopher Reeve, 1952–2004*

It happened in a split second: Christopher Reeve, the tall, athletic actor, who achieved worldwide recognition in his role as "Superman," was thrown from his horse. The accident shattered both atlas and axis, the first two cervical vertebrae, and Reeve became a ventilator-dependent quadriplegic. Although he spent the rest of his life in a wheelchair, Reeve redefined the recovery process for spinal cord injury with tremendous courage and determination.

Spinal cord injuries are usually caused by compression or bruising to one side of the cord, and often result from accidents or trauma. Injured cells in the center of the cord die, leaving a rim of functional cells. Depending on the nerve pathways that are interrupted, a variety of effects can result from the damage. Injury between the first thoracic vertebra (T1) and the second lumbar vertebra (L2) causes paralysis of the lower body, or **paraplegia.** If the injury is between the cervical vertebrae and the first thoracic vertebra (T1), the entire trunk is affected, resulting in **quadriplegia.** At the same time, sensation will be lost in all or parts of the body. Other serious consequences of spinal cord injury include loss of breathing ability, pain, absent or exaggerated reflexes, skeletal muscle spasms, and loss of bowel and bladder control.

After his injury, Reeve was determined to recover, though experts called his case hopeless. He pioneered *functional electrical stimulation* (FES), using computer-directed muscle activation to exercise his limbs. FES therapy is now used to improve muscle tone and bone density is spinal cord-injured patients, and studies have further shown that this exercise seems to stimulate spinal cord recovery. In Reeve's case, it worked: his was the *first* documented case of significant improvement in a quadriplegic occurring more than two years after injury. Reeve regained movement of his fingers and toes, as well as sensation over most of his body. He was also able to finally breathe independently for several hours at a time. Ironically, Reeve died in 2004 of complications resulting from an infected *decubitus ulcer,* or bedsore (see page 85), a common affliction of quadriplegics.

Reeve's foundation, the Christopher Reeve Paralysis Foundation, and other foundations dedicated to spinal cord research, have made significant advances. Emergency care of the spinal cord-injured patient has improved, and new drugs help injured spinal cord nerves to survive. Prosthetics for the spinal cord-injured include stair-climbing and voice-activated wheelchairs, and electronic devices that allow independent breathing and restore control of bladder and bowel. Studies using grafts of neural tissue, neural stem cells, and nerve growth factors—hormone-like chemicals that stimulate growth—have shown to be promising for regenerating spinal cord tissue. The goal of the Christopher Reeve Paralysis Foundation (www.christopherreeve.org) is to one day produce a cure for spinal cord paralysis.

The Cerebrum

Let's begin our study of the brain with the **cerebrum,** the largest portion of the brain in humans. The cerebrum is the last center to receive sensory input and carry out integration before commanding voluntary motor responses. It communicates with and coordinates the activities of the other parts of the brain. The cerebrum carries out the higher thought processes required for learning and memory and for language and speech.

The Cerebral Hemispheres The cerebrum has two halves called the left and right **cerebral hemispheres** (Fig. 8.8*b*). A deep groove, the longitudinal *fissure,* separates the left and right cerebral hemispheres. Still, the two cerebral hemispheres are connected by a bridge of white matter called the **corpus callosum.**

Ridges called *gyri* are separated by shallow grooves called *sulci* (sing., *sulcus*). Specific sulci divide each hemisphere into lobes (Fig. 8.9). Note that each lobe of the brain is located

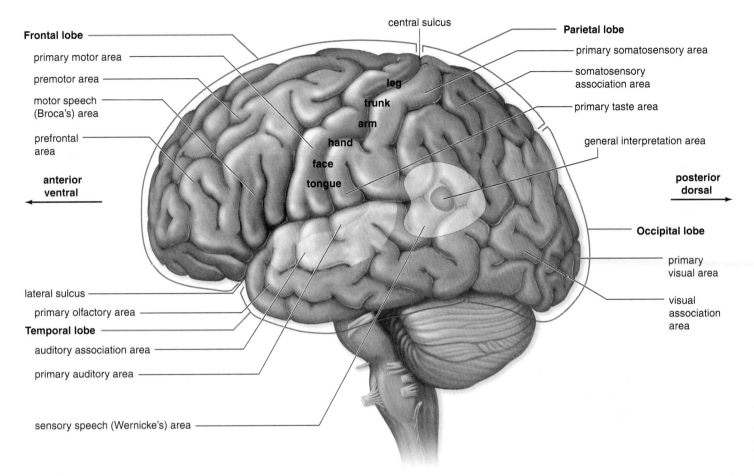

Figure 8.9 **The lobes of a cerebral hemisphere.** Each cerebral hemisphere is divided into four lobes: frontal, parietal, temporal, and occipital. These lobes contain centers for reasoning and movement (frontal lobe), somatic sensing including taste (parietal lobe), hearing (temporal lobe), and vision (occipital lobe).

underneath the skull bone that shares its name. The *frontal lobe* lies under the frontal bone, anterior to the *parietal lobe* and bone. The *occipital lobe* is deep to the occipital bone, in the posterior area of the cranial vault. The *temporal lobe* is the lateral portion of the cerebral hemisphere. A fifth, very small lobe called the *insula* (not shown) lies directly deep to the lateral sulcus.

The **cerebral cortex** is a thin but highly convoluted outer layer of gray matter that covers the cerebral hemispheres. The cerebral cortex contains over one billion cell bodies and is the region of the brain that accounts for sensation, voluntary movement, and all the thought processes we associate with consciousness.

Motor and Sensory Areas of the Cortex The **primary motor area** is in the frontal lobe just anterior to the central sulcus (Fig. 8.9). Voluntary commands to skeletal muscles begin in the primary motor area, and each part of the body is controlled by a certain section (see Fig. 8.10*a*). The right primary motor area controls the left side of the body and vice-versa.

The **primary somatosensory area** is just posterior to the central sulcus in the parietal lobe. Sensory information from the skin and skeletal muscles arrives here, where each part of the body is sequentially represented (see Fig. 8.10*b*). As you study

Figure 8.10, notice that the areas of the body with the greatest voluntary control—the face and hands—have the largest area of motor cortex in the brain dedicated to them. Similarly, the face and hands are among the most sensitive areas of the body, and the large area of the sensory cortex receiving information from them corresponds to that fact. As with the brain's motor areas, the left hemisphere's somatosensory area receives information from the right side of the body, and vice-versa.

A primary taste area, located within adjacent areas of the parietal lobe and insula, accounts for taste sensations. A primary visual area in the occipital lobe receives information from our eyes, and a primary auditory area in the temporal lobe receives information from our ears.

Begin Thinking Clinically

? Suppose that a woman slips on ice in the winter and falls backward, striking the back of her head on the pavement. The accident results in a concussion to the woman's occipital lobe. Which sense might be affected?

Answer and discussion in Appendix C, page 463.

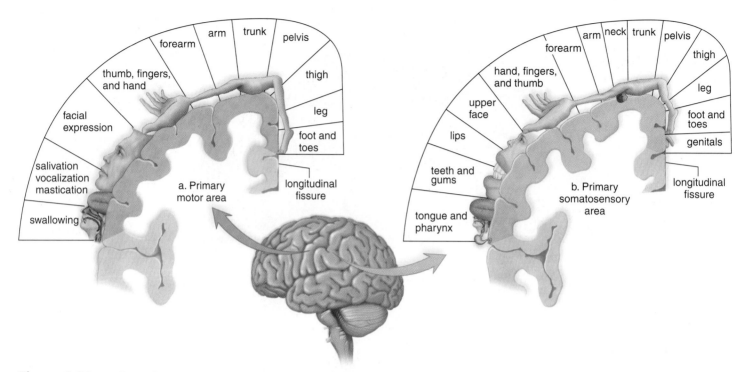

Figure 8.10 **Portions of the body controlled by the primary motor area and the primary somatosensory area of the cerebrum.** Notice that the size of the body part in the diagram reflects the amount of cerebral cortex devoted to that body part.

Association Areas Association areas are places where integration occurs and where memories are stored. Anterior to the primary motor area is a premotor area. The premotor area organizes motor functions for skilled motor activities, and then the primary motor area sends signals to the cerebellum and the basal nuclei, which integrate them. A momentary lack of oxygen during birth can damage the motor areas of the cerebral cortex so that **cerebral palsy,** a condition characterized by a spastic weakness of the arms and legs, develops.

The somatosensory association area, located just posterior to the primary somatosensory area, processes and analyzes sensory information from the skin and muscles. The visual association area associates new visual information with memories of previously received visual information. It might "decide," for example, whether we have seen this face, tool, or whatever before. The auditory association area performs the same functions with regard to sounds.

Processing Centers There are a few areas of the cortex that receive information from the other association areas and perform higher-level analytical functions. The **prefrontal area,** a processing area in the frontal lobe, receives information from the other association area and uses this information to reason and plan our actions. Integration in this area accounts for our most cherished human abilities to think critically and to formulate appropriate behaviors.

The unique ability of humans to speak is partially dependent upon the **motor speech area,** also called **Broca's area,** a processing area usually located in the left frontal lobe. Signals originating here pass to the premotor area before

reaching the primary motor area. Damage to this area can interfere with a person's ability to control the muscles of the face and neck that allow speech.

Wernicke's area, also called the **general interpretive area,** receives information from all of the other sensory association areas. Like Broca's area, Wernicke's area is usually located in the brain's left hemisphere. Damage to this area hinders a person's ability to interpret written and spoken messages, even though the spoken words may be a part of the person's vocabulary. Wernicke's area and Broca's area cooperate to allow human communication.

Central White Matter Much of the rest of the cerebrum beneath the cerebral cortex is composed of white matter. Tracts within the cerebrum take information between the different sensory, motor, and association areas pictured in Figure 8.9. The corpus callosum, previously mentioned, contains tracts that join the two cerebral hemispheres. Descending tracts from the primary motor area communicate with various parts of the brain, and transmit instructions to the opposite side of the brainstem and spinal cord. Ascending tracts from lower brain centers send sensory information up to the primary somatosensory area (Fig. 8.10).

Basal Nuclei While the bulk of the cerebrum is composed of tracts, there are masses of gray matter located deep within the white matter. These so-called **basal nuclei** (formerly termed *basal ganglia*) integrate motor commands, ensuring that proper muscle groups are activated or inhibited. Huntington disease and Parkinson's disease, which are both characterized by uncontrollable movements, are believed to be due to an imbalance of neurotransmitters in the basal nuclei.

The Diencephalon

The hypothalamus and the thalamus are both in the **diencephalon,** a region that encircles the third ventricle (see Fig. 8.8*a*). The **hypothalamus** forms the floor of the third ventricle. The hypothalamus is an integrating center that helps maintain homeostasis by regulating hunger, sleep, thirst, body temperature, and water balance. The hypothalamus produces the hormones secreted by the posterior pituitary gland and secretes hormones that control the anterior pituitary. Therefore, it is a link between the nervous and endocrine systems.

The **thalamus** consists of two masses of gray matter located in the sides and roof of the third ventricle. The thalamus is on the receiving end for all sensory input except smell; it functions as a sensory "relay center." Visual, auditory, and somatosensory information arrive at the thalamus via the cranial nerves and tracts from the spinal cord. The thalamus integrates this information and sends it on to the appropriate portions of the cerebrum. The thalamus is involved in arousal of the cerebrum, and it also participates in higher mental functions such as memory and emotions.

The pineal gland, which secretes the hormone melatonin and regulates our body's daily rhythms, is located in the diencephalon.

Limbic System The **limbic system** is a collection of structures from both the cerebrum and the diencephalon. It lies just inferior to the cerebral cortex and contains neural pathways that connect portions of the cerebral cortex and the temporal lobes with the thalamus and the hypothalamus:

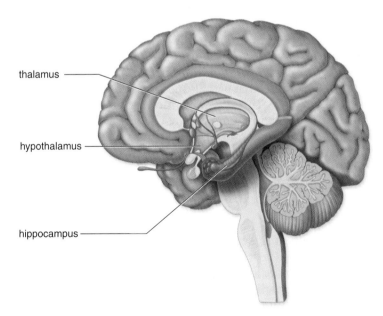

Stimulation of different areas of the limbic system causes the subject to experience rage, pain, pleasure, or sorrow. By causing pleasant or unpleasant feelings about experiences, the limbic system apparently guides the individual into behavior that is likely to increase the chance of survival.

The limbic system is also involved in learning and memory. In particular, the most inferior structure of the limbic system, the **hippocampus**, is vital in processing of short-term memory to become long-term memory. Learning requires memory and memory is stored in the sensory regions of the cerebrum, but just what permits memory development is not definitely known. The involvement of the limbic system in memory explains why emotionally charged events result in our most vivid memories. The fact that the limbic system communicates with the sensory areas for touch, smell, vision, and so forth accounts for the ability of any particular sensory stimulus to awaken a complex memory.

The Cerebellum

The **cerebellum** is separated from the brain stem by the fourth ventricle (see Fig. 8.8*a*). The cerebellum has two hemispheres, which are joined by a narrow median portion. Each portion is primarily composed of white matter, which in longitudinal section has a treelike pattern. Overlying the white matter is a thin layer of gray matter that forms a series of complex folds.

The cerebellum receives sensory input from the eyes, ears, joints, and muscles about the present position of body parts. It also receives motor output from the cerebral cortex about where these parts should be located. After integrating this in-

formation, the cerebellum sends motor impulses by way of the brain stem to the skeletal muscles. In this way, the cerebellum maintains posture and balance. It also ensures that all of the muscles work together to produce smooth, coordinated voluntary movements. In addition, the cerebellum assists the learning of new motor skills, such as playing the piano or hitting a baseball.

The Brain Stem

The brain stem is the structure that joins the upper brain regions to the spinal cord. The **brain stem** contains the midbrain, the pons, the medulla oblongata, and the reticular formation (see Fig. 8.8*a*). The **midbrain** acts as a relay station for tracts passing between the cerebrum and the spinal cord or cerebellum. It also has reflex centers for visual, auditory, and tactile responses. The word *pons* means "bridge" in Latin, and true to its name, the **pons** contains bundles of axons traveling between the cerebellum and the rest of the CNS. In addition, the pons functions with the medulla oblongata to regulate breathing rate and has reflex centers concerned with head movements in response to visual and auditory stimuli.

The **medulla oblongata** contains a number of reflex centers for regulating heartbeat, breathing, and vasoconstriction. It also contains the reflex centers for vomiting, coughing, sneezing, hiccuping, and swallowing. The medulla oblongata lies just superior to the spinal cord, and it contains tracts that ascend or descend between the spinal cord and higher brain centers.

The **reticular formation** assists the cerebellum in maintaining muscle tone; it also assists the pons and medulla in regulating respiration, heart rate, and blood pressure. The sensory component of the reticular formation processes sensory stimuli—sounds, sights, touch—and uses these signals to keep us mentally alert. Additionally, this portion of the brain helps to rouse a sleeping person. Damage to the reticular formation can result in **coma,** a state of temporary (sometimes permanent) unconciousness.

8.3 Peripheral Nervous System

The peripheral nervous system (PNS) lies outside the central nervous system and is composed of nerves and ganglia. Nerves are bundles of axons, both myelinated and unmyelinated, that travel together. Ganglia (sing., **ganglion**) are swellings associated with nerves that contain collections of cell bodies. As with muscles, connective tissue separates axons at various levels of organization:

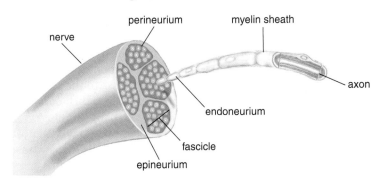

The PNS is subdivided into the **afferent,** or **sensory, system** and the **efferent,** or **motor, system** (see Fig. 8.1). The **somatic sensory system** serves the skin, skeletal muscles,

joints, and tendons. The special senses (vision, hearing, taste, smell) are also part of the somatic sensory system. The **visceral sensory system** supplies the internal organs. Nerves from both sensory systems take information from peripheral sensory receptors to the CNS.

The **somatic motor system** carries commands away from the CNS to the skeletal muscles. The **autonomic motor system,** with a few exceptions, regulates the activity of cardiac and smooth muscles and glands.

Types of Nerves

The **cranial nerves** are attached to the brain, and the **spinal nerves** are attached to the spinal cord.

Cranial Nerves

Humans have 12 pairs of cranial nerves (Table 8.1). By convention, the pairs of cranial nerves are referred to by Roman numerals (Fig. 8.11*a*). Some cranial nerves are sensory nerves—that is, they contain only sensory fibers; some are *motor nerves*, containing only motor fibers; and others are *mixed nerves*, so called because they contain both sensory and motor fibers. Cranial nerves are largely concerned with the head, neck, and facial regions of the body. However, the *vagus nerve* (cranial nerve X), has sensory and motor branches to the face and most of the internal organs. This nerve, whose name aptly derives from a Latin word that means "traveler," contains both somatic and visceral sensory nerves. Its efferent nerves belong to the somatic and the autonomic motor systems.

Spinal Nerves

Humans have 31 pairs of spinal nerves; one of each pair is on either side of the spinal cord (Fig. 8.11*b*). The spinal nerves are grouped as shown in Table 8.2 because they are at either the cervical, thoracic, lumbar, or sacral regions of the vertebral column. The spinal nerves are designated according to their location in relation to the vertebrae because each passes through an intervertebral foramen as it leaves the spinal cord. This organizational principle is illustrated in Figure 8.12.

Spinal nerves are called mixed nerves because they contain both sensory fibers that conduct impulses to the spinal cord from sensory receptors and motor fibers that conduct impulses away from the cord to effectors. The sensory fibers enter the cord via the posterior root, and the motor fibers exit by way of the anterior root. The cell body of a sensory neuron is in a **posterior (dorsal) root ganglion.** Each spinal nerve serves the particular region of the body in which it is located (Table 8.2).

Somatic Motor Nervous System and Reflexes

Most actions in the somatic motor nervous system are voluntary. These actions, such as when we decide to move a limb, always originate in the motor cortex. Recall that the motor

TABLE 8.1 Cranial Nerves

Nerve	Type		Brain Pathway	Transmits Nerve Impulses to (Motor) or from (Sensory)
Olfactory (I)	Sensory		I: Mucous membrane of nose to olfactory bulbs	Olfactory receptors for sense of smell
Optic (II)	Sensory		II: Retina → optic nerve → thalamus → occipital lobe	Retina for sense of sight
Oculomotor (III)	Motor		III: Midbrain → eye and eyelid	Eye muscles (including eyelids and lens); pupil (parasympathetic division)
Trochlear (IV)	Motor		IV: Midbrain → eye	Eye muscles
Trigeminal (V)	Mixed	Sensory	V: Sensory: Teeth, eye, skin, Tongue → pons	Teeth, eyes, skin, and tongue
		Motor	Motor: Pons → jaw muscles	Jaw muscles (chewing)
Abducens (VI)	Motor		VI: Pons → eye	Eye muscles
Facial (VII)	Mixed	Sensory	VII: Sensory: Tongue → pons	Taste buds of anterior tongue
		Motor	Motor: Pons → facial muscles, Salivary glands, tear glands	Facial muscles (facial expression) and glands (tear and salivary)
Vestibulocochlear (VIII) (also called auditory; acoustic)	Sensory		VIII: Inner ear → pons and medulla	Inner ear for sense of balance and hearing
Glossopharyngeal (IX)	Mixed	Sensory	IX: Sensory: Tongue, throat → pons	Pharynx
		Motor	Motor: Pons → Salivary gland, Throat muscles	Pharyngeal muscles (swallowing), salivary glands
Vagus (X)	Sensory Motor		X: Sensory: Eardrum, ear canal, throat, heart, lungs, abdominal organs → medulla	Internal organs, external ear canal, eardrum, back of throat
			Motor: Medulla → throat and larynx, heart, lungs, abdominal organs	Internal organs (parasympathetic division), throat muscles (somatic motor division)
Spinal accessory (XI)	Motor		XI: Medulla → muscles of throat, neck, shoulder	Neck and back muscles
Hypoglossal (XII)	Motor		XII: Medulla → tongue muscles	Tongue muscles

cortex is in the posterior part of the frontal lobe. Other actions in the somatic motor nervous system are due to **reflexes**—automatic involuntary responses to changes occurring inside or outside the body. A reflex occurs quickly; we don't even have to think about it. Reflexes are protective mechanisms that are essential to homeostasis. They keep the internal organs functioning within normal bounds and protect the body from external harm.

Some reflexes, called *cranial reflexes*, involve the brain, as when we automatically blink our eyes when an object nears the eye suddenly. Figure 8.13 illustrates the path of a reflex within the somatic motor nervous system that involves only the spinal cord (called a *spinal reflex*). If your skin is poked with a sharp pin, a pain sensory receptor in the skin generates action potentials that move along a sensory fiber through the posterior root ganglia toward the spinal cord. Sensory neurons enter the cord posteriorly and pass signals on to many interneurons. Some of these interneurons synapse with motor neurons whose short dendrites and cell bodies are in the spinal cord. Nerve action potentials travel along a motor fiber to an effector, which brings about a response to the stimulus. In this case, the effector is a skeletal muscle, which contracts so that you withdraw your hand from the pain stimulus.

Various other reactions are also possible—you will most likely look at the pin, wince, and cry out in pain. This whole series of responses occurs because certain interneurons carry nerve impulses to the brain via tracts in the spinal cord and brain. The brain makes you aware of the stimulus and directs your other reactions to it. You don't feel pain until the brain receives the information and interprets it.

a. **Nerve**

cranial
nerves

PNS

spinal
nerves

artery
and vein

single nerve fiber

bundle of
nerve fibers

b.

blood vessels

SEM 200×

c.

Cranial Nerves

I from olfactory receptors

II from retina of eyes

III to eyo muscles

IV to eye muscles

V from mouth and to jaw muscles

VI to eye muscles

VII from taste buds and to facial muscles and glands

VIII from inner ear

IX from pharynx and to pharyngeal muscles

XII to tongue muscles

X from and to internal organs

XI to neck and back muscles

Figure 8.11 Cranial and spinal nerves. a. A single nerve is composed of many individual fibers. **b.** Cross-section scanning electron micrograph of a single spinal nerve. **c.** Inferior surface of the brain showing the attachment of the 12 pairs of cranial nerves.

TABLE 8.2 Spinal Nerves		
Name	**Spinal Nerves Involved***	**Function**
Musculocutaneous nerves	$C_5–T_1$	Supply muscles of the arms on the anterior sides, and skin of the forearms
Radial nerves	$C_5–T_1$	Supply muscles of the arms on the posterior sides, and skin of the forearms and hands
Median nerves	$C_5–T_1$	Supply muscles of the forearms, and muscles and skin of the hands
Ulnar nerves	$C_5–T_1$	Supply muscles of the forearms and hands, and skin of the hands
Phrenic nerves	$C_3–C_5$	Supply the diaphragm
Intercostal nerves	$T_2–T_{12}$	Supply intercostal muscles, abdominal muscles, and skin of the trunk
Femoral nerves	$L_2–L_4$	Supply muscles and skin of the anterior thighs and legs
Sciatic nerves	$L_4–S_3$	Supply muscles and skin of the posterior thighs, legs, and feet

*C = cervical; T = thoracic; L = lumbar

musculocutaneous nerve

phrenic nerve

radial nerve

median nerve

ulnar nerve

intercostal nerves

femoral nerve

sciatic nerve

C1
C2
C3
C4
C5
C6
C7
C8

8 pairs of cervical nerves

T1
T2
T3
T4
T5
T6
T7
T8
T9
T10
T11
T12

12 pairs of thoracic nerves

L1
L2
L3
L4
L5

5 pairs of lumbar nerves

S1
S2
S3
S4
S5
Co

5 pairs of sacral nerves

1 pair of coccygeal nerves

Posterior view

Figure 8.12 Spinal nerves. The number and kinds of spinal nerves are given on the right. The location of major peripheral nerves is given on the left. Table 8.2 lists the functions of these nerves.

Reflexes can also be used to determine if the nervous system is reacting properly. Two of these types of reflexes are:

knee-jerk reflex (patellar reflex), initiated by striking the patellar ligament just below the patella. The response is contraction of the quadriceps femoris muscles, which causes the lower leg to extend;

ankle-jerk reflex, initiated by tapping the Achilles tendon just above its insertion on the calcaneus. The response is plantar flexion due to contraction of the gastrocnemius and soleus muscles.

Some reflexes are important for avoiding injury, but the knee-jerk and ankle-jerk reflexes are important for normal physiological functions. For example, the knee-jerk reflex helps a person stand erect. If the knee begins to bend slightly when a person stands still, the quadriceps femoris is stretched, and the leg straightens.

Autonomic Motor Nervous System and Visceral Reflexes

The autonomic motor nervous system (ANS) is composed of the sympathetic and parasympathetic divisions (Fig. 8.14). These two divisions have several features in common: (1) They function automatically and usually in an involuntary manner; (2) they innervate all internal organs; and (3) they utilize two motor neurons and one ganglion to transmit an action potential. (By contrast, a somatic motor neuron travels directly to its effector, without synapsing at a ganglion.) The first neuron has a cell body within the CNS and a preganglionic axon fiber. The second neuron has a cell body within the ganglion and a postganglionic axon fiber.

Visceral reflex actions, such as those that regulate blood pressure and breathing rate, are especially important to maintenance of homeostasis. These reflexes begin when the sensory neurons in contact with internal organs send messages via spinal nerves to the CNS. They are completed when motor neurons within the autonomic system stimulate smooth muscle, cardiac muscle, or a gland. These structures are also effectors.

Sympathetic Division: "Fight or Flight"

Most preganglionic fibers of the **sympathetic division** arise from the middle, or thoracic-lumbar, portion of the spinal cord and almost immediately terminate in ganglia that lie near the cord. Therefore, in this division, the preganglionic fiber is short, but the postganglionic fiber that makes contact with an organ is long:

CNS   ganglion   organ

NE

preganglionic fiber (axon)   postganglionic fiber (axon)

The sympathetic division is especially important during emergency situations when a person might be required to fight or take flight. It accelerates the heartbeat and dilates the bronchi—active muscles, after all, require a

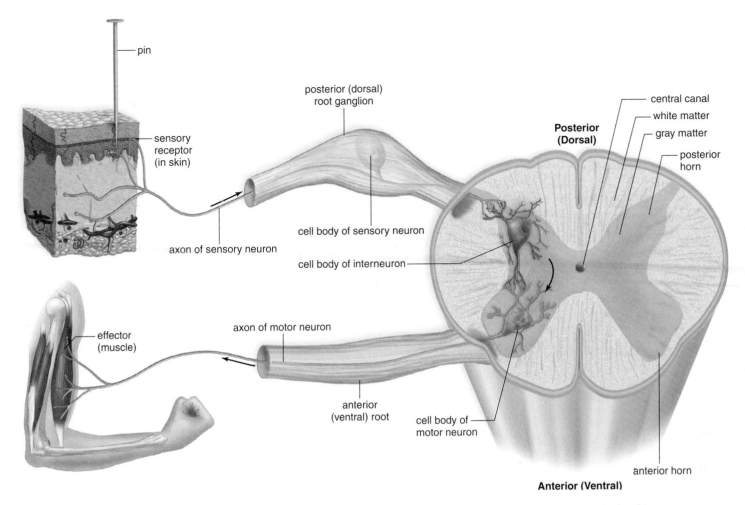

Figure 8.13 A reflex arc showing the path of a spinal reflex. A stimulus (e.g., a pinprick) causes sensory receptors in the skin to generate nerve signals that travel in sensory axons to the spinal cord. Interneurons integrate data from sensory neurons and then relay signals to motor neurons. Motor axons convey nerve signals from the spinal cord to a skeletal muscle, which contracts. Movement of the hand away from the pin is the response to the stimulus.

ready supply of glucose and oxygen. On the other hand, the sympathetic division inhibits the digestive tract—digestion is not an immediate necessity if you are under attack. The neurotransmitter released by the postganglionic axon is primarily norepinephrine (NE). The structure of NE is like that of epinephrine (adrenaline), an adrenal medulla hormone that usually increases heart rate and contractility.

Parasympathetic Division: "Rest and Digest"

The **parasympathetic division** includes several cranial nerves (e.g., the vagus nerve) as well as fibers that arise from the sacral (bottom) portion of the spinal cord. Therefore, this division is often referred to as the craniosacral portion of the autonomic system. In the parasympathetic division, the preganglionic fiber (axon) is long, and the

postganglionic fiber (axon) is short because the ganglia lie near or within the organ:

The parasympathetic division, sometimes called the "housekeeper" division, promotes all of the internal responses we associate with a relaxed state. Parasympathetic function can be referred to as "rest and digest." For example, parasympathetic nerves cause the pupil of the eye to contract, promote digestion of food, slow heart rate, and decrease the strength of cardiac contraction. The neurotransmitter utilized by the parasympathetic division is acetylcholine (ACh).

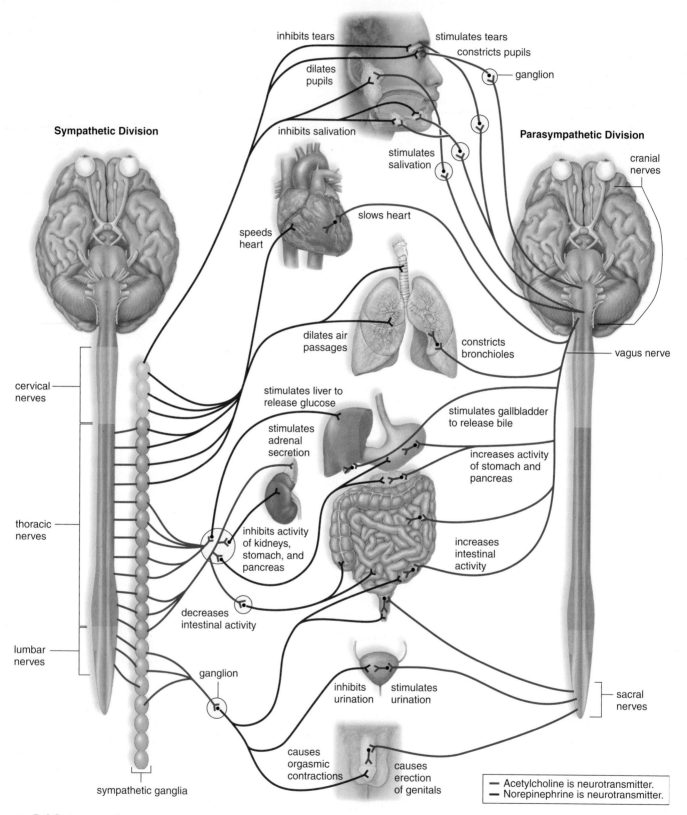

Sympathetic Division

inhibits tears

stimulates tears

constricts pupils

ganglion

dilates pupils

inhibits salivation

stimulates salivation

Parasympathetic Division

cranial nerves

speeds heart

slows heart

cervical nerves

dilates air passages

constricts bronchioles

vagus nerve

stimulates liver to release glucose

stimulates gallbladder to release bile

stimulates adrenal secretion

increases activity of stomach and pancreas

thoracic nerves

inhibits activity of kidneys, stomach, and pancreas

increases intestinal activity

lumbar nerves

decreases intestinal activity

ganglion

inhibits urination

stimulates urination

sacral nerves

causes orgasmic contractions

causes erection of genitals

sympathetic ganglia

— Acetylcholine is neurotransmitter.
— Norepinephrine is neurotransmitter.

Figure 8.14 Autonomic system structure and function. Sympathetic preganglionic fibers (*left*) arise from the cervical, thoracic, and lumbar portions of the spinal cord; parasympathetic preganglionic fibers (*right*) arise from the cranial and sacral portions of the spinal cord. Each system innervates the same organs but has contrary effects.

TABLE 8.3 Autonomic Motor Pathways

	Sympathetic	Parasympathetic
Type of control	Involuntary	Involuntary
Number of neurons per message	Two (preganglionic *shorter* than postganglionic)	Two (preganglionic *longer* than postganglionic)
Location of motor fiber	Thoracolumbar spinal nerves	Cranial (e.g., vagus) and sacral spinal nerves
Neurotransmitter	Norepinephrine	Acetylcholine
Effectors	Smooth and cardiac muscle, glands	Smooth and cardiac muscle, glands

Table 8.3 contrasts the two divisions of the autonomic system.

Content CHECK-UP!

7. Which of the following cranial nerves transmits sensory signals about hearing and balance?

   a. oculomotor   c. trochlear

   b. abducens   d. vestibulocochlear

8. Which spinal nerve supplies the diaphragm?

   a. radial   c. median

   b. phrenic   d. intercostal

9. Put the following parts of a reflex in the correct order:

   1. effector   2. sensory receptor   3. motor nerve
   4. sensory nerve   5. interneuron

   a. 2—5—4—3—1   c. 4—2—5—3—1

   b. 2—4—5—3—1   d. 5—2—4—1—3

Answers in Appendix C, page 463.

8.4 Effects of Aging

After age 60, brain mass begins to decrease. By age 80, the brain weighs about 10% less than when the person was a young adult. The cerebral cortex shrinks more than other areas of the brain, losing as much as 45% of its cells. Therefore, such mental activities as learning, memory, and reasoning decline.

Neurotransmitter production also decreases, resulting in slower synaptic transmission. As a person ages, thought processing and translating a thought into action take longer. This partly explains why younger athletes tend to outshine older athletes in sports. However, it is important to note that although structural changes occur, mental impairment is *not* an automatic consequence of getting older. Maintaining the health of the cardiovascular system is fundamental to retaining mental function—after all, the heart and blood vessels supply oxygen and nutrients to brain cells. In addition, older individuals can stay mentally alert by challenging the brain: taking courses, reading, solving puzzles, etc. Avoiding depression is important because it is a contributor to mental impairment; thus, the elderly should try to maintain a social network and interact with others in meaningful ways. Exercise is also an important ingredient in good health—including mental health—at every age.

Neurological disorders, especially Alzheimer disease and Parkinson's disease, are more apt to occur in the elderly. The Medical Focus readings on Alzheimer Disease and Parkinson's disease on pages 160 and 179 will give you more information about these conditions.

8.5 Homeostasis

The nervous system detects, interprets, and responds to changes in internal and external conditions to keep the internal environment relatively constant. Together with the endocrine system, it coordinates and regulates the functioning of the other systems in the body to maintain homeostasis.

The everyday regulation of internal organs that maintains the composition of blood and tissue fluid usually takes place below the level of consciousness. Subconscious control is dependent on reflex actions that involve the hypothalamus and the medulla oblongata. The hypothalamus and the medulla oblongata act through the autonomic motor nervous system to control such important parameters as the heart rate, the constriction of the blood vessels, and the breathing rate.

The illustration in Human Systems Work Together on page 178 shows how the nervous system works with other systems in the body to maintain homeostasis. The hypothalamus works closely with the endocrine system and even produces the hormone ADH, which causes the kidneys to reabsorb water.

Because the nervous system stimulates skeletal muscles to contract, it controls the major movements of the body. When we are in a "fight-or-flight" mode, the nervous system stimulates the adrenal glands and voluntarily controls the skeletal muscles to keep us from danger. On a daily basis, you might think that voluntary movements don't play a role in homeostasis, but actually we usually take all necessary actions to stay in as moderate an environment as possible. Otherwise, we are testing the ability of the nervous system to maintain homeostasis despite extreme conditions.

Human Systems Work Together

Integumentary System

Brain controls nerves that regulate size of cutaneous blood vessels; activates sweat glands and arrector pili muscles.

Skin protects nerves,* helps regulate body temperature; skin receptors send sensory input to brain.

Skeletal System

Receptors send sensory input from bones and joints to brain.

Bones protect sense organs, brain, and spinal cord; store Ca^{2+} for nerve function.

Muscular System

Brain controls nerves that innervate muscles; receptors send sensory input from muscles to brain.

Muscle contraction moves eyes, permits speech, and creates facial expressions.

Endocrine System

Hypothalamus is part of endocrine system; nerves innervate certain glands of secretion.

Sex hormones affect development of brain.

Cardiovascular System

Brain controls nerves that regulate the heart and dilation of blood vessels.

Blood vessels deliver nutrients and oxygen to neurons; carry away wastes.

How the Nervous System works with other body systems.

brain

spinal cord

Lymphatic System/Immunity

Microglial cells engulf and destroy pathogens.

Lymphatic vessels pick up excess tissue fluid; immune system protects against infections of nerves.

Respiratory System

Respiratory centers in brain regulate breathing rate.

Lungs provide oxygen for neurons and rid the body of carbon dioxide produced by neurons.

Digestive System

Brain controls nerves that innervate smooth muscle and permit digestive tract movements.

Digestive tract provides nutrients for growth, maintenance, and repair of neurons and neuroglial cells.

Urinary System

Brain controls nerves that innervate muscles that permit urination.

Kidneys maintain blood levels of Na^+, K^+, and Ca^{2+}, which are needed for nerve conduction.

Reproductive System

Brain controls onset of puberty; nerves are involved in erection of penis and clitoris, contraction of ducts that carry gametes, and contraction of uterus.

Sex hormones masculinize or feminize the brain, exert feedback control over the hypothalamus, and influence sexual behavior.

Parkinson's Disease

"My body is completely out of control. That's the hardest thing about this disease. Sometimes I can't move at all, or I move so slowly that it takes forever just to cross the room. Next thing you know, I'm jerking around like a puppet."

Your patient has just described the classic symptoms of **Parkinson's disease,** a progressive central nervous system disorder. The Parkinson's patient is usually a person age 60 or older. However, the disease is seen increasingly in younger people as well. Actor Michael J. Fox was just 38 years old when he announced publicly in 1999 that he suffers from Parkinson's disease.

If the patient's facial muscles are involved, his or her face may not be able to show emotion, resulting in a fixed, masklike appearance. Routine tasks such as dressing and bathing become very difficult. The sufferer has an increased risk of falling and injuring himself because balance and coordination are also affected. The disease takes its toll on the patient psychologically; most suffer depression as their activities and independence become more and more limited.

Parkinson's disease is caused by destruction of specific areas of the brain called the basal nuclei (see page 168). Researchers have determined that these basal nuclei nerve cells produce the neurotransmitter dopamine. The lack of this neurotransmitter seems to cause the signs and symptoms of the disorder. Treatment for the disease has, until recently, focused on ways to replace dopamine in the brain. Drug treatment produces temporary dopamine

replacement and relieves the symptoms completely for a few weeks to months. However, as the disease progresses, patients need increasingly stronger medications in higher dosages to relieve the symptoms. These stronger medications produce undesirable side effects, such as dizziness, sleepiness, and memory loss.

Implants of dopamine-producing cells have also been placed into the brain. These implants have had low to moderate success rates in relieving symptoms. Because the cells are often obtained from human embryos, scientists have also raised ethical concerns about the source of the implanted cells.

Increasingly, therapy for Parkinson's disease involves directly treating brain tissue. In **deep-brain stimulation (DES),** a set of electrodes is implanted into precise centers in the brain. The electrodes are connected to an electrical neurostimulator implanted into the chest near the clavicle, or collarbone. The neurostimulator delivers continuous electrical signals into the patient's brain, blocking the signals that cause Parkinsonian movement. Once implanted, the stimulator can be adjusted from outside the body to achieve maximum symptom relief. **Pallidotomy** is performed on an awake patient. In this procedure, a fine probe is placed into the brain tissue. Next, electrical signals are applied to pinpoint the area for treatment, and the patient describes the signal's effects to the surgeon. Based on these descriptions, the surgeon can burn a tiny hole in the brain tissue, permanently stopping the signals that cause abnormal movements.

SELECTED NEW TERMS

Basic Key Terms

acetylcholine (as″ē-til-ko′lēn), p. 162

acetylcholinesterase (as″ē-til-ko″lin-es′ter-ās), p. 162

action potential (ak′-shun po-ten′shul), p. 159

arachnoid mater (uh-rak′noyd mă′tər), p. 163

association area (uh-so″se-a′shun a′re-uh), p. 168

autonomic motor system (aw″to-nom′ik mō′tər sis′tem), p. 171

axon (ak′son), p. 157

basal nuclei (bas′al nu′kle-i), p. 168

brain stem (brān′ stĕm′), p. 170

Broca's area (brŏ′kəz âr′ē-ə), p. 168

cell body (sel bod′e), p. 157

central canal (sĕn′trəl kə-năl′), p. 163

central nervous system (sen′tral ner′vus sis′tem), p. 156

cerebellum (sĕr″ĕ-bel′um), p. 170

cerebral cortex (sĕr′ĕ-bral kor′teks), p. 166

cerebral hemisphere (sĕr″ĕ-bral hem′ĭ-sfēr), p. 166

cerebrospinal fluid (sĕr′ĕ-bro-spi′nal flu′id), p. 163

cerebrum (sĕr′ĕ-brum), p. 166

corpus callosum (kôr′pəs kə-lō′səm), p. 166

cranial nerve (kra′ne-al nerv), p. 171

dendrite (den′drīt), p. 157

depolarization (dē-pō′lə-rīz-ā-shən), p. 159

diencephalon (di″en-sef′uh-lon), p. 169

dura mater (dŏŏr′əmă′tər), p. 162

ganglion (gang′gle-on), p. 171

graded potential (grād-ed pə-tĕn′shəl), p. 162

gray matter (grā mat′er), p. 162

herniated disk (hûr′nē-āt-ed dĭsk), p. 163

hippocampus (hĭp′ə-kăm′pəs), p. 170

hypothalamus (hi″po-thal′uh-mus), p. 169

interneuron (in″ter-nu′ron), p. 157

limbic system (lim′bik sis′tem), p. 169

medulla oblongata (mĭ-dŭl′ə ŏb′lông-gä′tə), p. 170

midbrain (mid′brān), p. 170

nerve (nerv), p. 157

neurotransmitter (nu″ro-trans′mit-er), p. 160

norepinephrine (nor″ep-ĭ-nef′rin), p. 162

parasympathetic division (păr″uh-sim″puh-thet′ik dĭ-vizh′un), p. 175

peripheral nervous system (pĕ-rif′er-al ner′vus sis′tem), p. 156

pia mater (pī′ə mā′tər), p. 163

pons (ponz), p. 170

posterior-root ganglion (pos-tēr′e-or-rut gang′gle-on), p. 171

prefrontal area (prē-frŭn′tl âr′ē-ə), p. 168

primary motor area (pri′ma-re mo′tor a′re-uh), p. 166

primary somatosensory area (pri′ma-re so″măto-sen′so-re a′re-uh), p. 167

receptor potential (rĭ-sĕp′tər pə-tĕn′shəl), p. 162

reflex (re′fleks), p. 172

refractory period (rĭ-frăk′tə-rē pîr′ē-əd), p. 160

repolarization (rē-pō′lər-ĭ-zā′shən), p. 159

resting potential (rĕst-ing pə-tĕn′shəl), p. 159

reticular formation (rĭ-tĭk′yə-lər fôr-mā′-shən), p. 171

saltatory conduction (săl′tə-tôr′ē kən-dŭk′shən), p. 159

somatic sensory system (so-mat′ik sĕn′sə-rē sis′tem), p. 171

spinal cord (spi′nal kord), p. 163

spinal nerve (spi′nal nerv), p. 171

stimulus (stĭm′yə-ləs), p. 159

sympathetic division (sim″puh-thet′ik dĭ-vizh′un), p. 174

synapse (sin′aps), p. 160

synaptic cleft (sĭ-nap′tik kleft), p. 160

thalamus (thal′uh-mus), p. 169

tract (trakt), p. 157

ventricle (ven′trī-kl), p. 163

Wernicke's area (ver′nĭ-kez âr′ē-ə), p. 168

white matter (whīt mat′er), p. 162

Clinical Key Terms

Alzheimer disease (altz′hi-mer dĭ-zēz′), p. 160

ankle-jerk reflex (an′kl-jerk re′fleks), p. 174

cerebral palsy (sĕr′ē-bral pal′ze), p. 168

coma (kō′-ma), p. 171

deep-brain stimulation (DES) (dēp brān stĭm-ū-lā′ shŭn), p. 179

electroencephalogram(e-lek″tro-en-sef′uh-lo-gram), p. 165

hydrocephalus (hi″dro-sĕ′fuh-lus), p. 163

knee-jerk reflex (ne′jerk re′fleks), p. 174

pallidotomy (păl-ĭ-dot′ o-mē), p. 179

paraplegia (par-uh-ple′je-uh), p. 166

Parkinson's disease (par′kin-sun dĭ-zēz′), p. 179

quadriplegia (kwah-druh-ple′je-uh), p. 166

SUMMARY

8.1 Nervous System

A. The nervous system permits sensory input, performs integration, and stimulates motor output.

B. The nervous system is divided into the central nervous system (brain and spinal cord) and the peripheral nervous system (afferent and efferent nervous systems). The CNS lies in the midline of the body, and the PNS is located peripherally to the CNS.

C. Nervous tissue contains neurons and neuroglia. Each type of neuron (motor, sensory, and interneuron) has three parts (dendrites, cell body, and axon). Neuroglia support, protect, and nourish the neurons.

D. All axons transmit the same type of nerve signal: a change in polarity (called an action potential) that moves along the membrane of a nerve fiber. Saltatory conduction in myelinated axons is a faster type of conduction.

E. Transmission of a nerve signal across a synapse is dependent on the release of a neurotransmitter into a synaptic cleft.

8.2 Central Nervous System

A. The CNS, consisting of the spinal cord and brain, is protected by the meninges and the cerebrospinal fluid.

B. The spinal cord, located in the vertebral column, is composed of white matter and gray matter. White matter contains bundles of nerve fibers, called tracts, which conduct nerve impulses to and from the higher centers of the brain. Gray matter is mainly made up of short fibers and cell bodies. The spinal cord is a center for reflex action and allows communication between the brain and the peripheral nerves leaving the spinal cord.

C. The brain has four ventricles. The lateral ventricles are found in the left and right cerebral hemispheres. The third ventricle is found in the diencephalon. The fourth ventricle is found in the brain stem.

D. The cerebrum is divided into the left and right hemispheres. The cerebral cortex, a thin layer of gray matter, has four lobes in each hemisphere. The frontal lobe initiates motor output. The parietal lobe is the final receptor

for sensory input from the skin and muscles. The other lobes receive specific sensory input.

Various association areas integrate sensory data. Processing centers integrate data from other association areas: The prefrontal area carries out higher mental processes; Broca's area and Wernicke's area are concerned with speech.

E. The limbic system includes portions of the cerebrum, the thalamus, and the hypothalamus. It is involved in learning and memory and in causing the emotions that guide behavior.

F. The hypothalamus helps control the functioning of most internal organs and controls the secretions of the pituitary gland. The thalamus receives sensory impulses from all parts of the body and channels them to the cerebrum.

G. The cerebellum controls balance and complex muscular movements.

H. The brain stem contains the medulla oblongata, pons, midbrain, and reticular formation. The medulla oblongata contains vital centers for regulating heartbeat, breathing, and blood pressure. The pons assists the medulla oblongata in regulating the breathing rate. The midbrain contains tracts that conduct impulses to and from the higher parts of the brain. The reticular formation maintains alertness and assists other brain stem centers.

8.3 Peripheral Nervous System
A. A nerve contains bundles of long fibers covered by fibrous connective tissue layers.
B. Cranial nerves take impulses to and/or from the brain. Spinal nerves take impulses to and from the spinal cord.
C. Reflexes (automatic reactions to internal and external stimuli) depend on the reflex arc. Some reflexes are important for avoiding injury, and others are necessary for normal physiological functions.
D. The autonomic nervous system controls the functioning of internal organs.
1. The divisions of the autonomic nervous system: (1) function automatically and usually subconsciously in an involuntary manner; (2) innervate all internal organs; and (3) utilize two motor neurons and one ganglion for each impulse.
2. The sympathetic division brings about the responses associated with the "fight or flight" response.
3. The parasympathetic division "rest and digest" brings about the responses associated with normally restful activities.

8.4 Effects of Aging
A. The brain loses nerve cells, and this affects learning, memory, and reasoning. Mental impairment is not an automatic consequence of aging.
B. Alzheimer disease is more often seen among the elderly.

8.5 Homeostasis
A. The nervous system, along with the endocrine system, regulates and coordinates the other systems to maintain homeostasis.
B. Skeletal muscle contraction also plays a role because movement helps us take precautions or stay in a moderate environment.

STUDY QUESTIONS

1. What are the functions of the nervous system? (p. 156)
2. What are the two main divisions of the nervous system? How are these divisions subdivided? (p. 156)
3. What is the general structure of a neuron, and what are the functions of three different types of neurons? (pp. 156–157)
4. What constitutes a nerve impulse (action potential)? Describe the resting potential. Why do myelinated fibers have a faster speed of conduction? (pp. 157–160)
5. How is the nerve impulse transmitted across a synapse? Name two well-known neurotransmitters. (pp. 160–162)
6. Name the meninges, and describe their locations. Where do you find cerebrospinal fluid? (pp. 162–163)
7. Describe the structure and function of the spinal cord. (pp. 163–165)
8. What is the difference between the cerebrum and the cerebral cortex? Name the lobes of the cerebral cortex, and state their function. Describe the primary motor area and the primary somatosensory area. (pp. 165–168)
9. What is the limbic system, and what is its function? (pp. 169–170)
10. Name the other parts of the brain, and give a location and function for each part. (pp. 168–171)
11. Describe the structure of a nerve. In general, discuss the location and function of the cranial nerves and the spinal nerves. (pp. 171–172)
12. Describe a spinal reflex, including the role played by a sensory nerve fiber, interneurons, and a motor fiber. (pp. 172–173)
13. Contrast the actions of the sympathetic and the parasympathetic divisions of the autonomic system. (pp. 174–176)
14. What role does the nervous system play in homeostasis? (pp. 177–179)

LEARNING OBJECTIVE QUESTIONS

Fill in the blanks.
1. A(n) _____ carries nerve impulses away from the cell body.
2. During the depolarization portion of an action potential, _____ ions are moving to the _____ of the nerve fiber.
3. The space between the axon ending of one neuron and the dendrite of another is called the _____.
4. ACh is broken down by the enzyme _____ after it has initiated an action potential on a neighboring neuron.
5. Motor nerves stimulate _____.
6. In a simple reflex pathway, only the _____ is found entirely in the CNS.
7. The _____ is the part of the brain responsible for coordinating body movements.
8. The _____ is the part of the brain responsible for consciousness.
9. The brain and spinal cord are covered by protective layers called _____.

10. The vagus nerve is a _____ nerve that controls _____.

11. Whereas the central nervous system is composed of the _____ and _____ , the peripheral nervous system is composed of the _____ and _____.

12. The limbic system records emotions and also is involved in _____ and _____.

13. While the _____ division of the autonomic nervous system brings about organ responses that are part of the "fight-or-flight" response, the _____ division brings about responses associated with normal restful conditions.

14. The electrical activity of the brain can be recorded in the form of a(n) _____.

15. Label the following diagram.

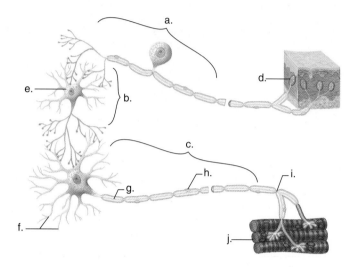

MEDICAL TERMINOLOGY EXERCISE

Consult Appendix B for help in pronouncing and analyzing the meaning of the terms that follow.

1. neuropathogenesis (nu″ro-path″o-jen″ĕ-sis)
2. anesthesia (an″es-the′ze-uh)
3. encephalomyeloneuropathy (en-sef″uh-lo-mi″ĕ-lo-nu-rop′uh-the)
4. hemiplegia (hem″ĭ-ple′je-uh)
5. glioblastoma (gli″o-blas-to′muh)
6. subdural hemorrhage (sub-du′ral hem′or-ij)
7. cephalometer (sef″uh-lom′ĕ-ter)
8. meningoencephalocele (me-ning″go-en-sef″ uh-lo-sēl)
9. neurorrhaphy (nu-rōr′uh-fe)
10. ataxiaphasia (uh-tak″se-uh-fa′ze-uh)
11. cerebrovascular accident (sĕr′-e-bro′-vas-kyū-ler ak′suh-dent)
12. duraplasty (du′ruh-plas-te)
13. brachycephalic (brak′e-sef-al′ik)
14. arachnoiditis (uh-rak″noy-di′tis)

WEBSITE LINK

Visit this text's website at <u>http://www.mhhe.com/longenbaker7</u> for additional quizzes, interactive learning exercises, and other study tools.

The Sensory System

CHAPTER 9

Once again, you hear that really annoying, repetitive song you've heard a million times before, and its lyrics replay over and over in your mind. You're affected by an *earworm*—a phenomenon that is well recognized by the public, but poorly understood by scientists. Researchers propose that earworms are auditory, or sound, memories stored in the temporal lobe, which you'll read about on page 199. Replaying the memories seems to allow the brain to "idle" (much as a car engine idles when it's standing still) while staying prepared for more important tasks. Want to get rid of a particular earworm? Activate your brain's frontal lobe by concentrating on a complex problem, or eat something very spicy to activate the taste areas. Or you can replace one earworm with another, just by thinking of a song more irritating than the first!

Chapter Outline & Learning Objectives

After you have studied this chapter, you should be able to:

When a sensory receptor is stimulated, it generates nerve signals that travel to your brain. Interpretation of these signals is the function of the brain, which has a special region for receiving information from each of the sense organs. Signals arriving at a particular sensory area of the brain can be interpreted in only one way; for example, those arriving at the olfactory area result in smell sensation, and those arriving at the visual area result in sight sensation.

Specialized sensory receptors throughout the body start signal transmission to the brain by using a system called a **receptor potential.** Receptor potentials begin with a stimulus (for example, a light stimulus for receptors in the eye). Unlike an action potential (which is an all-or-nothing event), receptor potentials can be weak or strong. Like the signals that occur when nerves synapse, receptor potentials can add together. Although receptors do not generate action potentials, they are parts of neurons or they synapse with other neurons that do create action potentials. In this way, sensory information is transmitted to the brain. The brain integrates data from various sensory receptors in order to perceive whatever caused the stimulation of olfactory and visual receptors—for example, a flower.

Sensory receptors may be categorized into five types based on their stimuli:

Mechanoreceptors, such as pressure receptors in the skin and proprioceptors (specialized stretch receptors) in muscle, are stimulated by changes in pressure or body movement. Other mechanoreceptors in the inner ear also detect body movements; these will be discussed in section 9.5.

Thermoreceptors in the skin and in the internal organs are stimulated by changes in the external or internal temperature.

Pain receptors, such as those in the skin, are stimulated by damage or oxygen deprivation to the tissues.

Chemoreceptors are stimulated by changes in the chemical concentration of substances. Taste buds in the tongue and olfactory (smell) receptors in the nose are chemoreceptors. Other chemoreceptors can sense the concentration of oxygen, carbon dioxide, and hydrogen ions in the blood.

Photoreceptors, which are located only in the eye, are stimulated by light energy.

9.1 General Senses

Sensory receptors in the muscles, joints, tendons, some internal organs, and skin send action potentials to the spinal cord. From there, the action potentials travel up the spinal cord in tracts to the **thalamus,** the sensory relay center of the brain. The information from these receptors is then relayed to the somatosensory areas of the cerebral cortex, located in the parietal lobe of the brain. These general sensory receptors can be categorized into three types: proprioceptors, cutaneous receptors, and pain receptors.

Proprioceptors

Proprioceptors are mechanoreceptors involved in reflex actions that maintain muscle tone and thereby the body's equilibrium and posture. They help us know the position of our limbs in

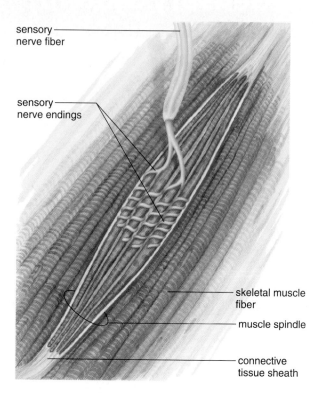

Figure 9.1 Muscle spindle. When a muscle is stretched, a muscle spindle sends sensory nerve signals to the spinal cord. Motor nerve signals from the spinal cord result in muscle fiber contraction so that muscle tone is maintained.

space by detecting the degree of muscle relaxation, the stretch of tendons, and the movement of ligaments. Muscle spindles act to increase the degree of muscle contraction, and Golgi tendon organs act to decrease it. The result is a muscle that has the proper length and tension, or **muscle tone.**

Figure 9.1 illustrates the activity of a muscle spindle. In a muscle spindle, sensory nerve endings are wrapped around thin muscle cells within a connective tissue sheath. When the muscle relaxes and undue stretching of the muscle spindle occurs, nerve action potentials are generated. The rapidity of the action potentials generated by the muscle spindle is proportional to the stretching of a muscle. A reflex action then occurs, which results in contraction of muscle fibers adjoining the muscle spindle. The knee-jerk reflex, which involves muscle spindles, offers an opportunity for physicians to test a reflex action. By contrast, a Golgi tendon organ, which is built into the tough connective tissue of a tendon, is activated by excessive muscle contraction. When Golgi tendon organs generate action potentials, inhibitory signals are sent to the muscle. The muscle is prevented from contracting too forcefully, which could possibly injure both muscle and tendon.

The information sent by muscle spindles and Golgi tendon organs to the CNS is used to maintain the body's equilibrium and posture despite the force of gravity always acting upon the skeleton and muscles.

free nerve endings
(pain, heat, cold)

Merkel disks (touch)

Krause end
bulbs (touch)

root hair
plexus (touch)

epidermis

Meissner
corpuscles (touch)

Pacinian corpuscles
(pressure)

Ruffini endings
(pressure)

dermis

Figure 9.2 **Sensory receptors in human skin.**

Cutaneous Receptors

The skin is composed of two layers: the epidermis and the dermis. In Figure 9.2, the artist has dramatically indicated these two layers by separating the epidermis from the dermis in one location. The epidermis is stratified squamous epithelium in which cells become keratinized as they rise to the surface. Once at the surface, cells are sloughed off. The dermis is a thick connective tissue layer. The deepest layer of epidermis, and the entire dermis layer, contain **cutaneous receptors,** which make the skin sensitive to touch, pressure, pain, and temperature (warmth and cold). The dermis is a mosaic of these tiny sensory receptors, as you can determine by slowly passing a metal probe over your skin. At certain points, you will feel touch or pressure, and at others, you will feel heat or cold (depending on the probe's temperature).

Three types of cutaneous receptors are sensitive to fine touch. *Meissner corpuscles* are concentrated in the fingertips, the palms, the lips, the tongue, the nipples, the penis, and the clitoris. *Merkel disks* are found in the deepest epidermal layer, where the epidermis meets the dermis. A *root hair plexus* winds around the base of a hair follicle and signals if the hair is touched.

The three different types of cutaneous receptors that are sensitive to pressure are Pacinian corpuscles, Ruffini endings, and Krause end bulbs. *Pacinian corpuscles* are onion-shaped sensory receptors that lie deep inside the dermis. *Ruffini endings* and *Krause end bulbs* are encapsulated by sheaths of connective tissue and contain lacy networks of nerve fibers.

Temperature receptors are simply free nerve endings in the superficial dermis and epidermis. Some free nerve endings

are responsive to cold; others are responsive to warmth. Cold receptors are far more numerous than warmth receptors, but the two types have no known structural differences.

As you've discovered, each type of receptor described in this section has a well-defined function. However, research continues to show that receptors may respond to more than one sensation. For example, microscopic examination of the skin of the ear shows only free nerve endings, which function as pain receptors. But as you know from everyday life, the skin of the ear responds to all sensations: temperature, fine touch, pressure, and so on. Therefore, it appears that the skin receptors are somewhat, but not completely, specialized.

Pain Receptors

The skin and many (but not all) internal organs have pain receptors, also called **nociceptors.** *Somatic nociceptors* from the skin and skeletal muscles respond to mechanical, thermal, electrical, or chemical damage to these tissues. Skinning a knee, burning a finger, or straining a muscle all stimulate somatic nociceptors. *Visceral nociceptors* react in response to excessive stretching of the internal organs, oxygen deprivation when blood supply is reduced, or chemicals released by damaged tissues. Visceral nociceptors create the pain sensation when the stomach is too full, as well as the crushing pain of a heart attack when blood supply to the heart is reduced.

Stimulation of internal pain receptors is sometimes felt as pain from the skin as well as the internal organs; this is called **referred pain.** Referred pain occurs because the somatic pain nociceptors travel in the same spinal cord pathway as the

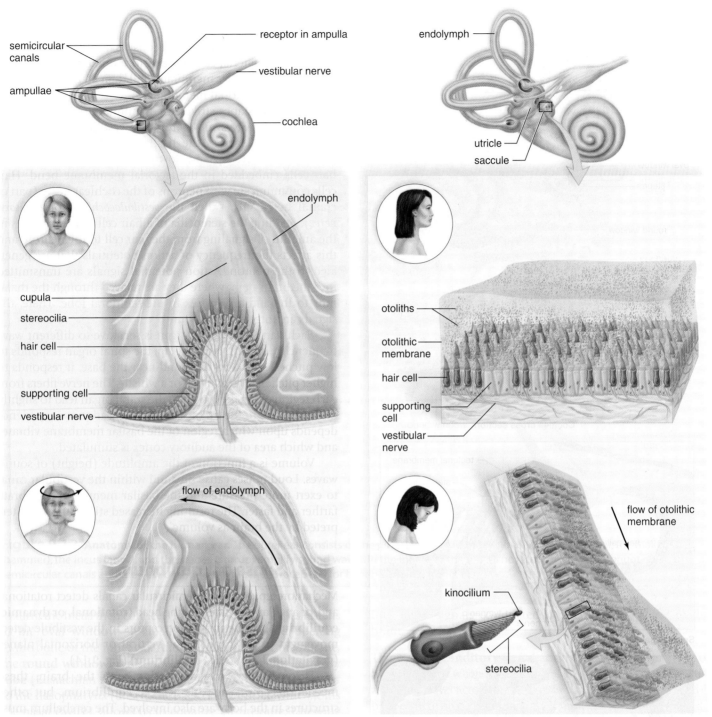

a. Rotational equilibrium: receptors in ampullae of semicircular canal

b. Gravitational equilibrium: receptors in utricle and saccule of vestibule

Figure 9.14 Mechanoreceptors for equilibrium. a. Rotational equilibrium. The ampullae of the semicircular canals contain hair cells with stereocilia embedded in a cupula. When the head rotates, the cupula is displaced, bending the stereocilia. Thereafter, nerve signals travel in the vestibular nerve to the brain. **b.** Gravitational equilibrium. The utricle and the saccule contain hair cells with stereocilia embedded in an otolithic membrane. When the head bends, otoliths are displaced, causing the membrane to sag and the stereocilia to bend. If the stereocilia bend toward the kinocilium, the longest of the stereocilia, nerve signals increase in the vestibular nerve. If the stereocilia bend away from the kinocilium, nerve signals decrease in the vestibular nerve. The difference tells the brain in which direction the head moved.

over and displaces a cupula, the stereocilia of the hair cells bend, and the pattern of signals carried by the vestibular nerve to the brain changes. The cerebellum and other brain centers use information from the hair cells within the ampullae of the semicircular canals to maintain balance. The sensory input from

semicircular canals tells the brain that you are rotating; motor output to various skeletal muscles can right your present position in space as need be (so that you don't fall).

Sometimes data regarding rotational equilibrium bring about unfortunate results. For example, continuous move-

TABLE 9.2 Functions of the Parts of the Ear

Part	Medium	Function	Mechanoreceptor
OUTER EAR	**AIR**		
Pinna		Collects sound waves	—
Auditory canal		Filters air; directs sound waves to tympanic membrane	—
MIDDLE EAR	**AIR**		
Tympanic membrane and ossicles		Amplify sound waves	—
Auditory tube (Eustachian tube)		Equalizes air pressure	—
INNER EAR	**FLUID**		
Cochlea (contains spiral organ)		Hearing	Stereocilia embedded in tectorial membrane
Semicircular canals		Rotational equilibrium	Stereocilia embedded in cupula
Vestibule (contains utricle and saccule)		Gravitational equilibrium	Stereocilia embedded in otolithic membrane

ment of fluid in the semicircular canals causes one form of motion sickness. **Vertigo** is dizziness and a sensation of rotation. It is possible to simulate a feeling of vertigo by spinning rapidly and stopping suddenly. Inflammation of cranial nerve VIII, the vestibulocochlear (or auditory) nerve from the ear can cause vertigo as well. Motion sickness is also possible if the sensory input from the inner ear is different from visual sensation. For example, a person standing *inside* a large ship on rough seas will sense movement with the inner ear (as the ship tosses back and forth on the waves), but the visual input will register no movement (because the ship's walls don't move relative to the person's body). The person is *seasick*, with severe nausea and vomiting. People who are seasick sometimes find relief if they stand outdoors and focus on the horizon, so that both visual and inner ear inputs to the brain signal movement.

Gravitational Equilibrium Pathway

Gravitational equilibrium depends on the **utricle** and **saccule,** two membranous sacs located in the vestibule. Both of these sacs contain little hair cells, whose stereocilia are embedded within a gelatinous material called an otolithic membrane. Calcium carbonate ($CaCO_3$) granules, or **otoliths,** rest on this membrane. The utricle is especially sensitive to horizontal (back-forth) movements and the bending of the head, while the saccule responds best to vertical (up-down) movements.

When the body is still, the otoliths in the utricle and the saccule rest on the otolithic membrane above the hair cells. When the head bends or the body moves in the horizontal and vertical planes, the otoliths are displaced and the otolithic membrane sags, bending the stereocilia of the hair cells beneath. If the stereocilia move toward the largest stereocilium, called the *kinocilium*, nerve signals increase in the

vestibular nerve. If the stereocilia move away from the kinocilium, nerve signals decrease in the vestibular nerve (see Fig. 9.14*b*). If you are upside down, nerve signals in the vestibular nerve cease. These data tell the brain the direction of the movement of the head at the moment. The cerebellum and other brain centers use this sensory information to maintain gravitational equilibrium and balance. Appropriate motor output from the brain to various skeletal muscles can right our present position in space.

Table 9.2 summarizes the functions of the parts of the ear.

Content CHECK-UP!

10. Choose the pathway for a sound wave to travel to the inner ear:
    a. auditory canal; oval window; malleus, stapes, incus; tympanic membrane
    b. auditory canal; tympanic membrane; stapes, incus, malleus; auditory tube
    c. auditory canal; tympanic membrane; malleus, incus, stapes; oval window
    d. auditory canal; tympanic membrane; incus, stapes, malleus; oval window

11. Which structure allows air pressure to equalize in the middle ear?
    a. auditory canal   c. cochlear canal
    b. auditory tube   d. semicircular canal

12. Which of the following receptors are found in the semicircular canals, utricle, and saccule?
    a. mechanoreceptors   c. chemoreceptors
    b. pressure receptors   d. nociceptors

Answers in Appendix C, page 463.

MEDICAL FOCUS

Hearing Damage and Deafness

There are two major types of deafness: conduction deafness and sensorineural, or nerve, deafness. **Conduction deafness** occurs when a mechanical blockage keeps sound waves from reaching the oval window, the membrane at the beginning of the inner ear. Conductive deafness can result from the presence of foreign objects in the external ear canal, impacted ear wax, or cancerous tumors in the external ear canal or middle ear. Conduction deafness can also be due to repeated infections or **otosclerosis.** With otosclerosis, the normal bone of the middle ear is replaced by vascular, spongy bone.

Nerve deafness most often occurs when cilia on the receptors within the cochlea have worn away. Because this may happen with normal aging, older people are more likely to have trouble hearing. However, studies also suggest that age-associated hearing loss can be prevented if ears are protected from loud noises, starting even during infancy. Hospitals are now aware of the problem and are taking steps to ensure that neonatal intensive care units and nurseries are as quiet as possible. Neural deafness may also result from congenital defects, particularly when a pregnant woman contracts German measles (rubella) during the first trimester of pregnancy. For this reason, every female should be immunized against rubella before her childbearing years.

In today's society, exposure to the types of noises listed in Table 9A is common. Everyone should consider three aspects of noise to prevent hearing loss: (1) how loud is the noise, (2) how long is the noise heard, and (3) how close is the noise to the ear. Loudness is measured in decibels, and any level above 80 decibels could damage the hair cells of the organ of Corti. Exposure to intense sounds of short duration, such as a burst of gunfire, can result in an immediate hearing loss. Since the butt of a rifle

offers some protection, hunters may have a significant hearing reduction in the ear opposite the shoulder they use for support while firing their gun. Even listening to city traffic for extended periods can damage hearing, and frequent attendance at rock concerts and constant listening to loud music from a stereo are obviously dangerous. Noisy indoor or outdoor equipment, such as a rug-cleaning machine or a chain saw, is also troublesome. Even motorcycles and recreational vehicles, such as snowmobiles and motocross bikes, can contribute to a gradual hearing loss.

The first hint of a problem could be temporary hearing loss, a "full" feeling in the ears, muffled hearing, or **tinnitus** (ringing in the ears). If you have any of these symptoms, modify your listening habits immediately to prevent further damage. If exposure to noise is unavoidable, use specially designed noise-reduction earmuffs or purchase earplugs made from a compressible, sponge-like material at a drugstore or sporting goods store. These earplugs are not the same as those worn for swimming, and they should not be used interchangeably.

Finally, people need to be aware that some medicines are **ototoxic** (damaging to any of the elements of hearing or balance). Anticancer drugs—most notably, cisplatin—and certain antibiotics (for example, streptomycin, kanamycin, and gentamicin) make the ears especially susceptible to a hearing loss. People taking such medications should protect their ears from any excessive noises.

Cochlear implants, which directly stimulate the auditory nerve, are available for persons with nerve deafness. Earlier models of cochlear implants were costly, and people wearing these electronic devices reported that the speech they heard was like that of a robot. Newer models produce better sound quality at a much lower cost.

TABLE 9A Sound Intensity and Hearing Damage

Type of Noise	Sound Level (decibels)	Effect
Rock concert, shotgun, jet engine	Over 125	Beyond threshold of pain; potential for hearing loss is high.
Nightclub, thunderclap, loud music from "ear bud" headphones	Over 120	Hearing loss is likely.
Chain saw, pneumatic drill, jackhammer, symphony orchestra, snowmobile, garbage truck, cement mixer	100–200	Regular exposure of longer than 1 minute risks permanent hearing loss.
Farm tractor, newspaper press, subway, motorcycle	90–100	Fifteen minutes of unprotected exposure is potentially harmful.
Lawnmower, food blender	85–90	Continuous daily exposure for more than 8 hours can cause hearing damage.
Diesel truck, average city traffic noise	80–85	Annoying; constant exposure may cause hearing damage.

Source: National Institute on Deafness and Other Communication Disorders, National Institutes of Health, January 1990.

9.6 Effects of Aging

As we age, we will likely need assistance to improve our sight and hearing. The lens of the eye in older individuals does not accommodate as well—a condition called **presbyopia**. Therefore, we will probably need eyeglasses, contact lenses, or corrective surgery to improve our vision.

Three serious visual disorders are seen more frequently in older persons: (1) Possibly due to exposure to the sun, the lens is subject to cataracts. (2) Age-related macular degeneration is the most frequent cause of blindness in older people. (3) Glaucoma is more likely to develop because of a tendency for eye pressure to increase in older individuals. These three disorders are described in the Medical Focus on page 195.

The need for a hearing aid also increases with age. Atrophy of the organ of Corti can lead to **presbycusis** (age-related hearing decline). First, people tend to lose the ability to detect high-frequency tones, and later the lower tones are affected. Eventually, they can hear speech but cannot understand the words being said.

Otosclerosis, an overgrowth of bone that causes the stapes to adhere to the oval window, is the most frequent cause of conduction deafness in adults (see the Medical Focus on page 202). The condition actually begins during youth but may not become evident until later in life. Dizziness and the inability to maintain balance may also occur in older people due to changes in the inner ear.

SELECTED NEW TERMS

Basic Key Terms

accommodation, p.190

ampulla (am-pul'uh), p. 199

aqueous humor (a'kwe-us hyū'mer), p. 191

auditory canal (aw'dĭ-to"re kuh-nal'), p. 197

auditory tube (aw'dĭ-to"re tūb), p. 198

blind spot (blīnd spot), p. 196

chemoreceptor (ké'mō-rĭ-sĕp'tar), p. 184

choroid (ko'royd), p. 190

ciliary body (sĭl'ē-ĕr'ē bŏd'ē), p. 190

ciliary muscle (sil'e-ĕr"e mus'l), p. 190

cochlea (kōk'le-uh), p. 198

cochlear canal (kōk'le-er kuh-nal'), p. 198

cone cell (kōn sel), p. 192

conjunctiva (kŏn'jŭngk-tī'və), p. 189

cornea (kor'ne-uh), p. 190

fovea centralis (fō'vē-əsĕn-trā'lĭs), p. 191

incus (ing'kus), p. 197

iris (i'ris), p. 190

lacrimal apparatus (lak'rĭ-mul ap"uh-rā'tus), p. 189

lens (lenz), p. 190

malleus (mal'e-us), p. 197

mechanoreceptor (mĕk'ə-nō-rĭ-sĕp'tər), p. 184

nociceptor (no"sē-sep'tor), p. 185

olfactory cell (ol-fak'to-re sel), p. 187

optic chiasma (ŏp'tĭk kī-ăz'mə), p. 196

optic nerve (op'tik nerv), p. 191

ossicle (os'ĭ-kl), p. 197

otolith (ō'tō-lith), p. 201

oval window (ŏ'vəl wĭn'dō), p. 198

pain receptors (pān rĭ-sĕp'tər), p. 184

photoreceptor (fō'tō-rĭ-sĕp'tər), p. 184

pinna (pin'uh), p. 197

proprioceptor (pro"pre-o-sep'tor), p. 184

pupil (pyū'pl), p. 190

receptor potential (rĭ-sĕp'tər pə-tĕn'shəl), p. 184

retina (rĕt'n-ə), p. 191

retinal (rĕt'n-ôl'), p. 192

rhodospin (rō-dŏp'sĭn), p. 192

rod cell (rod sel), p. 192

round window (round wĭn'dō), p. 198

saccule (sak'yūl), p. 201

sclera (sklĕr'uh), p. 190

semicircular canal (sem"e-ser'kyū-ler kuh-nal'), p. 198

spiral organ (spi'rul or'gun), p. 198

stapes (sta'pēz), p. 197

taste bud (tāst bud), p. 186

thermoreceptor (thûr'mō-rĭ-sĕp'tər), p. 184

tympanic membrane (tim-pan'ik mem'brān), p. 197

utricle (u'trĭ-kl), p. 201

vestibule (vĕs'tə-byōōl), p. 198

vitreous body (vit're-us bôd-ē), p. 191

vitreous humor (vit're-us hyū'mor), p. 191

Clinical Key Terms

Alzheimer cataract (alts'hi-mer kat' uh-rakt), p. 192

astigmatism (ə-stĭg'mə-tĭz'əm), p. 194

cataract (kat'uh-rakt), p. 195

cochlear implant (kōk'le-er im'plant), p. 202

color blindness (kŭl'ər-blīnd nəss), p. 192

conduction deafness (kon-duk'shun def'nes), p. 202

diabetic retinopathy (dī-ā-bet'ik ret-ĭ-nŏp'ă-thē), p. 195

glaucoma (glaw-ko'muh), p. 195

hyperopia (hi"per-o'pe-uh), p. 194

macular degeneration (mă'kyū-ler de"jen-er-a'shun), p. 195

myopia (mi-o'pe-uh), p. 194

nerve deafness (nerv def'nes), p. 202

otosclerosis (ō"tō-sklē-ro'sis), p. 202

ototoxic (ō"tō-tok'sik), p. 202

presbycusis (prez"be-ku'sis), p. 203

presbyopia (prez"be-oh'-pē-uh), p. 203

referred pain (rĭ-fŭr-əd pān), p. 185

sty (sti), p. 188

tinnitus (tĭn'ĭ-təs), p. 202

uncinate fit (uhn' sin-āte fĭt), p. 188

vertigo (vûr'tĭ-gō), p. 201

SUMMARY

9.1 General Senses
Each type of sensory receptor detects a particular kind of stimulus. When stimulation occurs, sensory receptors initiate graded potentials. Action potentials are generated and transmitted to the spinal cord and/or brain. Sensation occurs when nerve signals reach the cerebral cortex. Perception is an interpretation of the meaning of sensations.

9.2 Senses of Taste and Smell

A. Taste and smell are due to chemoreceptors that are stimulated by molecules in the environment. The taste buds contain taste cells that communicate with sensory fibers, while the chemoreceptors for smell are olfactory neurons.

B. After molecules bind to plasma membrane receptor proteins on the microvilli of taste cells and the cilia of olfactory cells, nerve action potentials eventually reach the cerebral cortex, which determines the taste and odor according to the pattern of stimulation.

9.3 Sense of Vision

A. Vision is dependent on the eye, the optic nerves, and the visual areas of the cerebral cortex. The eye has three layers. The outer layer, the sclera, can be seen as the white of the eye; it also becomes the transparent bulge in the front of the eye called the cornea. The middle pigmented layer, called the choroid, absorbs stray light rays. The rod cells (sensory receptors for dim light) and the cone cells (sensory receptors for bright light and color) are located in the retina, the inner layer of the eyeball. The cornea, the humors, and especially the lens bring the light rays to focus on the retina. To see a close object, accommodation occurs as the lens rounds up.

B. When light strikes rhodopsin within the membranous disks of rod cells, rhodopsin splits into opsin and retinal. A cascade of reactions leads to the closing of ion channels in a rod cell's plasma membrane. Neurotransmitter molecules are no longer released. Nerve cells, which synapse with photoreceptors, transmit action potentials, which are carried in the optic nerve to the brain.

C. Integration occurs in the retina, which is composed of three layers of cells: the rod and cone layer, the bipolar cell layer, and the ganglion cell layer. Integration also occurs in the brain. The visual field is taken apart by the optic chiasma and by the primary visual area in the cerebral cortex, which parcels out signals for color, form, and motion to the visual association area. Then the cortex rebuilds the field.

9.4 Sense of Hearing

A. Hearing is dependent on the ear, the cochlear nerve, and the auditory areas of the cerebral cortex.

B. The ear is divided into three parts: outer, middle, and inner. The outer ear consists of the pinna and the auditory canal, which direct sound waves to the middle ear. The middle ear begins with the tympanic membrane and contains the ossicles (malleus, incus, and stapes). The malleus is attached to the tympanic membrane, and the stapes is attached to the oval window, which is covered by a membrane. The inner ear contains the cochlea and the semicircular canals, plus the utricle and the saccule.

C. Hearing begins when the outer ear receives and the middle ear amplifies the sound waves that then strike the oval window membrane. Its vibrations set up pressure waves across the cochlear canal, which contains the spiral organ, consisting of hair cells whose stereocilia are embedded within the tectorial membrane. When the basilar membrane vibrates, the stereocilia of the hair cells bend. Nerve signals begin in the cochlear nerve and are carried to the brain.

9.5 Sense of Equilibrium

The ear also contains mechano-receptors for our sense of equilibrium. Rotational equilibrium is dependent on the stimulation of hair cells within the ampullae of the semicircular canals. Gravitational equilibrium relies on the stimulation of hair cells within the utricle and the saccule.

9.6 Effects of Aging

As we age, we will likely need assistance to improve our failing senses of sight and hearing. Three more serious visual disorders—cataracts, age-related macular degeneration, and glaucoma—may occur, making medical intervention necessary.

STUDY QUESTIONS

1. Name the three different types of general sensory receptors. (p. 184)
2. Name the receptors for the senses of taste and smell. Which brain areas receive their sensory information? (pp. 184, 186)
3. Describe the relationship between taste and smell. (p. 188)
4. Explain the accommodation process. Describe the shape of the lens when a person is viewing a faraway object; then when he or she is viewing a near object. (p. 191)
5. Describe sight in dim light. What chemical reaction is responsible for vision in dim light? (p. 192)
6. Compare and contrast rods and cones. Explain what role each plays in vision; what chemical each contains; where each is located on the retina; and how each reacts to light. (pp. 192–193, 196)
7. Place the structures of the visual pathway (optic radiations, optic nerve, optic chiasma, optic tracts, thalamus, occipital lobes) in the correct order as an action potential signal travels from the eye to the brain. How does the retina integrate visual information? How does the brain process visual information? (pp. 196–197)

8. Place the structures of the outer, middle, and inner ears in the correct order as a sound wave travels through the ear. How does the spiral organ (organ of Corti) in the cochlea translate a sound wave into an action potential? Which cranial nerve carries the action potential into the brain? (pp. 197–199, 201)

9. Compare and contrast the utricle and saccule and the semicircular canals. How are they alike? How are they different? (pp. 199–201)

10. Define conduction deafness and sensorineural deafness. Why do young people frequently suffer hearing loss? (p. 202)

LEARNING OBJECTIVE QUESTIONS

Fill in the blanks.

1. The sensory organs for position and movement are called _____.

2. Taste buds and olfactory cells are termed _____ because they are sensitive to chemicals in the air and food.

3. The sensory receptors for sight, the _____ and _____, are located in the _____, the inner layer of the eye.

4. The cones give us _____ vision and work best in _____ light.

5. The lens _____ for viewing close objects.

6. People who are nearsighted cannot see objects that are _____. A _____ lens will restore this ability.

7. The ossicles are the _____, _____, and _____.

8. The semicircular canals are involved in the sense of _____.

9. The spiral organ is located in the _____ canal of the _____.

10. Vision, hearing, taste, and smell do not occur unless nerve signals reach the proper portion of the _____.

MEDICAL TERMINOLOGY EXERCISE

Consult Appendix B for help in pronouncing and analyzing the meaning of the terms that follow.

1. ophthalmologist (of″thal-mol′o-jist)
2. presbyopia (pres″be-o′pe-uh)
3. blepharoptosis (blef″uh-ro-to′sis)
4. keratoplasty (ker′uh-to-plas″te)
5. optometrist (op-tom′ĕ-trist)
6. lacrimator (lak′rĭ-ma″tor)
7. otitis media (o-ti′tis me′de-uh)
8. tympanocentesis (tim″puh-no-sen-te′sis)
9. microtia (mi″kro′she-uh)
10. myringotome (mi-ring′go-tōm)
11. iridomalacia (ir′ĭ-do-muh′la′she-uh)
12. hypogeusia (hi-po′go′se-uh)

WEBSITE LINK

Visit this textbook's website at http://www.mhhe.com/longenbaker7 for additional quizzes, interactive learning exercises, and other study tools.

The Endocrine System

John F. Kennedy, the youngest man to be elected president, appeared healthy, vigorous, and active throughout his entire political career. Photos of the president showed a handsome, tanned sailor at the family estate in Massachusetts; others showed the Kennedy family playing football. However, throughout Kennedy's entire political life, he and his staff took great pains to hide his many ailments—in particular, the fact that Kennedy suffered from Addison's disease. This rare illness is caused by a deficiency of adrenal cortex hormones, and you can read about it on page 217. (Ironically, one symptom of Addison's disease is skin that looks like JFK's signature suntan.) Although he was hospitalized many times, Kennedy repeatedly denied having the disease or any other serious health problems. Yet, medical records tell a different story—the president was in almost constant pain, in part because his vertebrae had been destroyed by the medication for Addison's disease. At autopsy, his adrenal glands were shrunken and nonfunctional. Still, there is no evidence that Kennedy's illness affected his performance as president, and he left a legacy that includes the space program and the Peace Corps.

Chapter Outline & Learning Objectives   *After you have studied this chapter, you should be able to:*

10.1 Endocrine Glands

The endocrine system consists of glands and tissues that secrete hormones. This chapter will give many examples of the close association between the endocrine and nervous systems. Like the nervous system, the endocrine system is intimately involved in homeostasis.

Hormones are chemical signals that affect the behavior of other glands or tissues. Hormones influence the metabolism of cells, the growth and development of body parts, and homeostasis. **Endocrine glands** are ductless; they secrete their hormones directly into tissue fluid. From there, the hormones diffuse into the bloodstream for distribution throughout the body. Endocrine glands can be contrasted with exocrine glands, which have ducts and secrete their products into these ducts. For example, the salivary glands send saliva into the mouth by way of the salivary ducts.

Each type of hormone has a unique composition. Even so, hormones can be categorized as either a **peptide hormone** (which include proteins, glycoproteins, and modified amino acids) or a **steroid hormone.** All steroid hormones are lipids that have the same four-carbon ring complex, but each has different side chains. The majority of hormones are peptides.

Figure 10.1 depicts the locations of the major endocrine glands in the body and Table 10.1 lists the hormones they release. It's important for you to remember that other organs produce hormones, too. These additional hormones and their actions will be described in later chapters. Further, hormones aren't the only type of chemical signal. Neurotransmitters (which you learned about in Chapter 9) are one example of signaling molecules that allow direct communication from cell to cell, and you'll learn about others here as well.

How Hormones Function

Along with fundamental differences in structure, peptide and steroid hormones also function differently. Most peptide hormones bind to a receptor protein in the plasma membrane and activate a "second messenger" system (Fig. 10.2). The "second messenger" causes the cellular changes for which the hormone is credited. As an analogy, suppose you are the person in charge of a crew assigned to redecorate a room. As such, you stand outside the room and direct the workers inside the room. The workers clean, paint, apply wallpaper, etc. Like the "boss" in this analogy, the peptide hormone stays outside the cell and directs activities within. The peptide hormone, or "first messenger," activates a "second messenger"—the crew workers inside the cell. Common second messengers found in many body cells include **cyclic AMP** (made from ATP, and abbreviated cAMP) and calcium. The second messenger sets in motion an enzyme cascade, so-called because each enzyme in turn activates several others, and so on. These intracellular enzymes cause the changes in the cell that are associated with the hormone. Because of the second messenger system, the binding of a single peptide hormone can result in as much as a thousandfold response.

The cellular response can be a change in cellular behavior or the formation of an end product that leaves the cell. For example, by activating a second messenger, insulin causes the facilitated diffusion of glucose into body cells, while thyroid-stimulating hormone causes thyroxine release from the thyroid gland.

Because steroid hormones are lipids, they diffuse across the plasma membrane and other cellular membranes (Fig. 10.3). Inside the cell, steroid hormones such as estrogen and progesterone bind to receptor proteins. The hormone-receptor complex then binds to DNA, activating particular genes. Activation leads to production of a cellular enzyme in varying quantities. Again, it is largely intracellular enzymes that cause the cellular changes for which the hormone receives credit. For example, estrogen directs cellular enzymes that cause the growth of axillary and pubic hair in an adolescent female.

When protein hormones such as insulin are used for medical purposes, they must be administered by injection. If these hormones were taken orally, they would be acted on by digestive enzymes. Once digested, insulin cannot carry out its functions. Steroid hormones, such as those in birth control pills, can be taken orally because they're water-insoluble lipids and poorly digested. Steroids can pass through the digestive tract largely undigested, and then diffuse through the plasma membrane into the cell.

Hormone Control

The release of hormones is usually controlled by one or more of three mechanisms: (1) the concentration of dissolved molecules or ions in the blood, referred to as *humoral* control; (2) by the actions of other hormones; and/or (3) by the nervous system.

It's important for ions or molecules in the body to be kept within a normal range to maintain homeostasis. Thus, humoral control determines the secretion of many body hormones. For example, when the blood glucose level rises following a meal, the pancreas secretes insulin. Insulin causes the liver to store glucose and the cells to take up glucose, and blood glucose is lowered back to normal. Once the blood glucose concentration is corrected, the pancreas stops producing insulin. In much the same way, a low level of calcium ions in the blood stimulates secretion of parathyroid hormone (PTH) from the parathyroid gland. When blood calcium rises to a normal level, secretion of PTH stops. These examples of humoral control illustrate regulation by negative feedback.

Hormone release can also be controlled by specific stimulating or inhibiting hormones. Thyroid-stimulating hormone (TSH) from the pituitary gland (also called **thyrotropin**) does exactly what its name implies—it stimulates the thyroid gland to produce thyroid hormone. By contrast, the release of insulin is inhibited by the production of glucagon by the pancreas. Insulin lowers the blood glucose level, whereas glucagon raises it. In subsequent sections of this chapter, you'll learn about other instances in which pairs of hormones work opposite to one another, and thereby bring about the regulation of a substance in the blood.

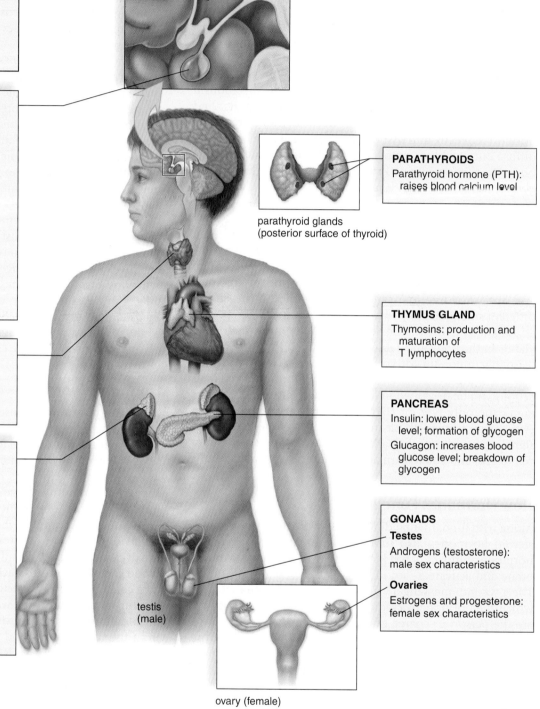

HYPOTHALAMUS

Releasing and inhibiting hormones:
  regulate the anterior pituitary
Antidiuretic (ADH):
  water reabsorption by kidneys
Oxytocin: stimulates uterine
  contraction and milk letdown

PITUITARY GLAND

Posterior Pituitary

Release ADH and oxytocin produced
  by the hypothalamus

Anterior Pituitary

Thyroid stimulating (TSH):
  stimulates thyroid
Adrenocorticotropic (ACTH):
  stimulates adrenal cortex
Gonadotropic (FSH, LH): egg and
  sperm production; sex hormone
  production
Prolactin (PL): milk production
Growth (GH): bone growth, protein
  synthesis, and cell division

THYROID GLAND

Thyroxine (T_4) and triiodothyronine
  (T_3): increase metabolic rate;
  regulates growth and development
Calcitonin: lowers blood calcium level

ADRENAL GLAND

Adrenal cortex

Glucocorticoids (cortisol):
  raises blood glucose level;
  stimulates breakdown of protein
Mineralocorticoids (aldosterone):
  reabsorption of sodium and
  excretion of potassium
Sex hormones: reproductive organs
  and bring about sex characteristics

Adrenal medulla

Epinephrine and norepinephrine:
  active in emergency situations;
  raise blood glucose level

PARATHYROIDS

Parathyroid hormone (PTH):
  raises blood calcium level

parathyroid glands
(posterior surface of thyroid)

THYMUS GLAND

Thymosins: production and
  maturation of
  T lymphocytes

PANCREAS

Insulin: lowers blood glucose
  level; formation of glycogen
Glucagon: increases blood
  glucose level; breakdown of
  glycogen

GONADS

Testes

Androgens (testosterone):
  male sex characteristics

Ovaries

Estrogens and progesterone:
  female sex characteristics

testis
(male)

ovary (female)

Figure 10.1 Anatomical location of major endocrine glands in the body. The hypothalamus and pituitary gland are in the brain, the thyroid and parathyroids are in the neck, and the adrenal glands and pancreas are in the pelvic cavity. The gonads include the ovaries in females, located in the pelvic cavity, and the testes in males, located outside this cavity in the scrotum. Not shown is the pineal gland, located in the brain. The thymus gland lies within the thoracic cavity.

TABLE 10.1 Principal Endocrine Glands and Hormones

Endocrine Gland	Hormone Released	Chemical Class	Target Tissues/Organs	Chief Function(s) of Hormone
Hypothalamus	Hypothalamic-releasing and inhibiting hormones	Peptide	Anterior pituitary	Regulate anterior pituitary hormones
Produced by hypothalamus, released from posterior pituitary	Antidiuretic (ADH)	Peptide	Kidneys	Stimulates water reabsorption by kidneys
	Oxytocin	Peptide	Uterus, mammary glands	Stimulates uterine muscle contraction, release of milk by mammary glands
Anterior pituitary	Thyroid-stimulating (TSH)	Glycoprotein	Thyroid	Stimulates thyroid
	Adrenocorticotropic (ACTH)	Peptide	Adrenal cortex	Stimulates adrenal cortex
	Gonadotropic Follicle-stimulating (FSH) Luteinizing (LH)	Glycoprotein	Gonads	Egg and sperm production; sex hormone production
	Prolactin (PRL)	Protein	Mammary glands	Milk production
	Growth (GH)	Protein	Soft tissues, bones	Cell division, protein synthesis, and bone growth
	Melanocyte-stimulating (MSH)	Peptide	Melanocytes in skin	Unknown function in humans; regulates skin color in lower vertebrates
Thyroid	Thyroxine (T_4) and triiodothyronine (T_3)	Iodinated amino acid	All tissues	Increases metabolic rate; regulates growth and development
	Calcitonin	Peptide	Bones, kidneys, intestine	Lowers blood calcium level
Parathyroids	Parathyroid (PTH)	Peptide	Bones, kidneys, intestine	Raises blood calcium level
Adrenal gland				
Adrenal cortex	Glucocorticoids (cortisol)	Steroid	All tissues	Raise blood glucose level; stimulate breakdown of protein
	Mineralocorticoids (aldosterone)	Steroid	Kidneys	Reabsorb sodium and excrete potassium
	Sex hormones	Steroid	Gonads, skin, muscles, bones	Stimulate reproductive organs and bring about sex characteristics
Adrenal medulla	Epinephrine and norepinephrine	Modified amino acid	Cardiac and other muscles	Released in emergency situations; raise blood glucose level
Pancreas	Insulin	Protein	Liver, muscles, adipose tissue	Lowers blood glucose level; promotes formation of glycogen
	Glucagon	Protein	Liver, muscles, adipose tissue	Raises blood glucose level
Gonads				
Testes	Androgens (testosterone)	Steroid	Gonads, skin, muscles, bones	Stimulate male sex characteristics
Ovaries	Estrogens and progesterone	Steroid	Gonads, skin, muscles, bones	Stimulate female sex characteristics
Thymus	Thymosins	Peptide	T lymphocytes	Stimulate production and maturation of T lymphocytes
Pineal gland	Melatonin	Modified amino acid	Brain	Controls circadian and circannual rhythms; possibly involved in maturation of sexual organs

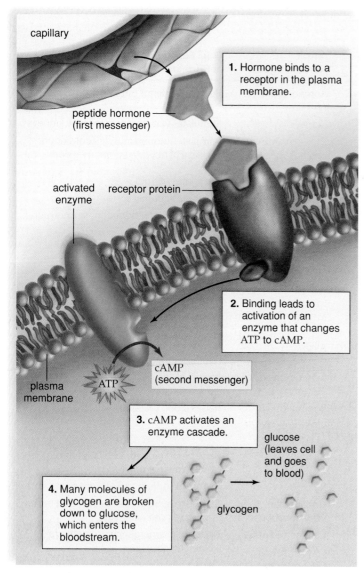

Figure 10.2 **Action of a peptide hormone.** The binding of a peptide hormone leads to cAMP and then to activation of an enzyme cascade. In this example, the hormone causes glycogen to be broken down to glucose.

Labels in figure (left):
- capillary
- 1. Hormone binds to a receptor in the plasma membrane.
- peptide hormone (first messenger)
- activated enzyme
- receptor protein
- 2. Binding leads to activation of an enzyme that changes ATP to cAMP.
- ATP
- cAMP (second messenger)
- plasma membrane
- 3. cAMP activates an enzyme cascade.
- glucose (leaves cell and goes to blood)
- 4. Many molecules of glycogen are broken down to glucose, which enters the bloodstream.
- glycogen

Figure 10.3 **Action of a steroid hormone.** A steroid hormone results in a hormone-receptor complex that activates DNA and protein synthesis.

Labels in figure (right):
- 1. Hormone diffuses through plasma membrane because it is lipid soluble.
- steroid hormone
- plasma membrane
- cytoplasm
- nucleus
- 2. Hormone binds to receptor inside nucleus.
- DNA
- receptor protein
- mRNA
- protein
- ribosome
- mRNA
- 3. Hormone-receptor complex activates gene and synthesis of a specific mRNA molecule follows.
- 4. mRNA moves to ribosomes, and protein synthesis occurs.

The nervous system is an important controller of the endocrine system. Upon receiving sensory information from the body, the brain can make appropriate adjustments to hormone secretion to ensure homeostasis. For example, while you eat a meal, sensory information is relayed to the brain. In turn, the brain signals parasympathetic motor neurons to release insulin from the pancreas. (Recall that the parasympathetic neurons control "rest and digest" functions.) Insulin will allow body cells to take up glucose from digested food.

It's important to stress that many hormones are influenced by more than one control mechanism. In the previous examples, you can see that insulin release is influenced by all three controllers: humoral, hormonal, and neural control. For the majority of hormones, control is regulated by negative feedback. As you know from Chapter 1, in a negative feedback system, a stimulus causes a body response. The body response, in turn, corrects the initial stimulus. The result is that the activity of the hormone is maintained within normal limits and homeostasis is ensured. However, in a few instances, positive feedback controls the release of a hormone—for example, release of oxytocin during labor and delivery (discussed on page 211).

1. Antidiuretic hormone (ADH), a peptide hormone, works by:
   a. binding to a receptor outside the cell and activating a second messenger.
   b. diffusing into the cell, binding to a receptor inside the cell, and activating a second messenger.
   c. diffusing into the cell, binding to a receptor inside the cell, and activating genes in DNA.

2. Testosterone, a steroid hormone, works by:
   a. binding to a receptor outside the cell and activating a second messenger.
   b. diffusing into the cell, binding to a receptor inside the cell, and activating a second messenger.
   c. diffusing into the cell, binding to a receptor inside the cell, and activating genes in DNA.

3. Antidiuretic hormone stimulates the kidneys to reabsorb water and return it to the blood plasma. ADH release is controlled by a negative feedback system. Which action causes ADH to be released?
   a. drinking a big bottle of water
   b. finishing a marathon race and becoming dehydrated

Answers in Appendix C, page 463.

10.2 Hypothalamus and Pituitary Gland

The **hypothalamus** regulates the internal environment. For example, through the autonomic system, the hypothalamus helps control heartbeat, body temperature, and water balance (by creating thirst). The hypothalamus also controls the glandular secretions of the **pituitary gland (hypophysis).** The pituitary, a small gland about 1 cm in diameter, is connected to the hypothalamus by a stalklike structure. The pituitary has two portions: the **posterior pituitary (neurohypophysis)** and the **anterior pituitary (adrenohypophysis).**

Posterior Pituitary

Neurons in the hypothalamus called neurosecretory cells produce the hormones antidiuretic hormone (ADH) and oxytocin (Fig. 10.4, *left*). These hormones pass through axons into the posterior pituitary (neurohypophysis) where they are stored in axon endings. Thus, the hypothalamic hormones antidiuretic hormone and oxytocin are produced in the hypothalamus, but are released into the bloodstream from the posterior pituitary.

Antidiuretic Hormone and Oxytocin

Certain neurons in the hypothalamus are sensitive to the water-salt balance of the blood. When these cells determine that the blood is too concentrated, **antidiuretic hormone (ADH,** also called *vasopressin*) is released from the posterior pituitary. In your body, blood becomes concentrated if you have just finished exercising heavily and body water has been lost as sweat. Upon reaching the kidneys, ADH causes more water to be reabsorbed into kidney capillaries. As the blood becomes dilute once again, ADH

is no longer released. This is an example of control by negative feedback because the effect of the hormone (to dilute blood) acts to shut down the release of the hormone. An additional effect of ADH is to raise blood pressure, by vasoconstriction of blood vessels throughout the body (hence, the hormone's additional name of vasopressin). This mechanism also illustrates negative feedback: Blood pressure falls because body water is lost as sweat (stimulus); vasopressin is released (response); blood vessels constrict and blood pressure rises to normal (stimulus corrected). Negative feedback maintains stable conditions and homeostasis.

Inability to produce ADH causes **diabetes insipidus** (watery urine), in which a person produces copious amounts of urine with a resultant loss of ions from the blood. The condition can be corrected by the administration of ADH.

It's interesting to note that alcohol suppresses ADH production and release. When ADH is absent, the kidneys don't reabsorb as much water. The person drinking alcohol urinates more often and may become dehydrated as a result. The symptoms of the drinker's "hangover"—headache, nausea, dizziness—are largely due to dehydration.

Oxytocin, the other hormone made in the hypothalamus, causes uterine contraction during childbirth and milk letdown when a baby is nursing. The more the uterus contracts during labor, the more nerve impulses reach the hypothalamus, causing oxytocin to be released. Similarly, the more a baby suckles, the more oxytocin is released. In both instances, the release of oxytocin from the posterior pituitary is controlled by **positive feedback**—that is, the stimulus continues to bring about an effect that ever increases in intensity. Positive feedback is not the best way to maintain stable conditions and homeostasis. However, it works during childbirth and nursing because external mechanisms interrupt the process. In childbirth, the delivery of the baby and afterbirth (the placenta and membranes surrounding the baby) eventually stops oxytocin secretion. When a baby with a full tummy stops nursing, that, too, halts oxytocin secretion. For a nursing mother, the letdown response to oxytocin becomes automatic over time. Thus, it is referred to as a *neuroendocrine reflex.* When the baby begins to suckle, the pressure and touch sensations signal the hypothalamus, oxytocin is released, and the stream of breast milk begins. Similarly, hearing a baby cry—even someone else's baby—will cause a nursing mother's breast milk to flow.

Anterior Pituitary

A portal system lies between the hypothalamus and the anterior pituitary (Fig. 10.4, *center*). In this example, the term **portal system** is used to describe the following unique pattern of circulation:

capillaries → vein → capillaries → vein

The hypothalamus controls the anterior pituitary by producing **hypothalamic-releasing hormones** and **hypothalamic-inhibiting hormones.** For example, there is a thyrotropin-releasing hormone (TRH) and a prolactin-inhibiting hormone (PIH). TRH stimulates the anterior pituitary to secrete thyroid-stimulating hormone, and PIH inhibits the pituitary from secreting prolactin.

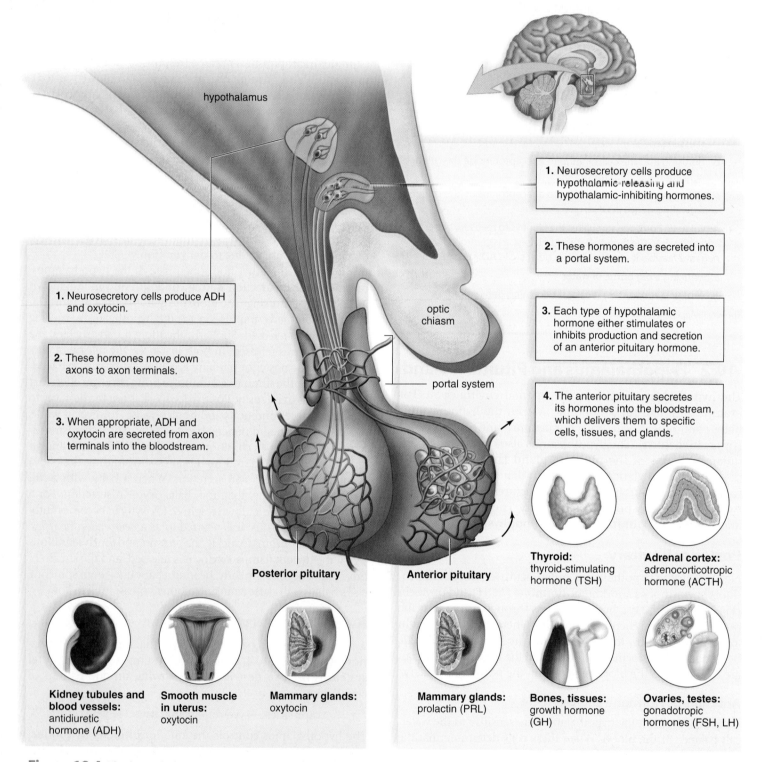

hypothalamus

1. Neurosecretory cells produce hypothalamic-releasing and hypothalamic-inhibiting hormones.

2. These hormones are secreted into a portal system.

optic chiasm

1. Neurosecretory cells produce ADH and oxytocin.

3. Each type of hypothalamic hormone either stimulates or inhibits production and secretion of an anterior pituitary hormone.

2. These hormones move down axons to axon terminals.

portal system

3. When appropriate, ADH and oxytocin are secreted from axon terminals into the bloodstream.

4. The anterior pituitary secretes its hormones into the bloodstream, which delivers them to specific cells, tissues, and glands.

Thyroid: thyroid-stimulating hormone (TSH)

Adrenal cortex: adrenocorticotropic hormone (ACTH)

Posterior pituitary

Anterior pituitary

Kidney tubules and blood vessels: antidiuretic hormone (ADH)

Smooth muscle in uterus: oxytocin

Mammary glands: oxytocin

Mammary glands: prolactin (PRL)

Bones, tissues: growth hormone (GH)

Ovaries, testes: gonadotropic hormones (FSH, LH)

Figure 10.4 The hypothalamus and the pituitary. *Left:* The hypothalamus produces two hormones, ADH and oxytocin, which are stored and secreted by the posterior pituitary. *Right:* The hypothalamus controls the secretions of the anterior pituitary, and the anterior pituitary controls the secretions of the thyroid, adrenal cortex, and gonads, which are also endocrine glands. It also secretes growth hormone and prolactin.

Hormones That Affect Other Glands

Four of the hormones produced by the anterior pituitary have an effect on other glands: **Thyroid-stimulating hormone (TSH)** stimulates the thyroid to produce the thyroid hormones; **adrenocorticotropic hormone (ACTH)** stimulates the adrenal cortex to produce its hormones; and **gonadotropic hormones** (follicle-stimulating hormone—FSH and luteinizing hormone—LH) stimulate the gonads—the testes in males and the ovaries in females—to produce gametes and sex hormones. The hypothalamus, the anterior pituitary, and other glands controlled by the anterior pituitary are all involved in self-regulating negative feedback mechanisms that maintain stable conditions. In each instance, the blood level of the last hormone in the sequence exerts negative feedback control over the secretion of the first two hormones:

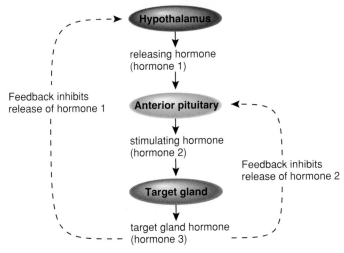

Effects of Other Hormones

Other hormones produced by the anterior pituitary do not affect other endocrine glands. **Prolactin (PRL)** is produced beginning at about the fifth month of pregnancy, and is produced in quantity after childbirth. It causes the mammary glands in the breasts to develop and produce milk. It also plays a role in carbohydrate and fat metabolism.

Growth hormone (GH), or somatotropic hormone, stimulates protein synthesis within cartilage, bone, and muscle. It stimulates the rate at which amino acids enter cells and protein synthesis occurs. It also promotes fat metabolism as opposed to glucose metabolism.

Begin Thinking Clinically

?

An *adenoma*, which is one type of pituitary gland tumor, can affect the production of one or more pituitary hormones. What would be the effect of a prolactin adenoma in a man?

Answer and discussion in Appendix C, page 463.

Effects of Growth Hormone

The amount of GH produced by the anterior pituitary affects the height of the individual. The quantity of GH produced is

Figure 10.5 Effect of growth hormone. The amount of growth hormone produced by the anterior pituitary during childhood affects the height of an individual. Too much growth hormone can lead to giantism, while an insufficient amount results in limited stature and even pituitary dwarfism.

greatest during childhood and adolescence, when most body growth is occurring (Fig. 10.5). If too little GH is produced during childhood, the individual has **pituitary dwarfism,** characterized by perfect proportions but small stature. If too much GH is secreted, a person can become a giant (Fig. 10.5). Giants usually have poor health, primarily because GH has a secondary effect on the blood sugar level, promoting an illness called diabetes mellitus (see page 220).

On occasion, GH is overproduced in the adult and a condition called **acromegaly** results. Because long bone growth is no longer possible in adults, the feet, hands, and face (particularly the chin, nose, and eyebrow ridges) can respond, and these portions of the body become overly large (Fig. 10.6).

Content CHECK-UP!

4. Name three anterior pituitary hormones that cause the release of another hormone(s).

5. Oxytocin release from the hypothalamus during labor and delivery is a mechanism that works by:
   a. positive feedback.   b. negative feedback.

6. Which of the following is an effect of growth hormone?
   a. It promotes fat metabolism.   c. It causes a person to grow taller.
   b. It stimulates protein synthesis in bone and cartilage.   d. all of the above

Answers in Appendix C, page 463.

Age 9   Age 16   Age 33   Age 52

Figure 10.6 Acromegaly. Acromegaly is caused by overproduction of GH in the adult. It is characterized by enlargement of the bones in the face, the fingers, and the toes as a person ages.

10.3 Thyroid and Parathyroid Glands

The **thyroid gland** is a large gland located in the neck, where it is attached to the trachea just below the larynx (see Fig. 10.1). The parathyroid glands are embedded in the posterior surface of the thyroid gland.

Thyroid Gland

The thyroid gland is composed of a large number of follicles, each a small spherical structure made of thyroid cells filled with **triiodothyronine (T_3)**, which contains three iodine atoms, and **thyroxine (T_4)**, which contains four iodine atoms. These are the two forms of thyroid hormone; T_3 is thought to have the greatest effect on the body.

Effects of Thyroid Hormones

To produce triiodothyronine and thyroxine, the thyroid gland actively acquires iodine. The concentration of iodine in the thyroid gland can increase to as much as 25 times that of the blood. If iodine is lacking in the diet, the thyroid gland is unable to produce the thyroid hormones. In response to constant stimulation by the anterior pituitary, the thyroid enlarges, resulting in a **simple,** or **endemic, goiter** (Fig. 10.7). Some years ago, it was discovered that the use of iodized salt allows the thyroid to produce the thyroid hormones, and therefore helps prevent simple goiter.

Thyroid hormones increase the metabolic rate. They do not have a single target organ; instead, they stimulate all cells of the body to metabolize at a faster rate. More glucose is broken down, and more energy is utilized.

If the thyroid fails to develop properly, a condition called **congenital hypothyroidism** results (Fig. 10.8). Individuals

Figure 10.7 Simple goiter. An enlarged thyroid gland is often caused by a lack of iodine in the diet. Without iodine, the thyroid is unable to produce its hormones, and continued anterior pituitary stimulation causes the gland to enlarge.

with this condition are short and stocky and have had extreme hypothyroidism (undersecretion of thyroid hormone) since infancy or childhood. Thyroid hormone therapy can initiate growth, but unless treatment is begun within the first two months of life, severe developmental delay results. The occurrence of hypothyroidism in adults produces the condition known as **myxedema,** which is characterized by lethargy, weight gain, loss of hair, slower pulse rate, lowered body

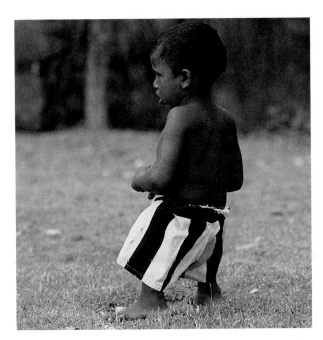

Figure 10.8 Congenital hypothyroidism. Individuals who have hypothyroidism since infancy or childhood do not grow and develop as others do. Unless medical treatment is begun, the body is short and stocky. Developmental delay is also likely.

temperature, and thickness and puffiness of the skin. The administration of adequate doses of thyroid hormones restores normal function and appearance.

In the case of hyperthyroidism (oversecretion of thyroid hormone), as seen in **Graves' disease,** the thyroid gland is overactive, and a goiter forms. This type of goiter is called **exophthalmic goiter.** The eyes protrude because of edema in eye socket tissues and swelling of the muscles that move the eyes. The patient usually becomes hyperactive, nervous, and irritable, and suffers from insomnia. Removal or destruction of a portion of the thyroid by means of radioactive iodine is generally effective in curing the condition. Hyperthyroidism can also be caused by a thyroid tumor, which is usually detected as a lump during physical examination. Again, the treatment is surgery in combination with administration of radioactive iodine. The prognosis for most patients is excellent.

Calcitonin

Calcium (Ca^{2+}) plays a significant role in both nervous conduction and muscle contraction. It is also necessary for coagulation (clotting) of blood. The blood calcium level is regulated in part by **calcitonin,** a hormone secreted by the thyroid gland when the blood calcium level rises (Fig. 10.9). The primary effect of calcitonin is to bring about the deposit of calcium in the bones. It does this by temporarily reducing the activity and number of osteoclasts. Recall from Chapter 6 that these cells break down bone. When the blood calcium lowers to normal, the release of calcitonin by the thyroid is inhibited, but a low calcium level stimulates the release of parathyroid

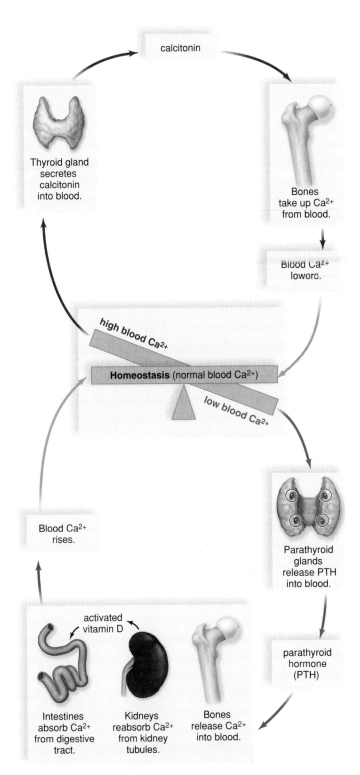

Figure 10.9 Regulation of blood calcium level. *Top:* When the blood calcium (Ca^{2+}) level is high, the thyroid gland secretes calcitonin. Calcitonin promotes the uptake of Ca^{2+} by the bones, and therefore the blood Ca^{2+} level returns to normal. *Bottom:* When the blood Ca^{2+} level is low, the parathyroid glands release parathyroid hormone (PTH). PTH causes the bones to release Ca^{2+}. It also causes the kidneys to reabsorb Ca^{2+} and activate vitamin D; thereafter, the intestines absorb Ca^{2+}. Therefore, the blood Ca^{2+} level returns to normal.

hormone (PTH) by the parathyroid glands. Calcitonin is an important hormone in children, whose skeleton is undergoing rapid growth. It is of minor importance in adults. Parathyroid hormone is the major controller of calcium homeostasis. However, calcitonin can be used therapeutically in adults to reduce the effects of osteoporosis (see the Medical Focus on page 102).

Parathyroid Glands

Parathyroid hormone (PTH), the hormone produced by the **parathyroid glands,** causes the blood phosphate (HPO_4^{2-}) level to decrease and the ionic blood calcium (Ca^{2+}) level to increase. The antagonistic actions of calcitonin, from the thyroid gland, and parathyroid hormone, from the parathyroid glands, maintain the blood calcium level within normal limits.

Note in Figure 10.9 that after a low blood calcium level stimulates the release of PTH, it promotes release of calcium from the bones. (It does this by promoting the activity of osteoclasts.) PTH promotes the reabsorption of calcium by the kidneys, where it also activates vitamin D. Vitamin D, in turn, stimulates the absorption of calcium from the intestine. These effects bring the blood calcium level back to the normal range so that the parathyroid glands no longer secrete PTH.

Many years ago, the four parathyroid glands were sometimes mistakenly removed during thyroid surgery because of their size and location in the thyroid. When insufficient parathyroid hormone production leads to a dramatic drop in the blood calcium level, hypocalcemic tetany results. In **tetany,** the body shakes from continuous muscle contraction. This effect is brought about by increased excitability of the nerves, which initiate nerve impulses spontaneously and without rest. In severe cases, hypocalcemic tetany is fatal.

Content **CHECK-UP!**

7. Which of the following conditions is caused by excessive thyroid hormone?

  a. Graves' disease

  b. cretinism (congenital hypothyroidism)

  c. simple goiter

  d. myxedema

8. Which mineral is necessary to manufacture thyroid hormones?

  a. sodium   c. calcium

  b. iron   d. iodine

9. The target organs for parathyroid hormone are:

  a. kidney, liver, stomach.

  b. kidney, bone, liver.

  c. kidney, bone, small intestine.

  d. bone, liver, small intestine.

Answers in Appendix C, page 463.

10.4 Adrenal Glands

The **adrenal glands** sit atop the kidneys (see Fig. 10.1). Each adrenal gland consists of an inner portion called the **adrenal medulla** and an outer portion called the **adrenal cortex.** These portions, like the anterior pituitary and the posterior pituitary, have no physiological connection with one another. The adrenal medulla is under nervous control, and the adrenal cortex is under the control of ACTH (also called **corticotropin**), an anterior pituitary hormone. Stress of all types, including emotional and physical trauma, prompts the hypothalamus to stimulate the adrenal glands (Fig. 10.10).

Adrenal Medulla

The hypothalamus initiates nerve impulses that travel by way of the brain stem, spinal cord, and sympathetic nerve fibers to the adrenal medulla, which then secretes its hormones.

Epinephrine (adrenaline) and **norepinephrine** (noradrenaline) produced by the adrenal medulla rapidly bring about all the body changes that occur when an individual reacts to an emergency situation. The release of epinephrine and norepinephrine achieves the same results as sympathetic stimulation—the "fight-or-flight" responses: increased heart rate, rapid respiration, dilation of the pupils, etc. Thus, these hormones assist sympathetic nerves in providing a short-term response to stress.

Adrenal Cortex

There are three layers in the adrenal cortex, and each produces a different set of steroid hormones. The hormones produced by the adrenal cortex provide a long-term response to stress (Fig. 10.10). The two major types of hormones produced by the adrenal cortex are the mineralocorticoids and the glucocorticoids. The **mineralocorticoids** regulate salt and water balance, leading to increases in blood volume and blood pressure. The **glucocorticoids** regulate carbohydrate, protein, and fat metabolism, leading to an increase in blood glucose level. Cortisone, the medication often administered for inflammation of joints, is a glucocorticoid.

The adrenal cortex also secretes small amounts of both male and female sex hormones—regardless of one's gender. Both male and female sex hormones promote skeletal growth in adolescents. The male hormones from the adrenal gland stimulate the growth of axillary and pubic hair at puberty. In addition, male hormones help to sustain the sex drive, or *libido,* in both men and women.

Glucocorticoids

Cortisol is a biologically significant glucocorticoid produced by the adrenal cortex. Cortisol raises the blood glucose level in at least two ways: (1) It promotes the breakdown of muscle proteins to amino acids, which are taken up by the liver from the bloodstream. The liver then converts these excess

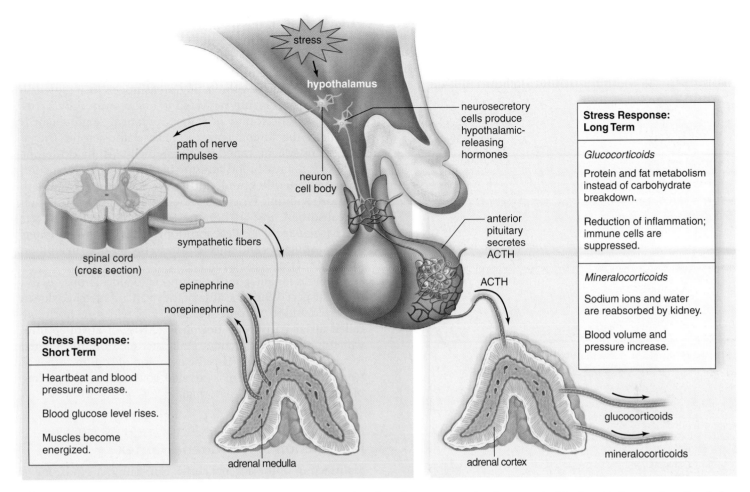

Figure 10.10 Adrenal glands. Both the adrenal medulla and the adrenal cortex are under the control of the hypothalamus when they help us respond to stress. *Left:* The adrenal medulla provides a rapid, but short-term stress response. *Right:* The adrenal cortex provides a slower, but long-term stress response.

amino acids to glucose, which enters the blood. (2) Cortisol promotes the metabolism of fatty acids rather than carbohydrates, and this spares glucose for the brain.

Glucocorticoid Therapy

Cortisol and other forms of glucocorticoids suppress the body's normal reaction to disease—the inflammatory reaction (see Fig. 13.4) and the immune process. Cortisone is the glucocorticoid that is used as a medication. Because it reduces inflammation, cortisone reduces swelling and pain in joint disorders such as tendonitis and osteoarthritis. Clinicians also treat autommimmune disorders, such as rheumatoid arthritis, organ transplant rejection, allergies, and severe asthma by suppressing the immune response with cortisone therapy. However, cortisone should be used for the minimum time possible because long-term administration for therapeutic purposes causes some degree of Cushing's syndrome (see page 218). Further, impaired immunity resulting from cortisone use predisposes the individual to infection and in-

creased cancer risk. In addition, sudden withdrawal from cortisone therapy causes symptoms of diminished secretory activity by the adrenal cortex. This occurs because cortisone medication suppresses the release of adrenocorticotropic hormone (ACTH) by the anterior pituitary, leading to a decrease in natural glucocorticoid production by the adrenal cortex. Therefore, withdrawal of cortisone following long-term use must be tapered. During an alternate-day schedule, the dosage is gradually reduced and then finally discontinued as the patient's adrenal cortex resumes activity.

Mineralocorticoids

Aldosterone is the most important of the mineralocorticoids. Aldosterone primarily targets the kidney where it promotes renal absorption of sodium (Na^+) and water, and renal excretion of potassium (K^+).

As one might expect, secretion of mineralocorticoids from the adrenal cortex is influenced by ACTH (adrenocorticotropic hormone or corticotropin) from the pituitary

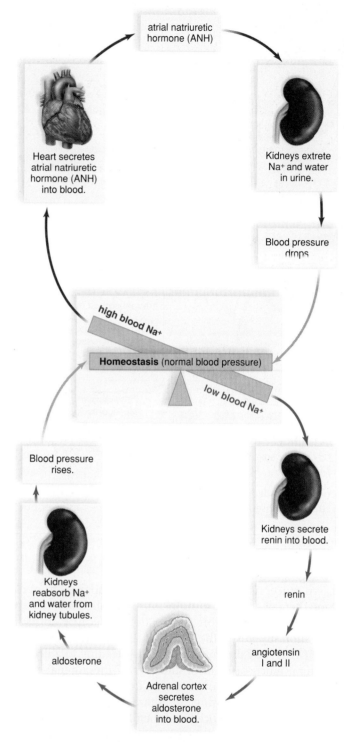

gland. However, the pituitary hormone is not the primary controller for aldosterone secretion. When the blood sodium level and therefore the blood pressure are low, the kidneys secrete **renin** (Fig. 10.11). Renin is an enzyme that converts the plasma protein angiotensinogen to angiotensin I. Angiotensin I is fully activated to angiotensin II by a converting enzyme found in lung capillaries. Angiotensin II stimulates the adrenal cortex to release aldosterone. The effect of this system, called the renin-angiotensin-aldosterone system, is to raise blood pressure in two ways: Angiotensin II constricts arterioles, and aldosterone causes the kidneys to reabsorb sodium. When the blood sodium level rises, water is reabsorbed, in part because the hypothalamus secretes ADH (see page 211). Reabsorption means that water enters kidney capillaries and thus returns to the blood. Then blood pressure increases to normal.

There is an antagonistic hormone to aldosterone, as you might suspect. When the atria of the heart are stretched due to increased blood volume, cardiac cells release a hormone called **atrial natriuretic hormone (ANH, or atriopeptide)**, which inhibits the secretion of aldosterone from the adrenal cortex. The effect of ANH is the excretion of sodium—that is, *natriuresis*. When sodium is excreted, so is water, and therefore blood pressure lowers to normal.

Malfunction of the Adrenal Cortex

Malfunction of the adrenal cortex can lead to a **syndrome**, a set of symptoms that occur together. The syndromes commonly associated with the adrenal cortex are Addison's disease and Cushing's syndrome.

Addison's Disease and Cushing's Syndrome

When the level of adrenal cortex hormones is low due to hyposecretion, a person develops **Addison's disease.** The presence of excessive but ineffective ACTH causes a bronzing of the skin because ACTH can lead to a buildup of melanin (Fig. 10.12). Without cortisol, glucose cannot be replenished when a stressful situation arises. Even a mild infection can lead to death. The lack of aldosterone results in a loss of sodium and water, the development of low blood pressure, and possibly severe dehydration. Left untreated, Addison's disease can be fatal.

When the level of adrenal cortex hormones is high due to hypersecretion, a person develops **Cushing's syndrome** (Fig. 10.13). The excess cortisol results in a tendency toward diabetes mellitus as muscle protein is metabolized and subcutaneous fat is deposited in the midsection. The trunk is obese, while the arms and legs remain a normal size. An excess of aldosterone and reabsorption of sodium and water by the kidneys leads to a basic blood pH and hypertension. The face is moon-shaped due to edema. Masculinization may occur in women because of excess adrenal male sex hormones.

Figure 10.11 Regulation of blood pressure and volume. *Bottom:* When the blood sodium (Na$^+$) level is low, low blood pressure causes the kidneys to secrete renin. Renin leads to the secretion of aldosterone from the adrenal cortex. Aldosterone causes the kidneys to reabsorb Na$^+$, and water follows, so that blood volume and pressure return to normal. *Top:* When the blood Na$^+$ is high, a high blood volume causes the heart to secrete atrial natriuretic hormone (ANH). ANH causes the kidneys to excrete Na$^+$, and water follows. The blood volume and pressure return to normal.

a.

b.

Figure 10.12 Addison's disease. Addison's disease is characterized by a peculiar bronzing of the skin, particularly noticeable in these light-skinned individuals. Note the color of **(a)** the face and **(b)** the hands compared to the hand of an individual without the disease.

a.

b.

Figure 10.13 Cushing's syndrome. Cushing's syndrome results from hypersecretion of adrenal cortex hormones. **a.** Patient first diagnosed with Cushing's syndrome. **b.** Four months later, after therapy.

Content CHECK-UP!

10. From the following list of hormones of the adrenal cortex and their corresponding effects, choose the pair, or pairs, that are correct.

   a. male hormones → stimulate sex drive

   b. aldosterone → increases sodium concentration in the blood

   c. female hormones → promote long bone growth in adolescents

   d. pairs b and c are correct.

   e. all are correct.

11. Which hormone opposes the effect of aldosterone in the body?

   a. renin

   b. angiotensin I

   c. atrial natriuretic hormone

   d. cortisol

12. Aldosterone returns blood pressure to normal by causing the kidneys to reabsorb water and sodium. Because it works by a negative feedback mechanism, which of the following actions could cause aldosterone to be released?

   a. giving a unit of blood

   b. drinking a big bottle of a sports drink

   c. running a marathon and becoming dehydrated

   d. both a and c are correct

Answers in Appendix C, page 463.

10.5 Pancreas

The **pancreas** is a long organ that lies transversely in the abdomen between the kidneys and near the duodenum of the small intestine. It is composed of two types of tissue. Exocrine tissue produces and secretes digestive juices that go by way of ducts to the small intestine. Endocrine tissue, called the **pancreatic islets** (islets of Langerhans), produces and secretes the hormones **insulin** and **glucagon** directly into the blood (Fig. 10.14).

The two antagonistic hormones, insulin and glucagon, both produced by the pancreas, help maintain the normal level of glucose in the blood. Insulin is secreted when the blood glucose level is high, which usually occurs just after eating. Insulin stimulates the uptake of glucose by most body cells. Insulin is not necessary for the transport of glucose into brain or red blood cells, but muscle cells and adipose tissue cells require insulin for glucose transport. In liver and muscle cells, insulin stimulates enzymes that promote the storage of glucose as glycogen. In muscle cells, the glucose supplies energy for muscle contraction, and in fat cells, glucose enters the metabolic pool and thereby supplies glycerol for the formation of fat. In these ways, insulin lowers the blood glucose level.

Glucagon is secreted from the pancreas, usually between meals, when the blood glucose level is low. The major target tissues of glucagon are the liver and adipose tissue. Glucagon stimulates the liver to break down glycogen to glucose and to use fat and protein in preference to glucose as energy sources. Adipose tissue cells break down fat to glycerol and fatty acids. The liver takes these up and uses them as substrates for glucose formation. In these ways, glucagon raises the blood glucose level.

Diabetes Mellitus

Diabetes mellitus is a fairly common hormonal disease in which insulin-sensitive body cells are unable to take up and/or metabolize glucose. Therefore, the blood glucose level is elevated—a condition called **hyperglycemia.** Because body cells cannot access glucose, starvation occurs at the cell level. The person becomes extremely hungry—a condition called **polyphagia.** As the blood glucose level rises, glucose will be lost in the urine (**glycosuria**). Glucose in the urine causes excessive water loss through urination (**polyuria**). The loss of water in this way causes the diabetic to be extremely thirsty (**polydipsia**). Glucose is not being metabolized, so the body turns to the breakdown of protein and fat for energy. Fat metabolism leads to the buildup of ketones in the blood, and excretion of ketones in the urine (**ketonuria**). Because ketones are acidic, their buildup in the blood causes **acidosis** (acid blood), which can lead to coma and death.

We now know that diabetes mellitus exists in two forms. In type I, more often called **insulin-dependent dia-**

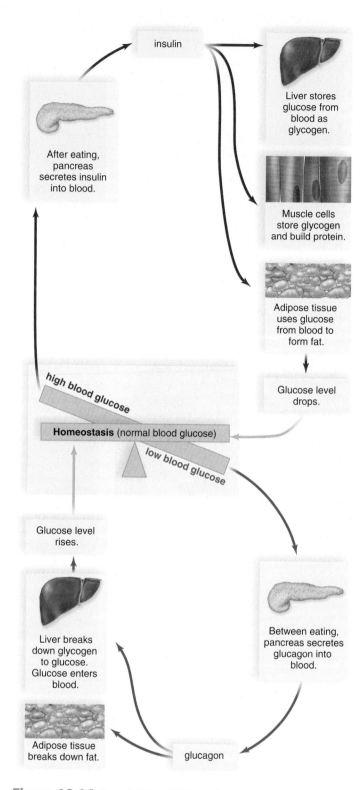

Figure 10.14 Regulation of blood glucose level. *Top:* When the blood glucose level is high, the pancreas secretes insulin. Insulin promotes the storage of glucose as glycogen in the liver and muscles and the use of glucose to form fat in adipose tissue. Therefore, insulin lowers the blood glucose level. *Bottom:* When the blood glucose level is low, the pancreas secretes glucagon. Glucagon acts opposite to insulin; therefore, glucagon raises the blood glucose level to normal.

I.C.E. — IN CASE OF EMERGENCY

Insulin Shock and Diabetic Ketoacidosis

If you're someone with *diabetes mellitus*, the disorder involving the hormone insulin, you already know the importance of maintaining a stable blood glucose level to ensure your long-term health. If your roommate, friend, or loved one is a diabetic, you need to be informed too, because a diabetic's possible problems don't always take a long time to develop. **Insulin shock** and **diabetic ketoacidosis** (DKA) can develop very rapidly, and both can be fatal or result in permanent brain damage. It's important for diabetics and their friends and family to recognize the symptoms of insulin shock and DKA, and know how to use a glucometer to measure blood glucose and how to give an injection.

Insulin shock (also called an *insulin reaction*) results when blood glucose falls to critically low levels—a condition called **hypoglycemia**. It often results when the diabetic patient accidentally injects too much insulin, or takes her insulin but misses a meal. The patient is likely to feel anxious, sweat profusely, and complain of a headache. She may become hyperactive, confused, and even psychotic as the condition worsens. Eventually, she'll lose consciousness and lapse into a so-called **diabetic coma.**

It's critical for first responders to try to raise the patient's blood glucose as quickly as possible. If she is conscious and alert, she can quickly drink milk, juice, or soda, or eat something sweet. She may be able to self-inject with glucagon, the hormone that raises blood glucose. If the blood glucose isn't too low, the insulin shock can be corrected at home. However, one should never attempt to give food or drinks to someone who is semiconscious or unconscious—she could easily choke. Instead, the inside of her cheeks can be smeared with glucose gel, honey, syrup, or frosting, which will melt and be swallowed. If the patient doesn't quickly become alert enough to eat or drink, she must be transported to an emergency room. There, an injection of glucagon or an intravenous solution will quickly raise the blood glucose level.

Diabetic ketoacidosis is caused by blood glucose that is too high. It commonly results when the diabetic eats a meal, but forgets to inject insulin. Infection, injury, or extreme stress can also lead to DKA. The symptoms of DKA are increased thirst, frequent urination, nausea, and vomiting. The patient breathes rapidly, and his breath smells like fruit-flavored gum. His pulse is very fast, but his blood pressure is low. Without an identifying bracelet or tag, he could easily be mistaken as someone who's had too much to drink: sluggish, lethargic, and increasingly sleepy. If untreated, he'll eventually fall into a diabetic coma. Fortunately, under a physician's direction, a trained paramedic can start an intravenous solution to help dilute the patient's blood. As he is being transported to the emergency room, EMS personnel can then measure the glucose and ketones in his blood to provide a complete history, in preparation for more complete treatment in the hospital.

---

betes mellitus (IDDM), the pancreas does not produce insulin. This condition is believed to be brought on, at least in part, by exposure to an environmental agent. This agent—very likely a virus—causes an extreme immune response, and immune cells destroy the pancreatic islets. As a result, the individual must have daily insulin injections. Daily injections control the diabetic symptoms, but diabetics can still experience life-threatening problems, as described in I.C.E. box above.

Of the 16 million people who now have diabetes in the United States, most have type II, more often called **noninsulin-dependent diabetes (NIDDM).** This type of diabetes mellitus usually occurs in people of any age who tend to be obese. Researchers theorize that perhaps adipose tissue produces a substance that interferes with the transport of glucose into cells. The amount of insulin in the blood of these patients is normal or elevated, but the insulin receptors on the cells do not respond to it. It is possible to prevent, or at least control, type II diabetes by adhering to a low-fat, low-sugar diet, maintaining a healthy weight, and exercising regularly. If these attempts fail, oral drugs are available to stimulate the pancreas to secrete more insulin. Other oral medications enhance the metabolism of glucose in the liver and muscle cells. It is projected that as many as 7 million Americans may have type II diabetes without being aware of it. Yet, the effects of untreated type II diabetes are as serious as those of type I diabetes. In addition, without stringent control the NIDDM diabetic will ultimately require insulin injections, thus becoming insulin-dependent.

"I can remember getting sick with the flu when I was 11. I missed two or three days of school, and I just never got my strength back. I ate and drank constantly because I was thirsty and hungry all the time. I was always in the bathroom, and I started wetting the bed—can you imagine, at age 11? I fell asleep in school, and the teacher could barely get me to wake up. That's when my doctor diagnosed my diabetes for the first time."

The patient, age 25, is a typical type I, juvenile-onset, or insulin-dependent, diabetic. Her symptoms are typical of insulin-dependent diabetes mellitus (IDDM) (see pages 220–221). As you know, in insulin-dependent diabetes, the body's own immune cells destroy the islet cells. To treat their illness, type I diabetic patients must inject insulin three or more times daily or use an insulin pump device continuously. Patients use blood tests to check their blood glucose levels. Further, diabetics must also monitor their diet, activity, and stress levels. Regular exercise is also a must.

To provide treatment options that are less painful, and potentially more effective and reliable, researchers are actively exploring alternative delivery systems for obtaining insulin without injections. Scientists at the University of Texas at Austin have recently published research on an insulin pill. Currently, diabetics can't swallow their insulin because harsh stomach acid destroys the protein, digesting it before it can be absorbed. A newly developed gel that can be used in pill form successfully shields the insulin as it passes through the stomach. The gel then attaches to the wall of the intestine, allowing insulin to be successfully taken into the bloodstream. In addition, skin patches and insulin aerosol sprays are already in clinical trials on human patients. Successful development of these products could help diabetics maintain stable levels of blood glucose, eliminate injections, and lead more normal lives.

Pancreatic islet cell transplantation shows great promise as a permanent cure for type I diabetes. Unlike a whole pancreatic transplant, which is major surgery, islet cell transplant is a relatively simple procedure. The cells are first harvested from a cadaver pancreas, then directly injected into the liver where they form colonies and produce insulin. Initially, an islet cell transplant required approximately 700,000 cells to produce enough insulin. New research has refined the combination of anti-rejection drugs, and more transplanted cells now survive. Now, fewer than 300,000 cells are needed, and the large majority of research participants no longer require insulin injections. Currently, both animal and living human donor cells are being investigated as potential sources for the large quantities of islet cells used for transplant (Fig. 10A). Investigators are also looking for ways to entirely eliminate the need for anti-rejection drugs. At present, all transplant patients must have lifelong anti-rejection medications, which have severe side effects (recurring infections, increased cancer risk, kidney damage, etc.). A technique called *microencapsulation* shields donor cells from being rejected by enclosing them in tiny membrane capsules. If these studies are successful, patients won't need anti-rejection medication or insulin injections. Researchers are optimistic that microencapsulated islet cells might soon be available for transplant.

Figure 10 A Encapsulated insulin-producing pancreatic islet cells from pigs can be transplanted into patients without the need for immune system-suppressing drugs.

Long-term complications of both types of diabetes are blindness, kidney disease, and circulatory disorders, including atherosclerosis, heart disease, stroke, and reduced circulation. The latter can lead to gangrene in the arms and legs. Pregnancy carries an increased risk of diabetic coma, and the child of a diabetic is somewhat more likely to be stillborn or to die shortly after birth. However, these complications of diabetes are not expected to appear if the mother's blood glucose level is carefully regulated and kept within normal limits during the pregnancy.

13. Insulin-sensitive cells in the human body include:

    a. muscle cells.

    b. adipose tissue cells.

    c. brain and nerve cells.

    d. a and b.

    e. all of the above.

14. Which of the following is an effect of glucagon?

    a. causes the liver to break down stored glycogen

    b. causes adipose tissue to store fat

    c. lowers blood glucose level

    d. stimulates the liver to store glucose as glycogen

15. Glucagon release is controlled by a negative feedback system. Which action causes glucagon to be released?

    a. skipping breakfast and going to morning classes with an empty stomach

    b. eating a big holiday meal

    c. running a marathon race for several hours without pausing for food

    d. a and c

    e. b and c

Answers in Appendix C, page 463.

10.6 Other Endocrine Glands

The body has a number of other endocrine glands, including the **gonads** (testes in males and the ovaries in females). Other lesser-known glands, such as the thymus gland and the pineal gland, also produce hormones. Some tissues within organs produce hormones and/or growth factors. Individual body cells produce local messenger chemicals termed prostaglandins.

Testes and Ovaries

The **testes** are located in the scrotum, and the **ovaries** are located in the pelvic cavity. The testes produce **androgens** (e.g., **testosterone**), which are the male sex hormones, and the ovaries produce **estrogens** and **progesterone**, the female sex hormones. The hypothalamus and the pituitary gland control the hormonal secretions of these organs in the manner previously described on page 212.

Androgens

Puberty is the time of life when sexual maturation occurs. Greatly increased testosterone secretion during puberty stimulates the growth of the penis and the testes. Testosterone also brings about and maintains the male secondary sex characteristics that develop during puberty, including the growth of a beard, axillary (underarm) hair, and pubic hair. It prompts the larynx and the vocal cords to enlarge, causing the voice to change. It is partially responsible for the muscular strength of males. This is why some athletes take supple-

mental amounts of **anabolic steroids,** which are either testosterone or related chemicals. The contraindications of taking anabolic steroids are discussed in the Medical Focus on page 224. Testosterone also stimulates oil and sweat glands in the skin; therefore, it is largely responsible for acne and body odor. Another side effect of testosterone is baldness. Genes for baldness are probably inherited by both sexes, but baldness is seen more often in males because of the presence of testosterone.

Estrogen and Progesterone

The female sex hormones, estrogens and progesterone, have many effects on the body. In particular, estrogens secreted during puberty stimulate the growth of the uterus and the vagina. Estrogen is necessary for ovum maturation and is largely responsible for the secondary sex characteristics in females, including female body hair and fat distribution. In general, females have a more rounded appearance than males because of a greater accumulation of fat beneath the skin. Also, the pelvic girdle is wider in females than in males, resulting in a larger pelvic cavity. Both estrogen and progesterone are required for breast development and for regulation of the uterine cycle, which includes monthly menstruation (discharge of blood and mucosal tissues from the uterus).

Thymus Gland

The lobular **thymus gland,** which lies just beneath the sternum (see Fig. 10.1), reaches its largest size and is most active during childhood. **Lymphocytes** are white blood cells that originate in the bone marrow and are responsible for specific defenses against a particular invader. When lymphocytes complete development in the thymus, they are transformed into **thymus-derived lymphocytes,** or **T lymphocytes.** The lobules of the thymus are lined by epithelial cells that secrete hormones called **thymosins.** These hormones aid in the differentiation of lymphocytes packed inside the lobules. Although the hormones secreted by the thymus ordinarily work only in the thymus, researchers hope that these hormones could be injected into AIDS or cancer patients where they would enhance T-lymphocyte function.

Pineal Gland

The **pineal gland,** which is located in the brain, produces the hormone **melatonin,** primarily at night. Melatonin is involved in our daily sleep-wake cycle; normally we grow sleepy at night when melatonin levels increase and awaken once daylight returns and melatonin levels are low (Fig. 10.15). Daily 24-hour cycles such as this are called **circadian rhythms.** Circadian rhythms are controlled by an internal timing mechanism called a biological clock.

Based on animal research, it appears that melatonin also regulates sexual development. It has also been noted that children whose pineal gland has been destroyed due to a brain tumor experience early puberty.

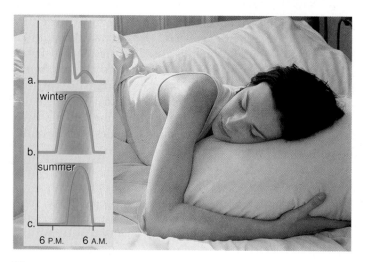

Figure 10.15 Melatonin production. Melatonin production is greatest at night when we are sleeping. Light suppresses melatonin production **(a)** so its duration is longer in the winter **(b)** than in the summer **(c).**

Hormones from Other Tissues

We have already mentioned that the heart produces atrial natriuretic hormone (see page 218). The kidney also influences cardiovascular system function by producing the hormone erythropoietin (EPO), which stimulates red blood cell production by the bone marrow. And you will see in Chapter 15 that organs and tissues of the digestive system produce hormones as well.

Leptin

Leptin is a protein hormone produced by adipose tissue. Leptin acts on the hypothalamus, where it signals satiety—that the individual has had enough to eat. Strange to say, the blood of obese individuals may be rich in leptin. It is possible that the leptin they produce is ineffective because of a genetic mutation, or else their hypothalamic cells lack a suitable number of receptors for leptin.

MEDICAL FOCUS

Side Effects of Anabolic Steroids

They're called "performance-enhancing steroids," and their use is alleged to be widespread in athletics, both amateur and professional. Whether the sport is baseball, football, professional cycling, or track and field events, no activity seems to be safe from drug abuse. Recent steroid abuse admitted by Marion Jones has forever changed Olympic history. Jones was the first female athlete to win five medals for track and field events during the 2000 Sydney Olympics. In 2008, she was stripped of all medals she had earned, as well as disqualified from a fifth-place finish in the 2004 Athens games. Future Olympic record books will not include her name, and the records of her teammates in the relay events have also been tainted.

Baseball records will also likely require revisions. The exciting slugfest between Mark McGwire and Sammy Sosa in the summer of 1998 was largely credited with reviving national interest in baseball. Yet, the great home-run competition drew unwanted attention to the darker side of professional sports, when it was alleged that both McGwire and Sosa were both using anabolic steroids at the time. Similar charges of drug abuse may prevent baseball great Roger Clemens from entering the Baseball Hall of Fame, despite his record for the number of Cy Young awards he holds. Likewise, because controversy continues to surround baseball legend Barry Bonds, this talented athlete may never achieve Hall of Fame status. Although he scored a record 715 home runs and won more Most Valuable Player awards than anyone in history, Bonds remains accused of steroid abuse. Lou Canseco of the Oakland Athletics and Jason Giambi of the New York Yankees have both admitted using performance-enhancing drugs.

Most athletes and officials continue to deny use of anabolic steroids in professional sports. However, many people from both inside and outside the industry maintain that such abuse has been going on for many years, and that it continues despite the negative publicity. Congress continues to investigate the controversy, yet the finger-pointing and accusations increase. Of tremendous concern to lawmakers, educators, and parents is the increased use of steroids by teens wishing to bulk up quickly, perhaps seeking to be just like the sports figures they admire.

Anabolic steroids are synthetic forms of the male sex hormone testosterone. Taking doses 10 to 100 times the amount prescribed by doctors for various illnesses promotes larger muscles when the person also exercises. Trainers may have been the first to acquire anabolic steroids for weight lifters, bodybuilders, and other athletes, such as professional baseball players. However, being a steroid user can have serious detrimental effects. Men often experience decreased sperm counts and decreased sexual desire due to atrophy of the testicles. Some develop an enlarged prostate gland or grow breasts. On the other hand, women can develop male sexual characteristics. They grow hair on their chests and faces, and lose hair from their heads; many experience abnormal enlargement of the clitoris. Some cease ovulating or menstruating, sometimes permanently.

Some researchers predict that two or three months of high-dosage use of anabolic steroids as a teen can cause death by age 30 or 40. Steroids have even been linked to heart disease in both sexes and implicated in the deaths of young athletes from liver cancer and one type of kidney tumor. Steroids can cause the body to retain fluid, which results in increased blood pressure. Users then try to get rid of

Growth Factors

A number of different types of organs and cells produce peptide **growth factors,** which stimulate cell division and mitosis. Some, such as lymphokines, are released into the blood; others diffuse to nearby cells. Growth factors of particular interest are the following:

Granulocyte and macrophage colony-stimulating factor (GM-CSF) is secreted by many different tissues. GM-CSF causes bone marrow stem cells to form either granulocyte or macrophage cells (both are forms of white blood cells, or leukocytes), depending on whether the concentration is low or high.

Platelet-derived growth factor is released from platelets and from many other cell types. It helps in wound healing and causes an increase in the number of fibroblasts, smooth muscle cells, and certain cells of the nervous system.

Epidermal growth factor and *nerve growth factor* stimulate the cells indicated by their names, as well as many others. These growth factors are also important in wound healing.

Tumor angiogenesis factor stimulates the formation of capillary networks and is released by tumor cells. One treatment for cancer is to prevent the activity of this growth factor.

Prostaglandins

Prostaglandins are potent chemical signals produced within cells from arachidonate, a fatty acid. Prostaglandins are not distributed in the blood; instead, they act locally, quite close

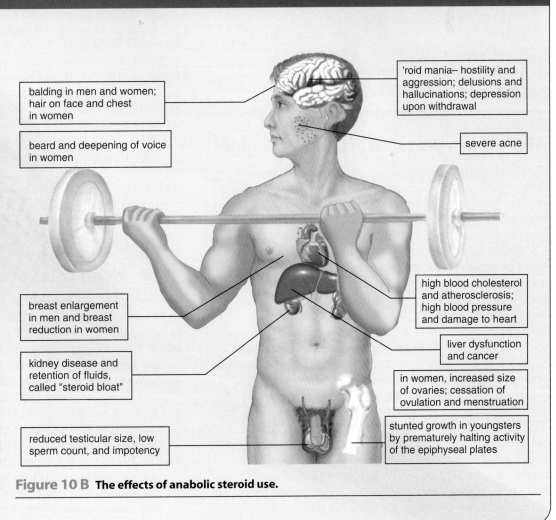

"steroid bloat" by taking large doses of diuretics. A young California weight lifter had a fatal heart attack after using steroids, and the postmortem showed a lack of **electrolytes,** salts that help regulate the heart. Finally, steroid abuse has psychological effects, including depression, hostility, aggression, and eating disorders. Unfortunately, these drugs make a person feel invincible. One abuser even had his friend videotape him as he drove his car at 40 miles an hour into a tree!

The many harmful effects of anabolic steroids are given in Figure 10B. The Federal Food and Drug Administration now bans most steroids, and steroid use has also been banned by the National Collegiate Athletic Association (NCAA), the National Football League (NFL), and the International Olympic Committee (IOC).

balding in men and women; hair on face and chest in women

beard and deepening of voice in women

breast enlargement in men and breast reduction in women

kidney disease and retention of fluids, called "steroid bloat"

reduced testicular size, low sperm count, and impotency

'roid mania– hostility and aggression; delusions and hallucinations; depression upon withdrawal

severe acne

high blood cholesterol and atherosclerosis; high blood pressure and damage to heart

liver dysfunction and cancer

in women, increased size of ovaries; cessation of ovulation and menstruation

stunted growth in youngsters by prematurely halting activity of the epiphyseal plates

Figure 10 B **The effects of anabolic steroid use.**

to where they were produced. In the uterus, prostaglandins cause muscles to contract and may be involved in the pain and discomfort of menstruation. Also, prostaglandins mediate the effects of pyrogens, chemicals that are believed to reset the temperature regulatory center in the brain. For example, aspirin reduces body temperature and controls pain because of its effect on prostaglandins.

Certain prostaglandins reduce gastric secretion and have been used to treat ulcers; others lower blood pressure and have been used to treat hypertension; and still others inhibit platelet aggregation and have been used to prevent thrombosis (the formation of stationary clots in blood vessels). However, different prostaglandins have contrary effects, and it has been very difficult to successfully standardize their use.

Content CHECK-UP!

16. From the following list of endocrine glands and their hormones, choose the pair that is correct:

    a. ovaries → androgens   c. kidney → aldosterone

    b. thymus → insulin   d. adipose tissue → leptin

17. Which of the following is a local tissue messenger that stimulates nearby cells?

    a. leptin   c. melatonin

    b. prostaglandin   d. thymosin

18. Describe three functions of the female sex hormones.

Answers in Appendix C, page 463.

10.7 The Importance of Chemical Signals

Chemical signals are molecules that affect the behavior of those cells that have receptor proteins to receive them. For example, a hormone that binds to a receptor protein affects the metabolism of the cell. Cells, organs, and even individuals communicate with one another by using chemical signals.

We are most familiar with chemical signals that are produced by organs some distance from one another in the body. For example, hormones produced by the anterior pituitary influence the function of numerous organs throughout the body. Insulin, produced by the pancreas, is transported in blood to muscle, adipose, and other insulin-sensitive cells. The nervous system at times utilizes chemical signals that are produced by an organ distant from the one being affected, as when the hypothalamus produces releasing hormones. As you know, these releasing hormones then travel in a portal system to the anterior pituitary gland.

Many chemical signals act locally—that is, from cell to cell. Prostaglandins are *local hormones,* and certain neurotransmitter substances released by one neuron affect a neuron nearby. Growth factors, which fall into this category, are very important regulators of cell division. Some growth factors are being used as medicines to promote the production of blood cells in AIDS and cancer patients. When a tumor develops, cell

division occurs even when no detectable stimulatory growth factor has been received. And the tumor produces a growth factor called tumor angiogenesis factor, which promotes the formation of capillary networks to service its cells.

Chemical Signals Between Individuals

Chemical signals that act between individuals are called *pheromones.* Pheromones are well exemplified in other animals, but they may also be effective between people. Humans produce airborne chemicals from a variety of areas, including the scalp, oral cavity, armpits, genital areas, and feet. For example, the armpit secretions of one woman could possibly affect the menstrual cycle of another woman. Women who live in the same household often have menstrual cycles in synchrony. Also, the cycle length becomes more normal when women with irregular cycles are exposed to extracts of male armpit secretions.

10.8 Effects of Aging

Thyroid disorders and diabetes are the most significant endocrine problems affecting health and function as we age. Both hypothyroidism and hyperthyroidism are seen in the elderly. Graves' disease is an autoimmune disease that targets the thyroid, resulting in symptoms of cardiovascular disease, increased body temperature, and fatigue. In addition, a patient may experience weight loss of as much as 20 pounds, depression, and mental confusion. Hypothyroidism (myxedema) may fail to be diagnosed because the symptoms of hair loss, skin changes, and mental deterioration are attributed simply to the process of aging.

The true incidence of IDDM diabetes among the elderly is unknown. Its symptoms can be confused with those of other medical conditions that are present. As in all adults, NIDDM diabetes is associated with being overweight and often can be controlled by proper diet.

The effect of age on the sex organs is discussed in Chapter 17.

10.9 Homeostasis

The endocrine system and the nervous system work together to regulate the organs of the body and thereby maintain homeostasis. It is clear from reviewing the Human Systems Work Together illustration on page 227 that the endocrine system particularly influences the digestive, cardiovascular, and urinary systems in a way that maintains homeostasis.

The endocrine system helps regulate digestion. The digestive system adds nutrients to the blood, and hormones produced by the digestive system influence the gallbladder and pancreas to send their secretions to the digestive tract. Another hormone, gastrin, promotes the digestion of protein by the stomach. Through its influence on the digestive process, the endocrine system promotes the presence of nutrients in the blood.

The endocrine system helps regulate fuel metabolism. Controlling the level of glucose in the blood is the function of insulin

Human Systems Work Together

Integumentary System

Androgens activate sebaceous glands and help regulate hair growth.

Skin provides sensory input that results in the activation of certain endocrine glands.

Skeletal System

Growth hormone regulates bone development; parathyroid hormone and calcitonin regulate Ca^{2+} content.

Bones provide protection for glands; store Ca^{2+} used as second messenger.

Muscular System

Growth harmone and androgens promote growth of skeletal muscle; epinephrine stimulates heart and constricts blood vessels.

Muscles help protect glands.

Nervous System

Sex hormones affect development of brain.

Hypothalamus is part of endocrine system; nerves innervate glands of secretion.

Cardiovascular System

Epinephrine increases blood pressure; ADH, aldosterone, and atrial natriuretic hormone help regulate blood volume; growth factors control blood cell formation.

Blood vessels transport hormones from glands; blood services glands; heart produces atrial natriuretic hormones.

How the Endocrine System works with other body systems.

hypothalamus
pineal body
pituitary

thyroid

thymus

adrenals
pancreas
kidney

testes (male)

ovaries (female)

parathyroids
(posterior part
of thyroid)

Lymphatic System/Immunity

Thymus is necessary for maturity of T lymphocytes.

Lymphatic vessels pick up excess tissue fluid; immune system protects against infections.

Respiratory System

Epinephrine promotes ventilation by dilating bronchioles; growth factors control production of red blood cells that carry oxygen.

Gas exchange in lungs provides oxygen and rids body of carbon dioxide.

Digestive System

Hormones help control secretion of digestive glands and accessory organs; insulin and glucagon regulate glucose storage in liver.

Stomach and small intestine produce hormones.

Urinary System

ADH, aldosterone, and atrial natriuretic hormone regulate reabsorption of water and Na^+ by kidneys.

Kidneys keep blood values within normal limits so that transport of hormones continues.

Reproductive System

Hypothalamic, pituitary, and sex hormones control sex characteristics and regulate reproductive processes.

Gonads produce sex hormones.

and glucagon. Just after eating, insulin encourages the uptake of glucose by cells and the storage of glucose as glycogen in the liver and muscles. In between eating, glucagon stimulates the liver to break down glycogen to glucose so that the blood glucose level stays constant. Adrenaline (epinephrine) from the adrenal medulla also stimulates the liver to release glucose. Glucagon (from the pancreas) and cortisol (from the adrenal cortex) promote the breakdown of protein to amino acids, which can be converted to glucose by the liver. They also promote the metabolism of fatty acids to conserve glucose, a process called **glucose sparing.** Finally, the thyroid hormones thyroxine and triiodothyronine set the body's metabolic rate, and thus are the hormones that ultimately regulate fuel metabolism.

The endocrine system helps regulate blood pressure and volume. ADH produced by the hypothalamus (but secreted by the posterior pituitary) promotes reabsorption of water by the kidneys, especially when we have not been drinking water that day. Aldosterone produced by the adrenal cortex causes the kidneys to reabsorb sodium, and when the level of sodium rises, water is automatically reabsorbed so that blood volume and pressure rise. Regulation by the endocrine system often involves antagonistic hormones; in this case, ANH (atriopeptide) produced by the heart causes sodium excretion.

The endocrine system helps regulate calcium balance. The concentration of calcium (Ca^{2+}) in the blood is critical because this ion is important to nervous conduction, muscle contraction, and the action of hormones. As you know from studying Chapter 6, the bones serve as a reservoir for calcium. When the blood calcium concentration lowers, parathyroid hormone promotes the breakdown of bone and the reabsorption of calcium by the kidneys, and the absorption of calcium by the intestines. Opposing the action of parathyroid hormone, calcitonin secreted by the thyroid brings about the deposit of calcium in the bones (although this function of calcitonin is more important in growing children than in adults).

The endocrine system helps regulate response to the external environment. In "fight-or-flight" situations, the nervous system stimulates the adrenal medulla to release epinephrine (adrenaline), which has a powerful effect on various organs. This, too, is important to homeostasis because it allows us to behave in a way that keeps us alive. Any damage due to stress is then repaired by the action of other hormones, including cortisol. Glucocorticoid (e.g., cortisone) therapy is useful for its antiinflammatory and immunosuppressive effects.

SELECTED NEW TERMS

Basic Key Terms

adrenal cortex (uh-dre'nul kor'teks), p. 216

adrenal gland (uh-dre'nul gland), p. 216

adrenal medulla
(uh-dre'nul mĕ-dūl'uh), p. 216

adrenocorticotropic hormone (uh-dre'no-kor″ti-kō-trōp'ik hor'mōn), p. 213

aldosterone (al″dos'ter-ōn), p. 217

anabolic steroid (an″uh-bol'ik stĕ'royd), p. 223

androgen (an'dro-jen), p. 223

anterior pituitary (an-tēr'ē-or pǐ-tu'ǐ-tār″e), p. 211

antidiuretic hormone (an″tǐ-dǐ″yū-ret'ik hor'mōn), p. 211

atrial natriuretic hormone (ā'trē-al nā″trē-yū-ret'ik hor'mōn), p. 218

atriopeptide (ā-tre-yo-pĕp-tīd), p. 218

calcitonin (kal″sǐ-to'nin), p. 215

circadian rhythm (ser″ka'de-an rǐ'thm), p. 223

corticotropin (kor'tǐ-ko-troh-pǐn), p. 216

cortisol (kor'tǐ-sol), p. 216

cyclic AMP (sik'lik AMP), p. 207

endocrine gland (en'do-krin gland), p. 207

epinephrine (ep″ǐ-nef'rin), p. 216

estrogen (es'tro-jen), p. 223

glucagon (glu'kuh-gon), p. 220

glucocorticoid (glu'ko-kor'tǐ-koyd), p. 216

gonad (go'nad), p. 223

gonadotropic hormone (go″nad-o-trōp'ik hor'mōn), p. 213

growth factor (grōth fak'tor), p. 224

growth hormone (grōth hor'mōn), p. 213

hormone (hor'mōn), p. 207

hypothalamic-inhibiting hormone (hi″po-thĕlam'ik-in-hib'it-ing hor'mōn), p. 211

hypothalamic-releasing hormone (hi″po-thĕlam'ik-re-lēs'-ing hor'mōn), p. 211

hypothalamus (hǐ'pō-thăl'ə-məs), p. 211

insulin (in'suh-lin), p. 220

leptin (lep'tin), p. 224

melatonin (mel″uh-to'nin), p. 223

mineralocorticoids (min″er-al-o-kor'tǐ-koyds), p. 216

norepinephrine (nôr'ĕp-nĕf'rǐn), p. 216

ovary (o'var-e), p. 223

oxytocin (ok″si-to'sin), p. 211

pancreas (pan'kre-us), p. 220

pancreatic islets (of Langerhans) (pan″kre-at'ik ī'lets ov lahng'er-hanz), p. 220

parathyroid gland (pār″uh-thi'royd gland), p. 216

parathyroid hormone
(pār″uh-thi'royd hor'mōn), p. 216

peptide hormone (pep'tid hor'mōn), p. 207

pineal gland (pin'e-ul gland), p. 223

pituitary gland (pǐ-tu'ǐ-tār″ē gland), p. 211

portal system (por'tul sis' tem), p. 211

posterior pituitary (pōs-tēr'e-or pǐ-tū'-ǐ-tār″ē), p. 211

progesterone (pro-jes'ter- ōn), p. 223

prolactin (pro-lak'tin), p. 213

prostaglandins (pros″tuh-glan'dinz), p. 225

renin (re'nin), p. 218

steroid hormone (stēr'oyd hor'mōn), p. 207

testis (tes'tis), p. 223

testosterone (tes-tos'tĕ-rōn), p. 223

thymosin (thi'mo-sin), p. 223

thymus gland (thi'mus gland), p. 223

thyroid gland (thi'royd gland), p. 214

thyroid-stimulating hormone (thi'royd stim'yū-lāt-ing hor'mōn), p. 211

thyrotropin (thǐ'ro-trō-'pǐn), p. 207

thyroxine (thī-rok'sin), p. 214

Clinical Key Terms

acidosis (ăs'ǐ-dō'sǐs), p. 220

acromegaly (ak″ro-meg'uh-le), p. 213

Addison's disease (ă'dǐ-sons dǐ-zēz'), p. 218

congenital hypothyroidism (kon-gen'i-tul hi″po-thi'-roy-dizm), p. 214

Cushing's syndrome (koosh'ings sin'drōm), p. 218

diabetic coma (dī″uh-bĕ tik kō'-muh), p. 221

diabetes insipidus (dī″uh-be'tēz in-sipǐ-dus), p. 211

diabetic ketoacidosis (dī″uh-bĕ tik kē'tō'-as-i-dō'sus), p. 221

exophthalmic goiter (ek″sof-thal'mik goy'ter), p. 215

glycosuria (gli'ko-sūr'e-uh), p. 220

Graves' disease (grāvz dǐ-zēz'), p. 215

hyperglycemia (hi"per-gli-se'me-uh), p. 220

hypoglycemia (hi"po-gli-se'me-uh), p. 221

insulin-dependent diabetes mellitus (in'sul-in-de-pen'dent di"uh-be'tēz mē-li'tus), p. 220

insulin shock (in'sul-in shŏk), p. 221

ketonuria (ke"to-nū're-uh), p. 220

myxedema (mik"sĕ-de'muh), p. 214

noninsulin-dependent diabetes (non'in'sūl-in-de-pen'dent di"uh-bē'tēz), p. 221

pituitary dwarfism (pĭ-tuĭ-tār"e dwarf'-izm), p. 213

polydipsia (pol"e-dip'se-uh), p. 220

polyphagia (pol"e-fa-je-uh), p. 220

polyuria (pol"e-yū-re-uh), p. 220

simple (endemic) goiter (sim'pl ĕn-dĕm'ĭk goy'ter), p. 214

syndrome (sin'drōm), p. 218

tetany (tet'uh-ne), p. 216

SUMMARY

10.1 Endocrine Glands
A. Endocrine glands secrete hormones into the bloodstream, and from there they are distributed to target organs or tissues.
B. Hormones are either peptides or steroids. Reception of a peptide hormone at the plasma membrane activates an enzyme cascade inside the cell. Steroid hormones combine with a receptor in the cell, and the complex attaches to and activates DNA. Protein synthesis follows. The major endocrine glands and hormones are listed in Table 10.1. Neural mechanisms, hormonal mechanisms, and/or negative feedback control the effects of hormones.

10.2 Hypothalamus and Pituitary Gland
A. Neurosecretory cells in the hypothalamus produce antidiuretic hormone (ADH) and oxytocin, which are stored in axon endings in the posterior pituitary until they are released.
B. The hypothalamus produces hypothalamic-releasing and hypothalamic-inhibiting hormones, which pass to the anterior pituitary by way of a portal system. The anterior pituitary produces at least six types of hormones, and some of these stimulate other hormonal glands to secrete hormones.

10.3 Thyroid and Parathyroid Glands
The thyroid gland requires iodine to produce triiodothyronine (T_3) and thyroxine (T_4), which increase the metabolic rate. If iodine is available

in limited quantities, a simple goiter develops; if the thyroid is overactive, an exophthalmic goiter develops. The thyroid gland also produces calcitonin, which helps lower the blood calcium level. The parathyroid glands secrete parathyroid hormone, which raises the blood calcium and decreases the blood phosphate levels.

10.4 Adrenal Glands
The adrenal glands respond to stress: Immediately, the adrenal medulla secretes epinephrine and norepinephrine, which bring about responses we associate with emergency situations. On a long-term basis, the adrenal cortex produces the glucocorticoids (e.g., cortisol) and the mineralocorticoids (e.g., aldosterone). Cortisol stimulates hydrolysis of proteins to amino acids that are converted to glucose; in this way, it raises the blood glucose level. Aldosterone causes the kidneys to reabsorb sodium ions (Na^+) and to excrete potassium ions (K^+). Addison's disease develops when the adrenal cortex is underactive, and Cushing's syndrome develops when the adrenal cortex is overactive.

10.5 Pancreas
The pancreatic islets secrete insulin, which lowers the blood glucose level, and glucagon, which has the opposite effect. The most common illness caused by hormonal imbalance is diabetes mellitus, which is due to the failure of the pancreas to produce insulin and/or the failure of the cells to take it up.

10.6 Other Endocrine Glands
A. The gonads produce the sex hormones. The thymus secretes thymosins, which stimulate T-lymphocyte production and maturation. The pineal gland produces melatonin, which is involved in circadian rhythms and may affect development of the reproductive organs.
B. Tissues also produce hormones. Adipose tissue produces leptin, which acts on the hypothalamus, and various tissues produce growth factors. Prostaglandins are produced and act locally.

10.7 The Importance of Chemical Signals
In the human body, some chemical signals, such as traditional endocrine hormones and secretions of neurosecretory cells, act at a distance. Others, such as prostaglandins, growth factors, and neurotransmitters, act locally. Whether humans have pheromones is under study.

10.8 Effects of Aging
Two concerns often seen in the elderly are thyroid malfunctioning and diabetes mellitus.

10.9 Homeostasis
Hormones particularly help maintain homeostasis in several ways: Hormones help maintain the level of nutrients (e.g., amino acids and glucose in blood); help maintain blood volume and pressure by regulating the sodium content of the blood; help maintain the blood calcium level; help regulate fuel metabolism; and help regulate our response to the external environment.

STUDY QUESTIONS

1. Explain how peptide hormones and steroid hormones affect the metabolism of the cell. (p. 207)
2. Contrast hormonal and neural signals, and show that there is an overlap

between the mode of operation of the nervous system and that of the endocrine system. (pp. 209–210)
3. Explain the relationship of the hypothalamus to the posterior pituitary

gland and to the anterior pituitary gland. List the hormones secreted by the posterior and anterior pituitary glands. (p. 211)

4. Give an example of the negative feedback relationship among the hypothalamus, the anterior pituitary, and other endocrine glands. (pp. 211–213)
5. Discuss the effect of growth hormone on the body and the result of having too much or too little growth hormone when a young person is growing. What is the result if the anterior pituitary produces growth hormone in an adult? (p. 213)
6. What types of goiters are associated with a malfunctioning thyroid? Explain each type. (pp. 214–215)

7. How do the thyroid and the parathyroid work together to control the blood calcium level? (pp. 215–216)
8. How do the adrenal glands respond to stress? What hormones are secreted by the adrenal medulla, and what effects do these hormones have? (pp. 216–218)
9. Name the most significant glucocorticoid and mineralocorticoid, and discuss their functions. Explain the symptoms of Addison's disease and Cushing's syndrome. (pp. 216–218)
10. Draw a diagram to explain how insulin and glucagon maintain the blood glucose level. Use your diagram to

explain the major symptoms of type I diabetes mellitus. (pp. 220–221)
11. Name the other endocrine glands discussed in this chapter, and discuss the functions of the hormones they secrete. (pp. 223–224)
12. What are leptin, growth factors, and prostaglandins? How do these substances act? (pp. 224–225)
13. Discuss five ways the endocrine system helps maintain homeotasis. (pp. 227–228)

LEARNING OBJECTIVE QUESTIONS

Fill in the blanks.

1. Generally, hormone production is self-regulated by a _____ mechanism.
2. The hypothalamus _____ the hormones _____ and _____, released by the posterior pituitary.
3. The _____ secreted by the hypothalamus control the anterior pituitary.
4. Growth hormone is produced by the _____ pituitary.
5. Simple goiter occurs when the thyroid is producing (too much or too little) _____ because it has (too much or too little) _____.

6. Parathyroid hormone increases the level of _____ in the blood.
7. Adrenocorticotropic hormone (ACTH), produced by the anterior pituitary, stimulates the _____ of the adrenal glands.
8. An overproductive adrenal cortex results in the condition called _____.
9. Type I diabetes mellitus is due to a malfunctioning _____ but type II diabetes is due to malfunctioning _____.
10. Prostaglandins are not carried in the _____ as are hormones secreted by the endocrine glands.

11. Whereas _____ hormones are lipid soluble and bind to receptor proteins within the cytoplasm of target cells, _____ hormones bind to membrane-bound receptors, thereby activating second messengers.
12. Whereas the adrenal _____ is under the control of the autonomic nervous system, the adrenal _____ secretes its hormones in response to _____ from the anterior pituitary gland.

MEDICAL TERMINOLOGY EXERCISE

Consult Appendix B for help in pronouncing and analyzing the meaning of the terms that follow.

1. antidiuretic (an″tĭ-di″yŭ-ret′ik)
2. hypophysectomy (hi-pof″ĭ-sek′to-me)
3. gonadotropic (go″nad-o-trōp′ik)

4. hypokalemia (hi″po-kal″e′me-uh)
5. lactogenic (lak″to-jen′ik)
6. adrenopathy (ad″ren-op′uh-the)
7. adenomalacia (ad″ĕ-no-muh-la′she-uh)
8. parathyroidectomy (pār″uh-thi″roy-dek′to-me)

9. polydipsia (pol″e-dip′se-uh)
10. dyspituitarism (dis-pĭ-tu′ĭ-ter′izm)
11. thyroiditis (thi-roy-di′tis)
12. glucosuria (glu-co-su′re-uh)
13. microsomia (mi′kro-so′me-uh)

WEBSITE LINK

Visit this text's website at http://www.mhhe.com/longenbaker7 for additional quizzes, interactive learning exercises, and other study tools.

Blood

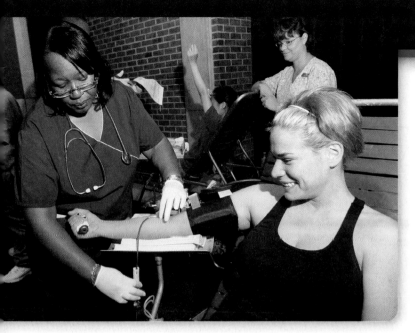

If you know about ABO blood types, and know your own, maybe you've wondered which of the four possible types is the most common worldwide. The answer depends on the genetic inheritance of a particular population. Worldwide, the most prevalent ABO type is type O. However, type A blood is most common in several European countries, including the Scandinavian countries. In certain geographic areas, virtually the entire population has the same blood type. For example, certain native Central and South American tribes are nearly all type O. The least common ABO type is type AB. If you also know your Rh type (positive or negative), you'll be interested to know that approximately 85% of humans are Rh positive. In addition to ABO and Rh, there are over 25 known blood typing systems, and more will doubtless be discovered. You can read more about blood types and blood compatibility on page 242.

Chapter Outline & Learning Objectives   *After you have studied this chapter, you should be able to:*

11.1 The Composition and Functions of Blood

When a blood sample is first prevented from clotting and then spun in a centrifuge tube, it separates into two layers (Fig. 11.1). The lower layer consists of the **formed elements**, which make up about 45% of the total volume of whole blood. Red blood cells are the heaviest cells due to their iron-containing hemoglobin, and can be found on the bottom of the tube. The volume of a blood sample that consists of red blood cells is that sample's **hematocrit**. The white blood cells and the blood platelets settle on top of the red blood cells in a thin, shiny layer called the **buffy coat**. The upper layer is **plasma**, a light-yellow fluid that contains a variety of inorganic and organic molecules dissolved or suspended in water. Plasma accounts for about 55% of the total volume of whole blood.

Functions of Blood

The functions of blood fall into three categories: transport, defense, and regulation.

Transport

Blood moves from the heart to all the various organs, where exchange with tissues takes place across thin capillary walls. Blood picks up oxygen from the lungs and nutrients from the digestive tract and transports these to the tissues. It also picks up and transports cellular wastes, including carbon dioxide, away from the tissues to exchange surfaces, such as the lungs and kidneys. We will see that capillary exchanges keep the composition of tissue fluid within normal limits.

Various organs and tissues secrete hormones and other messenger chemicals into the blood, and blood transports these to other organs and tissues, where they serve as signals that influence cellular metabolism.

Defense

Because blood contains the white blood cells, it defends the body against invasion by **pathogens** (microscopic infectious agents, such as bacteria and viruses). The white blood cells also remove dead and dying cells, clearing room for growth of healthy cells. Mutated cells—which could potentially develop into cancer—are also destroyed by white blood cells. Certain white blood cells are capable of engulfing and destroying pathogens or cancer cells; others produce and secrete antibodies into the blood. Antibodies incapacitate pathogens, making them subject to destruction by other white blood cells. This function of the blood will be discussed completely in Chapter 13.

When an injury occurs, blood forms a clot, and this prevents blood loss. Blood clotting involves platelets and the plasma proteins prothrombin and fibrinogen. Without blood clotting, we could bleed to death even from a small cut.

Regulation

Blood helps regulate body temperature by picking up heat, mostly from active muscles, and transporting it about the body. If the blood is too warm, the heat dissipates to the environment from dilated blood vessels in the skin.

The salts and plasma proteins in blood act to keep the osmotic pressure in the normal range. Recall that osmotic pressure causes diffusion of water (review section 3.2). Thus, the osmotic pressure of blood causes diffusion of water into the blood. In this way, blood plays a role in maintaining its own water-salt balance.

Because blood contains buffers, it also helps regulate body pH and keeps it relatively constant.

Content CHECK-UP!

1. Name the two layers in a sample of blood, then list three types of formed elements you will find in the sample.

2. Which of the following are defense mechanisms of the blood?
   a. Blood cells can engulf and destroy bacteria and viruses.
   b. Antibodies target pathogens for destruction.
   c. The blood clotting mechanism prevents blood loss.
   d. All of the above are defense mechanisms of the blood.

3. How does blood regulate body temperature? osmotic pressure? body pH?

Answers in Appendix C, page 463.

11.2 Components of Blood

Plasma

Plasma is the liquid portion of blood, and about 92% of plasma is water. The remaining 8% of plasma is composed of various salts (ions) and organic molecules (Table 11.1). The salts, which are simply dissolved in plasma, help maintain the osmotic pressure and pH of the blood. Small organic molecules such as glucose, amino acids, and urea are also dissolved

TABLE 11.1 Blood Plasma Solutes	
Plasma proteins	Albumin, globulins, fibrinogen
Inorganic ions (salts)	Na^+, Ca^{2+}, K^+, Mg^{2+}, Cl^-, HCO_3^-, HPO_4^{2-}, SO_4^{2-}
Gases	O_2, CO_2
Organic nutrients	Glucose, fats, phospholipids, amino acids, etc.
Nitrogenous waste products	Urea, ammonia, uric acid
Regulatory substances	Hormones, enzymes, other messenger chemicals

FORMED ELEMENTS	Function and Description	Source
Red Blood Cells (erythrocytes)	Transport O_2 and help transport CO_2.	Red bone marrow
4 million–6 million per mm^3 blood	7–8 μm in diameter; bright-red to dark-purple biconcave disks without nuclei.	
White Blood Cells (leukocytes) 5,000–11,000 per mm^3 blood	Fight infection. Remove dead/dying cells. Destroy cancer cells.	Red bone marrow
Granular leukocytes		
• Neutrophils 40–70%	Phagocytize pathogens. 10–14 μm in diameter; spherical cells with multilobed nuclei; fine, lilac granules in cytoplasm if stained.	
• Eosinophils 1–4%	Phagocytize antigen-antibody complexes and allergens. 10–14 μm in diameter; spherical cells with bilobed nuclei; coarse, deep-red, uniformly sized granules in cytoplasm if stained.	
• Basophils 0–1%	Release histamine and heparin, which promote blood flow to injured tissues. 10–12 μm in diameter; spherical cells with lobed nuclei; large, irregularly shaped, deep-blue granules in cytoplasm if stained.	
Agranular leukocytes		
• Lymphocytes 20–45%	Responsible for specific immunity. 5–17 μm in diameter (average 9–10 μm); spherical cells with large, round nuclei.	
• Monocytes 4–8%	Become macrophages that phagocytize pathogens and cellular debris. 10–24 μm in diameter; large, spherical cells with kidney-shaped, round, or lobed nuclei.	
Platelets 150,000–300,000 per mm^3 blood	Aid hemostasis. 2–4 μm in diameter; disk-shaped cell fragments with no nuclei; purple granules in cytoplasm.	Red bone marrow

Plasma 55%

Buffy coat

Formed elements 45%

PLASMA	Function	Source
Water (90–92% of plasma)	Maintains blood volume; transports molecules.	Absorbed from intestine
Plasma proteins (7–8% of plasma)	Maintain blood osmotic pressure and pH.	Liver
Albumins	Maintain blood volume and pressure, transport.	Select white blood cells
Globulins	Transport; fight infection.	
Fibrinogen	Coagulation.	
Salts (less than 1% of plasma)	Maintain blood osmotic pressure and pH; aid metabolism.	Absorbed from intestine
Gases		
Oxygen	Cellular respiration	Lungs
Carbon dioxide	End product of metabolism.	Tissues
Nutrients	Food for cells.	Absorbed from intestine
Lipids		
Glucose		
Amino acids		
Nitrogenous wastes	Excretion by kidneys.	Liver
Uric acid		
Urea		
Other		
Hormones, vitamins, etc.	Aid metabolism.	Varied

Figure 11.1 Composition of blood. When a blood sample is prevented from clotting and spun in a centrifuge tube, it forms two layers. The lucent, yellow top layer is plasma, the liquid portion of blood. The formed elements are in the bottom layer. This table describes these components in detail.

in plasma. Glucose and amino acids are nutrients for cells; urea is a nitrogenous waste product on its way to the kidneys for excretion. The large organic molecules in plasma include hormones and the plasma proteins.

Three major types of plasma proteins are the albumins, the globulins, and fibrinogen. Most plasma proteins are made in the liver. One exception is the antibodies produced by B lymphocytes, which function in immunity. Certain hormones are also plasma proteins made by various glands.

The plasma proteins have many functions that help maintain homeostasis. They are able to take up and release hydrogen ions; therefore, the plasma proteins help buffer the blood and keep its pH around 7.40. **Osmotic pressure** is the force caused by a difference in solute concentration on either side of a membrane. As mentioned previously, plasma proteins, particularly the **albumins,** contribute to the osmotic pressure. Their role will be described completely in Chapter 12.

There are three types of **globulins,** designated alpha, beta, and gamma globulins. The alpha and beta globulins, produced by the liver, bind to metal ions and to fat-soluble vitamins. They also bind to lipids, forming the **lipoproteins.** Antibodies, proteins made by white blood cells which help fight infections, are gamma globulins.

Both albumins and globulins combine with and transport large organic molecules. For example, albumin transports the molecule bilirubin, a breakdown product of hemoglobin. Lipoproteins, whose protein portion is a globulin, transport cholesterol.

The other plasma proteins, fibrinogen and prothrombin, are necessary to coagulation (blood clotting), which is discussed on page 239.

Formed Elements

The formed elements consist of blood cells and platelets. (Platelets are discussed in detail on page 238.) In the adult, the formed elements are produced continuously in the red bone marrow of the skull, ribs, and vertebrae, the iliac crests, and the ends of long bones.

The process by which formed elements are made is called **hematopoiesis** (Fig. 11.2). **Multipotent stem cells** are the red bone marrow cells that mature into all the various types of blood cells. At the top of Figure 11.2 is a multipotent stem cell that first replicates itself by mitosis. Each cell then *differentiates*, producing two other types of stem cells: myeloid and lymphatic stem cells. (When a stem cell differentiates, it commits itself to follow a single pathway in development.) The myeloid stem cell further differentiates to give rise to the cells that become red blood cells, granular leukocytes, monocytes, and megakaryocytes, the parent cell for platelets. The lymphatic stem cell differentiates to produce the lymphocytes.

Many scientists are very interested in developing ways to use blood stem cells, as well as stem cells from other adult tissues, to regenerate the body's tissues in the laboratory.

Red Blood Cells

Red blood cells (RBCs, or **erythrocytes)** are small, biconcave disks that lack a nucleus when mature. They occur in great quantity; there are 4 to 6 million red blood cells per mm^3 of whole blood.

Red blood cells transport oxygen, and each contains about 200 million molecules of **hemoglobin,** the respiratory pigment. If this much hemoglobin were suspended within the plasma rather than enclosed within the cells, blood would be so viscous that the heart would have difficulty pumping it.

Hemoglobin

In a molecule of hemoglobin, each of four polypeptide chains making up globin has an iron-containing heme group in the center. Oxygen combines loosely with iron when hemoglobin is oxygenated:

$$\text{Hb} + \text{O}_2 \underset{\text{tissues}}{\overset{\text{lungs}}{\rightleftharpoons}} \text{HbO}_2$$

In this equation, the hemoglobin on the right, which is combined with oxygen, is called oxyhemoglobin. Oxyhemoglobin forms in lung capillaries, and has a bright red color. The hemoglobin on the left, which is not combined with oxygen, is called deoxyhemoglobin. Deoxyhemoglobin forms in tissue capillaries, and has a dark maroon color.

Hemoglobin is remarkably adapted to its function of picking up oxygen in lung capillaries and releasing it in the tissues. As discussed in What's New on page 237, chemically altered hemoglobin may soon be available as a blood substitute. The higher concentration of oxygen, plus the slightly cooler temperature and slightly higher pH within lung capillaries, causes hemoglobin to take up oxygen. By contrast, in tissues the lower concentration of oxygen, plus the slightly warmer temperature and slightly lower pH within tissue capillaries, causes hemoglobin to give up its oxygen.

Production of Red Blood Cells

Erythrocytes are formed from red bone marrow stem cells (see Fig. 11.2): A multipotent stem cell descendant, called a myeloid stem cell, gives rise to erythroblasts, which divide many times. As they mature, erythroblasts gain many molecules of hemoglobin and lose their nucleus and most of their organelles. Possibly because mature red blood cells lack a nucleus, they live only about 120 days. It is estimated that about 2 million red blood cells are destroyed per second; therefore, an equal number must be produced to keep the red blood cell count in balance.

Whenever blood carries a reduced amount of oxygen, more red blood cells must be produced to maintain homeostasis. Lowered blood oxygen occurs when an individual

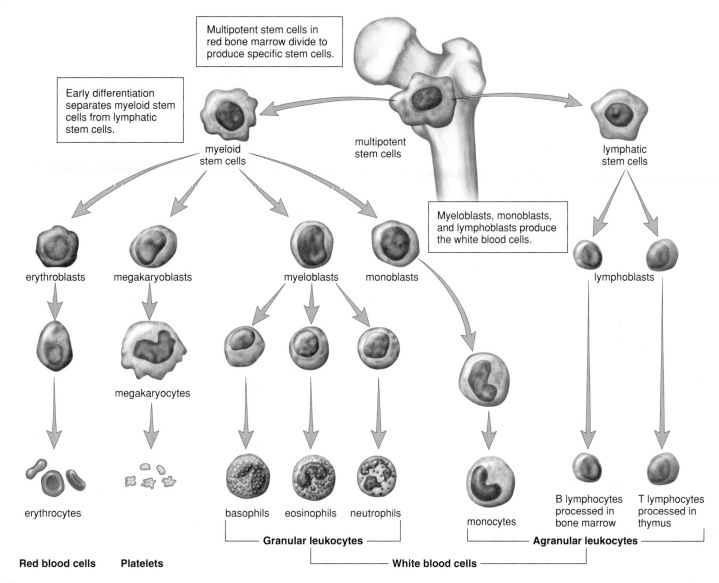

Multipotent stem cells in red bone marrow divide to produce specific stem cells.

Early differentiation separates myeloid stem cells from lymphatic stem cells.

myeloid stem cells

multipotent stem cells

lymphatic stem cells

Myeloblasts, monoblasts, and lymphoblasts produce the white blood cells.

erythroblasts

megakaryoblasts

myeloblasts

monoblasts

lymphoblasts

megakaryocytes

erythrocytes

basophils   eosinophils   neutrophils

monocytes

B lymphocytes processed in bone marrow

T lymphocytes processed in thymus

Granular leukocytes

Agranular leukocytes

Red blood cells   Platelets

White blood cells

Figure 11.2 Hematopoiesis. Multipotent stem cells give rise to two specialized stem cells. The myeloid stem cell gives rise to still other cells, which become red blood cells, platelets, and all the whole blood cells except lymphocytes. The lymphatic stem cell gives rise to lymphoblasts, which become lymphocytes.

first takes up residence at a high altitude, or after a hemorrhage when a person loses red blood cells, or in a patient with impaired lung function. These are stimuli for the kidneys to accelerate their release of **erythropoietin** (EPO) (Fig. 11.3). This hormone stimulates mitosis of stem cells and speeds up the maturation of red blood cells. The liver and other tissues also produce erythropoietin. Erythropoi-

etin, now mass-produced through biotechnology, is often prescribed for cancer patients who are anemic (low red blood cell count) due to chemotherapy or radiation therapy. Unfortunately, erythropoietin is sometimes abused by athletes in order to raise their red blood cell counts and thereby increase the oxygen-carrying capacity of their blood.

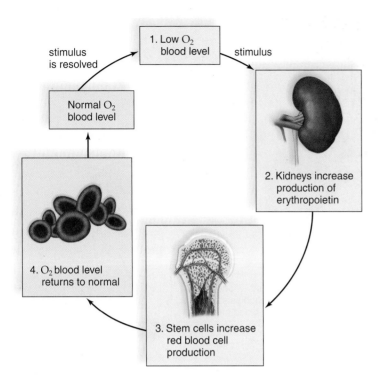

Figure 11.3 Action of erythropoietin. The kidneys release increased amounts of erythropoietin whenever the oxygen capacity of the blood is reduced. Erythropoietin stimulates the red bone marrow to speed up its production of red blood cells, which carry oxygen. Once the oxygen-carrying capacity of the blood is sufficient to support normal cellular activity, the kidneys cut back on their production of erythropoietin.

Begin Thinking Clinically

Polycythemia is a condition caused by excessive numbers of red blood cells in the blood. In this disorder, blood is excessively thick and unable to flow properly, and often forms unnecessary, dangerous clots. What do you think might cause the condition, and what effect could it have on the patient's heart? How might polycythemia be treated?

Answer and discussion in Appendix C, page 464.

Destruction of Red Blood Cells

With age, red blood cells are destroyed in the liver and spleen, where they are engulfed by macrophages. When red blood cells are broken down, hemoglobin is released. The globin portion of the hemoglobin is broken down into its component amino acids, which are recycled by the body. The iron is recovered and returned to the bone marrow for reuse. The heme portion of the molecule undergoes chemical degradation and is excreted as **bile pigments** by the liver into the bile. These bile pigments are bilirubin and biliverdin, which contribute to the color of feces. Chemical breakdown of heme is

Figure 11.4 Mobility of white blood cells. When injury or infection occurs, white blood cells can squeeze between the cells of a capillary wall and enter the tissues of the body.

also what causes a bruise on the skin to change color from red/purple to green to yellow.

Abnormal Red Blood Cell Counts

As discussed in the Medical Focus on page 239, anemia is an illness in which the patient has a tired, run-down feeling. The cells are not getting enough oxygen due to a reduction in the amount of functional hemoglobin or the number of red blood cells. **Hemolysis** (bursting of red blood cells) can also cause anemia.

White Blood Cells

White blood cells (WBCs, or **leukocytes**) differ from red blood cells in that they are usually larger, have a nucleus, lack hemoglobin, and are translucent unless stained. White blood cells are not as numerous as red blood cells; there are only 5,000–11,000 per mm³ of blood. White blood cells fight infection, destroy dead or dying body cells, and recognize and kill cancerous cells. Thus, they are important contributors to homeostasis. This function of white blood cells is discussed at greater length in Chapter 13, which concerns immunity.

White blood cells are derived from stem cells in the red bone marrow, and they, too, undergo several maturation stages (see Fig. 11.2). Each type of white blood cell is apparently capable of producing a specific growth factor that circulates back to the bone marrow to stimulate its own production.

Red blood cells are confined to the blood, but white blood cells are able to squeeze through pores in the capillary wall (Fig. 11.4). Therefore, they are found in tissue fluid and lymph (the fluid within lymphatic vessels) and in lymphatic organs. When an infection is present, white blood cells greatly increase in number. Many white blood cells live only a few days—they probably die while fighting pathogens. Others live months or even years.

Types of White Blood Cells

White blood cells are classified into the **granular leukocytes** and the **agranular leukocytes**. Both types of cells have granules

Improvements in Transfusion Technology

If you work in an emergency room (ER) setting, having a supply of donor blood for transfusion is a problem you and your co-workers will face every day. Your patient may be a victim of a car accident, a shooting, or a stabbing. Perhaps she's a pregnant woman who is hemorrhaging, or a leukemia sufferer whose hematocrit (see page 232) has dropped dangerously low because of chemotherapy. Without an immediate blood transfusion, tissue cells will die from lack of oxygen. If the emergency room is a major trauma center in a large, city hospital, donor blood for transfusion is usually available, and the ready supply of O-negative blood can theoretically be donated to anyone.

But what if the transfusion is needed in a remote area where whole blood isn't available, such as a wartime field hospital or the accident scene on an isolated stretch of highway? Likewise, small regional hospitals face regular shortages of donor blood for transfusion. Further, although the majority of blood transfusions are safe, there is always a small chance of a transfusion reaction, in which the recipient's immune system rejects the donor blood. Moreover, even though donated blood is carefully tested for viruses and other infectious agents, infection from donated blood is still a very slight possibility.

Recent developments show promise in providing solutions for the problems of availability and safety. Scientists are investigating ways to chemically alter red blood cell membranes so that all donated blood can be given safely to anyone needing a transfusion. One strategy involves antigen camouflage, in which the cell membrane proteins are covered with a plastic-like "coat." Once coated, the proteins can't trigger a transfusion reaction in the recipient. A second procedure uses bacterial-derived enzymes to "scrub" the red blood cells of A, B, and Rh antigens, essentially converting every donated unit into O-negative blood. Treated cells are called Enzyme Converted group O (ECO) cells, and blood containing these cells has already been safely transfused. These two techniques are currently in clinical trials, and may help to alleviate whole blood shortages.

Blood substitutes may also solve some of the blood transfusion problems. As you know, hemoglobin is the oxygen-transporting molecule contained in red blood cells (see page 234 and Fig. 11A). Untreated hemoglobin is useless as a blood substitute because it breaks down immediately into smaller molecules that are toxic to nerve, liver, and kidney cells. However, hemoglobin that is chemically altered to prevent breakdown can be safely eliminated from the body, and it transports oxygen even more effectively than does hemoglobin inside cells. What's more, adequate supplies are readily available because human hemoglobin can be produced by genetically engineered bacteria (also the source of the human insulin injected by diabetics). A second type of blood substitute replaces hemoglobin entirely with compounds called *perfluorocarbons*, which transport oxygen up to 90 times better than hemoglobin. Unlike whole blood, these types

Figure 11A **Hemoglobin contains four polypeptide chains.** There is an iron-containing heme group in the center of each chain. Oxygen combines loosely with iron when hemoglobin is oxygenated. Oxyhemoglobin is bright red, and deoxyhemoglobin is a dark maroon color.

of blood substitutes are always free of disease-causing contaminants, can be stored at room temperature, and won't cause a transfusion reaction because they lack the red blood cell membrane. This makes them a potential "one size fits all" substance for transfusion, and perhaps the ideal solution for critical-care emergencies. Blood substitutes are also currently in widespread clinical trials in the United States and Europe.

Wherever possible, physicians avoid human-to-human blood transfusion entirely by using the patient's own blood, a procedure called autologous transfusion or **autotransfusion**. For example, patients can have plasma or whole blood stored before an elective surgery. Then, the patient receives a transfusion using his or her own blood if needed. Intraoperative blood salvage, using an apparatus termed a "Cell Saver®," allows the emergency room physician or surgeon to suction blood from the patient's body. The suctioned blood is immediately mixed with an anticoagulant, washed with a saline solution to remove any contaminants, mixed with more saline, and then re-introduced into a vein just like a traditional transfusion. The procedure is quite fast: To prepare an entire unit of blood to re-infuse into the patient takes only 5 minutes. Autotransfusion completely eliminates the possibility of transfusion reaction.

Until these technologies are perfected, there will always be a critical need for blood donation. To find out how you can help, visit the Red Cross blood donor website at http://www.redcross.org/donate/give

in the cytoplasm surrounding the nucleus, but the granules are more visible upon staining in granular leukocytes. (The white cells shown in Figures 11.1 and 11.2 have been stained with a dye called Wright stain.) The granules contain various enzymes and proteins, which help white blood cells defend the body. There are four types of granular leukocytes and two types of agranular leukocytes. They differ somewhat by the size of the cell and the shape of the nucleus (see Fig. 11.1), and they also differ in their functions.

Granular Leukocytes

Neutrophils (see Fig. 13.3a) are the most abundant of the white blood cells. They have a multilobed nucleus joined by nuclear threads; therefore, they are also called polymorphonuclear ("many-shaped nuclei"). Some of their granules take up acid stain, and some take up basic stain (creating an overall lilac color). Neutrophils are the first type of white blood cell to respond to an infection, and they engulf pathogens during phagocytosis.

Eosinophils (see Fig. 13.3b) have a bilobed nucleus, and their large, abundant granules take up the dye called eosin and become a red color. (This accounts for their name, *eosinophil*.) Among several functions, they increase in number in the event of a parasitic worm infection. Eosinophils also lessen an allergic reaction by phagocytizing antigen-antibody complexes involved in an allergic attack.

Basophils (see Fig. 13.3c) have a U-shaped or lobed nucleus. Their granules take up the basic stain and become dark blue in color. (This accounts for their name, *basophil*.) In the connective tissues, basophils, as well as a similar type of cells called mast cells, release histamine and heparin. Histamine, which is associated with allergic reactions, dilates blood vessels and causes contraction of smooth muscle. Heparin prevents clotting and promotes blood flow. Natural killer cells (often abbreviated NK cells) are large, granular leukocytes that are similar in structure to lymphocytes. Their function will be detailed in Chapter 13.

Agranular Leukocytes The agranular leukocytes include lymphocytes, which have a spherical nucleus, and monocytes, which have a kidney-shaped nucleus. **Lymphocytes** are responsible for specific immunity to particular pathogens and their toxins (poisonous substances), as well as recognizing and destroying cancer cells. Lymphocytes (see Fig. 13.3d) are of two types, B lymphocytes and T lymphocytes. Pathogens have **antigens,** surface molecules that the immune system can recognize as foreign. When an antigen is recognized as foreign, B lymphocytes will form antibodies against it. **Antibodies** are proteins that neutralize antigens. T lymphocytes, on the other hand, directly attack and destroy any cell, such as a pathogen that has foreign antigens. B lymphocytes and T lymphocytes are discussed more fully in Chapter 13.

Monocytes (see Fig. 13.3e) are the largest of the white blood cells, and after taking up residence in the tissues, they differentiate into even larger macrophages. **Macrophages** phagocytize pathogens, old cells, and cellular debris. They

also stimulate other white blood cells, including lymphocytes, to defend the body.

Abnormal White Cell Counts

Abnormal white blood cell counts are discussed in the Medical Focus on page 239. Because specific white blood cells increase with particular infections, a differential white cell count, also discussed in the Medical Focus, can be quite helpful in diagnosing the cause of a particular illness.

Content **CHECK-UP!**

4. Which of the following might stimulate the release of erythropoietin (EPO)?

   a. hyperventilation   c. living at high attitude

   b. blood loss   d. both b and c can be stimuli

5. Which of the following is/are bile pigments?

   a. hemoglobin   d. b and c

   b. biliverdin   e. all of the above

   c. bilirubin

6. Granular white blood cells include all of the following except:

   a. eosinophils.   c. neutrophils.

   b. macrophages.   d. basophils.

Answers in Appendix C, page 464.

11.3 Platelets and Hemostasis

Platelets (thrombocytes) are formed elements necessary to the process of **hemostasis,** the cessation of bleeding.

Platelets

Platelets result from fragmentation of certain large cells, called **megakaryocytes,** that develop in red bone marrow. It is important to note that although platelets are called thrombocytes, they are *not* cellular; rather, they are cell *fragments.* Platelets are produced at a rate of 200 billion per day, and the blood contains 150,000–300,000 per mm^3. Because platelets have no nucleus, they last at most ten days, assuming they are not used sooner than that in hemostasis.

Hemostasis

Hemostasis is divided into three events: vascular spasm, platelet plug formation, and coagulation (Fig. 11.5).

Vascular spasm, the constriction of the smooth muscle layer in a broken blood vessel, is the immediate response to blood vessel injury. Platelets release serotonin, a chemical that prolongs smooth muscle contraction.

Platelet plug formation is the next event in hemostasis. Platelets don't normally adhere to undamaged blood vessel walls, but when the lining of a blood vessel breaks, connective

red blood cell

fibrin threads

blood clot

Figure 11.5 Process of hemostasis. Hemostasis requires three events: vascular spasm, platelet plug formation, and coagulation. Coagulation is further broken down into four steps.

tissue, including collagen fibers, is exposed. Platelets adhere to collagen fibers and release a number of substances, including one that promotes platelet aggregation so that a so-called *platelet plug* forms. As a part of normal activities, small blood vessels often break, and a platelet plug is usually sufficient to stop the bleeding.

Coagulation, also called blood clotting, is the last event to bring about hemostasis. As you will see, two plasma proteins, called **fibrinogen** and **prothrombin,** participate in blood clotting. Vitamin K, found in green vegetables and also formed by intestinal bacteria, is necessary for the production of prothrombin. If, by chance, vitamin K is missing from the diet, **hemorrhagic bleeding** disorders develop.

Coagulation

Coagulation requires many protein clotting factors which are constantly present in the blood in an inactive state. Most of these clotting factors are produced by the liver, although several are released "on demand" by platelets if tissues or blood vessels are damaged. There are two mechanisms for activation of clotting. The **intrinsic mechanism,** so-named thus because all clotting factors are *intrinsic* to the blood, is the slower pathway. The intrinsic mechanism explains why blood will coagulate if placed in a test tube that has not been treated with an anticoagulant. In the body, the intrinsic mechanism is initiated by contact with the exposed collagen in a torn blood vessel. The **extrinsic mechanism** is activated when damaged tissues release a substance called **tissue thromboplastin** (thus, its activation is *extrinsic* to blood itself). Tissue thromboplastin will interact with platelets, other clotting factors, and calcium ions (Ca^{2+}).

Continued—

Figure 11B Sickle-shaped red blood cells, as seen by a scanning electron microscope.

where malaria is common. Malaria is a disease caused by a parasite and spread by mosquitoes that infect humans when they bite. Having sickle-cell trait gives the person a better chance of surviving malaria, though it doesn't protect against the actual infection or subsequent illness. Scientists aren't sure why the trait enables some people to survive, but when survivors have children, the sickle-cell gene passes on to the next generation. In the United States, sickle-cell disease is most common in African Americans whose ancestors are from equatorial Africa, but it occurs in people of any race.

Dietary anemias occur because the patient's diet lacks substances needed for red blood cell development. In **iron deficiency anemia**, a common type of anemia, the person's diet does not contain enough iron, or there is excessive iron loss from the body. The hemoglobin count is low, red blood cells are small and pale in color, and the individual feels tired and run-down. Including iron supplements in the diet can help prevent this type of anemia. **Pernicious anemia** is another form of dietary anemia. The digestive tract is unable to absorb enough vitamin B_{12}, which is essential to red cell development. Without vitamin B_{12}, immature red cells tend to accumulate in the bone marrow. A special diet, vitamin supplements, and/or injections of vitamin B_{12} are effective treatments.

In **aplastic anemia,** the red bone marrow has been damaged due to radiation or chemicals, and red blood cells, white blood cells, and platelets are all deficient. Bone marrow transplant is one option to treat this condition.

Hemorrhagic anemia is decreased red blood cell count following a hemorrhage. Transfusions may be administered to increase red blood cell count. In addition, alternatives to transfusion are being developed, as detailed in "What's New" on page 237.

Certain viral illnesses, such as influenza, measles, and mumps, cause the white blood cell count to decrease. **Leukopenia** is a total white blood cell count below 5,000 per cubic millimeter. Other illnesses, including appendicitis and bacterial infections, cause the white blood cell count to increase dramatically. **Leukocytosis** is a white blood cell count above 10,000 per cubic millimeter.

Illness often causes an increase in a particular type of white blood cell. For this reason, a **differential white blood cell count** may help with diagnosis. In this microscopic procedure, 100 white blood cells are counted, and numbers of each type of cell are recorded. For example, the characteristic finding in the viral disease **mononucleosis** is a great number of large, dark-staining abnormal lymphocytes. This condition takes its name from the fact that the abnormal lymphocytes resemble monocytes (although monocytes are much larger than lymphocytes).

Leukemia is a form of cancer characterized by uncontrolled production of abnormal white blood cells. These cells accumulate in the bone marrow, lymph nodes, spleen, and liver so that these organs are unable to function properly. **Acute lymphoblastic leukemia (ALL),** which represents over 80% of the acute leukemias in children, also occurs in adults. Chemotherapy is used to destroy abnormal cells and restore normal blood cell production. Intraspinal injection of drugs and craniospinal irradiation are measures that prevent leukemic cells from infiltrating the central nervous system. In general, the prognosis is more favorable for children between the ages of 2 and 10 years, and females fare better because leukemia recurs in the testes of 8–16% of males. Remission occurs in 78% of adult patients after chemotherapy, and the median period of remission is 20 months. With chemotherapy, 50–60% of children survive past five years, and of those among this group who do not have a relapse, 85% are considered cured.

In the body, both clotting pathways are activated simultaneously; their common end point is production of prothrombin activator. Figure 11.5 breaks down the subsequent clotting process into four steps: ① **prothrombin activator** is formed. ② Prothrombin activator then converts prothrombin to **thrombin.** ③ Thrombin, in turn, severs two short

amino acid chains from each fibrinogen molecule, and these activated fragments join end-to-end, forming long threads of **fibrin.** ④ Fibrin threads wind around the platelet plug in the damaged area of the blood vessel and provide the framework for the clot. Red blood cells also are trapped within the fibrin threads. These cells make a clot appear red. Once clotting is

Hemorrhage

Anyone who drives a car might be involved in a minor fender-bender type of accident. Thankfully, in those situations only the car gets "hurt." However, as you know, a car can be a deadly piece of equipment. Car accidents are the leading cause of death for people ages 16–26, and the fifth-leading cause of death overall, according to the Centers for Disease Control (CDC). Tragically, many of these deaths are caused by severe bleeding, or **hemorrhage**, and might have been prevented.

When you're the first person at an accident scene, your help could make the difference between life or death for a hemorrhaging patient. It's important to react appropriately—first get yourself and the victim out of danger, and then call for assistance. Next, immediately check the patient's ABCs: Airway—ensure that the airway is intact and open; Breathing—observe regular respiration; and Circulation—check for heartbeat and pulse. Once you've taken these steps, act quickly to stop or slow any serious bleeding. Excessive blood loss can cause blood pressure to fall, a condition called **hypovolemic shock**. Shock causes organ failure and death if not treated.

Experts agree that the best way to stop a hemorrhage is to quickly apply direct pressure to the injury site, using gauze, towels, clothing, or even your bare hand if you must. Not only does pressing on a bleeding wound stop or slow the bleeding, it will also start the blood clotting process. If the cloth becomes soaked, don't remove it; add more layers. A clot may have begun to form in the fabric, and peeling it away might cause bleeding to start again. In addition, elevate the victim's legs, and keep him or her warm. Taking these steps will help prevent shock.

At the accident scene, EMS providers can begin an IV while transporting the patient to the emergency room. In addition, paramedic crews, hospital emergency rooms, and military medics now carry prepared dressings that are designed to stop bleeding, even in deep wounds. All work by quickly drawing water out of blood plasma, concentrating the clotting factors at the site of the wound. This promotes rapid clot formation, sometimes stopping even a severe hemorrhage within minutes.

---

initiated, the clotting process will be self-limiting and confined to just the area of injury. This important safeguard prevents the clot from becoming too large, possibly blocking blood flow to other body areas. Limiting the clotting process to only the injured area also prevents excessive consumption of the available clotting factors in the blood.

Clot retraction follows, and the clot gets smaller as platelets contract. A fluid called **serum** (plasma minus fibrinogen and prothrombin) is squeezed from the clot. A fibrin clot is present only temporarily. As soon as blood vessel repair is initiated, an enzyme called *plasmin* destroys the fibrin network and restores the fluidity of the plasma.

Naturally, it is vital to activate clotting when tissues have been injured. However, it is just as important that clotting *not* be activated if there is no injury. Normally, the smooth endothelial lining of an intact blood vessel prevents clots from forming in the blood vessel. Anticoagulants, such as heparin produced by basophils and mast cells, also help to prevent accidental clotting.

Disorders of Hemostasis

Both impaired clotting and excessive clotting can be life-threatening or fatal. **Thrombocytopenia**, a low platelet count, is one cause of impaired clotting. It occurs when one's own antibodies attack platelets, or if megakaryocytes in red bone marrow (parent cells that fragment to form platelets) are destroyed. The **hemophilias** are inherited clotting disorders caused by deficiencies of clotting factors. (The most severe form, hemophilia A, is due to the lack of clotting factor VIII, a step in the intrinsic clotting pathway.) In these clotting impairments, the slightest bump can cause bleeding into the tissues. Bleeding into a joint causes cartilage degeneration, and resorption of underlying bone can follow. Bleeding into muscles can lead to nerve damage and muscular atrophy. The most frequent cause of death is bleeding into the brain with accompanying neurological damage.

Undesirable, excessive clotting may block blood flow to tissues. Despite the presence of anticoagulants in the blood, sometimes a clot forms in an unbroken blood vessel. Such a clot is called a **thrombus** if it remains stationary. If the clot dislodges and travels in the blood, it is called an **embolus**. If this embolus blocks a blood vessel, it is a **thromboembolism**. Pulmonary thromboembolism (blockage of the pulmonary arteries by a clot) will seriously inhibit the oxygenation of the blood; cerebral thromboembolism will cause a **cerebrovascular accident** or **stroke**. Coronary thrombosis or thromboembolism causes a heart attack, as discussed in Chapter 12.

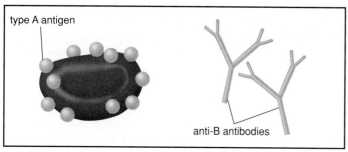

Type A blood. Red blood cells have type A surface antigens. Plasma has anti-B antibodies.

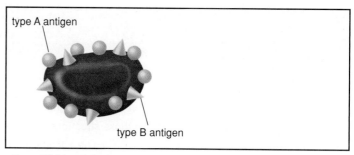

Type AB blood. Red blood cells have type A and type B surface antigens. Plasma has neither anti-A nor anti-B antibodies.

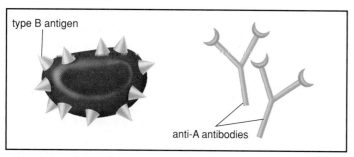

Type B blood. Red blood cells have type B surface antigens. Plasma has anti-A antibodies.

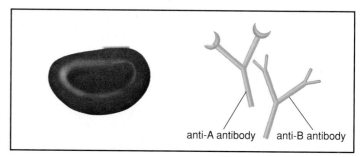

Type O blood. Red blood cells have neither type A nor type B surface antigens. Plasma has both anti-A and anti-B antibodies.

Figure 11.6 Types of blood. In the ABO system, blood type depends on the presence or absence of antigens A and B on the surface of red blood cells. In these drawings, A and B antigens are represented by different shapes on the red blood cells. The possible anti-A and anti-B antibodies in the plasma are shown for each blood type. Notice that an anti-B antibody cannot bind to an A antigen, and vice versa.

Content CHECK-UP!

7. The parent cell that fragments to form thrombocytes is called a:

   a. stem cell.   c. megalocyte.

   b. monocyte.   d. megakaryocyte.

8. Choose the correct sequence for hemostasis (first step to last):

   a. coagulation → vascular spasm → platelet plug formation

   b. vascular spasm → coagulation → platelet plug formation

   c. vascular spasm → platelet plug formation → coagulation

   d. coagulation → platelet plug formation → vascular spasm

9. Which vitamin is necessary to form prothrombin?

   a. B_{12} c. C

   b. folic acid   d. K

Answers in Appendix C, page 464.

11.4 Blood Typing and Transfusions

A **blood transfusion** is the transfer of blood from one individual into the blood of another. In order for transfusions to be done safely, it is necessary for blood to be typed so that **agglutination** (clumping of red blood cells) does not occur. Blood typing usually involves determining the ABO blood group and whether the individual is Rh⁻ (negative) or Rh⁺ (positive).

ABO Blood Groups

ABO blood typing is based on the presence or absence of two possible antigens, called type A antigen and type B antigen, on the surface of red blood cells. Whether these antigens are present or not depends on the particular inheritance of the individual.

A person with type A antigen on the surface of the red blood cells has type A blood; one with type B blood has type B antigen on the surface of the red blood cells. What antigens would be present on the surface of red blood cells if the person has type AB blood or type O blood? Notice in Figure 11.6 that a person with type AB blood has both antigens. A person with type O blood has no AB antigens on the surface of the red blood cell, though other protein antigens will be present.

It so happens that an individual with type A blood has anti-B antibodies in the plasma; a person with type B blood has anti-A antibodies in the plasma; and a person with type O blood has both antibodies in the plasma (Fig. 11.6). These antibodies are not present at birth, but they appear over the course of several months after birth.

Blood compatibility is very important when transfusions are done. The antibodies in the plasma must not combine with the antigens on the surface of the red blood cells, or else agglutination occurs. With agglutination, anti-A antibodies have combined with type A antigens, or anti-B antibodies have combined with type B antigens, or both types of binding have occurred. Therefore, agglutination is expected if the donor has type A blood and the recipient has type B blood (Fig. 11.7).

a. No agglutination

type A blood of donor + anti-B antibody of type A recipient → no binding

No clumping seen. Successful blood type match.

b. Agglutination

type A blood of donor + anti-A antibody of type B recipient → binding

Clumping seen. Hemolysis occurs. Unsuccessful blood type match.

Figure 11.7 Cross matching before a blood transfusion. No agglutination **(a)** versus agglutination **(b)** is determined by whether antibodies are present that can combine with antigens.

TABLE 11.2 Transfusion	
If your blood type is:	**You can safely receive blood of this type***
A	A or O
B	B or O
AB	A, B, AB, or O
O	O

*Regardless of ABO type, a person who is Rh⁻ can never be transfused with Rh⁺ blood.

What about other combinations of blood types? Table 11.2 shows when a blood transfusion is most likely safe. Type O blood is sometimes called the *universal donor* because it has no antigens on the red blood cells, and type AB blood is sometimes called the *universal recipient* because this blood type has no antibodies in the plasma. In practice, however, there are other possible blood groups, aside from ABO blood groups, so it is necessary to physically put the donor's blood on a slide with the recipient's blood and observe whether the blood types match (no agglutination occurs) before blood can be safely given from one person to another. This is called cross-matching blood.

As explained in the What's New reading on page 237, new technology may help to eliminate the problems of matching blood types.

Rh Blood Groups

The designation of blood type usually also includes whether the person has or does not have the Rh factor on the red blood cell. Rh⁻ individuals normally do not have antibodies to the Rh factor, but they make them when exposed to the Rh factor.

If a mother is Rh⁻ and the father is Rh⁺, the fetus conceived can be Rh⁺. The Rh⁺ red blood cells may begin leaking across the placenta into the mother's cardiovascular system, as placental tissues normally break down before and at birth. The presence of these Rh⁺ antigens causes the mother to produce anti-Rh antibodies. In a subsequent pregnancy with another Rh⁺ baby, the anti-Rh antibodies may cross the placenta and destroy the child's red blood cells. This is called hemolytic disease of the newborn (HDN) because hemolysis continues after the baby is born (Fig. 11.8). Due to red blood cell destruction the baby can be severely anemic at birth. Excess bilirubin in the blood can lead to brain damage and developmental delay, or even death.

The Rh problem is prevented by giving Rh⁻ women an Rh immunoglobulin injection at 28 weeks of pregnancy, followed by a second injection no later than 72 hours after giving birth to an Rh⁺ child. This injection, called Rho-Gam, contains anti-Rh antibodies that attack any of the baby's red blood cells in the mother's blood before these cells can stimulate her immune system to produce her own antibodies. This injection is not beneficial if the woman has already begun to produce antibodies; therefore, the timing of the injection is most important.

Content **CHECK-UP!**

10. Which of the following correctly describes type AB blood?

   a. A antigen and B antigen on the red cell membrane, anti-A and anti-B antibodies in plasma

   b. A antigen and B antigen on the red cell membrane, no antibodies in plasma

   c. A antigen on the red cell membrane, anti-B antibodies in plasma

   d. B antigen on the red cell membrane, anti-A antibodies in plasma

11. Which of the following transfusions would most likely be safe to administer?

   a. type A blood to a type B blood recipient

   b. type AB blood to a type B blood recipient

   c. type B blood to a type AB recipient

   d. type AB blood to a type O recipient

12. Hemolytic disease of the newborn occurs in which situation?

   a. Rh⁺ mother, Rh⁻ father, Rh⁺ fetus

   b. Rh⁻ mother, Rh⁻ father, Rh⁺ fetus

   c. Rh⁻ mother, Rh⁺ father, Rh⁺ fetus

   d. Rh⁺ mother, Rh⁻ father, Rh⁻ fetus

Answers in Appendix C, page 464.

red blood cell

anti-Rh antibody

Child is Rh positive; mother is Rh negative.

Red blood cells leak across placenta.

Mother makes anti-Rh antibodies.

Antibodies attack Rh-positive red blood cells in child.

Figure 11.8 Hemolytic disease of the newborn. Due to a pregnancy in which the child is Rh positive, an Rh-negative mother can begin to produce antibodies against Rh-positive red blood cells. In a subsequent pregnancy, these antibodies can cross the placenta and cause hemolysis of an Rh-positive child's red blood cells.

FOCUS on FORENSICS

Blood at the Crime Scene

In cases of violent crime such as homicide, assault, or sexual assault, the presence of blood at the crime scene is often the most important evidence. Blood evidence can tell an entire story about the crime—how a victim was attacked, whether the victim fought back or tried to escape, the type of weapon used, and perhaps even the identity of the attacker. Extensive research on the chemistry and properties of blood evidence, along with detailed examination of the crime scene, can enable a *forensic serologist* to help solve violent crime cases. (Forensic serologists are scientists trained in the examination of blood and body fluid at a crime scene.)

Upon arrival at the scene, investigators must first carry out a careful visual inspection for the presence of blood. Because blood hemoglobin is a protein molecule (globin) containing an iron pigment (heme), its stains are very difficult to remove completely from contaminated areas. As a result, a criminal who tries to clean up blood will almost always leave traces behind. Investigators sample between floorboards, under baseboards, in carpet padding, and on all surface areas. Tests can be done to confirm that a stain is due to human blood and not some other reddish-brown chemical. One test involves spraying the stain with luminol, a chemical that binds with blood and then glows in the dark. The Takayama and Teichmann tests cause crystals to form from hemoglobin, confirming the presence of blood. Furthermore, by testing with anti-human antibodies (protein molecules from rabbits), researchers can prove that a sample is in fact human in origin.

Careful documentation and analysis of the location and shape of any drops, spatters, or smears follows next. This type of evidence is carefully recorded and can also help investigators to piece together the events of the crime—the position of the suspect and the victim, whether a struggle occurred, if the victim moved or was dragged, etc. For example, research has shown that the shape of blood drops can yield useful information about a crime. If sprayed from a short distance, a blood drop will form a perfect circle on a wall or floor; however, blood that travels a greater distance will form a "sunburst" pattern. Droplet shape also yields information about the speed of travel from a person's body. When a droplet travels slowly from the body to an object (for example, a wall), it will form a circular drop. Blood that travels at a fast speed from the body to the object will form an elongated, "teardrop" shape. Blood flying in a trail of droplets from a knife or a club will leave a trail of spatters called a *castoff pattern*. Smeared blood may indicate that the body has been dragged, or that the victim attempted to escape.

If the suspect was wounded in a struggle, the suspect's blood may be present at the crime scene along with that of the victim. Blood chemistry is as individual as a fingerprint, and can positively confirm the identity of the suspect. ABO and Rh blood typing, studies of unique blood proteins, and DNA analysis (see page 59) are carried out on all blood samples. Other body fluids, if present, may also need to be tested because approximately 80% of the human population is comprised of *secretors*—individuals whose blood proteins and enzymes are present in saliva, tears, urine, semen, and skin, as well as in their blood. The data from these tests identify the source of blood samples, which can be statistically narrowed down to one person in several billion. Thus, for all practical purposes, a positive match against a suspect's blood can be made. Evidence such as this can prove that a suspect was at the crime scene and result in a conviction. Just as important, blood analysis may serve to acquit an innocent person.

11.5 Effects of Aging

Anemias, leukemias, and clotting disorders increase in frequency with age. As with other disorders, good health habits can help prevent these conditions from appearing.

Iron deficiency anemia most frequently results from an iron-deficient diet, but it can also result from a gradual and sometimes undetectable loss of blood. It is vital that the bleeding source be discovered as soon as possible. For example, iron deficiency anemia is often a first symptom of colon cancer in its early stage. The blood is often not visible in a person's stool because it has oxidized and is concealed in the stool. An *occult blood test* (test for oxidized blood present in the stool) can detect this source of blood loss.

Pernicious anemia most often signals that the digestive tract is unable to absorb enough vitamin B_{12}, but it can also be due to a diet deficient in B_{12}. Diets that completely exclude animal-derived protein sources (called *vegan* diets) are often B_{12} deficient. A vitamin supplement containing B_{12} will prevent pernicious anemia in vegan dieters.

Leukemia is a form of cancer that generally increases in frequency with age because of both intrinsic (genetic) and extrinsic (environmental) reasons.

Thromboembolism, a clotting disorder, may be associated with the progressive development of atherosclerosis in an elderly person. When arteries develop plaque (see Fig. 12A, p. 254), thromboembolism often follows. For many people, atherosclerosis can be controlled by diet and exercise, as discussed in the Chapter 12 Medical Focus, "Preventing Cardiovascular Disease."

SELECTED NEW TERMS

Basic Key Terms

agglutination (uh-glu"tĭ-na'shun), p. 242

agranular leukocyte (a-gran'yū-ler lu'ko-sīt), p. 236

albumin (al-byū'min), p. 234

antibody (an'tĭ-bod"e), p. 238

antigen (an'tĭ-jen), p. 238

basophil (ba'so-fil), p. 238

bile pigments (bīl pig'-měnts), p. 236

buffy coat (bŭ'fē kōt), p. 232

coagulation (ko-ag"yū-la'shun), p. 239

eosinophil (e"o-sin'o-fil), p. 238

erythrocyte (ĕ-rith"ro-sīt'), p. 234

erythropoietin (ĕ-rith"ro-poy'ĕ-tin), p. 235

fibrin (fi'brin), p. 240

fibrinogen (fi-brin'o-jen), p. 239

formed element (formd el'ĕ-ment), p. 232

globulin (glob'u-lin), p. 234

granular leukocyte (gran'u-ler lu'ko-sīt), p. 236

hematocrit (he-mat'o-krit), p. 232

hematopoiesis (hē'mə-tō-poi-ē'sĭs), p. 234

hemoglobin (he"mo-glo'bin), p. 234

hemolysis (he-mol'ĭ-sis), p. 236

hemostasis (hē'mə-stā'sĭs), p. 238

leukocytes (loō'kə-sīt's), p. 236

lipoproteins (lĭp'ō-prō'tēn's), p. 234

lymphocyte (lim'fo-sīt), p. 238

macrophage (mak'rō-fāj), p. 238

megakaryocyte (meg'uh-kār'e-o-sīt), p. 238

monocyte (mon'o-sīt), p. 238

multipotent stem cell (mŭl'tĭ-pōt'nt stĕm sĕl), p. 234

neutrophil (nu'tro-fil), p. 238

osmotic pressure (oz-mot'ik presh'ur), p. 234

pathogen (path'o-jen), p. 232

plasma (plaz'muh), p. 232

platelet (plāt'let), p. 238

platelet plug (plāt'let plŭg), p. 238

prothrombin (pro-throm'bin), p. 239

prothrombin activator (pro-throm'bin ak'tĭ-va"tor), p. 240

red blood cell (red blud sel), p. 234

thrombin (throm'bin), p. 240

thrombocyte (throm'bō-sīt), p. 238

white blood cell (whīt blud sel), p. 236

Clinical Key Terms

acute lymphoblastic leukemia (uh-kyūt' lim-fo-blas'tik lu-ke'me-uh), p. 240

anemia (uh-ne'me-uh), p. 239

autotransfusion (ăh'tō-trănz-fū"shun) p. 237

aplastic anemia (a-plas'tik uh-ne'me-uh), p. 240

blood transfusion (blud trans-fyū'zhun), p. 242

cerebrovascular accident (sĕr'ə-brō-văs'kyl-ər ăk'sĭ-dənt), p. 241

differential white blood cell count (dif"er-en'shul whīt blud sel kownt), p. 240

embolus (em'bo-lus), p. 241

hemolytic anemia (he-mo-lĭt'ik uh-ne'me-uh), p. 239

hemolytic disease of the newborn (he-mo-lĭt'ik dĭ-zēz' ov thah nu'born), p. 239

hemophilia (he-mo-fil'e-uh), p. 241

hemorrhage (hĕm'ō-rĭj), p. 241

hemorrhagic anemia (hem'o-raj-ik uh-ne'me-uh), p. 240

hemorrhagic bleeding (hem'o-raj-ik ble'ding), p. 239

hypovolemic shock (hĭ'-pō-vō-lĕ'mĭk shŏk), p. 241

iron deficiency anemia (i'ern dĭ-fi'shun-se uh-ne'me-uh), p. 240

leukemia (lu-ke'me-uh), p. 240

leukocytosis (lu"ko-si-to'-sis), p. 240

leukopenia (lu"ko-pe'ne-uh), p. 240

mononucleosis (mon"o-nu"kle-o'sis), p. 240

pernicious anemia (per-nĭ'shus uh-ne'me-uh), p. 240

polycythemia (pol"e-si-the'me-uh), p. 236

sickle-cell disease (sĭ'kl sel dĭ-z ēz'), p. 239

sickle-cell trait, p. 239

stroke (strōk), p. 241

thrombocytopenia (throm"bo-si-to-pe'ne-uh), p. 241

thromboembolism (throm"bo-em'bo-lizm), p. 241

thrombus (throm'bus), p. 241

SUMMARY

11.1 The Composition and Functions of Blood

A. Blood, which is composed of formed elements and plasma, has several functions. It transports hormones, oxygen, and nutrients to the cells and carbon dioxide and other wastes away from the cells. It fights infections. It regulates body temperature, and keeps the pH of body fluids within normal limits. All of these functions help maintain homeostasis.

11.2 Components of Blood

A. Plasma is mostly water (92%) and solutes (8%). Small organic molecules such as glucose and amino acids are dissolved in plasma and serve as nutrients for cells; urea is a waste product. Large organic molecules include the plasma proteins. The plasma proteins, most of which are produced by the liver, occur in three categories: albumins, globulins, and fibrinogen. The plasma proteins maintain osmotic pressure, help regulate pH, and transport molecules. Some plasma proteins have specific functions: The gamma globulins, which are antibodies produced by B lymphocytes, function in immunity; fibrinogen and prothrombin are necessary to blood clotting.

B. All blood cells, including red blood cells, are produced within red bone marrow from stem cells, which are ever capable of dividing and producing new cells.

C. Red blood cells are small, biconcave disks that lack a nucleus. They contain hemoglobin, the respiratory pigment, which combines with oxygen and transports it to the tissues. Red blood cells live about 120 days and are destroyed in the liver and spleen when they are old or abnormal. The production of red blood cells is controlled by the oxygen concentration of the blood. When the oxygen concentration decreases, the kidneys increase their production of erythropoietin, and more red blood cells are produced.

D. White blood cells are larger than red blood cells, have a nucleus, and are translucent unless stained. Like red blood cells, they are produced in the red bone marrow. White blood cells are divided into the granular leukocytes and the agranular leukocytes. The granular leukocytes have conspicuous granules; in eosinophils, granules are red when stained with eosin, and in basophils, granules are blue when stained with a basic dye. In neutrophils, some of the granules take up eosin, and others take up the basic dye, giving them a lilac color. Neutrophils are the most plentiful of the white blood cells, and they are able to phagocytize pathogens. Many neutrophils die within a few days when they are fighting an infection. The agranulocytes include the lymphocytes and the monocytes, which function in specific immunity. On occasion, the monocytes become large phagocytic cells of great significance. They engulf worn-out red blood cells and pathogens at a ferocious rate.

11.3 Platelets and Hemostasis

A. The extremely plentiful platelets result from fragmentation of megakaryocytes.

B. The three events of hemostasis are vascular spasm, platelet plug formation, and coagulation. Extrinsic and intrinsic pathways that activate the clotting mechanism both cause formation of prothrombin activator, which breaks down prothrombin to thrombin. Thrombin changes fibrinogen to fibrin threads, entrapping cells. The fluid that escapes from a clot is called serum and consists of plasma minus fibrinogen and prothrombin.

11.4 Blood Typing and Transfusions

A. ABO typing is the most common blood typing system used. Type A, type B, both type A and B, or no antigens can be on the surface of red blood cells. In the plasma, there are two possible antibodies: anti-A or anti-B. If the corresponding antigen and antibody are put together during a transfusion, agglutination occurs. Therefore, it is necessary to determine an individual's blood type before a transfusion is given.

B. Another important antigen is the Rh antigen. This particular antigen must also be considered in transfusing blood, and it is important during pregnancy because an Rh^- mother may form antibodies to the Rh antigen when giving birth to an Rh^+ child. These antibodies can cross the placenta and destroy the red blood cells of any subsequent Rh^+ child.

11.5 Effects of Aging

As we age, anemias, leukemias, and clotting disorders increase in frequency.

STUDY QUESTIONS

1. Name the two main components of blood, and describe the functions of blood. (p. 232)
2. List and discuss the major components of plasma. Name several plasma proteins, and give a function for each. (pp. 232–234)
3. What is hemoglobin, and how does it function? (p. 234)
4. Describe the life cycle of red blood cells, and tell how the production of red blood cells is regulated. (pp. 234–236)

5. Name the five types of white blood cells. Describe the structure and give a function for each type. (pp. 236, 238)
6. Name the steps that take place when blood clots. Which substances are present in blood at all times, and which appear during the clotting process? (pp. 238–241)
7. What are the four ABO blood types? For each, state the antigen(s) on the red blood cells and the antibody(ies) in the plasma. (p. 242)
8. Explain why a person with type O blood cannot receive a transfusion of type A blood. (pp. 242–243)
9. Problems can arise if the mother is which Rh type and the father is which Rh type? Explain why this is so. (pp. 243–244)

LEARNING OBJECTIVE QUESTIONS

I. Fill in the blanks.

1. The liquid part of blood is called _____.
2. Red blood cells carry _____, and white blood cells _____.
3. Hemoglobin that is carrying oxygen is called _____.
4. Human red blood cells lack a _____ and only live about _____ days.
5. The most common granular leukocyte is the _____, a phagocytic white blood cell.
6. B lymphocytes produce _____, and T lymphocytes attack and _____ pathogens.
7. When a blood clot occurs, fibrinogen has been converted to _____ threads.
8. AB blood has the antigens _____ and _____ on the red blood cells and _____ of these antibodies in the plasma.
9. Hemolytic disease of the newborn can occur when the mother is _____ and the father is _____.

II. Match the terms in the key to the descriptions in questions 10–13.

Key:
   a. hematocrit
   b. red blood cell count
   c. white blood cell count
   d. hemoglobin

10. 5,000 to 11,000 per cubic millimeter
11. 4 to 6 million per cubic millimeter in males
12. Just under 45% of blood volume
13. 200 million molecules in one red blood cell

MEDICAL TERMINOLOGY EXERCISE

Consult Appendix B for help in pronouncing and analyzing the meaning of the terms that follow.

1. hematemesis (hem″uh-tem′ĕ-sis)
2. erythrocytometry (ĕ-rith″ro-si-tom′ĕ-tre)
3. leukocytogenesis (lu″ko-si″to-jen′ĕ-sis)
4. hemophobia (he″mo-fo′be-uh)
5. afibrinogenemia (uh-fi″brin-o-jĕ-ne′me-uh)
6. lymphosarcoma (lim″fo-sar-ko′muh)
7. phagocytosis (fag″o-si-to′sis)
8. phlebotomy (flĕ-bot′o-me)
9. hemocytoblast (he′mo-si′to-blast)
10. megaloblastic anemia (meg′uh-lo-blas′tik uh-ne′me-uh)
11. microcytic hypochromic anemia (mi′kro-sit′ik hi″po-kro′mik uh-ne′me-uh)
12. hematology (he′muh-tol′o-je)
13. lymphedema (limf′uh-de′muh)
14. antithrombin (an″te-throm″bin)

WEBSITE LINK

Visit this text's website at http://www.mhhe.com/longenbaker7 for additional quizzes, interactive learning exercises, and other study tools.

The Cardiovascular System

As you'll recall from Chapter 5, tattoos have been part of human cultural practice for centuries. However, it may surprise you to hear that tattoos once had an interesting medical application as well: ensuring the health of the original Mercury 7 astronauts before their space voyages in the early 1960s. John Glenn, pictured here, was one of the original pioneer astronauts, and the first man to orbit the Earth. He and his fellow astronauts underwent hundreds of electrocardiograms throughout the years of the program. An electrocardiogram is a record of the heart's electrical activity, and is an important indicator of heart health. Cardiologists (physicians who specialize in heart care) tattooed Glenn's chest so that when each repeated electrocardiogram was taken, the recording pads (called electrodes) would always be placed in the same position. By creating these permanent markings, the doctors hoped to ensure the accuracy of the recordings. Glenn was a pioneer once again, as the oldest space traveler—at age 77, he flew on the space shuttle *Discovery*, launched on October 29, 1998. This time, however, improved ECG technology eliminated the need for his tattoos. Instead, computers provided constant, real-time information about every aspect of his heart's performance: electrocardiogram, heart rate, strength of contraction, blood pressure, etc. You can find out more about how electrocardiograms are recorded and used in the Medical Focus on page 256.

Chapter Outline & Learning Objectives

After you have studied this chapter, you should be able to:

Chapter 11 described the components of blood and detailed the role that blood has in transport, defense, and regulation. In this chapter, we will first study the anatomy and physiology of the heart and of the blood vessels. Then, we will take a look at various branches of the circulatory system. A crucial function of circulation is to connect the body's trillions of cells to the organs of exchange: lungs, small intestine, and kidney. In the lungs, oxygen enters and carbon dioxide exits the blood. The small intestine absorbs nutrient molecules into the blood, and the kidneys allow metabolic wastes to exit the blood. The circulatory system assures homeostasis by carrying nutrients and metabolites to and from these organs.

The cardiovascular system consists of two components: (1) the heart, which pumps blood so that it flows to tissue capillaries and lung capillaries, and (2) the blood vessels through which the blood flows. As you can see in Figure 12.1, the cardiovascular system is divided into two functional systems. The right side of the heart and its blood vessels form the **pulmonary circuit,** which pumps blood to the lungs. The left side of the heart and its vessels form the **systemic circuit,** which supplies blood to the entire body.

12.1 Anatomy of the Heart

The heart is located in the thoracic cavity within the mediastinum, a serous membrane sac between the lungs. It is a hollow, cone-shaped, muscular organ. To approximate the size of your heart, make a fist and then clasp the fist with your opposite hand. Figure 12.1 shows that the base (the widest part) of the heart is superior to its apex (the pointed tip), which rests on the diaphragm. Also, the heart lies on a slant; the base is directed toward the right shoulder, and the apex points to the left hip. The base is deep to the second rib, and the apex is at the level of the fifth intercostal space.

As the heart pumps the blood through the pulmonary and systemic vessels, it performs these functions:

1. keeps O_2-poor blood separate from O_2-rich blood;
2. keeps the blood flowing in one direction—blood flows away from and then back to the heart in each circuit;
3. creates blood pressure, which moves the blood through the circuits;
4. regulates the blood supply based on the current needs of the body;
5. serves as an endocrine gland, producing atrial natriuretic peptide (ANP) for regulation of blood pressure.

The Wall and Coverings of the Heart

The heart is a hollow organ composed of three layers, as shown in Figure 12.2. The innermost layer, the **endocardium,** is a single layer of simple squamous epithelium, called endothelium. Endothelium not only lines the heart but it also continues into and lines the blood vessels. Its smooth nature helps prevent blood from clotting unnecessarily. The central **myocardium** is the thickest part of the heart wall and is made

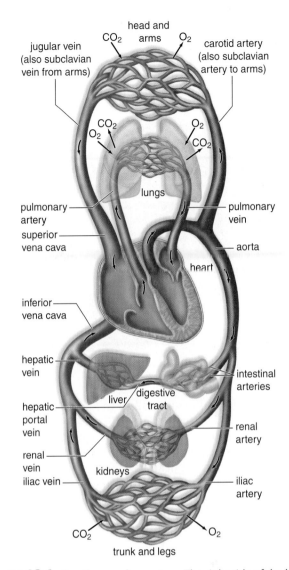

Figure 12.1 Cardiovascular system. The right side of the heart pumps blood through vessels of the pulmonary circuit. The left side of the heart pumps blood through vessels of the systemic circuit. Gas exchange occurs as blood passes through lung (pulmonary) capillaries. Gas exchange and nutrient-for-waste exchange occur as blood passes through tissue (systemic) capillaries. In this illustration, red vessels carry O_2-rich blood, and blue vessels carry O_2-poor blood.

up of cardiac muscle (see Fig. 4.15). When cardiac muscle fibers contract, the heart beats. The outermost layer is the **epicardium,** which is also called the **visceral serous pericardium** (the term *visceral* means organ, and refers to the fact that this layer covers the heart). After covering the heart, the visceral pericardium folds back over the heart, creating the parietal serous pericardium. The two serous membranes (epicardium and parietal pericardium) secrete pericardial fluid (a fluid similar to plasma). The pericardial fluid reduces friction as the heart beats. The parietal pericardium is fused to the outermost fibrous pericardium. The fibrous pericardium is a thick layer of fibrous connective tissue that adheres to the great blood vessels at the heart's base and anchors the heart to the diaphragm and the mediastinal wall. The coverings of the

simple squamous
epithelium

areolar connective
tissue and fat

epicardium
(visceral layer of
serous pericardium)

myocardium (cardiac muscle)

areolar connective tissue

endothelium

endocardium

Figure 12.2 The coverings and wall of the heart. The heart wall has three layers, from deep to superficial: endocardium, myocardium, and epicardium.

heart protect the heart, confine it to its location, and prevent it from overfilling, while still allowing the heart to contract and carry out its function of pumping the blood.

A layer of the heart can become inflamed due to infection, cancer, injury, or a complication of surgery. The suffix "itis" added to the name of a heart condition tells which layer is affected. For example, pericarditis refers to inflammation of the pericardium, and endocarditis refers to inflammation of the endocardium.

Chambers of the Heart

The heart has four hollow chambers: two superior **atria** (sing., **atrium**) and two inferior **ventricles** (Fig. 12.3). Each atrium has a wrinkled anterior pouch called an auricle. Internally, the atria are separated by the **interatrial septum,** and the ventricles are separated by the **interventricular septum.** Therefore, the heart's pulmonary circuit (its right side) is completely separated from its systemic circuit (the left side) by the septum.

The thickness of each chamber's myocardium is suited to its function. The atria have thin walls and each pumps blood into the ventricle below. The ventricles are thicker and they pump blood into blood vessels that travel to other parts of the body. The thinner myocardium of the right ventricle is suited for pumping blood to the lungs, which are nearby in the thoracic cavity. The left ventricle has a thicker wall than the right ventricle. Thicker myocardium enables the left ventricle to pump its blood to all other parts of the body.

Right Atrium

At its posterior wall, the **right atrium** receives O_2-poor blood from three veins: the *superior vena cava, the coronary sinus,* and the *inferior vena cava.* Venous blood passes from

the right atrium into the right ventricle through an **atrioventricular (AV) valve.** This valve, like the other heart valves, directs the flow of blood and prevents any backflow. The AV valve on the right side of the heart is specifically called the **tricuspid valve** because it has three *cusps,* or flaps.

Right Ventricle

In the **right ventricle,** the cusps of the tricuspid valve are connected to fibrous cords, called the **chordae tendineae** (meaning "heart strings"). The chordae tendineae in turn are connected to the **papillary muscles,** which are conical extensions of the myocardium.

Blood from the right ventricle passes through a **semilunar valve** into the *pulmonary trunk.* Semilunar valves are so called because their cusps are thought to resemble half-moons. This particular semilunar valve, called the **pulmonary semilunar valve,** prevents blood from flowing back into the right ventricle.

Note in Figure 12.3 that the pulmonary trunk divides into the left and right pulmonary arteries. For help in remembering how blood flows through the heart, trace the path of O_2-poor blood from the vena cava to the pulmonary arteries that take blood to the lungs (see Fig. 12.1).

Left Atrium

At its posterior wall, the **left atrium** receives O_2-rich blood from four *pulmonary veins.* Two pulmonary veins come from each lung. Blood passes from the left atrium into the left ventricle through an AV valve. The AV valve on the left side is specifically called the **bicuspid valve** because it has two cusps. (In the United States, the bicuspid valve is more commonly

left subclavian artery

left common carotid artery

brachiocephalic artery

superior vena cava

aorta

left pulmonary artery

pulmonary trunk

left pulmonary veins

right pulmonary artery

right pulmonary veins

pulmonary semilunar valve

left atrium

right atrium

atrioventricular (bicuspid) valve

atrioventricular (tricuspid) valve

chordae tendineae

papillary muscles

right ventricle

septum

left ventricle

inferior vena cava

a.

cardiac muscle cell

mitochondrion

intercalated disk

b.

Figure 12.3 Internal heart anatomy. a. The heart has four valves. The two atrioventricular valves allow blood to pass from the atria to the ventricles, and the two semilunar valves allow blood to pass out of the heart (the aortic semilunar valve lies deep to the pulmonary semilunar valve in this view). **b.** A cardiac muscle cell showing junctions called intercalated disks (see section 4.3 for review).

referred to as the **mitral valve,** so called because the valve is similar in shape to a bishop's hat, or mitre.)

Left Ventricle

The **left ventricle** forms the apex of the heart. The papillary muscles in the left ventricle are quite large, and the chordae tendineae attached to the AV valve are thicker and stronger than those in the right ventricle. As mentioned, the AV valve on the left side is also called the bicuspid (or mitral) valve.

Blood passes from the left ventricle through a semilunar valve into the aorta. This semilunar valve is appropriately called the **aortic semilunar valve.** The semilunar cusps of this valve are larger and thicker than those of the pulmonary semilunar valve.

Just beyond the aortic semilunar valve lie the first branches from the aorta. These are the **coronary arteries**—blood vessels that lie on and nourish the heart itself. The rest of the blood stays in the aorta, which continues as the arch of the aorta and then the descending aorta.

Operation of the Heart Valves and Heart Sounds

Let's take a look at how the valves of the heart operate to direct a one-way flow of blood from the atria to the ventricles to the arteries. The AV valves (tricuspid and mitral valve) are normally open. When a ventricle contracts, however, the increasing pressure of the blood inside the ventricle forces the cusps of the AV valve to slam shut. The force of the blood can be compared to a strong wind that can blow a door (the valve cusps) shut. However, when the ventricle contracts, the papillary muscles also contract, causing the chordae tendinae to tighten and pull on the valve. Thus, AV valves in the normal heart are prevented from inverting back up into the atrium.

The semilunar valves (pulmonary and aortic) are normally closed. However, the contraction of the ventricles pushes blood at high pressure against the valve cusps, forcing the valves open. Then, when the ventricle relaxes, the blood in the artery pushes backward, closing the valve once again.

A heartbeat produces the familiar "LUB-DUP" sounds as the chambers contract and the valves close. The first heart sound, "lub," is heard when the ventricles begin to contract and the atrioventricular (tricuspid and mitral) valves close. This sound lasts the longest and has a lower pitch. The second heart sound, "dup," is heard when the relaxation of the ventricles allows the semilunar (pulmonary and aortic) valves to close.

Like mechanical valves, the heart valves sometimes leak—in the language of medicine, the valves are *incompetent*. When valves don't close properly there is a backflow of blood. Blood leaks from the ventricle to the atrium (if the atrioventricular valve is incompetent) or from the pulmonary artery or aorta back into the ventricle (if the semilunar valve fails). Most often, the valves of the systemic circulation—bicuspid (or mitral) valves and the aortic semilunar valve—are affected. The backflow causes a heart murmur, which is typically a clicking or swishing sound heard after the first heart sound ("lub"). A trained physician or health professional can diagnose heart murmurs from their sound and timing. A person can be born with a valve deformity (called a *congenital valve defect*), or the valve can be damaged by bacterial infection. For example, rheumatic fever begins in the throat, and then spreads throughout the body. The bacteria attack connective tissue in the heart valves as well as other organs. Further, antibodies produced by the infected individual can also damage the valves. Fortunately, defective heart valves can be repaired surgically, or replaced with a synthetic valve or one taken from a pig's heart.

Coronary Circulation

Cardiac muscle fibers and the other types of cells in the wall of the heart are not nourished by the blood in the chambers; diffusion of oxygen and nutrients from this blood to all the cells that make up the heart would be too slow. Instead, these cells receive nutrients and rid themselves of wastes at capillaries embedded in the heart wall.

As mentioned previously, the right and left coronary arteries branch from the aorta just beyond the aortic semilunar valve (Fig. 12.4). Each of these arteries branches and then

left subclavian artery
left common carotid artery
brachiocephalic artery

superior vena cava
aorta

left pulmonary artery
pulmonary trunk
left pulmonary veins

right pulmonary artery

right pulmonary veins

left atrium
great cardiac vein
right atrium
right coronary artery

left ventricle
right ventricle

left anterior descending coronary artery

inferior vena cava

apex

Figure 12.4 Anterior view of exterior heart anatomy. The great vessels (venae cavae, pulmonary trunk, pulmonary arteries, and aorta) are attached to the base of the heart. The right ventricle forms most of the anterior surface of the heart, and the left ventricle forms most of the posterior surface. The coronary arteries bring oxygen and nutrients to cardiac cells, which derive little from the blood coursing through the heart.

rebranches, until the heart is encircled by small arterial blood vessels. Some of these join so that there are several routes to reach any particular capillary bed in the heart. Alternate routes are helpful if an obstruction should occur along the path of blood reaching cardiac muscle cells. The Medical Focus on p. 254 describes the cause, and some treatment options, for obstruction of the coronary arteries.

After blood has passed through cardiac capillaries, it is taken up by vessels that join to form veins. The coronary veins are specifically called **cardiac veins.** The cardiac veins enter a coronary sinus, which is essentially a thin-walled vein. The coronary sinus enters the right ventricle.

Content CHECK-UP!

1. Which valve is located between the right atrium and the right ventricle?
   a. pulmonary semilunar
   b. bicuspid (mitral)
   c. aortic semilunar
   d. tricuspid

2. The first heart sound is caused by the closure of the:
   a. AV valves (tricuspid and bicuspid).
   b. semilunar valves (aortic and pulmonary).

3. Conical extensions of the myocardium are called the:
   a. chordae tendinae.
   b. cusps.
   c. papillary muscles.
   d. valve leaflets.

Answers in Appendix C, page 464.

12.2 Physiology of the Heart

The physiology of the heart pertains to its pumping action—that is, the heartbeat. It is estimated that the heart beats two-and-a-half billion times in a lifetime, continuously recycling some 5 liters (L) of blood to keep us alive. In this section, we will consider what causes the heartbeat, what it consists of, and its consequences.

Conduction System of the Heart

The **conduction system of the heart** is a route of specialized cardiac muscle fibers that initiate and stimulate contraction of the atria and ventricles. The conduction system is said to be *intrinsic,* meaning that the heart beats automatically without the need for external nervous stimulation. The conduction system coordinates the contraction of the atria and ventricles so that the heart is an effective pump. Without this conduction system, the atria and ventricles would contract at different rates.

Nodal Tissue

The heartbeat is controlled by nodal tissue, which has both muscular and nervous characteristics. This unique type of cardiac muscle is located in two regions of the heart: The **SA (sinoatrial)**

node is located in the upper posterior wall of the right atrium. The **AV (atrioventricular) node** is located in the base of the right atrium very near the interatrial septum (Fig. 12.5).

The SA node initiates the heartbeat and automatically sends out an excitation impulse every 0.85 second. The SA node normally functions as the pacemaker for the entire heart because its intrinsic rate is the fastest in the system. From the SA node, impulses spread out over the atria, causing them to contract.

When the impulses reach the AV node, there is a slight delay that allows the atria to finish their contraction before the ventricles begin their contraction. The signal for the ventricles to contract travels from the AV node through the two branches of the **atrioventricular bundle (AV bundle)** before reaching the numerous and smaller **Purkinje fibers.** The AV bundle, its branches, and the Purkinje fibers consist of specialized cardiac muscle fibers that efficiently spread an electrical signal throughout the ventricles. Recall from Chapter 4 (p. 77) that cardiac muscle cells are connected to each other by specialized **gap junctions** called intercalated discs. These specialized intercellular connections allow electrical current to flow from cell to cell. Once stimulated electrically, the ventricular muscle contracts purposefully to pump blood.

The SA node **pacemaker** usually keeps the heartbeat regular. If the SA node fails to work properly, the ventricles still

Figure 12.5 Conduction system of the heart. ① The SA node sends out a stimulus, which causes the atria to contract. ② When this stimulus reaches the AV node, it passes through the atrial wall to the AV bundle. ③ Impulses pass down the two branches of the atrioventricular bundle. ④ Numerous small Purkinje fibers carry the electrical signal throughout the ventricle wall. Thereafter, the ventricles contract.

Arteriosclerosis, Atherosclerosis, and Coronary Artery Disease

The leading cause of heart attack and stroke, and the number-one killer in North America and Western Europe, is arteriosclerosis. **Arteriosclerosis** is the generalized term for abnormal thickening and hardening of the arterial wall over time. The most common form of arteriosclerosis is **atherosclerosis**. Scientists agree that atherosclerosis begins with injury to the arterial wall. Research has suggested several possible causes for injury: smoking, high blood pressure (called **hypertension**), and elevated levels of blood lipids, cholesterol, and homocysteine (a by-product of protein metabolism). Diabetics (especially those with type II or non-insulin-dependent diabetes mellitus) are at increased risk for atherosclerosis, probably because their disease causes high levels of blood lipids and cholesterol.

Research also indicates that low-level bacterial or viral infection that spreads to the blood may injure arterial walls and start the atherosclerosis process. This infection may originate with gum disease, or it can be caused by a bacterium called *Helicobacter pylori* (the microbe that also causes ulcers in the stomach). Antibodies specific to these microbes are found in people with atherosclerosis. In addition, a blood protein called C-reactive protein, or CRP, is an important piece of evidence suggesting that atherosclerosis may be caused by infection. For example, CRP levels rise in your blood if you suffer from a cold or are recovering from a wound. High blood levels of CRP in an otherwise healthy person imply that the arteries are damaged. Indeed, recent studies show that people with the highest blood levels of CRP have double the risk of heart attack.

Once the arterial wall is injured, the body's defense mechanisms respond. White blood cells called **macrophages** invade the injured area and stick to the arterial wall. These macrophages ingest low-density lipoprotein (LDL; often called "bad cholesterol"). Macrophages filled with LDL are called *foam cells* (because mixing fat with the cell's watery cytoplasm creates a foamy appearance). A collection of foam cells creates a **fatty streak**. Sadly, post-mortem studies on the arteries of young people have shown that these fatty streaks begin to develop during the early teenage years.

Over time, the artery's smooth muscle cells migrate to cover the fatty streak. Finally, fibroblasts and scar tissue will cover the smooth muscle cells. Calcium ions will invade the tissue, causing it to harden into an **atherosclerotic plaque.** Atherosclerotic plaques can grow so large that they completely block arterial blood flow, causing the tissue supplied by the artery to die. Because the plaque's surface is very rough compared to the smooth endothelium, the plaque may also trigger the clotting mechanism and cause a stationary blood clot, or thrombus, to form inside a blood vessel. Thrombi may also form if the surface of the plaque ulcerates (cracks open and bleeds). As mentioned in Chapter 11, **thromboembolism** occurs when a blood clot breaks away from its place of origin and is carried to a new location. Further, the plaque may cause the arterial wall to become brittle and harden, which might result in an **aneurysm.** Aneurysms are weakened areas in the arterial wall, which balloon outward and may even rupture.

Coronary artery disease is the term for atherosclerosis of the coronary arteries (Fig. 12A). If the coronary artery is partially occluded (blocked) by atherosclerosis, the individual may suffer from **ischemic heart disease.** Although enough oxygen normally reaches the resting heart, the person's heart is oxygen-deprived during periods of exercise or stress. This can lead to **angina pec-**

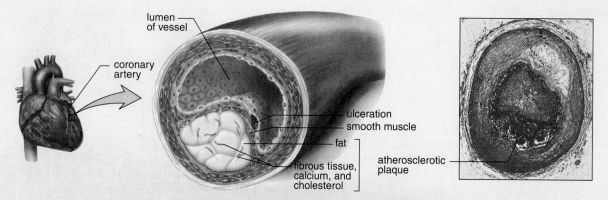

lumen of vessel
coronary artery
ulceration
smooth muscle
fat
fibrous tissue, calcium, and cholesterol
atherosclerotic plaque

Figure 12A Coronary artery disease. Atherosclerosis begins with injury to an artery; a fatty streak develops at the site and smooth muscle grows over the lesion. Fibrous tissue and ionic calcium enlarge and stiffen the atherosclerotic plaque. When plaque is present in a coronary artery, restricted blood flow may result in a heart attack.

toris, chest pain that is often accompanied by a radiating pain in the left arm. Angina pain is a warning sign of reduced blood flow to the heart and must not be ignored. Should the coronary artery become completely blocked by atherosclerotic plaque or thrombus, a portion of the heart will die from lack of oxygen. Dead tissue is called an **infarct** and, therefore, a heart muscle attack is termed a **myocardial infarction**.

Surgical Repair

Two surgical procedures can re-open occluded coronary arteries. In balloon angioplasty, a plastic tube is threaded into an artery of an arm or leg, then guided through a major blood vessel toward the heart. Once the tube reaches a blockage, a balloon attached to the end of the tube can be inflated to break up a blood clot or open a vessel clogged with plaque (Fig. 12B). Often, a small metal-mesh cylinder called a vascular **stent** is inserted into a blood vessel during balloon angioplasty. The stent holds the vessels open and decreases the risk of future occlusion. Stent devices currently in use have built-in medication, which is slowly released in the artery. These medications prevent blood clots and additional scar tissue from forming, thus helping to keep the stent open and blood flowing. In a similar fashion, a stent can be placed into an artery weakened by an aneurysm (Fig. 12C). This type of stent supports the arterial wall and can be monitored externally, allowing a physician to ensure that it is working properly.

In a **coronary bypass operation**, a portion of a blood vessel from another body part (usually one of the mammary arteries from the chest) is sutured from the aorta to the coronary artery, past the obstruction point. Blood can then flow normally again from the aorta to the heart muscle. Figure 12D shows a triple bypass in which three blood vessels connect the aorta to the coronary artery, restoring blood flow to the myocardium.

stent and inflated balloon

Figure 12B   Balloon angioplasty. A balloon inserted in an artery can be inflated to open up a clogged coronary blood vessel.

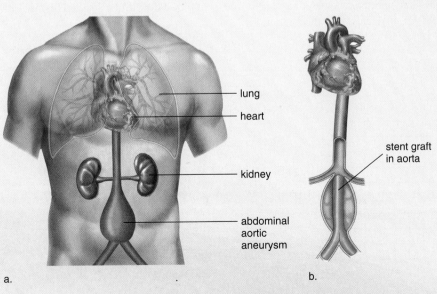

lung

heart

kidney

abdominal aortic aneurysm

stent graft in aorta

a.

b.

Figure 12C   Abdominal aortic aneurysm. a. before and **b.** after stent placement.

grafted vessels carry arterial blood

blocked vessels

Figure 12D   Coronary bypass surgery. During this operation, the surgeon grafts segments of another vessel between the aorta and the coronary vessels, bypassing areas of blockage.

beat due to impulses generated by the AV node. But the beat is slower (40 to 60 beats per minute). This condition is referred to as a **heart block.** To correct this condition, it is possible to implant an artificial pacemaker, which automatically gives an electrical stimulus to the heart every 0.85 second.

An area other than the SA node can become the pacemaker when it develops a rate of contraction that is faster than the SA node. This site, called an **ectopic pacemaker,** may cause an extra beat, if it operates only occasionally, or it can even pace the heart for a while. Caffeine and nicotine are two substances that can stimulate an ectopic pacemaker.

Electrocardiogram

With the contraction of any muscle, including the myocardium, electrolyte changes occur that can be detected by electrical recording devices. These changes occur as a muscle action potential sweeps over the cardiac muscle fibers. The resulting record, called an electrocardiogram, helps a physician detect and possibly diagnose the cause of an irregular heartbeat. There are many types of irregular heartbeats, called **arrhythmias.** The Medical Focus below discusses the electrocardiogram and some types of arrhythmias.

MEDICAL FOCUS

The Electrocardiogram

A graph that records the electrical activity of the myocardium during a cardiac cycle is called an **electrocardiogram,** or **ECG.*** An ECG is obtained by placing several electrodes on the patient's skin, then wiring the electrodes to a voltmeter (an instrument for measuring voltage). As the heart's chambers contract and then relax, the change in polarity is measured in millivolts.

An ECG consists of a set of waves: the P wave, a QRS complex, and a T wave (Fig. 12E). The P wave represents depolarization of the atria as an impulse started by the SA node travels throughout the atria. The P wave signals that the atria are going to be in systole and that the atrial myocardium is about to contract. The QRS complex represents depolarization of the ventricles following excitation of the Purkinje fibers. It signals that the ventricles are going to be in systole and that the ventricular myocardium is about to contract. The QRS complex shows greater voltage changes than the P wave because the ventricles have more muscle mass than the atria. The T wave represents repolarization of the ventricles. It signals

that the ventricles are going to be in diastole and that the ventricular myocardium is about to relax. Atrial diastole (depolarization) does not show up on an ECG as an independent event because the voltage changes are masked by the QRS complex.

An ECG records the duration of electrical activity and therefore can be used to detect arrhythmia, an irregular or abnormal heartbeat. A rate of fewer than 60 heartbeats per minute is called **bradycardia,** and more than 100 heartbeats per minute is called **tachycardia.** Another type of arrhythmia is **fibrillation,** in which the heart beats rapidly but the contractions are uncoordinated. The heart can sometimes be defibrillated by briefly applying a strong electrical current to the chest.

It is important to understand that an ECG only supplies information about the heart's electrical activity. To be used in diagnosis, an ECG must be coupled with other information, including X rays, studies of blood flow, and a detailed history from the patient.

Figure 12E Electrocardiogram. a. A portion of an electrocardiogram. **b.** An enlarged normal cycle.

Cardiac Cycle

A **cardiac cycle** includes all the events that occur during one heartbeat. On average, the heart beats about 70 times a minute, although a normal adult heart rate can vary from 60 to 100 beats per minute. After tracing the path of blood through the heart, it might seem that the right and left sides of the heart beat independently of one another, but actually, they contract together. First the two atria contract simultaneously; then the two ventricles contract together. The term **systole** refers to contraction of heart muscle, and the term **diastole** refers to relaxation of heart muscle. During the cardiac cycle, *atrial systole* is followed by *ventricular systole*.

As shown in Figure 12.6, the three phases of the cardiac cycle are:

Phase 1: Atrial Systole. Time = 0.15 sec. During this phase, both atria are in systole (contracted), while the ventricles are in diastole (relaxed). Rising blood pressure in the atria forces the blood to enter the two ventricles through the AV valves. At this time, both atrioventricular valves are open, and the semilunar valves are closed. Atrial systole ends when the atrioventricular valves (tricuspid and bicuspid/mitral) slam shut. Closure of the AV valves is caused by the rising pressure of blood filling the ventricle. Remember that closure of the AV valves causes the first heart sound, "lub" (page 252).

Phase 2: Ventricular Systole. Time = 0.30 sec. During this phase, both ventricles are in systole (contracted), while the atria are in diastole (relaxed). Rising blood pressure in the ventricles forces the semilunar valves (aortic and

Time	Atria	Ventricles	Atrioventricular Valves (tricuspid, bicuspid)	Semilunar Valves (pulmonary, aortic)
0.15 sec	Systole	Diastole	open	closed
0.30 sec	Diastole	Systole	closed	open
0.40 sec	Diastole	Diastole	open	closed

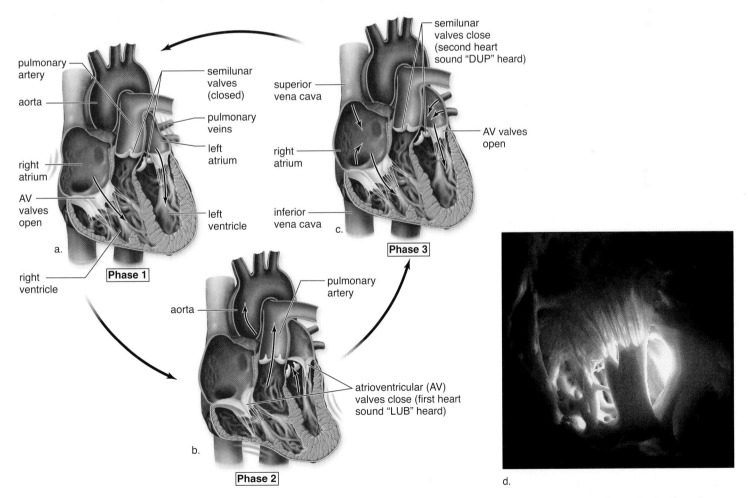

Figure 12.6 **Stages in the cardiac cycle. a.** Phase 1: atrial systole. **b.** Phase 2: ventricular systole. **c.** Phase 3: atrial and ventricular diastole. **d.** Photomicrograph of a heart valve, showing the fibrous chordae tendinae that attach the valve to the heart wall.

pulmonary) to open. Blood in the right ventricle exits through the pulmonary artery trunk to the right and left pulmonary arteries. Simultaneously, blood in the left ventricle exits into the aorta. During ventricular systole, both semilunar valves are open, and the atrioventricular valves are closed. Ventricular systole ends as the ventricles complete their pumping job; recall that backflow of blood in the pulmonary artery and aorta forces the semilunar valves to slam shut once more (page 252). Closure of the semilunar valves causes the second heart sound "dup."

Phase 3: Atrial and Ventricular Diastole. Time = 0.40 sec. During this period, both atria and both ventricles are in diastole (relaxed). At this point, pressure in all the heart chambers is low. Blood returning to the heart from the superior and inferior venae cavae, the coronary sinus, and the pulmonary veins fills the right and left atria and flows passively into the ventricles. At this time, both atrioventricular valves are open and the semilunar valves are closed.

Cardiac Output

Cardiac output (CO) is the volume of blood pumped out of a ventricle in one minute. (The same amount of blood is pumped out of each ventricle in one minute.) Cardiac output is dependent on two factors:

- *heart rate* (*HR*) = beats per minute
- *stroke volume* (*SV*) = amount of blood pumped by a ventricle each time it contracts

Thus, cardiac output = HR × SV.

The CO of an average human is 5,250 ml (or 5.25 L) per minute, which equates to about the total volume of blood in the human body. Each minute, the right ventricle pumps about 5.25 L through the pulmonary circuit, while the left ventricle pumps about 5.25 L through the systemic circuit. And this is only the resting cardiac output!

Cardiac output can vary because stroke volume and heart rate can vary, as discussed next. In this way, the heart regulates the blood supply, dependent on the body's needs.

Heart Rate

A **cardioregulatory center** in the medulla oblongata of the brain can alter the heart rate by way of the autonomic nervous system (Fig. 12.7). Parasympathetic motor signals conducted by the vagus nerve cause the heart rate to slow, and sympathetic motor signals conducted by sympathetic motor fibers cause the heart rate to increase.

The cardioregulatory center receives sensory input from receptors within the cardiovascular system. For example, *baroreceptors* are present in the aorta just after it leaves the heart, and also in the carotid arteries, which take blood from the aorta to the brain. If blood pressure falls, as it sometimes does when we stand up quickly, the baroreceptors signal the cardioregulatory center. Thereafter, sympathetic motor signals to the heart cause the heart rate to increase. Once blood pressure begins to rise above normal, nerve signals from the cardioregulatory center cause the heart rate to decrease. Such reflexes help control cardiac

Regulation of heart rate:

① Baroreceptors in the aortic arch and carotid arteries monitor blood pressure.

② Nerve signals from the baroreceptors signal the cardioregulatory center.

③ Increased parasympathetic signals decrease heart rate.

④ Increased sympathetic signals increase heart rate.

Regulation of blood pressure:

⑤ Increased sympathetic signals cause blood vessels to constrict.

carotid artery baroreceptors

① aortic arch baroreceptors

cardioregulatory and vasomotor centers in the medulla oblongata

vagus nerve (parasympathetic)

sympathetic nerves

sympathetic chain

⑤ blood vessels

Figure 12.7 Control of heart activity. The cardioregulatory center regulates the heart rate and the vasomotor center regulates constriction of blood vessels, according to input received from baroreceptors in the carotid artery and aortic arch.

output and, therefore, blood pressure, as discussed in section 12.4.

The cardioregulatory center is under the influence of the cerebrum and the hypothalamus. Therefore, when we feel anxious, the sympathetic motor nerves are activated. In addition, the adrenal medulla releases the hormones norepinephrine and epinephrine. The result is an increase in heart rate. On the other hand, activities such as yoga and meditation lead to activation of the vagus nerve, which slows the heart rate.

Other factors affect the heart rate as well. For example, a low body temperature slows the rate. Also, the proper electrolyte concentrations are needed to keep the heart rate regular.

Stroke Volume

Stroke volume, which is the amount of blood that leaves a ventricle, depends on the strength of heart contraction. As with heart rate, the autonomic nervous system helps to determine contraction force and stroke volume. Sympathetic activity strengthens each contraction and increases stroke volume. By contrast, parasympathetic activity decreases contractile force and blood volume. Further, the degree of contraction also depends on the correct blood electrolyte concentration. Recall from Chapter 8 that the proper concentration of ions, or electrolytes, is essential to create cardiac muscle action potentials. Without these electrolytes, cardiac conduction and contraction are impaired and stroke volume decreases. Two additional factors, venous return and difference in blood pressure, also influence the strength of contraction.

Venous Return   Venous return is the amount of blood entering the heart by way of the venae cavae (right side of heart) or pulmonary veins (left side of heart). The heart adjusts the strength of its own contraction beat by beat, based upon venous return. This principle is called the Frank-Starling law. The more blood returned to the heart before a given beat, the more strongly the heart will contract with that next beat. Thus, any event that increases the volume of blood entering the heart will increase the stroke volume leaving the heart. For example, exercise increases the strength of cardiac contraction because skeletal muscle contraction squeezes the veins within muscles and increases venous return. The opposite is also true: If venous return decreases, stroke volume decreases for the next beat. A low venous return, as might happen if there is blood loss, decreases the strength of cardiac contraction.

Difference in Blood Pressure   The strength of ventricular contraction has to be strong enough to oppose the blood pressure within the attached arteries. If a person has hypertension or atherosclerosis, the opposing arterial pressure may reduce the effectiveness of contraction and the stroke volume.

4. The SA node is the pacemaker of the heart and controls heart rate because:
   a. it sits at the top of the heart.
   b. it has the fastest intrinsic rate.
   c. it has the largest myocardial cells.
   d. it has the slowest intrinsic rate.
5. The semilunar valves are open during which phase of the cardiac cycle?
   a. atrial systole
   b. ventricular systole
   c. atrial and ventricular diastole
6. Cardiac output is equal to:
   a. heart rate plus stroke volume.
   b. stroke volume divided by heart rate.
   c. heart rate divided by stroke volume.
   d. heart rate times stroke volume.

Answers in Appendix C, page 464.

12.3 Anatomy of Blood Vessels

Blood vessels are of three types: arteries, capillaries, and veins (Fig. 12.8). These vessels function to:

1. transport blood and its contents (see page 249);
2. carry out exchange of gases in the pulmonary capillaries and exchange of gases plus nutrients for waste at the systemic capillaries;
3. regulate blood pressure; and
4. direct blood flow to those systemic tissues that most require it at the moment.

Arteries and Arterioles

Arteries (Fig. 12.8a) transport blood away from the heart. They have thick, strong walls composed of three layers: (1) The **tunica interna** is an endothelium layer with a basement membrane. (2) The **tunica media** is a thick middle layer of smooth muscle and elastic fibers. (3) The **tunica externa** is an outer connective tissue layer composed principally of elastic and collagen fibers. Arterial walls are sometimes so thick that they are supplied with blood vessels. The radius of an artery allows the blood to flow rapidly and the elasticity of an artery allows it to expand when the heart contracts and recoil when the heart rests. This means that blood continues to flow in an artery even when the heart is in diastole.

Arterioles are small arteries just visible to the naked eye. The middle layer of these vessels has some elastic tissue but is composed mostly of smooth muscle whose fibers encircle the arteriole. If the muscle fibers contract, the lumen (cavity) of the arteriole decreases; if the fibers relax, the lumen of the arteriole enlarges. Whether arterioles are constricted or dilated affects

v = vein; a = artery

Figure 12.8 **Blood vessels.** The walls of arteries and veins have three layers: tunica intima, tunica media, and tunica externa. Arteries have a thicker muscle layer and a thick wall, while veins have valves and the larger diameter. A capillary bed lies between an arteriole and a venule, and flow through it is controlled by a precapillary sphincter. When the sphincter relaxes, blood passes through the capillary bed and gives up its oxygen to the tissues. Thus, blood goes from being oxygen-rich in the arteriole (red color) to oxygen-poor in the venule (purple color). When the sphincter contracts, blood can bypass the capillary bed through an arteriovenous shunt.

blood distribution and blood pressure. When a muscle is actively contracting, for example, the arterioles in the vicinity dilate and additional blood flows to the muscle. In this way, the needs of the muscle for oxygen and glucose are met. As we shall see, the autonomic nervous system helps control blood pressure by regulating the number of arterioles that are contracted. The greater the number of vessels contracted, the higher the resistance to blood flow, and hence, the higher the blood pressure. The greater the number of vessels dilated, the lower the resistance to blood flow, and hence, the lower the blood pressure.

Capillaries

Arterioles branch into **capillaries** (Fig. 12.8*b*), which are extremely narrow, microscopic blood vessels. Their walls are composed of only one layer of endothelial cells connected by tight junctions. Capillaries are extremely numerous: The body probably contains a billion, and their combined surface area is estimated at 6,300 square meters. Capillary beds (networks of many capillaries) are present in all regions of the body, and each supplies the needs of neighboring cells. Therefore, most of the body's cells are near a capillary, and the heart and other vessels of the cardiovascular system can be considered a means for conducting blood to and from the capillaries.

Not all capillary beds are open or in use at the same time. For instance, after a meal, the capillary beds of the digestive tract are usually open, while during muscular exercise, the capillary beds of the skeletal muscles are open. Most capillary beds have a shunt that allows blood to move directly from an arteriole to a venule (a small vessel leading to a vein) when the capillary bed is closed. Sphincter muscles, called *precapillary sphincters*, encircle the entrance to each capillary (Fig. 12.9). When the capillary sphincters are constricted, the capillary bed is closed, preventing blood from entering the capillaries. Conversely, when the capillary sphincters are relaxed, the capillary bed is open. As would be expected, the larger the number of open capillary beds, the lower the blood pressure in the body.

Capillary Exchange

In the tissues of the body, metabolically active cells require oxygen and nutrients and give off wastes, including carbon dioxide. During capillary exchange between tissue capillaries and body cells, oxygen and nutrients leave a capillary. Cellular wastes, including carbon dioxide, enter a capillary. For this reason, systemic arterial blood contains more oxygen and nutrients than does systemic venous blood, and venous blood contains more wastes than does arterial blood. In pulmonary capillaries—the

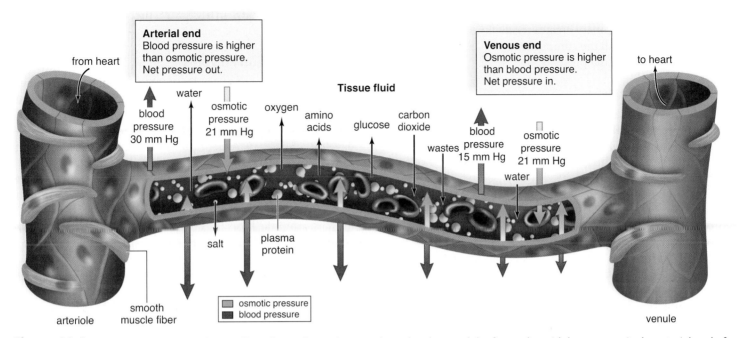

Figure 12.9 Capillary exchange. The capillary shows the exchanges that take place and the forces that aid the process. At the arterial end of a capillary, the blood pressure is higher than the osmotic pressure. Tissue fluid tends to leave the bloodstream. In the midsection, solutes, including oxygen (O_2) and carbon dioxide (CO_2), diffuse from high to low concentration. Carbon dioxide and wastes diffuse into the capillary while nutrients and oxygen enter the tissues. At the venous end of a capillary, the osmotic pressure is higher than the blood pressure. Tissue fluid tends to re-enter the bloodstream. The red blood cells and the plasma proteins are too large to exit a capillary.

capillaries supplying the lung—the exchange is reversed: Oxygen enters the blood and carbon dioxide leaves.

The internal environment of the body consists of blood and tissue fluid. **Tissue fluid** is simply the fluid that surrounds the cells of the body. In other words, substances that leave a capillary pass through tissue fluid before entering the body's cells, and substances that leave the body's cells pass through tissue fluid before entering a capillary. Water and other small molecules can cross through the cells of a capillary wall or through tiny clefts that occur between the cells. Large molecules in plasma, such as the plasma proteins, are too large to pass through capillary walls. Yet, the composition of tissue fluid stays relatively constant because of capillary exchange. Tissue fluid is a water-based solution that contains sodium chloride, other electrolytes, and scant protein. Any excess tissue fluid is collected by lymphatic capillaries, which are always found near blood capillaries.

Three processes influence capillary exchange—blood pressure, diffusion, and osmotic pressure:

Blood pressure, which is created by the pumping of the heart, pushes the blood through the capillary. Blood pressure also pushes blood against a vessel's (e.g., capillary) walls.

Diffusion, as you know, is simply the movement of substances from the area of higher concentration to the area of lower concentration.

Osmotic pressure is a force caused by a difference in solute concentration on either side of a membrane.

To understand osmotic pressure, consider that water will diffuse across a membrane toward the side that has the greater concentration of solutes, and the accumulation of this water results in a pressure. The presence of the plasma proteins, and also salts to some degree, means that blood has a greater osmotic pressure than does tissue fluid. Therefore, the osmotic pressure of blood pulls water into and retains water inside a capillary.

Notice in Figure 12.9 that a capillary has an arterial end (contains arterial blood) and a venous end (contains venous blood). In between, a capillary has a midsection. We will now consider the exchange of molecules across capillary walls at each of these locations.

Arterial End of Capillary

When arterial blood enters tissue capillaries, it is bright red because the hemoglobin in red blood cells is carrying oxygen. Blood is also rich in nutrients, which are dissolved in plasma.

At the arterial end of a capillary, blood pressure, an outward force, is higher than osmotic pressure, an inward force. Pressure is measured in terms of mm Hg (mercury). In this case, blood pressure is 30 mm Hg, and osmotic pressure is 21 mm Hg. Because blood pressure is higher than osmotic pressure at the arterial end of a capillary, water and other small molecules (e.g., glucose and amino acids) filter out of a capillary at its arterial end.

Red blood cells and a large proportion of the plasma proteins generally remain in a capillary because they are too large to pass through its wall. The exit of water and other small molecules from a capillary creates tissue fluid. Therefore, tissue fluid consists of all the components of plasma, except that it contains fewer plasma proteins.

Midsection of Capillary

Diffusion takes place along the length of the capillary, as small molecules follow their concentration gradient by moving from the area of higher to the area of lower concentration. In the tissues, the area of higher concentration of oxygen and nutrients is always blood, because after these molecules have passed into tissue fluid, they are taken up and metabolized by cells. The cells use oxygen and glucose in the process of cellular respiration, and they use amino acids for protein synthesis.

As a result of metabolism, tissue cells give off carbon dioxide and other wastes. Because tissue fluid is always the area of greater concentration for waste materials, they diffuse from tissue fluid into a capillary.

Venous End of Capillary

At the venous end of the capillary, blood pressure is much reduced to only about 15 mm Hg, as shown in Figure 12.9. Blood pressure is reduced at the venous end because capillaries have a greater cross-sectional area at their venous end than their arterial end. However, there is no reduction in osmotic pressure, which remains at 21 mm Hg and is now higher than blood pressure. Therefore, water tends to diffuse into a capillary at the venous end. As water enters a capillary, it brings with it additional waste molecules. Blood that leaves the capillaries to drain into veins is deep maroon in color because red blood cells now contain reduced hemoglobin—hemoglobin that has given up its oxygen and taken on hydrogen ions.

In the end, about 85% of the water that left a capillary at the arterial end returns to it at the venous end. Therefore, retrieving fluid by means of osmotic pressure is not completely effective. The body has an auxiliary means of collecting tissue fluid; any excess usually enters lymphatic capillaries.

Veins and Venules

Veins and smaller vessels called **venules** return blood from the capillary beds to the heart. The venules first drain the blood from the capillaries and then join together to form a vein. The wall of a vein is much thinner than that of an artery because the middle layer of muscle and elastic fibers is thinner (see Fig. 12.8a). Within some veins, especially the major veins of the arms and legs, **valves** allow blood to flow only toward the heart when they are open and prevent the backward flow of blood when they are closed.

At any given time, more than half of the total blood volume is found in the veins and venules. If blood is lost due to, for example, hemorrhaging, sympathetic nervous stimulation causes the veins to constrict, providing more blood to the rest of the body. In this way, the veins act as a blood reservoir.

Varicose Veins and Phlebitis

Varicose veins are abnormal and irregular dilations in superficial (near the surface) veins, particularly those in the lower legs. Varicose veins in the rectum, however, are commonly called piles, or more properly, **hemorrhoids.** Varicose veins develop when the valves of the veins become weak and ineffective due to backward pressure of the blood.

Phlebitis, or inflammation of a vein, is a more serious condition because thromboembolism can occur. In this instance, the embolus may eventually come to rest in a pulmonary arteriole, blocking circulation through the lungs. This condition, termed **pulmonary embolism,** can be fatal.

Content CHECK-UP!

7. The order for the three layers of a blood vessel, from superficial to deep, is:
    a. tunica externa—tunica intima—tunica media
    b. tunica media—tunica externa—tunica intima
    c. tunica externa—tunica media—tunica intima

8. If you described the differences between arteries and veins, you could say:
    a. Arteries have a thicker wall than veins.
    b. Veins have valves but arteries do not.
    c. Veins return blood to the heart; arteries carry blood away from the heart.
    d. All of these statements are correct.

9. Blood pressure is greatest at:
    a. the venous end of a capillary bed.
    b. the arterial end of a capillary bed.
    c. the center of a capillary bed.
    d. no single area; it is equal throughout the capillary bed.

Answers in Appendix C, page 464.

12.4 Physiology of Circulation

Circulation is the movement of blood through blood vessels, from the heart to the body and then back to the heart. In this section, we discuss various factors affecting circulation.

Velocity of Blood Flow

The velocity of blood flow is slowest in the capillaries. What might account for this? Consider that the aorta branches into the other arteries, and these in turn branch into the arterioles, and so forth until blood finally flows into the capillaries. Each time an artery branches, the total cross-sectional area of the blood vessels increases, reaching the maximum cross-sectional

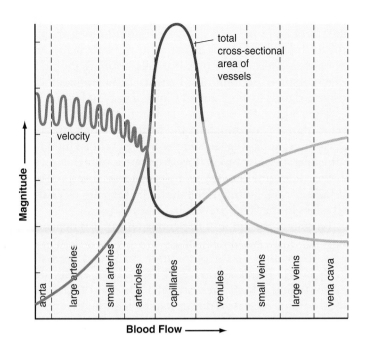

Figure 12.10 **Velocity of blood flow changes throughout the systemic circuit.** Velocity changes according to the total cross-sectional area of vessels.

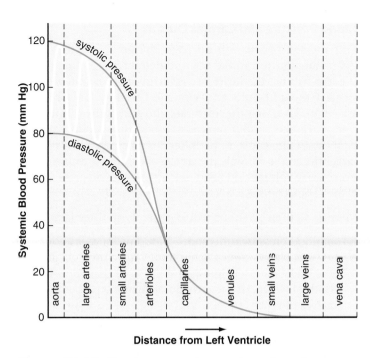

Figure 12.11 **Blood pressure changes throughout the systemic circuit.** Blood pressure decreases with distance from the left ventricle.

area in the capillaries (Fig. 12.10). The slow rate of blood flow in the capillaries is beneficial because it allows time for the exchange of gases in pulmonary capillaries and for the exchange of gases and nutrients for wastes in systemic capillaries (see Fig. 12.9).

Conversely, blood velocity increases as venules combine to form larger and larger veins. Thus, velocity is greatest in the venae cavae, which are the largest veins. However, even though the cross-sectional area of the two venae cavae is more than twice that of the aorta, the velocity of venous blood flow returning to the heart is always lower than that of arterial blood leaving the heart. Contractions of the powerful left ventricle generate a greater velocity for arterial blood. The velocity of the arterial and venous systems working together is very high—in a resting individual, it takes only about a minute for a drop of blood to go from the heart to the foot and back again to the heart!

Blood Pressure

Blood pressure is the force of blood against the walls of blood vessels. You would expect arterial blood pressure to be highest in the aorta. Why? Because the pumping action of the powerful, thick-muscled left ventricle forces blood into the aorta. Further, Figure 12.11 shows that systemic blood pressure decreases progressively with distance from the left ventricle. Blood pressure is lowest in the venae cavae because they are farthest from the left ventricle.

Note also in Figure 12.11 that blood pressure fluctuates in the arterial system between systolic blood pressure and diastolic blood pressure. Certainly, we can correlate this with the action

of the heart. During systole, the left ventricle is pumping blood out of the heart, and during diastole the left ventricle is resting.

More important than the systolic and diastolic pressure is the *mean arterial blood pressure (MABP)*, the pressure in the arterial system averaged over time. It is important to note that MABP is not determined by taking the average of systolic and diastolic pressures. Rather, MABP is the product of *cardiac output (CO)* times *peripheral resistance*. (Recall that cardiac output equals heart rate times stroke volume; see page 258). To put it as a simple math equation, MABP = CO × PR. According to this equation, increasing CO will also increase MABP. In other words, the greater the amount of blood leaving the left ventricle, the greater the pressure of blood against the wall of an artery.

Another factor that determines blood pressure is peripheral resistance, which is the resistance to flow between blood and the walls of a blood vessel. All things being equal, the smaller the blood vessel diameter, the greater the resistance and the higher the blood pressure. As an analogy, imagine a skinny, 1-inch-diameter garden hose (high resistance) compared to a firefighter's 12-inch canvas hose (low resistance). Similarly, total blood vessel length increases blood pressure because a longer vessel offers greater resistance. For this reason, an obese person is apt to have high blood pressure because about 200 miles of additional blood vessels develop for each extra pound of fat.

Let's summarize our discussion so far. The two factors that affect blood pressure are:

Cardiac output	Peripheral resistance
Heart rate	Arterial diameter and length
Stroke volume	

Blood Pressure and Cardiac Output

Our previous discussion on page 258 emphasized that the heart rate and the stroke volume determine cardiac output. We learned that the heart rate is intrinsic but is under extrinsic (nervous and endocrine) control. Therefore, heart rate can speed up. The faster the heart rate, the greater the cardiac output. As cardiac output increases, blood pressure increases as well (assuming constant peripheral resistance). Similarly, the larger the stroke volume, the greater the blood pressure. However, stroke volume and heart rate increase blood pressure *only* if the venous return is adequate.

Venous Return Venous return depends on three factors:

1. a blood pressure difference—blood pressure is about 16 mm Hg in venules versus 0 mm Hg in the right atrium;
2. the skeletal muscle pump and the respiratory pump, both of which are effective because of the presence of valves in veins;
3. total blood volume in the cardiovascular system.

The **skeletal muscle pump** works like this: When skeletal muscles contract, they compress the weak walls of the veins. This causes the blood to move past a valve (Fig. 12.12). Once past the valve, backward pressure of blood closes the valve and prevents its return. Blood in veins will always return to the heart.

As you might suspect, gravity can assist the return of venous blood from the head to the heart but not the return of blood from the extremities and trunk to the heart. The importance of the skeletal muscle pump in maintaining CO and blood pressure can be demonstrated by forcing a person to stand rigidly still for a number of hours. Frequently, the person faints because blood collects in the limbs, limiting venous return to the heart. As venous return decreases, cardiac output decreases as well. Blood pressure falls, robbing the brain of oxygen. In this case, fainting is beneficial. The horizontal body position caused by the faint aids in getting blood to the brain.

The **respiratory pump** works like this: When inhalation occurs, thoracic pressure falls and abdominal pressure rises as the chest expands. This aids in the flow of venous blood back to the heart because blood flows from areas of higher pressure (in the abdominal cavity) to areas of lower pressure (in the thoracic cavity). During exhalation, the pressure reverses, but the valves in the veins prevent backward flow.

As stated, the amount of venous return also depends on the total blood volume in the cardiovascular system. As you know, this volume in the pulmonary circuit and the systemic circuit is 5 L. If this amount of blood decreases, say due to hemorrhaging, blood pressure falls. On the other hand, if blood volume increases (due to water retention, for example), blood pressure rises.

Blood Pressure and Peripheral Resistance

The nervous system and the endocrine system both affect peripheral resistance.

Neural Regulation of Peripheral Resistance A **vasomotor center** in the medulla oblongata controls vasoconstriction. This center is under the control of the cardioregulatory center. As mentioned on page 258, if blood pressure falls, baroreceptors in the blood vessels signal the cardioregulatory center. The cardioregulatory center will activate its vasomotor center. The vasomotor center then stimulates sympathetic nerve fibers, which cause the heart rate to increase *and* the arterioles to constrict. Increasing heart rate increases cardiac output; constricting the arterioles increases peripheral resistance. The result is a rise in blood pressure. What factors lead to a reduction in blood pressure? If blood pressure rises above normal, the baroreceptors signal the cardioregulatory center in the medulla oblongata. Subsequently, the heart rate decreases *and* the arterioles dilate.

Nervous control of blood vessels also causes blood to be shunted from one area of the body to another. During exercise, arteries in the viscera and skin are more constricted than those in the muscles. Therefore, blood flow to the muscles increases. Also, dilation of the precapillary sphincters in muscles means that blood will flow to the muscles and not to the viscera.

Hormonal Regulation of Peripheral Resistance Certain hormones cause blood pressure to rise. Epinephrine and norepinephrine, the hormones from the adrenal medulla, increase the heart rate. As heart rate increases, so too does cardiac output and blood pressure. Epinephrine and norepinephrine also constrict arterioles in the capillary beds supplying the skin, abdominal viscera, and kidneys. Arteriolar vasoconstriction increases blood pressure by increasing peripheral resistance.

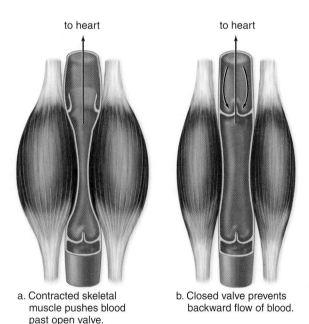

to heart   to heart

a. Contracted skeletal muscle pushes blood past open valve.

b. Closed valve prevents backward flow of blood.

Figure 12.12 Skeletal muscle pump. a. When skeletal muscles contract and compress a vein, blood is squeezed past a valve. **b.** When muscles relax, the valve closes.

When the blood volume and blood sodium level are low, the kidneys secrete the enzyme **renin.** Renin converts the plasma protein angiotensinogen to angiotensin I, which is changed to angiotensin II by a converting enzyme found in the lungs. Angiotensin II stimulates the adrenal cortex to release aldosterone. The effect of this system, called the renin-angiotensin-aldosterone system, is to raise the blood volume and pressure in two ways. First, angiotensin II constricts the arterioles directly. Second, aldosterone causes the kidneys to reabsorb sodium. When the blood sodium level rises, water is reabsorbed, and blood volume and pressure are maintained.

Two other hormones play a role in the homeostatic maintenance of blood volume and blood pressure. As discussed in Chapter 10, antidiuretic hormone (ADH) helps increase blood volume by causing the kidneys to reabsorb water. ADH also causes vasoconstriction of smooth muscle in arteries and veins throughout the body. Thus, ADH boosts blood pressure by simultaneously increasing blood volume (which increases cardiac output) *and* by causing vasoconstriction (which increases peripheral resistance).

The hormonal mechanism for decreasing blood pressure involves an endocrine hormone secreted by the heart. When the atria of the heart are stretched due to increased blood volume, cardiac cells release a hormone called **atrial natriuretic peptide (ANP),** which inhibits renin secretion by the kidneys and aldosterone secretion by the adrenal cortex. The effect of ANP, therefore, is to cause sodium excretion—that is, *natriuresis* (a term that means "salty urination"). When sodium is excreted, so is water, and therefore blood volume and blood pressure decrease (Fig. 12.13).

Evaluating Circulation

Taking a patient's pulse and blood pressure are two ways to evaluate circulation.

Pulse

The surge of blood entering the arteries causes their elastic walls to stretch, but then they almost immediately recoil. This alternating expansion and recoil of an arterial wall can be felt as a **pulse** in any artery that runs close to the body's surface. These superficial arteries are called pulse points (Fig. 12.14). It is customary to feel the pulse by placing several fingers on the radial artery, which lies near the outer border of the palm side of a wrist. The common carotid artery, on either side of the trachea in the neck, is another accessible location for feeling the pulse. Normally, the pulse rate indicates the rate of the heartbeat because the arterial walls pulse whenever the left ventricle contracts. The pulse is usually 70 times per minute, but can vary between 60 and 80 times per minute.

Figure 12.13 Blood volume maintenance. Normal blood volume is maintained by ADH (antidiuretic hormone) and aldosterone, whose actions raise blood volume, and by ANH (atrial natriuretic hormone), whose actions lower blood volume.

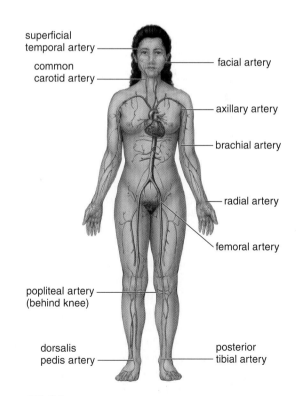

Figure 12.14 The pulse rate. Pulse points are the locations where the pulse can be taken. Each pulse point is named after the appropriate artery.

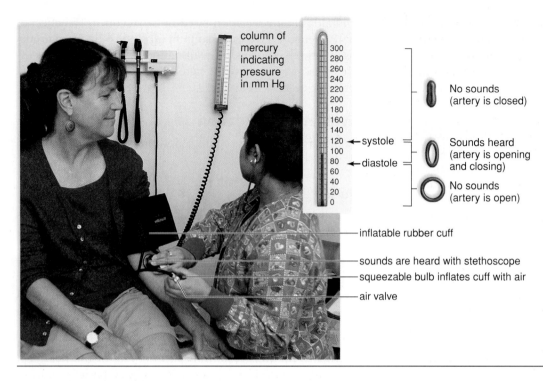

column of mercury indicating pressure in mm Hg

300
280
260
240
220
200
180
160
140
120 ← systole
100
80 ← diastole
60
40
20
0

No sounds (artery is closed)

Sounds heard (artery is opening and closing)

No sounds (artery is open)

inflatable rubber cuff

sounds are heard with stethoscope

squeezable bulb inflates cuff with air

air valve

Figure 12.15 Use of a sphygmomanometer. The technician inflates the cuff with air, gradually reduces the pressure, and listens with a stethoscope for the sounds that indicate blood is moving past the cuff in an artery. This is systolic blood pressure. The pressure in the cuff is further reduced until no sound is heard, indicating that blood is flowing freely through the artery. This is diastolic pressure.

Blood Pressure

Blood pressure is usually measured in the brachial artery with a sphygmomanometer, an instrument that records changes in terms of millimeters (mm) of mercury (Fig. 12.15). A blood pressure cuff connected to the sphygmomanometer is wrapped around the patient's arm, and a stethoscope is placed over the brachial artery. The blood pressure cuff is inflated until no blood flows through it; therefore, no sounds can be heard through the stethoscope. The cuff pressure is then gradually lowered. As soon as the cuff pressure declines below systolic pressure, blood flows through the brachial artery each time the left ventricle contracts. The blood flow is turbulent below the cuff. This turbulence produces vibrations in the blood and surrounding tissues that can be heard through the stethoscope. These sounds are called *Korotkoff sounds,* and the cuff pressure at which the Korotkoff sounds are heard the first time is the systolic pressure. As the pressure in the cuff is lowered still more, the Korotkoff sounds change tone and loudness. When the cuff pressure no longer constricts the brachial artery, no sound is heard. The cuff pressure at which the Korotkoff sounds disappear is the diastolic pressure.

Normal resting blood pressure for a young adult is 120/80. The higher number is the *systolic pressure,* the pressure recorded in an artery when the left ventricle contracts. The lower number is the *diastolic pressure,* the pressure recorded in an artery when the left ventricle relaxes.

It is estimated that about 20% of all Americans suffer from hypertension, or high blood pressure. Hypertension is present when the systolic blood pressure is 140 or greater, or the diastolic blood pressure is 90 or greater. Hypertension is sometimes called a silent killer because it may not be detected until a stroke or heart attack occurs. It has long been thought that a certain genetic makeup might account for the development of hypertension. Now researchers have discovered two genes that may be involved in some individuals. One gene codes for angiotensinogen, the plasma protein mentioned previously (see page 265). Angiotensinogen is converted to a powerful vasoconstrictor in part by the product of the second gene. Persons with hypertension due to overactivity of these genes might one day be cured by gene therapy.

At present, however, the best safeguard against developing hypertension is to have regular blood pressure checks and to adopt a lifestyle that lowers the risk of hypertension as described in the Medical Focus on page 274.

Content CHECK-UP!

10. Blood flow is slowest through which blood vessels?
    a. arterioles   b. veins   c. capillaries   d. venules

11. Mean arterial blood pressure is equal to:
    a. cardiac output times peripheral resistance.
    b. cardiac output plus peripheral resistance.
    c. cardiac output/peripheral resistance.
    d. cardiac output minus peripheral resistance.

12. Which of the following would increase mean arterial blood pressure?
    a. increased venous return to the heart
    b. increased constriction of arteries and arterioles throughout the body
    c. release of epinephrine from the adrenal medulla
    d. Any of these would increase mean arterial blood pressure

Answers in Appendix C, page 464.

Cardiopulmonary Resuscitation and Automated External Defibrillators

How many people's lives have been saved during cardiac arrest by cardiopulmonary resuscitation (CPR)? The exact number would probably be impossible to track, according to the American Heart Association (AHA). Regrettably, it's easier to track the tragedies that occur without it: Fewer than 30% of people who suffer cardiac arrest ever receive bystander help. Onlookers simply don't know what to do, or perhaps are afraid that they might do something wrong. For this reason, the AHA has published the *Chain of Survival*, a new set of four guidelines that make it easier to help a person in cardiac arrest:

- Recognize that you're in an emergency, and immediately call for help. Call 9-1-1, or use the EMS service in your area.
- Start CPR. The traditional method alternates chest compressions with mouth-to-mouth breathing, also known as rescue breathing. If you're not sure how to do rescue breathing, simply pump fast (100 or more times a minute) and hard on the center of the person's chest. Recent research from the AHA has shown that simple manual chest compression can be as effective as traditional CPR (chest compression accompanied by rescue breathing).

- If you have access to an Automatic External Defibrillator (AED), use it. This computerized device is available in many public places, such as airports and shopping malls, and is prominently labeled. An AED will explain, step by step, how to check for the person's breathing and a pulse first. Next, it will describe how to apply pads to the victim's chest so that the computer can analyze the heart activity. If a shock is needed to restart the victim's heart, the AED delivers a burst of intense electrical current to the chest. The rescuer simply moves back and pushes a button when prompted. Voice instructions will also detail how to do CPR until paramedics arrive.
- Transfer to advanced care. Clinicians refer to the first hour after admission to an emergency room as the "golden hour." It's not surprising that research has shown that patients who receive the best care during this critical time are the most likely to survive.

The AHA recommends that everyone learn both CPR and how to use an AED. The Red Cross and many hospitals regularly offer introductory and refresher classes in traditional classrooms and online. With training, you just might be able to save a person's life!

12.5 Circulatory Routes

Blood vessels belong to either the pulmonary circuit or the systemic circuit. The path of blood through the pulmonary circuit can be traced as follows: Blood from all regions of the body first collects in the right atrium and then passes into the right ventricle, which pumps it into the pulmonary trunk.

The pulmonary trunk divides into the **pulmonary arteries,** which in turn divide into the arterioles of the lungs. The arterioles then take blood to the pulmonary capillaries where carbon dioxide and oxygen are exchanged. The blood then enters the pulmonary venules and flows through the **pulmonary veins** back to the left atrium. Because the blood in the pulmonary arteries is O_2-poor but the blood in the pulmonary veins is O_2-rich, it is not correct to say that all arteries carry blood that is high in oxygen and that all veins carry blood that is low in oxygen. In fact, just the reverse is true in the pulmonary circuit.

The systemic circuit includes all of the other arteries and veins of the body. The largest artery in the systemic circuit is the aorta, and the largest veins are the **superior vena cava** and **inferior vena cava.** The superior vena cava collects blood from the head, chest, shoulders, and upper limbs, and the inferior vena cava collects blood from the lower body regions. Both venae cavae enter the right atrium. The aorta and venae cavae are the major pathways for blood in the systemic system.

The path of systemic blood to any organ in the body begins in the left ventricle, which pumps blood into the **aorta.** Branches from the aorta go to the major body regions and organs. Tracing the path of blood to any organ in the body requires mentioning only the aorta, the proper branch of the aorta, the organ, and the returning vein to the vena cava. In many instances, the artery and vein that serve the same organ have the same name. For example, the path of blood to and from the kidneys is: left ventricle; aorta; renal artery; arterioles, capillaries, venules; renal vein; inferior vena cava; right atrium. In the systemic circuit, unlike the pulmonary circuit, arteries contain O_2-rich blood and appear bright red, while veins contain O_2-poor blood and appear dark maroon.

The Major Systemic Arteries

After the aorta leaves the heart, it divides into the *ascending aorta*, the *aortic arch*, and the *descending aorta* (Fig. 12.16). The left and right coronary arteries, which supply blood to the heart, branch off the ascending aorta (Table. 12.1).

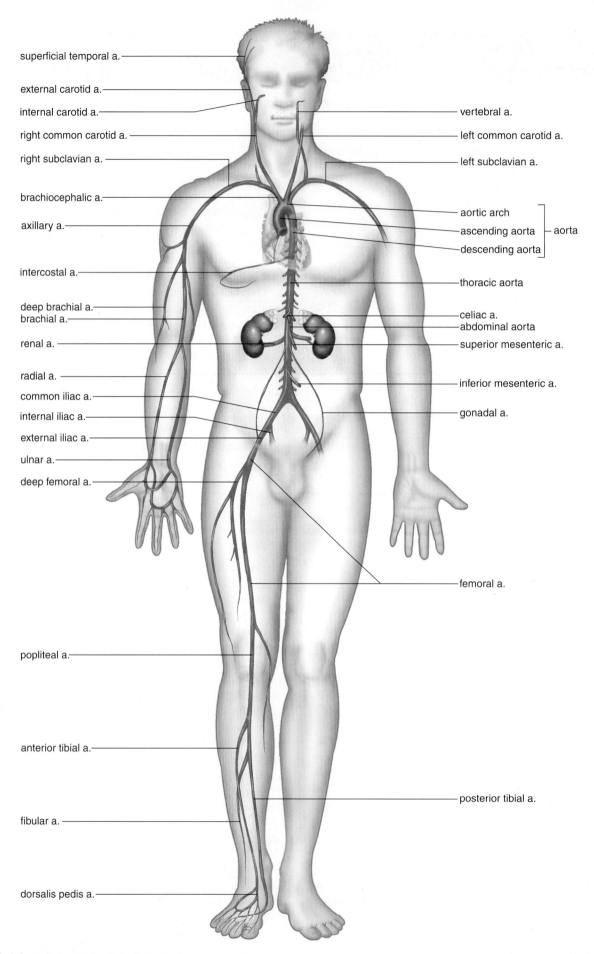

superficial temporal a.

external carotid a.

internal carotid a.

right common carotid a.

right subclavian a.

brachiocephalic a.

axillary a.

intercostal a.

deep brachial a.

brachial a.

renal a.

radial a.

common iliac a.

internal iliac a.

external iliac a.

ulnar a.

deep femoral a.

popliteal a.

anterior tibial a.

fibular a.

dorsalis pedis a.

vertebral a.

left common carotid a.

left subclavian a.

aortic arch

ascending aorta

descending aorta

aorta

thoracic aorta

celiac a.

abdominal aorta

superior mesenteric a.

inferior mesenteric a.

gonadal a.

femoral a.

posterior tibial a.

Figure 12.16 Major arteries (a.) of the body.

TABLE 12.1 The Aorta and Its Principal Branches

Portion of Aorta	Major Branch	Regions Supplied
Ascending aorta	Left and right coronary arteries	Heart
Aortic arch	Brachiocephalic artery	
	Right common carotid	Right side of head
	Right subclavian	Right shoulder and upper limb
	Left common carotid artery	Left side of head
	Left subclavian artery	Left shoulder and upper limb
Descending aorta		
Thoracic aorta	Intercostal artery	Thoracic wall
Abdominal aorta	Celiac artery	Stomach, spleen, and liver
	Superior mesenteric artery	Small and large intestines (ascending and transverse colons)
	Renal artery	Kidney
	Gonadal artery	Ovary or testis
	Inferior mesenteric artery	Lower digestive system (transverse and descending colons, and rectum)
	Common iliac artery	Pelvic organs and lower limbs

TABLE 12.2 Principal Veins That Join the Venae Cavae

Vein	Region Drained	Vena Cava
Right and left brachiocephalic veins	Head, neck, and upper limbs	Form superior vena cava
Right and left common iliac veins	Lower limbs	Form inferior vena cava
Right and left renal veins	Kidneys	Enters inferior vena cava
Right and left hepatic veins	Liver, digestive tract, and spleen	Enters inferior vena cava

Three major arteries branch off the aortic arch: the **brachiocephalic artery,** the **left common carotid artery,** and the **left subclavian artery.** The brachiocephalic artery divides into the **right common carotid** and the **right subclavian arteries.** These blood vessels serve the head (right and left common carotids) and the shoulders and upper limbs (right and left subclavians).

The descending aorta is divided into the *thoracic aorta,* which branches off to the organs within the thoracic cavity, and the *abdominal aorta,* which branches off to the organs in the abdominal cavity.

The descending aorta ends when it divides into the **common iliac arteries** that branch into the **internal iliac artery** and the **external iliac artery.** The internal iliac artery serves the pelvic organs, and the external iliac artery serves the legs. These and other arteries are shown in Figure 12.16.

The Major Systemic Veins

Figure 12.17 shows the major veins of the body. The **external** and **internal jugular veins** drain blood from the brain, head, and neck. An external jugular vein enters a **subclavian vein** that, along with an internal jugular vein, enters a **brachiocephalic vein.** Right and left brachiocephalic veins merge, giving rise to the superior vena cava.

In the abdominal cavity, as discussed in more detail later, the **hepatic portal vein** receives blood from the stomach, spleen, intestines, and other abdominal organs, and then enters the liver. Emerging from the liver, the **hepatic veins** enter the inferior vena cava.

In the pelvic region, veins from the various organs enter the **internal iliac** veins, while the veins from the lower limbs enter the **external iliac veins.** The internal and external iliac veins become the **common iliac veins** that merge, forming the inferior vena cava. Table 12.2 lists the principal veins that enter the venae cavae.

Begin Thinking Clinically

?

Sitting still for long periods of time, as you might have to do on a long plane flight, may cause you to develop a *deep vein thrombosis,* a clot in one of your leg veins. As you study Figure 12.17, imagine that the clot is in a deep branch of the inferior part of the great saphenous vein. Could blood still return to your heart and, if so, how? What could you do on the flight to prevent the thrombosis from forming?

Answer and discussion in Appendix C, page 464.

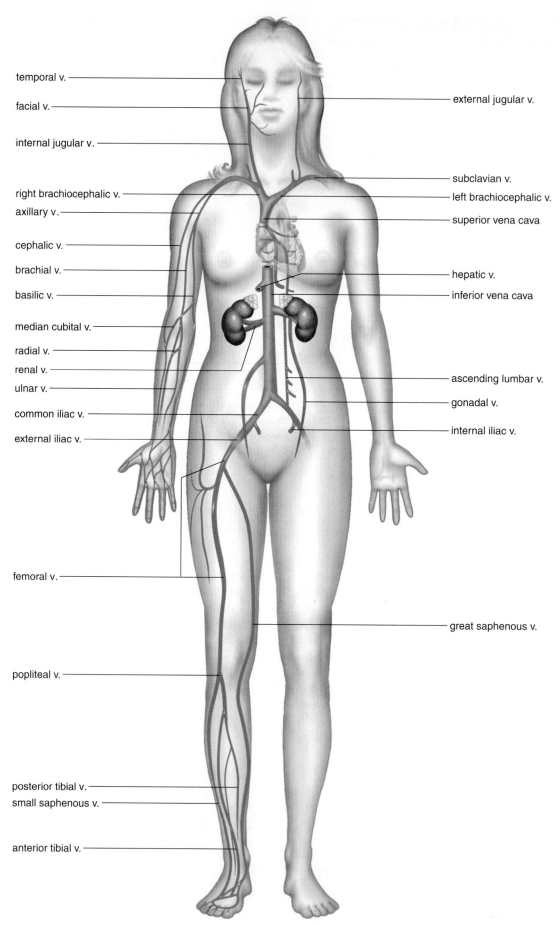

temporal v.

facial v.

internal jugular v.

right brachiocephalic v.

axillary v.

cephalic v.

brachial v.

basilic v.

median cubital v.

radial v.

renal v.

ulnar v.

common iliac v.

external iliac v.

femoral v.

popliteal v.

posterior tibial v.

small saphenous v.

anterior tibial v.

external jugular v.

subclavian v.

left brachiocephalic v.

superior vena cava

hepatic v.

inferior vena cava

ascending lumbar v.

gonadal v.

internal iliac v.

great saphenous v.

Figure 12.17 **Major veins (v.) of the body.**

Special Systemic Circulations

Hepatic Portal System

The **hepatic portal system** (Fig. 12.18) carries blood from the stomach, intestines, and other organs to the liver. The term portal system is used to describe the following unique pattern of circulation:

capillaries → vein → capillaries → vein

Capillaries of the digestive tract drain into the superior mesenteric vein and the splenic vein, which join to form the hepatic portal vein. The gastric veins empty directly into the hepatic portal vein. The hepatic portal vein carries blood to capillaries in the liver. The hepatic capillaries allow nutrients and wastes to diffuse into liver cells for further processing. Then, hepatic capillaries join to form venules that enter a hepatic vein. The hepatic veins enter the inferior vena cava.

In addition to receiving venous blood from the intestine, the liver also receives arterial blood via the hepatic artery. The hepatic artery is not a part of the hepatic portal system.

Hypothalamus-Pituitary Portal System

The hypothalamus-pituitary portal system (illustrated in the Visual Focus, Fig. 10.4) is an important endocrine portal system. This portal venous arrangement links the hypothalamus to the anterior pituitary. Through it, the hypothalamus sends releasing hormones to the pituitary.

Blood Supply to the Brain

The brain is supplied with O_2-rich blood by the anterior and posterior cerebral arteries and the carotid arteries. These arteries give off branches that join to form the **cerebral arterial circle** (circle of Willis), a vascular route in the region of the pituitary gland (Fig. 12.19). Because the blood vessels form a circle, alternate routes are available for bringing arterial blood to the brain and thus supplying the brain with oxygen. The presence of the cerebral arterial circle also equalizes blood pressure in the brain's blood supply.

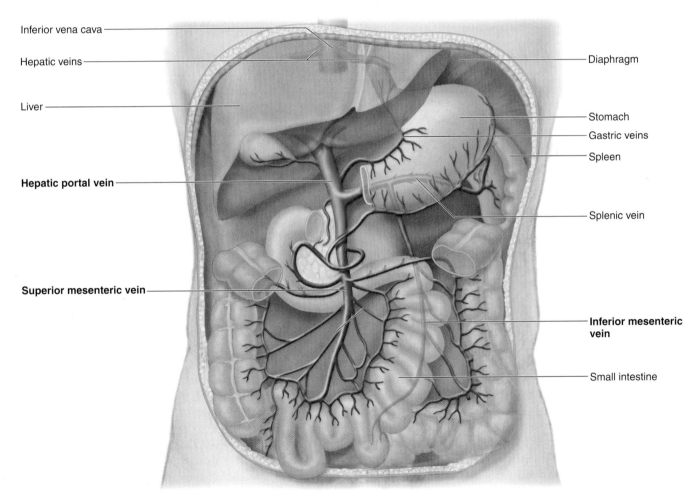

Figure 12.18 Hepatic portal system. This system provides venous drainage of the digestive organs and takes venous blood to the liver. (v. = vein)

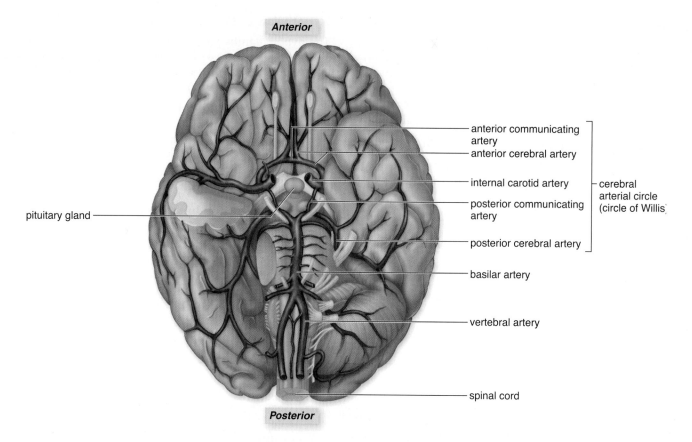

anterior communicating artery

anterior cerebral artery

internal carotid artery

posterior communicating artery

posterior cerebral artery

cerebral arterial circle (circle of Willis)

basilar artery

vertebral artery

spinal cord

pituitary gland

Anterior

Posterior

Figure 12.19 **Cerebral arterial circle.** The arteries that supply blood to the brain form the cerebral arterial circle (circle of Willis).

MEDICAL FOCUS

Stroke: Beware of "Brain Attack"!

Heart attack, or myocardial infarction, and **cerebrovascular accident,** or stroke, are two disease processes with a lot in common. Both are leading causes of death and disability in North America. Both most often result from atherosclerosis, the hardening and narrowing of arteries that results from injury and inflammation (see the Medical Focus on page 254). Just as a myocardial infarction causes death to heart muscle, a cerebrovascular accident causes death of nerve cells. Thus, a stroke could be termed a "brain attack."

An atherosclerotic plaque formed inside an artery often triggers the clotting cascade when blood clotting factors contact the roughened surface of the plaque, or if the plaque itself cracks open and bleeds. The subsequent clot causes a **thrombotic stroke** if it blocks arteries supplying the brain. A clot that forms elsewhere in the body, but breaks off to lodge in a cerebral vessel, will cause an **embolic stroke. Hemorrhagic stroke** occurs when intracranial blood vessels burst; this form of stroke is much less common (approximately 10% of strokes occur due to hemorrhage). Again, atherosclerosis and hypertension cause most hem-

orrhagic strokes because both can weaken the wall of an artery to the point that an aneurysm develops. An aneurysm is a weakened area in the arterial wall; rupture of the aneurysm causes the hemorrhage. Very rarely, an aneurysm results from congenital malformation of the cranial blood vessels during fetal development.

Just as a patient with angina pectoris is warned by his chest pain that blood flow through his coronary arteries is reduced, a potential stroke victim may be warned by a **transient ischemic attack (TIA).** Numbness or tingling in the hands or face, difficulty in speaking or understanding speech, and visual disturbances are all symptoms of a transient ischemic attack to the brain. Prompt medical attention during a TIA may prevent an actual stroke from occurring.

Heart attack and stroke can be prevented by minimizing the risk factors they also share in common: smoking; hypertension; elevated blood lipid, cholesterol, and homocysteine levels; and uncontrolled diabetes mellitus. See the Medical Focus on page 274 for a complete discussion on preventing cardiovascular disease.

Fetal Circulation

As Figure 12.20 shows, the fetus has four circulatory features that are not present in adult circulation:

1. The **foramen ovale,** or *oval window,* is an opening between the two atria. This window is covered by a flap of tissue that acts as a valve.
2. The **ductus arteriosus,** or *arterial duct,* is a connection between the pulmonary artery and the aorta.
3. The **umbilical arteries** and **vein** are vessels that travel to and from the placenta, leaving waste and receiving nutrients.

4. The **ductus venosus,** or *venous duct,* is a connection between the umbilical vein and the inferior vena cava.

All of these features can be related to the fact that the fetus does not use its lungs for gas exchange—it receives oxygen and nutrients from the mother's blood at the placenta. During development, the lungs receive only enough blood to supply their developmental need for oxygen and nutrients.

The path of blood in the fetus can be traced, beginning from the right atrium (Fig. 12.20). Most of the blood that enters the right atrium passes directly into the left atrium by way of the foramen ovale because the blood pressure in the

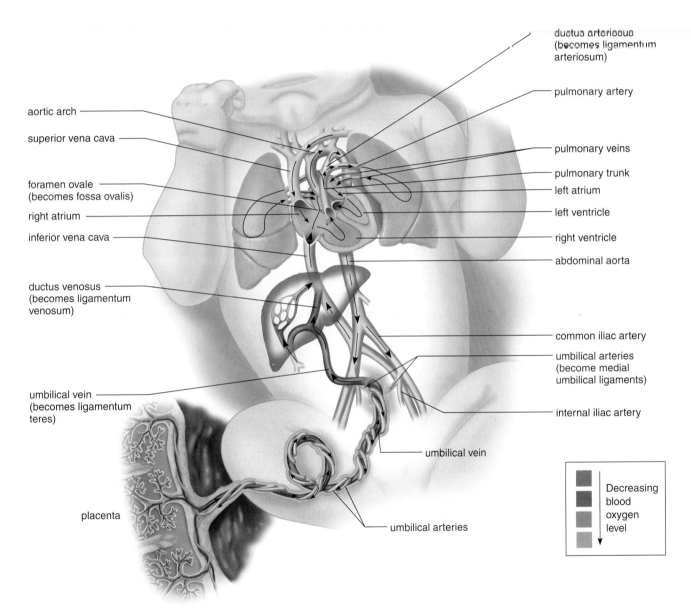

Figure 12.20 Fetal circulation. Arrows indicate the direction of blood flow. The lungs are not functional in the fetus, but they are developing. The blood passes directly from the right atrium to the left atrium via the foramen ovale or from the right ventricle to the aorta via the pulmonary trunk and ductus arteriosus. The umbilical arteries take fetal blood to the placenta where exchange of molecules between fetal and maternal blood takes place. Oxygen and nutrient molecules diffuse into the fetal blood, and carbon dioxide and urea diffuse from the fetal blood. The umbilical vein returns blood from the placenta to the fetus.

right atrium is somewhat greater than that in the left atrium. The rest of the fetal blood entering the right atrium passes into the right ventricle and out through the pulmonary trunk. However, because of the ductus arteriosus, most pulmonary trunk blood passes directly into the aortic arch. Notice that whatever route blood takes, most of it reaches the aortic arch instead of the pulmonary circuit vessels.

Blood within the aorta travels to the various branches, including the iliac arteries, which connect to the umbilical arteries leading to the placenta. Exchange between maternal and fetal blood takes place at the placenta. Blood in the umbilical arteries is O_2-poor, but blood in the umbilical vein, which travels from the placenta, is O_2-rich. The umbilical vein enters the ductus venosus, which passes directly through the liver. The ductus venosus then joins with the inferior vena cava, a vessel that contains O_2-poor blood. The vena cava returns this mixture to the right atrium.

Changes at Birth   Sectioning and tying the umbilical cord permanently separates the newborn from the placenta. The first breath inflates the lungs and oxygen enters the blood at the lungs instead of the placenta. O_2-rich blood returning from the lungs to the left side of the heart usually causes a flap on the left side of the interatrial septum to close the foramen ovale. What remains is a depression called the *fossa ovalis*. Incomplete closure occurs in nearly one out of four individuals, but even so, blood rarely passes from the right atrium to the left atrium because either the opening is small or it closes when the atria contract. In a small number of cases, the passage of O_2-poor blood from the right side to the left side of the heart is sufficient to cause **cyanosis**, a bluish cast to the skin. This condition can now be corrected by open-heart surgery.

The fetal blood vessels and shunts constrict and become fibrous connective tissue called ligamentums in all cases except the distal portions of the umbilical arteries, which become the medial umbilical ligaments. Regardless, these structures run between internal organs. For example, the ligamentum teres (which is the remnant of the umbilical vein) attaches the umbilicus to the liver.

density lipoprotein). LDL (often referred to as "bad cholesterol" or "bad lipoprotein" by physicians) takes cholesterol from the liver to the tissues. Recall that LDL is oxidized by macrophages to form foam cells, and later, a fatty streak that begins an atherosclerotic plaque. HDL ("good" cholesterol or lipoprotein) transports cholesterol out of the tissues to the liver, where cholesterol is metabolized. When the LDL level in the blood is abnormally high or the HDL level is abnormally low, cholesterol accumulates in artery walls.

It is recommended that everyone know his or her blood cholesterol level. Individuals with a high blood cholesterol level (240 mg/100 ml) should be further tested to determine their LDL cholesterol level. Blood tests can also determine high levels of homocysteine and C-reactive protein, two important indicators that atherosclerosis is occurring. Levels of these two markers, along with LDL cholesterol level, must be considered together with other cardiovascular risk factors such as age, family history, general health, and whether the patient smokes. A person with a moderate risk of cardiovascular disease may have success with dietary therapy to lower LDL. Cholesterol-lowering drugs are reserved for high-risk patients.

Evidence is mounting to suggest a role for antioxidant vitamins (A, E, and C) in the prevention of cardiovascular disease. Antioxidants protect the body from free radicals that may damage HDL cholesterol through oxidation or damage the lining of an artery, leading to a blood clot that can block the vessel. Nutritionists believe that consuming at least five servings of fruit and vegetables a day may protect against cardiovascular disease.

Proper Dental Care

Periodontal disease, the inflammation of the gums caused by poor dental hygiene, has been suggested as a cause for atherosclerosis. Scientists suspect that people with tooth decay and gum disease may have a low-level bacterial infection in the blood—not severe enough to cause illness, but enough to injure the endothelial lining and start the formation of atherosclerotic plaques. Proper care of the teeth and gums, along with regular visits to the dentist, just might prevent cardiovascular disease.

Exercise

People who exercise are less apt to have cardiovascular disease. One study found that moderately active men who spent an average of 48 minutes a day on a leisure-time activity such as gardening, bowling, or dancing had one-third fewer heart attacks than their peers who spent an average of only 16 minutes each day on such activities. Exercise helps keep weight under control, may help minimize stress, and reduces hypertension. The heart beats faster when exercising, but exercise slowly increases the heart's capacity. This means that the heart can beat more slowly when we are at rest and still do the same amount of work. One physician recommends that his cardiovascular patients walk for one hour, three times a week. In addition, they are to practice meditation and yoga-like stretching and breathing exercises to reduce stress.

Content CHECK-UP!

13. The superior and inferior venae cavae return their blood to the:
    a. right atrium.
    c. pulmonary artery.
    b. left atrium.
    d. pulmonary veins.

14. In an adult heart, the oxygen content is highest in blood found in the:
    a. pulmonary arteries.
    c. pulmonary veins.
    b. pulmonary trunk.
    d. right ventricle.

15. From the following list of pairs of major arteries and the smaller arteries that branch from them, choose the correct pair(s):
    a. brachiocephalic—right common carotid
    b. abdominal aorta—intercostal artery
    c. aortic arch—left subclavian
    d. a and c are correct.
    e. all are correct.

Answers in Appendix C, page 464.

12.6 Effects of Aging

The heart generally grows larger with age, primarily because of fat deposition in the epicardium and myocardium. In many middle-aged people, the heart is covered by a layer of fat, and the number of collagenous fibers in the endocardium increases. With age, the valves, particularly the aortic semilunar valve, become thicker and more rigid.

As a person ages, the myocardium loses some of its contractile power and some of its ability to relax. The resting heart rate decreases throughout life, and the maximum possible rate during exercise also decreases. With age, the contractions become less forceful; the heart loses about 1% of its reserve pumping capacity each year after age 30.

In the elderly, arterial walls tend to thicken with plaque and become inelastic, signaling that atherosclerosis and arteriosclerosis are present. Increased blood pressure was once believed to be inevitable with age, but now hypertension is known to result from other conditions, such as kidney disease and atherosclerosis. The Medical Focus on page 274 describes how diet and exercise in particular can help prevent atherosclerosis.

Myocardial infarction (described in the Medical Focus box on page 254) and other diseases related to atherosclerosis increase in frequency as a person ages. Congestive heart failure can result from myocardial infarction.

In **congestive heart failure,** a damaged left side of the heart fails to pump adequate blood, and blood backs up in the pulmonary circuit. Therefore, pulmonary blood vessels have become *congested.* The congested vessels leak fluid into tissue spaces, causing pulmonary edema. The result is shortness of breath, fatigue, and a constant cough with pink, frothy sputum. Treatment consists of the three Ds: diuretics (to increase urinary output), digoxin (to increase the heart's contractile force), and dilators (to relax the blood vessels). If necessary, a heart transplant is done.

The occurrence of varicose veins increases with age, particularly in people who are required to stand for long periods. Thromboembolism as a result of varicose veins can lead to death if a blood clot settles in a major branch of a pulmonary artery. (This disorder is called pulmonary embolism.)

12.7 Homeostasis

Homeostasis is possible only if the cardiovascular system delivers oxygen and nutrients to and takes metabolic wastes from the tissue fluid surrounding the cells. Human Systems Work Together on page 276 tells how the cardiovascular system works with other systems of the body to maintain homeostasis.

Maintaining Blood Composition, pH, and Temperature

As you study Human Systems Work Together, it's important to understand that the composition of the blood is maintained by the other systems of the body. The red bone marrow of the skeletal system contributes red blood cells for O_2 and CO_2 transport. The respiratory system is responsible for gas exchange. Both the respiratory and urinary systems cooperate to maintain proper pH. The digestive system supplies nutrients, and the liver detoxifies waste. Skeletal muscle contraction supplies heat to keep us warm, and the integumentary system allows us to eliminate any excess heat. Thus, blood's pH, temperature, O_2-CO_2 level, and nutrient concentration remain relatively constant.

Maintaining Blood Pressure

The pumping of the heart is critical to creating the blood pressure that moves blood throughout the body. The importance of the heart to survival can be seen in the speed with which it develops during prenatal life. Long before other major organs, the heart and its vessels have taken shape and are ready to function.

The body has multiple ways to maintain blood pressure. Blood vessel sensory receptors signal the brain when blood pressure falls. The brain's regulatory center subsequently increases heartbeat and constricts blood vessels, and blood pressure is restored. The lymphatic system collects excess tissue fluid, returning it to the cardiovascular system. In this way, the lymphatic system makes an important contribution to regulating blood volume and pressure.

The endocrine system assists the nervous system in regulating blood pressure. Epinephrine and norepinephrine cause constriction of arterioles. Aldosterone, ADH, and ANP regulate urine excretion. Retention of water helps to raise blood pressure, while water excretion helps to lower blood pressure.

Venous return from the capillaries to the heart is assisted by the muscular and respiratory systems. Skeletal muscle contraction pushes blood past the vein valves, and breathing movements encourage the flow of blood toward the heart. Smooth muscle in the walls of arterioles constricts and help raise blood pressure. Platelets are necessary to blood clotting, preventing excessive bleeding and loss of pressure.

An individual who loses more than 10% of his or her blood will suffer a sudden drop in blood pressure and go into shock. The decreased pressure triggers the body's last defense: A powerful wave of sympathetic impulses constricts the veins and arterioles throughout the body to slow the drop in blood pressure. Heart rate soars as high as 200 beats a minute to maintain blood flow, especially to the brain and heart itself. Because of this reflex, you can lose as much as 40% of your total blood volume and still live.

Human Systems Work Together

Integumentary System

Blood vessels deliver nutrients and oxygen to skin; carry away wastes; blood clots if skin is broken.

Skin prevents water loss; helps regulate body temperature; protects blood vessels.

Skeletal System

Blood vessels deliver nutrients and oxygen to bones; carry away wastes.

Rib cage protects heart; red bone marrow produces blood cells; bones store Ca^{2+} for blood clotting.

Muscular System

Blood vessels deliver nutrients and oxygen to muscles; carry away wastes.

Muscle contraction keeps blood moving in heart and blood vessels.

Nervous System

Blood vessels deliver nutrients and oxygen to neurons; carry away wastes.

Brain controls nerves that regulate the heart and dilation of blood vessels.

Endocrine System

Blood vessels transport hormones from glands; blood services glands; heart produces atrial natriuretic hormone.

Epinephrine increases blood pressure; ADH, aldosterone, and atrial natriuretic hormone factors help regulate blood volume; growth factors control blood cell formation.

How the Cardiovascular System works with other body systems.

Lymphatic System/Immunity

Blood vessels transport leukocytes and antibodies; blood services lymphatic organs and is source of tissue fluid that becomes lymph.

Lymphatic vessels collect excess tissue fluid and return it to blood vessels; lymphatic organs store lymphocytes; lymph nodes filter lymph, and the spleen filters blood.

Respiratory System

Blood vessels transport gases to and from lungs; blood services respiratory organs.

Gas exchange in lungs rids body of carbon dioxide, helping to regulate the pH of blood; breathing aids venous return.

Digestive System

Blood vessels transport nutrients from digestive tract to body; blood services digestive organs.

Digestive tract provides nutrients for plasma protein formation and blood cell formation; liver detoxifies blood, makes plasma proteins, destroys old red blood cells.

Urinary System

Blood vessels deliver wastes to be excreted; blood pressure aids kidney function; blood services urinary organs.

Kidneys filter blood and excrete wastes; maintain blood volume, pressure, and pH; produce renin and erythropoietin.

Reproductive System

Blood vessels transport sex hormones; vasodilation causes genitals to become erect; blood services reproductive organs.

Sex hormones influence cardiovascular health; sexual activities stimulate cardiovascular system.

Capillaries

Heart

Vein

Artery

SELECTED NEW TERMS

Basic Key Terms

aorta (a-or'tuh), p. 267

arteriole (ar-te're-ōl), p. 259

artery (ar'ter-e), p. 259

atrial natriuretic peptide (ANP)
(ā'trē-əl ney-truh-yoo-ret-ik pep-tahyd), p. 265

atrioventricular (AV) node
(a"tre-o-ven-trik'yū-ler nōd), p. 253

atrioventricular bundle
(a"tre-o-ven-trik'yū-ler bundle), p. 253

atrioventricular valve
(a"tre-o-ven-trik'yū-ler valv), p. 250

atrium (a'tre-um), p. 250

bicuspid valve (bi-kus'pid valv), p. 250

capillary (kap'ĭ-lār"e), p. 260

cardioregulatory center
(kar"de-o-reg'yū-luh-tor-e sen'ter), p. 258

cerebral arterial circle
(sĕr'ĕ-bral ar-te're'al ser'kl), p. 271

chordae tendinae (kor'de ten'dĭ-nēa), p. 250

coronary artery (kor'ŏ-na-re ar'ter-e), p. 251

diastole (di-as'to-le), p. 257

ductus arteriosus
(duk'tus ar-tēr-e-o'sus), p. 272

ductus venosus (duk'tus vē-no'sus), p. 272

endocardium (en"do-kar'de-um), p. 249

epicardium (ep-ĭ-kăr'dē-yŭm), p. 249

foramen ovale (fo-ra'men o-vah'le), p. 272

hepatic portal system
(hĕ-pat'ik por'tal sis'tem), p. 271

inferior vena cava
(in-fēr'e-or ve'nuh ka'vuh), p. 267

interatrial septum
(in"ter-a'tre-al sep'tum), p. 250

interventricular septum
(in"ter-ven-trik'yū-ler sep'tum), p. 250

macrophage (măk'rō-făj), p. 254

mitral valve (mī'trəl vălv), p. 251

myocardium (mi"o-kar'de-um), p. 249

pacemaker (pās'mā'kər), p. 253

papillary muscles (păp'ə-lĕr'ē mŭs'ə-l-s), p. 250

pulmonary artery
(pul'mo-nĕr"e ar'ter-e), p. 267

pulmonary circuit
(pul'mo-nĕr"e ser"kut), p. 249

pulmonary vein (pul'mo-nĕr"e vān), p. 267

pulse (puls), p. 265

Purkinje fiber (per-kin'je fi'ber), p. 253

renin (rē'nĭn), p. 265

semilunar valve (sem"e-lu'ner valv), p. 250

sinoatrial (SA) node (si"no-a'tre-ul nōd), p. 253

superior vena cava
(su-pēr'e-or ve'nuh ka'vuh), p. 267

systemic circuit (sis-tem'ik ser"kut), p. 249

systole (sis'to-le), p. 257

tricuspid valve (tri-kus'pid valv), p. 250

tunica externa (tōō'nĭ-kə ēk'stŭrnə), p. 259

tunica interna (tōō'nĭ-kə ĭn-tûr'nə), p. 259

tunica media (tōō'nĭ-kə mē'dē-ə), p. 259

umbilical artery and vein
(um-bil'ĭ-kl ar'ter-e and vān), p. 272

vasomotor center (vā'zō-mō'tər sĕn'tər), p. 264

vein (vān), p. 262

ventricles (vĕn'trĭ-kəlz), p. 250

venule (ven'ūl), p. 262

visceral serous pericardium,
(vis'ĕr-ăl sĕr'us per-ĭ-kăr'dē-ŭm), p. 249

Clinical Key Terms

aneurysm (an'yer'ĭzm), p. 254

angina pectoris (an-ji'nuh pek'to-ris), p. 254

arrhythmias (ə-rĭth'mē-əz), p. 256

arteriosclerosis (ar-te"re-o-sklĕ-ro'sis), p. 254

atherosclerosis (ath"er-o"sklĕ-ro'sis), p. 254

atherosclerotic plaque
(ath'er-ō-sklĕr-ŏt-ik plăk), p. 254

bradycardia (brad"e-kar'de-uh), p. 256

cerebrovascular accident
(sĕr"e-bro-vas'kyū-ler ak'si-dent), p. 274

congestive heart failure
(kon-jes'tiv hart fāl'yer), p. 276

coronary artery disease
(kôr'ə-nĕr'ē ăr'tə-rē dĭ-zēz), p. 254

coronary bypass operation
(kor'ŏ-na-re bi'pas op-er-a'shun), p. 255

cyanosis (si"uh-no'sis), p. 274

ectopic pacemaker
(ek-top'ik pās'ma-ker), p. 256

electrocardiogram
(e-lek"tro-kar'de-o-gram"), p. 256

fatty streak (fat'ē strēk) p. 254

fibrillation (fi"brĭ-la'shun), p. 256

heart block (hart blok), p. 256

hemorrhoid (hem'royd), p. 262

hypertension (hī'pər-tĕn'shən), p. 254

infarct (in'fărkt), p. 255

ischemic heart disease
(is-kem'ik hart dĭ-zēz'), p. 254

myocardial infarction
(mi"o-kar'de-ul in-fark'shun), p. 255

phlebitis (flĭ-bi'tus), p. 262

pulmonary embolism
(pul'mo-nĕr"e em'bo-lizm), p. 262

stent (stēnt), p. 255

tachycardia (tak'ĭ kar'de-uh), p. 256

thromboembolism
(thrŏm'bō-em'bə-lĭz'əm), p. 254

transient ischemic attack (TIA)
(trăn-zē-ənt is-kem'ik' ə-tăk'), p. 274

varicose vein (văr'ĭ-kōs vān), p. 262

SUMMARY

12.1 Anatomy of the Heart

A. The heart keeps O_2-poor blood separate from O_2-rich blood and blood flowing in one direction. It creates blood pressure and regulates the supply of blood to meet current needs.

B. The heart is covered by the pericardium. The visceral pericardium is equal to the epicardium of the heart wall. Myocardium is cardiac muscle, and endocardium is its lining.

C. The heart has a right and left side and four chambers, consisting of two atria and two ventricles. The heart valves are the tricuspid valve, the pulmonary semilunar valve, the bicuspid valve, and the aortic semilunar valve. The first heart sound, "lub," is caused by AV valve closure. The second heart sound, "dup," occurs when the semilunar valves close.

D. The right side of the heart pumps blood to the lungs (pulmonary circuit), and the left side pumps blood to the tissues (systemic circuit). The myocardium is serviced by blood in the coronary circuit.

12.2 Physiology of the Heart

A. The conduction system of the heart includes the SA node, the AV node, the AV bundle, the bundle branches, and the Purkinje fibers. The SA node causes the atria to contract. The AV node and the rest of the conduction system cause the ventricles to contract.

B. The heartbeat (cardiac cycle) is divided into three phases: (1) In atrial systole, the atria contract; (2) in ventricular systole, the ventricles contract; and (3) in atrial and ventricular diastole, both the atria and the ventricles rest.

C. The cardiac output (amount of blood discharged by the heart in one minute) is the product of stroke volume and heart rate. The heart rate is regulated largely by the cardioregulatory center and the autonomic nervous system.

12.3 Anatomy of Blood Vessels

A. Blood vessels transport blood; carry out exchange in pulmonary capillaries and systemic capillaries; regulate blood pressure; and direct blood flow.

B. Arteries and arterioles carry blood away from the heart; veins and venules carry blood to the heart; and capillaries join arterioles to venules.

C. Capillary exchange occurs via a balance of hydrostatic and osmotic pressures. At the arterial end of a capillary, outward hydrostatic pressure is greater and tissue fluid accumulates. At the venous end, osmotic pressure is greater and tissue fluid is reabsorbed.

12.4 Physiology of Circulation

A. Velocity of blood flow varies according to total cross-sectional area; therefore, blood flow is slowest in the capillaries.

B. Blood pressure decreases with distance from the left ventricle. Cardiac output (CO) and resistance to flow determine blood pressure. Venous return affects CO. The skeletal muscle pump and the respiratory pump assist venous return. A vasomotor center regulates peripheral resistance. Neural regulation of peripheral resistance is via a vasomotor center in the medulla that is under the control of the cardioregulatory center. Several different hormones regulate blood pressure through their influence over kidney reabsorption of water.

C. To evaluate a person's circulation, it is customary to take the pulse and blood pressure.

12.5 Circulatory Routes

A. The pulmonary arteries transport O_2-poor blood to the pulmonary capillaries, and the pulmonary veins return O_2-rich blood to the heart. Congestive heart failure occurs if blood backs up in the pulmonary vessels. In the systemic circuit, blood travels from the left ventricle to the aorta, systemic arteries, arterioles, and capillaries, and then from the capillaries to the venules and veins to the right atrium of the heart. The systemic circuit serves the body proper.

B. The hepatic portal system carries blood from the stomach and intestines to the liver.

C. Circulation to the brain includes the cerebral arterial circle, which protects all regions of the brain from reduced blood supply.

D. Fetal circulation includes four unique features: (1) the foramen ovale, (2) the ductus arteriosus, (3) the umbilical arteries and vein, and (4) the ductus venosus. These features are necessary because the fetus does not use its lungs for gas exchange.

12.6 Effects of Aging

As we age, the cardiovascular system is more apt to suffer from all the disorders discussed in this chapter.

12.7 Homeostasis

The cardiovascular system is essential to homeostasis because it functions to assure exchange at the pulmonary capillaries and the systemic capillaries. There are many examples of the interaction of the cardiovascular system with other systems. For example, the endocrine system is dependent on the cardiovascular system to transport its hormones; and hormones help maintain blood pressure. Blood vessels deliver wastes to the kidneys and the kidneys help maintain blood pressure. The respiratory system is dependent on the cardiovascular system to transport gases to and from cells, and the respiratory system assists venous return.

STUDY QUESTIONS

1. State the location and functions of the heart. (p. 249)
2. Describe the wall and coverings of the heart. (pp. 249–250)
3. Name the chambers and valves of the heart. Trace the path of blood through the heart. (pp. 250–252)
4. Describe the coronary circuit, and discuss several coronary circuit disorders. (pp. 252–255)
5. Describe the conduction system of the heart and an electrocardiogram. (pp. 253, 256)
6. Describe the cardiac cycle (using the terms systole and diastole), and explain the heart sounds. (pp. 257–258)
7. What is cardiac output (CO)? What two factors determine CO? How are these factors regulated? (pp. 258–259)
8. What types of blood vessels are in the body? Discuss their structure and function. (pp. 259–262)
9. What factors determine velocity of blood flow? Blood pressure? In what vessel is blood pressure highest? Lowest? (pp. 262–264)
10. What mechanisms assist venous return to the heart? Discuss nervous and hormonal control of blood pressure. (pp. 264–265)

11. What is pulse? How do you take a person's pulse? How do you take a person's blood pressure? What does a blood pressure of 120/80 mean? (pp. 265–266)
12. What are hypertension, stroke, aneurysm, and congestive heart failure? (pp. 253–254, 272)
13. Trace the path of blood from the superior mesenteric artery back to the aorta, indicating which of the vessels are in the systemic circuit and which are in the pulmonary circuit. (pp. 267–269)
14. Give examples to show that the cardiovascular system functions to maintain homeostasis and that interactions with other systems help it and the other systems maintain homeostasis. (pp. 276–277)

LEARNING OBJECTIVE QUESTIONS

Fill in the blanks.

1. When the left ventricle contracts, blood enters the _____.
2. The right side of the heart pumps blood to the _____.
3. The _____ node is known as the pacemaker.
4. Arteries are blood vessels that take blood _____ the heart.
5. The blood vessels that serve the heart are the _____ arteries and veins.
6. The major blood vessels taking blood to and from the shoulders and upper limbs are the _____ arteries and veins.

Those taking blood to and from the legs are the _____ arteries and veins.
7. Blood vessels to the brain end in a circular path known as the _____.
8. The human body contains a hepatic portal system that takes blood from the _____ to the _____.
9. The force of blood against the walls of a artery is termed _____.
10. Blood moves in arteries due to _____ and in veins movement is assisted by _____.
11. The blood pressure recorded when the left ventricle contracts is called the _____ pressure, and the pressure recorded when the left ventricle relaxes is called the _____ pressure.
12. The two factors that affect blood pressure are _____ and _____.
13. In the fetus, the opening between the two atria is called the _____ and the connection between the pulmonary artery and the aorta is called the _____.
14. The valve between the left atrium and left ventricle is the _____, or mitral, valve.

MEDICAL TERMINOLOGY EXERCISE

Consult Appendix B for help in pronouncing and analyzing the meaning of the terms that follow.

1. cryocardioplegia (kri-o-kar″de-o-ple′je-uh)
2. echocardiography (ek″o-kar″de-og′ruh-fe)
3. percutaneous transluminal coronary angioplasty (per″kyū-ta′ne-us trans″lu′mǐ-nal kor′ō-nā-re an′je-o-plas″te)
4. vasoconstriction (vas″o-kon-strik′shun)
5. valvuloplasty (val′vu-lo-plas″te)
6. antihypertensive (an″tǐ-hi″per-ten′siv)
7. thromboendarterectomy (throm″bo-end″ar-ter-ek′to-me)
8. cardiovalvulitis (kar′de-o-val-yū-li′tis)
9. vasospasm (va′-so-spazm)
10. pericardiocentesis (pěr-ǐ-kar′de-o-sen-te′sis)
11. ventriculotomy (ven-trik-yū-lot′o-me)
12. phlebectasia (fleb-ek-ta′ze-uh)
13. myocardiorrhaphy (mi′o-kar-de-or′uh-fe)

WEBSITE LINK

Visit this text's website at http://www.mhhe.com/longenbaker7 **for additional quizzes, interactive learning exercises, and other study tools.**

The Lymphatic System and Body Defenses

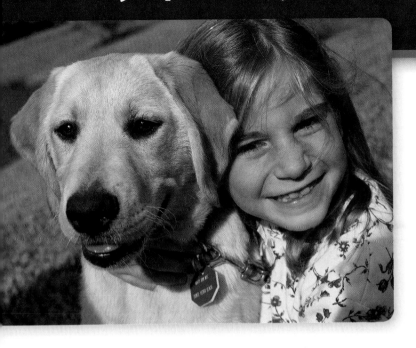

CHAPTER 13

As you'll discover in this chapter about the immune system, an allergy is an exaggerated immune response to ordinary, everyday substances in the environment. Mold, pollen, animal hair and dander, dust, bee stings, antibiotics, food—the list of things that can cause an allergy is a long one (and based on your own experience, you might be able to add to the list). Peanut allergy is one of the most prevalent and dangerous food allergies, and the leading cause of food-related death. For a person with this allergy, the world is a minefield of peanut products, in one form or another—nuts, oil, flour, empty peanut shells. Even peanut dust can contaminate other food: For example, baked goods produced in a factory that also makes peanut butter cookies are unsafe. Public places such as schools, restaurants, hotels, stores, and even public transportation are dangerous too, because all might have some peanut contamination. Enter man's best friend: a peanut-sniffing dog. Like a seeing eye dog is for a blind person, so a peanut-sniffer is for a person with this deadly allergy. The dogs are trained to detect even minute amounts of peanut substances and then lead their owners away. Equipped with a sniffing dog, the allergic person can navigate a peanut-filled world in relative safety. You can read more about allergies on page 300.

Chapter Outline & Learning Objectives
After you have studied this chapter, you should be able to:

13.1 Lymphatic System

The **lymphatic system** consists of lymphatic vessels and the lymphatic organs. This system, which is closely associated with the cardiovascular system, has three main functions that contribute to homeostasis:

1. **Fluid balance.** The lymphatic system takes up excess tissue fluid and returns it to the bloodstream. Recall that lymphatic capillaries lie very near blood capillaries, and they serve as an auxiliary way to take up fluid that has exited the blood capillaries.
2. **Fat absorption.** The lymphatic system absorbs fats from the digestive tract and transports them to the bloodstream. Special lymphatic capillaries called **lacteals** are located in the intestinal villi (see Fig. 15.7). This function ensures the absorption of dietary lipids as well as lipid-soluble vitamins.
3. **Defense.** The lymphatic system helps defend the body against disease. This function is carried out by the white blood cells present in lymphatic vessels and lymphatic organs, as well as those present in the blood and scattered between tissue cells. In addition to destroying foreign invaders, white blood cells are important in the destruction of dead and dying tissue cells. White blood cells also defend against body cells that have mutated to become cancerous.

Lymphatic Vessels

Lymphatic vessels form a one-way system that begins with lymphatic capillaries. Most regions of the body are richly supplied with lymphatic capillaries. These are tiny, closed-ended vessels consisting of a single layer of simple squamous epithelium, whose walls have large pores to allow particles to enter (Fig. 13.1). Lymphatic capillaries take up excess tissue fluid. Tissue fluid is mostly water, but it also contains solutes (e.g., nutrients, electrolytes, and oxygen) derived from plasma and cellular products (i.e., hormones, enzymes, and wastes) secreted by cells. Once inside lymphatic vessels, this tissue fluid is called **lymph**.

The lymphatic capillaries join to form lymphatic vessels that merge before entering one of two ducts: the **thoracic duct** or the **right lymphatic duct**. The larger thoracic duct returns lymph collected from the body below the thorax and the left arm and left side of the head and neck into the left subclavian vein. The right lymphatic duct returns lymph from the right arm and right side of the head and neck into the right subclavian vein.

The construction of the larger lymphatic vessels is similar to that of cardiovascular veins, including the presence of valves. The movement of lymph within lymphatic capillaries is largely dependent upon skeletal muscle contraction. Lymph forced through lymphatic vessels as a result of muscular compression is prevented from flowing backward by one-way valves (in the same way that vein valves prevent backflow). Failure of the lymphatic system to properly collect and return excessive tissue fluid to the bloodstream results in **edema,** as described in the Medical Focus on page 285.

Content CHECK-UP!

1. If you wanted to describe the lymphatic system, which statements would you use?
   a. It takes up excess tissue fluid and returns it to the circulation.
   b. It houses white blood cells.
   c. Lacteals in the intestines absorb dietary lipids.
   d. All of these statements are correct.

2. Imagine that lymph is draining from your right arm to return to the heart. What is the correct pathway for it to take?
   a. lymphatic capillaries → larger lymphatic vessels → right lymphatic duct → right subclavian vein → superior vena cava
   b. lymphatic capillaries → larger lymphatic vessels → thoracic duct → right subclavian vein → superior vena cava
   c. lymphatic capillaries → larger lymphatic vessels → thoracic duct → left subclavian vein → superior vena cava
   d. lymphatic capillaries → larger lymphatic vessels → cervical duct → left subclavian vein → superior vena cava

3. Large lymphatic vessels are similar in structure to:
   a. capillaries in the cardiovascular system.
   b. arteries in the cardiovascular system.
   c. veins in the cardiovascular system.

Answers in Appendix C, page 464.

13.2 Organs, Tissues, and Cells of the Immune System

The **immune system,** which plays an important role in keeping us healthy, consists of a network of lymphatic organs, tissues, and cells as well as products of these cells, including antibodies and regulatory agents. **Immunity** is the ability to react to antigens so that the body remains free of disease. Disease results from a failure of homeostasis. It can be due to infection by foreign microbes, and/or to the failure of the immune system to function properly.

Primary Lymphatic Organs

Lymphatic (lymphoid) organs contain large numbers of lymphocytes, the type of white blood cell that plays a pivotal role in immunity. The *primary lymphatic organs* are the red bone marrow and the thymus gland (Fig. 13.1, *bottom*). Lymphocytes originate and/or mature in these organs.

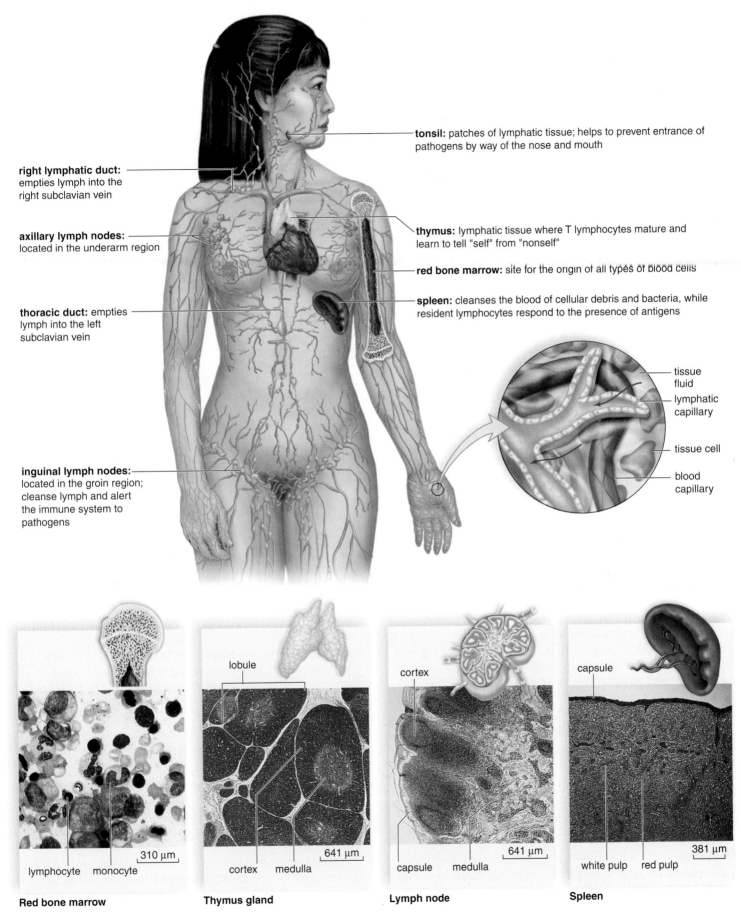

tonsil: patches of lymphatic tissue; helps to prevent entrance of pathogens by way of the nose and mouth

right lymphatic duct: empties lymph into the right subclavian vein

axillary lymph nodes: located in the underarm region

thoracic duct: empties lymph into the left subclavian vein

inguinal lymph nodes: located in the groin region; cleanse lymph and alert the immune system to pathogens

thymus: lymphatic tissue where T lymphocytes mature and learn to tell "self" from "nonself"

red bone marrow: site for the origin of all types of blood cells

spleen: cleanses the blood of cellular debris and bacteria, while resident lymphocytes respond to the presence of antigens

tissue fluid
lymphatic capillary
tissue cell
blood capillary

lymphocyte   monocyte
Red bone marrow
310 µm

lobule
cortex   medulla
Thymus gland
641 µm

cortex
capsule   medulla
Lymph node
641 µm

capsule
white pulp   red pulp
Spleen
381 µm

Figure 13.1 The lymphatic system. Lymphatic vessels drain excess fluid from the tissues and return it to the cardiovascular system. The enlargement shows a lymphatic vessel. The red bone marrow, thymus gland, lymph nodes, and spleen are among those lymphatic organs that assist immunity.

Red Bone Marrow

Red bone marrow is the site of stem cells that are ever capable of dividing and producing blood cells. Some of these cells become the various types of white blood cells: neutrophils, eosinophils, basophils, lymphocytes, and monocytes (Fig. 13.2).

In a child, most bones have red bone marrow, but in an adult it is limited to the sternum, vertebrae, ribs, the skull, part of the pelvic girdle, and the proximal heads of the humerus and femur.

The red bone marrow consists of a network of reticular tissue fibers, which support the stem cells and their progeny. They are packed around thin-walled sinuses filled with venous blood. Differentiated blood cells enter the bloodstream at these sinuses.

Lymphocytes differentiate into the B lymphocytes and the T lymphocytes. Bone marrow is not only the source of B lymphocytes, but also the place where B lymphocytes mature. T lymphocytes mature in the thymus.

Thymus Gland

The soft, bilobed **thymus gland** is located in the thoracic cavity between the trachea and the sternum superior to the heart. The thymus varies in size, but it is largest in children and shrinks as we get older. Connective tissue divides the thymus into lobules, which are filled with lymphocytes. The thymus gland produces thymic hormones, such as thymosin, that are thought to aid in the maturation of T lymphocytes. Thymosin may also have other functions in immunity.

Immature T lymphocytes migrate from the bone marrow through the bloodstream to the thymus, where they mature. Only about 5% of these cells ever leave the thymus. These T lymphocytes have survived a critical test: If any show the ability to react with "self" cells (i.e., one's own tissue cells), they die. If they have potential to attack a foreign cell, they leave the thymus.

The thymus is absolutely critical to immunity; without a thymus, the body does not reject foreign tissues, blood lymphocyte levels are drastically reduced, and the body's response to most antigens is poor or absent.

Secondary Lymphatic Organs

The secondary lymphatic organs are the spleen, the lymph nodes, and other organs, such as the tonsils, Peyer patches, and the appendix. All the secondary organs are places where lymphocytes encounter and bind with antigens, after which they proliferate and become actively engaged cells.

Spleen

The **spleen,** the largest lymphatic organ, is located in the upper left region of the abdominal cavity posterior to the stomach. Connective tissue divides the spleen into partial compartments, each of which contains tissue known as white pulp

a. Neutrophil
40–70%
Phagocytizes primarily bacteria

b. Eosinophil
1–4%
Phagocytizes and destroys antigen-antibody complexes and parasitic organisms

c. Basophil
0–1%
Releases histamine and other chemicals when stimulated

d. Lymphocyte
20–45%
B type produces antibodies in blood and lymph; T type kills virus-containing cells and cancer cells

e. Monocyte
4–8%
Becomes macrophage—phagocytizes bacteria and viruses

Figure 13.2 **The five types of white blood cells.** These cell types differ according to structure and function. The frequency of each type of cell is given as a percentage of the total found in circulating blood. (See Chapter 11 for a detailed description of these cells.)

and red pulp (Fig. 13.1, *bottom*). The white pulp contains a concentration of lymphocytes. The red pulp, which surrounds venous sinuses, is involved in filtering the blood. Blood entering the spleen must pass through the sinuses before exiting. Lymphocytes and macrophages react to pathogens, and macrophages engulf debris and also remove any old, worn-out red blood cells.

The spleen's outer capsule is relatively thin, and an infection or a blow can cause the spleen to burst. Although the spleen's functions are replaced by other organs, a person without a spleen is often slightly more susceptible to infections and may have to receive antibiotic therapy indefinitely.

Lymph Nodes

Lymph nodes, which are small, ovoid structures, occur along lymphatic vessels. Connective tissue forms the capsule of a lymph node and also divides the organ into compartments (see Fig. 13.1). Each compartment contains a nodule packed with B lymphocytes and a sinus that increases in size toward the center of the node. As lymph courses through the sinuses, it is filtered by macrophages (phagocytic white blood cells), which engulf pathogens and debris. T lymphocytes, also present in sinuses, fight infections and attack cancer cells.

Each portion of the anterior cavity (see Fig. 1.5) contains superficial and deep lymph nodes, named for their location. For example, inguinal nodes are in the groin, and axillary nodes are in the armpits. Physicians often examine for the presence of swollen, tender lymph nodes as evidence that the body is fighting an infection. This is a noninvasive, preliminary way to help make such a diagnosis.

Begin Thinking Clinically

Why do surgeons remove lymph nodes at the same time that a cancer is taken out? What is a sentinel node?

Answer and discussion in Appendix C, page 464.

Lymphatic Nodules

Lymphatic nodules are concentrations of lymphatic tissue not surrounded by a capsule. The **tonsils** are lymphatic nodules located in the posterior pharynx (see Fig. 14.2). A single **pharyngeal tonsil** (also referred to as adenoids) is in the nasopharynx, whereas **lingual tonsils** sit at the base of the tongue. **Palatine tonsils** are visible in the posterior oral cavity. The tonsils perform the same functions as lymph nodes, but because of their location, they are the first to encounter pathogens and antigens that enter the body by way of the nose and mouth.

MEDICAL FOCUS

The Lymphatic System and Illness

As the "next-door neighbor" to the cardiovascular system, the parallel lymphatic system defends against infectious disease and cancer, and regulates fluid balance in the tissues. Failure of the lymphatic system to maintain homeostatic balance in these critical areas results in disease.

The internal structure of a lymph node is designed to filter out and destroy any foreign material, including microbes and cancer cells, from lymph. An infection that causes swelling and tenderness of nearby lymph nodes is called **lymphadenitis.** If the infection is not contained, **lymphangitis,** an infection of the lymphatic vessels, may result. Red streaks can be seen through the skin, indicating that the infection may have spread to the bloodstream.

Like microbes, cancer cells sometimes enter lymphatic vessels and move undetected to other regions of the body, where they produce secondary tumors. In this way, the lymphatic system sometimes assists metastasis, the spread of cancer far from its place of origin. Because of this potential for metastasis, regional lymph nodes are usually removed for examination whenever surgery is used to diagnose or treat cancer. The presence or absence of tumor cells in the nodes can be used to determine how far the cancer has spread. Lymph node *biopsy* (the microscopic study of the tissue) also aids in making decisions concerning additional treatment, such as radiation or chemotherapy.

Failure of the lymphatic vessels to properly remove tissue fluid results in a condition called edema. Edema forms either because too much tissue fluid is made, and/or not enough tissue fluid is drained away. Excessive tissue fluid can compress the blood vessels within a tissue, preventing the flow of blood,

oxygen and nutrients to tissue cells. Thus, edema can lead to tissue damage and eventual death, illustrating the importance of tissue fluid collection by the lymphatic system.

Blockage of the lymphatic vessels prevents proper tissue fluid drainage, resulting in edema. As mentioned previously, surgical removal of lymph nodes is a common treatment for cancer. However, node removal often blocks lymphatic drainage, producing a painful and debilitating **lymphedema** for the patient. A dramatic example of edema caused by lymphatic obstruction occurs when a parasitic roundworm clogs the lymphatic vessels. Tremendous swelling of the arm, leg, or external genitals results, causing a condition called **elephantiasis.**

Edema can also be due to a low osmotic pressure of the blood, as when plasma proteins are excreted by the kidneys instead of being retained in the blood. Extra tissue fluid forms and lymphatic vessels may not be able to absorb it all. **Pulmonary edema** is a life-threatening condition associated with congestive heart failure. When the heart muscle is damaged and unable to effectively empty the heart, blood backs up in the pulmonary circulation. Pulmonary capillary blood pressure increases, which leads to excess fluid in lung tissue. Congested lung tissue cannot properly exchange oxygen and carbon dioxide with the blood, and the patient may suffocate.

Cancer of lymphatic tissue is called **lymphoma.** In **Hodgkin disease,** billions of lymphoma cells create swollen lymph nodes throughout the body. The lymphoma cells can migrate and grow in the spleen, liver, and bone marrow. The prognosis is good, however, if Hodgkin disease is diagnosed early.

Peyer patches are located in the intestinal wall and the walls of the **appendix.** These structures encounter pathogens that enter the body by way of the intestinal tract.

Content CHECK-UP!

4. Which lymphatic organ functions as a blood reservoir and removes dead and dying red blood cells?

    a. thymus   c. appendix

    b. Peyer patches   d. spleen

5. List and describe two functions of the thymus gland.

6. Which of the following structures has a connective tissue capsule?

    a. axillary lymph nodes   c. Peyer patches

    b. palatine tonsils   d. adenoids (pharyngeal tonsil)

Answers in Appendix C, page 464.

13.3 Nonspecific and Specific Defenses

Immunity includes nonspecific defenses and specific defenses. The five types of nonspecific defenses—barriers to entry, the inflammatory reaction, nonspecific phagocytic white blood cells, natural killer cells, and protective proteins—are effective against many types of infectious agents. Specific defenses are effective against a particular infectious agent.

Nonspecific Defenses

Barriers to Entry

The body has built-in barriers, both physical and chemical, that help to prevent infection by microbes. The intact skin is generally a very effective physical barrier that prevents infection. Mucous membranes lining the respiratory, digestive, reproductive, and urinary tracts are also physical barriers to entry by pathogens. (However, it must be noted that their effectiveness is limited, especially in the oral and nasal cavities, because most cold viruses enter the body by crossing these membranes.) The ciliated cells that line the upper respiratory tract sweep mucus and trapped particles up into the throat, where they can be swallowed or expectorated (spit out).

Chemical barriers to infection include the secretions of sebaceous (oil) glands in the skin, which contain chemicals that weaken or kill certain bacteria on the skin. As perspiration evaporates, the concentrated salt on the skin helps prevent infection. Further, perspiration, saliva, and tears contain an antibacterial enzyme called *lysozyme.* Saliva also helps to wash microbes off the teeth and tongue. Similarly, as urine is voided from the body, it flushes bacteria from the urinary tract. The acidic pH of the stomach inhibits bacterial growth or kills many types of bacteria. Finally, a significant chemical barrier to infection is created by the *normal flora*—microbes that normally reside in the mouth, intestine, and other areas. By using up available nutrients and releasing their own waste, these normal germs prevent potential pathogens from taking up residence. For this reason, abusing antibiotics can make a person susceptible to pathogenic infection by killing off the normal flora.

Inflammatory Reaction

Whenever tissue is damaged by physical or chemical agents or by pathogens, a series of events occurs that is known as the **inflammatory reaction.** The inflammatory reaction has four outward signs: redness, swelling, heat, and pain. All of these signs are due to capillary changes in the damaged area. Figure 13.3 illustrates the five basic steps of an inflammatory response:

1. Chemical mediators, such as **histamine,** cause capillaries to dilate and become more permeable. These tissue chemicals are released by damaged tissue cells and by **mast cells** found in the tissue. Mast cells resemble basophils, one of the types of white blood cells (see section 11.2 to review the types of white blood cells). Excess blood flow to the inflamed area causes the skin to redden and become warm. Elevated temperature tends to inhibit growth of some pathogens, and heightened blood flow brings more defensive white blood cells to the area.

2. Increased capillary permeability allows fluids and proteins, including blood clotting factors, to escape into the tissues. Escaped fluid causes swelling in the affected areas, putting pressure on nearby pain neurons and causing pain. Further, the inflammatory chemicals themselves can also stimulate the pain sensation. Although we can all agree that pain is a big nuisance, it's also a warning that tissue has been injured and we need to take corrective measures.

3. Edema and clots that form in the injured area help to "wall off" the area from the rest of the body. Thus, if pathogens have entered the body during injury, they will not be able to cause a systemic infection.

4. Migration of phagocytes, primarily neutrophils and monocytes, also occurs during the inflammatory reaction. Neutrophils and monocytes are amoeboid and can change shape to squeeze through capillary walls and enter tissue fluid. There, they phagocytize pathogens.

5. After monocytes appear on the scene, they differentiate into **macrophages,** large phagocytic cells that are able to devour as many as 100 pathogens and still survive. Some tissues, particularly connective tissue, have resident macrophages that routinely act as scavengers—devouring old blood cells, bits of dead tissue, and other debris. Macrophages also release colony-stimulating factors, which pass by way of the blood to the red bone marrow. There, the factors stimulate the production and the release of white blood cells, primarily neutrophils.

During the process of phagocytosis, endocytic vesicles form when neutrophils and macrophages engulf pathogens (see section 3.2, page 52). When the vesicle combines with a lysosome, a cellular organelle, the pathogen is destroyed by hydrolytic enzymes. As the infection is being overcome, some

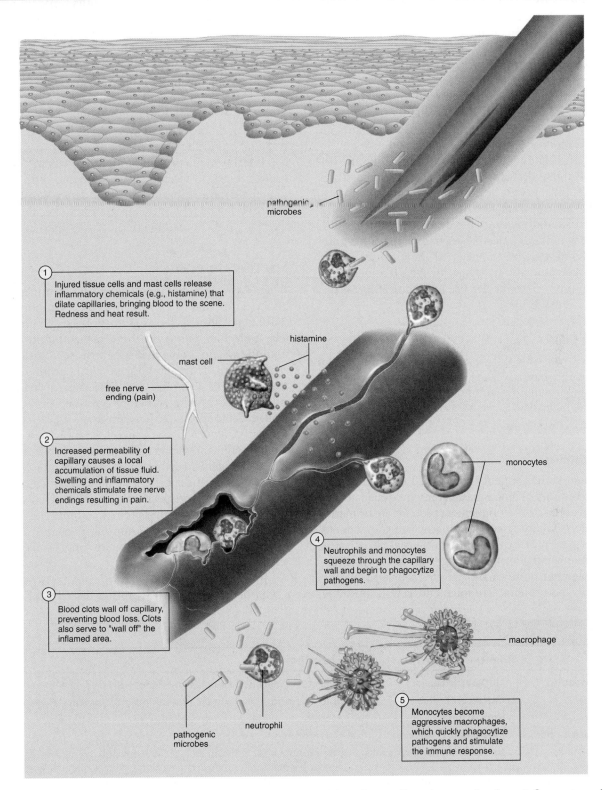

pathogenic microbes

1 Injured tissue cells and mast cells release inflammatory chemicals (e.g., histamine) that dilate capillaries, bringing blood to the scene. Redness and heat result.

histamine

mast cell

free nerve ending (pain)

2 Increased permeability of capillary causes a local accumulation of tissue fluid. Swelling and inflammatory chemicals stimulate free nerve endings resulting in pain.

monocytes

4 Neutrophils and monocytes squeeze through the capillary wall and begin to phagocytize pathogens.

3 Blood clots wall off capillary, preventing blood loss. Clots also serve to "wall off" the inflamed area.

macrophage

5 Monocytes become aggressive macrophages, which quickly phagocytize pathogens and stimulate the immune response.

neutrophil

pathogenic microbes

Figure 13.3 Steps of the inflammatory reaction. (1) When there is an injury, tissue cells and mast cells release inflammatory chemicals (such as histamine) that dilate blood vessels and cause them to leak. More blood flows to the area, and the skin appears red and feels warm. (2) Because capillaries are more permeable, edema forms. Swelling and inflammatory chemicals stimulate pain receptors (free nerve endings). (3) Blood clotting seals off the capillary, preventing blood loss and walling off the inflamed area. (4) Monocytes (which become macrophages) and neutrophils squeeze through the capillary wall, and phagocytosis begins. (5) Monocytes transform to macrophages, which stimulate specific immune responses in addition to destroying pathogens.

phagocytes die. These—along with dead tissue cells, dead bacteria, living white blood cells, and possibly living bacteria—form **pus,** a whitish material. The presence of pus indicates that the body is destroying dead tissue cells, and/or trying to overcome an infection.

Sometimes an inflammation persists, and the result is chronic inflammation that is often treated by administering anti-inflammatory agents such as aspirin, ibuprofen, or cortisone. These medications act against the chemical mediators released by the white blood cells in the damaged area and also decrease the phagocytic response by white blood cells.

Although an inflammatory reaction is a nonspecific defense mechanism that occurs whenever tissue is injured, it can be accompanied by a specific, directed response to the injury. As infectious microbes are destroyed by neutrophils and macrophages, bits and pieces of the dead cells can cross the barrier created by inflammation. These fragments, called **antigens,** can travel along with the released chemical mediators through the tissue fluid and lymph to the lymph nodes. Now lymphocytes mount a specific defense to the infection, as described in the next section.

Natural Killer Cells

Natural killer (NK) cells kill virus-infected cells and tumor cells by cell-to-cell contact. They are large, granular leukocytes, similar in structure to lymphocytes, but with no specificity and no memory. Their number is not increased by prior exposure to any kind of cell.

Protective Proteins

The **complement system,** often simply called complement, is composed of a number of blood plasma proteins designated by the letter C and a subscript. Complement proteins are primarily made by the liver. A limited amount of activated complement protein is needed because a cascade effect occurs: Each activated protein in a series is capable of activating many other proteins.

The complement proteins are activated when pathogens enter the body. The proteins "complement" certain immune responses, which accounts for their name. For example, they are involved in and amplify the inflammatory response because complement proteins attract phagocytes to the scene. Some complement proteins bind to the surface of pathogens already coated with antibodies (specific antibacterial protein molecules), which ensures that the pathogens will be phagocytized by a neutrophil or macrophage.

Certain other complement proteins join to form a membrane attack complex that punches holes in the walls and plasma membranes of bacteria. Fluids and salts then enter the bacterial cell to the point that it bursts.

Interferon is a protein produced by virus-infected cells. Interferon binds to receptors of noninfected cells, causing them to prepare for possible attack by producing substances that interfere with viral replication. Interferon is specific to the species; therefore, only human interferon can be used in humans.

Specific Defenses

When creating a specific defense, the immune system lymphocytes respond to foreign antigens. It's important to understand that normal cells also have antigens, in the form of protein molecules found on the cell membrane. However, the immune system is able to distinguish "self" antigens from foreign antigens. Thus, lymphocytes don't ordinarily attack normal body cells. Lymphocytes are capable of recognizing alien antigens because they have antigen receptors—plasma membrane receptor proteins that combine with a specific "nonself" antigen. Antigens are typically large molecules like proteins, though fragments of bacteria, viruses, molds, or parasites can also be antigenic. Further, abnormal cell membrane proteins produced by cancer cells may also be antigens. Table 13.1 summarizes the nonspecific and specific defenses.

Immunity usually lasts for some time. For example, once a child has been properly immunized against measles virus, the immune system confers a lifelong protection (see the Medical Focus on page 296).

Immunity is primarily the result of the action of the **B lymphocytes** and the **T lymphocytes.** B lymphocytes mature

TABLE 13.1 Three Lines of Defense		
	Types	**How They Work**
First Line: Nonspecific	**Mechanical barriers**	Intact skin, mucous membranes, respiratory cilia prevent entry
	Chemical barriers	Stomach acid, saliva, tears, etc. destroy and/or wash away pathogens
	Normal flora	Create a hostile environment for pathogens
Second Line: Nonspecific	**Inflammatory reaction**	Walls off infection site, draws white cells to area, signals a pain response
	Nonspecific phagocytic cells	Neutrophils, eosinophils, monocytes/macrophages engulf and destroy invaders; basophils/mast cells produce chemicals to stimulate immune response
	Natural killer cells	Kill pathogens by cell-to-cell contact
	Protective proteins	Complement participates, enhances defensive response Interferon interferes with viral replication
Third Line: Specific	**B lymphocytes**	Antibody response
	T lymphocytes	Produce messenger cytokines, stimulate B cells, direct cell destruction

in the bone marrow,[1] and T lymphocytes mature in the *thymus gland*. B lymphocytes, also called B cells, give rise to **plasma cells,** which produce antibodies. **Antibodies** combine with and neutralize foreign antigens. Each antibody protein is shaped like a specific antigen, and combines with that antigen in a precise, "lock-and-key" fit (Fig. 13.4). These antibodies are secreted into the blood, lymph, and other body fluids. In contrast, T lymphocytes, also called T cells, do not produce antibodies. Instead, certain T cells directly attack cells that bear nonself proteins. Other T cells regulate the immune response.

B Cells and Antibody-Mediated Immunity

There are millions of individual groups, or **clones,** of immature B lymphocytes in an adult. Each clone is pre-programmed to produce a particular type of antibody if exposed to a specific antigen. The **clonal selection theory** states that one particular lymphocyte in a clone is *selected* to respond because its antigen receptors combine with the antigen in the precise "lock-and-key" fit. Next, the selected lymphocyte goes on to reproduce multiple copies of itself. Each of the identical cells in the resulting group is a clone of the original cell. The majority of these B lymphocytes will mature to form plasma cells, all bearing the same type of antigen receptor.

Notice in Figure 13.5 that different types of antigen receptors are represented by their shaped ends: circular, triangular, square, and so on. The B cell with curved, circular receptors undergoes clonal expansion because a specific antigen (green dots) is present and binds to its receptors. B cells are stimulated to divide and become plasma cells by helper T-cell secretions called cytokines, as discussed later in this section. Some members of the clone become memory cells, which are the means by which long-term immunity is possible. If the same antigen enters the system again, **memory B cells** quickly divide and give rise to more lymphocytes capable of quickly producing antibodies. A second exposure to the same antigen produces a stronger, faster immune response.

[1]Historically, the B stands for bursa of Fabricius, an organ in the chicken where these cells were first identified.

Once the threat of an infection has passed, the development of new plasma cells ceases, and those present undergo apoptosis. **Apoptosis** is a process of programmed cell death involving a cascade of specific cellular events leading to the death and destruction of the cell. It is important that these cells die once they are no longer needed. Otherwise they could mistakenly destroy body cells in an **autoimmune response.**

Defense by B cells is called **antibody-mediated immunity** because the various types of B cells produce antibodies. It is also called **humoral immunity** because these antibodies are present in blood and lymph. A humor is any fluid normally occurring in the body.

Structure and Function of Antibodies

The basic unit that composes antibody molecules is a Y-shaped protein molecule with two arms. Each arm has a "heavy" (long) polypeptide chain and a "light" (short) polypeptide chain (Fig. 13.4). These chains have constant regions, located at the trunk of the Y, where the sequence of amino acids is set. The class of antibody for each individual molecule is determined by the structure of the antibody's constant region. The variable regions form an antigen-binding site, and their shape is specific to a particular antigen. The antigen combines with the antibody at the antigen-binding site in a lock-and-key manner. Antibodies may consist of single Y-shaped molecules called *monomers*, or they can be paired together in a molecule called a *dimer*. Very large antibodies belonging to the M class are *pentamers*—clusters of five Y-shaped molecules linked together.

The antigen-antibody reaction can take several forms. Antibodies react with viruses and toxins (the poisons made by bacteria) by coating them completely—a process called *neutralization*. Often the reaction produces *an immune complex*—a clump of antigens combined with antibodies. Clustering antigens into an immune complex marks the antigens for destruction. For example, an antigen-antibody complex may be engulfed by neutrophils or macrophages, or it may activate complement. Complement makes pathogens more susceptible to phagocytosis, as discussed previously.

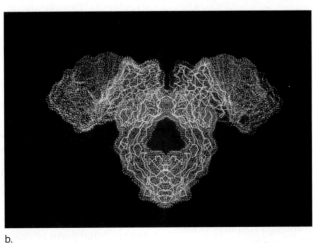

Figure 13.4 Structure of a single antibody molecule (monomer). **a.** Monomers contain two heavy (long) polypeptide chains and two light (short) chains arranged so that there are two variable regions, where a particular antigen is capable of binding with an antibody (V = Variable region, C = constant region). **b.** Computer model of an antibody molecule. The antigen combines with the two side branches.

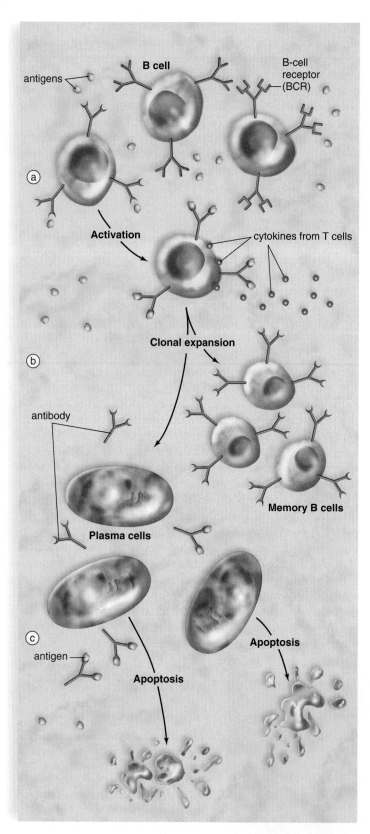

Figure 13.5 **Clonal selection theory as it applies to B cells.**
a. Each B cell has different antigen receptors. **b.** An antigen combines with only one B-cell receptor. Cytokines cause the B cell to multiply.
c. The chosen B cell forms plasma cells, which undergo apoptosis when the immune response is complete. Memory B cells remain in the body.

TABLE 13.2 Functional Classes of Antibodies		
Classes	**Presence**	**Function**
IgG	Main antibody type in circulation; crosses the placenta from mother to fetus	Binds to pathogens, activates complement proteins, and enhances phagocytosis
IgM	Antibody type found in circulation; largest antibody; first antibody formed by a newborn; first antibody formed with any new infection	Activates complement proteins; clumps cells
IgA	Main antibody type in secretions such as saliva and milk	Prevents pathogens from attaching to epithelial cells in digestive and respiratory tract
IgD	Antibody type found on surface of immature B cells	Presence signifies readiness of B cell
IgE	Antibody type found as antigen receptors on basophils in blood and on mast cells in tissues	Responsible for immediate allergic response and protection against certain parasitic infections

Classes of Antibodies

There are five different classes of circulating antibody proteins, or immunoglobulins (Igs) (Table 13.2). IgG antibodies are the major type in blood, and lesser amounts are also found in lymph and tissue fluid. IgG antibodies bind to pathogens and their toxins. IgG antibodies can cross the placenta from a mother to her fetus, so that the newborn has a temporary, partial immune response. As mentioned previously, IgM antibodies are pentamers. These antibodies are "first responders": the first antibodies produced by a newborn's body. Subsequently, IgM antibodies are the first to appear in blood soon after an infection begins and the first to disappear before the infection is over. They are good activators of the complement system. IgA antibodies are monomers or dimers containing two Y-shaped structures. They are the main type of antibody found in body secretions: saliva, tears, mucus, and breast milk. IgA molecules bind to pathogens and prevent them from reaching the bloodstream. The main function of IgD molecules seems to be to serve as antigen receptors on immature B cells. IgE antibodies are responsible for prevention of parasitic infections, but they can also cause immediate allergic responses.

T Cells and Cell-Mediated Immunity

When T cells leave the thymus, they have unique antigen receptors just as B cells do (Fig. 13.6a) Unlike B cells, however, T cells are unable to recognize an antigen present in lymph, blood, or the tissues without help. The antigen must be presented to them by an **antigen-presenting cell (APC)**, which is a specialized form of macrophage cell. When an APC presents a viral or cancer cell antigen to the T cell, the antigen is first

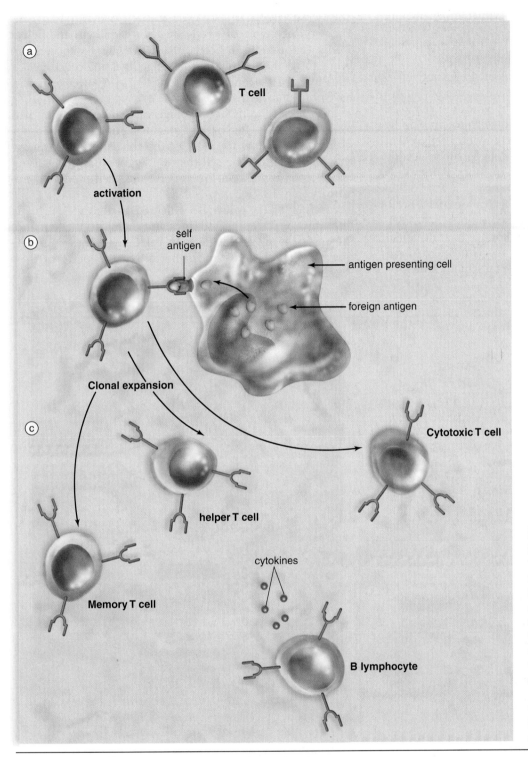

Figure 13.6 Clonal selection theory as it applies to T cells. a. Each T cell has different antigen receptors that combine with only one antigen. **b.** The macrophage presents this antigen in the groove of an HLA molecule, and the T cell that can combine with the antigen-HLA pair is stimulated to divide. **c.** Clonal expansion gives rise to three types of T cells. Helper T cells produce cytokines that stimulate immune responses, including B-cell response. Cytotoxic T cells attack cells having the same antigen. Memory cells remain after the immune response is complete.

Labels in figure:
a
T cell
activation
b
self antigen
antigen presenting cell
foreign antigen
Clonal expansion
c
Cytotoxic T cell
helper T cell
Memory T cell
cytokines
B lymphocyte

linked to a **major histocompatibility complex (MHC)** protein in the APC cell's plasma membrane (Fig 13.6*b*).

Human MHC proteins are called **HLA (human leukocyte antigens)**. These proteins are found on all of the body's cells. More than 50 different proteins have been identified, and each individual human being has a unique combination—no two sets are exactly alike. An exception is the set belonging to monozygotic, or identical twins. Because identical twins arise

from division of a single zygote, their HLA proteins are identical. Because HLA antigens mark the cell as belonging to a particular individual, they are *self* proteins. The importance of self proteins in plasma membranes was first recognized when it was discovered that they contribute to the specificity of tissues and make it difficult to transplant tissue from one human to another. Comparison studies of HLA antigens must always be carried out before a transplant is attempted. The more of the 50-plus proteins that the donor and recipient share in common, the better the tissue match. In other words, when the donor and the recipient are histo (tissue)-compatible, a transplant is more likely to be successful.

When an antigen-presenting cell links a foreign antigen to the self protein on its plasma membrane, it carries out an important safeguard for the rest of the body. Now the T cell to be activated can compare the antigen and self protein side by side. The activated T cell and all of the daughter cells it will form can recognize "foreign" from "self," and go on to destroy cells carrying foreign antigens while leaving normal body cells unharmed (Fig. 13.6b).

Presentation of the antigen leads to activation of the T cell. An activated T cell undergoes clonal expansion. (Fig. 13.6c). Many copies of the activated T cell are produced during clonal expansion. They destroy any cell, such as a virus-infected cell or a cancer cell, that displays the antigen presented earlier.

Types of T Cells

The two main types of T cells are cytotoxic T cells and helper T cells. **Cytotoxic T (T_c) cells,** also called CD_8 leukocytes, can bring about the destruction of antigen-bearing cells, such as virus-infected or cancer cells (Fig. 13.7a). Cancer cells also have nonself proteins.

Cytotoxic T cells have storage vacuoles containing perforin molecules. **Perforin** molecules punch holes into a plasma membrane, forming a pore that allows destructive enzymes to enter. The cell then undergoes apoptosis. Cytotoxic T cells are responsible for so-called **cell-mediated immunity** (Fig. 13.7b).

Helper T (T_h) cells, also called CD_4 leukocytes, regulate immunity by secreting **cytokines,** the chemicals that enhance the response of other immune cells. Cytokines stimulate macrophages, B cells, and other T cells to perform their

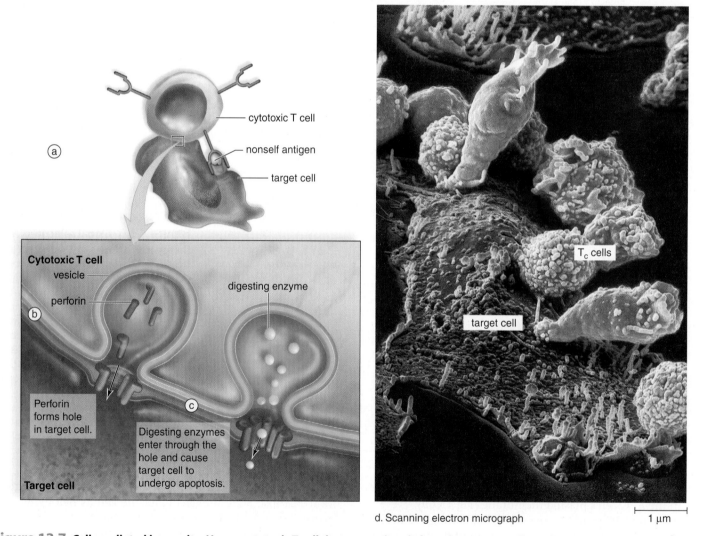

d. Scanning electron micrograph   1 µm

Figure 13.7 Cell-mediated immunity: How a cytotoxic T cell destroys a virus-infected or cancer cell. a. The activated T cell binds with a nonself antigen presented by the foreign cell. **b.** The T cell releases perforin molecules, which form pores in the target cell membrane. **c.** Digesting enzymes enter through the hole, triggering apoptosis in the target cell. **d.** This scanning electron micrograph shows cytotoxic cells destroying a cancer cell.

functions. Because HIV, the virus that causes AIDS, infects helper T cells and certain other cells of the immune system, it inactivates the immune response. AIDS is described in the Medical Focus, page 294.

Notice in Figure 13.6c that a few of the clonally expanded T cells are **memory T cells.** They remain in the body and can jump-start an immune reaction to an antigen previously present in the body.

When a person's immune response has successfully overcome the illness, ultimately that response must be shut down. Now, the activated T cells become susceptible to apoptosis. As mentioned previously, apoptosis is programmed cell death that contributes to homeostasis by regulating the number of cells present in an organ, or in this case, in the immune system. When apoptosis does not occur as it should, the potential exists for self-destruction (called an autoimmune response). Occasionally, T-cell cancers (i.e., lymphomas and leukemias) can result.

Apoptosis also occurs in the thymus as T cells are maturing. Any T cell that has the potential to destroy the body's own cells undergoes suicide.

Cytokines in Cancer Chemotherapy

Whenever cancer develops, it is possible that cytotoxic T cells have not been activated. With this possibility in mind, cytokines have been used as immunotherapeutic drugs to enhance the ability of T cells to fight cancer. Interferon, discussed on page 288, and also **interleukins,** which are cytokines produced by various white blood cells, are also being administered for this purpose.

Content **CHECK-UP!**

7. Which type of cell produces the chemical mediators that cause the four outward signs of an inflammatory response?

   a. macrophage   c. neutrophil

   b. mast cell   d. natural killer cells

8. Which sign of inflammation occurs when tissue capillaries become more permeable and leak protein and tissue fluid into the tissue spaces?

   a. swelling   c. heat

   b. redness   d. formation of pus

9. Which protein produced by virus-infected cells prevents other cells from being infected?

   a. complement   c. cytokine

   b. immunoglobulin   d. interferon

Answers in Appendix C, page 464.

13.4 Creating an Immune Response

Immunity occurs naturally or is brought about artificially (induced) by medical intervention. The two types of immune responses are active and passive. In **active immunity,** the individual alone produces antibodies against an antigen; in **passive immunity,** the individual receives prepared antibodies from another person.

Active Immunity

Active immunity often develops naturally after a person is infected with a pathogen. Once recovery from illness is complete, the person will have a *natural* active immunity thanks to the B and T memory cells that remain after the infection. However, active immunity can also be induced while a person is well, so that future infection will not take place. To prevent infections, people are immunized against them. Successful immunization provides *artificial* active immunity, because the individual produces his or her own antibodies against the antigen given. The United States is committed to immunizing all children against the common types of childhood disease, as discussed in the Medical Focus on page 296.

Immunization involves the use of **vaccines,** substances that contain an antigen to which the immune system responds. Traditionally, vaccines are the pathogens themselves, or their products, that have been treated so they are no longer virulent (able to cause disease). Today, it's possible to genetically engineer bacteria to mass-produce a protein from pathogens, and this protein can be used as a vaccine. This method has now produced a vaccine against hepatitis B, a viral-induced disease, and is being used to prepare a vaccine against malaria, a protozoan-induced disease.

After a vaccine is given, it's possible to follow an immune response by determining the amount of antibody present in a sample of plasma. This is called the **antibody titer.** After the first exposure to a vaccine, a primary response occurs. For a period of several days, no antibodies are present; then the titer rises slowly, levels off, and gradually declines as the antibodies bind to the antigen or simply break down (Fig. 13.8). After a second exposure to the vaccine, a secondary response is expected. The titer rises rapidly to a level much greater than

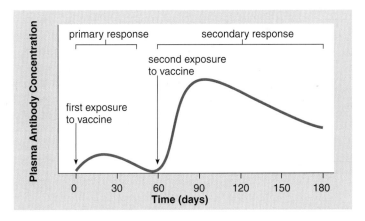

Figure 13.8 Vaccine responses. During immunization, the primary response, after the first exposure to a vaccine, is minimal, but the secondary response, which may occur after the second exposure, shows a dramatic rise in the amount of antibody present in plasma.

AIDS Epidemic

Acquired immunodeficiency syndrome (AIDS) is caused by a group of related retroviruses known as HIV (human immunodeficiency viruses). In the United States, AIDS is usually caused by HIV-1, which enters a host by attaching itself to a cell membrane receptor called a CD_4 receptor. HIV-1 infects helper T cells, the type of lymphocyte that stimulates B cells to produce antibodies and cytotoxic T cells to destroy virus-infected cells. Macrophages, which present antigens to helper T cells and thereby stimulate them, are also under attack. Other cells that have CD_4 receptors can also be infected and later destroyed.

HIV is a **retrovirus,** meaning that its genetic material consists of RNA instead of DNA. Once inside the host cell, HIV uses a special enzyme called *reverse transcriptase* to make a DNA copy (called cDNA) of its genetic material. Now cDNA integrates into a host chromosome, where it directs the production of more viral RNA. Each strand of viral RNA brings about synthesis of an outer protein coat called a **capsid**. The viral enzyme **protease** is necessary to the formation of capsids. Capsids assemble with RNA strands to form viruses, which bud from the host cell and spread to other cells.

Transmission of AIDS

HIV infection spreads when infected cells in body secretions, such as semen, and in blood are passed to another individual. To date, as many as 100 million people worldwide may have contracted HIV, and almost 31 million have died. A new infection is believed to occur every 15 seconds, the majority in heterosexuals. HIV infections are not distributed equally throughout the world. Most infected people live in Africa (66%) where the infection first began, but new infections are now occurring at the fastest rate in Southeast Asia and the Indian subcontinent.

HIV is transmitted by sexual contact with an infected person, including vaginal or rectal intercourse and oral/genital contact. Also, needle-sharing among intravenous drug users is high-risk behavior. Babies born to HIV-infected women may become infected before or during birth, or through breast-feeding after birth. The rise in the incidence of AIDS among women of reproductive age is paralleled by a rise in the incidence of AIDS in children younger than age 13.

Phases of an HIV Infection

The Centers for Disease Control and Prevention recognize three stages of an HIV-1 infection, called categories A, B, and C. During the category A stage, the helper T-lymphocyte count is 500 per mm^3 or greater (Fig. 13A). For a period of time after the initial infection with HIV, people don't usually have any symptoms at all. A few (1–2%) do have mononucleosis-like symptoms that may include fever, chills, aches, swollen lymph nodes, and an itchy rash. These symptoms disappear, however, and no other symptoms appear for quite some time. Although there are no symptoms, the person is highly infectious. Despite the presence of a large number of viruses in the plasma, the HIV blood test is not yet positive because it tests for the presence of antibodies and not for the presence of HIV itself. This means that HIV can still be transmitted before the HIV blood test is positive.

Several months to several years after a nontreated infection, the individual will probably progress to category B, in which the helper T-lymphocyte count is 200 to 499 per mm^3. During this stage, the patient may experience swollen lymph nodes in the neck, armpits, or groin that persist for three months or more. Other symptoms that indicate category B are severe fatigue not related to exercise or drug use; unexplained persistent or recurrent fevers, often with night sweats; persistent cough not associated with smoking, a cold, or the flu; and persistent diarrhea.

The development of non-life-threatening but recurrent infections is a signal that the disease is progressing. One possible infection is thrush, a fungal infection that is identified by the presence of white spots and ulcers on the tongue and inside the mouth. The fungus may also spread to the vagina, resulting in a chronic infection there. Another frequent infection is herpes simplex, with painful and persistent sores on the skin surrounding the anus, the genital area, and/or the mouth.

Previously, the majority of infected persons proceeded to category C, in which the helper T-lymphocyte count is below 200 per mm^3 and the lymph nodes degenerate. The patient is now suffering from AIDS, characterized by severe weight loss and weakness due to persistent diarrhea and coughing, and will most likely contract an opportunistic infection. An **opportunistic infection** is one that only has the opportunity to occur because the immune system is severely weakened. Such infections are extremely rare in healthy individuals. Persons with AIDS die from one or more opportunistic diseases, such as *Pneumocystis carinii* pneumonia, *Mycobacterium tuberculosis*, toxoplasmic encephalitis, Kaposi's sarcoma, or invasive cervical cancer. This last condition has been added to the list because the incidence of AIDS has now increased in women.

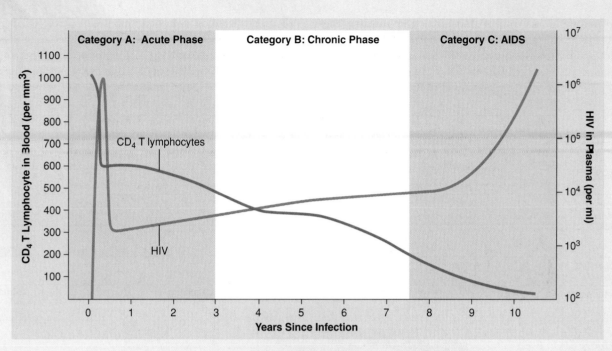

Figure 13A **Stages of an HIV infection.** In category A individuals, the number of HIV in plasma rises upon infection and then falls. The number of CD_4 T lymphocytes falls, but stays above 400 per mm^3. In category B individuals, the number of HIV in plasma is slowly rising, and the number of T lymphocytes is decreasing. In category C individuals, the number of HIV in plasma rises dramatically as the number of T lymphocytes falls below 200 per mm^3.

Treatment for AIDS

Therapy usually consists of combining two drugs that inhibit reverse transcriptase with another that inhibits protease, an enzyme needed for formation of a viral capsid. This multidrug therapy, when taken according to the manner prescribed, usually seems to slow the rate of mutation of the virus to a resistant strain. The sooner drug therapy begins after infection, the better the chances that the immune system will not be destroyed by HIV. Also, medication must be continued indefinitely. The medications all have serious side effects as well, including bone and joint destruction. Unfortunately, an HIV strain resistant to all known drugs has been reported, and persons who become infected with this strain have no drug therapy available to them.

The likelihood of transmission from mother to child at birth can be lessened if the mother takes an inhibitor of reverse transcriptase called AZT and if the child is delivered by cesarean section. The mother also must not breast-feed her child because the virus is transmissible in breast milk.

Many investigators are working on a vaccine for AIDS. Some are trying to develop a vaccine in the traditional way. Others are working on subunit vaccines that utilize just a single HIV protein as the vaccine. So far, no method has resulted in sufficient antibodies to keep an infection at bay. After many clinical trials, none too successful, most investigators now agree that a combination of various vaccines may be the best strategy to bring about a response in both B lymphocytes and cytotoxic T cells.

MEDICAL FOCUS

Immunization: The Great Protector

Immunization protects children and adults from diseases. The success of immunization is witnessed by the fact that the smallpox vaccination is no longer required because the disease has been eradicated. However, parents today often fail to get their children immunized because they don't realize the importance of immunizations, fear potential side effects (which occur rarely), or can't bear the expense. Newspaper accounts of an outbreak of measles at a U.S. college or hospital, therefore, are not uncommon because many adults were not immunized as children.

	Birth	1 month	2 mos.	4 mos.	6 mos.	12 mos.	15 mos.	18 mos.	24 mos.	4-6 yrs.	11-12 yrs.	13-18 yrs.	Adult
Hepatitis B	1st Dose	Booster 1			Booster 2				Catch Up*				
Rotavirus			1st Dose	2nd Dose	Catch Up*								
Diphtheria, Pertussis, Tetanus			1st Dose	2nd Dose	3rd Dose		Booster 1			Booster 2	Tetanus, Diphtheria booster		Tetanus, Diphtheria booster every 10 years
Haemophilus influenzae			1st Dose	2nd Dose	3rd Dose**	Final Dose							
Pneumococcal			1st Dose	2nd Dose	3rd Dose	Final Dose			Catch Up*				Booster after age 60
Polio			1st Dose	2nd Dose	Final Dose	Catch Up*				Booster			
Measles, Mumps, Rubella						1st Dose				2nd Dose	Catch Up*		1-2 Doses
Varicella (chickenpox)						1st Dose	2nd Dose	Final Dose	Catch Up***	Booster			Booster after age 60
Hepatitis A						1st Dose		2nd Dose	Catch Up*				
Human Papillomavirus											3 doses for females ages 9-18		
Meningococcal											1st Dose	Catch Up*	
Influenza					Annually	Annually	Annually	Annually	Annually	Annually	Annually	Annually	Annually

*for children or adolescents not previously immunized as infants
**optional, not necessary if doses 1 and 2 have been administered previously
***optional, not necessary if three doses administered previously

Figure 13B **Suggested immunization schedule for infants, young children, and adolescents.** Children who are immune-compromised may need additional immunizations. Always consult with your health care provider for up-to-date information on recommended immunizations for your area.

Source: Centers for Disease Control and Prevention. Recommended immunization schedules for persons aged 0–18 years—United States, 2008. MMWR 2007;56(51&52):Q1–Q4.

before; then it slowly declines. The second exposure is called a "booster" because it boosts the antibody titer to a high level. The high antibody titer now is expected to help prevent disease symptoms even if the individual is exposed to the disease-causing antigen.

As mentioned previously, active immunity is dependent upon the presence of memory B cells and memory T cells that are capable of responding to lower doses of antigen. Active immunity is usually long-lasting, although a booster may be required every few years.

Figure 13B shows a recommended immunization schedule for children and adults. The United States is now committed to the goal of immunizing all children against the common types of childhood diseases listed. Diphtheria, whooping cough, *Haemophilus influenzae,* and *Pneumococcus* infection are all life-threatening respiratory diseases. Tetanus is characterized by muscular rigidity, including a locked jaw. These extremely serious infections are all caused by bacteria; the rest of the diseases listed are caused by viruses. *Varicella* virus causes chickenpox. Polio is a type of paralysis; measles and rubella, sometimes called German measles, are characterized by high fever and skin rashes; and mumps is characterized by fever and enlarged parotid and other salivary glands.

Hepatitis B virus (HBV) is a blood-borne pathogen that is spread in the United States mainly by sexual contact and intravenous drug use. Health-care workers who are exposed to blood or blood products are also at risk, and maternal-neonatal transmission is a possibility as well. Recovery from an initial bout of hepatitis (inflammation of the liver) can lead to chronic hepatitis and then cancer of the liver. Fortunately, a hepatitis B vaccine is now available. It is rare for a child to come into contact with hepatitis B virus. However, physicians feel that the potential for liver damage from exposure to the virus warrants vaccination

early in life. Infants now receive their first dose of vaccine before leaving the hospital, with booster immunizations at approximately 3 months and 12 months of age.

Cervical cancer has recently been linked to the occurrence of genital warts, a sexually transmitted disease caused by human papillomavirus (HPV). Therefore, a new vaccine for papillomavirus type 16, the most frequent cause of genital warts, has been developed. Currently, the U.S. Centers for Disease Control recommends vaccination for girls and young women ages 11 to 26. Perhaps more vaccines for sexually transmitted diseases will one day become available.

It's important to note that the need for immunizations and boosters doesn't stop with adolescence. Regular boosters for tetanus and diptheria are recommended every 10 years for adults. Further, everyone should get a yearly flu shot, beginning at age 6 months.

Even though bacterial infections (e.g., tetanus) can be cured by antibiotic therapy, it's better to be immunized. Some patients are allergic to antibiotics, and their reaction to them can be fatal. In addition, antibiotics not only kill off disease-causing bacteria, but they also reduce the number of beneficial bacteria in the intestinal tract and elsewhere. These beneficial bacteria may have checked the spread of pathogens that now are free to multiply and to invade the body. This is why antibiotic therapy is often followed by a secondary infection, such as a vaginal yeast infection in women.

Abuse of antibiotic therapy also leads to resistant bacterial strains that are difficult to cure, even with antibiotics. For example, a once-harmless skin bacterium called *Staphylococcus aureus* now causes thousands of life-threatening infections and deaths every year. Post-surgical patients are particularly at risk, but anyone who is hospitalized can become infected. After decades of antibiotic exposure, this microbe (abbreviated MRSA, for *m*ethicillin-*r*esistant *S*taphylococcus *a*ureus) and others have become resistant to all but the strongest antibiotics.

Therefore, everyone should take advantage of appropriate vaccinations. Preventing a disease by becoming actively immune to it is preferable to becoming ill and needing antibiotic therapy to be cured.

Passive Immunity

Passive immunity occurs when an individual is given prepared antibodies (immunoglobulins) to combat a disease. Since these antibodies are not produced by the individual's own plasma cells, passive immunity is temporary. For example, newborn infants are passively immune to some diseases because antibodies have crossed the placenta from the mother's blood. This type of immunity is called natural passive immunity. These antibodies soon disappear, however, so that within a few months, infants become more susceptible to infections. Breast-feeding prolongs the natural passive immunity an infant receives from the mother because antibodies are present in the mother's milk (Fig. 13.9).

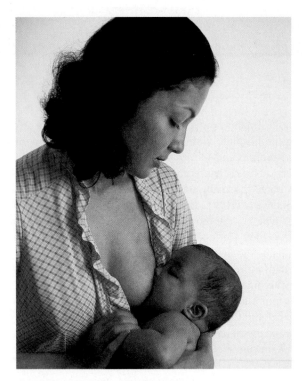

Figure 13.9 Passive immunity. Breast-feeding is believed to prolong the passive immunity an infant receives from the mother because antibodies are present in the mother's milk.

Even though passive immunity does not last, it is sometimes used to prevent illness in a patient who has been unexpectedly exposed to an infectious disease. Usually, the patient receives a **gamma globulin** injection (serum that contains IgG class antibodies), perhaps taken from individuals who have recovered from the illness. For example, a health care worker who suffers an accidental needle stick may come into contact with the blood from a patient infected with hepatitis B virus. Immediate treatment with an antiviral gamma globulin injection (along with simultaneous vaccination against the virus) can typically prevent the virus from causing infection in the health care worker. The gamma globulin injection imparts an artificial *passive* immunity. The vaccination will cause the worker to develop her own antibodies, thus creating artificial *active* immunity as well.

Monoclonal Antibodies

Every plasma cell derived from the same B cell secretes antibodies against one specific antigen. These are **monoclonal antibodies** because all of them are the same type and because they are produced by plasma cells derived from the same B cell. One method of producing monoclonal antibodies *in vitro* (outside the body in a laboratory) is depicted in Figure 13.10. B lymphocytes are removed from an animal (today, mice are usually used) and are exposed to a particular antigen. The resulting plasma cells are fused with myeloma cells (malignant plasma cells that live and divide indefinitely). The

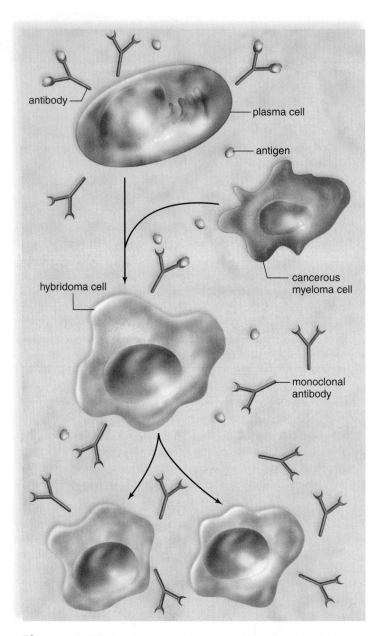

Figure 13.10 Production of monoclonal antibodies. Plasma cells of the same type (derived from immunized mice) are fused with myeloma (cancerous) cells, producing hybridoma cells that are "immortal." Hybridoma cells divide and continue to produce the same type of antibody, called monoclonal antibodies.

fused cells are called hybridomas—*hybrid-* because they result from the fusion of two different cells, and *-oma* because one of the cells is a cancer cell. It is important to note that the antibodies from these cells *cannot* cause cancer, however.

At present, monoclonal antibodies are being used for quick and certain diagnosis of various conditions. For example, a particular hormone is present in the urine of a pregnant woman. A monoclonal antibody can be used to detect this hormone; if it is present, the woman knows she is pregnant. Monoclonal antibodies are also used to identify infections. And because they can distinguish between cancerous and normal

Swine Flu—The Global Pandemic

*"Over there, over there, Send the word, send the word over there—
That the Yanks are coming, The Yanks are coming . . . And we won't
come back till it's over, Over there!"*
—"Over There," George M. Cohan

In April 1918, American soldiers who proudly marched off to World War I battlefields "over there" (referring to Europe, in the words of the popular patriotic ballad of the time), took an unwanted hitchhiker with them—a particularly virulent strain of influenza A, nicknamed the "Spanish Flu" because of the huge numbers of victims in that country. It quickly became a worldwide epidemic, or **pandemic,** causing an estimated 50 million deaths. Among the soldiers, Spanish flu caused twice as many deaths as those that resulted from combat. This particularly savage virus is once again a concern because of characteristics it shares with a modern-day virus: influenza A (H1N1), the swine flu virus.

Influenza viruses, like all viruses, have two basic parts. The core of the virus is its genetic material, which can be either DNA or RNA. Influenza A viruses are RNA viruses (Fig. 13C). Their genetic material is covered by a protective protein coat called a capsid. An additional third piece, called the *lipid envelope*, contains protein spikes that allow the virus to attach to its own specific kind of host cell and not another non-matching kind. The two types of protein spikes, abbreviated "H" (for hemagglutinin) and "N" (for neuraminidase), allow each type of influenza virus to be categorized. There are 16 known types of H proteins, and 9 known types of N proteins. Thus, many different combinations are possible, meaning there are many different ways for viruses to attach to different kinds of host cells. Each different combination of H and N represents a new form of flu virus. Avian influenza A, the "bird flu," which emerged in 2006, is designated "H5N1." The original Spanish flu virus is denoted "H1N1," and the current swine flu virus is its molecular "grandchild."

How do these new viruses emerge? It's well known that the genetic material of influenza virus has an extremely high mutation rate. Some of these mutations affect the structure of the protein spikes, so that a virus that previously could only infect a particular animal can "jump species." As the virus moves from one animal to the next, it can also swap genetic material and create new combinations. Scientists studying the ancestor H1N1 from 1918 have noted that it probably originated as a "bird flu" virus, which then mutated to infect human beings. Swine flu virus was first thought to be a virus common to pigs in North America, but this has been proven to be incorrect. Further studies have shown that this virus contains bird, human, and swine DNA—but the swine DNA is from Europe and Asia. Scientists call this four-way mutant a "quadruple reassortant" virus. The ever-changing nature of this and other influenza viruses explains why each year a new and different flu shot is needed.

Because the human population hasn't had ongoing exposure to the swine flu virus, little or no natural immunity is thought to exist among humans. In addition, public health officials are particularly concerned because the illness caused by this virus is eerily reminiscent of the 1918 Spanish flu epidemic. Like today, the original pandemic first began in the spring, not in the winter like most new viral illnesses. Both Spanish flu and swine flu have killed otherwise healthy young adults and children, and not just the elderly and infirmed. During its first outbreak in Mexico, the virus rapidly became lethal after the first symptoms appeared—just as it did in 1918. Moreover, back then the virus seemed to disappear for a time, then reappeared several more times, with ever-increasing severity. Scientists are particularly concerned that this particular pattern could repeat itself once again.

In June 2009, the World Health Organization declared the swine flu outbreak to be a global pandemic, with a strength rating of 6/6. This designation isn't intended to describe how dangerous the virus is. Rather, it's meant to alert public health officials so that plans can be made to cope with the virus. Fortunately, thanks to modern technology, we'll have a few more weapons in our antiviral arsenal than our great-grandparents had back in 1918. Vaccinations, drugs that prevent viral replication, antibiotics for secondary infections such as pneumonia, as well as communication tools like radio, television, and the Internet—all will hopefully create a different outcome during this pandemic. In the meantime, your best approach to avoid influenza is to use common sense. Get your flu shot every fall because each new vaccination is tailored to that season's viruses. Wash your hands often and well, don't touch your nose or eyes (because virus can enter your body when you do), and stay away from sick people. To keep from infecting others if you do get sick, don't go out in public. Cough or sneeze into tissues or your sleeve, and wash your hands before you touch anything. Further, if you do develop symptoms of swine flu, seek prompt medical attention. Antiviral drugs have the best chance to successfully treat you if you seek help quickly.

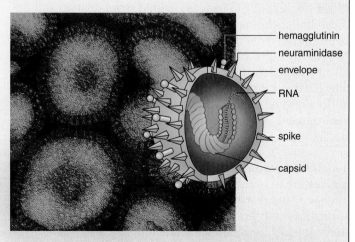

Figure 13C Influenza virus is an RNA virus. It has protein spikes on its outer envelope that make it specific to one kind of host cell.

tissue cells, they are used to carry radioactive isotopes or toxic drugs to tumors, which can then be selectively destroyed.

Hypersensitivity Reactions

The immune system usually protects us from disease because it can distinguish self from nonself. Sometimes, however, it responds in a manner that harms the body, as when individuals develop allergies, suffer tissue rejection, or have an autoimmune disease. These are all forms of hypersensitivity reactions.

Allergies

Allergies are hypersensitivities to substances such as pollen or animal hair that ordinarily would do no harm to the body. The response to these antigens, called **allergens,** usually includes some degree of tissue damage. There are four types of allergic responses, but we will consider only two of them: IgE-mediated allergic response and T-cell mediated allergic response.

IgE-Mediated Allergic Response An **IgE-mediated allergic response** is often referred to as an **immediate allergic response** because it can occur within seconds of contact with an antigen. The response is caused by antibodies known as IgE (see Table 13.2). IgE antibodies are attached to the plasma membrane of mast cells in the tissues and also to basophils in the blood. When an allergen attaches to the IgE antibodies on these cells, mast cells release histamine and other substances that bring about the allergic symptoms. When pollen is an allergen, histamine stimulates the mucosal membranes of the nose and eyes to release fluid, causing the runny nose and watery eyes typical of **hay fever.** If a person has **asthma,** the airways leading to the lungs constrict, resulting in difficult breathing accompanied by wheezing. When food contains an allergen, nausea, vomiting, and diarrhea result.

Anaphylactic shock is an IgE-mediated allergic response that occurs because the allergen has entered the bloodstream. Bee stings and penicillin shots are known to cause this reaction because both inject the allergen into the blood. Anaphylactic shock is characterized by a sudden and life-threatening drop in blood pressure due to increased blood vessel dilation and permeability of the capillaries, caused by histamine. Taking epinephrine can delay this reaction until medical help is available.

Allergy shots sometimes prevent the onset of IgE-mediated allergic responses. It has been suggested that injections of the allergen may cause the body to build up high quantities of IgG antibodies, and these combine with allergens received from the environment before they have a chance to reach the IgE antibodies located in the membrane of mast cells and basophils.

T-Cell Mediated Allergic Response A **T-cell mediated allergic response** is initiated by memory T cells at the site of allergen contact in the body. T-cell mediated responses are sometimes referred to as *delayed* allergic responses, because they develop more slowly than IgE-mediated (*immediate*) responses. The allergic response is regulated by the cytokines secreted by both T cells and macrophages.

A classic example of a T-cell mediated allergic response is the skin test for tuberculosis (TB). When the test result is positive, the tissue where the antigen was injected becomes red and hardened. This shows that there was prior exposure to tubercle bacilli, the cause of TB. Contact dermatitis, which occurs when a person is allergic to poison ivy, jewelry, cosmetics, and many other substances that touch the skin, is also an example of a T-cell mediated allergic response.

Tissue Rejection

Certain organs, such as skin, the heart, and the kidneys, could be transplanted easily from one person to another if the body did not attempt to reject them. Rejection of transplanted tissue results because the recipient's immune system recognizes that the transplanted tissue is not "self." Cytotoxic T cells respond by causing disintegration of the transplanted tissue.

Organ rejection can be controlled by carefully selecting the organ to be transplanted and administering **immunosuppressive** drugs. It is best if the transplanted organ has the same type of HLA antigens as those of the recipient, because T_c cells recognize foreign HLA antigens. Two well-known immunosuppressive drugs, cyclosporine and tacrolimus, both act by inhibiting the response of T cells to cytokines.

Researchers hope that tissue engineering, including the production of organs that lack antigens or that can be protected in some way from the immune system, will one day do away with the problem of rejection.

Autoimmune Disease

When a person has an **autoimmune disease,** cytotoxic T cells or antibodies mistakenly attack the body's own cells as if they bear foreign antigens. Exactly what causes autoimmune diseases is not known. However, sometimes they occur after an individual has recovered from an infection.

Virtually any tissue of the body can be a target for autoimmune disease. In **myasthenia gravis,** the synaptic junctions between motor nerves and skeletal muscles are destroyed. Neuromuscular junctions do not work properly and muscular weakness results. In **multiple sclerosis (MS),** the myelin sheath of nerve fibers is attacked. This can cause a wide variety of neurological and neuromuscular symptoms: muscle weakness, paralysis, blurred vision, dizziness, and deafness. A person with **systemic lupus erythematosus (SLE)** has an extremely serious disease characterized by antibodies directed against multiple self antigens, including her own DNA molecules. Symptoms can include facial rashes, arthritis, anemia, and kidney damage. In **rheumatoid arthritis,** the joints are affected. Researchers suggest that heart damage following rheumatic fever, and type I diabetes are also autoimmune illnesses. As yet, there are no cures for autoimmune diseases, but they can be managed with medication.

Immune Deficiency

When a person has an immune deficiency, the immune system is unable to protect the body against disease. AIDS (see the Medical Focus on page 294) is an example of an acquired immune deficiency. As a result of a weakened immune system, AIDS patients show a greater susceptibility to a variety of diseases, and also have a higher risk of cancer. Immune deficiency may also be genetic (that is, inherited from one's parents) or congenital (due to failure of lymphatic tissue to develop). For example, in some infants the thymus gland fails to develop, producing a child with severe immune deficiency. Infrequently, a child may be born with an impaired B- or T-cell system caused by a defect in lymphocyte development. In **severe combined immunodeficiency disease (SCID)**, a genetic disorder, both antibody and cell-mediated immunity are lacking or inadequate. Without treatment, even common infections can be fatal. Gene therapy has been successful in SCID patients (see page 439).

Content CHECK-UP!

10. Which of the following is a form of active immunity?
    a. transfer of antibodies from mother to fetus across the placenta
    b. transfer of antibodies from mother to baby in breast milk
    c. immunization for hepatitis B virus
    d. gamma globulin injection for hepatitis B virus

11. Which of the following statements about passive immunity is true?
    a. Passive immunity's effects are permanent.
    b. It requires activated T cells to produce antibodies.
    c. Plasma cells produce antibodies, which are transferred from one person to another.
    d. Antibody molecules giving passive immunity can't be passed from a pregnant woman to her fetus.

12. Which term describes the amount of antibody present in a blood plasma sample?
    a. booster   c. hematocrit
    b. vaccine   d. titer

Answers in Appendix C, page 464.

13.5 Effects of Aging

With advancing age, people become more susceptible to all types of infections and disorders because the immune system exhibits lower levels of function. One reason is that the thymus gland degenerates. Having reached its maximum size in early childhood, it begins to shrink after puberty and has virtually disappeared by old age. As the gland decreases in size, so does the number of T cells. The T cells remaining do not respond to foreign antigens; therefore, the chance of having cancer increases with age.

Among the elderly, the B cells sometimes fail to form clones. Or, when they do form clones, the antibodies released may not function well. Therefore, infections are more common among the elderly. Also, the antibodies are more likely to attack the body's own tissues, increasing the incidence of autoimmune diseases.

The response of elderly individuals to vaccines is decreased. However, considering that their overall level of immune response is low, it is better that these people be vaccinated than not. For this reason, elderly individuals are encouraged to get an influenza (flu) vaccination each year.

13.6 Homeostasis

The three functions of the lymphatic system listed on page 282 assist homeostasis. The lymphatic system helps the digestive system by absorbing fats. In the process of absorbing dietary fats, lacteals also absorb fat-soluble vitamins. The lymphatic system assists the cardiovascular system by absorbing excess tissue fluid. The lymphatic vessels return excess tissue fluid as lymph to cardiovascular veins in the thorax, thus helping to maintain the blood volume and pressure needed for capillary exchange.

The lymphatic organs and the immune system protect us from infectious diseases. Nonspecific defenses precede specific immunity. The skin and the mucous membranes of the respiratory, digestive, reproductive, and urinary systems all resist infection. If a pathogen should enter the body, the infection is localized as much as possible by an inflammatory reaction. Phagocytic white blood cells engulf as many pathogens as possible. If the pathogens do enter the blood, complement proteins work in diverse ways to keep the blood free of disease-causing organisms and their toxins.

Specific defenses depend on blood cells; the lymphocytes and macrophages play central roles. B and T cells have antigen receptors and can distinguish self from nonself. Antigen binding selects which B or T cells will undergo clonal expansion. B cells recognize an antigen directly, but T cells must have the antigen displayed by an APC in the groove of an HLA antigen. Plasma cells (mature B cells) produce antibodies, but T cells kill virus-infected and cancer cells outright.

The lymphatic organs play a central role in immunity. White blood cells are made in the red bone marrow where B cells also mature. T cells mature in the thymus. The spleen filters the blood directly. Clonal expansion of lymphocytes occurs in the lymph nodes, which also filter the lymph.

A strong connection exists between the immune, nervous, and endocrine systems. Lymphocytes have receptors for a wide variety of hormones and the thymus produces hormones that influence the immune response. Cytokines produced by T lymphocytes affect the brain's temperature control center. The high body temperature of a fever is thought to create an unfavorable environment for the foreign invaders. Further, cytokines promote behaviors that tend to make us take care of ourselves by causing sleepiness and appetite loss. A close connection between the immune and endocrine systems is illustrated by the ability of cortisone to decrease the inflammatory reaction in the joints.

Integumentary System

Lymphatic vessels pick up excess tissue fluid; immune system protects against skin infections.

Skin serves as a barrier to pathogen invasion; Langerhans cells phagocytize pathogens; protects lymphatic vessels.

Skeletal System

Lymphatic vessels pick up excess tissue fluid; immune system protects against infections.

Red bone marrow produces leukocytes involved in immunity.

Muscular System

Lymphatic vessels pick up excess tissue fluid; immune system protects against infections.

Skeletal muscle contraction moves lymph; physical exercise enhances immunity.

Nervous System

Lymphatic vessels pick up excess tissue fluid; immune system protects against infections of nerves.

Microglia engulf and destroy pathogens.

Endocrine System

Lymphatic vessels pick up excess tissue fluid; immune system protects against infections.

Thymus is necessary to maturity of T lymphocytes.

How the Lymphatic System works with other body systems.

tonsils

cervical lymph nodes

thymus

axillary lymph nodes

thoracic duct

spleen

inguinal lymph nodes

popliteal lymph node

lymphatic vessel

Cardiovascular System

Lymphatic organs produce and store formed elements; lymphatic vessels transport leukocytes and return tissue fluid to blood vessels; spleen serves as blood reservoir, filters blood.

Blood vessels transport leukocytes and antibodies; blood services lymphatic organs and is source of tissue fluid that becomes lymph.

Respiratory System

Lymphatic vessels pick up excess tissue fluid; immune system protects against respiratory tract and lung infections.

Tonsils and adenoids occur along respiratory tract; breathing aids lymph flow.

Digestive System

Lacteals absorb fats; Peyer patches prevent invasion of pathogens; appendix contains lymphatic tissue.

Digestive tract provides nutrients for lymphatic organs; stomach acidity prevents pathogen invasion of body.

Urinary System

Lymphatic system picks up excess tissue fluid, helping to maintain blood pressure for kidneys to function; immune system protects against infections.

Kidneys control volume of body fluids, including lymph.

Reproductive System

Female immune system does not attack sperm or fetus, even though they are foreign to the body.

Sex hormones influence immune functioning; acidity of vagina helps prevent pathogen invasion of body; milk passes antibodies to newborn.

Basic Key Terms

active immunity (ak′tiv ĭ–myū′nĭ-te), p. 293

antibodies (ăn′tĭ -bŏd′ez), p. 289

antibody-mediated immunity (an″tĭ-bod″ e-me′de-āt-ed ĭ-myū′nĭ-te), p. 289

antigens (ăn′tĭ-jən), p. 288

antigen-presenting cell (ăn′tĭ-jən prĭ-zĕnt-ing sĕl`), p. 290

apoptosis (ăp′ō-tō′sis), p. 289

appendix (ă-pen′-diks), p. 286

autoimmune response (ô′tō-ĭmyo͞onrĭ-spŏns), p. 289

B lymphocytes (B lim′fo-sītz), p. 288

capsid (kap′sid), p. 294

cell-mediated immunity (sel me″de-āt′ed ĭ-myū′nĭ-te), p. 292

clone (klōn), p. 289

complement system (kom′plĕ-ment sis′tem), p. 288

cytokines (si′to-kīnz), p. 292

cytotoxic T cells (si′to-tok′sik T selz), p. 292

helper T cells (help′er T selz), p. 292

histamine (his′tuh-min), p. 286

HLA (human leukocyte) antigens (hyū′mun lu′ko-sīt an′tĭ-jenz), p. 291

immunity (ĭ-myū-nĭ-te), p. 282

inflammatory reaction (in-flam′uh-to″re re-ak′shun), p. 286

lacteal (lak′tē-ăl), p. 282

lymph (lĭmf), p. 282

lymphatic organ (lim-fat′ik or′gan), p. 282

lymphatic system (lim-fat′ik sis′tem), p. 282

lymphatic vessels (lĭm-făt′ĭk vĕs′əlz), p. 282

lymph node (limf nōd), p. 285

macrophages (măk′ra-făj-ez), p. 286

mast cells (măst sĕlz), p. 286

major histocompatibility complex, (mā′jōr his′tō.kŏm-păt-ĭ-bĭl′ĭ-tē kŏm′plĕks), p. 291

memory B cells (mem′o-re B selz), p. 289

memory T cells (mem′o-re T selz), p. 293

monoclonal antibodies (mon″o-klōn′al an″tĭ-bod′ēz), p. 298

natural killer cells (nat′u-ral kil′er selz), p. 288

passive immunity (păs′-iv ĭ-myū′nĭ-te), p. 293

perforin (pĕr′for-ə n), p. 292

Peyer patches (pi′er pach′ēz), p. 286

plasma cells (plaz′muh selz), p. 289

protease (prō′tē-āse), p. 294

pus (pus), p. 288

red bone marrow (rĕd bōn măr′ō), p. 284

retrovirus (rĕt′rō-vī′rəs), p. 294

right lymphatic duct (rīt lĭm făt′ ĭk dŭkt), p. 282

spleen (splēn), p. 284

thoracic duct (thō-răs′ĭk dŭkt), p. 282

thymus gland (thĭ′mə s glănd), p. 284

T lymphocytes (T lim′fo-sītz), p. 288

tonsils (ton′silz), p. 285

Clinical Key Terms

AIDS (acquired immunodeficiency syndrome) (uh-kwīr′d ĭ-myŭ″no-dĭ-fi′shun-se sin′ drōm), p. 294

allergens (al′er-jenz), p. 300

allergies (al′er-jēz), p. 300

anaphylactic shock (an″uh-fi-lak′tik shok), p. 300

antibody titer (an″ti-bod″e ti′ter), p. 293

asthma (az′muh), p. 300

autoimmune disease (aw″to-ĭ-myūn′ dĭ-zez′), p. 300

edema (ĕ-de′muh), p. 282

elephantiasis (ĕ″luh-fun-ti′uh-sis), p. 285

gamma globulin (găm-uh glŏb′yoo-lin), p. 298

hay fever (hā fē′vĕr), p. 300

Hodgkin disease (hoj″kin dĭ-zĕz′), p. 285

immediate allergic response (ĭm′-mē′dē-ĭt o-lūr′jik rĭ-spŏns′), p. 300

IgE-mediated allergic response (eye-gē ēē mē′dē-at′ĕd əlûr′jik rĭ-spŏns′), p. 300

immunization (ĭ-myū-nĭ-za′shun), p. 293

immunosuppressive drugs (ĭ-myū″no-sŭ-pres′iv drugz), p. 300

interferon (in″ter-fēr′on), p. 288

interleukin, (in-ter-loo′kin), p. 293

lymphadenitis (lim-fad″ĕ-ni′tis), p. 285

lymphangitis (lim″fan-jī′tis), p. 285

lymphedema (lim″fe-de′mah), p. 285

lymphoma (lĭm-fō′mə), p. 285

multiple sclerosis (mul′tĭ-pl skler-o′sis), p. 300

opportunistic infection (op″er-tu-nis′tik in-fek′shun), p. 294

pulmonary edema (pul′mo-nĕr″e ĕ-de′muh), p. 285

severe combined immunodeficiency disease (sĕ-vĕr′ kum-bīnd′ ĭ-myu″no-dĭ-fi′shun-se di-zēz′), p. 301

systemic lupus erythematosus (sis-tem′ik lu′pus er-ĭ-the-muh-to′sus), p. 300

T-cell-mediated allergic response (tē sel me′dē-at′ĕd ə-lûr′jik rĭ-spŏns′), p. 300

vaccines (vak-sēnz′), p. 293

SUMMARY

13.1 Lymphatic System

A. The lymphatic system consists of lymphatic vessels and lymphatic organs. The lymphatic vessels return excess tissue fluid to the bloodstream, absorb fats at intestinal villi, and help the immune system defend the body against disease.

B. Lymphatic capillaries have thin walls. Larger vessels are structured the same as cardiovascular veins, with valves that prevent backward flow.

13.2 Organs, Tissues, and Cells of the Immune System

Lymphocytes are produced and accumulate in the lymphatic organs (primary organs: red bone marrow, thymus gland; secondary organs: lymph nodes, spleen, and other lymphatic tissues). Lymph is cleansed of pathogens and/or their toxins in lymph nodes and blood is cleansed of pathogens in the spleen. T lymphocytes mature in the thymus, while B lymphocytes mature in the red bone marrow where all blood cells are produced. White blood cells are necessary for nonspecific and specific defenses.

13.3 Nonspecific and Specific Defenses

A. Immunity involves nonspecific and specific defenses. Nonspecific defenses include barriers to entry, the inflammatory reaction, nonspecific white blood cells, natural killer cells, and protective proteins.

B. Specific defenses require B lymphocytes and T lymphocytes, also called B cells and T cells. B cells undergo clonal selection with production of plasma cells and memory B cells after their antigen receptors combine with a specific antigen. Plasma cells secrete antibodies and eventually undergo apoptosis. Plasma cells are responsible for antibody-mediated immunity. Antibodies are Y-shaped molecules with two binding sites for a specific antigen. Memory B cells remain in the body and produce antibodies if the same antigen enters the body at a later date.

C. T cells are responsible for cell-mediated immunity. The two main types of T cells are cytotoxic T cells and helper T cells. Cytotoxic T cells kill virus-infected or cancer cells on contact because they bear a nonself antigen. Helper T cells produce cytokines and stimulate other immune cells. Like B cells, each T cell bears antigen receptors. However, for a T cell to recognize an antigen, the antigen must be presented by an antigen-presenting cell (APC), usually a macrophage, in the groove of an HLA (human leukocyte-associated antigen). Thereafter, the activated T cell undergoes clonal expansion until the illness has been stemmed. Then most of the activated T cells undergo apoptosis. A few cells remain, however, as memory T cells. Cytokines, including interferon and interleukins, are used in an attempt to promote the body's ability to recover from cancer and to treat AIDS.

13.4 **Creating an Immune Response**
A. Immunity can be induced in various ways. Vaccines are available to induce long-lasting, active immunity. Antibodies sometimes are available to provide an individual with temporary, passive immunity. Monoclonal antibodies are produced in the laboratory and used for diagnosis and treatment purposes.

B. Allergic responses occur when the immune system reacts vigorously to substances not normally recognized as foreign. IgE-mediated or immediate allergic responses, usually consisting of coldlike symptoms, are due to the activity of antibodies. T-cell mediated allergic responses, such as contact dermatitis, are due to the activity of T cells.

13.5 **Effects of Aging**
The thymus gets smaller as we age, and fewer antibodies are produced. The elderly are at great risk of infections, cancer, and autoimmune diseases.

13.6 **Homeostasis**
The lymphatic system assists the cardiovascular system by returning excess tissue fluid to the bloodstream. It assists the digestive system by absorbing fats from the intestinal tract, and it assists the immune system through the functioning of its lymphatic organs.

STUDY QUESTIONS

1. What is the lymphatic system, and what are its three functions? (p. 282)
2. Describe the structure and the function of red bone marrow, the thymus, the spleen, lymph nodes, and the tonsils. (pp. 282–286)
3. What are the body's nonspecific defense mechanisms? (pp. 286–282)
4. Describe the inflammatory reaction and give a role for each type of cell and molecule that participates in the reaction. (pp. 286–288)
5. What is the clonal selection theory as it applies to B cells? B cells are responsible for which type of immunity? (pp. 289–290)
6. Describe the structure of an antibody, and define the terms variable regions and constant regions. (pp. 289–290)
7. Describe the clonal selection theory as it applies to T cells. (pp. 290–292)
8. Name the two main types of T cells and state their functions. (pp. 292–293)
9. What are cytokines and how are they used in immunotherapy? (p. 293)
10. How is active immunity artificially achieved? How is passive immunity achieved? (pp. 293, 296–298)
11. How are monoclonal antibodies produced, and what are their applications? (pp. 298–300)
12. Discuss allergies, tissue rejection, and autoimmune diseases as they relate to the immune system. (pp. 300–301)
13. How do the lymphatic and immune systems help maintain homeostasis? (pp. 301–302)
14. How does the skeletal system assist the immune system in maintaining homeostasis? (p. 302)

LEARNING OBJECTIVE QUESTIONS

Fill in the blanks.
1. Lymphatic vessels contain _____, which close, preventing lymph from flowing backward.
2. _____ and _____ are two types of white blood cells produced and stored in the lymphatic organs.
3. Lymph nodes cleanse the _____ while the spleen cleanses the _____.
4. _____ and _____ are phagocytic white blood cells.
5. T lymphocytes have matured in the _____.
6. A stimulated B cell divides and differentiates into antibody-secreting _____ cells and also into _____ cells that are ready to produce the same type of antibody at a later time.
7. B cells are responsible for _____ mediated immunity.

8. Cytotoxic T cells are responsible for _____ mediated immunity.
9. Immunization with _____ brings about active immunity.
10. Allergic reactions are associated with the release of _____ from mast cells.

11. Whereas _____ immunity occurs when an individual is given antibodies to combat a disease, _____ immunity occurs when an individual develops the ability to produce antibodies against a specific antigen.

12. Barriers to entry, protective proteins, and the inflammatory reaction are all examples of _____ defenses.
13. Proteins that function to form holes in bacterial cell walls comprise the _____ system.

MEDICAL TERMINOLOGY EXERCISE

Consult Appendix B for help in pronouncing and analyzing the meaning of the terms that follow.

1. metastasis (mě-tas'tuh-sis)
2. allergist (al'er-jist)
3. immunosuppressant (i-myū"no-sŭ-pres'ant)

4. immunotherapy (i-myū"no-thĕr'uh-pe)
5. splenorrhagia (sple"no-ra'je-uh)
6. lymphadenopathy (lim-fad"ĕ-nop'uh-the)
7. lymphangiography (lim-fan"je-og'ruh-fe)
8. eosinophilia (e'oh-sin'o-fil'e-uh)
9. thymectomy (thi-mek'to-me)

10. lymphopenia (limf'o-pe'ne-uh)
11. agammaglobulinemia (ā-gam'uh-glob'yu-lĭ-ne'me-uh)
12. pyemia (pi-ē'me-uh)
13. tonsillectomy (tŏn' sə-lek'tə-mē)
14. hypersensitivity (hi'per-sen-sĭ-tiv'ĭ-te)

WEBSITE LINK

Visit this text's website at http://www.mhhe.com/longenbaker7 **for additional quizzes, interactive learning exercises, and other study tools.**

The Respiratory System

Hold your breath, drink from a glass (sideways), eat a spoonful of sugar, have someone startle or scare you, pull hard on your tongue—all of these are "cures" for **singultus,** also known as the hiccups (and any of them might work). Technically, hiccups are spasms of the respiratory diaphragm muscle, caused by irritation of the phrenic nerve which controls it. Everyone gets hiccups, even babies in the womb. Common causes include excessive alcohol and nicotine, taking a bite of food that's too big, and excessive stress (for example, if you've been crying). Normally, hiccups last only a few minutes at most, and they're harmless. However, they can continue for a long time—the longest-recorded bout lasted six months—and might be a sign of much more serious conditions such as a heart attack or a tumor in the chest. If your hiccups last more than three hours, and you're in pain, too, it's time to see a doctor. You can read more about the diaphragm and phrenic nerve on pages 310 and 316. And if you want to give up smoking to stop your hiccups, see the Medical Focus on page 321.

Chapter Outline & Learning Objectives

After you have studied this chapter, you should be able to:

14.1 The Respiratory System

The primary function of the respiratory system is gas exchange—allowing oxygen from the air to enter the blood and carbon dioxide from the blood to exit into the air. During **inspiration,** or inhalation (breathing in), and **expiration,** or exhalation (breathing out), air is conducted toward or away from the lungs by a series of cavities, tubes, and openings, illustrated in Figure 14.1.

The respiratory system also works with the cardiovascular system to accomplish these four respiratory events:

1. pulmonary ventilation (breathing): the entrance and exit of air into and out of lungs
2. external respiration: the exchange of gases (oxygen and carbon dioxide) between air and blood
3. internal respiration: the exchange of gases between blood and tissue fluid
4. transport of gases to and from the lungs and the tissues.

Cellular respiration, which produces ATP, uses the oxygen and produces the carbon dioxide that makes gas exchange with the environment necessary. Without a continuous supply of ATP, the cells cease to function. The four events listed here allow cellular respiration to continue.

The Respiratory Tract

Table 14.1 traces the path of air from the nose to the lungs. As air moves in along the airways, it is cleansed, warmed, and moistened by the mucous membrane lining of the nasal cavity. Cleansing is accomplished by coarse hairs just inside the nostrils and by cilia and mucus in the nasal cavities and the other airways of the respiratory tract. Nasal hairs, cilia, and mucus act as screening devices for dust, dirt, pollen, fungal spores, etc., in inhaled air. Cilia inside the nasal cavity beat backwards, sweeping the mucous sheet into the throat where it can be swallowed. Lysozyme in the mucus helps to kill bacteria, while

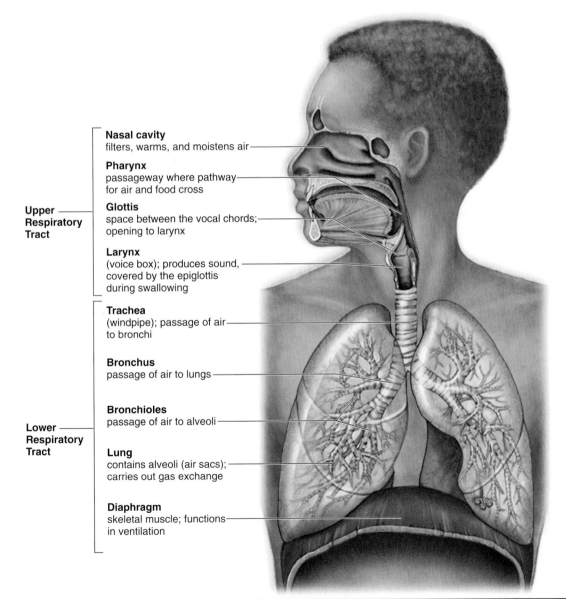

Upper Respiratory Tract

Nasal cavity
filters, warms, and moistens air

Pharynx
passageway where pathway for air and food cross

Glottis
space between the vocal chords; opening to larynx

Larynx
(voice box); produces sound, covered by the epiglottis during swallowing

Lower Respiratory Tract

Trachea
(windpipe); passage of air to bronchi

Bronchus
passage of air to lungs

Bronchioles
passage of air to alveoli

Lung
contains alveoli (air sacs); carries out gas exchange

Diaphragm
skeletal muscle; functions in ventilation

Figure 14.1 The respiratory tract. Note the structures of the upper respiratory tract and the lower respiratory tract. The tract extends from the nasal cavities to the lungs.

TABLE 14.1 Path of Air

Structure	Description	Function
UPPER RESPIRATORY TRACT		
Nasal cavitities	Hollow spaces in nose	Filter, warms, and moisten air
Pharynx	Chamber posterior to oral cavity; lies between nasal cavity and larynx	Connection to surrounding regions
Glottis	Opening into larynx	Passage of air into larynx
Larynx	Cartilaginous organ that houses the vocal cords; voice box	Sound production
LOWER RESPIRATORY TRACT		
Trachea	Flexible tube that connects larynx with bronchi	Passage of air to bronchi
Bronchi	Paired tubes inferior to the trachea that enter the lungs	Passage of air to lungs
Bronchioles	Branched tubes that lead from bronchi to alveoli	Passage of air to each alveolus
Lungs	Soft, cone-shaped organs that occupy lateral portions of thoracic cavity	Gas exchange / Acid-base balance; conversion of angiotensin I to angiotensin II
Alveoli	Air sacs that branch from bronchioles	Passage of gases into and out of pulmonary capillaries
Pulmonary capillaries	Capillaries that cover surface of alveoli	Receive oxygen from alveoli; oxygenation of hemoglobin; deliver CO_2 from tissues to alveoli

lymphocytes populating underlying tissue create an immune response whenever pathogens are accidentally inhaled. In the trachea and other airways, the cilia beat upward, carrying mucus, dust, and other trapped potential contaminants into the pharynx. This important protective mechanism is called the **mucociliary escalator.** Once in the pharynx, the mucous accumulation can be swallowed (thus, destroyed by stomach acid and digestive enzymes) or spit out. Inhaled air is also warmed by heat given off by blood vessels lying close to the surface of the lining of the airways, and it is moistened by the wet surface of the mucous membrane.

Conversely, as air moves out during expiration, it cools and loses its moisture. As the air cools, it deposits its moisture on the lining of the trachea and the nose, and the nose may even drip as a result of this condensation. The air still retains so much moisture, however, that upon expiration on a cold day, it condenses and forms a small cloud.

The Nose

The **nose,** a prominent feature of the face, is the only external portion of the respiratory system. Air enters the nose through external openings called **nostrils.** The nose contains two **nasal cavities,** which are narrow canals separated from one another by a septum composed of bone and cartilage (Fig. 14.2). Mucous membrane lines the nasal cavities. The nasal **conchae** are bony ridges that project laterally into the nasal cavity. They increase the surface area for moistening and warming air during inhalation and for trapping water droplets during exhalation. Odor receptors are on the cilia of cells in the olfactory epithelium, located high in the recesses of the nasal cavities.

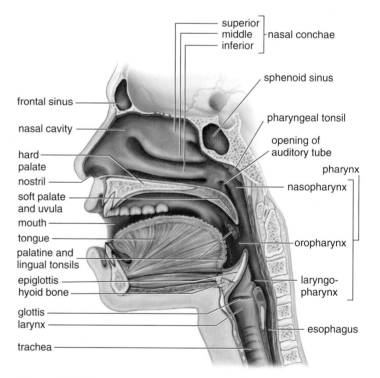

Figure 14.2 **The path of air.** This drawing shows the path of air from the nasal cavities to the trachea, which is a part of the lower respiratory tract. The other organs are in the upper respiratory tract.

The tear (lacrimal) glands drain into the nasal cavities by way of tear ducts. For this reason, crying produces a runny nose. The nasal cavities also communicate with the **paranasal sinuses,** air-filled spaces that reduce the weight of the skull and act as resonating chambers for the voice. If the ducts leading from the sinuses become inflamed, fluid may accumulate, causing a sinus headache. The nasal cavities are separated from the oral cavity by the palate, a partition that has two portions. Anteriorly, the hard palate is supported by the maxilla and palatine bones. Posteriorly, the soft palate is composed of muscle and glandular tissue.

The Pharynx

The **pharynx** is a funnel-shaped passageway that connects the nasal and oral cavities to the larynx. Consequently, the pharynx, commonly referred to as the "throat," has three parts: the nasopharynx, where the nasal cavities open posterior to the soft palate; the oropharynx, where the oral cavity joins the pharynx; and the laryngopharynx, which opens into the larynx. The soft palate has a soft extension called the *uvula* that can be seen projecting into the oropharynx.

The single pharyngeal tonsil (or *adenoids*) in the posterior nasal cavity is the primary lymphatic tissue defense for breathing. Its action is supported by the paired palatine tonsils at the rear of the oropharynx, as well as the lingual tonsils at the base of the tongue. The tonsils contain lymphocytes that protect against invasion of inhaled pathogens. Here, both B cells and T cells can respond to antigens that may subsequently invade internal tissues and fluids. Thus, the respiratory tract assists the immune system in maintaining homeostasis.

In the pharynx, the air and food passages cross because the larynx, which receives air, is anterior to the esophagus, which receives food. The larynx lies at the top of the trachea. The larynx and trachea are normally open, allowing air to pass, but the esophagus is normally collapsed and opens only when a person swallows.

The Larynx

The cartilaginous **larynx** serves as a passageway for air between the pharynx and the trachea. The larynx can be pictured as a triangular box whose apex, the Adam's apple, is located at the anterior neck. The larynx is called the voice box because it houses the vocal cords. The **vocal cords** are mucosal folds supported by elastic ligaments, and the slit between the vocal cords is called the **glottis** (Fig. 14.3). When air is expelled past the vocal cords through the glottis, the vocal cords vibrate, producing sound. At puberty, the growth of the larynx and the vocal cords is much faster and accentuated in males than females, causing males to have a more prominent Adam's apple and deeper voices. The voice "breaks" in the young male because he cannot control the longer vocal cords. These changes cause the lower pitch of the voice in males.

Voice pitch is regulated when we speak and sing by changing the tension on the vocal cords. The greater the tension, as when the glottis becomes more narrow, the higher the pitch. When the vocal cords relax, the glottis is wider, and the pitch is lower (Fig. 14.3b). The loudness, or intensity, of the voice depends upon the amplitude of the vibrations—that is, how much the vocal cords vibrate.

Coughing, sneezing, and hiccuping also involve the glottis. When we cough, the glottis closes, then suddenly opens as a blast of air is forced upward from the lower respiratory tract. Sneezing is much like coughing, except that the air blast is directed through the nasal passages. As you read in the chapter introduction, a hiccup is caused by spasm of the diaphragm muscle while the glottis is closed.

When food is swallowed, the larynx moves upward against the **epiglottis,** a flap of elastic cartilage that prevents food from passing through the glottis into the larynx. You can detect this movement by placing your hand gently on your larynx and swallowing.

The Trachea

The trachea, or windpipe, is a tube connecting the larynx to the primary bronchi. The trachea lies anterior to the esophagus and is held open by C-shaped cartilaginous rings. The open part of the C-shaped rings forms the anterior wall of the esophagus, and this allows the esophagus to expand when swallowing. The mucosa that lines the trachea has a layer of pseudostratified ciliated columnar epithelium. (Pseudostratified—"false layers"—means that while the epithelium appears to be layered, actually each cell touches the basement membrane.)

The cilia of cells lining the bronchial wall help keep the lungs clean by moving trappped particles.

larynx

Figure 14.3 Anatomy of the larynx. a. The larynx lies posterior to the thyroid cartilage of the neck (commonly called the Adam's apple). **b.** Sagittal view of the larynx, showing location of the vocal cords. **c.** Viewed from above, the vocal cords can be seen to stretch from anterior to posterior across the larynx. When air is forced past the vocal cords, they vibrate, producing sound. The vocal cords are taut when we produce a high-pitched sound (*top*), and they relax as the pitch deepens (*bottom*).

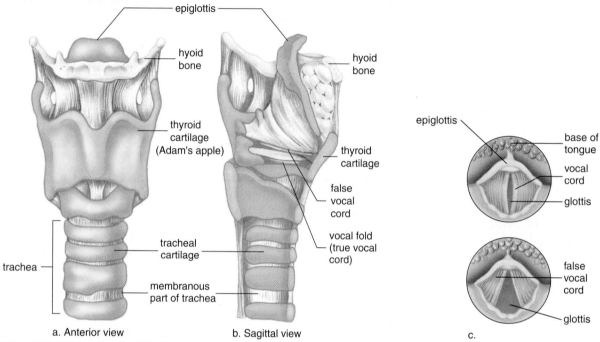

epiglottis

hyoid bone

thyroid cartilage (Adam's apple)

trachea

tracheal cartilage

membranous part of trachea

a. Anterior view

hyoid bone

thyroid cartilage

false vocal cord

vocal fold (true vocal cord)

b. Sagittal view

epiglottis

base of tongue

vocal cord

glottis

false vocal cord

glottis

c.

As mentioned previously, the epithelial cilia sweep cilia mucus produced by goblet cells, and debris toward the pharynx (a mechanism called the *mucociliary escalator*). Smoking is known to destroy these cilia, and consequently the soot in cigarette smoke collects in the lungs. Smoking is discussed more fully in the Medical Focus on page 321.

If the trachea is blocked because of illness or the accidental swallowing of a foreign object, it is possible to insert a breathing tube by way of an incision made in the trachea. This tube acts as an artificial air intake and exhaust duct. The operation is called a **tracheostomy.**

The Bronchial Tree

The trachea divides into right and left primary **bronchi** (sing., **bronchus**), which lead into the right and left lungs (see Fig. 14.1). The primary bronchi then branch into secondary bronchi: one for each lobe of the lung. Thus, there are three secondary bronchi for the right lung, which has three lobes. Two secondary bronchi supply the left lung, which has only two lobes in order to allow room for the heart. Each secondary bronchus then divides into smaller tertiary bronchi. These smaller bronchi are supported by smaller plates of cartilage, in place of the cartilage rings of the trachea. **Bronchioles** are

the smallest conducting airways. They lack cartilage support, but possess a ciliated epithelium and a well-developed smooth muscle layer. During an asthma attack, the smooth muscle of the bronchioles contracts, causing bronchiolar constriction and characteristic wheezing. Each bronchiole leads to an elongated space enclosed by a multitude of air pockets, or sacs, called **alveoli** (sing., **alveolus**). The components of the bronchial tree beyond the primary bronchi, including the alveoli, compose the lungs.

The Lungs

The **lungs** are paired, cone-shaped organs. Each fills its own pleural cavity inside the thoracic cavity, separated by the mediastinum. Recall that the mediastinum is the central membrane that separates the thoracic cavity, covering the primary bronchi, heart, thymus gland, trachea, and esophagus (see Chapter 1, page 6). The apex is the superior narrow portion of a lung, and the base is the inferior broad portion that curves to fit the dome-shaped diaphragm, the muscle of respiration that separates the thoracic cavity from the abdominal cavity. The lateral surfaces of the lungs follow the contours of the ribs in the thoracic cavity.

Each lobe of the lung is further divided into lobules, and each lobule has a bronchiole supplying many alveoli.

Lung Collapse

Imagine that you're a military medic who's called upon to respond when troops have been injured due to a bomb blast. As you arrive at the scene, two fallen soldiers need your attention. One has an open chest wound, caused by shrapnel cutting his chest. The second was nearby when the blast occurred, but has no obvious wounds. Yet, both have the same symptoms: sharp pain when they inhale, difficulty speaking, and a feeling of breathlessness. Both soldiers' blood pressure is low and pulse is rapid, indicating that they might slip into shock. You take a quick history from both victims and from their buddies.

Right away, you suspect each soldier has *atelectasis*—the technical term for a collapsed lung. As you'll recall, the lungs are held up against the chest wall by the attraction force of surface tension. If air enters the thorax, surface tension will fail, and the lungs will collapse. The first soldier's chest wound is allowing air from the atmosphere to enter the thorax. A section of the second soldier's lung burst as a result of the high pressure from the bomb blast, and air has filled his thorax from the hole in his lung. When the lungs collapse, the air filling the chest compresses the heart and prevents it from filling with blood. This is termed *tension pneumothorax*, or air in the thorax. You'll need to act fast, or both victims will slip into shock.

With the help of the first soldier's buddies, you put a special airtight pressure bandage over his open chest wound, which will prevent additional air from entering the wound and help stop bleeding. Next, you'll start his intravenous solution (IV). By listening to the second soldier's chest with your stethoscope, you'll be able to tell where the lung has collapsed because it will sound hollow. When you trained as a medic, you learned to do a *thoracocentesis*, and you'll rapidly insert a catheter between the soldier's ribs to let the trapped air out into the atmosphere. Now your patients are ready for their helicopter trip to a field hospital for more advanced care.

Figure 14.4 Gas exchange in the lungs. The lungs consist of portions of the bronchial tree leading to the alveoli, each of which is surrounded by an extensive capillary network. Notice that the pulmonary artery and arteriole carry O_2-poor blood (colored blue), and the pulmonary vein and venule carry O_2-rich blood (colored red).

Pulmonary arteries travel alongside the bronchi; likewise, pulmonary arterioles parallel the bronchioles. Each pulmonary arteriole then further branches to form **pulmonary capillaries.** Pulmonary capillaries surround and cover each alveolus of the lung. Elastic connective tissue binds the air passages to the blood vessels within each lung; this elastic tissue helps the lungs return to their resting position, or *recoil,* when a person exhales.

Each lung is enclosed by a double layer of serous membrane called the **pleurae** (sing., **pleura**). The visceral pleura adheres to the surface of the lung; the parietal pleura lines the inside of the thoracic cavity. The pleurae produce a lubricating serous fluid that reduces friction and allows the two layers to slide across one another. Serous fluid, a water-based solution, also creates **surface tension:** the tendency for water molecules to cling to each other (due to hydrogen bonding between the molecules) and to form a droplet. Surface tension holds the two pleural layers together, thus holding the lungs open against the chest wall.

The Alveoli

With each inhalation, air passes by way of the bronchial tree to the alveoli. An alveolar sac is made up of simple squamous epithelium surrounded by pulmonary capillaries. Gas exchange occurs between the air in the alveoli and the blood in the capillaries (Fig. 14.4). Oxygen diffuses across the alveolar and capillary walls to enter the bloodstream, while carbon dioxide diffuses from the blood across these walls to enter the alveoli.

Each alveolus is covered with an extremely thin layer of water-based tissue fluid. Gas exchange takes place across moist cellular membranes, and the attractive force created by the fluid's surface tension helps the distended lung tissue to

Figure 14.5 **The respiratory membrane.** The alveolar wall and the capillary wall allow gases to easily diffuse between the alveolus and the capillary.

return to its resting position when a person exhales. However, the alveoli must stay open to receive the inhaled air if gas exchange is to occur, and the surface tension of water lining the alveoli is capable of causing them to close up. (As an analogy, rinse a balloon several times with water, then empty the balloon completely. Note how the balloon's sides stick together and how hard it is to blow it up.) Normal alveoli are lined with **surfactant**—a film of lipoprotein that lowers surface tension to an acceptable level. Thus, in healthy lungs surface tension is high enough to help the lungs recoil (return to a resting position) yet low enough to prevent the alveoli from collapsing. The lungs collapse in some newborn babies, especially premature infants, who lack this surfactant film. The condition, called **infant respiratory distress syndrome,** is now treatable by surfactant replacement therapy.

Respiratory Membrane Gas exchange occurs very rapidly because of the characteristics of the so-called respiratory membrane (Fig. 14.5). The **respiratory membrane** consists of the juxtaposed alveolar epithelium and the capillary endothelium. Throughout most of the lungs, their basement membranes are fused, meaning that very little tissue fluid separates the two portions[1] of the respiratory membrane and they are indeed one membrane. This membrane is extremely thin—only 0.2–0.6 μm thick.

The total surface area of the respiratory membrane is the same as the area of the alveoli, namely 50–70 m^2. The blood that enters the many pulmonary capillaries spreads thin. The red blood cells within the capillaries are pressed up against the narrow capillary walls, and little plasma is present. This too facilitates the speed of gas exchange during external respiration.

Content CHECK-UP!

1. Which structure covers the larynx as swallowing occurs?
   a. epiglottis   c. uvula
   b. soft palate   d. hard palate
2. Put the following structures in order, from external to internal: trachea-alveoli-tertiary bronchi-primary bronchi secondary bronchi-bronchioles
3. Which membrane structure divides the thoracic cavity into two separate cavities?
   a. parietal pleura   c. parietal pericardium
   b. mediastinum   d. visceral pericardium

Answers in Appendix C, page 464.

14.2 Mechanism of Breathing

Ventilation

To understand **ventilation**—the manner in which air enters and exits the lungs—it is helpful to be aware of the following conditions:

1. *The lungs lie within the sealed-off thoracic cavity.* The rib cage, consisting of the ribs joined to the vertebral column posteriorly and to the sternum anteriorly, forms the top and sides of the thoracic cavity. The intercostal muscles lie between the ribs. The diaphragm and connective tissue form the floor of the thoracic cavity.
2. *The lungs adhere to the thoracic wall by way of the pleurae.* The visceral pleura covering the lungs attaches to the parietal pleura covering the chest wall, by utilizing the force of surface tension created by the fluid between them. Any space between the two layers of the pleura is minimal.

a. Inspiration

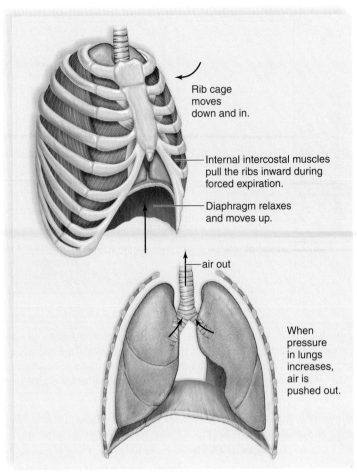

b. Expiration

Figure 14.6 **Inspiration versus expiration. a.** During inspiration, the thoracic cavity and lungs expand so that intrapulmonary pressure decreases. Now air flows into the lungs. **b.** During expiration, the thoracic wall and lungs recoil, assuming their original positions and pressures. Now air is forced out. The internal intercostal muscles only contract during forceful expiration.

(As an analogy, wet a clean glass microscope slide with a drop of water, then cover it with a second glass slide. Note the two pieces of glass will slide back and forth, but can't be easily pulled apart.) This surface tension force creates **intrapleural pressure**—the pressure between the pleurae. Intrapleural pressure is less than atmospheric pressure, and this also helps to keep the lungs inflated. The importance of a reduced intrapleural pressure is demonstrated when, by design or accident, atmospheric air enters the intrapleural space and the lung then collapses.

3. *A continuous column of air extends from the pharynx to the alveoli of the lungs.*

Inspiration

Inspiration is the active phase of ventilation because this is the phase in which the diaphragm and the external intercostal muscles contract (Fig. 14.6*a*). In its relaxed state, the diaphragm is dome-shaped; during inspiration, it contracts and becomes a flattened sheet of muscle. Simultaneously, the external intercostal muscles contract, and the rib cage moves upward and outward.

As the thoracic volume increases, the lungs increase in volume as well because the lung adheres to the wall of the thoracic cavity. As lung volume increases, the air pressure within the alveoli (called *intrapulmonary pressure*) decreases, creating a partial vacuum. In other words, alveolar pressure is now less than atmospheric pressure (air pressure outside the lungs). Because a continuous column of air fills the lungs, air will naturally flow from outside the body into the respiratory passages and into the alveoli. Air flow continues, until intrapulmonary pressure equals atmospheric pressure.

It's important to realize that air comes into the lungs because they have already opened up; air does not force the lungs open. This is why it's sometimes said that *humans breathe by subatmospheric pressure.* The creation of a partial vacuum in the alveoli causes air to enter the lungs. While inspiration is the active phase of breathing, the actual flow of air into the alveoli is passive.

Expiration

Usually, expiration is the passive phase of ventilation, and no muscular effort is required to bring it about. During expiration, the diaphragm and the external intercostal muscles relax.

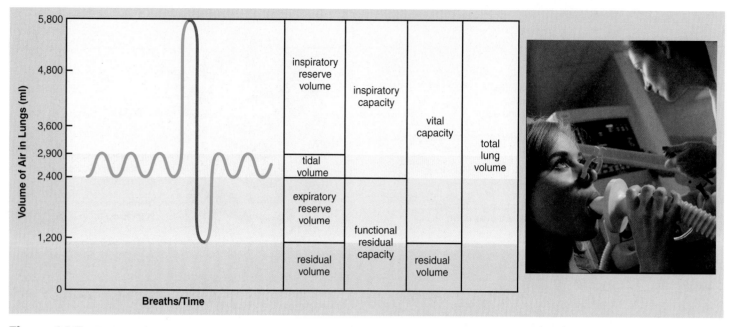

Figure 14.7 **Vital capacity.** A spirometer measures the amount of air inhaled and exhaled with each breath. During inspiration, the pen moves up, and during expiration, the pen moves down. The volume of one normal breath (tidal volume) multiplied by the number of breaths per minute is called the minute ventilation. A lower-than-normal minute ventilation can be a sign of pulmonary malfunction. Vital capacity (red) is the maximum amount of air a person can exhale after taking the deepest inhalation possible.

Therefore, the diaphragm resumes its dome shape and the rib cage moves down and in (Fig. 14.6*b*). As the volume of the thoracic cavity decreases, the lungs are free to recoil. Lung recoil occurs due to the elastic tissue built into the lungs' walls, and also due to the slight alveolar surface tension. As thoracic cavity volume and, thus, *lung* volume, decrease, the air pressure within the alveoli (intrapulmonary pressure) increases above atmospheric pressure. Because the airways are filled with a continuous column of air, some of that air will naturally flow out of the body. Air flow continues until intrapulmonary pressure equals atmospheric pressure.

What keeps the alveoli from collapsing as they decrease in size during expiration? Recall that the presence of surfactant lowers the surface tension within the alveoli. Also, as the lungs recoil, pressure between the two layers of pleura decreases, and this also helps the alveoli stay open.

Maximum Inspiratory Effort and Forced Expiration

If you recall the last time you exercised vigorously—perhaps running in a race or even just climbing all those stairs to your classroom—you'll probably remember that you were breathing a lot harder than normal during and immediately after that heavy exercise. Maximum inspiratory effort involves the *accessory muscles* of respiration: erector spinae muscles of the back, pectoralis minor (chest), and scalene and sternocleidomastoid muscles of the anterior neck. Their combined efforts can help to increase the size of the thoracic cavity larger than normal, thus allowing maximum expansion of the lungs.

While inspiration is always the active phase of breathing, expiration is usually passive—that is, the diaphragm and

external intercostal muscles are simply allowed to relax, the lungs recoil, and expiration occurs. However, expiration can also be forced. Forced expiration accompanies the maximum inspiratory efforts of heavy exercise. Forced expiration is also necessary to sing, blow air into a trumpet, or blow out birthday candles. Contraction of the internal intercostal muscles can force the rib cage to move downward and inward. Also, when the abdominal wall muscles contract, they push on the viscera, which push against the diaphragm, and the increased pressure in the thoracic cavity helps to expel air.

During breathing, air moves into the lungs from the nose or mouth (called inspiration, or inhalation), and then moves out of the lungs during expiration, or exhalation. A free flow of air from the nose or mouth to the lungs and from the lungs to the nose or mouth is vitally important. Therefore, a technique has been developed that allows physicians to determine if there is a medical problem that prevents the lungs from filling with air upon inspiration and releasing air from the body upon expiration. An instrument called a **spirometer** records the volume of air exchanged during normal breathing and during deep breathing. A spirogram (recording from a spirometer) shows the measurements recorded by a spirometer when a person breathes as directed by a technician (Fig. 14.7).

Respiratory Volumes

Normally when we are relaxed, only a small amount of air moves in and out with each breath. This amount of air, called the **tidal volume,** is only about 500 ml.

It is possible to increase the amount of air inhaled, and therefore the amount exhaled, by deep breathing. The maximum

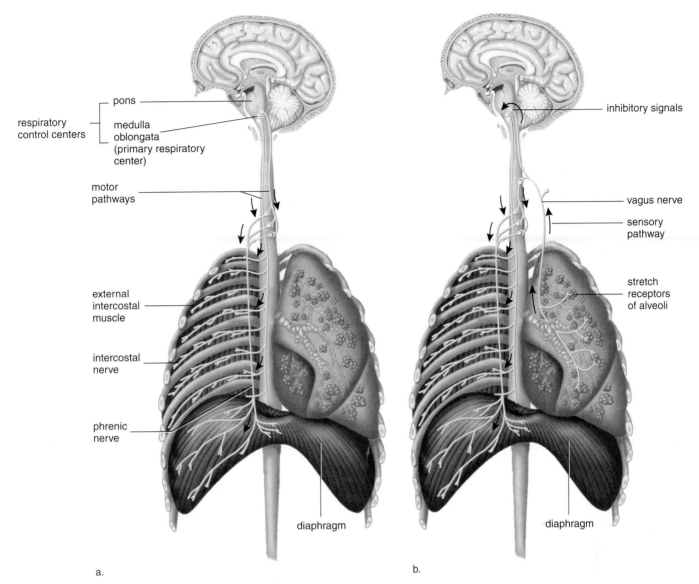

respiratory control centers
- pons
- medulla oblongata (primary respiratory center)

motor pathways

external intercostal muscle

intercostal nerve

phrenic nerve

diaphragm

a.

inhibitory signals

vagus nerve

sensory pathway

stretch receptors of alveoli

diaphragm

b.

Figure 14.8 Nervous control of breathing. a. During inspiration, respiratory control areas of the pons and medulla stimulate the intercostal and phrenic nerves. Intercostal nerves cause the external intercostal muscles to contract. The phrenic nerve stimulates the diaphragm. Expiration occurs rhythmically when respiratory control areas stop nervous stimulation and allow the muscles to relax. **b.** Hering-Breuer reflex. When tidal volume increases above 1.5 liters, alveolar stretch receptors signal the respiratory centers via the vagus nerve and inspiration ceases. This prevents overinflation of the lungs.

volume of air that can be moved in plus the maximum volume that can be moved out during a single breath is the **vital capacity.** It is called vital capacity because your life depends on breathing, and the more air you can move, the better off you are. A number of different illnesses, discussed in section 14.4, can decrease vital capacity.

Vital capacity varies by how much we can increase inspiration and expiration over the tidal volume amount. We can increase inspiration by using the accessory muscles of respiration (scalenes, sternocleidomastoid, pectoralis minor, etc.) to expand the chest, and also by lowering the diaphragm to the maximum extent possible. Forced inspiration usually increases the volume of air beyond the tidal volume by 2,900 ml, and that amount is called the **inspiratory reserve volume.**

We can increase the amount of air expired by contracting the abdominal and internal intercostal muscles. This so-called **expiratory reserve volume** is usually about 1,400 ml of air. You can see from Figure 14.6 that vital capacity is the sum of the tidal, inspiratory reserve, and expiratory reserve volumes.

It's a curious fact that some of the inhaled air never reaches the alveoli; instead, it fills the nasal cavities, trachea, bronchi, and bronchioles (see Fig. 14.1). In an average adult, some 70% of the tidal volume does reach the aveoli; but 30% remains in the airways. These passages are not used for gas exchange, and therefore they are said to contain **dead-space** air. To ensure that a large portion of inhaled air reaches the lungs, it is better to breathe slowly and deeply. Also, note in Figure 14.8 that even after a very deep exhalation, some air (about 1,000 ml) remains in the alveoli; this is called the **residual volume.** This air is not as useful for gas exchange because it contains a great deal of CO_2 and has been depleted of oxygen. In some lung diseases, such as emphysema, the residual volume increases because the individual has difficulty emptying the lungs. This means that the vital capacity is reduced because the lungs have more residual volume.

Control of Ventilation

Normally, adults have a breathing rate of 12 to 20 ventilations per minute. The basic rhythm of ventilation is controlled by a **primary respiratory center** located in the medulla oblongata of the brain.

The primary respiratory center automatically sends out motor nerve signals by way of the phrenic nerve to the diaphragm. Simultaneously, the intercostal nerves stimulate the external intercostal muscles of the rib cage (Fig. 14.6). When these muscles contract, the thoracic volume and lung volume increase, and the person inhales. When the respiratory center stops sending neuronal signals to the diaphragm and the rib cage, the diaphragm relaxes, resuming its dome shape. The rib cage moves down and in. Decreasing thoracic and lung volumes allow the person to exhale.

The primary respiratory center allows the *basic* pattern of inhalation and exhalation. However, if the medulla functions alone (as might occur in a person with a serious head injury, if nerves from higher centers are damaged), respiration is short and gasping. Breathing rhythmically at a normal rate and volume requires nervous input from the **pons,** the brain stem center immediately superior to the medulla. Functioning together, these two brain centers allow normal, quiet breathing, a pattern referred to as **eupnea:** smooth, sustained inspiration, followed by a smooth, sustained expiration.

Although the respiratory center controls the rate and depth of breathing, its activity can be influenced by nervous input and chemical input.

Nervous Input By observing our breathing during day-to-day living, it's evident that higher brain centers influence the rate and depth of respiration. Input from the cerebral cortex, limbic system, hypothalamus, and other brain centers cause increased respiration rate and depth if you're angry, frightened, or otherwise upset. Faster than normal respiration is referred to as **hyperpnea.** Conversely, respiration rate and depth decrease in the soundest stages of sleep. Conscious control of respiration allows a person to hold his breath for a time, or to voluntarily hyperventilate. Neural influence over the respiratory centers also helps increase respiratory rate and depth during exercise.

Nervous control over respiration also helps to protect delicate lung tissue, as illustrated by the Hering-Breuer reflex. During exercise, inspiratory depth increases, due to recruitment of muscle fibers in the diaphragm, intercostal muscles, and accessory muscles. Then, stretch receptors in the alveolar walls are stimulated, and they initiate inhibitory nerve signals that travel from the inflated lungs to the respiratory center. This causes the respiratory center to stop sending out nerve signals. This reflex helps support rhythmic respiratory movements by limiting inspiration, thus preventing over-expansion of lung tissue.

Chemical Input The respiratory center is directly sensitive to the levels of carbon dioxide (CO_2) and hydrogen ions (H^+). When they rise, due to increased cellular respiration (during exercise, for example), the respiratory center increases respiratory rate and depth. The center is not affected directly by low oxygen (O_2) levels. However, chemoreceptors in the **carotid bodies,** located in the carotid arteries, and in the **aortic bodies,** located in the aorta, are sensitive to blood oxygen levels. (Do not confuse the carotid and aortic bodies with the carotid and aortic baroreceptors, which monitor blood pressure.) When oxygen concentration decreases, these bodies communicate with the respiratory center, and the rate and depth of breathing increase.

A condition called Cheyne-Stokes respiration is characterized by alternating periods of deep, labored breathing, then shallow or absent breathing (apnea). In this condition, the respiratory center is largely controlled by chemical input. Breathing rate first increases when blood CO_2 and H^+ are high and O_2 is low, then decreases when blood CO_2 and H^+ are low and O_2 is high. Congestive heart failure and other disorders may cause this condition.

Content **CHECK-UP!**

4. Name the muscles that contract when a person inhales.

5. Which of the following has to be true for a person to inhale?

   a. Intrapulmonary pressure is greater than atmospheric pressure.

   b. Atmospheric pressure is greater than intrapulmonary pressure.

   c. The abdominal muscles must contract.

   d. a and c

6. Which of the following is the volume of air left over in the lungs after the maximum expiratory effort?

   a. inspiratory reserve volume   c. residual volume

   b. expiratory reserve volume   d. alveolar volume

Answers in Appendix C, page 464.

14.3   Gas Exchange and Transport

Respiration includes both exchange of gases in the lungs *and* in the tissues. Both are critical to homeostasis. Recall that diffusion is the movement of molecules from a higher concentration to a lower concentration. The principles of diffusion alone govern whether oxygen (O_2) or carbon dioxide (CO_2) enters or leaves the blood in the lungs and in the tissues.

External Respiration

External respiration is the exchange of gases in the lungs. Specifically, during external respiration, gases are exchanged between the alveolar air and the pulmonary capillary blood. Blood that enters the pulmonary capillaries is dark maroon because it is relatively O_2-poor.

Once inspiration has occurred, the alveoli have a higher concentration of O_2 than does blood entering the lungs. Therefore, O_2 diffuses from the alveoli into the blood. The reverse is true of CO_2. The alveoli have a lower concentration of CO_2 than does blood entering the lungs. Therefore, CO_2 diffuses out of the blood into the alveoli. This CO_2 exits the body during expiration.

Another way to explain gas exchange in the lungs is to consider the partial pressure of the gases involved. Gases exert pressure, and the amount of pressure each gas exerts is its partial pressure, symbolized as P_{O_2} and P_{CO_2}. Alveolar air has a much higher P_{O_2} than does blood. Therefore, O_2 diffuses into the blood from the alveoli. The pressure pattern is the reverse for CO_2. Blood entering the pulmonary capillaries has a higher P_{CO_2} than the air in the alveoli. Therefore, CO_2 diffuses out of the blood into the alveoli.

Internal Respiration

Internal respiration refers to the exchange of gases in the tissues. Specifically, during internal respiration, gases are exchanged between the blood in systemic capillaries and the tissue fluid. Blood that enters the systemic capillaries is a bright red color because the blood is O_2-rich. Tissue fluid, on the other hand, has a low concentration of O_2. Why? Because the cells are continually consuming O_2 during cellular respiration. Therefore, O_2 diffuses from the blood into the tissue fluid. Tissue fluid has a higher concentration of CO_2 than does the blood entering the tissues. Why? Because CO_2 is an end product of cellular respiration. Therefore, CO_2 diffuses from the tissue fluid into the blood. Figure 14.9 summarizes our discussion of gas exchange in the lungs and tissues and shows the differences in O_2 and CO_2 that lead to diffusion of these gases.

Again, we can explain exchange in the tissues by considering the partial pressure of the gases involved. In this case, oxygen diffuses out of the blood into the tissues because the P_{O_2} in tissue fluid is lower than that of the blood. Carbon dioxide diffuses into the blood from the tissues because the P_{CO_2} in tissue fluid is higher than that of the blood.

Gas Transport

The mode of transport of oxygen and carbon dioxide in the blood differs, although red blood cells are involved in transporting both of these gases.

Oxygen Transport

After O_2 enters the blood contained in pulmonary capillaries of the lungs, it enters red blood cells and combines with the iron portion of **hemoglobin,** the pigment in red blood cells. In addition, a small amount of oxygen is transported as a dissolved gas in the watery blood plasma. This dissolved oxygen amounts to only 2–3% of the body's oxygen at any given time because oxygen is not very soluble in water.

Hemoglobin is remarkably suited to the task of transporting oxygen because it can either combine with or release oxygen

Figure 14.9 External and internal respiration. During external respiration in the lungs, CO_2 leaves the blood and O_2 enters the blood passively by diffusion. During internal respiration in the tissues, O_2 leaves the blood and CO_2 enters the blood passively by diffusion.

(depending upon its surroundings). The higher concentration of oxygen in the alveoli, plus the slightly higher pH and slightly cooler temperature, causes hemoglobin to take up oxygen and become **oxyhemoglobin** (HbO_2). The lower concentration of oxygen in the tissues, plus the slightly lower pH and slightly warmer temperature in the tissues, causes hemoglobin to release oxygen and become deoxyhemoglobin (Hb). This equation summarizes our discussion of oxygen transport:

Carbon Dioxide Transport

Transport of CO_2 to the lungs involves a number of steps. After CO_2 diffuses into the blood at the tissues, it can be transported in one of three ways:

1. Approximately 10% of CO_2 is transported as a dissolved gas in blood plasma and in the cytoplasm of red blood cells. Carbon dioxide is much more soluble in water than is oxygen and, thus, three to five times as much CO_2 can be transported as a dissolved gas.
2. Roughly 30% of CO_2 molecules formed are taken up by the protein (globin) portion of hemoglobin, forming a compound called **carbaminohemoglobin.**
3. Most of the CO_2—60%—combines with water, forming carbonic acid (H_2CO_3). The carbonic acid dissociates to hydrogen ions (H^+) and bicarbonate ions (HCO_3^-). This reaction is assisted by an enzyme, **carbonic anhydrase,** which is present in all body cells but is especially abundant in red blood cells. The release of hydrogen ions from carbonic acid explains why the blood in systemic capillaries has a lower pH than the blood in pulmonary capillaries. However, the difference in pH is slight because the globin portion of hemoglobin combines with excess hydrogen ions and becomes **reduced hemoglobin (HHb).**

Bicarbonate ions are carried in the plasma because they diffuse out of red blood cells and go into the plasma. Most of the carbon dioxide in blood is carried as HCO_3^-, the **bicarbonate ion.** As bicarbonate ions diffuse out of red blood cells, chloride ions (Cl^-) diffuse into them. This so-called **chloride shift** maintains the electrical balance between the plasma and the red blood cells.

In pulmonary capillaries, a reverse reaction occurs. Bicarbonate combines with hydrogen ions to form carbonic acid, which this time splits into CO_2 and H_2O, and the CO_2 diffuses out of the blood into the alveoli. The following equation summarizes the reaction between carbon dioxide and water:

$$CO_2 + H_2O \underset{\text{lungs}}{\overset{\text{tissues}}{\rightleftharpoons}} H_2CO_3 \underset{\text{lungs}}{\overset{\text{tissues}}{\rightleftharpoons}} H^+ + HCO_3^-$$

By changing the breathing rate, the respiratory system uses this chemical relationship to help regulate blood pH.[2] As you know, tissue cells are constantly producing carbon dioxide as a waste product of metabolism. Thus, if breathing rate slows, the amount of CO_2 in the blood increases, and more carbonic acid is formed. As carbonic acid dissociates, free H^+ (hydrogen ions) are released. As mentioned previously, hydrogen ions will combine with hemoglobin. However, if enough H^+ are released, ultimately the blood pH decreases. By contrast, if respiratory rate increases, the blood carbon dioxide decreases as more is exhaled. Carbonic acid will reform, and fewer free H^+ will be found in the plasma. Now blood pH will increase. Breathing rate is constantly being "fine-tuned" in this way, so that blood pH remains at normal levels and homeostasis is maintained.

Many respiratory diseases interfere with normal excretion of carbon dioxide, causing excess carbonic acid to be formed. If blood pH falls below the normal range as a result, a condition termed **acidosis** develops. Other conditions may cause a person to hyperventilate, or breathe too rapidly. If too much CO_2 is eliminated in this fashion, blood pH will rise above normal level, causing **alkalosis.** Stress and anxiety can cause a person to hyperventilate. Disorders involving the respiratory control center (such as a brain tumor or brain damage near the brain stem) may cause a person to hyperventilate as well. Both acidosis and alkalosis can be fatal because they interfere with cell enzyme function.

Content CHECK-UP!

7. If you describe internal respiration, you can use this statement(s):
   a. Oxygen diffuses from the blood into the tissue fluid.
   b. Carbon dioxide diffuses from tissue fluid into the blood.
   c. Internal respiration takes place in systemic capillaries.
   d. All of these statements are correct.

8. Hemoglobin with carbon dioxide attached is termed:
   a. reduced hemoglobin.   c. oxyhemoglobin.
   b. carbaminohemoglobin.   d. dioxyhemoglobin.

9. List three mechanisms used to transport CO_2 in the blood.

Answers in Appendix C, page 464.

14.4 Respiration and Health

The respiratory tract is constantly exposed to environmental air. The quality of this air can affect our health. The presence of a disease means that homeostasis is threatened and if the condition is not brought under control, death is inevitable.

Upper Respiratory Tract Infections

The upper respiratory tract consists of the nasal cavities, the pharynx, and the larynx. Upper respiratory infections (URI) can spread from the nasal cavities to the sinuses, middle ears, and larynx. Viral infections sometimes lead to secondary bacterial infections. What we call "strep throat" is a primary bacterial infection caused by *Streptococcus pyogenes* that can lead to a generalized upper respiratory infection and even a systemic (affecting the body as a whole) infection. Although antibiotics have no effect on viral infections, they are successfully used to treat most bacterial infections, including strep throat. The symptoms of strep

---

[2]pH means the negative logarithm of free hydrogen ion concentration, so the smaller the pH number, the greater the number of free hydrogen ions. The pH of arterial blood should range from 7.35 to 7.45. Blood with pH 7.3 (acidosis) has *more* free hydrogen ions than blood with pH 4 (normal). See page 25 to review the concept of pH.

throat are severe sore throat, high fever, and white patches on a dark red throat.

Sinusitis

Sinusitis is an infection of the cranial sinuses, the cavities within the facial skeleton that drain into the nasal cavities. Only about 1–3% of upper respiratory infections are accompanied by sinusitis. Sinusitis develops when nasal congestion blocks the tiny openings leading to the sinuses. Symptoms include postnasal discharge as well as facial pain that worsens when the patient bends forward. Pain and tenderness usually occur over the lower forehead or over the cheeks. If the latter, toothache is also a complaint. Successful treatment depends on restoring proper drainage of the sinuses. Antibiotic therapy is necessary if the condition results from bacterial infection. Even a hot shower and sleeping upright can be helpful. Otherwise, spray decongestants are preferred over oral antihistamines, which thicken rather than liquefy the material trapped in the sinuses.

Otitis Media

Otitis media is a bacterial infection of the middle ear. The middle ear is not a part of the respiratory tract, but this infection is considered here because it is a complication often seen in children who have a nasal infection. Infection can spread by way of the **auditory (eustachian) tube** that leads from the nasopharynx to the middle ear. Pain is the primary symptom of a middle ear infection. A sense of fullness, hearing loss, vertigo (dizziness), and fever may also be present. Antibiotics almost always bring about a full recovery, and recurrence is probably due to a new infection. Tubes (called *tympanostomy tubes*) are sometimes placed in the eardrums of children with multiple recurrences to help promote fluid drainage, as well as to prevent the buildup of pressure in the middle ear and the possibility of hearing loss. Normally, the tubes fall out with time.

Tonsillitis

Tonsillitis occurs when the **tonsils,** masses of lymphatic tissue in the pharynx, become inflamed and enlarged. If tonsillitis occurs frequently and enlargement makes breathing difficult, the tonsils can be removed surgically in a **tonsillectomy.** Fewer tonsillectomies are performed today than in the past because we now know that the tonsils remove many of the pathogens that enter the pharynx; therefore, they are a first line of defense against invasion of the body.

Laryngitis

Laryngitis is an inflammation of the larynx, with accompanying hoarseness leading to the inability to talk in an audible voice. It can result from an upper respiratory infection, or simply from overuse of the larynx (if, for example, you've screamed for several hours at a sports event or concert). So long as the immune system is functioning properly, an upper respiratory infection usually resolves, with rest, over time. Likewise, laryngitis disappears over time as long as the larynx is rested. Persistent hoarseness without the presence of an upper respiratory infection is one of the warning signs of cancer, and therefore should be looked into by a physician.

Lower Respiratory Tract Disorders

Lower respiratory tract disorders include infections, restrictive pulmonary disorders, obstructive pulmonary disorders, and lung cancer.

Lower Respiratory Infections

Acute bronchitis, pneumonia, and tuberculosis are infections of the lower respiratory tract. **Acute bronchitis** is an infection of the primary and secondary bronchi. Usually, it is preceded by a viral URI that has led to a secondary bacterial infection. Most likely, a nonproductive cough has become a deep cough that expectorates mucus and perhaps pus.

Pneumonia is a viral or bacterial infection of the lungs in which the bronchi and alveoli fill with thick fluid (Fig. 14.10). Risk factors for pneumonia include advanced age (it is most common and most often fatal in the elderly), weakened immune system, smoking, and being immobilized. A common scenario for development of pneumonia is an elderly person who is immobilized due to a fractured hip. High fever, productive cough, difficulty in breathing, headache, and chest pain are symptoms of pneumonia. Rather than being a generalized lung infection, pneumonia may be localized in specific lobules of the lungs. Obviously, the more lobules involved, the more serious is the infection. Pneumonia can be caused by a bacterium that is usually held in check but has gained the upper hand due to stress and/or reduced immunity. AIDS patients are subject to a particularly rare form of pneumonia caused by the protozoan *Pneumocystis carinii*. Pneumonia of this type is almost never seen in individuals with a healthy immune system.

Pulmonary tuberculosis is caused by the tubercle bacillus, a type of bacterium. When tubercle bacilli invade the lung tissue, the cells build a protective capsule about the foreigners, isolating them from the rest of the body. This tiny capsule is called a tubercle. If the resistance of the body is high, the imprisoned organisms die, but if the resistance is low, the organisms eventually can be liberated. If a chest X ray detects active tubercles, the individual is put on appropriate antibiotic therapy to ensure the localization of the disease and the eventual destruction of any live bacteria. It is possible to tell if a person has ever been exposed to tuberculosis with a test in which a highly diluted extract of the bacillus is injected into the skin of the patient. A person who has never been in contact with the tubercle bacillus shows no reaction, but one who has had or is fighting an infection shows an area of inflammation that peaks in about 48 hours.

Restrictive Pulmonary Disorders

In restrictive pulmonary disorders, vital capacity is reduced because the lungs have lost their elasticity. Inhaling particles such as silica (sand), coal dust, asbestos, clay, cement, flour,

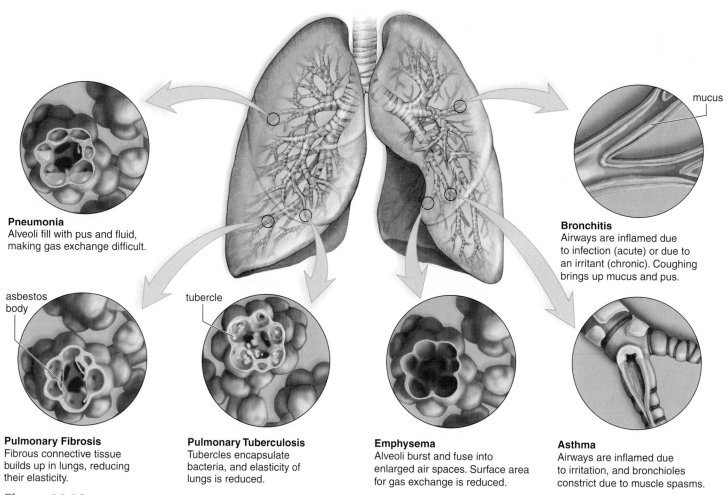

Pneumonia
Alveoli fill with pus and fluid, making gas exchange difficult.

Bronchitis
Airways are inflamed due to infection (acute) or due to an irritant (chronic). Coughing brings up mucus and pus.

mucus

asbestos body

Pulmonary Fibrosis
Fibrous connective tissue builds up in lungs, reducing their elasticity.

tubercle

Pulmonary Tuberculosis
Tubercles encapsulate bacteria, and elasticity of lungs is reduced.

Emphysema
Alveoli burst and fuse into enlarged air spaces. Surface area for gas exchange is reduced.

Asthma
Airways are inflamed due to irritation, and bronchioles constrict due to muscle spasms.

Figure 14.10 **Common bronchial and pulmonary diseases.** Exposure to infectious pathogens and/or polluted air, including tobacco smoke, causes the diseases and disorders shown here.

and fiberglass can lead to **pulmonary fibrosis,** a condition in which fibrous connective tissue builds up in the lungs. The lungs cannot inflate properly and are always tending toward deflation. Breathing asbestos is also associated with the development of cancer. Because asbestos was formerly used widely as a fireproofing and insulating agent, unwarranted exposure has occurred. It has been projected that two million deaths caused by asbestos exposure—mostly in the workplace—will occur in the United States between 1990 and 2020.

Obstructive Pulmonary Disorders

In obstructive pulmonary disorders, air does not flow freely in the airways, and the time it takes to inhale or exhale maximally is greatly increased. Chronic bronchitis and emphysema are referred to as **chronic obstructive pulmonary disorders (COPD)** because they develop slowly, over a long period of time, and are recurrent. Asthma is an obstructive disorder as well. However, asthma is generally an acute illness—that is, one that occurs in intermittent episodes that flare up and disappear quickly.

In **chronic bronchitis,** the airways are inflamed and filled with mucus. A cough that brings up mucus is common. The bronchi have undergone degenerative changes, including the loss of cilia and their normal cleansing action. Under these conditions, an infection is more likely to occur. Smoking cigarettes and cigars is the most frequent cause of chronic bronchitis. Exposure to other pollutants can also cause chronic bronchitis.

Emphysema is a chronic and incurable disorder in which the alveoli are distended and their walls damaged so that the surface area available for gas exchange is reduced. Emphysema is often preceded by chronic bronchitis. Air trapped in the lungs leads to alveolar damage and a noticeable ballooning of the chest. On an X ray, the diaphragm (which is normally curved when at rest; see page 307) appears flattened as the lungs increase in size. Breakdown in the elastic tissue of the lung diminishes the elastic recoil. Not only are the airways narrowed, but the driving force behind expiration is also reduced. The patient has a very difficult time exhaling, and residual volume increases. The victim often feels a breathless sensation and may have a cough. Because the surface area for

The Most-Often-Asked Questions About Tobacco and Health

Is there a safe way to smoke?

No. All forms of tobacco can cause damage, and smoking even a small amount is dangerous. Tobacco is perhaps the only legal product whose advertised and intended use—that is, smoking it—will hurt the body.

Does smoking cause cancer?

Yes, and not only lung cancer. Besides lung cancer, smoking a pipe, cigarettes, or cigars is also a major cause of cancers of the mouth, larynx (voice box), and esophagus. In addition, smoking increases the risk of cancer of the bladder, kidney, pancreas, stomach, and uterine cervix.

What are the chances of being cured of lung cancer?

Very low; the five-year survival rate is only 13%. However, it is important to remember that lung cancer, or **bronchogenic carcinoma** in technical terms, is very rare (although not unheard of) in nonsmokers. Thus, lung cancer is largely preventable: Don't smoke, and avoid exposure to smoke.

Does smoking cause other lung diseases?

Yes. Smoking leads to chronic bronchitis, a disease in which the airways produce excess mucus, forcing the smoker to cough frequently. Smoking is also the major cause of emphysema, a disease that slowly destroys a person's ability to breathe. Chronic bronchitis and pulmonary emphysema rates are higher in smokers than in nonsmokers.

Why do smokers have "smoker's cough"?

Remember the mucociliary escalator (page 308)? This is the protective mechanism for the lungs and the airways: a sheet of mucus present in the trachea, bronchi, and bronchioles. Normally, constantly beating cilia on the cells lining the airways move the sheet upward. Once at the pharynx, mucus can be swallowed or spit out. Smoking is a double assault on the mucociliary escalator: First, inhaled, nasty toxins and particles cling to the sticky mucus. Then, a big dose of nicotine paralyzes the cilia for 30 minutes or more. Long-term smokers will eliminate the delicate, ciliated epithelial cells entirely, replacing them with a sturdier epithelium better able to stand the heat of cigarette smoke. The result: There is no way to clear accumulated mucus (and the trapped contaminants in it) except to cough it out. The longer a person smokes, the more pronounced and productive the cough becomes.

If you smoke but don't inhale, is there any danger?

Yes. Wherever smoke touches living cells, it does harm. So, even if smokers of pipes, cigarettes, and cigars don't inhale, they are at an increased risk for lip, mouth, and tongue cancer.

Does smoking affect the heart?

Yes. Smoking increases the risk of heart disease, which is the number one killer in the United States. Smoking, high blood pressure, high cholesterol, and lack of exercise are all risk factors for heart disease. Smoking alone doubles the risk of heart disease.

Is there any risk for pregnant women and their babies?

Pregnant women who smoke endanger the health and life of their unborn babies. When a pregnant woman smokes, she really is smoking for two because the nicotine, carbon monoxide, and other dangerous chemicals in smoke enter her bloodstream and then pass into the baby's body. Smoking mothers have more stillbirths and babies of low birthweight than do nonsmoking mothers.

Does smoking cause any special health problems for women?

Yes. Women who smoke and use the birth control pill have an increased risk of stroke and blood clots in the legs. In addition, women who smoke increase their chances of getting cancer of the uterine cervix.

What are some of the short-term effects of smoking cigarettes?

Almost immediately, smoking can make it hard to breathe. Within a short time, it can also worsen asthma and allergies. Only seven seconds after a smoker takes a puff, nicotine reaches the brain, where it produces a morphinelike effect.

Are there any other risks to the smoker?

Yes, there are many more risks. Smoking is a cause of stroke, which is the third leading cause of death in the United States. Smokers are more likely to have and die from stomach ulcers than are nonsmokers. Smokers have a higher incidence of cancer in general. If a person smokes and is exposed to radon or asbestos, the risk for lung cancer increases dramatically.

What are the dangers of passive smoking?

Passive smoking causes lung cancer in healthy nonsmokers. Children whose parents smoke are more likely to suffer from

—Continued

Continued—

pneumonia or bronchitis in the first two years of life than are children who come from smoke-free households. Passive smokers have a 30% greater risk of developing lung cancer than do nonsmokers who live in a smoke-free house.

Are chewing tobacco and snuff safe alternatives to cigarette smoking?

No, they are not. Many people who use chewing tobacco or snuff believe it can't harm them because there is no smoke. Wrong. Smokeless tobacco contains nicotine, the same addicting drug found in cigarettes and cigars. Although not inhaled through the lungs, the juice from smokeless tobacco is absorbed through the lining of the mouth. There it can cause sores and white patches, which often lead to cancer of the mouth. Snuff dippers actually take in an average of over ten times more cancer-causing substances than cigarette smokers.

OK, OK, enough already! You've convinced me that I should quit. But how? I've tried before. Is there a better way to quit smoking?

The U.S. Office of the Surgeon General has done extensive research on the most effective ways to quit smoking. This research has determined that to be successful, a smoker should use the **Five Keys for Quitting:**

1. **Get ready to quit:** Set your date and plan ahead. Review what worked, or didn't, in the past. Throw out all smoking materials: cigarettes, pipes, ashtrays, lighters, etc. Inform your family, friends, and co-workers of your intention to quit, and ask them not to smoke around you. Once you quit, don't take another puff from a cigarette.
2. **Get support and encouragement:** See your health-care provider for help. Private and group smoking cessation counseling often helps and is available at a low cost or is free. Your local health department will have information about programs in your area. Avoid spending time with people who might undermine your efforts.
3. **Learn new behaviors that aren't associated with smoking:** When you first quit, change your routine completely so that subconscious cues to smoke don't keep appearing. Take a different route to work, eat breakfast in a different place, drink tea instead of coffee. When the urge to smoke recurs, distract yourself with physical or mental activity. Exercise or talk to someone to reduce stress. Each day, plan something fun and pleasurable for yourself.
4. **Use medication correctly:** Never be embarrassed to ask for, and use, anti-smoking medication. Nicotine comes in gum and patches (both sold over the counter) and in prescription inhaler and nasal spray forms. Using these products can help to overcome the cravings for nicotine that occur immediately after quitting. In addition, use of antidepressants (available by prescription) has been shown to significantly increase a patient's chances of quitting for good. Chantix®, a recently approved antidepressant, also reduces the effects of nicotine addiction by blocking nicotine receptors.
5. **If at first you don't succeed, try, try again!** The first 3–6 months after quitting will be the toughest, and that's when many people go back to smoking. Don't be discouraged if this happens to you. On average, most people try to quit at least twice before finally succeeding. To avoid a relapse, don't drink alcohol and/or hang around with other smokers. Studies have shown that both can lead to failure in the attempt to quit. Weight gain and increased irritability or depression can also be expected. Exercise and eating a healthy diet will minimize both of these side effects.

From *You Can Quit Smoking Consumer Guide,* available online at http://www.surgeongeneral.gov/tobacco/consquits.htm.

gas exchange in the lungs is reduced, less oxygen reaches the heart and the brain. Even so, the heart works furiously to force more blood through the lungs, and an increased workload on the heart can result. Lack of oxygen to the brain can make the person feel depressed, sluggish, and irritable. Before therapy can be effective, the patient must stop smoking. Medical treatment for emphysema is limited to drugs that dilate the bronchi and bronchioles by relaxing the smooth muscle, along with inhaled oxygen to improve oxygenation of the blood. Surgical options include lung volume reduction surgery (LVRS) and lung transplantation. In LVRS, the diseased upper portions of the lung are surgically removed, and the remaining lung tissue is sealed with a flap of pericardium. Removing the diseased portion of the lung allows the remaining healthy tissue to expand and fill the thoracic cavity. Lung transplantation is the other surgical option for emphysema, but it's rarely used because of the scarcity of donor organs. As with all transplant recipients, the patient must remain on lifelong anti-rejection therapy after a lung transplant.

Asthma is a disease of the bronchi and bronchioles that is marked by wheezing, breathlessness, and sometimes a cough and expectoration of mucus. The airways are unusually sensitive to specific irritants, which can include a wide

a.   b.

Figure 14.11 **Normal lung versus cancerous lung. a.** Normal lung with heart in place. Note the healthy red color. **b.** Lungs of a heavy smoker. Notice how black the lungs are except where cancerous tumors have formed.

trachea

primary
bronchus

heart

range of allergens such as pollen, animal dander, dust, cigarette smoke, and industrial fumes. Even cold air can be an irritant. When exposed to the irritant, the smooth muscle in the bronchioles undergoes spasms. It now appears that chemical mediators given off by immune cells in the bronchioles cause the spasms. Most asthma patients have some degree of bronchial inflammation that reduces the diameter of the airways and contributes to the seriousness of an attack. Asthma is not curable, but it is treatable. Special inhalers can control the inflammation and hopefully prevent an attack, while other types of inhalers can stop the muscle spasms should an attack occur.

Lung Cancer

Lung cancer used to be more prevalent in men than in women, but recently it has surpassed breast cancer as a cause of death in women. The recent increase in the incidence of lung cancer in women is directly correlated to increased numbers of women who smoke. Autopsies on smokers have revealed the progressive steps by which the most common form of lung cancer develops. The first event appears to be thickening and callusing of the cells lining the primary bronchi. (Callusing occurs whenever cells are exposed to irritants.) Then cilia are lost, making it impossible to prevent dust and dirt from settling in the lungs. Following this, cells with atypical nuclei appear in the callused lining. A tumor consisting of disordered cells with atypical nuclei is considered cancer in situ (at one location) (Fig. 14.11). A final step occurs when some of these cells break loose and penetrate other tissues, a process called **metastasis.** Now the cancer has spread. The original tumor may grow until a bronchus is blocked, cutting off the supply of air to that lung. The entire lung then collapses, the secretions trapped in the lung spaces become infected, and pneumonia or a lung abscess (localized area of pus) results. The only treatment that offers a possibility of cure is to remove a lobe or the whole lung before metastasis has had time to occur. This operation is called **pneumonectomy.** If the cancer has spread, chemotherapy and radiation are also required.

The Medical Focus on pages 321–322 lists the various illnesses, including cancer, that are apt to occur in persons who smoke. Current research indicates that passive smoking—exposure to smoke created by others who are smoking—can also cause lung cancer and other illnesses associated with smoking. If a person stops voluntary smoking and avoids passive smoking, and if the body tissues are not already cancerous, they may return to normal over time.

Content **CHECK-UP!**

10. Obstructive pulmonary diseases include:
    a. emphysema.   c. asthma.
    b. chronic bronchitis.   d. all of the above.
11. Which of the following increases in a person with emphysema?
    a. amount of surface area in the lung for gas exchange
    b. residual volume
    c. amount of elastic tissue in the lungs
    d. amount of oxygen delivered to the brain and heart
12. Antibiotics would be useful medication for:
    a. streptococcal infection.   d. a and b.
    b. tuberculosis.   e. all of the above.
    c. viral upper respiratory infection.

Answers in Appendix C, page 464.

14.5　Effects of Aging

Respiratory fitness decreases with age. Maximum breathing capacities decline, while the likelihood of fatigue increases. Inspiration and expiration are not as effective in older persons. With age, weakened intercostal muscles and increased inelasticity of the rib cage combine to reduce the inspiratory reserve volume, while the lungs' inability to recoil reduces the expiratory reserve volume. More residual air is found in the lungs of older people.

With age, gas exchange in the lungs becomes less efficient, not only due to changes in the lungs but also due to changes in the blood capillaries. The respiratory membrane thickens, and the gases cannot diffuse as rapidly as they once did.

In the elderly, the ciliated cells of the trachea are reduced in number, and those remaining are not as effective as they once were. Respiratory diseases, such as those discussed in section 14.4, are more prevalent in older people than in the general public. Pneumonia and other respiratory infections are among the leading causes of death in older persons.

14.6　Homeostasis

The illustration in Human Systems Work Together on page 325 describes how the respiratory system assists and depends on other body systems. The respiratory system contributes to homeostasis in many ways—in particular, by carrying on gas exchange and regulating blood pH.

Gas Exchange

First and foremost, the respiratory system performs gas exchange. Carbon dioxide, a waste molecule given off by cellular respiration, exits the body, and oxygen, a molecule needed for cellular respiration, enters the body at the lungs. Cellular respiration produces ATP, a molecule that allows the body to perform all sorts of work, including muscle contraction and nerve conduction. It is estimated that the brain uses 15–20% of the oxygen taken into the blood. Not surprisingly, a lack of oxygen affects the functioning of the brain before it affects other organs.

Regulation of Blood pH

The respiratory system can alter blood pH by changing blood carbon dioxide levels. In the tissues, carbon dioxide enters the blood and red blood cells where this reaction occurs. The bicarbonate ion (HCO_3^-) diffuses out of the red blood cells to be carried in the plasma:

$$CO_2 + H_2O \underset{lungs}{\overset{tissues}{\rightleftharpoons}} H_2CO_3 \underset{lungs}{\overset{tissues}{\rightleftharpoons}} H^+ + HCO_3^-$$

This reaction lowers blood pH because free H^+ (hydrogen ions) are formed during the reaction as carbonic acid dissociates. The greater the amount of CO_2 in the blood, the greater the carbonic acid formed, and the more free hydrogen ions are formed. Conversely, when carbon dioxide starts to diffuse out of the blood in the lungs, the reaction occurs in the reverse direction. Carbon dioxide concentration in blood decreases (because it is being exhaled), and carbonic acid concentration decreases, too. Now, the blood pH rises.

What happens to your blood pH if you hypoventilate—that is, breathe at a lower than normal rate? Acidosis results because carbon dioxide is not being exhaled at the normal rate. Too much carbonic acid is formed from the excess carbon dioxide. Any respiratory condition, such as emphysema, that hinders the passage of carbon dioxide out of the blood also results in acidosis. By contrast, what happens to your blood pH if you hyperventilate—that is, breathe at a higher than normal rate? Alkalosis (high blood pH) will develop because carbon dioxide is leaving the body at a high rate. This decreases the concentration of carbonic acid and the concentration of free H^+. Severe anxiety can cause a person to hyperventilate.

Control of Blood Pressure

The lungs play a role in the control of blood pressure by assisting in the renin-angiotensin-aldosterone pathway (see p. 217). The kidneys produce the enzyme renin whenever blood pressure falls. Renin activates **angiotensin** I (an inactive plasma protein produced by the liver). As blood that contains angiotensin I flows through the lungs, angiotensin I is further activated to form angiotensin II by angiotensin-converting enzyme (ACE), which is found in endothelial cells that compose the pulmonary capillaries. Once formed, angiotensin II causes arterioles to constrict, increasing blood vessel resistance. Angiotensin II also causes the release of aldosterone from the adrenal gland. In turn, aldosterone causes the kidneys to reabsorb salt and water. Through this mechanism, blood pressure returns to normal and homeostasis is maintained.

Defense

The respiratory tract assists defense against pathogens by preventing their entry into the body and by removing them from respiratory surfaces. For example, the cilia that line the trachea sweep impurities toward the throat. The respiratory tract also assists immunity. We now know that the tonsils serve as a location where T cells are presented with antigens before they enter the body as a whole. This action helps the body prepare to respond to an antigen before it enters the bloodstream.

Human Systems Work Together

Integumentary System

Gas exchange in lungs provides oxygen to skin and rids body of carbon dioxide from skin.

Skin helps protect respiratory organs and helps regulate body temperature.

Skeletal System

Gas exchange in lungs provides oxygen and rids body of carbon dioxide.

Rib cage protects lungs and assists breathing; bones provide attachment sites for muscles involved in breathing.

Muscular System

Lungs provide oxygen for contracting muscles and rid the body of carbon dioxide from contracting muscles.

Muscle contraction assists breathing; physical exercise increases respiratory capacity.

Nervous System

Lungs provide oxygen for neurons and rid the body of carbon dioxide produced by neurons.

Respiratory centers in brain regulate breathing rate.

Endocrine System

Gas exchange in lungs provides oxygen and rids body of carbon dioxide.

Epinephrine promotes ventilation by dilating bronchioles; growth factors control production of red blood cells that carry oxygen.

How the Respiratory System works with other body systems

Diagram labels:
- nasal cavity
- nose
- pharynx (throat)
- larynx
- trachea
- bronchi
- lungs

Cardiovascular System

Gas exchange in lungs rids body of carbon dioxide, helping to regulate the pH of blood; breathing aids venous return.

Blood vessels transport gases to and from lungs; blood services respiratory organs.

Lymphatic System/Immunity

Tonsils and adenoids occur along respiratory tract; breathing aids lymph flow; tracheal cilia sweep impurities toward throat.

Lymphatic vessels pick up excess tissue fluid; immune system protects against respiratory tract and lung infections.

Digestive System

Gas exchange in lungs provides oxygen to the digestive tract and excretes carbon dioxide from the digestive tract.

Breathing is possible through the mouth because digestive tract and respiratory tract share the pharynx.

Urinary System

Lungs excrete carbon dioxide, provide oxygen, and convert angiotensin I to angiotensin II, leading to kidney regulation.

Kidneys compensate for water lost through respiratory tract; work with lungs to maintain blood pH.

Reproductive System

Gas exchange increases during sexual activity.

Sexual activity increases breathing; pregnancy causes breathing rate and vital capacity to increase.

SELECTED NEW TERMS

Basic Key Terms

alveolus (al-ve'o-lus), p. 310

angiotensin (an-jē-o-ten'sin), p. 324

aortic body (a-or'tik bod'e), p. 316

auditory (eustachian) tube (ô'dĭ-tôr'ē (yōō-stā'shən) tōōb), p. 319

bicarbonate ion (bi-kar'bo-nāt i'on), p. 318

bronchiole (brong'ke-ōl), p. 310

bronchus (brong'kus), p. 310

carbaminohemoglobin (kar-buh-me"no-he"mo-glo'bin), p. 318

carbonic anhydrase (kär-bŏn'ĭk ăn-hī'drəs), p. 318

carotid body (kar-ah'tid bod'e), p. 316

chloride shift (klōr'ĭd shĭft), p. 318

concha (kong'kuh), p. 308

epiglottis (ĕ"pĭ-glot'is), p. 309

expiration (eks"pĭ-ra'shun), p. 307

expiratory reserve volume (ek-spi'ruh-to-re re-zerv' vol'yūm), p. 315

external respiration (eks-ter'nal res"pĭ-ra'shun), p. 316

glottis (glot'is), p. 309

hemoglobin (hē'mə-glō'bĭn), p. 317

inspiration (in"spī-ra'shun), p. 307

inspiratory reserve volume (in-spi'ruh-to-re re-zerv' vol'yūm), p. 315

internal respiration (in-ter'nal res"pĭ-ra'shun), p. 317

intrapleural pressure (in'trah-ploor'al prēsh'ər), p. 313

larynx (lār'inks), p. 309

lungs (lungz), p. 310

mucociliary escalator (mū-kō-sil'ē-ă-rē), p. 308

nasal cavities (na'zal kav'ĭ-tēz), p. 308

nose (nōz), p. 308

nostrils (nŏs'trĭls), p. 308

oxyhemoglobin (ok"se-he"mo-glo'bin), p. 317

paranasal sinus (pār-uh-na'zul si'nus), p. 308

pharynx (fār'inks), p. 309

pleurae (pleura) (ploŏr'-ā), p. 311

pons (poňz), p. 316

primary respiratory center (prī'mĕr'ē res'pĭ-ruh-tor-e sen'ter), p. 316

pulmonary capillaries (pŭl'-mōh-nār-ee kap'ĭl-lār-ee), p. 311

reduced hemoglobin (re-dūs'd he'mo-glo-bin), p. 318

residual volume (re-zid'yū-ul vol'yūm), p. 316

spirometer (spī-rŏm'-ĕ-ter), p. 314

surface tension (ser'fus ten'shun), p. 310

surfactant (sur-fak'tant), p. 312

tidal volume (ti'dul vol'yūm), p. 314

tonsils (tŏn'səlz), p. 319

trachea (tra'ke-uh), p. 309

ventilation (ven"tĭ-la'shun), p. 312

vital capacity (vi'tal kuh-pas'ĭ-te), p. 315

vocal cord (vo'kul kord), p. 309

Clinical Key Terms

acidosis (as-i-dō'sis), p. 318

acute bronchitis (uh-kyūt brong-ki'tis), p. 319

alkalosis (al-kă-lō'sis), p. 318

apnea (ap'ne-uh), p. 309

asthma (ăz'mə), p. 322

bronchogenic carcinoma (brong-ko-jen'ik kär'sə-nō'mə), p. 321

Cheyne-Stokes respiration (shān-stōks res"pĭ-ra'shun), p. 316

chronic bronchitis (kron'ik brong-ki'tis), p. 320

chronic obstructive pulmonary disorders (krŏn'ĭk əb-strŭkt'ive dĭs-or'dərs), p. 320

emphysema (em"fi-se'muh), p. 320

eupnea (yūp-ne'uh), p. 316

hyperpnea (hi"per-ne'uh), p. 316

infant respiratory distress syndrome (in'funt res'pĭ-ruh-tor-e dis-tres' sin'drōm), p. 312

laryngitis (lār-in-ji'tis), p. 319

lung cancer (lung kan'ser), p. 323

metastasis (mĕ-tas'-tă-sis), p. 323

obstructive sleep apnea (ob'struk-tĭv slēp ap'nē-uh), p. 309

otitis media (o-ti'tis me'-de-uh), p. 319

pneumonectomy (nu-mah-nek'tuh-me), p. 323

pneumonia (nu-mo'ne-uh), p. 319

pulmonary fibrosis (pul'mo-nĕr"e fi-bro'sis), p. 320

pulmonary tuberculosis (pul'mo-nĕr"e tū-ber"kyū-lo'sis), p. 319

singultus (sing-gŭl'tŭs), p. 306

sinusitis (si-nŭ-si'tis), p. 319

tonsillectomy (ton'sĭ-lek'to-me), p. 319

tonsillitis (ton-sil-ī'tis), p. 319

tracheostomy (tra"ke-ahs'to-me), p. 310

SUMMARY

14.1 The Respiratory System

A. The nasal cavities, which filter, warm, and humidify incoming air, open into the pharynx.

B. The food and air passages cross in the pharynx, which conducts air to the larynx and food to the esophagus.

C. The larynx is the voice box. It houses the vocal cords. The glottis, a slit between the vocal cords, is covered by the epiglottis when food is being swallowed.

D. The trachea and the primary bronchi are held open by cartilaginous rings. The rings gradually disappear as the primary bronchi branch into bronchioles, which enter the alveoli. The lungs are composed of the air tubes and alveoli beyond the primary bronchi and the alveoli.

E. The respiratory membrane is the juxtaposed alveolar wall and the capillary wall. The large surface area and thinness of the respiratory membrane allow rapid gas exchange.

14.2 Mechanism of Breathing

A. Ventilation is the movement of gases into and out of the lungs. The pleural membranes, which enclose the lungs, attach them to the thoracic wall. During inspiration, the thoracic cavity increases in size as the diaphragm lowers and the rib cage moves upward and outward. Therefore, the lungs expand, creating a partial vacuum, which causes air to rush in. During expiration, the diaphragm relaxes and resumes its dome shape. As the rib cage and lungs recoil, air is pushed out of the lungs. Expiration can be forceful when the internal intercostal muscles contract, causing the rib cage to move farther downward and inward. Also, contraction of the abdominal wall pushes the viscera against the diaphragm, and the increased pressure expels air.

B. Tidal volume is the amount of air inhaled and exhaled with each breath. Vital capacity is the total volume of air that can be moved in and out of the lungs during a single breath. Some air remains in the lungs after expiration. This is called the residual volume. Passages within the airways are called dead space because no gas exchange takes place in the airways.

C. The primary respiratory center in the medulla oblongata is assisted by the pons. It rhythmically controls the ventilation rate. The respiratory center increases the rate when CO_2 and H^+ levels increase, as detected by chemoreceptors in the respiratory center or the carotid and aortic bodies. The latter also detects a low O_2 level and stimulates the respiratory center, which then increases the ventilation rate.

14.3 Gas Exchange and Transport

A. Diffusion accounts for the movement of gases during external respiration and internal respiration. External respiration occurs when CO_2 moves from the area of higher concentration in the blood to lower concentration in the alveoli. O_2 moves from a higher concentration in the alveoli to a lower concentration in the blood. Internal respiration occurs when O_2 moves from a higher concentration in the blood to a lower concentration in the tissue fluid, and CO_2 moves from a higher concentration in the tissue fluid to a lower concentration in the blood. The occurrence of cellular respiration always causes gas transport.

B. Oxygen is transported to the tissues in combination with hemoglobin as oxyhemoglobin (HbO_2). Carbon dioxide is mainly carried to the lungs within the plasma as the bicarbonate ion (HCO_3^-). Hemoglobin combines with hydrogen ions and becomes reduced (HHb) and this helps maintain the pH level of the blood within normal limits.

C. By adjusting respiratory rate, the respiratory system helps to maintain pH balance. Breathing slowly decreases pH and breathing rapidly increases it.

14.4 Respiration and Health

A. A number of illnesses are associated with the respiratory tract. These disorders can be divided into those that affect the upper respiratory tract and those that affect the lower respiratory tract. Infections of the nasal cavities, sinuses, throat, tonsils, and larynx are all well known. In addition, infections can spread from the nasopharynx to the ears.

B. The lower respiratory tract is subject to infections such as acute bronchitis, pneumonia, and pulmonary tuberculosis. In restrictive pulmonary disorders, exemplified by pulmonary fibrosis, the lungs lose their elasticity. In obstructive pulmonary disorders, exemplified by chronic bronchitis, emphysema, and asthma, the bronchi (and bronchioles) do not effectively conduct air to and from the lungs. Smoking, which is associated with chronic bronchitis and emphysema, can eventually lead to lung cancer.

14.5 Effects of Aging
All aspects of respiration decline with age. The elderly often die from pulmonary infections.

14.6 Homeostasis

A. The respiratory system carries on two main functions: (1) gas exchange, which is essential to the process of cellular respiration, and (2) maintenance of blood pH. The respiratory system helps maintain blood pressure by converting angiotensin I to angiotensin II.

B. The respiratory system works with other systems of the body. The cardiovascular system transports gases, and breathing helps systemic venous blood return to the heart. The respiratory tract assists defense against pathogens by keeping the tract clean of debris. Also, the tonsils are lymphatic tissue where antigens are presented to T cells. The nervous system maintains rhythmic ventilation, and the sensory organs for olfaction are located in the nasal cavities. The respiratory center responds to the increased gas exchange needs of the muscular system when we exercise.

STUDY QUESTIONS

1. Name and explain the four events that comprise respiration. (p. 307)
2. What is the path of air from the nose to the lungs? Describe the structure and state the function of all the organs mentioned. (pp. 307–312)
3. What is the respiratory membrane, and how does its structure promote rapid gas exchange? (p. 312)
4. What three conditions should be borne in mind in order to understand ventilation? (pp. 312–313)
5. Explain the volume and pressure changes necessary to inspiration and expiration. How is ventilation controlled? (pp. 313, 316)
6. What is the difference between tidal volume and vital capacity? Of the air we inhale, some is not used for gas exchange. Why not? (pp. 314–316)
7. Contrast external respiration with internal respiration, and explain why there is a flow of gases during each of these. (pp. 316–317)
8. How are oxygen and carbon dioxide transported in the blood? What role does hemoglobin play in the transport of CO_2? (pp. 317–318)
9. Name and describe several upper and several lower respiratory tract disorders (other than cancer). If appropriate, explain why breathing is difficult with these conditions. (pp. 318–320, 322–323)
10. List the steps by which lung cancer develops. (p. 323)

LEARNING OBJECTIVE QUESTIONS

Fill in the blanks.

1. In tracing the path of air, the _____ immediately follows the pharynx.
2. The lungs contain air sacs called _____.
3. The breathing rate is primarily regulated by the amount of _____ in the blood.
4. Air enters the lungs after they have _____.
5. Gas exchange is dependent on the physical process of _____.
6. During external respiration, oxygen _____ the blood.
7. During internal respiration, carbon dioxide _____ the blood.
8. Carbon dioxide is carried in the blood as _____ ions.
9. The most likely cause of emphysema and chronic bronchitis is _____.
10. Most cases of lung cancer actually begin in the _____.
11. The amount of air moved in and out of the respiratory system with each normal breath is called the _____.
12. The total amount of air that can be moved in and out of the lungs during a single breath is called the _____.
13. The _____ closes the opening into the larynx during swallowing.
14. The respiratory membrane consists of the walls of the _____ and _____.
15. Label structures a. through i. in the diagram of the respiratory tract.

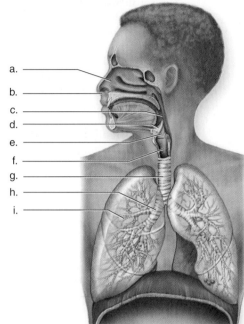

MEDICAL TERMINOLOGY EXERCISE

Consult Appendix B for help in pronouncing and analyzing the meaning of the terms that follow.

1. eupnea (yūp-ne′uh)
2. nasopharyngitis (na″zo-fār″in-ji′tis)
3. tracheostomy (tra″ke-ahs′to-me)
4. pneumonomelanosis (nu-mo″no-mel″uh-no′sis)
5. pleuropericarditis (pler″o-pĕr″ĭ-kar″di′tis)
6. bronchoscopy (brong-kos′kuh-pe)
7. dyspnea (disp′ne-uh)
8. laryngospasm (luh-ring′go-spazm)
9. hemothorax (he″mo-tho′raks)
10. otorhinolaryngology (o″to-ri″no-lār″in-gol′o-je)
11. hypoxemia (hi″pok-se′me-uh)
12. pulmonectomy (pul-mo-nek′to-me)
13. hypercapnea (hi-per-kap′ne-uh)
14. spirometry (spy-rom′uh-tre)
15. thoracentesis (thor′uh-sen-te′sis)

WEBSITE LINK

Visit this text's website at http://www.mhhe.com/longenbaker7 for additional quizzes, interactive learning exercises, and other study tools.

The Digestive System

In a time before modern, well-equipped laboratories existed, one man's misfortune turned into an unparalleled opportunity to study digestion. In 1822, Alexis St. Martin was injured by a gunshot wound. The bullet tore through his stomach, leaving a *fistula* (a hole in the body wall) that refused to heal. Dr. William Beaumont, one of St. Martin's physicians, took advantage of his patient's situation to study the stomach's digestive processes. After tying various forms of food to a string, Beaumont first pushed the food through the fistula into St. Martin's stomach. Next, Beaumont retrieved the food at specified intervals by pulling on the string. He then measured the remaining food morsel's mass and volume to track its digestion rate, and sampled the liquid in the stomach to analyze its composition. Much of the early knowledge regarding the stomach resulted from this unusual research. You can find out more about the stomach's role in digestion on page 335.

Chapter Outline & Learning Objectives

After you have studied this chapter, you should be able to:

15.1 Anatomy of the Digestive System

The organs of the digestive system are located within a tube called the **alimentary canal,** or **gastrointestinal tract.** The tube begins with the mouth and ends with the **anus** (Fig. 15.1). Although the term *digestion*, strictly speaking, means the breakdown of food by enzymatic action, we'll expand the definition to include both the physical and chemical processes that reduce food to small, soluble molecules that can be absorbed into the bloodstream.

The functions of the digestive system are to:

1. ingest the food;
2. break food down into small molecules that can cross plasma membranes;
3. absorb these nutrient molecules;
4. eliminate nondigestible wastes.

The Mouth

The **mouth,** also termed the *oral cavity*, receives food and begins physical and chemical digestion. The oral cavity is bounded externally by the lips and cheeks. The space between the teeth, the lips, and cheeks is the **vestibule.** The structure of human teeth is discussed in the Medical Focus on p. 331.

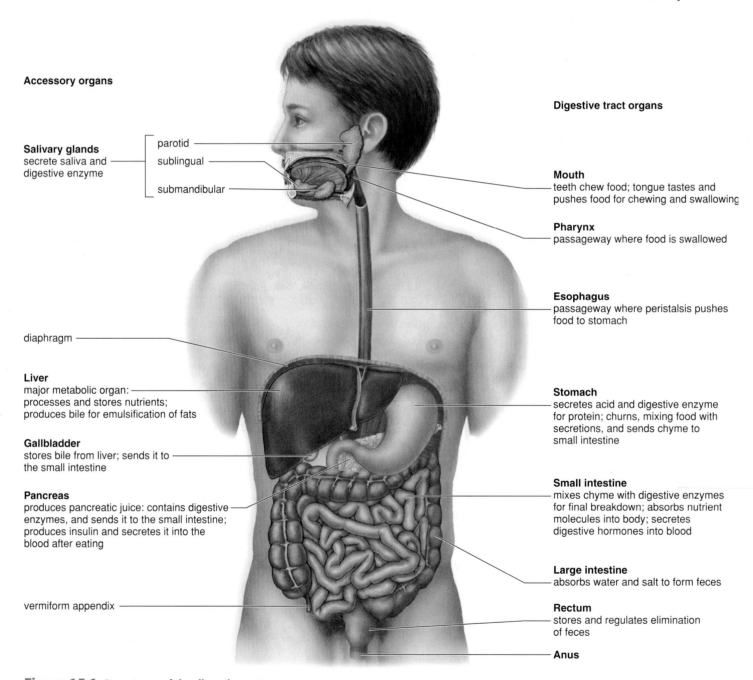

Accessory organs

Salivary glands
secrete saliva and
digestive enzyme

parotid

sublingual

submandibular

diaphragm

Liver
major metabolic organ:
processes and stores nutrients;
produces bile for emulsification of fats

Gallbladder
stores bile from liver; sends it to
the small intestine

Pancreas
produces pancreatic juice: contains digestive
enzymes, and sends it to the small intestine;
produces insulin and secretes it into the
blood after eating

vermiform appendix

Digestive tract organs

Mouth
teeth chew food; tongue tastes and
pushes food for chewing and swallowing

Pharynx
passageway where food is swallowed

Esophagus
passageway where peristalsis pushes
food to stomach

Stomach
secretes acid and digestive enzyme
for protein; churns, mixing food with
secretions, and sends chyme to
small intestine

Small intestine
mixes chyme with digestive enzymes
for final breakdown; absorbs nutrient
molecules into body; secretes
digestive hormones into blood

Large intestine
absorbs water and salt to form feces

Rectum
stores and regulates elimination
of feces

Anus

Figure 15.1 **Structures of the digestive system.**

The tongue is composed of skeletal muscle whose contraction changes the shape of the tongue. Muscles exterior to the tongue cause it to move about. Rough projections on the tongue, called *papillae*, help it handle food and also contain the sensory receptors called taste buds. A fold of mucous membrane, called the lingual frenulum, on the underside of the tongue attaches it to the floor of the mouth. If the frenulum is too short, the individual cannot speak clearly and is said to be tongue-tied. Posteriorly, the tongue is anchored to the hyoid bone.

MEDICAL FOCUS

Human Teeth

During the first two years of life, the 20 deciduous, or baby, teeth appear. Eventually, the deciduous teeth are replaced by the adult teeth. Normally, adults have 32 teeth (Fig. 15Aa). The maxilla (upper jaw bone) and the mandible (lower jaw bone) each contain teeth of four different types. Anteriorly, four chisel-shaped **incisors** function for biting. Flanking them on either side are two pointed canine teeth, or **cuspids**, which help tear food. Laterally, four flat premolar teeth, or **bicuspids**, are used for grinding food. The most posterior are the six **molar** teeth, designed for crushing and grinding food. The last molars, called the wisdom teeth, may fail to come in, or if they do, they may grow in crooked and be useless. Frequently, wisdom teeth are extracted.

Each tooth (Fig. 15Ab) has a crown and a root. The crown has a layer of enamel, an extremely hard outer covering of calcium compounds; dentin, a thick layer of bonelike material; and an inner pulp, which contains the nerves and blood vessels. Dentin and pulp are also in the root. **Caries** (tooth decay; commonly called *cavities*) occur when bacteria within the mouth break down sugar and give off acids that corrode the teeth. Once these acids dissolve the enamel and dentin, the pulp is compromised, triggering a toothache. Fluoride treatments, particularly in children, can make the enamel stronger and more resistant to decay.

Gum disease is more likely as we age. One example is inflammation of the gums, called **gingivitis**, that may spread to the periodontal membrane lining the tooth socket (Fig. 15Ab). When this occurs, the individual develops **periodontitis**, characterized by loss of bone and loosening of the teeth. Brushing and flossing your teeth after every meal cleans the teeth and stimulates the gums, preventing these conditions. Make sure you brush gently and away from the gumline, to keep delicate gum tissue from being destroyed. Taking good care of your teeth and gums is also important in avoiding atherosclerosis (see Chapter 12).

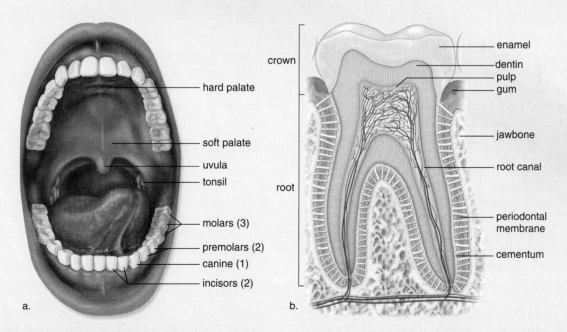

Figure 15A Human teeth. a. The chisel-shaped incisors bite; the pointed canines tear; the fairly flat premolars grind; and the flattened molars crush food. The last molar, called a wisdom tooth, may fail to erupt, or if it does, it is sometimes crooked and useless. Often dentists recommend the extraction of the wisdom teeth. **b.** Longitudinal section of a tooth. The crown is the portion that projects above the gumline and can be replaced by a dentist if damaged. When a "root canal" is done, the nerves are removed. When the periodontal membrane is inflamed, the tooth can loosen.

The mouth has a roof that separates it from the nasal cavities. The roof has two parts: an anterior (toward the front) **hard palate** and a posterior (toward the back) **soft palate** (see Fig. 15.2). The hard palate is formed by the maxilla and palatine bones; the soft palate is formed by muscle and glandular tissue. The soft palate ends in a finger-shaped projection called the **uvula.**

Three pairs of **salivary glands** send saliva by way of ducts to the mouth. The parotid glands lie anterior and somewhat inferior to the ears between the cheek and the masseter muscle. They have ducts that open on the inner surface of the cheek at the location of the second upper molar. The parotid glands swell when a person has the mumps, a disease caused by a viral infection. The sublingual glands are located beneath the tongue, and the submandibular glands are in the floor of the mouth on the inside surface of the lower jaw. The ducts from the sublingual and submandibular glands open under the tongue. You can locate the openings for the salivary glands if you use your tongue to feel for small flaps on the inside of your cheek and under your tongue. Saliva is a solution of mucus and water. It also contains bicarbonate and an enzyme called **salivary amylase,** which begins the process of digesting carbohydrate. Saliva moistens food and prepares it for swallowing. Saliva contains antibacterial lysozyme, as well as secretory antibodies, which help to protect the body. By constantly bathing the teeth, tongue, and oral mucous membrane, saliva removes microbes. Swallowed microbes can be destroyed by stomach acid and enzymes.

The Pharynx

Table 15.1 traces the path of food. From the mouth, food passes through the pharynx and esophagus to the stomach, small intestine, and large intestine. The food passage and the air passage cross in the **pharynx** because the trachea is anterior to the esophagus, a long muscular tube that takes food to the stomach (Fig. 15.2).

The pharynx has three parts: (1) The **nasopharynx,** posterior to the nasal cavity, serves as a passageway for air; (2) the **oropharynx,** posterior to the soft palate, is a passageway for both air and food; and (3) the **laryngopharynx,** just superior to the esophagus, is a passageway for food entering the esophagus.

The tonsils are embedded in the mucous membrane of the tongue and pharynx. The **palatine tonsils** are on either side of the tongue close to the soft palate, and the **lingual tonsils** sit at the base of the tongue. Together, they help to protect the body from infection caused by ingested and/or inhaled

TABLE 15.1 Path of Food in the Digestive Process

Organ	Function of Organ	Anatomic Features	Function of Anatomic Features
Oral cavity	Mechanical breakdown of food; swallowing; starch digestion	Teeth Tongue	Biting, tearing, chewing food Forms bolus; swallowing
Pharynx	Transport of food to esophagus	Soft palate and uvula Epiglottis	Prevents food from entering nasal cavity Prevents food from entering glottis
Esophagus	Transport of food to stomach	Four-layer construction	Muscularis layer carries out peristalsis
Stomach	Food storage; antibacterial; begins protein digestion; intrinsic factor production; slow release of chyme to small intestine	Rugae Parietal cells of gastric glands Chief cells of gastric glands Parietal cells of gastric glands Pyloric sphincter	Expand as stomach fills Produce hydrochloric acid Produce pepsin Produce intrinsic factor Controls release of chyme
Small intestine	Completes digestion of food; absorption of nutrients	Brush border enzymes Intestinal villi; microvilli	Complete digestive process Absorb nutrients
Large intestine	Absorption of water; storage of indigestible remains; elimination of fecal material	Epithelial cells Haustra Teniae coli and muscularis	Water absorption by osmosis Pouch-like structures store material Move feces toward anus

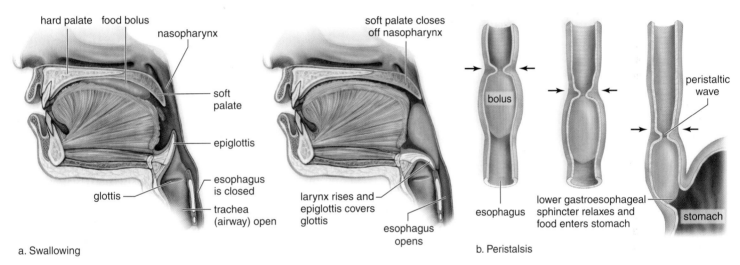

a. Swallowing

b. Peristalsis

Figure 15.2 Swallowing. a. When food is swallowed, the soft palate closes off the nasopharynx, and the epiglottis covers the glottis, forcing the bolus to pass down the esophagus. Therefore, a person does not breathe while swallowing. **b.** Peristalsis moves food through the gastroesophageal sphincter into the stomach.

microbes. The single **pharyngeal tonsil** (sometimes called the **adenoids**) sits in the posterior nasopharynx. As mentioned in Chapter 14, this tonsillar tissue defends against microbes in inhaled air. A person who has inflamed tonsils has tonsillitis. If the tonsillitis keeps recurring, the tonsils may be surgically removed (called a tonsillectomy).

Swallowing

During swallowing, food normally enters the esophagus because other possible avenues are blocked. Swallowing has a voluntary phase—from day-to-day living, you know that you can swallow voluntarily (try it). However, once food or drink is pushed back to the oropharynx, swallowing then becomes a reflex action performed automatically (without our willing it). When we swallow, the soft palate moves back to close off

the **nasopharynx**, and the trachea moves up under the epiglottis so that food is less likely to enter it. (We don't breathe when we swallow.) The tongue presses against the soft palate, sealing off the oral cavity, and the esophagus opens to receive a food **bolus** (Fig. 15.2). *Bolus* is the technical term for a morsel of chewed and swallowed food or drink.

Unfortunately, we've all had the unpleasant experience of having food "go the wrong way." The wrong way may be either into the nasal cavities or into the trachea. If it is the latter, coughing will most likely force the food up out of the trachea and into the pharynx again.

The Wall of the Digestive Tract

The entire digestive tract, from the esophagus to the rectum, is collectively referred to by the technical term *gut*. The gut is a

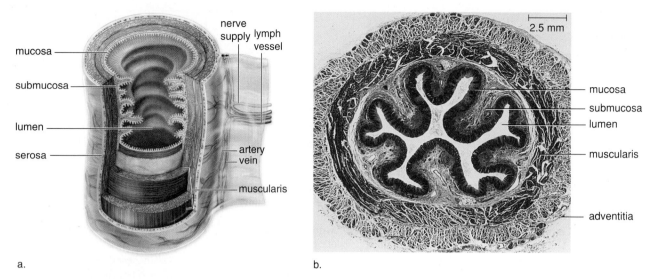

Figure 15.3 Wall of the alimentary canal. a. Several different types of tissues are found in the wall of the alimentary canal. Note the placement of circular muscle inside longitudinal muscle. **b.** Micrograph of the wall of the esophagus.

continuous tube composed of four layers (Fig. 15.3). From deepest to most superficial, the layers are:

Mucosa (mucous membrane layer) A layer of epithelium supported by connective tissue and smooth muscle lines the **lumen** (central cavity). This layer contains glandular epithelial cells that secrete digestive enzymes and goblet cells that secrete mucus.

Submucosa (submucosal layer) A broad band of loose connective tissue that contains blood vessels, lymphatic vessels, and nerves lies beneath the mucosa. The submucosa joins the mucosa to the muscularis layer. Lymph nodules, called Peyer patches, are scattered throughout the submucosa, of the small intestine. Like the tonsils, they help protect us from disease.

Muscularis (smooth muscle layer) Two layers of smooth muscle make up this section. The inner, circular layer encircles the gut; the outer, longitudinal layer lies in the same direction as the gut. The stomach also has oblique muscles.

Serosa (serous membrane layer) Most of the alimentary canal has a serosa, a very thin, outermost layer of squamous epithelium supported by connective tissue. The serosa secretes a serous fluid that keeps the outer surface of the intestines moist so that the organs of the abdominal cavity slide against one another. The esophagus has an outer layer composed only of loose connective tissue called the *adventitia*.

The structure of the digestive tract (a continuous tube from mouth to anus) allows study by a fascinating new device. The Medical Focus reading on page 339 describes a miniature camera that passes through the digestive system while photographing the internal organs.

The Esophagus

The **esophagus** is a muscular tube that passes from the pharynx through the thoracic cavity and diaphragm into the abdominal cavity where it joins the stomach. The esophagus is ordinarily collapsed but it opens and receives the bolus when swallowing occurs.

A rhythmic contraction called **peristalsis** pushes the food along the alimentary canal. In peristalsis, short segments of smooth muscle built into the wall of the digestive tract alternately contract and then relax. Peristalsis constantly moves food forward through the digestive tract. (As an analogy, think of squeezing a toothpaste tube, starting at the bottom, then continuing upward along the entire length of the tube. If done correctly, the toothpaste moves forward until the tube is empty.) Peristalsis begins in the esophagus and continues in all the organs of the alimentary canal. Occasionally, peristalsis begins even though there is no food in the esophagus. This produces the sensation of a lump in the throat.

The esophagus plays no role in the chemical digestion of food. Its sole purpose is to transport the food bolus from the mouth to the stomach. The entrance of the esophagus to the stomach is marked by a constriction, often called the esophageal **sphincter.** It's important to note that the esophageal sphincter isn't a true sphincter muscle. A true sphincter muscle (such as the pyloric sphincter, discussed with the stomach) is an actual ring of muscle that closes a tube when it contracts and opens the tube when it relaxes. Although not an actual ring of muscle, the esophageal sphincter functions as one because it allows the bolus to pass into the stomach when relaxed (Fig. 15.2). Likewise, the sphincter prevents the stomach's acidic contents from backing up into the esophagus when contracted.

Heartburn, which feels like a burning pain rising up into the throat, occurs during reflux when some of the stomach

esophagus

lower gastroesophageal sphincter

cardiac region

pylorus

pyloric sphincter

fundic region

body of stomach

muscularis layer has three layers of muscle

mucosa layer has rugae

a.

duodenum

gastric pit

gastric gland

parietal cells

chief cells

b. Gastric glands

gastric pit

gastric gland

c.

20 μm

Figure 15.4 Anatomy and histology of the stomach. a. The stomach has a thick wall with deep folds, rugae, that allow it to expand and fill with food. **b.** The stomach lining has gastric glands, which secrete mucus and protein-digesting gastric juice. **c.** Photomicrograph of the stomach lining.

contents escape into the esophagus. When vomiting occurs, a contraction of the abdominal muscles and diaphragm propels the contents of the stomach upward through the esophagus.

The Stomach

The **stomach** (Fig. 15.4) is a thick-walled, J-shaped organ that lies on the left side of the abdominal cavity inferior to the diaphragm and posterior to the liver. The stomach is continuous with the esophagus above and the duodenum of the small intestine below.

The length of the stomach remains at about 25 cm (10 in.) regardless of the amount of food it holds, but the diameter varies, depending on how full it is. As the stomach expands, deep folds in its wall, called **rugae**, stretch out and gradually disappear. When full, the stomach can hold about 4 liters (1 gallon). The stomach receives food from the esophagus, stores food, liquifies food by mixing food with its juices (thereby starting the digestion of proteins), and moves food into the small intestine.

Regions of the Stomach

The stomach has four regions. The **cardiac region**, which is near the heart, surrounds the lower esophageal sphincter

where food enters the stomach. The **fundic region**, which holds food temporarily, is an expanded portion superior to the cardiac region. The **body region**, which comes next, is the main part. The **pyloric region** narrows to become the pyloric canal leading to the pyloric sphincter. Food passes through this sphincter and enters the duodenum, the first part of the small intestine.

Digestive Functions of the Stomach

The stomach acts on food both chemically and physically. Its wall contains three muscle layers: One layer is longitudinal, another is circular, and the third is obliquely arranged. This muscular wall not only moves the food along, but it also churns, mixing the food with gastric juice and breaking it down to small pieces.

The term *gastric* always refers to the stomach. The columnar epithelial lining of the stomach has millions of **gastric pits,** which lead into **gastric glands** (Fig. 15.4*a, b*). Gastric glands contain four types of secretory cells: *chief cells, parietal cells, enteroendocrine cells,* and *mucous cells.* The gastric glands produce **gastric juice,** a watery solution that contains pepsinogen, hydrochloric acid (HCl), intrinsic factor, and mucus. Chief cells secrete pepsinogen, which becomes the protein-digesting enzyme **pepsin** when exposed to hydrochloric acid. Parietal cells produce the hydrochloric acid. The HCl causes the gastric juice in the stomach to have a high acidity with a pH of about 2; this is beneficial because it kills most of the bacteria present in food. Although HCl does not digest food, it does break down the connective tissue of meat and activates pepsin.

Parietal cells also produce **intrinsic factor,** a glycoprotein that binds to vitamin B$_{12}$, preventing this vitamin from being destroyed in the harsh, acidic environment of the stomach. If the stomach fails to produce intrinsic factor, or if the diet is deficient in vitamin B$_{12}$, a serious disorder called *pernicious anemia* will result. Lacking the vitamin, red blood cells will fail to develop.

Enteroendocrine cells of the gastric glands produce the hormone **gastrin.** This hormone enters stomach blood vessels and is circulated throughout the stomach. Gastrin regulates muscular contraction and secretion by the stomach.

The wall of the stomach is protected by the thick layer of mucus secreted by the *mucous cells.* If, by chance, HCl penetrates this mucus, the wall can begin to break down, and an ulcer results. An **ulcer** is an open sore in the wall caused by the gradual disintegration of tissue. It now appears that most ulcers are due to a bacterial infection *(Helicobacter pylori)* that impairs the ability of mucous cells to produce protective mucus.

Alcohol and water are absorbed in the stomach, but food substances are not. Normally, the stomach empties in about 2–6 hours. When food leaves the stomach, it is a thick, soupy liquid called **chyme.** Chyme enters the small intestine in squirts by way of the pyloric sphincter, dilating and opening, then contracting and closing repeatedly.

Peritoneum

The abdominal wall and the organs of the abdomen are covered by **peritoneum,** a serous membrane (Fig. 15.5). The portion of the peritoneum that lines the wall is called the parietal peritoneum. The portion that covers the organs is called the *visceral peritoneum*. In between the organs, the visceral peritoneum is a double-layered **mesentery** that supports the visceral organs, including the blood vessels, nerves, and lymphatic vessels.

Some portions of the mesentery have specific names. The **lesser omentum** is mesentery that runs between the stomach and the liver. The **greater omentum** is indeed "greater." It hangs down in front of the intestines like a large, double-layered apron. The greater omentum has several functions: It contains fat that cushions and insulates the abdominal cavity; it contains macrophages that can take up and rid the body of pathogens; and it can wall off portions of the alimentary wall that may be infected, keeping the infection from spreading to other parts of the peritoneal cavity.

The Small Intestine

The **small intestine** extends from the pyloric valve of the stomach to the *ileocecal valve* where it joins the large intestine. It is named for its small diameter (compared to that of the large intestine), but perhaps it should be called the long

intestine. The small intestine takes up a large portion of the abdominal cavity, averaging about 6 m (18 ft) in length.

All the contents of food—fats, proteins, and carbohydrates—are digested in the small intestine to molecules that can be absorbed. To this end, the small intestine receives secretions from the pancreas and liver and produces intestinal juices. Absorption of nutrients for the body's cells, such as amino acids and sugars, occurs in the small intestine. It also transports nondigestible remains to the large intestine.

Regions of the Small Intestine

The small intestine has the following regions (Fig. 15.6):

Duodenum The first 25 cm (10 in.) contain distinctive glands that secrete mucus and also receive the pancreatic secretions and the bile from the liver through a common duct. Folds and villi (Fig. 15.6) are more numerous at the end than at the beginning.

Jejunum The next 1 m (3 ft) contains folds and villi, more at the beginning than at the end.

Ileum The last 2 m (6–7 ft) contain fewer folds and villi than the jejunum. The ileum wall contains Peyer patches, aggregates of lymph nodules mentioned in Chapter 13.

Wall of the Small Intestine

It's been suggested that the surface area of the small intestine is approximately that of a tennis court. Three features

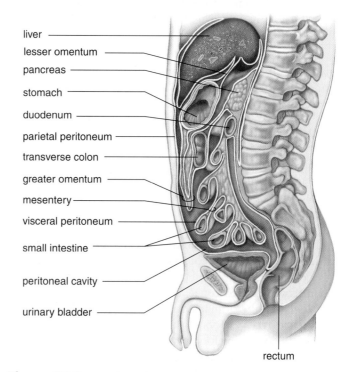

liver
lesser omentum
pancreas
stomach
duodenum
parietal peritoneum
transverse colon
greater omentum
mesentery
visceral peritoneum
small intestine
peritoneal cavity
urinary bladder
rectum

Figure 15.5 Function of mesentery. Mesentery formed by two layers of the peritoneal membrane supports the abdominal viscera. Deep folds of the peritoneal membrane, called the greater omentum, cover these organs anteriorly.

contribute to increasing its surface area: circular folds, villi, and microvilli (Fig. 15.7). The **circular folds** are permanent transverse folds involving the mucosa and submucosa of the small intestine. The **villi** (sing., *villus*) are fingerlike projections of the mucosa into the lumen of the small intestine. The villi are so numerous and closely packed that they

give the wall a velvetlike appearance. A villus has an outer layer of columnar epithelial cells, and each of these cells has thousands of microscopic extensions called **microvilli.** Collectively, in electron micrographs, microvilli give the villi a fuzzy border known as a "brush border" (Fig. 15.7*d*). Because the microvilli bear the intestinal enzymes, these enzymes are called *brush-border enzymes.*

Functions of the Small Intestine The digestive process is brought to completion in the small intestine. Ducts from the liver, gallbladder, and pancreas join to form one duct that enters the duodenum (Fig. 15.8). The small intestine receives bile from the gallbladder and pancreatic juice from the pancreas via this duct. **Bile,** which is produced by the liver but stored in the gallbladder, emulsifies fat—emulsification causes fat droplets to disperse in water. The intestine has a slightly basic pH because pancreatic juice contains sodium bicarbonate ($NaHCO_3$), which neutralizes the acidic chyme from the stomach. The enzymes in pancreatic juice and the enzymes produced by the intestinal wall complete the process of food digestion. The functions of the liver, gallbladder, and pancreas will be detailed in section 15.2.

The other primary function of the small intestine is *absorption of nutrients.* The tremendous increase in surface area created by the circular folds, villi, and microvilli makes this an efficient process—the greater the surface area, the greater the volume of intake in a given unit of time. Also, a villus contains a generous supply of blood capillaries and a small lymphatic capillary, called a **lacteal** (Fig. 15.7*c, d*). The lymphatic system, as you know, is an adjunct to the cardiovascular system; its vessels carry a fluid called *lymph* to the cardiovascular veins. Sugars (digested in part from carbohydrates) and amino

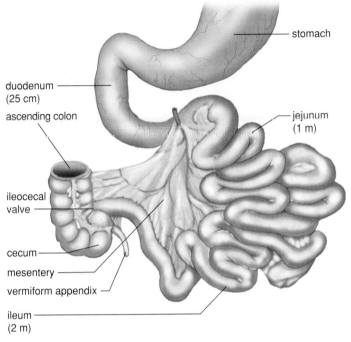

Figure 15.6 **Regions of the small intestine.** The duodenum is attached to the stomach. The jejunum leads to the ileum, which is attached to the large intestine.

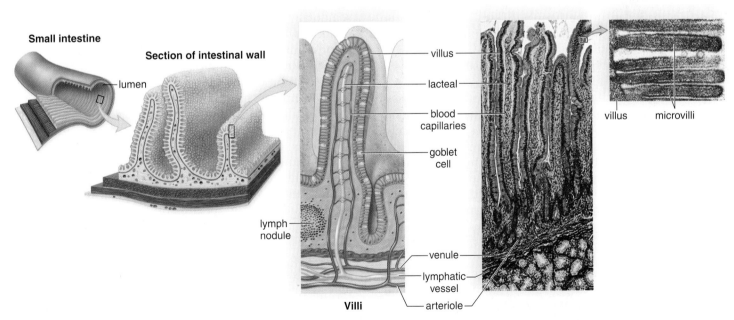

Figure 15.7 **Anatomy of the small intestine.** The wall of the small intestine has folds that bear fingerlike projections called *villi.* Villi in turn have projections called *microvilli.* The products of digestion are absorbed by microvilli and they enter the blood capillaries and the lacteals of the villi.

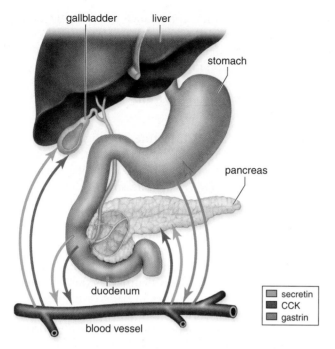

gallbladder   liver

stomach

pancreas

duodenum

blood vessel

- ☐ secretin
- ☐ CCK
- ☐ gastrin

Figure 15.8 Hormonal control of digestive gland secretions.
Gastrin (blue), produced by the lower part of the stomach, enters the bloodstream and thereafter stimulates the upper part of the stomach to produce more gastric juice. Secretin (green) and CCK (purple), produced by the duodenal wall, stimulate the pancreas to secrete its juice, the liver to produce bile, and the gallbladder to release stored bile.

acids (digested from protein) enter the blood capillaries of a villus. Glycerol and fatty acids (digested from fats) enter the epithelial cells of the villi, and within these cells are joined and packaged as lipoprotein droplets, which enter a lacteal. Thus, lipids first enter the lymphatic capillaries and are ultimately transported in the lymph to the lymphatic ducts. Recall that the lymphatic ducts return lymph to the bloodstream by way of the subclavian veins (p. 282). After nutrients are absorbed into the blood, they are eventually carried to all the cells of the body by the bloodstream.

As we noted previously, a third function of the small intestine is movement of nondigested remains to the large intestine. The wall of the small intestine has two types of movements: segmentation and peristalsis. Segmentation refers to localized contractions and constrictions that mix the chyme thoroughly with digestive juices. Segmentation also encourages absorption of nutrients into the bloodstream or lymph. Peristalsis then moves nondigested remains toward the large intestine.

Regulation of Contraction and Secretion in the Digestive Tract

A combination of three control mechanisms cooperates to regulate the muscular contractions needed to move food throughout the gut, as well as the secretion of digestive juices needed for breaking down the food as it travels along. The central nervous system allows us to respond to hunger signals (and for many, stimulating eating behavior when we're

stressed). The **enteric nervous system** is a nerve network built directly into the wall of the gut, allowing it to control its own movements and secretion. Hormones produced by cells in the stomach and small intestine participate as well.

Three phases occur as the control mechanisms become active. The first, the **cephalic phase** (remember from Chapter 1 that *cephalic* means "head"), begins when we look at, think about, smell, or taste food. Sensory signals are relayed to the hypothalamus, the brain center that stimulates hunger. In turn, the hypothalamus triggers autonomic nerves belonging to the parasympathetic nervous system to automatically stimulate secretion and muscle contraction. Filling the stomach helps to trigger the next regulatory mechanism: the **gastric phase** of control. The presence of food (particularly food high in amino acids) that stretches the stomach causes stimulation of the enteric nerve network. At the same time, the stomach hormone gastrin is also released. Gastrin is transported in the bloodstream to nearby smooth muscle and gastric glands. Gastric contraction and secretion by gastric glands are controlled internally by cooperation of these two mechanisms. Both muscle contraction and gastric gland secretion increase in response.

A similar nerve network and hormone set cooperate with parasympathetic nerve activity to trigger the third, or **intestinal phase** of secretion. When the duodenum is stretched by digested food that's rich in fatty acids and carbohydrate, the enteric nervous system responds to stimulate intestinal secretion and contraction. Simultaneously, cells of the duodenal wall produce three hormones that are of particular interest: **secretin, gastric inhibitory polypeptide** (GIP; sometimes referred to as glucose-dependent insulinotropic polypeptide), and **cholecystokinin** (CCK). Acid, especially hydrochloric acid (HCl) present in chyme, is a potent stimulator for secretin release, whereas partially digested protein and fat stimulate the release of CCK. Secretin stimulates secretion of bicarbonate solution by the pancreas, and cholecystokinin stimulates pancreatic enzyme secretion. Thus, when these hormones enter the bloodstream, the pancreas increases its output of pancreatic juice. Pancreatic juice buffers the stomach's acid chyme and helps to digest food. The liver and gallbladder also respond: The liver increases its production of bile in response to secretin and the gallbladder contracts to release stored bile in response to CCK. The third hormone produced by the duodenal wall, GIP, works opposite to gastrin. It inhibits gastric gland secretion and slows the contractions of the stomach. In this way, GIP prevents the stomach from emptying too quickly. Figure 15.8 summarizes the actions of gastrin, secretin, and CCK. As the figure illustrates, these digestive hormones must first enter the bloodstream to be carried to their target cells (just as all other endocrine hormones are transported in the blood).

It's important to note that these three regulatory methods—cephalic, gastric, and intestinal phases—occur in sequence only as your meal begins. As you continue eating, digesting, and absorbing your food, reflexes from each phase may be occurring at the same time.

Two additional hormones contribute to hunger (the desire for food) and appetite (the desire for specific foods that

MEDICAL FOCUS

Swallowing a Camera?

"I really just don't even want to eat anymore. My gut just hurts so badly, and then I can't even control the diarrhea. I moved out of the dorm last year because I was so embarrassed to always be in the bathroom." The patient, a 20-year-old college student, is anemic, pale, and underweight. She identifies the lower right abdominal quadrant as the source of her pain, which has persisted, off and on, for more than six months.

To diagnose digestive system problems, gastroenterologists (physicians who specialize in treatment of the digestive tract) traditionally use several techniques: endoscopy, colonoscopy, and barium X ray. In endoscopy, a long, flexible tube equipped with a video camera is used to examine the upper GI tract: the esophagus, stomach, and upper small intestine. Colonoscopy is a similar technique, except that the video camera tube is inserted through the anus to examine the lower GI tract: the rectum and colon. To examine the ileum (the intestinal portion found between the upper and lower GI tracts), patients drink a barium solution that coats the entire GI tract and makes the digestive organs visible, after which a series of X rays are taken.

A newer technique involves placing a video camera directly into the digestive tract, in the form of a camera pill. The patient swallows the pill (which is about the size of a large vitamin), then wears a receiver fitted into a belt pack for the next 8 to 12 hours. Images are transmitted by the camera as the pill passes through the patient's body. Once the camera has been passed, physicians study the video images from the belt pack receiver. In clinical studies, these video images have been extremely useful in diagnosis of disorders affecting the ileum: intestinal tumors, celiac disease (an allergic condition affecting the small intestine), and chronic inflammatory ileitis, or Crohn's disease.

In this young woman's case, video images taken by the camera pill reveal that she does have Crohn's disease, an autoimmune disorder. Although there is no cure for Crohn's disease, its symptoms can be controlled. Anti-inflammatory and anti-diarrheal medications can help to relieve the irritation, pain, blood loss, and diarrhea caused by the disease. Anti-cancer drugs have also proven effective to control Crohn's symptoms. Vitamins and nutritional supplements will help the young woman to regain lost weight and restore her red blood cell count. Over time, Crohn's disease symptoms usually resolve, but sufferers must learn to expect flare-ups.

The Crohn's and Colitis Foundation of America, www.ccfa.org, is a support network for people with Crohn's disease and similar disorders. It offers tips for managing this chronic illness.

---

one enjoys). These are **leptin,** produced by fat cells (and briefly discussed in Chapter 10), and **ghrelin,** produced by the stomach. Both hormones act on the hypothalamus. Leptin stimulates a feeling of fullness and contentment after eating and causes us to stop eating. In contrast, ghrelin triggers a hunger sensation and causes us to seek food. Scientists are actively studying both hormones to learn about their roles in the development of obesity.

The Large Intestine

The **large intestine,** which includes the cecum, the colon, the rectum, and the anal canal (Fig. 15.9), is larger in diameter than the small intestine (6.5 cm compared to 2.5 cm), but it is shorter in length (see Fig. 15.1). The large intestine absorbs water, salts, and some vitamins. It also stores indigestible material until it is eliminated at the anus.

The **cecum,** which lies below the junction with the small intestine, is the blind end of the large intestine. The cecum has a small projection called the **vermiform appendix** (*vermiform* means wormlike). In humans, the appendix also may play a role in fighting infections. This organ is subject to inflammation, a condition called *appendicitis*. If inflamed, the appendix should be removed before the fluid content rises to the point that the appendix bursts. A ruptured (burst) appendix will cause **peritonitis,** a generalized infection of the lining of the abdominal cavity. Peritonitis can lead to death.

The **colon** has four portions: the **ascending colon,** which goes up the right side of the body to the level of the liver; the **transverse colon,** which crosses the abdominal cavity just below the liver and the stomach; the **descending colon,** which passes down the left side of the body; and the **sigmoid colon,** which enters the **rectum,** the last 20 cm of the large intestine. The colon is characterized by two distinct anatomic features: pouches called **haustra** along its length, and a band of muscle called **taenia coli** on its surface (see Fig. 15.9). Haustra expand to store material in the colon. The taenia coli and the muscularis layer of the colon cause peristaltic movement called a *mass movement* several times daily.

Feces are 75% water and 25% solids. A breakdown product of bilirubin (see page 236) and the presence of oxidized iron cause the brown color of feces. Bacteria, shed intestinal cells, undigested cellulose fiber, and other indigestible materials are in the solid portion. Colon bacteria use undigested cellulose as an energy source, producing fatty acids, B-complex vitamins, and most of the vitamin K needed by our bodies. Bacteria also release hydrogen gas and sulfur-containing compounds that contribute to human flatulence (gas).

Although most intestinal bacteria are harmless, many cause disease. To ensure water quality, public health departments

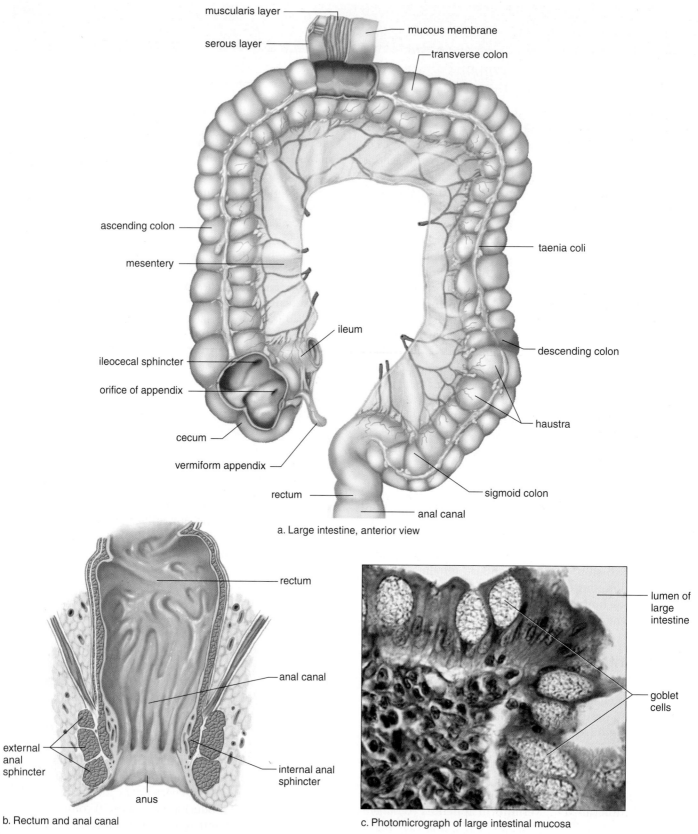

muscularis layer

serous layer

mucous membrane

transverse colon

ascending colon

mesentery

taenia coli

ileum

ileocecal sphincter

orifice of appendix

descending colon

cecum

haustra

vermiform appendix

rectum

sigmoid colon

anal canal

a. Large intestine, anterior view

rectum

anal canal

external anal sphincter

internal anal sphincter

anus

b. Rectum and anal canal

lumen of large intestine

goblet cells

c. Photomicrograph of large intestinal mucosa

Figure 15.9 The large intestine. a. The colon has four regions: the ascending colon, the transverse colon, the descending colon, and the sigmoid colon. **b.** The rectum and anal canal are at the distal end of the alimentary canal. **c.** The intestinal mucosa has many goblet cells.

constantly monitor water supplies for coliform (nonpathogenic intestinal) bacterial count. A high count indicates that a significant amount of fecal material has entered the water, and serves as a warning that disease-causing bacteria might also be present.

Water is removed from the nondigestible intestinal contents entering the ascending colon from the small intestine. Then, fecal material begins to form in the transverse colon. From there, feces are propelled down the descending colon toward the rectum by periodic, firm peristaltic contractions (see page 334 to review peristalsis). When sufficient feces are in the rectum, a reflex defecatory urge is felt. The involuntary defecation reflex contracts the rectal muscle and relaxes the internal anal sphincter, a ring of muscle that closes off the rectum. Then, feces move toward the anus. A pushing motion, along with the relaxation of the external anal sphincter, propels feces from the body. Because these activities are under voluntary control, it is possible to control defecation.

Defecation normally occurs from three times a week to three times a day; therefore some variation in occurrence is nothing to worry about. However, if the frequency of defecation declines and if defecation becomes difficult, constipation is present. When a person is constipated, the feces are dry and hard. Temporary **constipation** due to traveling, pregnancy, or medication can be relieved by increasing the intake of dietary fiber, drinking plenty of water, and getting moderate amounts of exercise. The use of oral laxatives (agents that aid emptying the colon) can provide temporary relief from constipation. However, laxative use must be limited because the person can become dependent on them for defecation. Very rarely, medical attention is required for constipation. Chronic constipation is associated with the development of **hemorrhoids**—enlarged and inflamed blood vessels at the anus.

A second complaint associated with the large intestine is **diarrhea**. The major causes of diarrhea are infection of the lower intestinal tract and excessive nervous stimulation. In the case of infection, such as food poisoning caused by eating contaminated food, the intestinal wall becomes irritated and peristalsis increases. Water is not absorbed and the diarrhea that results rids the body of the infectious organisms. In nervous diarrhea, the nervous system stimulates the intestinal wall excessively and diarrhea results. Prolonged diarrhea can lead to dehydration because of water loss. In severe cases, disturbances in the heart's contraction may occur due to an imbalance of salts in the blood.

Other Disorders of the Large Intestine

The colon is subject to the development of **polyps,** small growths arising from the epithelial lining. Polyps, whether benign or cancerous, can be removed surgically along with a portion of the colon if necessary. If colon cancer is detected while still confined to a polyp, the expected outcome is a complete cure. If the last portion of the rectum and the anal canal must be removed, then the intestine is sometimes attached to the abdominal wall through a procedure known as a **colostomy,** and the digestive remains are collected in a plastic bag fastened around the opening. Recently, the use of metal staples has permitted surgeons to join the colon to a piece of rectum that formerly was considered too short.

Some investigators believe that dietary fat increases the likelihood of colon cancer because dietary fat causes an increase in bile secretion. Researchers believe that intestinal bacteria convert bile salts to substances that promote the development of cancer. On the other hand, fiber in the diet seems to inhibit the development of colon cancer. Dietary fiber absorbs water and adds bulk, thereby diluting the concentration of bile salts and facilitating the movement of substances through the intestine. Regular elimination reduces the time that the colon wall is exposed to any cancer-promoting agents in feces.

Diverticulosis is characterized by the presence of diverticula, or saclike pouches, in the colon. Ordinarily, these pouches cause no problems. But about 15% of people with diverticulosis develop an inflammation known as diverticulitis. The symptoms of diverticulitis are similar to those of appendicitis—cramps or steady pain with local tenderness. Fever, loss of appetite, nausea, and vomiting may also occur. Today, high-fiber diets are recommended to prevent the development of these conditions and of cancer of the colon.

Begin Thinking Clinically

As you know, the large intestine, or colon, is filled with bacteria. This population is referred to as the *normal flora* of the gut. Like all bacteria, they're sensitive to antibiotics. If the normal flora organisms are destroyed by antibiotics, what would be the effect on the body?

Answer and discussion in Appendix C, page 465.

Content CHECK-UP!

1. The correct order of layering in the digestive tract wall, from deep to superficial, is:
   a. mucosa, muscularis, submucosa, serosa.
   b. serosa, mucosa, submucosa, muscularis.
   c. mucosa, submucosa, muscularis, serosa.
   d. submucosa, mucosa, muscularis, serosa.

2. Which hormone is produced by the stomach and controls stomach contraction and secretion?
   a. pepsin
   b. gastrin
   c. secretin
   d. cholecystokinin (CCK)

3. Which nutrient is absorbed by lymphatic lacteals of the small intestine?
   a. amino acids from protein digestion
   b. fatty acids and glycerol from fat digestion
   c. sugars

Answers in Appendix C, page 465.

15.2 Accessory Organs of Digestion

The salivary glands and even the teeth are accessory organs of digestion that were discussed earlier (see pages 330–332). The pancreas, liver, and gallbladder are also accessory digestive organs. Figure 15.10 shows how the pancreatic duct from the pancreas and the common bile duct from the liver and gallbladder join before entering the duodenum.

The Pancreas

The **pancreas** lies deep in the abdominal cavity, behind the peritoneum, resting on the posterior abdominal wall. Its broad end, called the head, more than fills the loop formed by the duodenum, and its tail extends in the opposite direction (Fig. 15.10). The pancreas has both an endocrine and an exocrine function. As you know from Chapter 10, pancreatic islets (formerly called islets of Langerhans) secrete insulin and glucagon, hormones that help keep the blood glucose level within normal limits.

In this chapter, however, we are interested in the exocrine function of the pancreas. Most pancreatic cells, called *pancreatic acinar cells*, produce pancreatic juice, which is secreted into tiny tubes that unite, forming ever-larger ones. Finally, a pancreatic duct extends the length of the pancreas to the duodenum.

Pancreatic Juice

Pancreatic juice contains sodium bicarbonate ($NaHCO_3$) and digestive enzymes for all types of food. Sodium bicarbonate neutralizes the acidic chyme from the stomach. This is important because pancreatic enzymes require a slightly basic pH to function optimally. **Pancreatic amylase** digests starch, and lipase digests fat. There are three protein-digesting enzymes in pancreatic juice: trypsin, chymotrypsin, and carboxypeptidase. Pancreatic juice also contains two nucleases, which are enzymes that break down nucleic acid molecules into nucleotides. Whereas the majority of pancreatic enzymes are already functional when they reach the small intestine, the three protein-digesting pancreatic enzymes are secreted in an inactive form. Once inside the small intestine, trypsin is activated by an intestinal enzyme, **enterokinase.** In turn, trypsin goes on to activate chymotrypsin and carboxypeptidase so protein digestion in the small intestine can start. It's important for the pancreas to secrete nonfunctional protein enzymes—if the enzymes were active, they could digest the pancreas itself!

In the genetic disorder *cystic fibrosis*, the production of abnormally thick mucus blocks the pancreatic duct. The patient must take supplemental pancreatic enzymes by mouth for proper digestion to occur.

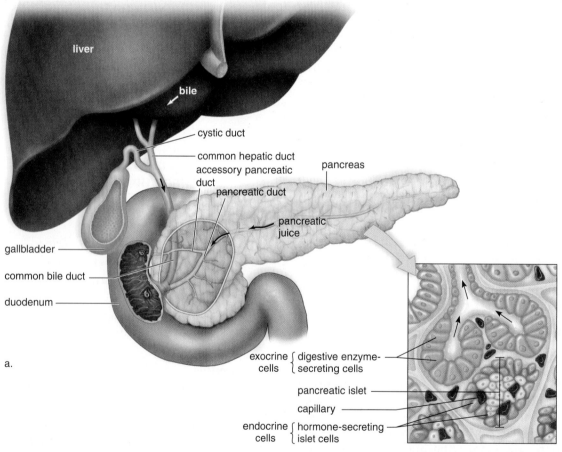

Figure 15.10 The pancreas. The pancreas is an exocrine gland when it secretes digestive enzymes into tubes that join to become the pancreatic duct. The pancreatic duct and the common bile duct empty into the duodenum of the small intestine. Pancreatic juice contains enzymes that digest all types of food: carbohydrates, fats, proteins, and nucleic acids.

The Liver

The **liver,** the largest organ in the body, lies mainly in the upper right section of the abdominal cavity, just inferior to the diaphragm (see Fig. 15.1).

Liver Structure

The liver has two main lobes, the right lobe and the smaller left lobe, separated by a ligament. Each lobe is divided into many **hepatic lobules** that serve as its structural and functional units (Fig. 15.11). A lobule consists of many hepatic cells arranged in longitudinal groups that radiate out from a central vein. Hepatic sinusoids separate the groups of cells from each other. Large, fixed phagocytic macrophages called *Kupffer cells* are attached to the lining of the hepatic sinusoids. They remove pathogens and debris that may have entered the hepatic portal vein at the small intestine.

Portal triads consisting of the following three structures are located between the lobules: a bile duct that takes bile away from the liver; a branch of the hepatic artery that brings O_2-rich blood to the liver; and a branch of the hepatic portal vein that transports nutrients from the intestines. The bile ducts merge to form the common hepatic duct.

The central veins from each of the lobules enter a hepatic vein. With the help of Figure 12.18, trace the path of blood from the intestines to the liver via the hepatic portal vein and from the liver to the inferior vena cava via the hepatic veins.

Liver Functions

As the blood from the hepatic portal vein passes through the liver, hepatic cells remove poisonous substances and detoxify them. The liver also removes and stores nutrients and works to keep the contents of the blood constant. It removes and stores iron and the fat-soluble vitamins A, D, E, and K; makes the plasma proteins from amino acids; and helps regulate the quantity of cholesterol in the blood.

The liver maintains the blood glucose level at about 100 mg/100 ml (0.1%), even though a person eats

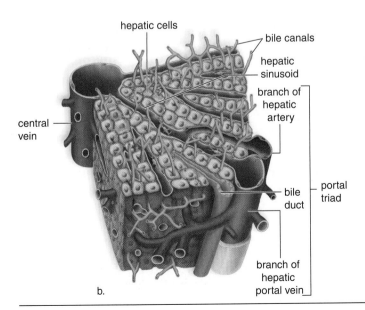

Figure 15.11 Macroscopic and microscopic anatomy of the liver. a. The liver has two lobes viewed anteriorly (*left*) and posteriorly from the visceral surface (*right*). **b.** Cross section of a hepatic lobule, illustrating microscopic structure.

intermittently. When insulin is present, any excess glucose in the blood is removed and stored by the liver as glycogen. Between meals, glycogen is broken down to glucose, which enters the hepatic veins. In this way, the blood glucose level remains constant.

If the supply of glycogen is depleted, the liver converts glycerol (from fats) and amino acids to glucose molecules. This process is called *gluconeogenesis*. To convert amino acids to glycerol for gluconeogenesis, the liver must first carry out *deamination*, a complex metabolic pathway. Through this process, amino groups are removed, converted to ammonia, and combined with carbon dioxide to form urea:

$$2\,NH_3 + CO_2 \longrightarrow H_2N - \overset{\displaystyle O}{\overset{\displaystyle \|}{C}} - NH_2$$

$$\text{ammonia} \quad \text{carbon dioxide} \qquad\qquad \text{urea}$$

Urea is the usual nitrogenous waste product from amino acid breakdown by the liver. Urea is normally excreted by the kidneys. However, if it accumulates excessively, it can be toxic to the body. It's important to drink plenty of water following a high-protein meal so that urea can be quickly and effectively eliminated.

The liver produces bile, which is stored in the gallbladder. Bile has a yellowish-green color because it contains the bile pigment **bilirubin,** which is derived from the breakdown of hemoglobin from red blood cells. Bilirubin from bile is excreted by the kidneys and accounts for urine's yellow color. Bile also contains **bile salts,** which are derived from cholesterol. Bile salts emulsify fat in the small intestine. When fat is emulsified, it disperses into droplets. Emulsification of fats provides a much larger surface area that can be more easily acted upon by pancreatic digestive enzymes.

Because the liver is the most important of the metabolic organs, it works closely with the cardiovascular system. A large amount of the body's blood constantly flows to the liver. Between meals, more than three-quarters of this supply reaches the liver by way of the hepatic portal vein, which drains the intestine. The remainder comes from the abdominal aorta, via the hepatic artery. Immediately after a meal, additional blood is diverted to the intestine to cope with the tasks of digestion and absorption, and the hepatic portal vein drains even more blood to the liver.

Altogether, the following are significant ways in which the liver helps maintain homeostasis:

1. Detoxifies blood by removing and metabolizing poisonous substances.
2. Stores iron (Fe^{2+}) and the fat-soluble vitamins A, D, E, and K.
3. Makes plasma proteins, such as albumins and fibrinogen, from amino acids.
4. Stores glucose as glycogen after a meal. Between meals, it breaks down glycogen to glucose to maintain the glucose concentration of blood. When blood glucose is depleted, the liver converts glycerol to glucose in gluconeogenesis.
5. Produces urea after breaking down amino acids.
6. Forms and secretes bile. Bile can be secreted directly into the duodenum or into the gallbladder for storage. Bile salts, formed from cholesterol, emulsify fats. Bilirubin, the bile pigment, is a product of the breakdown of hemoglobin from red blood cells.
7. Helps regulate the blood cholesterol level, converting some to bile salts.

Liver Disorders

When a person has a liver ailment, jaundice may occur. **Jaundice** is a yellowish tint to the whites of the eyes and also to the skin of light-pigmented persons. Bilirubin is deposited in the skin due to an abnormally large amount in the blood. In hemolytic jaundice, red blood cells have been broken down in abnormally large amounts; in obstructive jaundice, bile ducts are blocked, or liver cells are damaged.

Jaundice can also result from **hepatitis,** inflammation of the liver. Viral hepatitis occurs in several forms. Hepatitis A is usually acquired from sewage-contaminated drinking water. Hepatitis B, which is usually spread by sexual contact, can also be spread by blood transfusions or contaminated needles. The hepatitis B virus is more contagious than the AIDS virus, which is spread in the same way. Thankfully, however, a vaccine is now available for hepatitis B. Hepatitis C, which is usually acquired by contact with infected blood and for which there is no vaccine, can lead to chronic hepatitis, liver cancer, and death.

Cirrhosis is another chronic disease of the liver. First the organ becomes fatty, and then liver tissue is replaced by inactive fibrous scar tissue. Cirrhosis of the liver is often seen in alcoholics due to malnutrition and the excessive amounts of alcohol (a toxin) the liver is forced to break down.

Hepatitis and cirrhosis affect the entire liver and hinder its ability to repair itself. Therefore, they are life-threatening diseases. The liver has amazing regenerative powers and can recover if the rate of regeneration exceeds the rate of damage. During liver failure, however, there may not be enough time to let the liver heal itself. Liver transplantation is usually the preferred treatment for liver failure, but artificial livers have been developed and tried in a few cases. One type is a cartridge that contains liver cells. The patient's blood passes through the cellulose acetate tubing of the cartridge and is serviced in the same manner as with a normal liver. In the meantime, the patient's liver has a chance to recover.

The Gallbladder

The **gallbladder** is a pear-shaped, muscular sac located in a depression on the inferior surface of the liver (see Fig. 15.11a). About 1,000 ml of bile are produced by the liver each day, and any excess is stored in the gallbladder. Water is reabsorbed by the gallbladder so that bile becomes a thick, mucous-like material. When needed, bile leaves the gallbladder by way of the **cystic duct.** The cystic duct and the **common hepatic duct** join to form the common bile duct, which enters the duodenum.

Function of Bile Salts

Bile salts carry out emulsification; they break up masses of fat into droplets that can be acted on by enzymes that digest fat. Through their ability to make fats interact with water, they also enhance absorption of fatty acids, cholesterol, and the fat-soluble vitamins A, D, E, and K.

The cholesterol content of bile can come out of solution and form crystals. If the crystals grow in size, they form gallstones. The passage of the stones from the gallbladder may block the common bile duct and cause obstructive jaundice. Then the gallbladder may have to be removed.

Content **CHECK-UP!**

4. The protein-digesting enzyme produced by the pancreas is:
   a. pepsin.
   b. gastrin.
   c. amylase.
   d. trypsin.

5. Describe three functions of the liver.

6. From the following list of accessory organs of digestion and their function, choose the correct pair.
   a. pancreas → production of insulin and glucagon
   b. gallbladder → production of bile
   c. salivary glands → production of protein-digesting enzymes
   d. liver → production of nucleases

Answers in Appendix C, page 465.

15.3 Chemical Digestion

The digestive enzymes are **hydrolytic enzymes,** which break down substances by the introduction of water at specific bonds. Digestive enzymes have an optimum pH at which they work best.

In the mouth, saliva from the salivary glands has a neutral pH and contains salivary amylase, the first enzyme to act on carbohydrate:

$$\text{starch} + H_2O \xrightarrow{\text{salivary amylase}} \text{maltose}$$

Notice that the name of the enzyme is written above the arrow to indicate that it is not used up. In this case, the enzyme speeds the breakdown of starch to maltose, a disaccharide. Maltose molecules are too large to be absorbed by the alimentary canal; therefore, more digestion is required.

In the stomach, gastric juice secreted by gastric glands has a very low pH—about 2—because it contains hydrochloric acid (HCl). The precursor, pepsinogen, is converted to the enzyme pepsin when exposed to HCl. Pepsin acts on protein to produce peptides:

$$\text{protein} + H_2O \xrightarrow{\text{pepsin}} \text{peptides}$$

Peptides vary in length, but they are usually too large to be absorbed and must be broken down further.

In the small intestine, starch, proteins, nucleic acids, and fats are all enzymatically broken down. Pancreatic juice, which enters the duodenum, has a basic pH because it contains sodium bicarbonate ($NaHCO_3$). One pancreatic enzyme, pancreatic amylase, digests starch:

$$\text{starch} + H_2O \xrightarrow{\text{pancreatic amylase}} \text{maltose}$$

Recall that there are three pancreatic proteases, or enzymes that digest protein: trypsin, chymotrypsin, and carboxypeptidase. The following chemical equation summarizes their action:

$$\text{protein} + H_2O \xrightarrow{\text{pancreatic proteases}} \text{peptides and free amino acids}$$

Free amino acids are ready to be absorbed by the blood. Peptides are too large and must undergo further digestion in the small intestine. It is important to note that all three pancreatic proteases are secreted in an inactive form and cannot digest protein until they are inside the small intestine. This prevents them from accidentally digesting the pancreas itself.

Lipase, a third pancreatic enzyme, digests fat molecules in the fat droplets after they have been emulsified by bile salts:

$$\text{fat} \xrightarrow{\text{bile salts}} \text{fat droplets}$$

$$\text{fat droplets} + H_2O \xrightarrow{\text{lipase}} \text{glycerol} + \text{fatty acids}$$

As mentioned previously, glycerol and fatty acids enter the cells of the villi, and within these cells, they are rejoined and packaged as lipoprotein droplets before entering the lacteals.

Peptidases and **disaccharases (maltase, sucrase,** and **lactase**) are brush border enzymes produced by cells lining the small intestine. Intestinal peptidases complete the digestion of protein to amino acids. Intestinal disaccharases digest compound sugar molecules, called *disaccharides,* into single sugar molecules, or *monosaccharides.* Amino acids and monosaccharides are small molecules that are able to cross into the cells of the intestinal villi. Peptides, which result from the first step in protein digestion, are digested to amino acids by peptidases:

$$\text{peptides} + H_2O \xrightarrow{\text{peptidases}} \text{amino acids}$$

Maltose, a disaccharide that results from the first step in starch digestion, is digested to glucose by maltase:

$$\text{maltose} + H_2O \xrightarrow{\text{maltase}} \text{glucose} + \text{glucose}$$

Similarly, the enzyme sucrase digests table sugar (the disaccharide sucrose) and lactase digests milk sugar (the disaccharide lactose). The monosaccharides that result are absorbed by intestinal cells.

TABLE 15.2 Major Digestive Enzymes

Enzyme	Produced By	Site of Action	Optimum pH	Digestion Process
CARBOHYDRATE-DIGESTING ENZYMES				
Salivary amylase	Salivary glands	Mouth	Neutral	Starch + $H_2O \rightarrow$ maltose
Pancreatic amylase	Pancreas	Small intestine	Basic	Starch + $H_2O \rightarrow$ maltose
Maltase	Small intestine	Small intestine	Basic	Maltose $\rightarrow$ glucose + glucose
Sucrase	Small intestine	Small intestine	Basic	Sucrose $\rightarrow$ glucose + fructose
Lactase	Small intestine	Small intestine	Basic	Lactose $\rightarrow$ glucose + galactose
PROTEIN-DIGESTING ENZYMES				
Pepsin	Stomach	Stomach	Acidic	Protein + $H_2O \rightarrow$ peptides
Trypsin	Pancreas	Small intestine	Basic	Peptides + $H_2O \rightarrow$ smaller peptides + amino acids
Chymotrypsin	Pancreas	Small intestine	Basic	Peptides + $H_2O \rightarrow$ smaller peptides + amino acids
Carboxypeptidase	Pancreas	Small intestine	Basic	Peptides + $H_2O \rightarrow$ smaller peptides + amino acids
Peptidases	Small intestine	Small intestine	Basic	Small peptides $\rightarrow$ amino acids
LIPID-DIGESTING ENZYMES				
Pancreatic lipase	Pancreas	Small intestine	Basic	Fat droplet + $H_2O \rightarrow$ glycerol + fatty acids
NUCLEIC ACID-DIGESTING ENZYMES				
Nuclease	Pancreas	Small intestine	Basic	RNA or DNA + $H_2O \rightarrow$ nucleotides
Nucleotidase	Small intestine	Small intestine	Basic	Nucleotide + $H_2O \rightarrow$ base + sugar + phosphate

Digestion of the nucleic acids DNA and RNA into free nucleotides is accomplished by a pancreatic nuclease enzyme. The free nucleotides can be further broken down into their components—sugar, phosphate, and organic base—by a nucleotidase enzyme built into the brush border cells.

Table 15.2 lists some of the major digestive enzymes produced by the alimentary canal, salivary glands, or the pancreas. Each type of food is broken down by specific enzymes.

Content CHECK-UP!

7. Choose the correct enzyme to cause this reaction:

     starch + water → maltose

   a. trypsin
   c. lipase
   b. pancreatic amylase
   d. pepsin

8. Which of the following enzymes is not produced by the pancreas?

   a. trypsin
   c. nuclease
   b. lipase
   d. sucrase

9. Which of the following is not a brush border enzyme produced by the small intestine?

   a. nuclease
   c. sucrase
   b. maltase
   d. peptidase

Answers in Appendix C, page 465.

15.4 Effects of Aging

The incidence of gastrointestinal disorders increases with age. Periodontitis, which is common in elderly people, leads to the loss of teeth and the need for false teeth.

The esophagus, which rarely causes any difficulties in younger people, is more prone to disorders in the elderly. The portion of the esophagus normally found inferior to the diaphragm can protrude into the thoracic cavity, causing an esophageal hiatal hernia. In some cases, the lower esophageal sphincter opens inappropriately and allows chyme to regurgitate into the esophagus, causing heartburn. Or in some older persons, chest pain may occur when this sphincter fails to open and a bolus cannot enter the stomach. Eventually, the esophagus may develop a diverticulum that allows food to collect abnormally.

Peristalsis generally slows within the alimentary canal as the muscular wall loses tone. Peptic ulcers increase in frequency with age. The failure of older people to consume sufficient dietary fiber can result in diverticulosis and constipation. Constipation and hemorrhoids are frequent complaints among the elderly, as is fecal incontinence.

The liver shrinks with age and receives a smaller blood supply than in younger years. Notably, it needs more time to metabolize drugs and alcohol. With age, gallbladder difficulties occur; there is an increased incidence of gallstones and

Human Systems Work Together

Integumentary System

Digestive tract provides nutrients needed by skin.

Skin helps to protect digestive organs; helps to provide vitamin D for Ca^{2+} absorption.

Skeletal System

Digestive tract provides Ca^{2+} and other nutrients for bone growth and repair.

Bones provide support and protection; hyoid bone assists swallowing.

Muscular System

Digestive tract provides glucose for muscle activity; liver metabolizes lactic acid following anaerobic muscle activity.

Smooth muscle contraction accounts for peristalsis; skeletal muscles support and help protect abdominal organs.

Nervous System

Digestive tract provides nutrients for growth, maintenance, and repair of neurons and neuroglia.

Brain controls nerves, which innervate smooth muscle and permit tract movements.

Endocrine System

Stomach and small intestine produce hormones.

Hormones help control secretion of digestive glands and accessory organs; insulin and glucagon regulate glucose storage in liver.

How the Digestive System works with other body systems.

oral cavity (mouth)
salivary glands
pharynx (throat)
esophagus
liver
stomach
large intestine
small intestine

Cardiovascular System

Digestive tract provides nutrients for plasma protein formation and blood cell formation; liver detoxifies blood, makes plasma proteins, destroys old red blood cells.

Blood vessels transport nutrients from digestive tract to body; blood services digestive organs.

Lymphatic System/Immunity

Digestive tract provides nutrients for lymphatic organs; stomach acidity prevents pathogen invasion of body.

Lacteals absorb fats; Peyer patches prevent invasion by pathogens; appendix contains lymphatic tissue.

Respiratory System

Breathing is possible through the mouth because digestive tract and respiratory tract share the pharynx.

Gas exchange in lungs provides oxygen to digestive tract and excretes carbon dioxide from digestive tract.

Urinary System

Liver synthesizes urea; digestive tract excretes bile pigments from liver and provides nutrients.

Kidneys convert vitamin D to active form needed for Ca^{2+} absorption; compensate for any water loss by digestive tract.

Reproductive System

Digestive tract provides nutrients for growth and repair of organs and for development of fetus.

Pregnancy crowds digestive organs and promotes heartburn and constipation.

cancer of the gallbladder. In fact, cancer of the various organs of the gastrointestinal tract is seen more often among the elderly. For example, most cases of pancreatic cancer occur in people over the age of 60.

15.5 Homeostasis

Human Systems Work Together on page 347 tells how the digestive system works with other systems in the body to maintain homeostasis.

Within the alimentary canal, the food we eat is broken down to nutrients small enough to be absorbed by the villi of the small intestine. Digestive enzymes are produced by the salivary glands, gastric glands, and intestinal glands. Three accessory organs of digestion (the pancreas, the liver, and the gallbladder) also contribute secretions that help break down food. The liver produces bile (stored by the gallbladder), which emulsifies fat. The pancreas produces enzymes for the digestion of carbohydrates, proteins, and fat. Secretions from these glands, which are sent by ducts into the small intestine, are regulated by hormones such as secretin produced by the alimentary canal. Therefore, the alimentary canal is also a part of the endocrine system. Other accessory organs, such as the salivary glands and teeth, are also essential to digestion.

The nutrients absorbed by the alimentary canal are converted by the body into energy and used for physical activities and for the growth and repair of body tissue. As we'll see in section 15.6, carbohydrates and fats are used to fuel all the body's processes and functions, while protein is mainly used as a building material. Besides these three basic components, the body must also have vitamins and minerals. Vitamins are essential for normal growth and development, and because they cannot be manufactured in the body, they must be supplied ready-made in the diet or as supplements. Minerals assist in many body processes, such as normal nerve and muscle function, but are needed only in small quantities. By providing all other tissues with required nutrients, the digestive tract helps to ensure homeostasis. Its interactions with other body systems are summarized in the Human Systems Work Together figure.

15.6 Nutrition

Nutrition involves an interaction between food and the living organism. A nutrient is a substance that the body uses to maintain health. A balanced diet contains all the essential nutrients and includes a variety of foods, proportioned as shown in Figure 15.12.

Following digestion, nutrients enter the blood in the cardiovascular system, which distributes them to the tissues, where they are utilized by the body's cells. Mitochondria use glucose to produce a constant supply of ATP for the cell. In other words, glucose is the body's immediate energy source. Because the brain's preferred source of energy is glucose, it needs a constant supply.

The liver is able to chemically alter ingested fats to suit the body's needs, with the exception of linoleic acid and linolenic

Figure 15.12 **Newest dietary guidelines, published by the U.S. Department of Agriculture, and available at www.mypyramid.gov.** The website can be customized for an individual's age, gender, and activity level.

acid. Both fatty acids are required, for construction of plasma membranes and for synthesis of messenger chemicals. Thus, they are **essential fatty acids.** Essential molecules must be present in food because the body is unable to manufacture them. Other fats, especially saturated fats, should be restricted, as discussed in the Medical Focus on page 275.

If glucose isn't available, fats can be metabolized into smaller molecules, which are then used as an alternate energy source. Therefore, fats are said to be a long-term energy source. When adipose tissue cells store fats, the body increases in weight. Cells have the capability of converting excess sugar molecules into fats for storage, which accounts for the fact that carbohydrates can also contribute to weight gain.

Amino acids from protein digestion are used by the cells to construct their own proteins, including the enzymes that carry out metabolism. Protein formation requires 20 different types of amino acids. Of these, nine are required in the diet because the body is unable to produce them. These are termed the **essential amino acids.** The body produces the other 11 amino acids by simply transforming one type into another type. Some protein sources, such as meat, are complete in the sense that they provide all the different types of amino acids. Vegetables and grains supply the body with amino acids, but they are typically incomplete sources because at least one of the essential amino acids is absent. A combination of certain vegetables, however, can provide all of the essential amino acids. In addition, soy protein is a complete protein.

Vitamins

Vitamins are *vital* to life because they play essential roles in cellular metabolism. Because the body is unable to produce them, most vitamins must be present in the diet. Vitamins are organic molecules, but they differ radically from carbohydrates, fats,

TABLE 15.3 Vitamins: Their Role in the Body and Food Sources

Vitamins	Role in Body	Good Food Sources
FAT-SOLUBLE VITAMINS		
Vitamin A	Assists in the formation and maintenance of healthy skin, hair, and mucous membranes; aids in the ability to see in dim light (night vision); is essential for proper bone growth, tooth development, and reproduction	Deep yellow/orange and dark green vegetables and fruits (carrots, broccoli, spinach, cantaloupe, sweet potatoes); cheese, milk, and fortified margarine
Vitamin D	Aids in the formation and maintenance of bones and teeth; assists in the absorption and use of calcium and phosphorus	Milk fortified with vitamin D; tuna, salmon, or cod liver oil; also made in the skin when exposed to sunlight
Vitamin E	Protects vitamin A and essential fatty acids from oxidation; prevents plasma membrane damage	Vegetable oils and margarine; nuts; wheat germ and whole grain breads and cereals; green, leafy vegetables
Vitamin K	Aids in synthesis of substances needed for clotting of blood; helps maintain normal bone metabolism	Green, leafy vegetables, cabbage, and cauliflower; also made by bacteria in intestines of humans, except for newborns
WATER-SOLUBLE VITAMINS		
Vitamin C	Is important in forming collagen, a protein that gives structure to bones, cartilage, muscle, and vascular tissue; helps maintain capillaries, bones, and teeth; aids in absorption of iron; helps protect other vitamins from oxidation	Citrus fruits, berries, melons, dark green vegetables, tomatoes, green peppers, cabbage, potatoes
B-COMPLEX VITAMINS		
Thiamin	Helps in release of energy from carbohydrates; promotes normal functioning of nervous system	Whole-grain products, dried beans and peas, sunflower seeds, nuts
Riboflavin	Helps body transform carbohydrates, proteins, and fats into energy	Nuts, yogurt, milk, whole-grain products, cheese, poultry, leafy green vegetables
Niacin	Helps body transform carbohydrates, proteins, and fats into energy	Nuts, poultry, fish, whole-grain products, dried fruit, leafy greens, beans; can be formed in the body from tryptophan, an essential amino acid found in protein
Vitamin B_6	Aids in the use of fats and amino acids; aids in the formation of protein	Sunflower seeds, beans, poultry, nuts, bananas, dried fruit, leafy green vegetables
Folic acid	Aids in the formation of hemoglobin in red blood cells; aids in the formation of genetic material	Nuts, beans, whole-grain products, fruit juices, dark green leafy vegetables
Pantothenic acid	Aids in the formation of hormones and certain nerve-regulating substances; helps in the metabolism of carbohydrates, proteins, and fats	Nuts, beans, seeds, poultry, dried fruit, milk, dark green leafy vegetables
Biotin	Aids in the formation of fatty acids; helps in the release of energy from carbohydrates	Occurs widely in foods, especially eggs; made by bacteria in the human intestine
Vitamin B_{12}	Aids in the formation of red blood cells and genetic material; helps in the functioning of the nervous system; requires intrinsic factor from the stomach to be absorbed	Milk, yogurt, cheese, fish, poultry, eggs; not found in plant foods unless fortified (as in some breakfast cereals)

Source: From David C. Nieman, et al., *Nutrition,* Revised 1st ed. Copyright © 1992 Wm. C. Brown Communications, Inc., Dubuque, Iowa. Reprinted by permission of C. V. Mosby, St. Louis, MO.

and proteins. They are much smaller in size and are not broken down to be used as building blocks or as a source of energy. Instead, the body protects them and provides many of them with protein carriers that transport them in the blood to the cells. In the cells, vitamins become helpers in metabolic processes that break down or synthesize other organic molecules. Because vitamins can be used over and over again, they are required in very small amounts.

Vitamins fall into two groups: fat-soluble vitamins (vitamins A, D, E, and K) and water-soluble vitamins (the B complex vitamins and vitamin C; see Table 15.3). Most of the water-soluble vitamins are coenzymes, or enzyme helpers, that help speed up specific reactions. The functions of the fat-soluble vitamins, some of which have been previously discussed, are more specialized. Vitamin A, as noted in Chapter 9, is used to synthesize the visual pigments. Vitamin D is needed to produce

TABLE 15.4 Minerals: Their Role in the Body and Food Sources

Minerals	Role in Body	Good Food Sources
MACRONUTRIENTS		
Calcium	Is used for building bones and teeth and for maintaining bone strength; also involved in muscle contraction, blood clotting, and maintenance of plasma membranes	All dairy products; dark green, leafy vegetables; beans, nuts, sunflower seeds, dried fruit, molasses, canned fish
Phosphorus	Is used to build bones and teeth; to release energy from carbohydrates, proteins, and fats; and to form genetic material, plasma membranes, and many enzymes	Beans, sunflower seeds, milk, cheese, nuts, poultry, fish, lean meats
Magnesium	Is used to build bones, to produce proteins, to release energy from muscle carbohydrate stores (glycogen), and to regulate body temperature	Sunflower and pumpkin seeds, nuts, whole-grain products, beans, dark green vegetables, dried fruit, lean meats
Sodium	Regulates body-fluid volume and blood acidity; aids in transmission of nerve impulses	Most of the sodium in the U.S. diet is added to food as salt (sodium chloride) in cooking, at the table, or in commercial processing; animal products contain some natural sodium
Chloride	Is a component of gastric juice and aids in acid-base balance	Table salt, seafood, milk, eggs, meats
Potassium	Assists in muscle contraction, the maintenance of fluid and electrolyte balance in the cells, and the transmission of nerve impulses; also aids in the release of energy from carbohydrates, proteins, and fats	Widely distributed in foods, especially fruits and vegetables, beans, nuts, seeds, and lean meats
MICRONUTRIENTS (TRACE ELEMENTS)		
Iron	Is involved in the formation of hemoglobin in the red blood cells of the blood and myoglobin in muscles; also is a part of several enzymes and proteins	Molasses, seeds, whole-grain products, fortified breakfast cereals, nuts, dried fruits, beans, poultry, fish, lean meats
Zinc	Is involved in the formation of protein (growth of all tissues), in wound healing, and in prevention of anemia; is a component of many enzymes	Whole-grain products, seeds, nuts, poultry, fish, beans, lean meats
Iodine	Is an integral component of thyroid hormones	Table salt (fortified), dairy products, shellfish, and fish
Fluoride	Is involved in maintenance of bone and tooth structure	Fluoridated drinking water is the best source; also found in tea, fish, wheat germ, kale, cottage cheese, soybeans, almonds, onions, milk
Copper	Is vital to enzyme systems and in manufacturing red blood cells; is needed for utilization of iron	Nuts, oysters, seeds, crab, wheat germ, dried fruit, whole grains, legumes
Selenium	Functions in association with vitamin E; may assist in protecting tissues and plasma membranes from oxidative damage; may also aid in preventing cancer	Nuts, whole grains, lean pork, cottage cheese, milk, molasses, squash
Chromium	Is required for maintaining normal glucose metabolism; may assist insulin function	Nuts, prunes, vegetable oils, green peas, corn, whole grains, orange juice, dark green vegetables, legumes
Manganese	Is needed for normal bone structure, reproduction, and the normal functioning of the central nervous system; is a component of many enzyme systems	Whole grains, nuts, seeds, pineapple, berries, legumes, dark green vegetables, tea
Molybdenum	Is a component of enzymes; may help prevent dental caries	Tomatoes, wheat germ, lean pork, legumes, whole grains, strawberries, winter squash, milk, dark green vegetables, carrots

Source: From David C. Nieman, et al., *Nutrition,* Revised 1st ed. Copyright © 1992 Wm. C. Brown Communications, Inc., Dubuque, Iowa. Reprinted by permission of C. V. Mosby, St. Louis, MO.

a hormone that regulates calcium and phosphorus metabolism (see Chapter 5), and vitamin E is an antioxidant. Vitamin K is required to form *prothrombin,* a substance necessary for normal blood clotting (see Chapter 11).

Minerals

In contrast to vitamins, **minerals** are inorganic elements (Table 15.4). An element, you'll recall, is one of the basic substances of matter that cannot be broken down further into simpler substances. Minerals sometimes occur as a single atom, in contrast to vitamins, which contain many atoms. Minerals cannot lose their identity, no matter how they are handled. Because they are indestructible, no special precautions are needed to preserve them when cooking.

Minerals are divided into macronutrients, which are needed in gram amounts per day, and micronutrients (trace

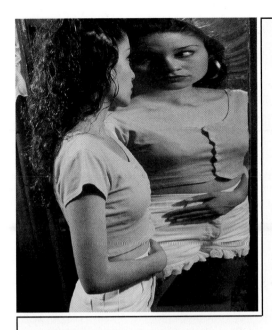

Persons with bulimia nervosa have

- recurrent episodes of binge eating characterized by consuming an amount of food much higher than normal for one sitting and a sense of lack of control over eating during the episode.
- an obsession about their body shape and weight.
- increase in fine body hair, halitosis, and gingivitis.

Body weight is regulated by

- a restrictive diet, excessive exercise.
- purging (self-induced vomiting or misuse of laxatives).

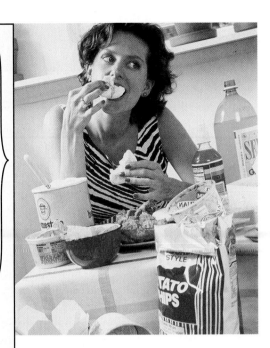

Persons with anorexia nervosa have

- a morbid fear of gaining weight; body weight no more than 85% normal.
- a distorted body image so that person feels fat even when emaciated.
- in females, an absence of a menstrual cycle for at least three months.

Body weight is kept too low by either/or

- a restrictive diet, often with excessive exercise.
- binge eating/purging (person engages in binge eating and then self-induces vomiting or misuses laxatives).

Figure 15.13 Recognizing anorexia nervosa and bulimia nervosa.

elements), which are needed in only microgram amounts per day. The macronutrients sodium, magnesium, phosphorus, chlorine, potassium, and calcium serve as constituents of cells and body fluids, and as structural components of tissues. The micronutrients have very specific functions, as noted in Table 15.4. As research continues, more elements will be added to the list of those considered essential.

Eating Disorders

Authorities recognize three primary eating disorders: obesity, bulimia nervosa, and anorexia nervosa. Although they exist in a continuum as far as body weight is concerned, all represent an inability to maintain normal body weight because of eating habits.

Obesity

Obesity is most often defined as a body weight 20% or more above the ideal weight for a person's height. By this standard, 33% of women and 30% of men in the United States are obese. Moderate obesity is 41–100% above ideal weight, and severe obesity is 100% or more above ideal weight.

Obesity is most likely caused by a combination of hormonal, metabolic, and social factors. It is known that obese individuals have more fat cells than normal. When an obese person loses weight, the fat cells simply get smaller; they don't

disappear. The social factors that cause obesity include the eating habits of other family members. Consistently eating fatty foods, for example, may cause you to gain weight. Sedentary activities, such as watching television instead of exercising, also determine how much body fat you have. The risk of heart disease is higher in obese individuals, and this alone tells us that excess body fat is not consistent with optimal health.

Treatment depends on the degree of obesity. Surgery to restrict stomach volume may be required for those who are moderately or greatly overweight (see the Medical Focus on page 352). But for most people, a knowledge of good eating habits along with behavior modification may be enough, particularly if a balanced diet is accompanied by a sensible exercise program. A lifelong commitment to a properly planned program is the best way to prevent a cycle of weight gain followed by weight loss. Such a cycle is not conducive to good health.

Bulimia Nervosa

Bulimia nervosa can coexist with either obesity or anorexia nervosa, which is discussed next. People with this condition have the habit of eating to excess (called binge eating) and then purging themselves by some artificial means, such as self-induced vomiting or the use of a laxative (Fig. 15.13). Bulimic individuals are overly concerned about their body shape and weight, and therefore they may be on a very restrictive diet.

Bariatric Surgery for Obesity

It's a final measure sought by increasing numbers of people: **bariatric surgery,** or surgical intervention with the specific goal of causing drastic weight loss. For many overweight people, years of dieting haven't worked. Name a diet plan, and they have tried it—sure, they lose weight only to regain it. Obesity is approaching epidemic levels in the United States, with more than 64 million adults classified as obese. In addition, pediatricians are especially concerned with the dramatic rise in childhood and teenage obesity.

Obesity is defined as a body mass index (BMI) greater than 30,[1] or a weight that is 41% or higher than the ideal weight for one's height. It's a physical and emotional challenge for the overweight patient. Obesity is a primary risk factor for hypertension, type II diabetes mellitus, atherosclerosis, stroke, coronary artery disease, and early heart attack. It has been linked to increased risk of breast, ovarian, uterine, and prostate cancer. Obese individuals suffer disrespect and ridicule from society and discrimination on the job. After years of struggling with the problem, many are willing to undergo surgery as a last-chance option.

However, reputable programs offering bariatric surgery have strict requirements for patients. To be admitted as a surgical candidate, the patient must be morbidly obese (BMI greater than 40; or greater than 100 pounds over ideal weight). Patients cannot have respiratory or cardiac problems that might complicate surgery. Most important, the patient must understand the risks of surgery as well as its benefits. Patients must also understand that lifestyle changes will be necessary even after successful surgery and recovery. Psychological and nutritional counseling is usually required before surgery to prepare the patient for new eating habits and ways of thinking about food. Follow-up counseling tracks the patient's progress in adapting.

The two most commonly used interventions are laparoscopic banding and the Roux-en-Y gastric bypass (Fig. 15B). A more conservative approach, laparoscopic banding, requires making a series of small incisions around the stomach and using an instrument called a laparoscope to illuminate structures in the abdomen. A band is placed around the stomach. Once tightened like a belt, the stomach is divided into a smaller upper chamber, which receives food, and a lower chamber that remains connected to the duodenum. The belt can later be tightened further, or removed if necessary. In Roux-en-Y gastric bypass, the top section of the stomach is cut free and stapled shut to make a pouch about the size of an egg. The small intestine is cut free between the duodenum and jejunum, and the jejunum is sewn to the end of the stomach pouch. Finally, the duodenal segment is sewn back to the jejunum (forming the Y-shaped branch for which the procedure is named). The lower stomach and duodenum remain healthy and continue to secrete digestive enzymes, but never receive food. Regardless of the approach used, the person will only be able to eat small amounts of food, but should feel full due to the small size of the stomach after surgery.

It's important to note that bariatric surgery comes with extremely serious potential complications. Postoperative bleeding and infection are risks of any surgical procedure. Blood clots in the legs can form during hospitalization, causing pulmonary embolism or stroke. After gastric bypass, staple lines in the stomach can leak. In rare cases, the connection between the stomach pouch and jejunum narrows, requiring additional corrective surgery. Worst of all, a small percentage of patients die during the surgery itself. After surgery, vitamin and mineral deficiencies are possible.

Further, bariatric surgery offers no guarantees of permanent weight loss—patients can in fact regain any weight that is lost, even with a drastically smaller stomach. However, with proper nutrition and behavioral changes, bariatric surgery can result in dramatic weight loss and improvement to health.

---

[1]To calculate BMI, use the following formula: $\dfrac{\text{weight in pounds}}{(\text{height in inches}) \times (\text{height in inches})} \times 703.$
BMI is not always accurate in determining obesity. Other factors such as percent body fat may also need to be used.

After laparoscopic banding   **Before surgery**

stomach

duodenum
jejunum

small intestine

After Roux-en-Y gastric bypass

stomach (bypassed)

pouch

jejunum

duodenum (bypassed)

Figure 15B (*far left figure*) A laparascopic band is placed around the stomach and then tightened to decrease stomach size. (*center*) The normal movement of food prior to surgery. (*right*) The flow of food after Roux-en-Y gastric bypass surgery.

A restrictive diet may bring on the desire to binge, and typically the person chooses to consume sweets, such as cakes, cookies, and ice cream. The amount of food consumed is far beyond the normal number of calories for one meal, and the person keeps on eating until every bit is gone. Then, a feeling of guilt most likely brings on the next phase, which is a purging of all the calories that have been taken in.

Bulimia is extremely dangerous to health. Blood composition is altered, leading to an abnormal heart rhythm, and damage to the kidneys can even result in death. At the very least, vomiting can lead to inflammation of the pharynx and esophagus, and stomach acids can cause the teeth to erode. The esophagus and stomach may even rupture and tear due to strong contractions during vomiting.

The most important aspect of treatment is to get the patient on a sensible and consistent diet. Again, behavioral modification is helpful, and so perhaps is psychotherapy to help the patient understand the emotional causes of the behavior. Medications, including antidepressants, have sometimes helped to reduce the bulimic cycle and restore normal appetite.

Anorexia Nervosa

In **anorexia nervosa**, a morbid fear of gaining weight causes the person to be on a very restrictive diet (see Fig. 15.13). Athletes such as distance runners, wrestlers, and dancers are at risk of anorexia nervosa because they believe that being thin gives them a competitive edge. In addition to eating only low-calorie foods, the person may induce vomiting and use laxatives to bring about further weight loss. No matter how thin they have become, people with anorexia nervosa think they are overweight. Such a distorted self-image may prevent recognition of the need for medical help.

Actually, the person is starving and has all the symptoms of starvation, including low blood pressure, irregular heartbeat, constipation, and constant chilliness. Bone density decreases, and stress fractures occur. The body begins to shut down; menstruation ceases in females; the internal organs, including the brain, don't function well; and the skin dries up. Impairment of the pancreas and alimentary canal means that any food consumed does not provide nourishment. Death may be imminent. If so, the only recourse may be hospitalization and force-feeding. Eventually, it is necessary to use behavior therapy and psychotherapy to enlist the cooperation of the person to eat properly. Family therapy may be necessary, because anorexia nervosa in children and teens is believed to be a way for them to gain some control over their lives.

SELECTED NEW TERMS

Basic Key Terms

alimentary canal (al-i-men′ter-ē kuh-nal), p. 330

anus (ā-nus), p. 330

ascending colon (uh-send′ing ko′lon), p. 339

bicuspid (bī-cus′-pīd), p. 331

bile (bīl), p. 337

bile salts (bīl sôlt), p. 344

bilirubin (bī′l′y-roo′bĭn), p. 344

body region (bod′é rē′jen), p. 335

bolus (bo′lus), p. 333

cardiac region (kăr′dē-ăk rē′jen), p. 335

cecum (se′kum), p. 339

cephalic phase (sə-fāl′ik fāz), p. 338

cholecystokinin (kō′lē-sĭs-tō-kī′-nin), p. 338

chyme (kīm), p. 336

colon (kō′lən), p. 339

cuspid (cŭs′-pid), p. 331

descending colon (de-send′ing ko′lon), p. 339

disaccharases (dī-săk′oh-rə-ses), p. 336

duodenum (du″o-de′num), p. 334

enteric nervous system (ĕn-tĕr′ik nĕr′vŭs sĭs′təm), p. 338

enterokinase (ĕn′tēr-ō-ki′nās), p. 342

esophagus (ĕ-sof′uh-gus), p. 334

essential amino acid (ĕ-sen′shul uh-me′no as′id), p. 348

essential fatty acid (ĕ-sen′shul fat′e as′id), p. 348

fundic region (fŭn′dik rē′jen), p. 335

gallbladder (gawl′blad-er), p. 344

gastric gland (gas′trik gland), p. 335

gastric juice (găs′trĭk jōōs), p. 335

gastric inhibitory polypeptide (găs′trik in-hib′i-torē pol-ē-pep′tīd), p. 338

gastric phase (găs′trik fāz), p. 338

gastric pits (găs′trik pits), p. 335

gastrin (găs′trĭn), p. 336

ghrelin (grē′lĭn), p. 339

hard palate (hard pal′ut), p. 332

haustra (hăw′stră), p. 339

hepatic lobule (hə-pat′ik lob′yool), p. 343

ileum (il′e-um), p. 336

incisor (ĭn-sī-zōr), p. 331

intestinal phase (in-tĕs′tĭn-ul fāz), p. 336

intrinsic factor (in-trĭn′zik făk′tər), p. 336

jejunum (jĕ-ju′num), p. 336

lacteal (lăk′tē-əl), p. 337

large intestine (larj in-tes′tin), p. 339

leptin (lĕp′ tĭn), p. 339

lipase (lĭp′as′), p. 345

liver (liv′er), p. 343

lumen (loo′mən), p. 334

mineral (min′er-al), p. 350

molar (mō′lăr), p. 330

mouth (mowth), p. 331

mucosa (mū-kó′sa), p. 334

muscularis (mŭs-kū-lā′-ris), p. 334

pancreas (păng′ krē-əs), p. 342

pancreatic amylase (păng′krē-ăt′ĭk am′ə-lăs′), p. 342

pepsin (pĕp′-sĭn), p. 336

peptidases (pĕp′tĭ-dās′), p. 345

peristalsis (per″ĭ-stal′sis), p. 334

pharynx (făr′ĭngks), p. 332

portal triads (pôr′tl trī′ăd), p. 343

pyloric region (pī-lō′ik rē′jĕn), p. 335

rectum (rek′tum), p. 339

rugae (roo′je), p. 335

salivary amylase (săl′ə-vēr′ē am′ə-lăs), p. 332

salivary gland (sal′ĭ-vĕr-e gland), p. 332

secretin (sĭ-krēt′n), p. 338

serosa (se-rō′să), p. 334

sigmoid colon (sig′moid ko′lon), p. 339

small intestine (smawl in-tes′tin), p. 336

soft palate (sawft pal′ut), p. 332

sphincter (sfingk′ter), p. 334

stomach (stum′ak), p. 335

submucosa (sŭb-moo-kō′să), p. 334

Clinical Key Terms

SUMMARY

15.1 Anatomy of the Digestive System

The alimentary canal consists of the mouth, pharynx, esophagus, stomach, small intestine, and large intestine. Only these structures actually contain food, while the salivary glands, liver, and pancreas supply substances that aid in the digestion of food.

A. The salivary glands send saliva into the mouth, where the teeth chew the food and the tongue forms a bolus for swallowing. Saliva contains salivary amylase, an enzyme that begins the digestion of starch.

B. The air passage and the food passage cross in the pharynx. When a person swallows, the air passage is normally blocked off, and food must enter the esophagus, where peristalsis begins.

C. The stomach expands and stores food. While food is in the stomach, the stomach churns, mixing food with the acidic gastric juices. Gastric juices contain pepsin, an enzyme that digests protein.

D. The duodenum of the small intestine receives bile from the liver and pancreatic juice from the pancreas. Bile, which is produced in the liver and stored in the gallbladder, emulsifies fat and readies it for digestion by lipase, an enzyme produced by the pancreas. The pancreas also produces enzymes that digest starch (pancreatic amylase) and protein (trypsin, chymotrypsin, and carboxypeptidase). The intestinal enzymes finish the process of chemical digestion.

E. The walls of the small intestine have fingerlike projections called villi. The villi have microvilli where small nutrient molecules are absorbed. Amino acids and glucose enter the blood vessels of a villus. Glycerol and fatty acids are joined and packaged as lipoproteins before entering lymphatic vessels called lacteals in a villus.

F. The large intestine consists of the cecum, the colon (including the ascending, transverse, descending, and sigmoid colon), and the rectum, which ends at the anus. The large intestine does not produce digestive enzymes; it does absorb water, salts, and some vitamins. Reduced water absorption results in diarrhea. The intake of water and fiber helps prevent constipation.

15.2 Accessory Organs of Digestion

A. Three accessory organs of digestion—the pancreas, liver, and gallbladder—send secretions to the duodenum via ducts. The pancreas produces pancreatic juice, which contains digestive enzymes for carbohydrate, protein, and fat.

B. The liver produces bile, which is stored in the gallbladder. The liver receives blood from the small intestine by way of the hepatic portal vein. It has numerous important functions, and any malfunction of the liver is a matter of considerable concern.

15.3 Chemical Digestion

A. Digestive enzymes are present in digestive juices and break down food into the nutrient molecules glucose, amino acids, fatty acids, and glycerol (see Table 15.2). Salivary amylase and pancreatic amylase begin the digestion of starch. Pepsin and pancreatic peptidases digest protein to peptides. Lipase digests fat to glycerol and fatty acids. Intestinal enzymes finish the digestion of starch and protein.

B. Digestive enzymes have the usual enzymatic properties. They are specific to their substrate and speed up specific reactions at optimum body temperature and pH.

15.4 Effects of Aging

The structure and function of the digestive system generally decline with age. The various illnesses associated with the digestive system are more likely to be seen among the elderly.

15.5 Homeostasis

The digestive system works with the other systems of the body in the ways described in Human Systems Work Together on page 347.

15.6 Nutrition

A. The nutrients released by the digestive process should provide us with an adequate amount of energy, essential amino acids and fatty acids, and all necessary vitamins and minerals.

B. The diet should be balanced and low in saturated fatty acids and cholesterol molecules, whose intake is linked to cardiovascular disease. Aside from carbohydrates, proteins, and fats, the body requires vitamins and minerals. The vitamins C, E, and A are antioxidants that protect cell contents from damage due to free radicals. The mineral calcium is needed for strong bones.

C. The reasons for eating disorders, including obesity, bulimia nervosa, and anorexia, are being explored in order to help people maintain a normal weight for their height.

STUDY QUESTIONS

1. List the organs of the alimentary canal, and state the contribution of each to the digestive process. (pp. 330–341)
2. Discuss the absorption of the products of digestion into the lymphatic and cardiovascular systems. (pp. 336–338)
3. Name and state the functions of the hormones that assist the nervous system in regulating digestive secretions. (pp. 336, 338–339)
4. Name the accessory digestive organs, and describe the part they play in the digestion of food. (pp. 342–345)
5. Name and discuss any three functions and two serious illnesses of the liver. (pp. 343–344)
6. Discuss the digestion of starch, protein, and fat, listing all the steps that occur with each. (pp. 345–346)
7. How does the digestive system help maintain homeostasis? (pp. 347–348)
8. What is the chief contribution of each of these constituents of the diet: (a) carbohydrates; (b) proteins; (c) fats; (d) fruits and vegetables? (p. 348)
9. What role do water-soluble vitamins usually play in the body? (p. 349)
10. Name and discuss three eating disorders. (pp. 351–353)

LEARNING OBJECTIVE QUESTIONS

Fill in the blanks.

1. In the mouth, salivary _____ digests starch.
2. When we swallow, the _____ covers the opening to the larynx.
3. The _____ takes food to the stomach, where _____ is primarily digested.
4. The gastric juices are _____, and therefore, they usually destroy any bacteria in the food.
5. The large intestine has a colon with four _____ and a(n) _____, which leads to the _____.
6. The pancreas transports digestive juices to the _____, the first part of the small intestine.
7. After a meal, the liver stores glucose as _____.
8. The gallbladder stores _____, a substance that _____ fat.
9. Pancreatic juice contains _____, _____, and _____ for digesting protein, _____ for digesting fat, and _____ for digesting starch.
10. The products of digestion are absorbed into the cells of the _____, fingerlike projections of the intestinal wall.

MEDICAL TERMINOLOGY EXERCISE

Consult Appendix B for help in pronouncing and analyzing the meaning of the terms that follow.

1. stomatoglossitis (sto″muh-to-glos-si′tis)
2. glossopharyngeal (glos″o-fah-rin′je-al)
3. esophagectasia (ĕ-sof″ah-jek-ta′se-uh)
4. gastroenteritis (gas″tro-en-ter-i′tis)
5. sublingual (sub-ling′gwal)
6. gingivoperiodontitis (jin″ji-vo-pĕr″e-o-don-ti′tis)
7. dentalgia (den-tal′je-uh)
8. pyloromyotomy (pi-lo″ro-mi-ot′o-me)
9. cholangiogram (ko-lan′je-o-gram)
10. cholecystolithotripsy (ko″le-sis″to-lith′o-trip″se)
11. proctosigmoidoscopy (prok″to-sig″moy-dos′kuh-pe)
12. colocentesis (ko-lo′sen-te′sis)
13. trichophagia (tri-ko-fāj′e-uh)
14. duodenorrhaphy (du-o-dĕ-nor′uh-fe)
15. ileocecal (il′e-o-se′kul)

WEBSITE LINK

Visit this text's website at http://www.mhhe.com/longenbaker7 for additional quizzes, interactive learning exercises, and other study tools.

The Urinary System and Excretion

C H A P T E R

16

Before you study this chapter about the urinary system, take a moment to reflect on your kidneys, shown here in an X ray. (The kidneys are colored light green, and the renal pelvis and ureters, through which urine passes, are colored red.) It's probably safe to say that most healthy people rarely, if ever, think about their kidneys, except during a visit to a restroom—and probably not even then. However, when kidneys are analyzed by the numbers, suddenly these little bean-shaped organs become much more impressive. Consider this: A healthy, average man's kidneys filter about 125 mL of blood every minute—that's 7.5 L per hour and 180 L per day. Keep going with the math: The same man's kidneys filter 65,700 L of blood per year. Now assume he lives for 75 years (the average lifespan for a man in the United States) and for 60 of those years he's filtering at the normal adult rate. That's almost 4 million L, or roughly 1 million gallons, of blood filtered by the kidneys during his lifetime (the figures are slightly less for women) and, of course, filtration also occurred as he grew to adult size. However, of that volume, only about 1% ever leaves the body as urine. The remainder is quickly reabsorbed and returned to the bloodstream. (Still, 10,000 gallons of urine—now that's a lot of trips to a restroom!)

Chapter Outline & Learning Objectives

After you have studied this chapter, you should be able to:

16.1 Urinary System

The kidneys are the primary organs of excretion. **Excretion** is the removal of metabolic wastes from the body. People sometimes confuse the terms excretion and defecation, but they do not refer to the same process. Defecation, the elimination of feces from the body, is a function of the digestive system. Excretion, on the other hand, is the elimination of metabolic wastes, which are the products of metabolism. For example, the undigested food and bacteria that make up feces have never been a part of the functioning of the body, while the substances excreted in urine were once metabolites in the body.

Functions of the Urinary System

The urinary system filters the blood to produce urine, then transports the urine out of the body. As the kidneys produce urine, they carry out four functions that are essential to homeostasis: excretion of metabolic wastes, preservation of water-salt balance in blood and body fluids, maintenance of blood pressure, and maintenance of acid-base balance. As we shall see, the kidneys assist the endocrine system; they also produce erythropoietin (EPO), their own endocrine hormone.

Excretion of Metabolic Wastes

The kidneys excrete metabolic wastes, notably nitrogenous wastes. Urea is the primary nitrogenous end product of metabolism in human beings, but humans also excrete some ammonium, creatinine, and uric acid.

Urea is a by-product of amino acid metabolism. The breakdown of amino acids in the liver releases ammonia, which the liver combines with carbon dioxide to produce urea. Ammonia is very toxic to cells, but urea is much less toxic. Because urea is less toxic than ammonia, it can be safely excreted in the relatively small amounts of water passed daily as urine. (If our bodies had to excrete ammonia instead of urea, the ammonia would have to be diluted in extremely large volumes of urine to avoid being toxic to body cells.)

Creatine phosphate is a high-energy phosphate reserve molecule in muscles. The metabolic breakdown of creatine phosphate results in **creatinine.**

The breakdown of nucleotides, such as those containing adenine and thymine, produces **uric acid.** Uric acid is rather insoluble. If too much uric acid is present in blood, crystals form and precipitate out. Crystals of uric acid sometimes collect in the joints producing a painful ailment called **gout.**

Urochrome is a waste product formed as the liver breaks down hemoglobin. As you know from reading Chapter 15, hemoglobin is first formed into bilirubin, which is excreted in the bile. Once in the intestine, bilirubin is further broken down into smaller compounds, including urochrome. It is urochrome that gives urine its yellow color.

Preservation of Water-Salt Balance

A principal function of the kidneys is to maintain the appropriate water-salt balance of the blood. As we shall see, blood volume is intimately associated with the salt balance of the body. Salts, such as NaCl, have the ability to cause osmosis, the diffusion of water—in this case, into the blood. The more salts there are in the blood, the greater the blood volume. The kidneys also maintain the appropriate level of other ions (electrolytes), such as potassium ions (K^+), bicarbonate ions (HCO_3^-), and calcium ions (Ca^{2+}), in the blood.

Maintenance of Blood Pressure

Recall that blood pressure is the product of cardiac output and peripheral resistance, and the kidneys influence both of these variables. By preserving salt and water balance, the kidneys maintain a normal blood volume. In turn, blood volume determines the heart's stroke volume and, thus, cardiac output. The kidneys also influence peripheral resistance by producing the enzyme renin, which activates the plasma protein angiotensin. Angiotensin constricts blood vessel smooth muscle, increasing resistance.

Maintenance of Acid-Base Balance

The kidneys regulate the acid-base balance of the blood. In order for a person to remain healthy, the blood pH should be just about 7.4. The kidneys monitor and control blood pH, mainly by excreting hydrogen ions (H^+) from the blood into the urine. At the same time, the kidneys reabsorb bicarbonate ions (HCO_3^-) and return them to the blood as needed to keep blood pH at about 7.4. Urine usually has a pH of 6 or lower because our diet often contains acidic foods.

Secretion of Hormones

The kidneys assist the endocrine system in hormone secretion. By releasing the enzyme renin whenever their own blood supply decreases, the kidneys activate angiotensin, which leads to the secretion of the hormone aldosterone. Aldosterone is produced by the adrenal cortex, the outer portion of the adrenal glands, which lie atop the kidneys. As described in section 16.3, aldosterone promotes the reabsorption of sodium ions (Na^+) and water by the kidneys.

Whenever the oxygen-carrying capacity of the blood is reduced, the kidneys secrete the hormone **erythropoietin (EPO)**, which stimulates red blood cell production.

The kidneys also help activate vitamin D from the skin. Vitamin D is the precursor of the hormone calcitriol, which promotes calcium (Ca^{2+}) absorption from the digestive tract.

Organs of the Urinary System

The urinary system consists of the kidneys, ureters, urinary bladder, and urethra. Figure 16.1 shows these organs and also traces the path of urine.

Kidneys

The **kidneys** are paired organs located near the small of the back in the lumbar region on either side of the vertebral

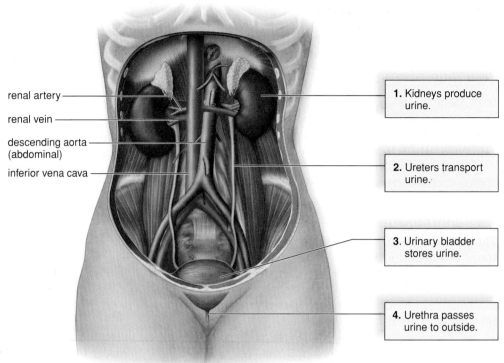

renal artery

renal vein

descending aorta
(abdominal)

inferior vena cava

1. Kidneys produce
   urine.

2. Ureters transport
   urine.

3. Urinary bladder
   stores urine.

4. Urethra passes
   urine to outside.

Figure 16.1 **The urinary system.** Urine is found only within the kidneys, the ureters, the urinary bladder, and the urethra.

column. The kidneys are *retroperitoneal*, which means that they are covered by the parietal peritoneum (the serous membrane that lines the abdominopelvic cavity). The kidneys lie in depressions against the deep muscles of the back, where they receive some protection from the lower rib cage. Each kidney is usually held in place by connective tissue, called renal fascia. Masses of adipose tissue adhere to each kidney. A sharp blow to the back can dislodge a kidney, which is then called a **floating kidney.**

The kidneys are bean-shaped and reddish-brown in color. The fist-sized organs are covered by a tough capsule of fibrous connective tissue, called a renal capsule. The concave side of a kidney has a depression called the hilum where a **renal artery** enters and a **renal vein** and a ureter exit the kidney.

Ureters

The **ureters,** which extend from the kidneys to the bladder, are small, muscular tubes about 25 cm long and 5 mm in diameter. Each descends behind the parietal peritoneum, from the hilum of a kidney, to enter the bladder posteriorly at its inferior surface.

The wall of a ureter has three layers. The inner layer is a mucosa (mucous membrane), the middle layer consists of smooth muscle, and the outer layer is a fibrous coat of connective tissue. Peristaltic contractions of the ureters cause urine to enter the bladder even if a person is lying down. Urine enters the bladder in spurts that occur at the rate of one to five per minute.

Urinary Bladder

The **urinary bladder** is located in the pelvic cavity, below the parietal peritoneum and just posterior to the pubic symphysis. In males, it is directly anterior to the rectum; in females, it is anterior to the vagina and inferior to the uterus. Its function is to store urine until it is expelled from the body. The bladder has three openings—two for the ureters and one for the urethra, which drains the bladder. The *trigone* is a smooth triangular area at the base of the bladder outlined by these three openings (Fig. 16.2).

Collectively, the muscle layers of the bladder wall are called the *detrusor muscle.* The wall contains a middle layer of circular fiber and two layers of longitudinal muscle, and it can expand. The transitional epithelium of the mucosa becomes thinner, and folds in the mucosa called *rugae* disappear as the bladder fills and enlarges.

The bladder has other features that allow it to retain urine. After urine enters the bladder from a ureter, small folds of bladder mucosa act like a valve to prevent backward flow.

Two sphincters in close proximity are found where the urethra exits the bladder. The internal sphincter occurs around the opening to the urethra. Inferior to the internal sphincter, the external sphincter is composed of skeletal muscle that can be voluntarily controlled.

Urethra

The **urethra** is a small tube that extends from the urinary bladder to an external opening. The urethra is a different

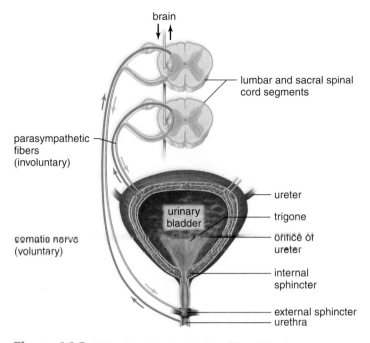

Figure 16.2 Urination. As the bladder fills with urine, sensory signals (blue arrows) go to the spinal cord and then to the brain. The brain can override the urge to urinate. When urination occurs, motor nerve signals cause the bladder to contract and an internal sphincter to open. Nerve signals also cause an external sphincter to open.

length in females than in males. In females, the urethra is only about 4 cm long. The short length of the female urethra makes bacterial invasion easier and helps explain why females are more prone to urinary tract infections than males. In males, the urethra averages 20 cm when the penis is flaccid (limp, nonerect). As the urethra leaves the male urinary bladder, it is encircled by the **prostate gland.** In older men, enlargement of the prostate gland can restrict urination. A surgical procedure can usually correct the condition and restore a normal flow of urine.

In females, the reproductive and urinary systems are not connected. In males, the urethra carries urine during urination and sperm during ejaculation. This double function of the urethra in males does not alter the path of urine.

Urination

When the urinary bladder fills to about 250 ml with urine, stretch receptors send sensory nerve signals to the spinal cord. Subsequently, motor nerve signals from the lumbar and sacral regions of the spinal cord cause the urinary bladder to contract and the sphincters to relax so that urination, also called **micturition,** is possible (Fig. 16.2). In older children and adults, the brain controls this reflex, delaying urination until a suitable time.

16.2 Anatomy of the Kidney and Excretion

A sagittal section of a kidney shows that many branches of the renal artery and renal vein reach inside a kidney (Fig. 16.3*a*). Removing the blood vessels shows that a kidney has three regions (Fig. 16.3*b*). The **renal cortex** is an outer, granulated layer that dips down in between a radially striated inner layer called the renal medulla. The **renal medulla** consists of cone-shaped tissue masses called renal pyramids. The **renal pelvis** is a central space, or cavity, that is continuous with the ureter.

Anatomy of a Nephron

Each kidney is composed of over one million **nephrons,** microscopic structures that are the functional units of the kidney (Fig. 16.3*c*). A nephron consists of two combined structures: a **renal corpuscle** and a **renal tubule.** The renal corpuscle is formed by the **glomerulus,** a tightly coiled capillary network, and the **glomerular capsule** (also called *Bowman's capsule*). The glomerular capsule is formed by the closed end of the renal tubule, which widens to form a cuplike structure that surrounds the glomerulus. The blood supply of a nephron is unusual because it contains two distinct capillary regions (Fig. 16.4). The nephron receives oxygenated arterial blood from an afferent arteriole, which originates from a smaller branch of the renal artery. The afferent arteriole leads to the glomerulus, the first capillary bed. Blood leaves the glomerulus through the efferent arteriole. The efferent arteriole takes blood to the second, or **peritubular capillary network,** which surrounds the rest of the renal tubule. A specialized section of the peritubular capillary, the vasa recta, helps

renal artery

renal vein

ureter

renal cortex

a. Blood vessels

nephrons

collecting duct

renal pelvis

renal cortex

renal medulla

d. Nephrons

renal cortex

renal pelvis

ureter

renal pyramid
in renal medulla

renal pyramid
in renal medulla

b. Gross anatomy, photograph

c. Gross anatomy, art

Figure 16.3 **Gross anatomy of the kidney. a.** A sagittal section of the kidney showing the blood supply. Note that the renal artery divides into smaller arteries, and these divide into arterioles. Venules join to form small veins, which join to form the renal vein. **b.** Sagittal section photograph of the kidney. The same section without the blood supply. Now it is easier to distinguish the renal cortex, the renal medulla, and the renal pelvis, which connects with the ureter. The renal medulla consists of the renal pyramids. **d.** An enlargement showing the placement of nephrons.

to maintain solute concentration in the renal medulla, and will be discussed in detail later. From the peritubular capillary, the blood goes into a venule. Venules join larger veins that ultimately empty onto the renal vein.

Parts of a Renal Tubule

Each renal tubule is made up of several parts, beginning with the glomerular capsule. This wide proximal end of the tubule surrounds the glomerulus and receives the fluid it filters. The inner layer of the glomerular capsule is composed of *podocytes*

that have long, cytoplasmic extensions. The podocytes cling to the capillary walls of the glomerulus and leave pores that allow easy passage of small molecules from the glomerulus to the inside of the glomerular capsule.

Next, there is a **proximal convoluted tubule (PCT)**, called "proximal" because it is near the glomerular capsule. The cuboidal epithelial cells lining this part of the nephron have numerous microvilli about 1 μm in length. These microvilli are tightly packed and form a brush border, increasing the surface area for reabsorption.

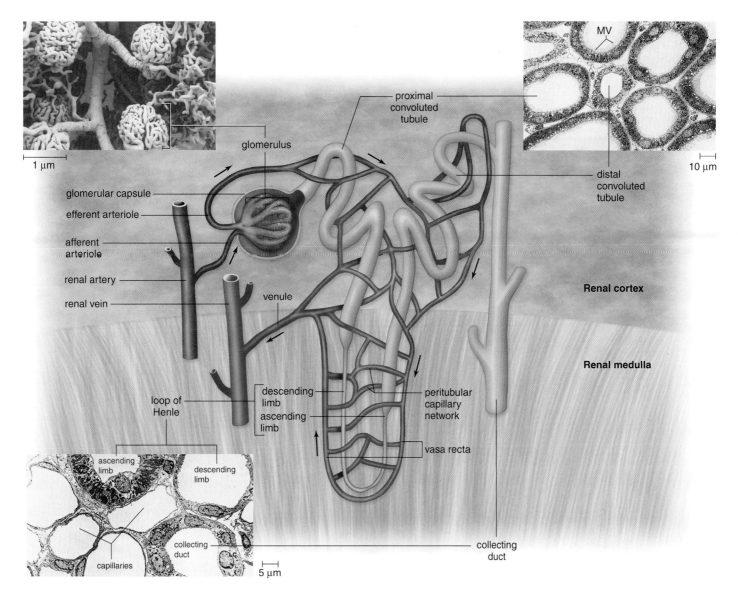

Figure 16.4 Nephron anatomy. A nephron is made up of a glomerular capsule, the proximal convoluted tubule, the loop of Henle, the distal convoluted tubule, and the collecting duct. The photomicrographs show these structures in cross section; MV = microvilli. Trace the path of blood about the nephron by following the arrows. *Source: Top left; R.G. Kessel and R.H. Kardon, Tissues and Organs: A Text Atlas of Scanning Electron Microscopy, 1979.*

Simple squamous epithelium appears as the tube narrows and makes a U-turn called the **loop of Henle**. Each loop consists of a descending limb and an ascending limb.

The cuboidal epithelial cells of the **distal convoluted tubule (DCT)** have numerous mitochondria, but they lack microvilli. This is consistent with the active role they play in moving molecules from the blood into the tubule, a process called tubular secretion. The distal convoluted tubules of several nephrons enter a single collecting duct. Many **collecting ducts** carry urine to the renal pelvis.

As shown in Figure 16.4, the glomerular capsule and the convoluted tubules always lie within the renal cortex. The loop of Henle dips down into the renal medulla; a few nephrons have a very long loop of Henle, which penetrates deep into the renal medulla. Collecting ducts are also located

in the renal medulla, and they give the renal pyramids their lined appearance.

Urine Formation

Figure 16.5 gives an overview of urine formation, which is divided into these steps: glomerular filtration, tubular reabsorption, and tubular secretion.

Glomerular Filtration

Glomerular filtration occurs when whole blood enters the afferent arteriole and the glomerulus. Due to glomerular blood pressure, water and small molecules are pushed from the glomerulus to the inside of the glomerular capsule. This is a filtration process because large molecules and formed

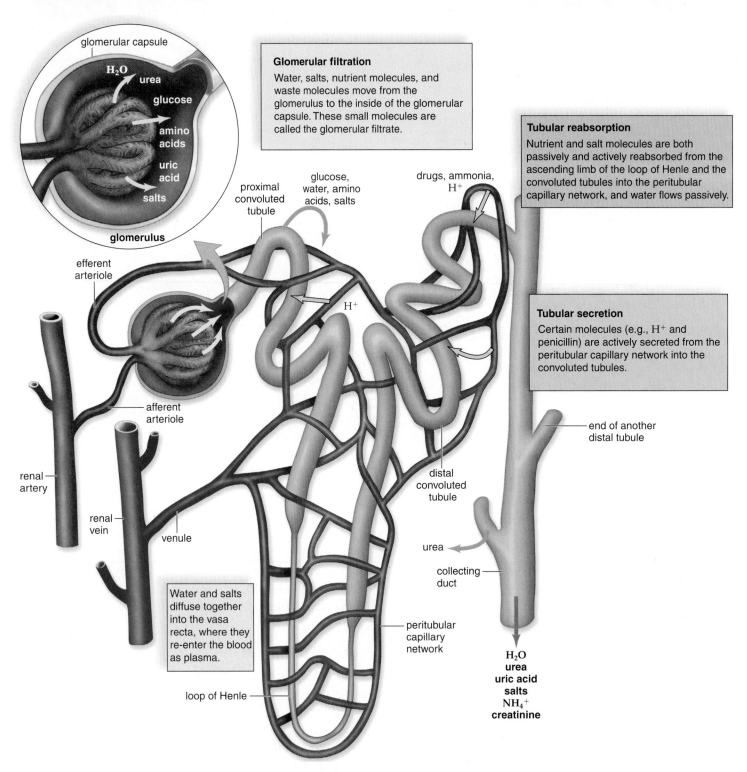

glomerular capsule

H_2O

urea

glucose

amino acids

uric acid

salts

glomerulus

Glomerular filtration

Water, salts, nutrient molecules, and waste molecules move from the glomerulus to the inside of the glomerular capsule. These small molecules are called the glomerular filtrate.

Tubular reabsorption

Nutrient and salt molecules are both passively and actively reabsorbed from the ascending limb of the loop of Henle and the convoluted tubules into the peritubular capillary network, and water flows passively.

proximal convoluted tubule

glucose, water, amino acids, salts

drugs, ammonia, H^+

efferent arteriole

H^+

Tubular secretion

Certain molecules (e.g., H^+ and penicillin) are actively secreted from the peritubular capillary network into the convoluted tubules.

afferent arteriole

end of another distal tubule

renal artery

distal convoluted tubule

renal vein

venule

urea

collecting duct

Water and salts diffuse together into the vasa recta, where they re-enter the blood as plasma.

peritubular capillary network

H_2O
urea
uric acid
salts
NH_4^+
creatinine

loop of Henle

Figure 16.5 Steps in urine formation. *Top:* The three main steps in urine formation are described in boxes that are color-coded to arrows showing the movement of molecules into or out of the nephron at specific locations. In the end, urine is composed of the substances within the collecting duct (see brown arrow, lower right).

TABLE 16.1 Reabsorption from Nephrons

Substance	Amount Filtered (per day)	Amount Excreted (per day)	Reabsorption (%)
Water (L)	180	1.8	99.0
Sodium (g)	630	3.2	99.5
Glucose (g)	180	0.0	100.0
Urea (g)	54	30.0	44.0

L = Liter; g = grams

elements are unable to pass through the capillary wall. In effect, then, blood in the glomerulus has two portions: the filterable components and the nonfilterable components:

Filterable Blood Components	Nonfilterable Blood Components
Water	Formed elements (blood cells and platelets)
Nitrogenous wastes	Large plasma proteins
Small molecules of nutrients	
Salts (ions)	
Creatinine and urochrome	

The **glomerular filtrate** contains small, dissolved molecules in approximately the same concentration as plasma. Small molecules that escape being filtered as well as the nonfilterable components leave the glomerulus by way of the efferent arteriole.

As indicated in Table 16.1, nephrons in the kidneys filter 180 liters of water per day, along with a considerable amount of small molecules (such as glucose) and ions (such as sodium). If the composition of the urine were the same as that of the glomerular filtrate, the body would constantly lose water, salts, and nutrients in the urine. This does not happen in a healthy person, because the composition of the original filtrate is altered significantly as this fluid passes through the remainder of the tubule.

It is important to stress that many small chemicals are freely filtered from the blood. Once in the filtrate, they cannot be reabsorbed and are passed in urine. Thus, urinalysis provides a history of chemicals recently present in the blood, and can be used for detection of illegal drug use, as described in Focus on Forensics (page 370.)

Tubular Reabsorption

Tubular reabsorption occurs as molecules and ions are both passively and actively reabsorbed from the nephron into the blood of the peritubular capillary network. The osmolarity of the blood is maintained by the presence of both plasma proteins and salt. When sodium ions (Na^+) are actively reabsorbed, chloride ions (Cl^-) follow passively. The reabsorption

of salt (NaCl) increases the osmolarity of the blood compared to the filtrate, and therefore water moves by osmosis from the tubule into the blood. About 67% of Na^+ is reabsorbed at the proximal convoluted tubule.

Nutrients such as glucose and amino acids also return to the blood at the proximal convoluted tubule. This is a selective process. Only molecules recognized by carrier proteins built into tubular cell membranes will be actively reabsorbed. Glucose is an example of a molecule that ordinarily is completely reabsorbed because there is a plentiful supply of carrier molecules for it. However, every substance has a maximum rate of transport. After all its carriers are in use, any excess in the filtrate will appear in the urine. For example, as reabsorbed levels of glucose approach 1.8–2 mg/ml plasma, the rest appears in the urine. Excess glucose occurs in the blood in untreated diabetes mellitus because of lack of insulin (type I diabetes), or because of failure of insulin receptors to respond normally (type II diabetes). Without insulin function, glucose cannot be transported into cells, and liver and muscle cannot store glucose as glycogen. Excess blood glucose is filtered into the tubular fluid and subsequently passes into the urine because the kidneys cannot reabsorb it all. The presence of glucose in the filtrate increases its osmolarity compared to that of the blood, and therefore less water is reabsorbed into the peritubular capillary network. The frequent urination and increased thirst experienced by untreated diabetics are due to the fact that water isn't being reabsorbed.

We've seen that the filtrate that enters the proximal convoluted tubule is divided into two portions: components that are reabsorbed from the tubule into the blood and components that are not reabsorbed and continue to pass through the nephron to be further processed into urine:

Reabsorbed Filtrate Components	Nonreabsorbed Filtrate Components
Most water	Some water
Nutrients	Much nitrogenous waste
Required salts (ions)	Excess salts (ions)
Some urea and uric acid	Creatinine and urochrome
	Molecules not recognized by carriers

The substances that are not reabsorbed become the tubular fluid, which enters the loop of Henle.

Tubular Secretion

Tubular secretion is a second way by which substances are removed from blood and actively transported into the tubular fluid. Hydrogen ions, potassium ions, creatinine, and drugs such as penicillin are some of the substances that are moved by active transport from the blood into the distal convoluted tubule. In the end, urine contains (1) substances that have undergone glomerular filtration but have not been reabsorbed, and (2) substances that have undergone tubular secretion.

4. Select the correct order of the blood supply for a nephron:

   a. afferent arteriole → efferent arteriole → glomerulus → peritubular capillary

   b. efferent arteriole → glomerulus → afferent arteriole → peritubular capillary

   c. afferent arteriole → glomerulus → efferent arteriole → peritubular capillary

   d. glomerulus → afferent arteriole → efferent arteriole → peritubular capillary

5. Select the correct order of the parts of a nephron:

   a. glomerular capsule → proximal convoluted tubule → loop of Henle → distal convoluted tubule

   b. glomerular capsule → proximal convoluted tubule → distal convoluted tubule → loop of Henle

   c. proximal convoluted tubule → glomerular capsule → loop of Henle → distal convoluted tubule

   d. proximal convoluted tubule → loop of Henle → distal convoluted tubule → glomerular capsule

6. Name two substances you would expect to find in the blood, but not the tubular fluid.

Answers in Appendix C, page 465.

16.3 Regulatory Functions of the Kidneys

The kidneys are involved in maintaining the fluid and electrolyte balance, and also the acid-base balance, of the blood. If the kidneys fail to carry out these vital functions, either hemodialysis or a kidney transplant is needed.

Fluid and Electrolyte Balance

The average adult male body is about 60% water by weight. The average adult female body is only about 50% water by weight because females generally have more subcutaneous adipose tissue, which contains less water. About two-thirds of this water is inside the cells (called *intracellular fluid*), and the rest is largely distributed in the plasma, tissue fluid, and lymph (called *extracellular fluid*). Water is also present in such fluids as cerebrospinal fluid and synovial fluid; in Figure 16.6, these fluids are referred to as "other" fluids.

For body fluids to be normal, it is necessary for the body to be in fluid balance. The total water intake should equal the total water loss. Table 16.2 shows how water enters the body—namely, in liquids we drink, in foods we eat, and as a by-product of metabolism. We drink water when the osmolarity of the blood rises as determined by the hypothalamus. Table 16.2 also shows how water exits the body—namely, in urine, sweat, exhaled air, and feces. Similar to the gain and loss of water, the body also gains and loses electrolytes. Despite these changes, the kidneys keep the fluid and electrolyte balance of the blood within normal limits. In this way, they also maintain the blood volume and blood pressure.

Reabsorption of Water

Because of the process of osmosis (see Chapter 3, page 50), the reabsorption of salt (NaCl) automatically leads to the reabsorption of water until the osmolarity is the same on both sides of a plasma membrane. Most of the salt, and therefore water, present in the filtrate are reabsorbed across

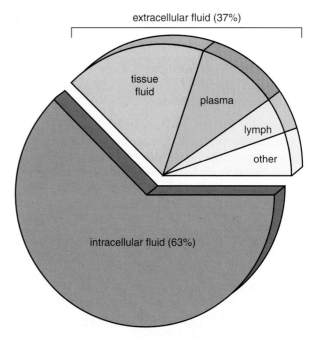

Figure 16.6 Location of fluids in the body. Most of the body's water is inside cells (intracellular fluid), and only about one-third is located outside cells (extracellular fluid).

TABLE 16.2 Fluid Balance			
Water Input	**Average ml/day and % of Total**	**Water Output**	**Average ml/day and % of Total**
In liquids	1,500; 60%	In urine	1,300; 52%
In moist food	750; 30%	In sweat	650; 26%
From metabolism	250; 10%	In exhaled air	450; 18%
		In feces	100; 4%
	Total 2,500; 100%		Total 2,500; 100%

the plasma membranes of the cells lining the proximal convoluted tubule. But the amount of salt and water reabsorbed is not sufficient to result in a hypertonic urine—one in which the osmolarity is higher than that of blood. How is it, then, that humans produce a hypertonic urine? We now know that the excretion of a hypertonic urine is dependent upon the reabsorption of water from the loop of Henle and the collecting duct.

Loop of Henle and Collecting Duct A loop of Henle has a descending limb (*descending* through the medulla toward the center of the kidney, in the area near the renal pelvis) and an ascending limb (*ascending* toward the renal cortex). The kidney's longest loops of Henle penetrate deep into the center of the renal medulla (just adjacent to the renal pelvis). The ascending limb has two portions that are structurally different from each other. Salt (NaCl) passively diffuses out of the narrow, first portion of the ascending loop of Henle because it has a thinner wall. The wide, thick-walled second portion actively transports salt into the tissue of the renal medulla (Fig. 16.7). As the second portion of the loop continues to transport salt, less and less salt is available in the fluid moving through it. Therefore, the concentration of salt is greater in the center of the medulla than in the outer rim of the medulla. (It's important to realize that water cannot leave the thick segment of the ascending limb because the ascending limb is impermeable to water.)

The large arrow on the left side of Figure 16.7 indicates that the central portion of the medulla (closest to the renal pelvis) has the highest concentration of solutes. You can see that this is due not only to the presence of salt, but also to the presence of urea. Urea is reabsorbed from the lower portion of the collecting duct, and it is this molecule that contributes to the high solute concentration of the lowest portion of the inner medulla.

Because of the osmotic gradient within the renal medulla, water leaves the descending limb along its entire length. There is a higher concentration of water in the first portion of the descending limb. Thus, less solute is needed for water to be able to diffuse into the renal medulla. The remaining fluid within the descending limb encounters an increasingly greater osmotic concentration of solute as it moves along, so that water continues to leave the descending limb along its entire length. Such a mechanism is called a *countercurrent mechanism*.

At the top of the ascending limb, any remaining water enters the distal convoluted tubule. Surprisingly, the fluid inside the nephron is still not hypertonic—the net effect of reabsorption of salt and water so far is the production of a fluid that has the same tonicity as blood plasma. However, the collecting duct also encounters the same osmotic gradient as did the descending limb of the loop of the nephron (Fig. 16.7). Therefore, water diffuses out of the collecting duct into the renal medulla, and the urine within the collecting duct becomes hypertonic to blood plasma.

As you can see in Figures 16.4 and 16.5, the renal tubule is surrounded by peritubular capillaries. As blood flows through these capillaries, water will naturally diffuse from high concentration (in the blood) to low concentration (in the renal

medulla) because the medulla has high urea and salt concentrations. How is it, then, that the high particle concentration found in the medulla isn't simply "washed away"? A specialized, U-shaped peritubular capillary, the **vasa recta**, parallels the loop of Henle and prevents this from happening. As it passes alongside the loop, the vasa recta collects both salt and water in equal amounts, forming plasma that is quickly transported out of the kidney and back to the circulation. This helps to minimize loss of solutes from the medulla.

Antidiuretic Hormone (ADH) ADH produced by the hypothalamus, and released by the posterior lobe of the pituitary, plays a role in water reabsorption at the distal convoluted tubule and collecting duct. In order to understand the action of this hormone, consider its name. *Diuresis* means flow of urine, and *antidiuresis* means against a flow of urine. In practical terms, if an individual does not drink much water on a given day, the hypothalamus will detect an increase in the osmotic pressure of the blood (i.e., the solutes in the blood will

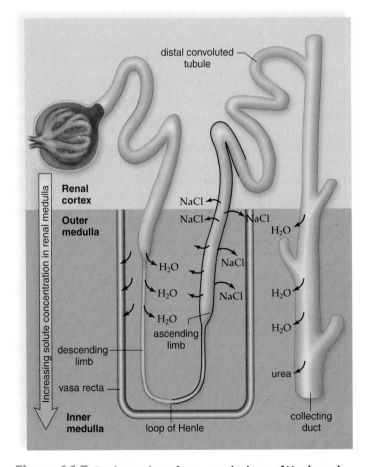

Figure 16.7 Reabsorption of water at the loop of Henle and the collecting duct. A hypertonic environment in the tissues of the medulla of a kidney draws water out of the descending limb and the collecting duct. This water is returned to the cardiovascular system. The thick black line means the ascending limb is impermeable to water. In the distal convoluted tubule and collecting duct, water permeability is hormone dependent.

become more concentrated). As an analogy, imagine leaving a glass of salt water on a windowsill for a week or more. The salt will become more concentrated over time because water is lost by evaporation. In the same way, blood solutes become more concentrated if the water lost daily isn't replaced by drinking. In response to increased blood osmolarity, the hypothalamus will produce ADH, which is then released from the posterior pituitary. ADH is transported to the cells of the collecting duct. There, it causes specialized cell membrane channels called *aquaporins* to open. Water diffuses from the collecting duct to the interstitial space of the kidney through these aquaporins. Thus, more water is reabsorbed into the blood, blood volume and blood pressure rise, and less urine is formed. On the other hand, if an individual drinks a large amount of water and does not perspire much, ADH is not released. Collecting duct aquaporins remain closed, more urine forms, and more water is excreted.

Reabsorption of Electrolytes

As previously discussed, the osmolarity of body fluids is dependent upon the concentration of particular electrolytes within the fluids. **Electrolytes** are compounds and molecules that are able to ionize and, thus, carry an electrical current. The kidneys regulate electrolyte excretion and therefore help control blood composition.

The Electrolytes The most common electrolytes in the plasma are sodium (Na^+), chloride (Cl^-), and bicarbonate (HCO_3^-). The most common electrolytes in the intracellular fluid are potassium (K^+) and phosphate (PO_4^{-3}). Na^+ and K^+ are termed *cations* because they are positively charged. Cl^-, HCO_3^-, and PO_4^{-3} are termed *anions* because they are negatively charged (see Chapter 2).

Sodium The movement of Na^+ across an axon membrane, you will recall, is necessary to the formation of a nerve action potential signal and also to muscle contraction. The concentration of Na^+ in the blood is also the best indicator of the blood's osmolarity.

Potassium The movement of K^+ across an axon membrane is also necessary to the formation of a nerve action potential and muscle contraction. Abnormally low K^+ concentrations in the blood, as might occur if diuretics are abused, can lead to cardiac arrest.

Bicarbonate ion HCO_3^- is the form in which carbon dioxide is carried in the blood. The bicarbonate ion has the very important function of helping maintain the pH of the blood, as will be discussed later in this section.

Other ions The plasma contains many other ions. For example, calcium ions (Ca^{2+}) and phosphate ions (HPO_4^{2-}) are important to bone formation and cellular metabolism. Their absorption from the intestine and excretion by the kidneys is regulated by hormones, as discussed in Chapter 10.

The Kidneys More than 99% of sodium (Na^+) filtered at the glomerulus is returned to the blood. Most sodium (67%) is reabsorbed at the proximal convoluted tubule, and a sizable amount (25%) is actively transported from the tubule by the ascending limb of the loop of Henle. The rest is reabsorbed from the distal convoluted tubule and collecting duct.

Aldosterone Hormones control the reabsorption of sodium and water in the distal portion of the renal tubule. **Aldosterone,** a hormone secreted by the adrenal cortex, promotes the excretion of potassium ions (K^+) and the reabsorption of sodium ions (Na^+). Water is subsequently reabsorbed by osmosis. Aldosterone acts on the ascending limb of the loop of Henle, the distal convoluted tubule, and the collecting ducts.

The release of aldosterone is set in motion by the kidneys themselves. The **juxtaglomerular apparatus** is a region of contact between the afferent arteriole and the distal convoluted tubule (Fig. 16.8). When blood volume, and therefore blood pressure, is not sufficient to promote glomerular filtration, the juxtaglomerular apparatus secretes renin. **Renin** is an enzyme that changes angiotensinogen (a large plasma protein produced by the liver) into angiotensin I. Later, angiotensin I is converted to angiotensin II, a powerful vasoconstrictor that also stimulates the adrenal cortex to release aldosterone. The reabsorption of sodium ions is followed by the reabsorption of water. Therefore, blood volume and blood pressure increase.

Atrial Natriuretic Peptide (ANP) ANP (sometimes referred to as atrial natriuretic hormone) is a hormone secreted by the atria of the heart when cardiac cells are stretched due to increased blood volume. ANP inhibits the secretion of renin by the juxtaglomerular apparatus and the secretion of aldosterone by the adrenal cortex. Its effect, therefore, is to promote the excretion of Na^+, called *natriuresis*. When Na^+ is excreted, so is water, and therefore blood volume and blood pressure decrease.

Diuretics

Diuretics are chemicals that increase the flow of urine. Drinking alcohol causes diuresis because it inhibits the secretion of ADH. The dehydration that follows is believed to contribute to the symptoms of a hangover. Caffeine is a diuretic because it increases the glomerular filtration rate and decreases the tubular reabsorption of Na^+. Diuretic drugs developed to counteract high blood pressure cause water reabsorption to diminish. Subsequently, blood pressure decreases because cardiac output is diminished.

Begin Thinking Clinically

? Two categories of diuretic drugs are separately classified as *loop diuretics* and *aldosterone antagonists*. Where in the kidney tubule would each of these function? How does each category work? Can you name a trade/generic version for each?

Answers and discussion in Appendix C, page 465.

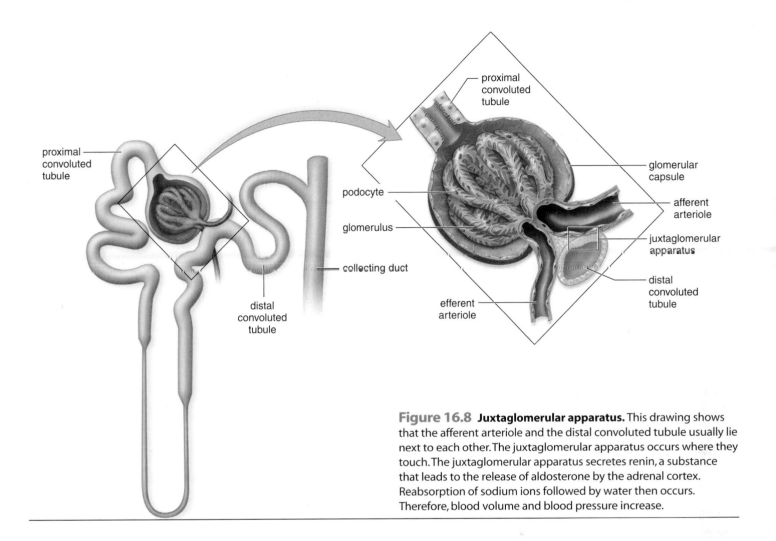

Figure 16.8 Juxtaglomerular apparatus. This drawing shows that the afferent arteriole and the distal convoluted tubule usually lie next to each other. The juxtaglomerular apparatus occurs where they touch. The juxtaglomerular apparatus secretes renin, a substance that leads to the release of aldosterone by the adrenal cortex. Reabsorption of sodium ions followed by water then occurs. Therefore, blood volume and blood pressure increase.

Acid-Base Balance

The pH scale, as discussed in Chapter 2, can be used to indicate the basicity (alkalinity) or the acidity of body fluids. A basic solution has a lower hydrogen ion concentration $[H^+]$ than the neutral pH of 7.0. An acidic solution has a greater $[H^+]$ than neutral pH. The normal pH for body fluids is about 7.4. This is the pH at which our proteins, such as cellular enzymes, function properly. If the blood pH rises above 7.4, a person is said to have **alkalosis,** and if the blood pH decreases below 7.4, a person is said to have **acidosis.** Alkalosis and acidosis are abnormal conditions that may need medical attention.

The foods we eat add basic or acidic substances to the blood, and so does metabolism. For example, cellular respiration adds carbon dioxide that combines with water to form carbonic acid, and fermentation adds lactic acid. The pH of body fluids stays at just about 7.4 via several mechanisms, primarily acid-base buffer systems, the respiratory center, and the kidneys.

Acid-Base Buffer Systems

The pH of the blood stays near 7.4 because the blood is buffered. A **buffer** is a chemical or a combination of chemicals that can take up excess hydrogen ions (H^+) or excess hydroxide ions (OH^-). Proteins are the primary buffers of both the intracellular and extracellular fluid. Their amino acids can combine with or release hydrogen ions as needed. One of the most important buffers in the blood is a combination of carbonic acid (H_2CO_3) and bicarbonate ions (HCO_3^-). Carbonic acid is a weak acid that minimally dissociates and re-forms in the following manner:

H_2CO_3	dissociates $\rightleftharpoons$ re-forms	H^+	+	HCO_3^-
carbonic acid		hydrogen ion		bicarbonate ion

When hydrogen ions (H^+) are added to blood, the following reaction occurs:

$$H^+ + HCO_3^- \rightarrow H_2CO_3$$

When hydroxide ions (OH^-) are added to blood, this reaction occurs:

$$OH^- + H_2CO_3 \rightarrow HCO_3^- + H_2O$$

These reactions temporarily prevent any significant change in blood pH. A blood buffer, however, can be overwhelmed unless some more permanent adjustment is made. The next adjustment to keep the pH of the blood constant occurs at pulmonary capillaries.

Respiratory Regulation of Acid-Base Balance

As discussed in Chapter 14, the respiratory center in the medulla oblongata increases the breathing rate, as well as the depth of respiration, if the hydrogen ion concentration of the blood rises. This change rids the body of hydrogen ions because the following reaction takes place in pulmonary capillaries:

$$H^+ + HCO_3^- \rightleftharpoons H_2CO_3 \rightleftharpoons H_2O + CO_2$$

In other words, when carbon dioxide is exhaled, hydrogen ions recombine with bicarbonate ions to form carbonic acid. Carbonic acid then dissociates to form the carbon dioxide that is eliminated from the body. Now the hydrogen ions are no longer free in solution, but tied up harmlessly with water. Increasing the respiratory rate provides the body with a rapid way to remove excess free hydrogen ions from the blood.

It is important to have the correct proportion of carbonic acid and bicarbonate ions in the blood. Breathing readjusts this proportion so that this particular acid-base buffer system can continue to absorb both H^+ and OH^- as needed.

Renal Regulation of Acid-Base Balance

As powerful as the acid-base buffer and the respiratory center mechanisms are, only the kidneys can rid the body of a wide range of acidic and basic substances and otherwise adjust the pH. The kidneys are slower acting than the other two mechanisms, but they have a more powerful effect on pH. For the sake of simplicity, we can think of the kidneys as reabsorbing bicarbonate ions and excreting hydrogen ions as needed to maintain the normal pH of the blood (Fig. 16.9). If the blood is acidic,

Figure 16.9 Acid-base balance. In the kidneys, bicarbonate ions (HCO_3^-) are reabsorbed and hydrogen ions (H^+) are excreted as needed to maintain the pH of the blood. Excess hydrogen ions are buffered, for example, by ammonia (NH_3), which becomes ammonium (NH_4^+). Ammonia is produced in tubule cells by the deamination of amino acids.

hydrogen ions are excreted and bicarbonate ions are reabsorbed. If the blood is basic, hydrogen ions are not excreted and bicarbonate ions are not reabsorbed. Because the urine is usually acidic, it follows that an excess of hydrogen ions is usually excreted. Ammonia (NH_3) provides a means for buffering these hydrogen ions in urine. Ammonia (whose presence is quite obvious in the diaper pail or kitty litter box) is produced by the deamination of amino acids in both liver and kidney tubule cells. Ammonia is a basic molecule that diffuses easily through tubule cells into the urine. Once there, it combines with hydrogen ions to form ammonium ions (NH_4^+) by this reaction:

$$NH_3 + H^+ \rightarrow NH_4^+$$

Ammonium ions cannot diffuse back out of the kidney tubules and pass out of the body in the urine. Phosphate provides another means of buffering hydrogen ions in urine.

The importance of the kidneys' ultimate control over the pH of the blood cannot be overemphasized. As mentioned, the enzymes of cells cannot continue to function if the internal environment does not have near-normal pH.

Content **CHECK-UP!**

7. Which of the following statements about body water is correct?
   a. Males have a higher percentage of body water by weight than females.
   b. Most of the body's water is found inside cells.
   c. Water can be lost from the body in exhaled air.
   d. A portion of daily water intake comes from water produced during metabolism.
   e. All of these statements are correct.

8. Which hormone is produced by the hypothalamus and released from the posterior pituitary?
   a. atrial natriuretic peptide (ANP)
   b. aldosterone
   c. antidiuretic hormone (ADH)
   d. angiotensin

9. Which hormone causes excretion of sodium and water from the body?
   a. atrial natriuretic peptide (ANP)
   b. aldosterone
   c. antidiuretic hormone (ADH)
   d. angiotensin

Answers in Appendix C, page 465.

16.4 Problems with Kidney Function

The composition of normal urine is given in Table 16.3. Water accounts for almost all of the volume of urine (95%). The remaining 5% consists of electrolytes and various solutes, including nitrogenous end products and substances derived from drugs. Notice that urine is typically free of proteins and blood cells because they are not filtered at the glomerulus.

Prostate Enlargement and Cancer

The prostate gland, which is part of the male reproductive system, surrounds the urethra at the point where the urethra leaves the urinary bladder. The prostate gland produces and adds a fluid to semen as semen passes through the urethra within the penis. At about age 50, the prostate gland often begins to enlarge, growing from the size of a walnut to that of a lime or even a lemon. This condition is called **benign prostatic hyperplasia (BPH).** As it enlarges, the prostate squeezes the urethra, causing urine to back up—first into the bladder, then into the ureters, and finally, perhaps, into the kidneys.

The treatment for BPH can involve (1) taking a drug that is expected to shrink the prostate and/or improve urine flow, or (2) undergoing a more invasive procedure to reduce the size of the prostate. Prostate tissue can be destroyed by applying microwaves to a specific portion of the prostate. In many cases, however, a physician may decide that prostate tissue should be removed surgically. In some cases, rather than performing abdominal surgery, which requires an incision in the abdomen, the physician gains access to the prostate via the urethra. This operation, called transurethral resection of the prostate (TURP), requires careful consideration because one study found that the death rate during the 5 years following TURP is much higher than that following abdominal surgery.

Prostate enlargement is due to a prostate enzyme (5α-reductase) that acts on the male sex hormone testosterone, converting it into a substance that promotes prostate growth. That growth is fine during puberty, but continued growth in an adult is undesirable. Two substances, one a nutrient supplement and the other a prescription drug, interfere with the action of this enzyme. Saw palmetto, which is sold in tablet form as an over-the-counter nutrient supplement, is derived from a plant of the same name. This drug should not be taken unless the need for it is confirmed by a physician, but it is particularly effective during the early stages of prostate enlargement. Finasteride, a prescription drug, is a more powerful inhibitor of the enzyme, but patients complain of erectile dysfunction and loss of libido while on the drug.

Two other medications have a different mode of action. Nafarelin prevents the release of luteinizing hormone, which leads to testosterone production. When it is administered, approximately half of the patients report relief of urinary symptoms even after drug treatment is halted. However, again, the patients experience erectile dysfunction and other side effects, such as hot flashes. The drug terazosin, which is on the market for hypertension because it relaxes arterial walls, also relaxes muscle tissue in the prostate. Improved urine flow was experienced by 70% of the patients taking this drug. However, the drug has no effect on the prostate's overall size.

Many men are concerned that BPH may be associated with prostate cancer, but the two conditions are not necessarily related. BPH occurs in the inner zone of the prostate, while cancer tends to develop in the outer area. If prostate cancer is suspected, blood tests and a biopsy, in which a tiny sample of prostate tissue is surgically removed, will confirm the diagnosis.

Although prostate cancer is the second most common cancer in men, it is not a major killer. Typically, prostate cancer is so slow growing that the survival rate is about 98% if the condition is detected early.

possibility of organ rejection. Receiving a kidney from a close relative has the highest chance of success. The current one-year survival rate is 97% if the kidney is received from a relative and 90% if it is received from a non-relative. Even with a successful surgery, the transplant patient must take antirejection medication for the rest of his or her life.

16.5 Effects of Aging

Urinary disorders are significant causes of illness and death among the elderly. Total renal function in an elderly individual may be only 50% of that of the young adult. With increasing age, the kidneys decrease in size and have significantly fewer nephrons. However, vascular changes may play a more significant role in declining renal efficiency than renal tissue loss. Microscopic examination shows many degenerate glomeruli through which blood no longer flows and many other glomeruli that are completely destroyed.

Kidney stones occur more frequently with age, possibly as a result of improper diet, inadequate fluid intake, and kidney infections. Infections of the urethra, bladder, ureters, and kidneys increase in frequency among the elderly. Enlargement of the prostate occurs in males and, as is discussed in the Medical Focus, this can lead to urine retention and kidney disease. Cancer of the prostate and bladder are the most common cancers of the urogenital system.

The involuntary loss of urine, called **incontinence,** increases with age. The bladder of an elderly person has a capacity of less than half that of a young adult and often contains residual urine. Therefore, urination is more urgent and frequent.

16.6 Homeostasis

The illustration in Human Systems Work Together on page 372 tells how the urinary system works with the other systems of the body to help maintain homeostasis.

Human Systems Work Together

Integumentary System

Kidneys compensate for water loss due to sweating; activate vitamin D precursor made by skin.

Skin helps regulate water loss; sweat glands carry on some excretion.

Skeletal System

Kidneys provide active vitamin D for Ca^{2+} absorption and help maintain blood level of Ca^{2+}, needed for bone growth and repair.

Bones provide support and protection.

Muscular System

Kidneys maintain blood levels of Na^+, K^+, and Ca^{2+}, which are needed for muscle innervation, and eliminate creatinine, a muscle waste.

Smooth muscular contraction assists voiding of urine; skeletal muscles support and help protect urinary organs.

Nervous System

Kidneys maintain blood levels of Na^+, K^+, and Ca^{2+}, which are needed for nerve conduction.

Brain controls nerves, which innervate muscles that permit urination.

Cardiovascular System

Kidneys keep blood values within normal limits so that transport of hormones continues.

ADH and aldosterone, and atrial natriuretic peptide regulate reabsorption of Na^+ by kidneys.

How the Urinary System works with other body systems.

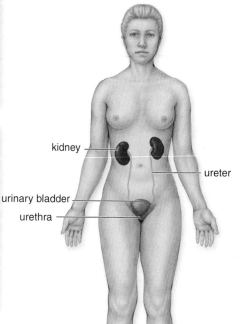

kidney —
— ureter
urinary bladder —
urethra —

Reproductive System

Semen is discharged through the urethra in males; kidneys excrete wastes and maintain electrolyte levels for mother and child.

Penis in males contains the urethra and performs urination; prostate enlargement hinders urination.

Cardiovascular System

Kidneys filter blood and excrete wastes; maintain blood volume, pressure, and pH; produce renin and erythropoietin.

Blood vessels deliver waste to be excreted; blood pressure aids kidney function; heart produces atrial natriuretic peptide.

Lymphatic System/Immunity

Kidneys control volume of body fluids, including lymph.

Lymphatic system picks up excess tissue fluid, helping to maintain blood pressure for kidneys to function; immune system protects against infections.

Respiratory System

Kidneys compensate for water lost through respiratory tract; work with lungs to maintain blood pH.

Lungs excrete carbon dioxide, provide oxygen, and convert angiotensin I to angiotensin II, leading to kidney regulation.

Digestive System

Kidneys convert vitamin D to active form needed for Ca^{2+} absorption; compensate for any water loss by digestive tract.

Liver synthesizes urea; digestive tract excretes bile pigments from liver and provides nutrients.

Recall that *excretion* means to rid the body of a metabolic waste. Using this definition, it is possible to classify three other organs in addition to the kidneys as excretory organs:

1. The sweat glands in the skin excrete perspiration, which is a solution of water, salt, and some urea. Although perspiration is a form of excretion, we perspire more to cool the body than to rid it of wastes. In times of kidney failure, urea is excreted by the sweat glands and forms a so-called uremic frost on the skin.

2. The liver breaks down hemoglobin and excretes bile pigments, which are derived from heme. Bile pigments are incorporated into bile, a substance stored in the gallbladder before it passes into the small intestine by way of ducts.

   The yellow pigment found in urine, called urochrome, also is derived from the breakdown of heme, but this pigment is deposited in blood and is subsequently excreted by the kidneys. As you know, the liver also produces urea—our main nitrogenous end product—which is excreted by the kidneys.

3. The lungs excrete carbon dioxide. The process of exhalation not only removes carbon dioxide, but also results in the loss of water. The air we exhale contains moisture, as demonstrated by breathing onto a cool mirror.

The kidneys are the primary organs of excretion. They excrete almost all of our nitrogenous wastes—namely urea, creatinine, and uric acid. The liver makes urea, and muscles make creatinine. Urea is the end product of protein metabolism, and creatinine is a breakdown product of creatine phosphate, a molecule that stores energy in muscles. Uric acid is produced by cells when they break down nucleotides. Nitrogenous wastes are carried by the cardiovascular system to the kidneys. In this way, the cardiovascular system and the kidneys work together to clear the blood of nitrogenous end products. The excretion of nitrogenous wastes may not be as critical as maintaining the water-salt and the acid-base balances, but it is still necessary because urea is a toxic substance.

The kidneys are primary organs of homeostasis because they maintain the water-salt (electrolyte) and the acid-base balances of the blood. If blood does not have the usual water-salt balance, blood volume and blood pressure are affected. Without adequate blood pressure, exchange across capillary walls cannot take place, nor is glomerular filtration possible in the kidneys themselves. The kidneys and the endocrine system work together to help maintain blood pressure. The production of renin by the kidneys and subsequently the renin-angiotensin-aldosterone sequence help ensure that the sodium (Na^+) concentration of the blood, and therefore osmolarity and blood pressure, stay normal. The lymphatic system assists the urinary system because it makes a significant contribution to blood pressure by picking up excess tissue fluid and returning it to the cardiovascular veins.

Aside from producing renin, the kidneys assist the endocrine system and also the cardiovascular system by producing erythropoietin. Erythropoietin stimulates red bone marrow to produce red blood cells. The kidneys assist the skeletal, nervous, and muscular systems by helping to regulate the amount of calcium ions (Ca^{2+}) in the blood. The kidneys convert vitamin D to its active form needed for Ca^{2+} absorption by the digestive tract, and they regulate the excretion of electrolytes, including Ca^{2+}. The kidneys also regulate the sodium (Na^+) and potassium (K^+) content of the blood. These ions are necessary to the contraction of the heart and other muscles in the body, and are also needed for nerve conduction.

We have already described how the kidneys work with the cardiovascular system and the respiratory system to maintain the acid-base balance (pages 367–368). This is a critical function to prevent the occurrence of alkalosis, or acidosis, both of which are life-threatening conditions. This function must be performed by a machine when people undergo hemodialysis. So, while we tend to remember that the kidneys excrete urea, we must also keep in mind all the other functions of the kidneys that are absolutely essential to homeostasis.

SELECTED NEW TERMS

Basic Key Terms

acidosis (ăs'ĭdō'sĭs), p. 367

aldosterone (ăl-dŏs'tə-rōn), p. 366

alkalosis (ăl'kə-lō'sĭs), p. 367

antidiuretic hormone (ADH), (an'tĭ-dī'u-ret'ik hôr'mōn'), p. 365

atrial natriuretic peptide (ANP), (a'tre-al na'tre-u-ret'ik pep'tīd), p. 366

buffer (bŭf'ər), p. 367

collecting duct (kuh-lek'ting dukt), p. 361

creatinine (kre-at'ĭ-nēn), p. 357

distal convoluted tubule (dis'tal kon'vo-lūt-ed tu'byūl), p. 361

electrolytes (ĭ-lĕk'trə-lit'), p. 366

erythropoietin (ĭ-rĭth'rō-poi-ē'tĭn), p. 357

excretion (ek-skre'shun), p. 357

glomerular capsule (glo-mĕr'yū-ler kap'sul), p. 359

glomerular filtrate (glo-mĕr'yū-ler fil'trāt), p. 363

glomerular filtration (glo-mĕr'yū-ler filtra'shun), p. 361

glomerulus (glo-mĕr'yū-lus), p. 359

gout (gout), p. 357

juxtaglomerular apparatus (juks"tuh-glo-mĕr'yū-ler ap"uh-ra'tus), p. 366

kidney (kid'ne), p. 357

loop of Henle (lōōp of Hĕn'lē), p. 361

micturition (mik"tu-rish'un), p. 359

nephron (nef'ron), p. 359

peritubular capillary network (per"ĭ-tu'byū-ler kap'ĭ-lār"e net'werk), p. 359

proximal convoluted tubule (prok'sĭ-mal kon'vo-lū-ted tu'byūl), p. 360

prostate gland (pros'tāt glănd), p. 359

renal corpuscle (rē'năl kŏr'pŭs-l), p. 359

renal cortex (re'nul kor'teks), p. 359

renal medulla (re'nul mĕ-du'luh), p. 359

renal pelvis (re'nul pel'vis), p. 359

renal tubule (rē'năl too'būl), p. 359

renin (rē″nĭn), p. 366

tubular reabsorption (tu′byū-ler re″ab-sorp′shun), p. 363

tubular secretion (tu′byū-ler sĕ-kre′shun), p. 363

urea (yoŏ-rē′ə), p. 357

ureters (yū-re′terz), p. 358

urethra (yū-re′thruh), p. 358

uric acid (yū-rik as′id), p. 357

urinary bladder (yūr′in-ār-e blad′er), p. 358

vasa recta (vā′sä rec′tă), p. 365

Clinical Key Terms

benign prostatic hyperplasia (bĭ-nīn′ prahs-tat′ik hi″per-pla′ze-uh), p. 371

cystitis (sis′ti′tis), p. 369

dialysate (dī-al′ĭ-sāt), p. 369

diuretic (di-yū-ret′ik), p. 366

floating kidney (flōt′ing kid′ne), p. 358

hemodialysis (he-mo-di-al′ĭ-sis), p. 369

incontinence (in-con′tin-ents), p. 371

pyelonephritis (pi-lo-nef-ri′tis), p. 369

uremia (yū-re′me-uh), p. 369

urethritis (yū-re-thri′tis), p. 369

urinalysis (yū-rĭ-nal′ĭ-sis), p. 369

SUMMARY

16.1 Urinary System

A. The kidneys excrete nitrogenous wastes, including urea, uric acid, and creatinine. They maintain the normal water-salt balance, blood pressure, and acid-base balance of the blood, as well as influencing the secretion of certain hormones.

B. The kidneys produce urine, which is conducted by the ureters to the bladder, where it is stored before being released by way of the urethra.

16.2 Anatomy of the Kidney and Excretion

A. Macroscopically, the kidneys are divided into the renal cortex, renal medulla, and renal pelvis. Microscopically, they contain the nephrons.

B. Each nephron has its own blood supply; the afferent arteriole approaches the glomerular capsule and divides to become the glomerulus, a knot of capillaries. The efferent arteriole leaves the capsule and immediately branches into the peritubular capillary network.

C. Each region of the nephron is anatomically suited to its task in urine formation. The spaces between the podocytes of the glomerular capsule allow small molecules to enter the capsule

from the glomerulus. The cuboidal epithelial cells of the proximal convoluted tubule have many mitochondria and microvilli to carry out active transport (following passive transport) from the tubule to the blood. In contrast, the cuboidal epithelial cells of the distal convoluted tubule have numerous mitochondria but lack microvilli. They carry out active transport from the blood to the tubule.

D. The steps in urine formation are glomerular filtration, tubular reabsorption, and tubular secretion.

16.3 Regulatory Functions of the Kidneys

A. The kidneys regulate the fluid and electrolyte balance of the body. Water is reabsorbed from certain parts of the tubule, and the loop of Henle establishes an osmotic gradient that draws water from the descending loop of the nephron and also from the collecting duct. The permeability of the collecting duct is under the control of the hormone ADH. The reabsorption of salt increases blood volume and pressure because more water is also reabsorbed. Two other hormones, aldosterone and ANH, control

the kidneys' reabsorption of sodium (Na^+) and water.

B. The kidneys regulate the acid-base balance of the blood. Before the work of the kidneys begins, the acid-base buffer systems of the blood have functioned to keep the pH temporarily under control; also, the respiratory center has regulated the breathing rate to control the excretion of carbon dioxide at pulmonary capillaries. The kidneys largely function by excreting hydrogen ions and reabsorbing bicarbonate ions as needed.

16.4 Problems with Kidney Function

Various types of problems, including repeated urinary infections, can lead to renal failure, which necessitates receiving a kidney from a donor or undergoing hemodialysis by utilizing a kidney machine or CAPD.

16.5 Effects of Aging

Kidney function declines with age. Also, kidney stones, infections, and urination problems are more common.

16.6 Homeostasis

The urinary system works with the other systems of the body to maintain homeostasis in the ways described in the Human Systems Work Together on page 372.

STUDY QUESTIONS

1. List and explain four functions of the urinary system. (p. 357)
2. Trace the path of urine, and describe the function of each organ mentioned. (pp. 357–359)

3. Explain how urination is controlled. (p. 359)
4. Describe the detailed anatomy of a kidney. (pp. 359–361)

5. Trace the path of blood about a nephron. (pp. 359–361)
6. Name the parts of a nephron, and tell how the structure of the convoluted

tubules suits their respective functions. (pp. 359–361)
7. State and describe the three steps of urine formation. (pp. 361–364)
8. Where in particular are salt and water reabsorbed along the length of the nephron? Describe the contribution of the loop of the nephron. (pp. 361–364)
9. Name and describe the action of antidiuretic hormone (ADH), the renin-aldosterone connection, and atrial natriuretic peptide (ANP). (pp. 365–366)
10. How do the kidneys maintain the pH of the blood within normal limits? (pp. 367–368)
11. Explain how the artificial kidney machine works. (pp. 369–370)

LEARNING OBJECTIVE QUESTIONS

Fill in the blanks.

1. The lungs are organs of excretion because they rid the body of _____.
2. The capillary tuft inside the glomerular capsule is called the _____.
3. Urine leaves the bladder in the _____.
4. _____ is a substance that is found in the filtrate, is reabsorbed, and is present in urine.
5. Tubular secretion takes place at these portions of the nephron_____.
6. The primary nitrogenous end product of humans is _____.
7. _____ is a substance that is found in the filtrate, is not reabsorbed, and is concentrated in urine.
8. In addition to excreting nitrogenous wastes, the kidneys adjust the _____, _____, and _____ balance of the blood.
9. Reabsorption of water from the collecting duct is regulated by the hormone _____.
10. A _____ is a chemical that can combine with either hydrogen ions or hydroxide ions, depending on the pH of the solution.
11. The accumulation of uric acid crystals in a joint cavity produces a condition called _____.
12. Urine is carried from the kidneys to the urinary bladder by a pair of organs called _____.
13. The outer granulated layer of the kidney is the renal _____, whereas the inner striated layer is the renal _____.

MEDICAL TERMINOLOGY EXERCISE

Consult Appendix B for help in pronouncing and analyzing the meaning of the terms that follow.

1. hematuria (hem″uh-tu-re′uh)
2. oliguria (ol″ĭ-gu′re-uh)
3. polyuria (pol″e-yŭ′re-uh)
4. extracorporeal shock wave lithotripsy (ESWL) (eks″truh-kor-po′re-al lith″o-trip′se)
5. antidiuretic (an″tĭ-di″yŭ-ret′ik)
6. urethratresia (yū-re″thruh-tre′ze-uh)
7. cystopyelonephritis (sis″to-pi″e-lo-nĕ-fri′tis)
8. nocturia (nok-tu′re-uh)
9. glomerulonephritis (glo-mĕr″yū-lo-nĕ-fri′tis)
10. ureterovesicostomy (yū-re″ter-o-ves″ĭ-kos′to-me)

WEBSITE LINK

Visit this text's website at http://www.mhhe.com/longenbaker7 for additional quizzes, interactive learning exercises, and other study tools.

The Reproductive System

" . . . No sooner met but they looked;
No sooner looked but they loved;
No sooner loved but they sighed;
No sooner sighed but they asked one another the reason;
No sooner knew the reason but they sought the remedy;
. . . they are in the very wrath of love . . . clubs cannot part
them . . . "

As You Like It, William Shakespeare

Do you believe in love at first sight? There's no doubt that human beings have a particularly acute visual sense, and so poets, authors, and lovers would answer *yes* to that question. But here is another question: Is there "love at first smell," influenced by human pheromones? As you recall from Chapter 10, pheromones are thought to be airborne messenger chemicals that influence human sexual behavior through the sense of smell. While you study the reproductive system in this chapter, the "love at first smell" question is interesting to think about. Pheromones clearly exist for animals, and behavioral studies seem to indicate that we humans have them, too. For example, it's well known that women who live or work together experience menstrual synchrony (menstrual cycles in unison); this effect seems to be caused by a component of axillary sweat, which shortens or lengthens women's cycles until eventually group members menstruate together. Men's axillary secretions have also been shown to influence women's menstrual cycles. Further, a woman's vagina produces fatty-acid compounds called *copulines*, which can raise a man's testosterone (the male sex hormone) level when he is exposed to them. Likewise, women who were first exposed to adrostenol, a testosterone-like compound found in male sweat, were subsequently more likely to find a man's photo attractive. Scientists, however, aren't completely convinced about the influence of human pheromones. The exact compounds responsible for these effects haven't been isolated and identified, nor have their precise receptors in the nose and brain been found. Research into the topic will undoubtedly continue. For now, though, since we don't have all the answers yet, you might not want to count on your pheromone cologne—no matter how good you smell—to bring you true love.

Chapter Outline & Learning Objectives

After you have studied this chapter, you should be able to:

17.1 Human Life Cycle

Unlike the other systems of the body, which are the same in males and females, the reproductive system is quite different in males and females. Development of a child into a sexually mature young adult is a sequence of events called **puberty.** The reproductive system does not begin to fully function until puberty is complete. Sexual maturity occurs between the ages of 11 and 13 in girls, and 14 and 16 in boys. At the completion of puberty, the individual is capable of producing children.

The reproductive organs (**genitals**) have the following functions:

1. Males produce sperm within testes, and females produce ova (eggs) within ovaries.
2. In males, sperm mature and are then transported in a duct system until they exit through the penis. In females, ova are transported in the uterine tubes to the uterus.
3. The male penis functions to deliver sperm to the female vagina, which functions to receive the sperm. The vagina also transports menstrual fluid to the exterior and is the birth canal.
4. The uterus of the female allows the fertilized egg to develop within her body. After birth, the female breast provides nourishment in the form of milk.
5. The testes and ovaries produce the sex hormones that maintain the testes and ovaries and have a profound effect on the body because they bring about masculinization and feminization of various features. In females, ovarian hormones allow a pregnancy to continue.

Meiosis

We studied the type of cell division called mitosis in Chapter 3 and learned that mitosis occurs during growth and repair of the body's tissues. Mitosis is *duplication division* (as an analogy, imagine the cell making exact copies of itself during mitosis, much like a copy machine does with a page of notes). As a result of mitosis, the chromosome number stays constant, and every cell in your body has an identical set of 46 chromosomes. These chromosomes are organized into pairs of homologous chromosomes. Homologous chromosomes are the same size and length. As you recall (from Chapter 2), chromosome genes are individual sections of DNA that determine particular traits such as hair or eye color. These specific genes are located in the same position on homologous chromosomes.

It might be helpful to review the cell cycle and mitosis (Chapter 3). The terminology used to describe DNA might seem confusing because the terms sound so similar. As you'll recall, *chromatin* is the term for unorganized DNA found in the nucleus of the cell, and a *chromosome* is formed when the chromatin condenses. There are 46 chromosomes in a normal human cell. A *chromatid* is the term used for each single

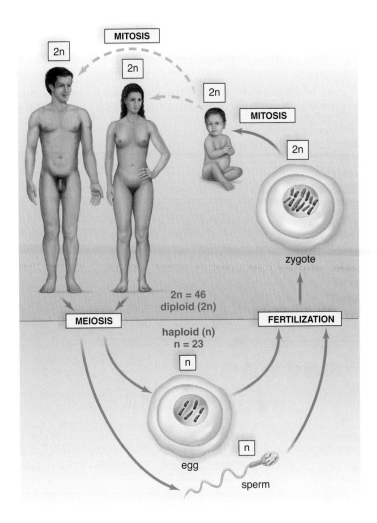

Figure 17.1 Human life cycle. The human life cycle has two types of cell divisions: mitosis, in which the chromosome number stays constant; and meiosis, in which the chromosome number is reduced. Meiosis only occurs in the testes of males during the production of sperm and in the ovaries of females during the production of ova (eggs). In human beings, the ovum (egg) and sperm have 23 chromosomes each, called the n number. Following fertilization, the new individual has 46 chromosomes, called the 2n number.

chromosome while it's attached to its duplicate copy at the centromere (following the S-phase of the cell cycle). Once the chromatids have separated, each is called a chromosome once more.

In addition to mitosis, the human life cycle (Fig. 17.1) includes a type of cell division called **meiosis,** which is *reduction division.* Meiosis only takes place in the testes of males during the production of sperm and in the ovaries of females during the production of ova. During meiosis, the chromosome number is reduced from the normal 46 chromosomes, called the **diploid** or **2n** number, down to 23 chromosomes, called the **haploid** or **n** number of chromosomes. This occurs in two successive divisions, called meiosis I and meiosis II (Fig. 17.2a). The stages of division in meiosis and mitosis are the same—prophase, metaphase, anaphase, and telophase.

To begin meiosis, the cell (called a primary cell) duplicates each chromatid (individual DNA strand), so there are 92 individual chromatids, which remain joined together (Fig. 17.2b, step 1). (This is the same process that occurs in body cells during the S-phase of the cell cycle, which precedes mitosis; see page 57). Next, chromosomes of the same shape and size, called homologous pairs of chromosomes, line up side-by-side in prophase 1 (Fig. 17.2b, steps 2, 3).

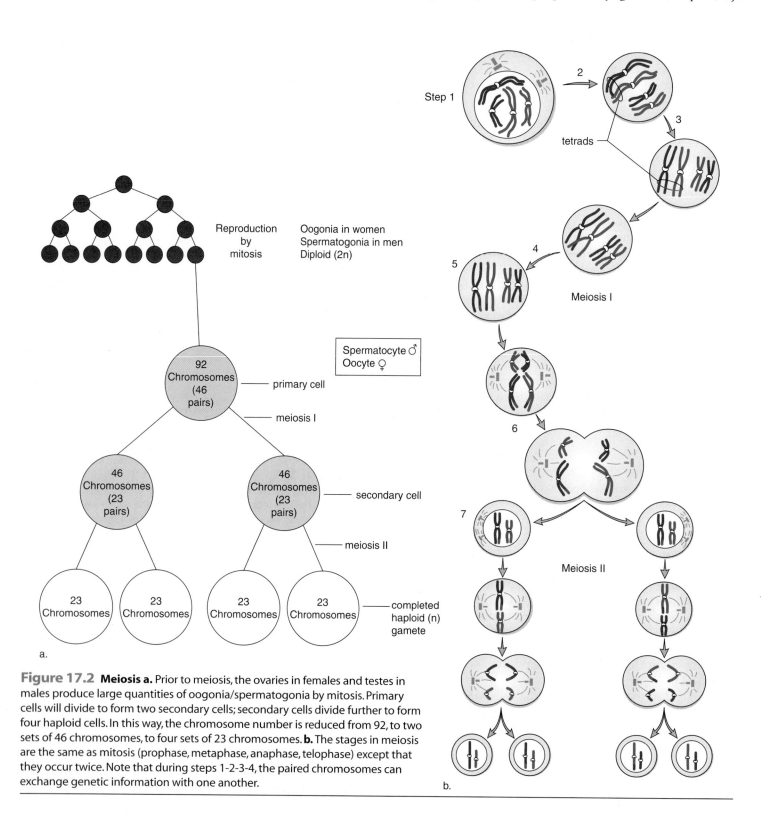

Figure 17.2 Meiosis a. Prior to meiosis, the ovaries in females and testes in males produce large quantities of oogonia/spermatogonia by mitosis. Primary cells will divide to form two secondary cells; secondary cells divide further to form four haploid cells. In this way, the chromosome number is reduced from 92, to two sets of 46 chromosomes, to four sets of 23 chromosomes. **b.** The stages in meiosis are the same as mitosis (prophase, metaphase, anaphase, telophase) except that they occur twice. Note that during steps 1-2-3-4, the paired chromosomes can exchange genetic information with one another.

Each set of four chromatids is called a tetrad. As they line up, each of the chromatids in a tetrad can swap bits and pieces of DNA with the others (Fig 17.2b, steps 3, 4). The effect of this "gene-swap" (the technical term is *crossing over*) is that one's **gametes** (ova or sperm) are *not* exact copies of body cells. Instead, new and different gene combinations are created and passed down to one's children. As meiosis I is completed, two secondary cells are created, each with 46 chromatids, which are still joined together (Fig 17.2b, steps 4, 5, 6). In meiosis II, only one member of each pair goes into the daughter cells (Fig. 17.2b, step 7). To summarize, each primary cell starts with 92 chromatids, which are divided during meiosis I into two cells with 46 chromatids, and then divided once more during meiosis II into four cells with 23 chromosomes.

A **zygote,** the first cell of a new human being, forms following *fertilization*, when a sperm joins with an egg. Because the sperm has 23 chromosomes and the egg has 23 chromosomes, the zygote has 23 pairs of homologous chromosomes, or 46 chromosomes altogether. Without meiosis, the chromosome number in each generation of human beings would double, and cells would no longer be able to function. The zygote undergoes mitosis during development to produce the many cells of a newborn, and mitosis also occurs as a child becomes an adult.

Content **CHECK-UP!**

1. If you wanted to describe meiosis, which statement(s) could you use?
   a. Meiosis occurs only in the ovaries and testes.
   b. Meiosis begins with 92 chromosomes, in sets called tetrads.
   c. Meiosis results in two identical cells that are haploid.
   d. Statements a and b are correct.
   e. All three statements are correct.
2. At the end of meiosis I, the two cells that are produced contain:
   a. 46 individual chromosomes.
   b. 46 chromosomes, paired in 23 sets.
   c. 23 individual chromosomes.
3. Which type of cell is haploid?
   a. sperm cell   d. sperm and ovum
   b. zygote   e. all of the above
   c. ovum

Answers in Appendix C, page 465.

17.2 **Male Reproductive System**

The male reproductive system includes the organs depicted in Figure 17.3. The *primary sex organs* of a male are the paired testes (sing, **testis**), which are suspended within the sacs of

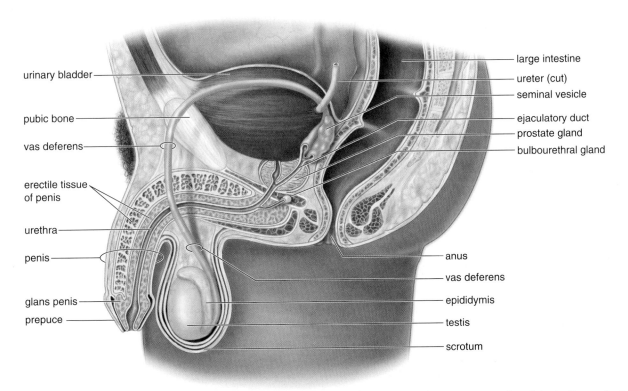

Figure 17.3 Male reproductive system. The testes produce sperm. The seminal vesicles, the prostate gland, and the bulbourethral glands provide a fluid medium for the sperm, which move from a testis to an epididymis to a vas deferens and through the ejaculatory duct to the urethra in the penis. The foreskin (prepuce) is removed when a penis is circumcised.

the **scrotum.** The testes are the primary sex organs because they produce sperm and the male sex hormones (**androgens**).

The other organs depicted in Figure 17.3 are the *accessory* (or secondary) *sex organs* of a male. Sperm produced by the testes are stored within the **epididymis** (pl., *epididymides*). Then they enter a **vas deferens** (pl., *vasa deferentia*), which transports them to an **ejaculatory duct.** The ejaculatory ducts enter the **urethra.** (The urethra in males is a part of both the urinary system and the reproductive system.) The urethra passes through the penis and transports sperm to outside the body.

Several important secretory structures are also accessory sex organs. The **seminal vesicles** lie lateral to the each of the two vas deferens. The duct of each seminal vesicle joins the duct of its lateral vas deferens. The shared ducts are termed *ejaculatory ducts* and each empties into the urethra. The **prostate gland** is a single, walnut-sized gland that surrounds the upper portion of the urethra just inferior to the bladder. **Bulbourethral glands** (also called Cowper glands) are pea-sized organs that lie inferior to the prostate on either side of the urethra. At the time of **ejaculation,** sperm leave the penis in a fluid celled **semen** (seminal fluid). The seminal vesicles and the prostate gland form the majority of the volume of semen. The bulbourethral glands produce a mucoid solution that neutralizes any acidic urine that may be present in the urethra.

Each component of seminal fluid seems to have a particular function. Sperm are more viable in a basic solution, and seminal fluid, which is milky in appearance, has a slightly basic pH (about 7.5). Swimming sperm require energy and seminal fluid contains the sugar, fructose, which presumably serves as an energy source. Semen also contains **prostaglandins,** chemicals that cause the uterus in females to contract. Uterine contractions help propel the sperm toward the egg.

The Testes

The testes, which produce sperm and also the male sex hormones, lie outside the abdominal cavity of the male within the scrotum. The testes begin their development inside the abdominal cavity but descend into the scrotal sacs during the last two months of fetal development. If, by chance, the testes do not descend (a condition called **cryptorchidism**), and the male is not operated on to place the testes in the scrotum, sterility—the inability to produce offspring—usually follows. This is because sperm require a temperature approximately 2°C lower than body temperature to form; thus, the internal temperature of the body is too high to produce viable sperm. A subcutaneous muscle along with an adjoining muscle raise the scrotum during sexual excitement and when a higher temperature is need to warm the testes.

Anatomy of a Testis

A sagittal section of a testis shows that it is enclosed by a tough, fibrous capsule. The connective tissue of the capsule extends into the testis, forming septa that divide the testis into compartments called lobules. Each lobule contains one to three tightly coiled **seminiferous tubules** (Fig. 17.4*a*). Altogether, these tubules have a combined length of approximately 250 m. A microscopic cross section of a seminiferous tubule reveals that it is packed with cells undergoing spermatogenesis (Fig. 17.4*b*), the production of sperm.

Delicate connective tissue surrounds the seminiferous tubules. Cells that secrete the male sex hormones, the androgens, are located here between the seminiferous tubules. Therefore, these endocrine cells are called **interstitial cells** (also called interstitial cells of Leydig or, simply, Leydig cells). The most important of the androgens is **testosterone,** whose functions are discussed later in this section.

Testicular cancer, or cancer of the testes, is one type of cancer that can be detected by self-examination, as explained in the Medical Focus on page 397.

Spermatogenesis

Spermatogenesis, the production of sperm, includes the process of meiosis as the sperm form. Before puberty, the testes, including the seminiferous tubules, are small and nonfunctioning. At the time of puberty, the interstitial cells become larger and start producing androgens. Then, the seminiferous tubules also enlarge, and they start producing sperm.

The seminiferous tubules contain two types of cells: **germ cells,** which are involved in spermatogenesis, and **sustentacular (Sertoli) cells.** Sustentacular cells are large; they extend from the capsule to the lumen of the seminiferous tubule. The sustentacular cells support, nourish, and regulate the development of cells undergoing spermatogenesis.

The germ cells near the capsule are called **spermatogonia.** The spermatogonia divide, producing more cells by mitosis. Some of these cells remain as spermatogonia, which continue to divide. These provide a constant source of sperm cells throughout a normal man's lifetime. Other spermatogonia replicate their DNA and become **primary spermatocytes,** containing 92 chromatids (strands of DNA) held together in 46 pairs (Fig. 17.5). Primary spermatocytes are termed *diploid*, or 2n cells. The primary spermatocytes start the process of meiosis, which requires two divisions. Following meiosis I, two **secondary spermatocytes** are formed, each containing 46 chromatids in 23 pairs. When meiosis II has been completed, there are four spermatids, cells that are termed *haploid* because they possess *half* the normal chromosome number. Spermatids then mature into sperm.

Mature **sperm,** or spermatozoa, have three distinct parts: a head, a middle piece, and a tail (see Fig. 17.4*c*). Mitochondria in the middle piece provide energy for the movement of the tail, which has the structure of a flagellum. The head contains a nucleus covered by a cap called the **acrosome,** which stores enzymes needed to penetrate the egg. Notice in Figure 17.4*b*, that the sperm are situated so that their tails project into the lumen of the seminiferous tubules.

When formed, the sperm are transported to the epididymis because the seminiferous tubules unite to form a complex network of channels that join, forming ducts. When the ducts join, an epididymis is formed.

a. Testis
(cut to show lobules)

vas deferens
epididymis
lobule
testis
scrotal sac

spermatogonia
sperm cells
seminiferous tubule
interstitial cells

b. Seminiferous tubules (micrograph)

acrosome
head
middle piece

tail

end piece

c. Sperm

Figure 17.4 **Testis and sperm. a.** The lobules of a testis contain seminiferous tubules. **b.** Micrograph showing a cross section of the seminiferous tubules, where spermatogenesis occurs. **c.** A sperm has a head, a middle piece, and a tail. The nucleus is in the head, which is capped by the enzyme-containing acrosome.

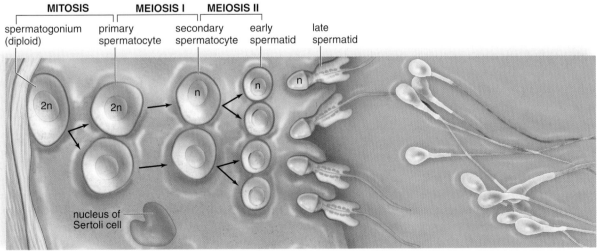

MITOSIS   **MEIOSIS I**   **MEIOSIS II**

spermatogonium
(diploid)
primary
spermatocyte
secondary
spermatocyte
early
spermatid
late
spermatid

nucleus of
Sertoli cell

Figure 17.5 Spermatogenesis.

TABLE 17.1 Male Internal Accessory Organs

Organ	Function
Epididymides	Ducts where sperm mature and some sperm are stored
Vas deferens	Transport and store sperm
Seminal vesicles	Contribute nutrients and fluid to semen
Ejaculatory ducts	Transport sperm
Prostate gland	Contributes basic fluid to semen
Urethra	Transports sperm
Bulbourethral glands	Produce mucoid fluid that neutralizes urine

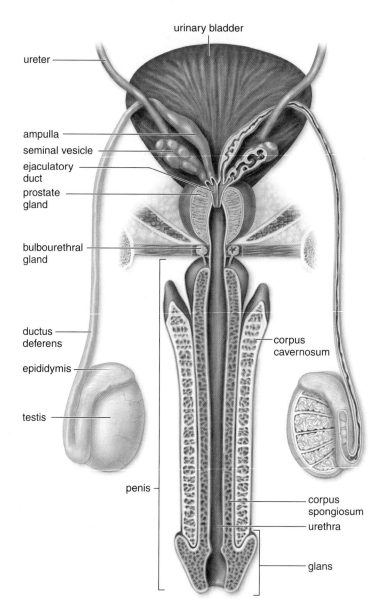

Figure 17.6 Male reproductive system, posterior view. This view shows the duct system that transports sperm from each testis to the urethra, which continues in the penis.

The ejaculated semen of a normal human male contains several hundred million sperm, but only one sperm normally enters an egg. Sperm typically survive in the female reproductive tract for approximately 48 hours after sexual intercourse. However, under optimal conditions sperm can survive for 4–6 days after intercourse.

Male Internal Accessory Organs

Table 17.1 lists and Figure 17.6 depicts the internal accessory organs, as well as the other reproductive organs, of the male. Sperm are transported to the urethra by a series of ducts. Along the way, various glands add secretions to seminal fluid.

Epididymides

Each epididymis is a tightly coiled, threadlike tube that would stretch about 6 meters if uncoiled. An epididymis runs posteriorly down along a testis and becomes a vas deferens that ascends a testis medially.

The lining of an epididymis consists of pseudostratified columnar epithelium with long cilia. Sperm are stored in the epididymides, and the lining secretes a fluid that supports them. The wall of an epididymis contains a thin layer of smooth muscle. Peristaltic contractions move the sperm along as they mature. By the time the sperm leave the epididymides, they are capable of fertilizing an egg even though they do not "swim" until they enter the vagina.

Vas Deferens

Each vas deferens is a continuation of the epididymis. Like the epididymis, the vas deferens is lined by pseudostratified columnar epithelium that is ciliated at the testicular end of the tube. The vas deferens is contained within a protective structure called the **spermatic cord.** The spermatic cord consists of connective tissue and muscle fibers, and contains the testicular artery, vein, and nervous supply in addition to the vas deferens. As each spermatic cord ascends into the abdomen, it passes through the inguinal canal. This is the passageway by which the testis descended from the abdomen

into the scrotum during fetal life. The inguinal canal remains a weak point in the abdominal wall. As such, it is frequently a site of **inguinal hernia.** (A **hernia** is an opening or separation of some part of the abdominal wall through which a portion of an internal organ, usually the intestine, protrudes. Hernias also can occur in the umbilical region, diaphragm, or elsewhere in the abdomen.)

After each vas deferens enters the abdomen, it passes over the superior surface of the bladder to reach the posterior side of the urinary bladder. Each vas deferens widens at its ampulla, located at the posterior base of the urinary bladder, then joins with the duct of a seminal vesicle to form an ejaculatory duct. The ejaculatory ducts pass through the prostate gland to join the urethra.

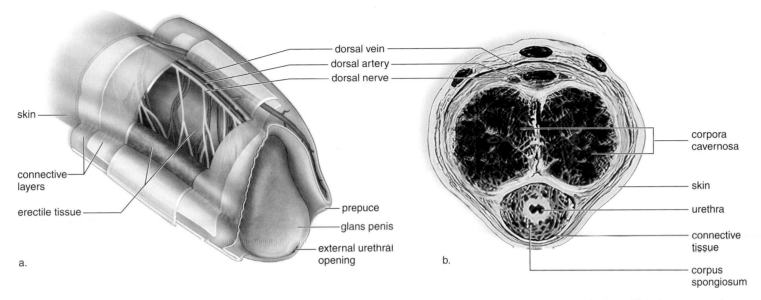

Figure 17.7 Penis anatomy. a. Beneath the skin and the connective tissue lies the urethra surrounded by erectile tissue. This tissue expands to form the glans penis, which in uncircumcised males is partially covered by the foreskin (prepuce). **b.** Two other columns of erectile tissue in the penis are located posteriorly.

Seminal Vesicles

The two seminal vesicles lie lateral to each vas deferens on the posterior side of the bladder. They are coiled, membranous pouches about 5 cm long. The glandular lining of the seminal vesicles secretes an alkaline fluid that contains fructose and prostaglandins into an ejaculatory duct. The pH of the fluid helps modify the pH of seminal fluid; the fructose provides energy for sperm; and the prostaglandins promote muscular contractions of the female genital tract that help move sperm along.

Prostate

The prostate gland encircles the urethra just inferior to the bladder. The walnut-sized gland is about 4 cm across, 2 cm thick, and 3 cm in length. The fibrous connective tissue of its capsule extends inward to divide the gland into lobes, each of which contains about 40 to 50 tubules. The epithelium lining the tubules secretes a fluid that is thin, milky, and alkaline. In addition to adjusting the pH of seminal fluid, prostatic fluid enhances the motility of sperm. The secretion of the prostate gland enters the urethra when the smooth muscle in its capsular wall contracts.

As discussed in the Medical Focus on page 397, the prostate gland is a frequent site for cancer.

Bulbourethral Glands

The bulbourethral glands (Cowper glands) are two small glands about the size of peas. They are located inferior to the prostate gland and enclosed by fibers of the external urethral sphincter. These glands also contain many tubules that secrete a mucuslike fluid. This fluid lubricates the end of the penis preparatory to sexual intercourse. It also neutralizes any acidic urine that may be in the urethra.

Male External Genitals

The **external genitals** are the sex organs that can be easily observed because they are located outside the body. The **penis** and the scrotum are the external genitals of the male. The penis is the male organ of sexual intercourse by which sperm are introduced into the female reproductive tract.

Penis

The penis has an internal root that anchors it to the body wall, and an external shaft (see Fig. 17.6). The tip of the penis is called the *glans*. At the glans penis, the skin folds back on itself to form the prepuce, or foreskin (Fig. 17.7a). This is the structure that is removed in the surgical procedure called **circumcision**. The penis contains three cylindrical bodies of erectile tissue. The two paired **corpora cavernosa** (sing., *corpus cavernosum*) are found on either side of the single **corpus spongiosum**. The urethra passes through the corpus spongiosum. These three columns are supported by fibrous connective tissue, and all are then covered with a thin, loose skin (Fig. 17.7b).

Male Sexual Response

A male's sexual response begins with an erection. The erection process is a spinal reflex; thus, it can even occur in male infants in the womb (see Chapter 8 for a review of reflexes). After puberty, sexual excitation in a male can be triggered by touch, smell, visual or emotional stimuli, or a combination of these.

The three erectile bodies contain distensible spaces that can fill with blood during the erection process.

During sexual arousal, autonomic nerves release nitric oxide, NO. This stimulus leads to the production of cGMP (cyclic guanosine monophosphate), causing the smooth muscle walls of incoming arteries to relax and the erectile tissue to fill with blood. The veins that take blood away from the penis are compressed, and the penis becomes erect. **Erectile dysfunction** (formerly called impotency) exists when the erectile tissue doesn't expand enough to compress the veins. Medications for treatment of erectile dysfunction inhibit an enzyme that breaks down cGMP, ensuring that a full erection will take place. However, individuals who take these medications may experience vision problems because the same enzyme occurs in the retina.

Orgasm (climax) in males is marked by ejaculation, which has two phases: emission and expulsion. During emission, sperm enter the urethra from each ejaculatory duct, and the prostate, seminal vesicles, and bulbourethral glands contribute secretions to the seminal fluid. Once seminal fluid is in the urethra, rhythmic muscle contractions cause seminal fluid to be expelled from the penis in spurts. During ejaculation, a sphincter closes off the bladder so that no urine enters the urethra. (Notice that the urethra carries either urine or semen at different times.)

Male orgasm includes expulsion of seminal fluid from the penis but also the physiological and psychological sensations that occur at the climax of sexual stimulation. The psychological sensation of pleasure is centered in the brain, but the physiological reactions involve the genital organs and associated muscles, as well as the entire body. Marked muscular tension is followed by contraction and relaxation.

Following ejaculation and/or loss of sexual arousal, the penis returns to its normal flaccid state. After ejaculation, a male typically experiences a period of time, called the refractory period, during which stimulation does not bring about an erection. The length of the refractory period increases with age.

There may be in excess of 400 million sperm in the 3.5 ml of semen expelled during ejaculation. The sperm count can be much lower than this, however, and fertilization of the egg by a sperm can still take place.

Regulation of Male Hormone Levels

At the time of puberty, the sex organs mature, and then changes occur in the physique of males. The cause of puberty is related to the level of sex hormones in the body, as regulated by the negative feedback system described in Figure 17.8. We now know that this feedback system functions long before puberty, but the level of hormones is low because the hypothalamus is supersensitive to feedback control. At the start of puberty, the hypothalamus becomes less sensitive to feedback control and begins to increase its production of gonadotropin-releasing hormone (GnRH), which stimulates the anterior pituitary to produce the gonadotropic hormones. Two gonadotropic hormones, **FSH (follicle-stimulating hormone)** and **LH (luteinizing hormone)**, are named for

their function in females but exist in both sexes, stimulating the appropriate organs in each. FSH promotes spermatogenesis in the seminiferous tubules, and LH promotes androgen (e.g., testosterone) production in the interstitial cells. LH in males is also called interstitial cell-stimulating hormone (ICSH).

Negative Feedback Mechanisms

As mentioned, the hypothalamus, anterior pituitary, and testes are involved in a negative feedback system. The system maintains testosterone production at a fairly constant level. When the amount of testosterone in the blood rises to a certain level, it causes the hypothalamus and anterior pituitary to decrease their respective secretion of GnRH and LH. As the level of testosterone begins to fall, the hypothalamus increases its secretion of GnRH, and the anterior pituitary increases its secretion of LH. Thus, stimulation of the interstitial cells occurs. Only minor fluctuations of the testosterone level occur in the male, and the feedback mechanism in this case acts to maintain testosterone at a normal level.

A similar feedback mechanism maintains the continuous production of sperm. The sustentacular cells in the wall of the seminiferous tubules produce a hormone called **inhibin**, which blocks GnRH and FSH secretion when appropriate.

Figure 17.8 Negative feedback. Regulation of testosterone secretion involves negative feedback (reverse arrows) by testosterone on GnRH and LH. Regulation of sperm production involves negative feedback by inhibin on GnRH and FSH.

Testosterone

The male sex hormone, testosterone, has many functions. It is essential for normal development and function of the sex organs in males. For example, greatly increased testosterone secretion at the time of puberty stimulates maturation of the penis and the testes.

Secondary Sex Characteristics Testosterone also brings about and maintains the male **secondary sex characteristics,** which develop at the time of puberty and visibly distinguish males from females. These characteristics include a pattern of male hair growth, activity of cutaneous glands, deeper pitch to the voice, and muscle strength.

At puberty, males experience growth of a beard, axillary (underarm) hair, and pubic hair. In males, pubic hair tapers toward the navel. A side effect of testosterone activity is baldness. Genes for baldness are probably inherited by both sexes, but baldness is seen more often in males because of the presence of testosterone. This makes some types of baldness gender-influenced traits.

Testosterone also causes oil and sweat glands in the skin to secrete, thereby contributing to acne and body odor. The larynx and vocal cords enlarge, causing the voice to change. The "Adam's apple" is a part of the larynx, and it is usually more prominent in males than in females.

Testosterone is responsible for the greater muscular strength of males, which is why some athletes take a supplemental anabolic steroid, such as testosterone or a related chemical. It's important to note that anabolic steroids will build muscle in both men and women. The disadvantages of anabolic steroid use are discussed in a Medical Focus in Chapter 10, page 224.

17.3 Female Reproductive System

The female reproductive system includes the organs depicted in Figure 17.9. The primary sex organs of a female are the paired **ovaries** that lie in shallow depressions, one on each side of the upper pelvic cavity. The ovaries are the primary sex

Figure 17.9 **The female reproductive system.** The ovaries release one egg per month. Fertilization occurs in the uterine tube and development occurs in the uterus. The vagina is the birth canal as well as the organ of sexual intercourse.

organs because they produce **ova** and the female sex hormones, estrogen and progesterone.

The other organs depicted in Figure 17.9 are the accessory (or secondary) sex organs of a female. When an **ovum** leaves an ovary, it is usually swept into a uterine (fallopian) tube by the combined action of the fimbriae (fingerlike projections of a uterine tube) and the beating of cilia that line the uterine tube.

Once in a uterine tube, the ovum is transported toward the uterus. Fertilization, and therefore zygote formation, usually takes place in the uterine tube. The developing embryo normally arrives at the **uterus** several days later, and then **implantation** occurs as the embryo embeds in the uterine lining, which has been prepared to receive it.

Development of the embryo and fetus normally takes place in the uterus. The lining of the uterus, called the **endometrium,** participates in the formation of the placenta (see Chapter 18, page 416), which supplies nutrients needed for embryonic and fetal development.

The uterine tubes join the uterus at its upper end, while at its lower end, the **cervix** enters the vagina nearly at a right angle. A small opening in the cervix leads from the uterus to the vagina.

The **vagina** is the birth canal and organ of sexual intercourse in females. The vagina also acts as an exit for menstrual flow. If fertilization and implantation do not occur, the endometrium is sloughed off during menstruation.

The external genital organs of the female are known collectively as the **vulva.** The vulva consists of the mons pubis, clitoris, labia majora, and labia minora. The area between the labia minora, which is termed the vestibule, contains the openings of the urethra and the vagina.

Notice that the urinary and reproductive systems in the female are entirely separate. For example, the urethra carries only urine, and the vagina has functions that pertain only to reproduction.

The Ovary

The ovaries are paired, oval bodies about 3 cm in length by 1 cm in width and less than 1 cm thick. They lie to either side of the uterus on the lateral walls of the pelvic cavity.

Several ligaments hold the ovaries in place (see Fig. 17.12). The largest of these, the broad ligament, is also attached to the uterine tubes and the uterus. The suspensory ligament holds the upper end of the ovary to the pelvic wall, and the ovarian ligament attaches the lower end of the ovary to the uterus.

A sagittal section through an ovary shows that it is made up of an outer cortex and an inner medulla. In the cortex are many **follicles,** each one containing an immature ovum, called an **oocyte** (Fig. 17.10). A female is born with as many as 2 million follicles, but the number is reduced to 300,000–400,000 by the time of puberty. Only a small number of follicles (about 400) ever become completely mature because a female usually produces only one ovum per month

during her reproductive years. Because oocytes are present at birth, they age as the woman ages. This may be one reason why older women are more likely to produce children with chromosome defects.

Cancer of an ovary, or ovarian cancer, which is discussed in the Medical Focus on page 389, causes more deaths than cervical and uterine cancer.

Oogenesis

Oogenesis, the production of an ovum, includes the process of meiosis. Similar to spermatogenesis, oogenesis begins with a primary oocyte that undergoes meiosis I to become a secondary oocyte having 23 paired chromosomes. The secondary oocyte undergoes meiosis II, but only if it is first fertilized by a sperm cell. If the secondary oocyte remains unfertilized, it never completes meiosis and will die shortly after being released from the ovary.

Oogenesis begins within a follicle. As the follicle matures, it develops from a primary follicle to a secondary follicle to a **vesicular (Graafian) follicle** (see Fig. 17.10). The epithelium of a primary follicle surrounds a primary oocyte. Pools of follicular fluid surround the oocyte in a secondary follicle. In a vesicular follicle, a fluid-filled cavity increases to the point that the follicle wall balloons out on the surface of the ovary.

Figure 17.11 traces the steps of oogenesis. As a follicle matures, the primary oocyte divides, producing two cells. One cell is a secondary oocyte, and the other is a polar body. A **polar body** is formed only during oogenesis, and its function is simply to hold the discarded chromosomes, as a sort of cellular "trash can." Because it is almost completely lacking in cytoplasm, the polar body can never be fertilized or develop further. The vesicular follicle bursts, releasing the secondary oocyte surrounded by a clear membrane and attached follicular cells. This process is referred to as **ovulation.**

The secondary oocyte, often called an egg, enters a uterine tube. If fertilization occurs, a sperm enters the secondary oocyte, which then completes meiosis II. A haploid ovum with 23 chromosomes and a second polar body result. Thus, formation of sperm cells (spermatogenesis) and formation of the ovum (oogenesis) both result in cells that are haploid, containing only 23 chromosomes. However, spermatogenesis produces four sperm cells for every primary cell. In oogenesis, only one haploid ovum is formed for each primary oocyte (the extra chromosomes are contained in polar bodies). When the sperm nucleus unites with the egg nucleus during conception, a diploid zygote with 46 chromosomes is produced.

A follicle that has lost its egg develops into a **corpus luteum,** a glandlike structure. If implantation does not occur, the corpus luteum begins to degenerate after about 10 days. The remains of a corpus luteum is a white scar called the **corpus albicans.** If implantation does occur, the corpus luteum continues to secrete for about six months. Hormones from the corpus luteum help keep the uterine lining intact and allow a pregnancy to continue.

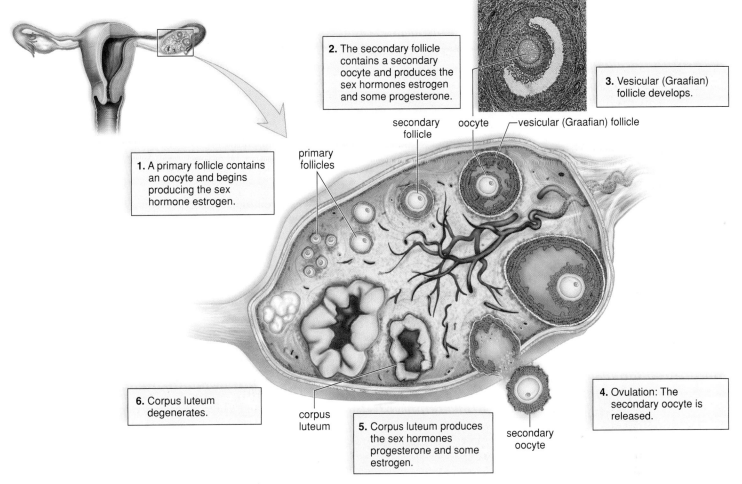

2. The secondary follicle contains a secondary oocyte and produces the sex hormones estrogen and some progesterone.

3. Vesicular (Graafian) follicle develops.

secondary follicle   oocyte   vesicular (Graafian) follicle

1. A primary follicle contains an oocyte and begins producing the sex hormone estrogen.

primary follicles

6. Corpus luteum degenerates.

corpus luteum

5. Corpus luteum produces the sex hormones progesterone and some estrogen.

secondary oocyte

4. Ovulation: The secondary oocyte is released.

Figure 17.10 Anatomy of ovary and follicle. As a follicle matures, the oocyte enlarges and is surrounded by layers of follicular cells and fluid. The micrograph shows the mature vesicular (Graafian) follicle. Eventually, ovulation occurs, the mature follicle ruptures, and the secondary oocyte is released. A single follicle actually goes through all the stages in one place within the ovary.

Although a number of follicles grow during each month, only one reaches full maturity and ruptures to release a secondary oocyte. Presumably the ovaries alternate in producing functional ova. The number of secondary oocytes produced by a female during her lifetime is minuscule compared to the number of sperm produced by a male.

Female Internal Accessory Organs

Table 17.2 lists and Figure 17.12 depicts the internal accessory organs, as well as the other reproductive organs, of a female.

TABLE 17.2 Female Internal Accessory Organs

Organ	Function
Uterine tubes (fallopian tubes, oviducts)	Transport ovum; location of fertilization
Uterus (womb)	Houses developing fetus
Cervix	Contains opening to uterus
Vagina	Receives penis during sexual intercourse; serves as birth canal and as an exit for menstrual flow

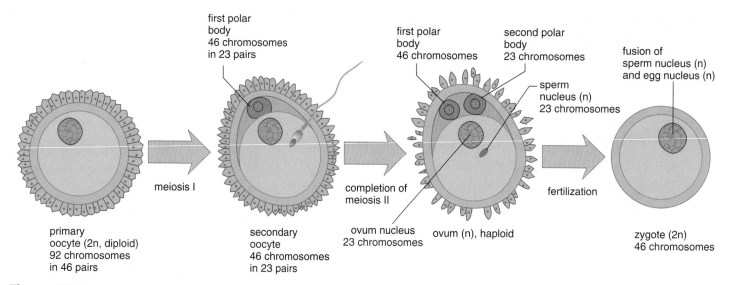

Figure 17.11 **Oogenesis in an ovary.** Oogenesis involves meiosis I, during which the chromosome number is reduced, and meiosis II, which results in a single egg. Meiosis II takes place after a sperm enters the secondary oocyte. At the end of oogenesis, there are also at least two polar bodies, nonfunctional cells that later disintegrate.

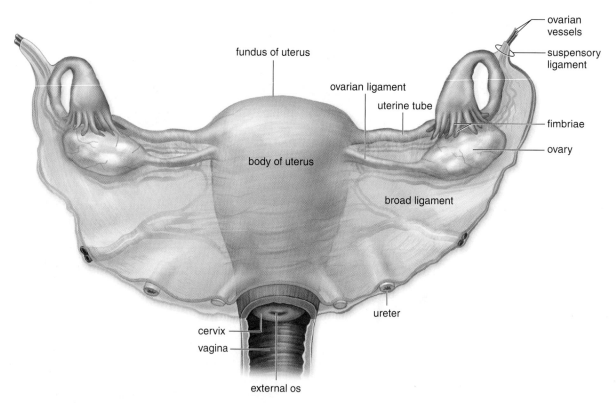

Figure 17.12 **Female reproductive system, posterior view.**

Uterine Tubes

The **uterine tubes,** also called fallopian tubes or oviducts, extend from the uterus to the ovaries. Usually the secondary oocyte enters a uterine tube because the **fimbriae** sweep over the ovary at the time of ovulation, and the beating of the cilia that line uterine tubes creates a suction effect. Once in the uterine tube, the egg is propelled slowly toward the uterus by action of the cilia and by muscular contractions in the wall of the uterine tubes.

Fertilization, the completion of meiosis, and zygote formation normally occur in the upper one-third of a uterine tube.

Ovarian cancer is often "silent," showing no obvious signs or symptoms until late in its development. The most common sign is enlargement of the abdomen, which is caused by the accumulation of fluid. Rarely is there abnormal vaginal bleeding. In women over 40, vague digestive disturbances (stomach discomfort, gas, distention) that persist and cannot be explained by any other cause may indicate the need for a thorough evaluation for ovarian cancer.

The risk for ovarian cancer increases with age. The highest rates are for women over age 60. Women who have never had children are twice as likely to develop ovarian cancer as those who have. Early age at first pregnancy, early menopause, and the use of oral contraceptives, which reduces ovulation frequency, appear to be protective against ovarian cancer. If a woman has had breast cancer, her chances of developing ovarian cancer double. Certain rare genetic disorders are associated with increased risk. The highest incidence rates are reported in the more industrialized countries, with the exception of Japan.

Early detection requires periodic, thorough pelvic examinations. The Pap smear, useful in detecting cervical cancer, does not reveal ovarian cancer. Women over age 40 should have a cancer-related checkup every year. Researchers are currently trying to develop a blood test that will allow for consistent diagnosis of early ovarian cancer. Testing for high levels of tumor marker CA-125, a protein antigen, is sometimes used to screen for ovarian cancer. However, CA-125 is unreliable because it gives false positive results for many other disorders, including benign ovarian cysts and pregnancy. A combination of blood testing and ultrasound examination of the ovaries is currently the most accurate way to diagnose ovarian cancer.

Surgery, radiation therapy, and drug therapy are treatment options. Surgery usually includes the removal of one or both ovaries (**oophorectomy**), the uterus (**hysterectomy**), and the uterine tubes (**salpingectomy**). In some very early tumors, only the involved ovary is removed, especially in young women. In advanced disease, an attempt is made to remove all intra-abdominal cancerous tissue to enhance the effect of chemotherapy.

The developing embryo usually does not arrive at the uterus for several days. Once in the uterus, the embryo embeds itself in the uterine lining, which has been prepared to receive it.

Occasionally, the embryo becomes embedded in the wall of a uterine tube, where it begins to develop. Tubular pregnancies cannot succeed and must be surgically removed because the tubes are not anatomically capable of allowing full development to occur. Any pregnancy that occurs outside the uterus is called an **ectopic pregnancy.**

Begin Thinking Clinically

?

Rarely, an ectopic pregnancy can occur outside of the womb, in the abdominal cavity. How could this occur? On which organs could a fertilized ovum construct a placenta? How would a full-term infant resulting from such a pregnancy be delivered?

Answers and discussion in Appendix C, page 465.

Uterus

The *uterus* is a thick-walled, muscular organ about the size and shape of an inverted pear. Normally, it lies above and is tipped over the urinary bladder. The uterus has three sections. The *fundus* is the region superior to the entrance of the uterine tubes. The *body* of the uterus is the major region. The *cervix* is

the narrow end of the uterus that projects into the vagina. A cervical orifice leads to the lumen of the vagina.

Development of the embryo normally takes place in the uterus. This organ, sometimes called the womb, is approximately 5 cm wide in its usual state but is capable of stretching to over 30 cm to accommodate the growing baby. The lining of the uterus, called the *endometrium*, participates in the formation of the placenta (see Chapter 18), which supplies nutrients needed for embryonic and fetal development. In the nonpregnant female, the endometrium varies in thickness during a monthly menstrual cycle, discussed later in this chapter.

Cancer of the cervix is a common form of cancer in women. Early detection is possible by means of a **Pap smear,** which entails the removal of a few cells from the region of the cervix for microscopic examination. If the cells are cancerous, a hysterectomy (the removal of the uterus) may be recommended. Removal of the ovaries in addition to the uterus is termed an **ovariohysterectomy.** Because the vagina remains intact, the woman can still engage in sexual intercourse.

Vagina

The vagina is a tube that makes a 45° angle with the small of the back. The mucosal lining of the vagina lies in folds that extend when the fibromuscular wall stretches. This capacity to extend is especially important when the vagina serves as the birth canal, and it can also facilitate intercourse, when the vagina receives the penis.

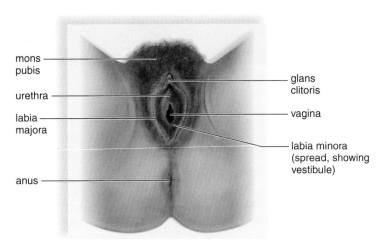

mons
pubis

urethra

labia
majora

anus

glans
clitoris

vagina

labia minora
(spread, showing
vestibule)

Figure 17.13 External genitals of the female. At birth, the opening of the vagina is partially blocked by a membrane called the hymen. Physical activities and sexual intercourse disrupt the hymen.

External Genitals

The female external genitals (Fig. 17.13) are known collectively as the vulva. The vulva includes two large, hair-covered folds of skin called the **labia majora** (sing., *labium majus*). They extend posteriorly from the **mons pubis,** a fatty prominence underlying the pubic hair. The **labia minora** (sing., *labium minus*) are two small folds of skin lying just inside the labia majora. They extend forward from the vaginal opening to encircle and form a foreskin for the **clitoris,** an organ that is homologous to the penis. Although quite small, the clitoris has a shaft of erectile tissue and is capped by a pea-shaped glans. The clitoris also has sensory receptors that allow it to function as a sexually sensitive organ.

The **vestibule,** a cleft between the labia minora, contains the orifices of the urethra and the vagina. The vagina can be partially closed by a ring of tissue called the hymen. The hymen ordinarily is ruptured by initial sexual intercourse; however, it can also be disrupted by other types of physical activities. If the hymen persists after sexual intercourse, it can be surgically ruptured.

The urinary and reproductive systems in the female are entirely separate: The urethra carries only urine, and the vagina serves only as the birth canal and as the organ for sexual intercourse.

Female Sexual Response

Upon sexual stimulation, the labia minora, the vaginal wall, and the clitoris become engorged with blood. The breasts also swell, and the nipples become erect.

The vagina expands and elongates. Blood vessels in the vaginal wall release small droplets of fluid that seep into the vagina and lubricate it. Mucus-secreting glands called **Bartholin glands,** beneath the labia minora on either side of the vagina, also provide lubrication for entry of the penis into the vagina. (Bartholin glands are similar to male bulbourethral

glands). Although the vagina is the organ of sexual intercourse in females, the clitoris plays a significant role in the female sexual response. The extremely sensitive clitoris can swell to two or three times its usual size. The thrusting of the penis and the pressure of the pubic symphyses of the partners act to stimulate the clitoris.

Orgasm occurs at the height of the sexual response. Blood pressure and pulse rate rise, breathing quickens, and the walls of the uterus and uterine tubes contract rhythmically. A sensation of intense pleasure is followed by relaxation when organs return to their normal size. Females have no refractory period, and multiple orgasms can occur during a single sexual experience.

Regulation of Female Hormone Levels

At the time of puberty in females, the hypothalamus increases its secretion of GnRH, and the anterior pituitary releases larger amounts of the gonadotropins, FSH and LH. These hormones stimulate the ovaries to produce and release ova. FSH and LH also cause elevated estrogen and progesterone production by the ovaries.

Estrogen and Progesterone

In particular, estrogen stimulates the growth of the uterus and the vagina. **Estrogen** is also necessary for maturation of the ovum and for the onset of the menstrual cycle. Development of the secondary sex characteristics in the female is caused by increased estrogen production during puberty. Along with sexual development, girls going through puberty experience an early increase in long bone growth that leads to increased height. However, estrogen also leads to early fusion of the growth plates in long bones, so that most young women have achieved their full adult height by about age 16.

Secondary Sex Characteristics These characteristics include the female pattern of body hair and fat distribution. In general, females have a more rounded appearance than males because of a greater accumulation of fat beneath the skin. Also, the pelvic girdle enlarges in females so that the pelvic cavity has a larger relative size compared to that of males; this means that females have wider hips. Both estrogen and **progesterone** are required for breast development, which is discussed on page 392.

Menstrual Cycle

The **menstrual cycle** is a monthly series of events that involve the ovaries and uterus plus the hormones already mentioned. The cycle is about 28 days long, but it can be as short as 18 days or as long as 40 days (Fig. 17.14).

Pre-Ovulation Events

Under the influence of follicle-stimulating hormone (FSH) from the anterior pituitary, several follicles begin developing in the ovary. Therefore, this period of time (days 1–14) is called the *follicular phase* of the ovary (Fig. 17.14). Although several

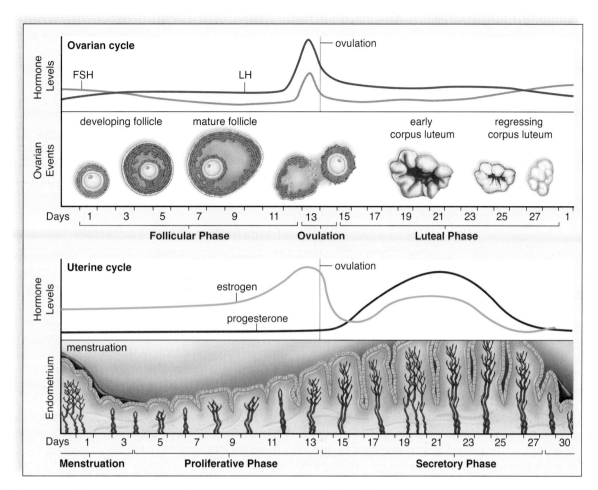

Figure 17.14 Events of the menstrual cycle. During the menstrual cycle, FSH and LH are released by the anterior pituitary. FSH promotes the maturation of a follicle in the ovary. The follicle produces increasing levels of estrogen, which cause the endometrium to thicken during the proliferative phase in the uterus. An LH surge causes ovulation. After ovulation, LH promotes the development of the corpus luteum. This structure produces increasing levels of progesterone, which causes the endometrial lining to become secretory. Menses due to the breakdown of the endometrium begins when progesterone production declines to a low level due to corpus luteum disintegration.

follicles begin growing, only one follicle continues developing, and it secretes increasing amounts of estrogen. This particular follicle becomes more and more sensitive to FSH and then LH. Eventually, the very high level of estrogen exerts *positive feedback control* over the hypothalamus so that it secretes ever greater amounts of GnRH. GnRH induces a surge in FSH and LH secretion by the pituitary. The LH level rises to a greater extent than does the FSH level. Ovulation is triggered by the very high level of hormonal stimulation, and in particular the spike in LH, called the LH surge. (It is interesting to note that over-the-counter ovulation prediction test kits work by detecting high levels of LH metabolites in the urine, combining the metabolites with monoclonal antibodies to produce a color change.)

While the ovary is experiencing its follicular phase, first menstruation and then the proliferative phase occur in the uterus. During menstruation (days 1–5), a low level of female sex hormones in the body causes the endometrial tissue to disintegrate and its blood vessels to rupture. A flow of blood and tissues, known as the **menses,** passes out of the vagina during menstruation, also called the menstrual period.

Under the influence of estrogen released by the new follicle, the endometrium thickens and becomes vascular and glandular. This is the *proliferative phase* of the uterus, which ends when ovulation occurs.

Post-Ovulation Events

Under the influence of LH from the anterior pituitary, the emptied ovarian follicle tissue becomes a hormone-secreting tissue called the corpus luteum. Therefore, this period of time (days 15–28) is known as the *luteal phase* of the ovary (Fig. 17.14). The corpus luteum secretes progesterone and some estrogen. As the blood level of progesterone rises, it exerts *negative feedback control* over the anterior pituitary's secretion of LH so that the corpus luteum in the ovary begins to degenerate. If fertilization of the egg does occur, the corpus luteum persists for reasons that will be discussed shortly.

Under the influence of progesterone secreted by the corpus luteum, a secretory phase (days 15–28) begins in the uterus. During the *secretory phase* of the uterus, the endometrium of the uterus doubles or even triples in thickness (from 1 mm to 2–3 mm), and the uterine glands mature, producing a thick, mucoid secretion. The endometrium is now prepared to receive the embryo. If implantation of an embryo does not take place, the corpus luteum disintegrates, and menstruation occurs. The disintegrated corpus luteum in the ovary becomes a patch of white scar tissue called the corpus albicans. With age, more of a woman's ovarian tissue becomes nonfunctional corpus albicans. This is why ovulation and menstruation eventually stop during menopause.

If fertilization occurs and is followed by implantation, the developing placenta produces **human chorionic gonadotropin (HCG),** which maintains the corpus luteum in the ovary until the placenta begins to produce progesterone and estrogen. The placental hormones shut down the anterior pituitary so that no new follicle in the ovaries matures. Placental hormones also maintain the endometrium so that the corpus luteum in the ovary is no longer needed. Usually, no menstruation occurs during pregnancy. After human chorionic gonadotropin is metabolized and excreted in the urine, it can be detected using prepared monoclonal antibodies. Over-the-counter home pregnancy test kits function in this fashion, as discussed in Chapter 18.

Menopause

Menopause, the period in a woman's life during which the menstrual cycle ceases, is likely to occur between ages 45 and 55. The ovaries are no longer responsive to the gonadotropic hormones produced by the anterior pituitary, and the ovaries secrete very low amounts of estrogen and progesterone. At the onset of menopause, the uterine cycle becomes irregular, but as long as menstruation occurs, it is still possible for a woman to conceive. Therefore, a woman is usually not considered to have completed menopause until menstruation has been absent for a year.

The hormonal changes during menopause often produce physical symptoms, such as "hot flashes" (caused by circulatory irregularities), dizziness, headaches, insomnia, sleepiness, and depression. These symptoms may be mild or even absent. If they are severe, medical attention should be sought. Women sometimes report an increased sex drive following menopause. It has been suggested that this may be due to androgen production by the adrenal cortex.

Female Breast and Lactation

Early growth of the female breasts during puberty is referred to as *budding* of the breasts. Budding is followed by development of lobes, the functional portions of the breast, and deposition of adipose tissue, which gives breasts their adult shape.

A breast contains 15 to 25 lobules, each with a milk duct that begins at the nipple. The nipple is surrounded by a pigmented area called the **areola.** Hair and glands are absent from the nipples and areola, but glands are present that secrete a saliva-resisting lubricant to protect the nipples, particularly during nursing. Smooth muscle fibers in the region of the areola may cause the nipple to become erect in response to sexual stimulation or cold.

Within each lobe, the mammary duct divides into numerous alveolar ducts that end in blind sacs called alveoli (Fig. 17.15). The alveoli are made up of the cells that can produce milk. Estrogen and progesterone are required for lobe

a. **Anteromedial view**

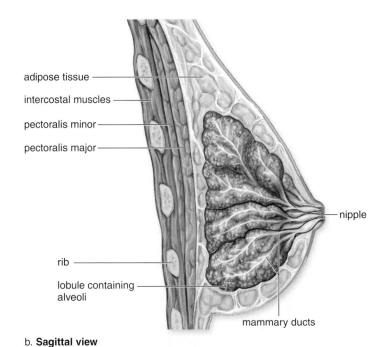

b. **Sagittal view**

Figure 17.15 **Structure of the female breast and mammary glands. a.** Anteromedial view. **b.** Sagittal section.

development. It is believed that estrogen causes proliferation of ducts and that both estrogen and progesterone bring about alveolar development. The abundance of these hormones during pregnancy means that the alveoli proliferate at this time. A nonlactating breast has ducts but few alveoli, while a lactating breast has many ducts and alveoli.

During pregnancy, the breasts enlarge as the ducts and alveoli increase in number and size. The same hormones that affect the mother's breasts can also affect the child's. Some newborns, including males, even secrete a small amount of milk for a few days.

Usually, no milk is produced during pregnancy. The hormone prolactin is needed for **lactation** to begin, and the production of this hormone is suppressed during pregnancy. Negative feedback control by high levels of estrogen and progesterone produced during pregnancy shuts down prolactin secretion by the anterior pituitary. Once the baby is delivered, however, the pituitary begins secreting prolactin. It takes a couple of days for milk production to begin, and in the meantime, the breasts produce **colostrum,** a thin, yellow, milky fluid rich in protein, including antibodies.

The continued production of milk requires a suckling child. When a breast is suckled, the nerve endings in the areola are stimulated, and a nerve impulse travels along neural pathways from the nipples to the hypothalamus, which directs the posterior pituitary gland to release the hormone oxytocin. When this hormone arrives at the breast, it causes the lobules to contract so that milk flows into the ducts (called milk letdown), where it may be drawn out of the nipple by the suckling child.

Breast cancer is one of the few types of cancer that can be detected by the female herself. The Medical Focus on page 397 tells how to do a shower check for breast cancer.

Content CHECK-UP!

7. Which structure in the ovary eventually releases the secondary oocyte or ovum?

   a. vesicular follicle   c. corpus albicans

   b. corpus luteum   d. polar body

8. The second stage of meiosis in an ovum occurs:

   a. in the ovary, before the ovum is released at ovulation.

   b. after the ovum is released and travels through the uterine tube.

   c. only if the ovum is fertilized by a sperm cell.

   d. just after the ovum is released from the ovary, but before it enters the uterine tube.

9. During which stage of the monthly cycle of the uterus does the endometrium thicken and become glandular and vascular?

   a. menstrual phase   c. proliferative phase

   b. secretory phase   d. ovulation phase

Answers in Appendix C, page 465.

17.4 Control of Reproduction and Sexually Transmitted Infections

Contraception

Today, there are more options than ever before for **contraception,** or the prevention of pregnancy. Although many contraceptives are available without prescription, women should consult a health-care provider before choosing a birth control method. Health-care providers can give patients advice on the best method to use based on lifestyle, and can also ensure that the patient uses the product correctly. Table 17.3 lists the means of birth control used in the United States and rates their effectiveness. For example, consider any of the prescription combined hormone methods, which are among the most effective methods listed in the table. Based on statistics, we expect that within a year, 98 or 99 out of 100, or 98–99%, of sexually active women will *not* get pregnant using one of these methods, while one or two women will get pregnant. (It is important to note that the failure rate for any contraception method might be a bit high—it also includes those users who don't follow directions and therefore use the product incorrectly.)

Figure 17.16 features some of the most effective and commonly used means of birth control. Combination hormone methods can be in the form of pills taken orally, and a new regimen of combined pill allows a woman to have only three or four menstrual cycles a year (Fig. 17.16a). In addition, combination hormones can be introduced into the blood via patches that allow hormone absorption through the skin (Fig. 17.16e), or by a plastic ring placed in the vagina. The ring slowly releases hormones into the bloodstream. Regardless of delivery method, the estrogen and progesterone in the combined hormone form effectively shut down the pituitary's production of both FSH and LH so that no follicle begins to develop in the ovary. Because ovulation does not occur, pregnancy cannot take place. Minor side effects of combined hormone methods include dizziness, nausea, weight gain, and changes in menstruation and mood. Rarely, serious complications can occur in women who use combined hormonal methods, including high blood pressure, cardiovascular disease, blood clots, heart attack, and strokes. Serious complications are more likely in women who smoke; therefore, smokers should use other birth control methods. Because of their possible side effects, these methods are offered by prescription only. Women using combined hormonal methods should see a physician or other health-care provider on a regular basis to monitor their use.

A second hormonal method of contraception involves use of only a synthetic progesterone, without estrogen. The oral form is commonly referred to as the "mini-pill" (because it contains only one hormone) and is taken daily. An injection, Depo-Provera, can be taken once every three months (Fig. 17.16f). Another form of the same hormone is an implantable rod that is inserted into the skin under the arm. All three work by inhibiting ovulation and making the mucus of the uterine cervix thick and sticky.

TABLE 17.3 Common Methods of Contraception

Name	Procedure	How Does It Work?	How Effective Is It?	Side Effects and Other Health Risks
NATURAL METHODS				
Abstinence	Refrain from sexual intercourse	No sperm in vagina	100%	None; also protects against sexually transmitted infections
Natural family planning	Determine day of ovulation by keeping records; various testing methods	Intercourse is avoided only during the time period while ovum is viable	80%	None
Withdrawal method	Penis withdrawn from vagina just before ejaculation	Ejaculation outside the woman's body; no sperm in vagina	75%	None
Douching	Vagina cleansed after intercourse	Washes sperm out of woman's body	Less than 70%	May cause infection or inflammation
NONPRESCRIPTION METHODS				
Male condom	Sheath of latex, polyurethane, or natural material fitted over erect penis	Traps sperm and prevents entry into vagina; latex and polyurethane forms protect against STIs	89%	With latex: latex allergy; natural material condoms give no protection against STIs
Female condom	Polyurethane liner fitted inside vagina	Traps sperm and prevents entry into vagina; some protection against STIs	79%	Possible allergy or irritation; urinary tract infection
Spermicide: jellies, foams, films, creams, tablets	Spermicidal products inserted into the vagina before intercourse	Spermicidal chemical nonoxynol-9 kills large numbers of sperm cells	50–80%	Irritation, inflammation; allergic reaction; urinary tract infection
Contraceptive sponges	Sponge containing spermicide inserted into vagina and placed against cervix	Spermicidal chemical nonoxynol-9 kills large numbers of sperm cells	72–86%	Irritation, inflammation; allergic reaction; urinary tract infection; toxic shock syndrome; limited availability
PRESCRIPTION HORMONAL METHODS				
Combined hormone, vaginal ring	Flexible plastic ring inserted into vagina; releases hormones that are absorbed into the bloodstream	Suppresses ovulation by the combined actions of the hormones estrogen and progestin	98%	Vaginal irritation, inflammation; dizziness, nausea; changes in menstruation, mood, and weight; rarely: cardiovascular disease, including high blood pressure, blood clots, heart attack, and strokes
Progesterone-only injection (Depo-Provera)	Injection of progestin once every three months	Inhibits ovulation; prevents sperm from reaching the egg and prevents the fertilized egg from implanting in the uterus	99%	Irregular bleeding; weight gain; breast tenderness; headaches; osteoporosis if used for extended period
Progesterone-only implantable rod (Implanon)	Single rod implanted into skin under the arm	Inhibits ovulation; prevents sperm from reaching the egg; prevents fertilized egg from implanting in the uterus	99%	Irregular bleeding, weight gain, increased risk of blood clots, heart attack, stroke
Emergency contraception	Must be taken within 72 hours after unprotected intercourse	Suppresses ovulation by the combined actions of the hormones estrogen and progestin; prevents implantation by embryo	80%	Nausea; vomiting; abdominal pain; fatigue; headache

PRESCRIPTION HORMONAL METHODS

Method		Effectiveness	
Combined hormone pill	Pills are swallowed daily; chewable form also available	Suppresses ovulation by the combined actions of the hormones estrogen and progestin	98%	Dizziness, nausea; changes in menstruation, mood, and weight; rarely: cardiovascular disease, including high blood pressure, blood clots, heart attack, and strokes
Progestin-only mini-pill	Pills are swallowed daily	Thickens cervical mucus, preventing sperm from contacting ovum	98%	Irregular bleeding; weight gain; breast tenderness
Combined hormone 91, daily regimen	Pills are swallowed daily; user has 3–4 menstrual periods a year	Suppresses ovulation by the combined actions of the hormones estrogen and progestin	98%	Dizziness, nausea; changes in menstruation, mood, and weight; rarely: cardiovascular disease, including high blood pressure, blood clots, heart attack, and strokes
Combined hormone patch	Patch is applied to skin and left in place for 1 week; new patch applied	Suppresses ovulation by the combined actions of the hormones estrogen and progestin	98%	Skin irritation, allergy; dizziness, nausea; changes in menstruation, mood, and weight; rarely: cardiovascular disease, including high blood pressure, blood clots, heart attack, and strokes

PRESCRIPTION BARRIER METHODS

Method		Effectiveness	
Diaphragm	Latex cup, placed into vagina to cover cervix before intercourse	Blocks entrance of sperm into uterus; spermicide kills sperm	90% with spermicide	Irritation, inflammation; allergic reaction; urinary tract infection; toxic shock syndrome
Cervical cap	Latex cap held over cervix	Blocks entrance of sperm into uterus; spermicide kills sperm	90% with spermicide	Irritation, inflammation; allergic reaction; toxic shock syndrome; abnormal Pap smear
Cervical shield	Latex cap placed in upper vagina, held in place by suction	Blocks entrance of sperm into uterus; spermicide kills sperm	90% with spermicide	Irritation, inflammation; allergic reaction; toxic shock syndrome; urinary tract infection
Intrauterine device, copper T	Placed in uterus	Causes cervical mucus to thicken; fertilized embryo cannot implant	99%	Cramps; bleeding; pelvic inflammatory disease; infertility; perforation of uterus; limited availability
Intrauterine device, progesterone-releasing type	Placed in uterus	Prevents ovulation; causes cervical mucus to thicken; fertilized embryo cannot implant	99%	Cramps; bleeding; pelvic inflammatory disease; infertility; perforation of uterus; limited availability

SURGICAL STERILIZATION

Method		Effectiveness	
Vasectomy	Vasa deferentia are cut and tied	No sperm in seminal fluid	Almost 100%	May not be reversible; minor risk of surgical complications such as infection or reaction to anesthetic
Tubal ligation: trans-abdominal surgery	Oviducts are cut and tied, or a clip is placed on the oviduct	Sperm cannot enter oviduct; ova cannot pass through oviduct	Almost 100%	May not be reversible; pain, bleeding, infection; other post-surgical complications; ectopic (tubal) pregnancy
Sterilization implant (Essure system)	Small metallic implant is placed into uterine tubes; inserted through the vagina using a catheter	Causes scar tissue to form, blocking uterine tubes and preventing conception	Almost 100%	May not be reversible; pain after placement; ectopic pregnancy

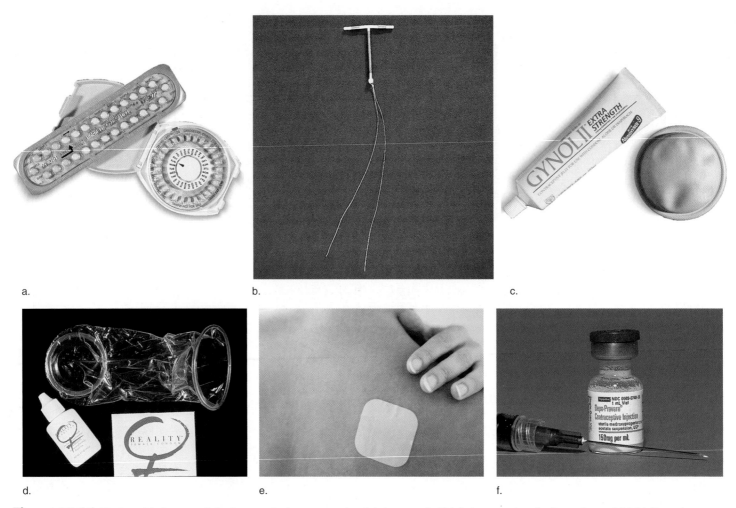

Figure 17.16 **Various birth control devices. a.** Oral contraception (birth control pills). **b.** Intrauterine device. **c.** Spermicidal jelly and diaphragm. **d.** Female condom. **e.** Contraceptive patch. **f.** Contraceptive injection.

Sperm cells cannot swim through the thick mucus, thus preventing fertilization of the ovum. If fertilization does occur, the uterine lining is changed by this medication so that an embryo will not be able to implant. As with other hormonal forms of contraception, progesterone-only forms are available by prescription only and do have some side effects and health risks (see Table 17.3).

An intrauterine device (IUD) is a small piece of molded plastic that is inserted into the uterus by a physician (Fig. 17.16b). IUDs are believed to alter the environment of the uterus and uterine tubes so that fertilization probably will not occur—but if fertilization does occur, implantation cannot take place. The type of IUD shown in Figure 17.16b has copper wire wrapped around the plastic. Other forms will slowly release progesterone over time into the bloodstream. IUDs are very effective. However, because of their possible side effects—cramps, bleeding, pelvic inflammatory disease, infertility, and a slight risk of perforation of uterus—intrauterine devices currently have a limited availability.

The diaphragm is a soft latex cup with a flexible rim (Fig. 17.16c) that lodges behind the pubic bone and fits over the cervix. The diaphragm can be inserted into the vagina no

more than two hours before sexual relations. Also, it must be used with spermicidal jelly or cream and should be left in place at least six hours after sexual relations. A cervical shield is smaller than a diaphragm, and the cervical cap is even smaller still. All three devices must be fitted by a physician or other health-care provider. Each carries a slight risk of urinary tract infection in the user. Rarely, incidents of a very serious illness—toxic shock syndrome—have been reported in users of these methods. Use of proper hygiene methods, including handwashing before inserting the device, and prompt removal of the device, can minimize the risk of serious illness.

There has been a revival of interest in barrier methods of birth control because these methods also offer some protection against sexually transmitted infections. A female condom, now available, consists of a large polyurethane tube with a flexible ring that fits onto the cervix (Fig. 17.16d). The open end of the tube has a ring that covers the external genitals. A male condom is most often a latex or polyurethane sheath that fits over the erect penis. The ejaculate is trapped inside the sheath and, thus, does not enter the vagina. When used in conjunction with a spermicide, protection is better than with the condom alone. So-called "natural" condoms made of

The American Cancer Society urges women to do a breast self-exam and men to do a testicle self-exam every month. Breast cancer and testicular cancer are far more curable if found early, and we must all take the responsibility of checking for one or the other.

Breast Self-Exam for Women

1. Check your breasts for any lumps, knots, or changes about one week after your period.
2. Place your right hand behind your head. Move your *left* hand over your *right* breast in a circle. Press firmly with the pads of your fingers (Fig. 17A). Also check the armpit.
3. Now place your left hand behind your head and check your *left* breast with your *right* hand in the same manner as before. Also check the armpit.
4. Check your breasts while standing in front of a mirror right after you do your shower check. First, put your hands on your hips and then raise your arms above your head (Fig. 17B). Look for any changes in the way your breasts look: dimpling of the skin, changes in the nipple, or redness or swelling.
5. If you find any changes during your shower or mirror check, see your doctor right away.

You should know that the best check for breast cancer is a mammogram. When your doctor checks your breasts, ask about getting a mammogram. Yearly mammograms are recommended for women over 40 years of age. Women with a personal or family history of breast cancer (mother, sister, or aunt who have or had the disease) may be advised to start such screening exams at an earlier age, and to have the exams more frequently. Additional exams such as ultrasound or MRI of the breast may also be advised.

Testicle Self-Exam for Men

1. Check your testicles once a month.
2. Roll each testicle between your thumb and finger as shown in Figure 17C. Feel for hard lumps or bumps.
3. If you notice a change or have aches or lumps, tell your doctor right away so he or she can recommend proper treatment.

Cancer of the testicles can be cured if you find it early. You should also know that prostate cancer is the most common cancer in men. Men over age 50 should have an annual health checkup that includes a prostate examination.

Information provided by the American Cancer Society. Used by permission.

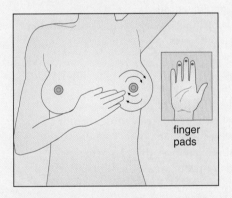

Figure 17A **Shower check for breast cancer.**

finger pads

Figure 17B **Mirror check for breast cancer.**

Figure 17C **Shower check for testicular cancer.**

lambskin or other material do not provide any protection against sexually transmitted infection.

Research continues into other forms of contraception, including contraceptive vaccines. One form, which immunizes against human chorionic gonadotropin, the hormone of pregnancy, was successful in one limited clinical trial. However, such vaccines have limited practical application currently because they have not been proven to be completely effective in some users. There is also a question as to whether

their effects are reversible. Current hormonal contraceptives designed for use in males cause undesirable side effects such as erectile dysfunction. More study needs to be done into alternative methods that will be effective, safe, and inexpensive.

Emergency Contraception

Emergency contraception, sometimes referred to as "morning-after pills," refers to a medication that will prevent pregnancy

Endocrine-Disrupting Contaminants

One of the first synthetic estrogens to be produced in a laboratory was diethylstilbestrol, or DES, created in 1938. In the ensuing years, DES was used to treat a variety of conditions such as prostate cancer in men and menopausal symptoms in women. For a short time, it was even used as an emergency contraceptive. Tragically, DES was also given to many pregnant women to prevent early miscarriage. Years later, those same women's children are suffering from its effects on their reproductive systems. Daughters of the original patients (called *DES daughters*) often have birth defects involving reproductive system organs, including the ovaries, uterus, and vagina. Further, these same women are 2.5 times as likely to have breast cancer after age 40. DES daughters are also more likely to suffer from cancers of the vagina and cervix. DES sons have an increased risk of genital abnormalities, and possibly prostate and testicular cancers as well.

The consequences of DES use have raised concern about the many chemicals created after 1938 that might also affect the endocrine system. Because endocrine hormones influence nearly all aspects of physiology and behavior in animals, synthetic compounds called *endocrine-disrupting contaminants* (EDCs) that show hormonal effects are being studied extensively. Environmental biologists first became aware of EDCs when wildlife—especially fish, amphibians, reptiles, and birds in polluted areas—began to show abnormalities in tissue differentiation, sexual development, and reproduction. Human studies suggest that these same pollutants lower sperm counts, reduce male and female fertility, and increase the rates of certain cancers (breast, ovarian, testicular, and prostate). Further, EDCs are thought to be responsible for early onset of menses in young girls, some of whom begin their periods as early as age 10. Additionally, some studies seem to indicate that EDCs contribute to learning deficits and behavioral problems in children.

Many EDCs are chemicals used as pesticides and herbicides in agriculture, and some are associated with the manufacture of various other synthetic organic compounds such as PCBs (poly-chlorinated biphenyls). Other chemicals shown to influences hormones are found in plastics, food additives, and personal hygiene products. Baby bottles, plastic cups, and toys recently have been discontinued from the market because compounds used for hardening their plastic are suspected EDCs. In mice, some plastic components affect neonatal development when present in extremely low concentrations (parts per *trillion*). EDCs have been found in nature at levels 1,000 times greater than this—even in amounts comparable to functional hormone levels in the human body. It's also important to recognize that hormones and hormonelike contaminants are concentrated in adipose tissue. Obesity, which is epidemic in the United States, increases the risk that EDCs might be found at toxic levels in humans.

Scientists and those representing industrial manufacturers continue to debate whether EDCs pose a health risk to humans. Currently, the EPA regulates the concentration of 23 inorganic chemicals including mercury, lead, and arsenic, and more than 30 organic chemicals also are regulated. Doubtless other inorganic and organic chemicals will be added to the list as research continues. In the meantime, there are steps you can take to minimize the dangers of EDCs:

- Maintain a healthy weight. Excess adipose tissue can contain high levels of EDCs that you may have been exposed to throughout your life.
- Use less plastic. Substitute reusable cloth bags, washable dinnerware, glass containers, etc. for the plastic products you might otherwise buy.
- Recycle plastic, so that EDCs that might be contained in it don't leak into the groundwater.
- Don't heat or reheat food in plastic containers.
- If you eat a lot of fish, be aware of those that may contain high levels of EDCs and other contaminants. The EPA maintains a database that can help you find out which fish in your area are safe to eat: http://www.epa.gov/waterscience/fish/

after unprotected intercourse. The name is a misnomer because the medication does not have to be taken the next morning—the woman can begin the medication one to several days after having unprotected intercourse.

One type of emergency contraception contains four synthetic estrogen-progesterone pills; a second type contains progesterone-only pills. Two pills are taken up to 72 hours after unprotected intercourse, and two more are taken 12 hours later. The medication upsets the normal uterine cycle, making it difficult for an embryo to implant in the endometrium. A recent study estimated that the medication was 80% effective in preventing unintended pregnancies. The medication should not be used for regular contraception. Side effects may include nausea, vomiting, abdominal pain, fatigue, and headache. Other, more serious complications include hypertension, cardiovascular disease, and stroke.

Mifepristone, better known as RU-486, does not prevent conception. Rather, this medication is presently used to cause the loss of an implanted embryo by blocking the progesterone receptor proteins of endometrial cells. Without functioning receptors for progesterone, the endometrium sloughs off, carrying the embryo with it. When taken in conjunction with a prostaglandin to induce uterine contractions, RU-486 is 95% effective in terminating an early pregnancy. However, cases of blood-borne infection and hemorrhage have occurred in women using the drug, and it has caused fatalities in Europe and the U.S. Thus, it is critical that patients using RU-486 be under the constant supervision of qualified health-care professionals.

Although it may seem an impossibly old-fashioned, outmoded ideal, it is still ultimately true—and therefore worth mentioning in any textbook dealing with human anatomy and physiology: The most reliable method of birth control is *still* abstinence—that is, not engaging in sexual intercourse. This form of birth control has the added advantage of preventing transmission of a sexually transmitted infection.

Infertility

Infertility is the failure of a couple to achieve pregnancy after one year of regular, unprotected intercourse. The American Medical Association estimates that 15% of all couples are infertile. The cause of infertility can be attributed to the male (40%), the female (40%), or both (20%).

Causes of Infertility

The most frequent cause of infertility in males is low sperm count and/or a large proportion of abnormal sperm. There are numerous causes for low sperm count. Herbicides and pesticides may cause feminizing effects in the man's body, as discussed in the previous Medical Focus. Chemicals used on the job, such as paints, varnishes, and degreasers, may cause infertility as well. Smoking, alcohol and drug abuse decrease the numbers and quality of a man's sperm cells. In particular, abuse of anabolic (body-building) steroids can shrink the testes and decrease sperm count. Exposing the testicles to overheating must be avoided, as it impairs sperm production and lowers sperm count. Common causes for overheating include obesity and/or remaining seated for very long periods (as in long-distance driving, for example). Frequent use of saunas or hot tubs, wearing extremely tight clothing such as biking shorts, and long periods of using a very warm laptop computer can also cause overheating and a subsequent low sperm count.

Body weight appears to be the most significant factor in causing female infertility. In women of normal weight, fat cells produce a hormone called leptin that stimulates the hypothalamus to release GnRH. In overweight women, the ovaries often contain many small follicles, and the woman fails to ovulate. Other causes of infertility in females are blocked uterine tubes due to pelvic inflammatory disease (see page 400) and endometriosis. **Endometriosis** is the presence of uterine tissue outside the uterus, particularly in the uterine tubes and on the abdominal organs. Backward flow of menstrual fluid allows living uterine cells to establish themselves in the abdominal cavity, where they go through the usual uterine cycle, causing pain and structural abnormalities that make it more difficult for a woman to conceive.

Sometimes the causes of infertility can be corrected by medical intervention so that couples can have children. If no obstruction is apparent and body weight is normal, it is possible to give females fertility drugs, which are gonadotropic hormones that stimulate the ovaries and bring about ovulation. Such hormone treatments may cause multiple ovulations and multiple births.

When reproduction does not occur in the usual manner, many couples adopt a child. Others sometimes try one of the assisted reproductive technologies discussed in the following paragraphs.

Assisted Reproductive Technologies

Assisted reproductive technologies (ARTs) consist of techniques used to increase the chances of pregnancy. Often, sperm and/or eggs are retrieved from the testes and ovaries, and fertilization takes place in a clinical or laboratory setting.

Artificial Insemination During **artificial insemination**, sperm are placed in the vagina by a physician. Sometimes a woman is artificially inseminated by her partner's sperm. This is especially helpful if the partner has a low sperm count, because the sperm can be collected over a period of time and concentrated so that the sperm count is sufficient to result in fertilization. Often, however, a woman is inseminated by sperm acquired from a donor who is a complete stranger to her. At times, a combination of partner and donor sperm is used.

During *intrauterine insemination (IUI)*, fertility drugs are given to stimulate the ovaries, and then the donor's sperm is placed in the uterus rather than in the vagina.

If the prospective parents wish, sperm can be sorted into those believed to be X-bearing or Y-bearing to increase the chances of having a child of the desired sex.

In Vitro Fertilization During **in vitro fertilization (IVF)**, conception occurs in the laboratory. Ultrasound machines can now spot follicles in the ovaries that hold immature eggs; therefore, the latest method is to forgo the administration of fertility drugs and retrieve immature eggs by using a needle. The immature eggs are then brought to maturity in the laboratory before concentrated sperm are added. After about two to four days, the embryos are ready to be transferred to the uterus of the woman, who is now in the secretory phase of her uterine cycle. If desired, preimplantation genetic analysis can be done and only embryos found to be free of genetic disorders are used. If implantation is successful, development is normal and continues to term.

Intracytoplasmic Sperm Injection In **intracytoplasmic sperm injection (ICSI)**, a highly sophisticated procedure, a single sperm is injected into an egg. This method is used effectively when a man has severe infertility problems.

Sexually Transmitted Infections

Sexually transmitted infections (STIs) are caused by organisms ranging from viruses to arthropods; however, we will discuss only certain STIs caused by viruses and bacteria. Unfortunately, for unknown reasons, humans cannot develop effective immunity to any STIs. Therefore, any person exposed to an STI should seek prompt medical treatment. To prevent the spread of STIs, a latex or polyurethane condom can be used; the concomitant use of a spermicide containing nonoxynol-9 gives added protection.

Among those STIs caused by viruses, treatment is available for AIDS and genital herpes. However, it is important to note that treatment for HIV/AIDS and genital herpes cannot presently eliminate the virus from the person's body. Drugs used for treatment can merely slow replication of the viruses. Thus, neither viral disease is presently curable. Further, antiviral drugs have serious, debilitating side effects on the body.

Only STIs caused by bacteria (e.g., chlamydia, gonorrhea, and syphilis) are curable with antibiotics. Bacteria that acquire antibiotic resistance may necessitate treatment with extremely strong drugs for an extended period to achieve a cure.

Genital Warts

Genital warts are caused by the human papillomaviruses (HPVs). Many times, carriers either do not have any sign of warts or merely have flat lesions. When present, the warts commonly are seen on the penis and foreskin of men and near the vaginal opening in women. A newborn can become infected while passing through the birth canal.

Individuals who are currently infected with visible growths may have those growths removed by surgery, freezing, or burning with lasers or acids. However, visible warts that are removed may recur. You'll recall (from Chapter 13) that a vaccine for the human papillomaviruses that most commonly cause genital warts is now available. The vaccine is an extremely important step in the prevention of cancer, as well as the warts themselves. Genital warts are associated with cancer of the cervix, as well as tumors of the vulva, vagina, anus, and penis. Researchers believe that these viruses may be involved in up to 95% of all cases of cancer of the cervix. Vaccination might make such cancers a thing of the past.

Genital Herpes

Genital herpes is caused by herpes simplex virus. Type 1 usually causes cold sores and fever blisters, while type 2 more often causes genital herpes.

Persons usually get infected with herpes simplex virus type 2 when they are adults. Some people exhibit no symptoms; others may experience a tingling or itching sensation before blisters appear on the genitals. Once the blisters rupture, they leave painful ulcers that may take as long as three weeks or as little as five days to heal. The blisters may be accompanied by fever, pain on urination, swollen lymph nodes in the groin, and in women, a copious discharge. At this time, the individual has an increased risk of acquiring an HIV infection.

After the ulcers heal, the disease is only latent, and blisters can recur, although usually at less frequent intervals and with milder symptoms. Fever, stress, sunlight, and menstruation are associated with recurrence of symptoms. Exposure to herpes in the birth canal can cause an infection in the newborn, which

leads to neurological disorders and even death. Birth by cesarean section prevents this possibility.

Hepatitis

Hepatitis infects the liver and can lead to liver failure, liver cancer, and death. The type of hepatitis and the virus that causes it are designated by the same letter. There are six known viruses that cause hepatitis, designated A-B-C-D-E-G. Hepatitis A is usually acquired from sewage-contaminated food or drinking water, but this infection can also be transmitted sexually through oral/anal contact. Hepatitis B is spread through sexual contact and by blood-borne transmission (accidental needle stick on the job, receiving a contaminated blood transfusion, a drug abuser sharing infected needles while injecting drugs, from mother to fetus, etc.). Simultaneous infection with hepatitis B and HIV is common because both share the same routes of transmission. Fortunately, a combined vaccine is available for hepatitis A and B; it is currently recommended that all children receive the vaccine to prevent infection (see pages 296–297). Hepatitis C (also called non-A, non-B hepatitis) causes most cases of post-transfusion hepatitis. Hepatitis D and hepatitis G are transmitted sexually, and hepatitis E is acquired from contaminated water. Screening of blood and blood products can prevent transmission of hepatitis viruses during a transfusion. Proper water treatment techniques can prevent contamination of drinking water.

Chlamydia

Chlamydia is named for the tiny bacterium that causes it (*Chlamydia trachomatis*). The incidence of new chlamydia infections has steadily increased since 1984.

Chlamydia infections of the lower reproductive tract are usually mild or asymptomatic, especially in women. About 18 to 21 days after infection, men may experience a mild burning sensation on urination and a mucoid discharge. Women may have a vaginal discharge along with the symptoms of a urinary tract infection. Chlamydia also causes cervical ulcerations, which increase the risk of acquiring HIV.

If the infection is misdiagnosed or if a woman does not seek medical help, there is a particular risk of the infection spreading from the cervix to the uterine tubes so that **pelvic inflammatory disease (PID)** results. This very painful condition can result in blockage of the uterine tubes with the possibility of sterility and infertility. If a baby comes in contact with chlamydia during birth, inflammation of the eyes or pneumonia can result.

Gonorrhea

Gonorrhea is caused by the bacterium *Neisseria gonorrhoeae.* Diagnosis in the male is not difficult, since typical symptoms are pain upon urination and a thick, greenish-yellow urethral discharge. In males and females, a latent infection leads to pelvic inflammatory disease (PID), which can also cause sterility in males. If a baby is exposed during birth, an eye

MEDICAL FOCUS

Preventing Transmission of STIs

It is wise to protect yourself from getting a sexually transmitted infection (STI). Some of the STIs, such as gonorrhea, syphilis, and chlamydia, can be cured by taking an antibiotic, but medication for the ones transmitted by viruses is much more problematic. In any case, it is best to prevent the passage of STIs from person to person so that treatment becomes unnecessary.

Sexual Activities Transmit STIs (Fig. 17D)

Abstain from sexual intercourse or develop a long-term monogamous (always the same partner) *sexual relationship* with a partner who is free of STIs.

Refrain from multiple sex partners or having relations with someone who has multiple sex partners. If you have sex with two other people and each of these has sex with two people and so forth, the number of people who are relating is quite large.

Remember that, although the prevalence of AIDS is presently higher among homosexuals and bisexuals, the highest rate of increase is now occurring among heterosexuals. The lining of the uterus is only one cell thick, and it does allow infected cells from a sexual partner to enter.

Be aware that having relations with an intravenous drug user is risky because the behavior of this group risks hepatitis and an HIV infection. Be aware that anyone who already has another sexually transmitted infection is more susceptible to an HIV infection.

Uncircumcised males are more likely to become infected with an STI than circumcised males because vaginal secretions can remain under the foreskin for a long period of time.

Avoid anal-rectal intercourse (in which the penis is inserted into the rectum) because the lining of the rectum is thin and cells infected with HIV can easily enter the body there.

Unsafe Sexual Practices Transmit STIs

Always use a latex condom during sexual intercourse if you do not know for certain that your partner has been free of STIs for some time. Be sure to follow the directions supplied by the manufacturer. Use of a water-based spermicide containing nonoxynol-9 in addition to the condom can offer further protection because nonoxynol-9 immobilizes viruses and virus-infected cells.

Avoid fellatio (kissing and insertion of the penis into a partner's mouth) *and cunnilingus* (kissing and insertion of the tongue into the vagina) because they may be a means of transmission. The mouth and gums often have cuts and sores that facilitate the entrance of infected cells.

Be cautious about using alcohol or any drug that may prevent you from being able to control your behavior.

Drug Use Transmits Hepatitis and HIV (Fig. 17E)

Stop, if necessary, or do not start the habit of injecting drugs into your veins. Be aware that hepatitis and HIV can be spread by blood-to-blood contact.

Always use a new, sterile needle for injection or one that has been cleaned in bleach if you are a drug user and cannot stop your behavior.

Figure 17D Sexual activities transmit STIs.

Figure 17E Sharing needles transmits STIs.

infection leading to blindness can result. All newborns are given eyedrops to prevent this possibility.

Gonorrhea proctitis, an infection of the anus characterized by anal pain and blood or pus in the feces, also occurs in patients. Oral/genital contact can cause infection of the mouth, throat, and tonsils. Gonorrhea can spread to internal parts of the body, causing heart damage or arthritis. If, by chance, the person touches infected genitals and then touches his or her eyes, a severe eye infection can result. Up to now, gonorrhea was curable by antibiotic therapy, but resistance to antibiotics is becoming more and more common, and 40% of all strains are now known to be resistant to therapy.

Syphilis

Syphilis is caused by a bacterium called *Treponema pallidum*. As with many other bacterial diseases, penicillin is an effective antibiotic. Syphilis has three stages, often separated by latent periods, during which the bacteria are resting before multiplying again. During the *primary stage*, a hard **chancre** (ulcerated sore with hard edges) indicates the site of infection. The chancre usually heals spontaneously, leaving little scarring. During the *secondary stage*, the victim breaks out in a rash that does not itch and is seen even on the palms of the hands and the soles of the feet. Hair loss and infectious gray patches on the mucous membranes may also occur. These symptoms disappear of their own accord.

During the *tertiary stage*, which lasts until the patient dies, syphilis may affect the cardiovascular system by causing aneurysms, particularly in the aorta. In other instances, the disease may affect the nervous system, resulting in psychological disturbances. Also **gummas**, large destructive ulcers, may develop on the skin or within the internal organs.

Congenital syphilis is caused by syphilitic bacteria crossing the placenta. The child is born blind and/or with numerous anatomical malformations. Control of syphilis depends on prompt and adequate treatment of all new cases; therefore, it is crucial for all sexual contacts to be traced so they can be treated. Diagnosis of syphilis can be made by blood tests or by microscopic examination of fluids from lesions.

Content CHECK-UP!

10. A combined hormonal method of contraception is available through the following delivery method(s):

    a. pills taken orally   c. injections received monthly

    b. patches applied to the skin   d. all of the above

11. Name three bacterial forms of sexually transmitted infection.

12. Which of the following is a characteristic of tertiary syphilis?

    a. large, destructive ulcers called gummas

    b. formation of a hard chancre

    c. a skin rash that does not itch

    d. hair loss

Answers in Appendix C, page 465.

17.5 Effects of Aging

Sex hormone levels decline with age in both men and women. Menopause, the period in a woman's life during which menstruation ceases, is likely to occur between the ages of 45 and 55. The ovaries are no longer responsive to pituitary gonadotropic hormones because after years of ovulating, much of functional ovarian tissue has been replaced by corpus albicans. After menopause, the ovaries stop producing ova and produce only minimal amounts of estrogen and progesterone. At the onset of menopause, the menstrual cycle becomes irregular and then eventually ceases. Hormonal imbalance often produces physical symptoms, such as dizziness, headaches, insomnia, sleepiness, depression, and "hot flashes" that are caused by circulatory irregularities. Menopausal symptoms vary greatly among women, and some symptoms may be absent altogether.

Following menopause, atrophy of the uterus, vagina, breasts, and external genitals is likely. The lack of estrogen also promotes changes in the skin (e.g., wrinkling; see Chapter 5) and in the skeleton (e.g., osteoporosis; see Chapter 6).

In men, testosterone production diminishes steadily after age 50, which may be responsible for the enlargement of the prostate gland. Sperm production declines with age, yet men can remain fertile well into old age. With age, however, the chance of erectile dysfunction due to degenerative vascular changes in the penis increases.

Sexual desire and activity need not decline with age, and many older men and women enjoy sexual relationships. Men are likely to experience reduced erection until close to ejaculation, and women may experience a drier vagina.

17.6 Homeostasis

Regulation of sex hormone blood level is an example of homeostatic control. Figure 17.8 shows how the blood level of testosterone is maintained within normal limits. Negative feedback results in a self-regulatory mechanism that maintains the appropriate level of these hormones in the blood.

The illustration in Human Systems Work Together on page 404 shows how the reproductive system works with the other systems of the body to maintain homeostasis. Usually we stress that the function of sex hormones produced not only by the gonads but also by the adrenal glands is to foster the maturation of the reproductive organs and to maintain the secondary sex characteristics. However, these functions of sex hormones have nothing to do with homeostasis. Why? Because homeostasis pertains to the constancy of the internal environment of cells. Other activities of the sex hormones do affect the internal environment. For example, estrogen promotes fat deposition, which serves as a source of energy for cells and also helps the body maintain a normal temperature because of its insulating effect.

In recent years, it's been discovered that the sex hormones perform other activities that affect homeostasis even more directly. We are just now beginning to discover the role that

FOCUS on FORENSICS

Rape

Rape is a crime of violence, and not a crime of sexuality. It happens to all manner of victims: women, men, and children; heterosexual and homosexual; even babies and the elderly. It is a crime that reportedly occurs once every 2 minutes—but that statistic may be much lower than the number of actual incidents. An estimated 25% of reported cases are committed by a complete stranger. Family members carry out approximately 12% of rapes, but the highest numbers of such crimes—63%—are committed by an acquaintance of the victim. Many rapes go unreported, their victims too afraid or too ashamed to seek medical and legal assistance. When rape victims do seek emergency medical care, it is imperative that their needs for physical, psychological, and legal assistance are met promptly and compassionately.

Many communities have formed Sexual Assault Response Teams (SARTs). A key member of such a team is a trained sexual assault nurse examiner (SANE). Forensic nursing is a subspecialty of nursing recognized by the American Nursing Association, which provides training for nurses desiring SANE certification. The team also includes EMS personnel, physicians, psychologists, social workers, and law enforcement officials.

The first step in care for the rape victim is taking a thorough medical history. The patient should be made as comfortable as possible, and may prefer to remain clothed during the initial assessment. A physical exam will follow during which a standardized rape kit may be used to collect evidence. The patient should disrobe while standing on a sheet of exam-table paper in order to catch any hairs from the suspect that might remain on the patient's body. All clothing should be placed in paper evidence bags, and any fluid on the clothing (blood, urine, semen, etc.) should be allowed to air-dry before the paper bags are sealed. This will prevent mold growth, which might destroy evidence. Bruises and abrasions on the patient's body must be documented and photographed. Vaginal exams of females and oral-rectal exams for either gender may also give evidence of injury. The pubic hair is combed through with a fine-tooth comb to capture any pubic hair from the perpetrator.

Fluids found on the patient's body should be collected on swabs for DNA and other chemical testing, and their presence recorded. The victim's blood, hair, and urine will also be sampled for comparison. A high-frequency ultraviolet lamp called a Wood's lamp can detect the presence of semen on the patient's body. Seminal fluid will fluoresce at specific wavelengths of ultraviolet light. Other chemical testing for the components of semen can be done as well. However, it is important to note that absence of semen does not mean that rape did not occur. The perpetrator may have had a vasectomy. Further, a study of assailants documented that 50% reported erectile dysfunction and inability to ejaculate while attempting a rape.

Medical treatment for the victim will include testing for sexually transmitted infection, including gonorrhea, syphilis, and HIV. Antibiotics and anti-retroviral drugs can be prescribed. If the patient is not immunized for hepatitis B, he or she should receive the vaccine. Testing for HIV and hepatitis should be repeated approximately 6 months after the incident. Female victims should also be tested for a pre-existing pregnancy. Nonpregnant women may elect to receive emergency contraception to prevent pregnancy. Such treatment can damage an existing pregnancy and should not be given to a pregnant woman. The patient should also receive a urinalysis, especially if he or she does not recall details of the rape. Testing can show evidence of so-called "date-rape" drugs, which induce short-term memory loss.

Psychological counseling is essential if a rape victim is to recover. A rape victim may initially feel anger, fear, or anxiety—although he or she may express no emotion at all. Many victims will internalize blame for the incident, believing that some past incident may have invited the assault. Flashbacks—painful, detailed memories of the incident—are common after rape. Anxiety disorders and post-traumatic stress disorder may occur years after the incident, and also require psychological counseling.

By properly documenting history and evidence, EMS, nursing, and medical professionals can enable successful prosecution of the rapist. Detailed evidence is particularly important in cases of nonstranger sexual assault, such as child, spousal, or elder abuse. Evidence from health-care professionals in these cases can support a victim's claims of lack of consent to sexual activity. Just as important, careful medical and psychological care can enable the victim to recover from his or her ordeal.

---

estrogen and androgens play in the metabolism of cells and therefore their role in homeostasis in general. Estrogen induces the liver to produce many types of proteins that transport substances in the blood. These include proteins that bind iron and copper and lipoproteins that transport cholesterol. Iron and copper are enzyme cofactors necessary to cellular metabolism. Although we associate cholesterol with cardio-vascular diseases, in fact, it is also a substance that contributes to the functioning of the plasma membrane.

Estrogen also induces synthesis of bone matrix proteins and counteracts the loss of bone mass. At menopause, when the rate of estrogen secretion is drastically reduced, osteoporosis (decrease in bone density) may develop. Nevertheless, long-term estrogen therapy is no longer recommended.

Human Systems Work Together

Integumentary System

Androgens activate oil gland; sex hormones stimulate fat deposition, affect hair distribution in males and females.

Skin receptors respond to touch; modified sweat glands produce milk; skin stretches to accommodate growing fetus.

Skeletal System

Sex hormones influence bone growth and density in males and females.

Bones provide support and protection of reproductive organs.

Muscular System

Androgens promote growth of skeletal muscle.

Muscle contraction occurs during orgasm and moves gametes; abdominal and uterine muscle contractions occur during childbirth.

Nervous System

Sex hormones masculinize or feminize the brain, exert feedback control over the hypothalamus, and influence sexual behavior.

Brain controls onset of puberty; nerves are involved in erection of penis and clitoris, movement of gametes along ducts, and contraction of uterus.

Endocrine System

Gonads produce the sex hormones.

Hypothalamic, pituitary, and sex hormones control sex characteristics and regulate reproductive processes.

How the Reproductive System works with other body systems.

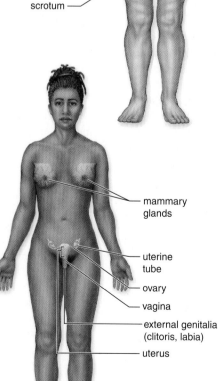

seminal vesicle
ductus deferens
prostate gland
urethra
epididymis
penis
testis
scrotum

mammary glands
uterine tube
ovary
vagina
external genitalia (clitoris, labia)
uterus

Cardiovascular System

Sex hormones influence cardiovascular health; sexual activities stimulate cardiovascular system.

Blood vessels transport sex hormones; vasodilation causes genitals to become erect; blood services the reproductive organs.

Lymphatic System/Immunity

Sex hormones influence immune functioning; acidity of vagina helps prevent pathogen invasion of body; milk passes antibodies to newborn.

Immune system does not typically attack sperm or fetus, even though they are foreign to the body.

Respiratory System

Sexual activity increases breathing; pregnancy causes breathing rate and vital capacity to increase.

Gas exchange increases during sexual activity.

Digestive System

Pregnancy crowds digestive organs and promotes heartburn and constipation.

Digestive tract provides nutrients for growth and repair of organs and for development of fetus.

Urinary System

Penis in males contains the urethra and performs urination; prostate enlargement hinders urination.

Semen is discharged through the urethra in males; kidneys excrete wastes and maintain electrolyte levels for mother and child.

Nervous System

The National Institutes of Health conducted a study of 16,608 healthy women who were taking both estrogen and progesterone—that is, hormone replacement therapy (HRT)—or a placebo. The study was halted after 5.2 years because physicians concluded that the risks for the group on HRT outweighed the benefits. The women on HRT had a small but significant increased risk for breast cancer, coronary heart disease, stroke, and blood clots. The benefits of HRT included lower risk of hip fractures and colon cancer.

Similarly, besides the action of androgens (e.g., testosterone) on the sexual organs and functions of males, androgens play a metabolic role in cells. They stimulate synthesis of structural proteins in skeletal muscles and bone, and they also affect the activity of various enzymes in the liver and kidneys. For example, in the kidney, androgens stimulate synthesis of erythropoietin, the protein that signals the bone marrow to increase the production of red blood cells.

SELECTED NEW TERMS

Basic Key Terms

acrosome (ak′ro-sōm), p. 380

androgens (ăn′drə-jən), p. 380

areola (ă-rē′ō-lă), p. 392

Bartholin glands (Băr′tō-lĭn glănds), p. 390

bulbourethral gland (bul″bo-yū-re′thral gland), p. 380

cervix (ser′viks), p. 386

clitoris (klĭ′to-ris), p. 390

colostrum (ko-los′trum), p. 393

corpus albicans (kor′pus al′bĭ-kanz), p. 386

corpus luteum (kor′pus lu′te-um), p. 386

corpora cavernosa (kōr′pōr-ă kăv-ĕr-noś-ə), p. 383

corpus spongiosum (kōr′pŭs spŭn-gē-ō′sŭm), p. 383

diploid (dĭp′loid′), p. 377

ejaculation (e-jak″yū-la′shun), p. 380

ejaculatory duct (e-jak′yū-luh-to″re dukt), p. 380

endometrium (en-do-me′tre-um), p. 386

epididymis (ep″ĭ-did′ĭ-mis), p. 380

estrogen (ĕs′trə-jən), p. 390

external genitals (eks-ter′nal jen′ĭ-talz), p. 383

fimbriae (fim′bre-e), p. 388

follicle (fol′ĭ-kl), p. 386

FSH (follicle-stimulating hormone) (fol′ĭ-kl stim′yū-la-ting hor′mōn), p. 384

gametes (găm′ēt′), p. 379

genitals (jen′i-tălz), p. 377

germ cells (jûrm sēl), p. 380

haploid (hăp′loid′), p. 377

human chorionic gonadotropin (hyū′man ko″re-on′ik go′nad-o-tro′pin), p. 392

implantation (im″plan-ta′shun), p. 386

inhibin (in-hib′in), p. 384

interstitial cell (in″ter-stish′ul sel), p. 380

labia majora (la′be-uh muh-jor′uh), p. 390

labia minora (la′be-uh mi-nor′uh), p. 390

lactation (lak-tā′-shun), p. 393

LH (luteinizing hormone) (lu′te-ĭ-nīz-ing hor′mōn), p. 384

meiosis (mi-o′sis), p. 377

menopause (men′o-pawz), p. 392

menses (měn′sēz′), p. 391

menstrual cycle (men″stru-ăl), p. 390

mons pubis (mŏnz pyoo′bis), p. 390

oocyte (o′o-sīt), p. 386

oogenesis (o″o-jen′ĕ-sis), p. 386

orgasm (ŏr-gazm′), p. 384

ova (ō-vŭh), p. 386

ovaries (ō′və-rēs), p. 385

ovulation (ov′yū-la′shun), p. 386

ovum (o-vŭm), p. 386

Pap smear (păp smîr), p. 389

penis (pe′nis), p. 383

polar body (pō′lĕr bŏd-ē), p. 386

progesterone (prŏ-jĕs′tə-rōn′), p. 390

prostaglandins (prŏs′tə-glăn′dĭn), p. 380

prostate gland (prŏs′tāt glănd), p. 380

puberty (pyu′ber-te), p. 377

scrotum (skro′tum), p. 380

semen (se′men), p. 380

seminal vesicle (sem′i-năl věs-i- kul), p. 380

seminiferous tubule (se″mĭ-nif′er-us tu′byūl), p. 380

sperm (sperm), p. 380

spermatic cord (sperm-ă′-tik kŏrd), p. 382

spermatogenesis (sper″muh-to-jen′ĕ-sis), p. 380

spermatogonia (spĕr′-mă-tō-gō-nē-ŭm), p. 380

sustentacular (Sertoli) cells (sus′tĕn-tăk-yoo-lĕr sells), p. 380

testis (tĕs′tĭs), p. 379

testosterone (tĕs-tŏs′tə-rōn′), p. 380

urethra (yo͞o-rē′thrə), p. 380

uterine tube (yū′ter-in tūb), p. 388

uterus (yoo′-tĕr-ŭs), p. 386

vagina (vuh-ji′nuh), p. 386

vas deferens (vas def′er-ens), p. 380

vesicular (Graafian) follicle (ves-ik′yū-ler [graf′e-un] fol′ĭ-kl), p. 386

vestibule (věs′tə-byo͞o-l′), p. 390

vulva (vul′vuh), p. 386

zygote (zi′gŏt), p. 379

Clinical Key Terms

artificial insemination (ar″tĭ-fĭ′shul in-sem-ĭ-na′shun), p. 399

assisted reproductive technologies (uh-sis′ted re″pro-duk′tiv tek-nah′lo-jēz), p. 399

chancre (shan′ker), p. 402

chlamydia (kluh-mĭ′de-uh), p. 400

circumcision (ser″kum-sizh′un), p. 383

contraception (kon-tră-sep′shun), p. 393

cryptorchidism (krĭp-tōr′ki-dizm), p. 380

ectopic pregnancy (ek-top′ik preg′nun-se), p. 389

endometriosis (en-do-me″tre-o′sis), p. 399

erectile dysfunction (e-rek′til dis-funk′shun), p. 384

genital herpes (jen′ĭ-tal her′pēz), p. 400

genital warts (jen′ĭ-tal worts), p. 400

gonorrhea (gon-o-re′uh), p. 400

gumma (gŭm′ă), p. 402

hepatitis (hĕp′ə-tī′tĭs), p. 400

hernia (her′ne-uh), p. 382

hysterectomy (his″ter-ek′to-me), p. 389

infertility (in-fer-til′ĭ-te), p. 399

inguinal hernia (ĭng′gwə-nel hûr′nē-ə), p. 382

in vitro fertilization (ĭn vē′trō fûr′tl-ĭ-zā′shən), p. 399

oophorectomy (o-ahf-or-ek′to-me), p. 389

ovarian cancer (o-vā′re-un can′ser), p. 389

ovariohysterectomy (o-vār″e-o-his-ter-ek′to-me), p. 389

pelvic inflammatory disease (pel′-vik-in-flăm-uh-tor-e dĭz-ēz), p. 400

salpingectomy (sal-pin-jek′to-me), p. 389

sexually transmitted infection (sek′shu-ah-le tranz-mit′ed ĭn′fēk-shŭn), p. 399

syphilis (sĭf-ŭh-lĭs), p. 402

SUMMARY

17.1 Human Life Cycle
The life cycle of humans requires two types of nuclear division: mitosis and meiosis. Meiosis is involved in sperm production in males and ova production in females.

17.2 Male Reproductive System
A. In males, spermatogenesis, occurring in seminiferous tubules of the testes, produces sperm that mature and are stored in the epididymides. Sperm may also be stored in the vasa deferentia before entering the urethra, along with secretions produced by the seminal vesicles, prostate gland, and bulbourethral glands. Sperm and these secretions are called semen, or seminal fluid.

B. The external genitals of males are the penis, the organ of sexual intercourse, and the scrotum, which contains the testes. Orgasm in males is a physical and emotional climax during sexual intercourse that results in ejaculation of semen from the penis.

C. Hormonal regulation, involving secretions from the hypothalamus, the anterior pituitary, and the testes, maintains testosterone, produced by the interstitial cells of the testes, at a fairly constant level. FSH from the anterior pituitary promotes spermatogenesis in the seminiferous tubules, and LH promotes testosterone production by the interstitial cells.

17.3 Female Reproductive System
A. In females, oogenesis occurring within the ovaries typically produces one mature follicle each month. This follicle balloons out of the ovary and bursts, releasing a secondary oocyte, which enters a uterine tube. The uterine tubes lead to the uterus, where implantation and development occur. The external genital area includes the vaginal opening, mons pubis, the clitoris, the labia minora, and the labia majora.

B. The vagina is the organ of sexual intercourse and the birth canal in females. The vagina and the external genitals, especially the clitoris, play an active role in orgasm, which culminates in uterine and uterine tube contractions.

C. Hormonal regulation in females involves the production of FSH and LH by the anterior pituitary and also production of estrogen and progesterone by the ovaries.

D. The menstrual cycle is a monthly series of events that can be divided into the pre-ovulation events and the post-ovulation events. Before ovulation, FSH from the anterior pituitary causes an estrogen-producing follicle to begin developing in the ovary. Meanwhile in the uterus, menstruation occurs before the proliferation phase occurs. During the proliferation phase, estrogen causes the uterine lining to thicken.

   Ovulation is caused by a positive feedback cycle in which an abundance of estrogen brings about an FSH and LH surge. After ovulation, the corpus luteum in the ovary secretes primarily progesterone that causes the uterine lining to become secretory. If the egg is fertilized, it implants itself in the uterine lining and the corpus luteum does not disintegrate.

E. If fertilization takes place, the embryo implants itself in the thickened endometrium. If fertilization and implantation occur, the corpus luteum in the ovary is maintained because of HCG production by the placenta, and therefore progesterone production does not cease. Menstruation usually does not occur during pregnancy.

17.4 Control of Reproduction and Sexually Transmitted Infections
A. Numerous contraceptive methods and devices, such as the birth control pill, diaphragm, and condom, are available for those who wish to prevent pregnancy.

   Emergency contraception can be taken before there is any indication of pregnancy. Mifepristone blocks progesterone receptors, causing an implanted embryo to be shed with the uterine lining.

B. Some couples are infertile, and if so, they may use assisted reproductive technologies in order to have a child. Artificial insemination and in vitro fertilization have been followed by more sophisticated techniques such as intracytoplasmic sperm injection.

17.5 Effects of Aging
Menopause occurs between the ages of 45 and 55 in women. Following menopause, atrophy of the genitals is likely. In men, testosterone production decreases after age 50, and the incidence of erectile dysfunction increases.

17.6 Homeostasis
The reproductive system works with the other systems of the body in the ways described in Human Systems Work Together on page 404.

STUDY QUESTIONS

1. Outline the path of sperm. What glands contribute fluids to semen? (pp. 379–383)
2. Discuss the anatomy and physiology of the testes. Describe the structure of sperm. (pp. 380–382)
3. Name the endocrine glands involved in maintaining the sex characteristics of males and the hormones produced by each. (pp. 384–385)
4. Describe the organs of the female genital tract. Where do fertilization and implantation occur? Name two functions of the vagina. (pp. 385–386)
5. Name and describe the external genitals in females. (p. 390)
6. Discuss the anatomy and the physiology of the ovaries. (pp. 386–387)

7. Describe the pre- and post-ovulation events of the menstrual cycle. In what way is menstruation prevented if pregnancy occurs? (pp. 390–392)
8. Name three functions of the female sex hormones. (pp. 390–392)
9. Discuss the various means of birth control and their relative effectiveness in preventing pregnancy. (pp. 393–399)
10. Describe how in vitro fertilization is carried out. (p. 399)

LEARNING OBJECTIVE QUESTIONS

Fill in the blanks.

1. In tracing the path of sperm, the structure that follows the epididymis is the _____.
2. The prostate gland, the bulbourethral glands, and the _____ all contribute to seminal fluid.
3. An erection is caused by the entrance of _____ into erectile tissue of the penis.
4. The primary male sex hormone is _____.

5. In the female reproductive system, the uterus lies between the uterine tubes and the _____.
6. The female sex hormones are _____ and _____.
7. In the menstrual cycle, once each month a _____ produces an ovum and the _____ is prepared to receive the embryo.
8. The most frequent causes of male infertility are _____ and _____.
9. Spermatogenesis occurs within the _____ of the testes.

10. Androgens are secreted by the _____ cells that lie between seminiferous tubules.
11. Once a vesicular follicle has released an egg, it develops into a glandlike structure called a _____.
12. The release of an egg from a vesicular follicle is called _____.
13. Whereas AIDS and genital herpes are caused by _____, gonorrhea and chlamydia are caused by _____.

MEDICAL TERMINOLOGY EXERCISE

Consult Appendix B for help in pronouncing and analyzing the meaning of the terms that follow.

1. orchidopexy (or″kĭ-do-pek′se)
2. transurethral resection of prostate (TURP) (trans″yū-re′thral re-sek′shun ov pros′tāt)
3. gonadotropic (go″nad-o-trōp′ik)

4. gynecomastia (jin″ĕ-ko-mas′te-uh)
5. hysterosalpingo-oophorectomy (his″ter-o-sal-ping′go-o″ahf-or-ek′to-me)
6. colporrhaphy (kol-por′uh-fe)
7. menometrorrhagia (men′o-me-tro-ra-je-uh)
8. multipara (mul-tip′uh-ruh)
9. balanitis (bal″uh-ni′tis)

10. seminoma (sem′ĭ-no′muh)
11. genitourinary (jen′ĭ-to-yū′rĭ-nār′e)
12. prostatic hypertrophy (pros-tat′ik hy′per-tro′fe)
13. azoospermia (ā-zo′o-sper′me-uh)

WEBSITE LINK

Visit this text's website at http://www.mhhe.com/longenbaker7 for additional quizzes, interactive learning exercises, and other study tools.

Human Development and Birth

The town of Mount Airy, North Carolina is most famous for its connections to the fictional town of "Mayberry," from the 1960's era television show, *The Andy Griffith Show*. Mount Airy is the birthplace of Andy Griffith, the show's star and namesake, and Mayberry was modeled after Mount Airy. Andy Griffith was not the town's only famous resident, however. There were two other Mount Airy citizens who were famous long before *The Andy Griffith Show* ever began—Chang and Eng Bunker. The Bunkers were the original "Siamese twins," a pair of conjoined brothers born in Siam (now modern-day Thailand) in 1811. As you'll discover in this chapter, conjoined twins like Chang and Eng are formed when the developing zygote begins separating but doesn't completely divide. Two individuals are formed but remain connected somewhere along their bodies. The two Bunker brothers were united at the chest by a tube of flesh, and modern-day surgeons speculate that the twins could easily have been separated with today's medical knowledge and surgical procedures. Separation was a medical impossibility then, however, and the two earned a handsome living for a time as human oddities, giving lectures and demonstrations throughout the United States and Europe. In 1839, they settled in on a farm in Mount Airy, then married sisters Adelaide and Sarah Yates, and between them, fathered 21 children. The twins died within hours of each other in 1874, at the age of 63.

Chapter Outline & Learning Objectives
After you have studied this chapter, you should be able to:

18.1 Fertilization

Fertilization of an ovum at conception requires that a single sperm penetrate the ovum. When the genetic materials from ovum and sperm are combined, the **zygote** is formed. This single cell is the foundation for a new, unique individual. Figure 18.1 shows the manner in which an ovum is fertilized by a sperm in humans.

Sperm and Ovum Anatomy

As you'll recall from Chapter 17, a sperm has three distinct parts: a head, a middle piece, and a tail. The tail is a flagellum, which allows the sperm to swim through the female reproductive tract toward the ovum. The middle piece contains energy-producing mitochondria. The head contains the sperm nucleus, which encloses the 23 chromosomes from the male. Capping the sperm head is a membrane-bound **acrosome,** which contains digestive enzymes. Very little cytoplasm is present in a sperm cell. Notice that only the nucleus from the sperm head fuses with the ovum nucleus. Therefore,

the zygote receives cytoplasm, organelles, nutrients, and structural components only from the mother.

The plasma membrane of the ovum is surrounded by a nonliving extracellular matrix termed the **zona pellucida.** In turn, the zona pellucida is surrounded by a few layers of adhering follicular cells, collectively called the **corona radiata.** These cells nourished the ovum when it was in a follicle of the ovary.

Steps of Fertilization

During fertilization, (1) several sperm penetrate the corona radiata, (2) several sperm attempt to penetrate the zona pellucida, and (3) one sperm enters the ovum. The acrosome plays a role in allowing sperm to penetrate the zona pellucida. After a sperm head binds tightly to the zona pellucida, the acrosome releases digestive enzymes that forge a pathway for the sperm through the zona pellucida. When a sperm binds to the ovum, their plasma membranes fuse, and this sperm (the head, the middle piece, and usually the tail) enters the ovum. Fusion of the sperm nucleus and the ovum nucleus follows.

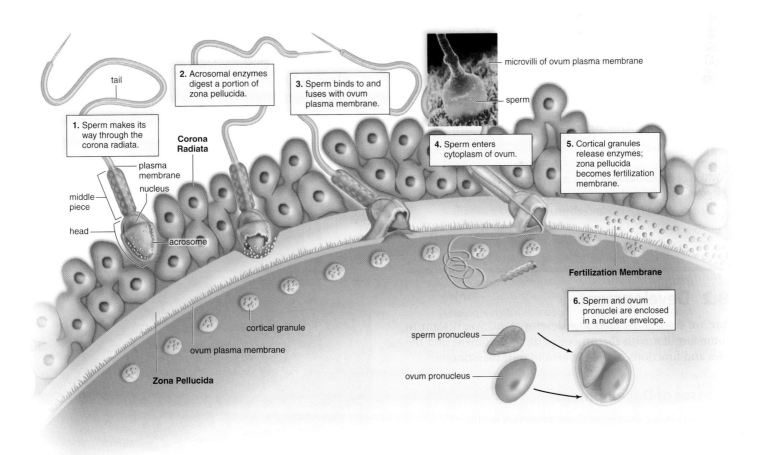

Figure 18.1 Fertilization. Fertilization is accomplished in a series of steps: **1.** sperm cells break aside the cells of the corona radiata, until a single sperm is able to penetrate and adhere to the zona pellucida; **2.** enzymes in the sperm's acrosome digest the zona pellucida; **3.** sperm plasma membrane binds to the ovum plasma membrane; **4.** entire sperm enters the ovum, and its nucleus is released; **5.** fertilization membrane is formed, polyspermy is prevented; and **6.** sperm and ovum pronuclei fuse together in a nuclear envelope.

Figure 18.3 Pre-embryonic development. At ovulation, the egg leaves the ovary. Fertilization occurs in the upper one-third of the uterine tube. The zygote is termed a pre-embryo when cell division (cleavage) begins. When the pre-embryo begins implanting itself into the endometrium, it becomes an embryo.

The many cells of the blastocyst arrange themselves so that there is an **inner cell mass** surrounded by a layer of cells, the **trophoblast.** The trophoblast will become the *chorion.* The early appearance of the chorion emphasizes the complete dependence of the developing embryo on this extraembryonic membrane. The inner cell mass will become the embryo.

Each cell within the morula and the inner cell mass of the blastocyst has the genetic capability of becoming any tissue. This recognition has recently led to a new procedure called therapeutic cloning, as discussed in the What's New reading on page 413. Sometimes during development, the cells of the morula separate, or the inner cell mass splits, and two pre-embryos are present rather than one. If each of the two

pre-embryos is able to complete development, the two babies formed will be *identical twins* because they have inherited exactly the same chromosomes. Should separation of the cell mass be incomplete, the identical twins formed are called *conjoined twins* (formerly referred to as Siamese twins). Conjoined twins can be joined at any part of the body. *Fraternal twins,* who arise when two different eggs are fertilized by two different sperm, do not have identical chromosomes.

Embryonic Development

Embryonic development begins with the second week and lasts until the end of the second month of development.

Therapeutic Cloning

The term *cloning* means making exact multiple copies of genes, a cell, or an organism. For example, identical twins are clones of a single zygote. The cloning of human beings may someday be possible. The procedure would begin as described in Figure 18A. In other words, the person being cloned need not contribute sperm or an egg to the process. Instead, a 2n (diploid) nucleus from, say, a fibroblast can be placed in an enucleated egg, and that egg begins developing. The pre-embryo (blastocyst) would be implanted in the uterus of a surrogate mother where it would develop until birth. Although subject to environmental influences, the clone would be expected to closely resemble its "parent."

Therapeutic cloning is not the same as cloning a human being because the cells of the pre-embryo are separated and treated to become particular tissues, which can be used to treat the patient. The separated cells of a pre-embryo are called stem cells because they divide repeatedly and can become various tissues, as shown in Figure 18A. The tissues resulting from this procedure will not be subject to rejection by the patient because they bear the same surface molecules as the patient's cells. However, there is another way to carry out therapeutic cloning. Fertility clinics store extra pre-embryos prepared by in vitro fertilization but not used. These pre-embryos could be a source of stem cells to make tissues that could be stored and used when needed by any patient, if they were stripped of rejection-causing surface molecules.

So far, therapeutic cloning is experimental and has not been perfected. (Claims of success by a Korean researcher in autumn 2005 were later proven to be a hoax.) However, one day it may provide insulin-secreting cells for diabetics, nerve cells for stroke patients or those with Parkinson's disease, cardiac cells for those with heart disease, and so forth. Yet, it is important to recognize that ethical concerns about therapeutic cloning remain—after all, any pre-embryo is potentially a living, breathing human being.

Anticipating intense interest in therapeutic cloning, the U.S. National Academy of Sciences proposed strict new guidelines for federally funded research in 2005. Among the committee's recommendations was the establishment of Embryonic Stem Cell Research Oversight, or ESCRO, committees, to approve the research before it is begun. An ESCRO committee review would be in addition to the current review required by an Institutional Review Board. ESCRO committees would include bioethicists and legal experts, as well as members of the general public.

Further, the Academy recommended that prior to beginning any research, informed consent from the donors of ova or sperm must be obtained. Stem cells created for therapeutic cloning would never be used for reproductive purposes under the proposed guidelines. The guidelines also would require that the embryos created could not be grown in culture for longer than 14 days. At that point, the primitive streak of the developing nervous system begins to form. Finally, the Academy called for the formation of a national agency charged with regularly reviewing all guidelines on stem cell research.

Because of the controversy over the cloning of stem cells derived from pre-embryos, some researchers have begun searching for other sources. It turns out that the adult body has not only blood stem cells, but also neural stem cells in the brain. It has even been possible to coax blood stem cells and neural stem cells to become other types of human tissues in the body. Another potential source of blood stem cells is a baby's umbilical cord and umbilical blood can now be stored for future use. Once researchers have the know-how, they may be able to use any type of stem cell to cure many of the disorders afflicting human beings.

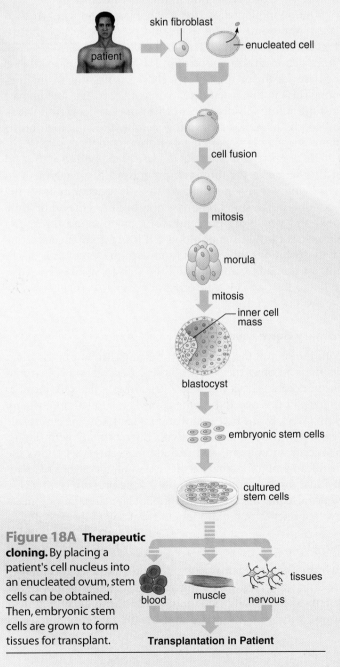

Figure 18A Therapeutic cloning. By placing a patient's cell nucleus into an enucleated ovum, stem cells can be obtained. Then, embryonic stem cells are grown to form tissues for transplant.

Second Week

At the end of the first week, the **embryo** usually begins the process of implanting itself in the wall of the uterus. If **implantation** is successful, the woman is clinically pregnant. On occasion, it happens that the embryo implants itself in a location other than the uterus—most likely, the uterine tube. Implantation of the embryo in any location outside the uterus is called extrauterine or **ectopic pregnancy.** An ectopic pregnancy in the uterine tube must be surgically removed, because the tube is too narrow to allow the continuing growth of the embryo. Should the pregnancy continue, the uterine tube will eventually rupture, causing a fatal hemorrhage to both mother and embryo. In rare circumstances, ectopic pregnancies elsewhere in the abdominal cavity (for example, on the outside of the uterine wall or on the wall of the urinary bladder) can continue to term, producing a live infant. Of course, birth is by Caesarian section in this circumstance.

During implantation, the trophoblast secretes enzymes to digest away some of the tissue and blood vessels of the endometrium of the uterus. The trophoblast also begins to secrete **human chorionic gonadotropin (HCG),** the hormone that is the basis for the pregnancy test. HCG acts like luteinizing hormone (LH) in that it serves to maintain the corpus luteum past the time it normally disintegrates. Because it is being stimulated, the corpus luteum secretes progesterone, the endometrium is maintained, and the expected menstruation does not occur.

The embryo is now about the size of the period at the end of this sentence. As the week progresses, further growth of cells below the inner cell mass causes it to separate from the trophoblast. The inner cell mass now becomes the **embryonic disk,** and two more extraembryonic membranes form. The yolk sac is the first site of blood cell formation. The amniotic cavity surrounds the embryo (and then the fetus) as it develops. In humans, amniotic fluid acts as an insulator against cold and heat and also absorbs shock, such as that caused by the mother exercising.

Primary Germ Layers

With the start of the major event called **gastrulation,** the inner cell mass becomes the embryonic disk. Gastrulation is an example of morphogenesis (see page 410) during which cells move or migrate, in this case to become tissue layers called the **primary germ layers.** By the time gastrulation is complete, the embryonic disk has become an embryo with three primary germ layers: **ectoderm, mesoderm,** and **endoderm.** Figure 18.4 shows the significance of the primary germ layers—all the organs of an individual can be traced back to one of the primary germ layers. Notice also that the original trophoblast layer and the mesoderm layer together form the chorion, the fetal half of the placenta.

Third Week

Two important organ systems make their appearance during the third week (Fig. 18.5). The nervous system is the first organ system to be visually evident. Growth of the nervous system begins when the ectoderm layer on the posterior surface of the embryonic disk becomes thickened. This thickened area is termed the **neural plate.** A visible trench, called the **neural groove,** next begins to form in the center of the neural plate. As the neural groove deepens, two ridges of tissue become evident on either side; these are the **neural folds.** When the neural folds meet in the middle, the hollow **neural tube** is formed. This tube will later develop into the brain and spinal cord. Complete closure of the neural tube is critical for normal development. Should this fail to occur for whatever reason, a **neural tube defect** will be the result. Failure of the inferior spinal cord area to close causes **spina bifida,** which can sometimes be surgically repaired. Failure of the superior neural tube to close causes a fatal condition called **anencephaly** (literally, "without brain"), resulting in a fetus whose brain and cranium never complete development. Multivitamin supplements containing the vitamin folic acid have been shown to help to prevent the occurrence of neural tube defects (see the Medical Focus on page 424).

Development of the heart begins in the third week and continues into the fourth week. At first, there are right and left heart tubes; when these fuse, the heart begins pumping blood, even though the chambers of the heart are not fully formed. The veins enter posteriorly, and the arteries exit anteriorly from this largely tubular heart, but later the heart twists so that all major blood vessels are located anteriorly.

Fourth and Fifth Weeks

At four weeks, the embryo is barely larger than the height of this print. A body stalk connects the caudal (tail) end of the embryo with the chorion. Branching projections of the chorion called **chorionic villi** extend into the endometrial wall of the uterus (see Fig. 18.5). Within each chorionic villus is a capillary network. Exchange of gases, nutrients, and wastes from mother to fetus can take place across this capillary network. The fourth extraembryonic membrane, the allantois, lies within the body stalk, and its blood vessels become the umbilical blood vessels. The head and the tail then lift up, and the body stalk moves anteriorly by constriction. Once this process is complete, the **umbilical cord,** which connects the developing embryo to the placenta, is fully formed (see Fig. 18.5d).

Little flippers called limb buds appear (Fig. 18.6); later, the arms and the legs develop from the limb buds, and even the hands and the feet become apparent. At the same time—during the fifth week—the head enlarges and the sense organs become more prominent. It is possible to make out the developing eyes and ears, and even the nose.

Sixth Through Eighth Weeks

During the sixth through eighth weeks of development, the embryo changes to a form that is easily recognized as a human being. Concurrent with brain development, the head achieves its normal relationship with the body as a neck region develops. The nervous system is developed well enough to permit reflex actions, such as a startle response to touch. At the end of this period, the embryo is about 38 mm (1.5 in.)

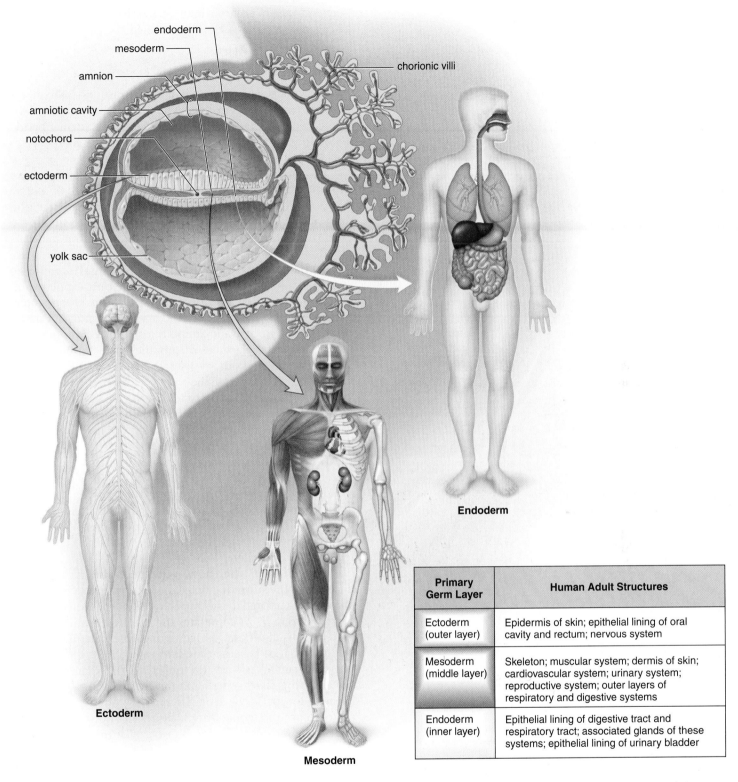

endoderm	
mesoderm	
amnion	
amniotic cavity	
notochord	
ectoderm	
	chorionic villi
yolk sac	

Ectoderm

Mesoderm

Endoderm

Primary Germ Layer	Human Adult Structures
Ectoderm (outer layer)	Epidermis of skin; epithelial lining of oral cavity and rectum; nervous system
Mesoderm (middle layer)	Skeleton; muscular system; dermis of skin; cardiovascular system; urinary system; reproductive system; outer layers of respiratory and digestive systems
Endoderm (inner layer)	Epithelial lining of digestive tract and respiratory tract; associated glands of these systems; epithelial lining of urinary bladder

Figure 18.4 Embryonic germ layers. An embryo has three primary germ layers: ectoderm, mesoderm, and endoderm. Organs and tissues can be traced back to a particular germ layer as indicated in this illustration.

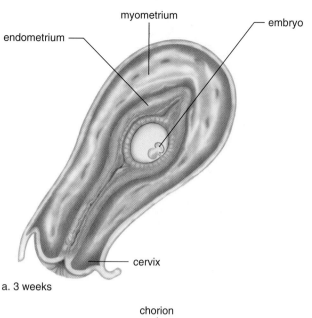

a. 3 weeks

myometrium
endometrium
embryo
cervix

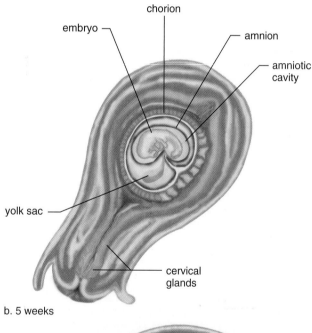

b. 5 weeks

chorion
embryo
amnion
amniotic cavity
yolk sac
cervical glands

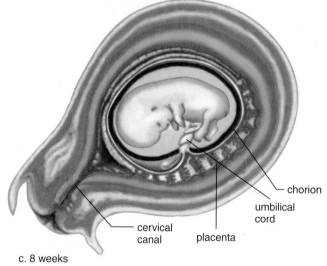

c. 8 weeks

chorion
umbilical cord
placenta
cervical canal

long and weighs no more than an aspirin tablet, even though all organ systems have been established.

Placenta

The placenta is shaped like a pancake, measuring 15 to 20 cm in diameter and 2.5 cm thick. The placenta is normally fully formed and functional by the end of the embryonic period and before the fetal period begins. The placenta is expelled as the **afterbirth** following the birth of an infant.

The placenta has two portions, a fetal portion composed of chorionic tissue and a maternal portion composed of uterine tissue. Chorionic villi cover the entire surface of the chorion until about the eighth week when they begin to disappear, except in one area. These villi are surrounded by maternal blood, and it is here that exchanges of materials take place across the placental membrane. The **placental membrane** consists of the epithelial wall of an embryonic capillary and the epithelial wall of a chorionic villus. Maternal blood rarely mingles with fetal blood. Instead, oxygen and nutrient molecules, such as glucose and amino acids, diffuse from maternal blood across the placental membrane into fetal blood, and carbon dioxide and other wastes, such as urea, diffuse out of fetal blood into maternal blood.

Note that the digestive system, lungs, and kidneys do not function in the fetus. The functions of these organs are not needed because the placenta supplies the fetus with its nutritional and excretory needs.

The umbilical cord transports fetal blood to and from the placenta (Fig. 18.7; see Fig. 12.19). The umbilical cord is the fetal lifeline because it contains the two umbilical arteries and a single umbilical vein, which transport waste molecules (carbon dioxide and urea) to the placenta for disposal and oxygen and nutrient molecules from the placenta to the rest of the fetal circulatory system.

As mentioned, the chorion and then the placenta produce HCG, the hormone detected by a pregnancy test. HCG prevents the normal degeneration of the corpus luteum of the ovary. Instead, HCG stimulates the corpus luteum to secrete even larger quantities of progesterone. Later, the placenta begins to produce progesterone and estrogen, and the corpus luteum degenerates—it is no longer needed. Placental estrogen and progesterone maintain the endometrium and have a negative feedback effect on the anterior pituitary so that it ceases to produce gonadotropic hormones during pregnancy. Menstruation ordinarily does not occur during the length of pregnancy.

Figure 18.5 Embryonic development within the uterus.
a. Three weeks after fertilization. **b.** Five weeks after fertilization, amnion and chorion are present, and the uterus is about the size of a hen's egg. **c.** Two months after fertilization, the placenta and umbilical cord are well formed.

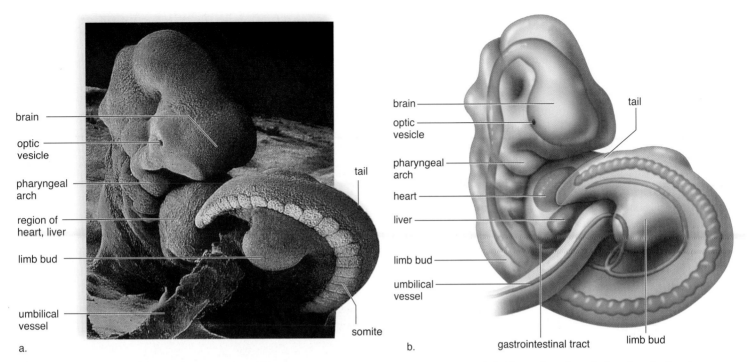

Figure 18.6 Human embryo at beginning of fifth week. a. Scanning electron micrograph. **b.** The embryo is curled so that the head touches the region of the heart and liver. The organs of the gastrointestinal tract are forming, and the arms and the legs develop from the bulges called limb buds. The tail is an evolutionary remnant; its bones regress and become those of the coccyx (tailbone).

Placenta

Figure 18.7 The placenta. Blood vessels within the umbilical cord lead to the placenta where exchange takes place between fetal blood and maternal blood. Note that the umbilical vein is colored red here because it is bringing oxygenated blood from the placenta to the fetus. Likewise, the umbilical arteries are colored blue to reflect the deoxygenated blood flowing from the fetus to the placenta for gas exchange with the maternal blood vessels.

Fetal Development and Birth

Fetal development includes the third through the ninth months of development. At this time, the fetus looks human (Fig. 18.8).

Third and Fourth Months

At the beginning of the third month, the fetal head is still very large, the nose is flat, the eyes are far apart, and the ears are well formed. Head growth now begins to slow down as the rest of the body increases in length. Epidermal refinements, such as fingernails, nipples, eyelashes, eyebrows, and hair on the head, appear.

Cartilage begins to be replaced by bone as ossification centers appear in most of the bones. Cartilage remains at the ends of the long bones, and ossification is not complete until age 18 or 20 years. The skull has six large membranous areas called **fontanels,** which permit a certain amount of flexibility as the head passes through the birth canal and allow rapid growth of the brain during infancy. Progressive fusion of the skull bones causes the fontanels to close, usually by 2 years of age.

Sometime during the third month, it is possible to distinguish males from females. Researchers have discovered a series of genes on the X and Y chromosomes that cause the differentiation of gonads into testes and ovaries. Once these have differentiated, they produce the sex hormones that influence the differentiation of the genital tract.

During the fourth month, the fetal heartbeat is loud enough to be heard when a physician applies a stethoscope to the mother's abdomen. By the end of this month, the fetus is about 152 mm (6 in.) in length and weighs about 171 g (6 oz).

Fifth Through Seventh Months

During the fifth through seventh months (Fig. 18.8), the mother begins to feel movement. At first, she experiences only a fluttering sensation, but as the fetal legs grow and develop, she feels kicks and jabs. The fetus remains tightly curled in the uterus, with the head bent down and in contact with the flexed knees. This posture is aptly termed the fetal position.

The wrinkled, translucent, pink-colored skin is covered by a fine down called **lanugo.** This in turn is coated with a white, greasy, cheeselike substance called **vernix caseosa,** which probably protects the delicate skin from the amniotic fluid. The eyelids are now fully open.

At the end of this period, the fetus's length has increased to about 300 mm (12 in.), and it weighs about 1,380 g (3 lb). It is possible that, if born now, the baby will survive.

Eighth Through Ninth Months

As the end of development approaches, the fetus usually rotates so that the head is pointed toward the cervix. However, if the fetus does not turn, a **breech birth** (rump first) is likely. It is very difficult for the cervix to expand enough to accommodate this form of birth, and asphyxiation of the baby is more likely to occur. Thus, a **cesarean section** may be prescribed for delivery of the fetus (incision through the abdominal and uterine walls).

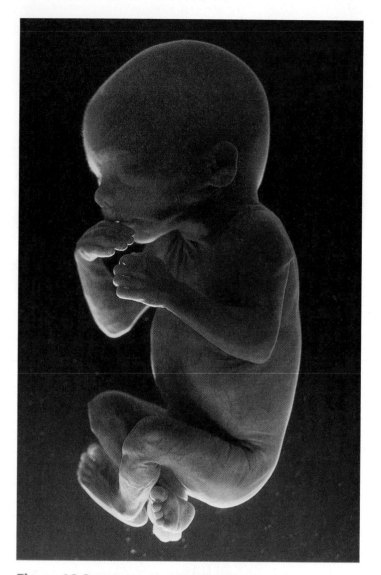

Figure 18.8 **Five- to seven-month-old fetus.**

At the end of nine months, the fetus is about 530 mm (20½ in.) long and weighs about 3,400 g (7½ lb). Weight gain is due largely to an accumulation of fat beneath the skin. Full-term babies have the best chance of survival; as discussed in the Medical Focus on page 419, premature babies are subject to various challenges.

Development of Male and Female Sex Organs

The gender of an individual is determined at the moment of fertilization. Both males and females have 23 pairs of chromosomes; in males, one of these pairs is an X and Y, while females have two X chromosomes. During the first several weeks of development, it is impossible to tell by external inspection whether the unborn child is a boy or a girl. Gonads don't start developing until the seventh week of development. The tissue that gives rise to the gonads is called *indifferent* because it can become testes or ovaries depending on the action

Premature Babies

In Memoriam: Brandon Nelson, 1982–1983

Each and every day in the United States, approximately 1,300 babies will be born too soon. **Premature birth** is defined as birth prior to 37 weeks gestation (normal gestation is approximately 40 weeks). Many other babies, although full-term, have a low birth weight (less than 5.5 pounds at birth). These babies face weeks to months in the neonatal intensive care unit, or NICU. Fortunately, advances in the care of these tiniest patients allow most of them to continue to grow and develop while slowly gaining weight. Most eventually go home. However, "premies" and term babies with low birth weight face many serious challenges to their survival:

Respiratory distress syndrome (hyaline membrane disease) The lungs do not produce enough of a chemical surfactant that helps the alveoli stay open. Therefore, the lungs tend to collapse, instead of expanding to be filled with air.

Retinopathy of prematurity The high level of oxygen needed to ensure adequate gas exchange by the immature lungs can lead to proliferation of blood vessels within the eyes, with ensuing blindness.

Intracranial hemorrhage The delicate blood vessels in the brain are apt to break, causing swelling and inflammation of the brain. If not fatal, this can lead to brain damage.

Jaundice The immature liver fails to excrete the waste product bilirubin, which instead builds up in the blood, possibly causing brain damage.

Infections The level of antibodies in the body is low, and the various medical procedures performed could possibly introduce germs. Also, bowel infection is common, along with perforation, bleeding, and shock.

Circulatory disorders Fetal circulation, discussed in Chapter 12, has two features: (1) the oval opening between the atria, and (2) the arterial duct that allows blood to bypass the lungs. If these features persist in the newborn, oxygen-rich blood will mix with oxygen-poor blood. Blood circulation will be impaired, perhaps leading to the delivery of a "blue baby"—that is, a baby with cyanosis, a bluish cast to the skin. Heart failure can also result from these conditions.

Risk for permanent disability Many premies grow up to be normal and healthy, but many others face permanent disability.

Developmental delays, cerebral palsy, learning disorders, chronic lung disease, blindness, and deafness are all potential consequences of premature birth. Further, pediatricians estimate that fully half of all neurological disabilities in children are related to premature birth.

Why do women deliver prematurely? Obstetricians (physicians who specialize in the treatment of women during and after pregnancy) have come up with four general causes for pre-term labor and delivery. The placenta may be disrupted by a bacterial infection; this triggers uterine contractions. Bleeding from the placenta or from the uterus itself may trigger labor. The uterus may be stretched excessively, as often occurs in multiple births. Finally, maternal or fetal stress may trigger a hormone cascade that leads to uterine contraction and pre-term delivery. Pregnant women who have already had a previous pre-term delivery are known to be at risk for subsequent pre-term delivery, as are women pregnant with twins, triplets, or more. Structural abnormalities of the uterus or the cervix may also lead to delivery of a premature baby.

How can a pregnant woman maximize her chances for a full-term pregnancy? First, she should take those precautions needed for all healthy pregnancies: continuing prenatal care throughout the pregnancy; abstaining from alcohol, smoking, and recreational drugs; and early antibiotic therapy for bacterial infections, especially sexually transmitted infections (see the Medical Focus on page 424). If she has had a previous pre-term delivery, or is carrying more than one fetus, evaluation by an obstetrician who specializes in high-risk pregnancy is an appropriate precaution. Stress in her life also needs to be kept to a minimum, if possible. Being in an abusive relationship, lacking a social network, and working long hours with long periods of standing have all been linked to an increased risk for pre-term labor and delivery.

Unfortunately, despite increased research and improvements in medical care, the rate of premature birth has increased by 31% in the past 20 years. Obviously, more needs to be done to prevent premature birth. The March of Dimes supports educational efforts and research into the causes of birth defects and prematurity. More information can be obtained at www.marchofdimes.com.

and concentration of hormones. Genes on the Y chromosome cause the production of testosterone, and then the indifferent tissue becomes testes.

In Figure 18.9*a*, notice that at six weeks both males and females have the same types of ducts. During this indifferent stage, an embryo has the potential to develop into a male or a female. If a Y chromosome is present, testosterone stimulates the Wolffian ducts to become male genital ducts. The Wolffian ducts enter the urethra, which belongs to both the urinary and reproductive systems in males. The testes secrete an anti-Müllerian hormone that causes the Müllerian ducts to regress.

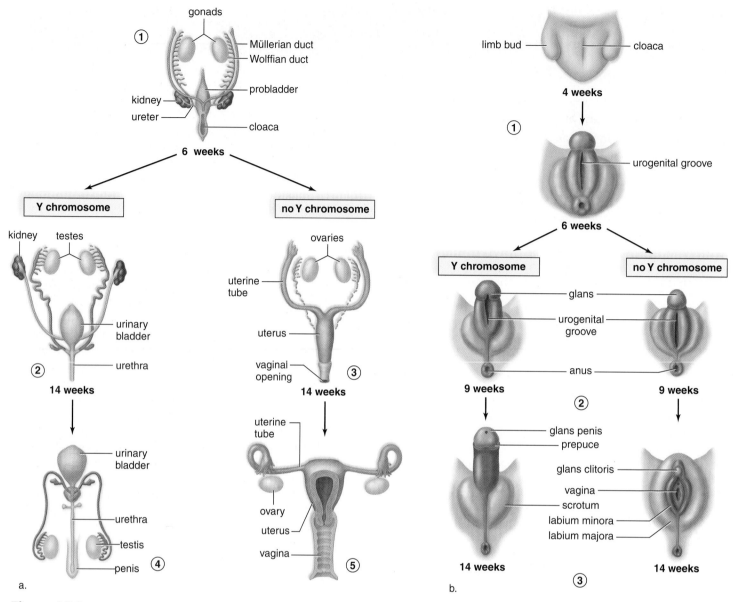

Figure 18.9 **Male and female organs. a.** Development of gonads and ducts. **b.** Development of external genitals.

In the absence of a Y chromosome, ovaries develop instead of testes from the same indifferent tissue. Now the Wolffian ducts regress, and the Müllerian ducts develop into the uterus and uterine tubes. A developing vagina also extends from the uterus. There is no connection between the urinary and genital systems in females.

At 14 weeks, both the primitive testes and ovaries are located deep inside the abdominal cavity. An inspection of the interior of the testes would show that sperm are even now starting to develop, and similarly, the ovaries already contain large numbers of tiny follicles, each having an ovum. At this time, either testes or ovaries are located within the abdominal cavity. Later, in the last trimester of fetal development, both testes and ovaries descend into the pelvis. As you know from Chapter 17, the ovaries remain in the pelvis. However, the testes descend out of the pelvis into the scrotal sacs (scrotum). Because the testes are normally found

in the scrotal sacs, testicular temperature is cooler than body temperature, and this allows sperm cells to develop. As you'll recall from Chapter 17, sometimes the testes fail to descend, a condition called cryptorchidism. In that case, an operation may be done later to place them in their proper location.

Figure 18.9*b* shows the development of the external genitals. These tissues are also indifferent at first—they can develop into either male or female genitals. At six weeks, a small bud appears between the legs; this can develop into the male penis or the female clitoris, depending on the presence or absence of the Y chromosome and testosterone. At nine weeks, a urogenital groove bordered by two swellings appears. By 14 weeks, this groove has disappeared in males, and the scrotum has formed from the original swellings. In females, the groove persists and becomes the vaginal opening. Labia majora and labia minora are present instead of a scrotum.

4. The site of red blood cell production in an embryo is the:

   a. bone marrow.   c. yolk sac.

   b. chorion.   d. allantois.

5. Which statement(s) accurately describe the umbilical cord?

   a. Blood vessels of the allantois become the umbilical blood vessels.

   b. A body stalk that connects the embryo to the chorion becomes the umbilical cord.

   c. There are two umbilical veins and two umbilical arteries in the cord.

   d. Statements a and b are both correct.

   e. All of the statements are correct.

6. If an embryo has a Y chromosome, which of the following will happen?

   a. The embryo will develop into a male.

   b. The Wolffian duct will develop fully, whereas the Müllerian duct regresses.

   c. The Müllerian duct will develop fully, whereas the Wolffian duct regresses.

   d. a and b

   e. a and c

Answers in Appendix C, page 466.

18.3 Birth

The uterus has contractions throughout pregnancy. At first, these are light, lasting about 20–30 seconds and occurring every 15–20 minutes. Near the end of pregnancy, the contractions may become stronger and more frequent so that a woman might think she is in labor. "False-labor" contractions are called **Braxton Hicks** contractions. However, the onset of true labor is marked by uterine contractions that occur regularly every 15–20 minutes (and more frequently in labor's final stages) and last for 40 seconds or longer.

A positive feedback mechanism can explain the onset and continuation of labor. Uterine contractions are induced by a stretching of the cervix, which also brings about the release of oxytocin, which is produced by the hypothalamus and released from the posterior pituitary. Oxytocin stimulates the uterine muscles, both directly and through the action of prostaglandins. Uterine contractions push the fetus downward, and the cervix stretches even more. This cycle keeps repeating itself until birth occurs.

Prior to or at the first stage of **parturition,** which is the process of giving birth to an offspring, there can be a "bloody show." This is caused by expulsion of the mucous plug from the cervix. During pregnancy, this thick, sticky mucus effectively sealed the uterine cervix, preventing bacteria and sperm from entering the uterus. Loss of the plug is often a first sign that the baby's birth is imminent.

Stage 1

During the first stage of labor, the uterine contractions of labor occur in such a way that the cervical canal slowly disappears as the lower part of the uterus is pulled upward toward the baby's head. This process is called effacement, or "taking up the cervix." With further contractions, the baby's head acts as a wedge to assist cervical dilation (Fig. 18.10b). If the amniotic membrane has not already ruptured, it is apt to do so during this stage, releasing the amniotic fluid, which leaks out the vagina (an event sometimes referred to as "breaking water"). The first stage of parturition ends once the cervix is dilated completely.

Stage 2

During the second stage of parturition, the uterine contractions occur every 1–2 minutes and last about one minute each. They are accompanied by a desire to push, or bear down. As the baby's head gradually descends into the vagina, the desire to push becomes greater. When the baby's head reaches the exterior, it turns so that the back of the head is uppermost (Fig. 18.10c). Because the vaginal orifice may not expand enough to allow passage of the head, an **episiotomy** is often performed to prevent the mother's tissues from tearing. This incision, which enlarges the opening, is sewn together later. As soon as the head is delivered, the baby's shoulders rotate so that the baby faces either to the right or the left. At this time, the physician may hold the head and guide it downward, while one shoulder and then the other emerges. The rest of the baby follows easily.

Once the baby is breathing normally, the umbilical cord is cut and tied, severing the child from the placenta. The stump of the cord shrivels and falls off to create a scar, which is the umbilicus.

Stage 3

The placenta, or afterbirth, is delivered during the third stage of parturition (Fig. 18.10d). About 15 minutes after delivery of the baby, uterine muscular contractions shrink the uterus and dislodge the placenta. The placenta then is expelled into the vagina. As soon as the placenta and its membranes are delivered, the third stage of parturition is complete.

> **Begin Thinking Clinically**
>
> **?**
>
> **During an actual delivery, a pregnant woman's uterine contractions will sometimes begin, seemingly strengthen, and then stop abruptly. What might cause this to occur? What should be done to correct this?**
>
> *Answers and discussion in Appendix C, page 466.*

placenta

pubic symphysis
urethra
urinary bladder
vagina
cervix
rectum

a. 9-month-old fetus

ruptured amniotic sac

b. First stage of birth: cervix dilates

placenta

c. Second stage of birth: baby emerges

placenta

uterus
umbilical cord

d. Third stage of birth: afterbirth is expelled

Baby has arrived

Figure 18.10 **Three stages of parturition (birth). a.** Position of fetus just before birth begins. **b.** Dilation of cervix. **c.** Birth of baby. **d.** Expulsion of afterbirth.

Effects of Pregnancy on the Mother

Major changes take place in the mother's body during pregnancy. When first pregnant, the mother may experience nausea and vomiting, loss of appetite, and fatigue. These symptoms subside, and some mothers report increased energy levels and a general sense of well-being despite an increase in weight. During pregnancy, the mother gains weight due to breast and uterine enlargement, weight of the fetus, amount of amniotic fluid, size of the placenta, her own increase in total body fluid; and an increase in storage of proteins, fats, and minerals. The increased weight can lead to lordosis (swayback) and lower back pain.

Aside from an increase in weight, many of the physiological changes in the mother are due to the presence of the placental hormones that support fetal development (Table 18.2). Progesterone decreases uterine motility by relaxing smooth muscle, including the smooth muscle in the walls of arteries. The arteries expand, and this leads to a low blood pressure. Hypotension causes the kidneys to release their enzyme, renin. The renin-angiotensin-aldosterone mechanism is thus activated, and is promoted by estrogen. Aldosterone activity promotes sodium and water retention, and blood volume increases until it reaches its peak sometime during weeks 28–32 of pregnancy. Altogether, blood volume increases from 5 liters to 7 liters—a 40% rise. An increase in the number of red blood cells follows. With the rise in blood volume, cardiac output increases by 20–30%. Blood flow to the kidneys, placenta, skin, and breasts rises significantly. Smooth muscle relaxation also explains the common gastrointestinal effects of pregnancy. The heartburn experienced by many is due to relaxation of the esophageal sphincter and reflux of stomach contents into the esophagus. Constipation is caused by a decrease in intestinal tract motility.

Of interest is the increase in pulmonary values in a pregnant woman. The bronchial tubes relax, but this alone cannot explain the typical 40% increase in vital capacity and tidal volume. The increasing size of the uterus from a nonpregnant weight of 60–80 g to 900–1,200 g contributes to an improvement in respiratory functions. The uterus comes to occupy most of the abdominal cavity, reaching nearly to the xiphoid process of the sternum. This increase in size not only pushes the intestines, liver, stomach, and diaphragm superiorly, but it also widens the thoracic cavity. Compared to nonpregnant values, the maternal oxygen level changes little, but blood carbon dioxide levels fall by 20%, creating a concentration gradient favorable to the flow of carbon dioxide from fetal blood to maternal blood at the placenta.

The enlargement of the uterus does result in some problems. In the pelvic cavity, compression of the ureters and urinary bladder can result in stress incontinence. Compression of the inferior vena cava, especially when lying down, decreases venous return, and the result is edema and varicose veins.

Aside from the steroid hormones progesterone and estrogen, the placenta also produces some peptide hormones. One of these makes cells resistant to insulin, and the result can be pregnancy-induced diabetes. Some of the integumentary changes observed during pregnancy are also due to placental hormones. **Striae gravidarum,** commonly called "stretch marks," typically form over the abdomen and lower breasts in response to increased steroid hormone levels rather than stretching of the skin. Melanocyte activity also increases during pregnancy. Darkening of the areolae, skin in the line from the navel to the pubis, areas of the face and neck, and vulva is common.

Changes in breast anatomy and the occurrence of lactation are discussed in Chapter 17, pages 392–393.

TABLE 18.2 Effect of Placental Hormones on Mother	
Hormone	**Chief Effects**
Progesterone	Relaxation of smooth muscle; reduced uterine motility; reduced maternal immune response to fetus
Estrogen	Increased uterine blood flow; increased renin-angiotensin-aldosterone activity; increased protein biosynthesis by the liver
Peptide hormones	Increased insulin resistance

Source: Moore, Thomas R., *Gestation Encyclopedia of Human Biology,* Vol. 7, 7th edition. Copyright © 1997 Academic Press.

Content **CHECK-UP!**

7. Which of the following statements about oxytocin is/are true?
   a. Oxytocin causes uterine smooth muscle contraction.
   b. Oxytocin is manufactured by the posterior pituitary.
   c. Stretching of the uterine cervix stimulates oxytocin production and release.
   d. Both a and b are correct.
   e. Both a and c are correct.

8. Effacement is the technical term for:
   a. leakage of amniotic fluid.
   b. false labor contractions.
   c. the process of gradually stretching the cervix until it slowly disappears.
   d. loss of the mucous plug that blocked the cervical canal.

9. Which of the following is an effect of placental estrogen?
   a. reduced uterine motility
   b. relaxation of smooth muscle
   c. increased insulin resistance
   d. increased uterine blood flow

Answers in Appendix C, page 466.

Preventing Birth Defects

Before the nursery room gets painted pink or blue, before the baby registry is completed at the department store, before deciding whether the baby gets named after Great-Aunt Adelaide or Grandma Gertrude, a couple that has decided to start (or add to) a family must take steps to ensure a healthy pregnancy for both baby and mother. One way to maximize the odds of having a healthy baby is to take precautions to prevent **congenital,** or **birth, defects.** Some birth defects are unavoidable—those caused by chromosomal or genetic abnormalities, for example. Tragically, however, many other birth defects are completely preventable.

For decades, the placenta was believed to be an effective barrier between mother and fetus. Continuing research has shown instead that the placenta is not a barrier at all; rather, it functions as a *filter* through which many substances can enter the baby's bloodstream from the mother's circulation. Thus, almost anything that mom eats, drinks, smokes, or injects—good and bad—is potentially capable of winding up in the baby's body. Chemicals, bacteria, viruses, parasites, antibodies from the mother's immune system (both harmful and beneficial) can all act as **teratogens**—agents capable of causing birth defects.

Ideally, preventing birth defects, as well as ensuring a healthy pregnancy for mom-to-be, is a process that should begin before a woman ever becomes pregnant. A complete physical exam is an important first step. At that time, she can receive any necessary immunization boosters, or "catch up" with immunizations she may have missed as a child (see page 296). It is especially important to immunize against **rubella (German measles).** If a pregnant woman contracts rubella, it can cause blindness, deafness, mental retardation, heart malformations, and other serious problems in an unborn child. A woman who has neither contracted chickenpox nor been immunized against it should be vaccinated at this time as well. Screening for sexually transmitted infections should also take place during this initial physical. Antibiotic treatment can cure bacterial infections (syphilis, gonorrhea, chlamydia, etc.) and prevent serious harm to the embryo. Although there are no cures for viral STIs, it is still important to know if a woman is infected. HIV (the causative agent for AIDS) can be transmitted across the placenta to infect the fetus. Anti-retroviral therapy can prevent this type of infection in many cases (see pages 294 and 295). During a pre-pregnancy physical, health-care providers will also recommend prenatal vitamin/mineral supplements. Studies have shown that taking vitamins with high levels of vitamin B_{12} and folate, before and after conception, minimizes the risk of neural tube defects.

Couples trying to conceive should also adapt their lifestyle, if necessary. If either or both partners smoke, there will never be a more perfect time to quit. It has been shown that the best way to quit smoking is by having supportive family and friends around (see pages 321 and 322). Males who smoke are more likely to have erectile dysfunction than nonsmoking males, making a pregnancy more difficult to achieve. Further, it will later be essential for both the pregnant mother and the fetus to be in a smoke-free environment at all times. Lifestyle changes must also include giving up any use of recreational drugs because *all* will decrease a man's sperm count. Use of recreational drugs by the pregnant woman can harm the fetus.

Because the placenta functions as a filter, if mom smokes or exposes herself to secondhand smoke, her baby "smokes" too. Cleft lip/cleft palate may occur because of exposure to toxins in cigarette smoke. Children of smoking mothers are more likely to be stillborn, or die shortly after birth. Death from sudden infant death syndrome (SIDS—unexplained death of an otherwise healthy baby) is more common in babies of smoking mothers. The occurrence of preterm babies with low birth weight is doubled for women who smoke.

Likewise, alcohol easily crosses the placenta. There is no safe dosage during pregnancy. Even one drink a day appears to increase the chances of a spontaneous abortion. The more alcohol consumed, the greater are the chances of physical abnormalities if the pregnancy continues. **Fetal alcohol syndrome (FAS)** is the term given to the collection of physical, mental, and behavioral abnormalities seen in babies born to heavy drinkers. Babies with FAS have decreased weight, height, and head size, with malformation of the head and face. Later, developmental delay is common, as are numerous other physical malformations. Babies born to heavy drinkers are apt to undergo an extremely painful withdrawal, called **delirium tremens,** after birth—shaking, vomiting, and extreme irritability.

Certainly, illegal drugs such as marijuana, cocaine, and heroin are teratogens that must be completely avoided during pregnancy. Cocaine babies now make up 60% of drug-affected babies. Cocaine use causes severe fluctuations in a mother's blood pressure that temporarily deprive the developing fetus's brain of oxygen. Cocaine babies have visual problems, lack coordination, and are often developmentally delayed.

Once pregnancy occurs, continuing prenatal medical care and good health habits are a must. Basic nutrients are required in adequate amounts to meet the demands of both fetus and mother. This includes increasing calories and protein intake. The pregnant woman of normal weight should expect to gain 25–35 pounds during her pregnancy. Without adequate weight gain during pregnancy, the woman risks giving birth to an infant with low birth weight. Low birth weight increases the risk of complications, such as respiratory problems, immediately after birth. Furthermore, growing numbers of studies confirm that newborns of low birth weight (less than 5 pounds, 8 ounces at birth) are more likely to develop certain chronic diseases, such as diabetes and high blood pressure, when they become adults.

It is important that mom continues to take prenatal vitamin/mineral supplements throughout her pregnancy. Calcium is needed for bone growth, iron for red blood cell formation, and vitamin B_6 for proper metabolism. Perhaps most important of all, pregnant women need more folate (folic acid) each day to meet the increased rate of cell division and DNA synthesis in their own bodies and that of the developing child. As mentioned previously, maternal deficiency of folate has been linked to development of neural tube defects in the fetus. These defects include spina bifida (spinal cord or spinal fluid bulge through the back) and anencephaly (absence of the brain). Perhaps as many as 75% of these defects could be avoided by adequate folate intake even before a pregnancy occurs. Consuming fortified breakfast cereals is a good way to meet folate needs because they contain a more absorbable form of folate.

Certain foods must be avoided during pregnancy, however. Game fish species, such as shark, mackerel, swordfish, and salmon, should be passed up because of potentially high levels of mercury that might be in these fish. Mom should also turn down unpasteurized milk or cheese/yogurt products made from unpasterized milk. Raw or undercooked shellfish, poultry, pork, beef, or eggs are also off the menu for 9 months (and longer if mom decides to breast-feed her baby). These foods may contain bacteria or parasites that could cause illness and, in some cases, also infect the unborn baby.

Mom has to be careful at home and on the job as well. Toxic chemicals, such as pesticides, and many organic industrial chemicals, such as vinyl chloride, formaldehyde, asbestos, and benzenes, are teratogens and can cross the placenta, resulting in fetal abnormalities. Cleaning the kitty litter box should become someone else's job during the mom's entire pregnancy—cat feces can cause her to contract **toxoplasmosis,** a dangerous parasitic infection. Lead (from old-fashioned, lead-based paint or pipes) circulating in a pregnant woman's blood can cause her child to be severely developmentally delayed. If mom works in a health-care setting (hospital, doctor or dentist office, etc.), she must avoid accidental X-ray exposure. It takes a lower amount of exposure to X rays to cause mutations in a developing embryo or fetus than in an adult. X rays and other ionizing radiation increase the baby's risk of developing leukemia after birth. If a woman must have diagnostic X rays during pregnancy (if she has broken a bone, for example), she should inform her physician that she is pregnant. Shielding can minimize or eliminate danger to the fetus.

The medicine cabinet can be a pretty dangerous spot if you're pregnant. Many medications, including herbal supplements, nonprescription medications, and prescription drugs, can be teratogenic. For example, the isotretinoins, a group of drugs chemically similar to vitamin A and prescribed for severe acne, are known to cause severe birth defects. A woman must have a negative pregnancy test and sign a contract agreeing to use two forms of birth control simultaneously before she can receive these drugs. As you'll recall from the Medical Focus box in Chapter 17, in the 1950s and 1960s, DES (diethylstilbestrol) was given to pregnant women to prevent cramps, bleeding, and threatened miscarriage. But in the 1970s and 1980s, some adolescent girls and young women whose mothers had been treated with DES showed various abnormalities of the reproductive organs and increased tendency toward cervical cancer. Other sex hormones, including birth control pills, can possibly cause abnormal fetal development, including abnormalities of the sex organs. The drug thalidomide was a popular tranquilizer during the 1950s and 1960s in many European countries and, to a degree, in the United States. The drug, which was taken to prevent nausea in pregnant women, arrested the development of arms and legs in some children and also damaged heart and blood vessels, ears, and the digestive tract. Some mothers of affected children report that they took the drug for only a few days. Because of these situations and others, physicians are generally very cautious about prescribing drugs during pregnancy. No pregnant woman should take any drug—including ordinary cold remedies and herbal supplements—without checking first with her physician.

As her big day draws nearer, mom must continue to be careful to prevent birth defects and illness in her newborn. When a mother has HIV, herpes, gonorrhea, or chlamydia, newborns can become infected as they pass through the birth canal. Blindness and other physical and mental defects may develop. Birth by cesarean section can prevent these occurrences. An Rh-negative woman who has given birth to an Rh-positive child should receive an Rh immunoglobulin injection within 72 hours after birth to prevent her body from producing Rh antibodies. She will start producing these antibodies when some of the child's Rh-positive red blood cells enter her bloodstream, possibly before, but particularly at, birth. Rh antibodies can cause nervous system and heart defects in a fetus. The first Rh-positive baby is not usually affected. But in subsequent pregnancies, antibodies created at the time of the first birth cross the placenta and begin to destroy the blood cells of the fetus, thereby causing anemia and other complications in the baby.

With proper prenatal care, a woman's chances of having a normal pregnancy and healthy baby are better now than ever before in human history. Now that physicians and laypeople are aware of the various ways birth defects can be prevented, it is hoped that the incidence of birth defects will decrease in the future. For more information on all aspects of pregnancy, consult www.marchofdimes.com.

SELECTED NEW TERMS

Basic Key Terms

acrosome (ak'ro-sōm), p. 409

afterbirth (af'ter-berth), p. 416

allantois (uh-lan'to-is), p. 411

amnion (am'ne-on), p. 411

blastocyst (blas'to-sist), p. 411

chorion (ko're-on), p. 410

chorionic villi (ko"re-on'ik vil'l), p. 414

cleavage (klĕv'ij), p. 410

corona radiata (kō-rō'nă rā dĕ ă-toh), p. 409

differentiation (dif'er-en"she-a'shun), p. 410

ectoderm (ek'tō-derm), p. 414

embryo (em'bre-o), p. 414

embryonic development (em"bre-on'ik de-vel'op-ment), p. 412

embryonic disk (em" bre-on'ik disk), p. 414

endoderm (en'dō-derm), p. 414

extraembryonic membrane (eks " truh-em "bre-on'ik mem'brān), p. 410

fertilization (fer'tĭ-lĭ-za'shun), p. 409

fetal development (fe'tal de-vel'op-ment), p. 418

fontanels (fŏn'tn-ĕl'), p. 418

gastrulation (gas'tru-la'shun), p. 414

gestation (jes-ta'shun), p. 411

growth (grōth), p. 410

human chorionic gonadotropin (HCG) (hyōō-măn'kôr'ē-ŏn'ik gō-năd'ə-trō'pĭn), p. 414

implantation (im'plăn-tăshən), p. 414

inner cell mass (in'er sel mas), p. 412

lanugo (luh-nu'go), p. 418

mesoderm (mez'o-derm), p. 414

morphogenesis (morf-o-jen'ĕ-sis), p. 410

morula (mor'u-luh), p. 411

parturition (par"tu-rish'un), p. 421

placenta (pluh-sen'tuh), p. 410

primary germ layers (prī'-mă-ree gĕrm lā'-yerz), p. 414

trophoblast (trof'o-blast), p. 412

umbilical cord (um-bil'ĭ-kl kord), p. 414

vernix caseosa (ver'niks ka"se-o'suh), p. 418

yolk sac (yŏk sak), p. 411

zona pellucida (zō'-nă pĕl-loo-sĭ-duh), p. 409

zygote (zī'gōt'), p. 409

Clinical Key Terms

anencephaly (an'en-sef'uh-le), p. 414

Braxton Hicks contraction (braks'ton hiks con-trak'shun), p. 421

breech birth (brĕch berth), p. 418

Cesarean section (sĭ-zār'e-un sek'shun), p. 418

congenital birth defect (kon-jen'ĭ-tul burth de'fekt), p. 424

delirium tremens (de-lēr'e-um tre'mens), p. 424

ectopic pregnancy (ĕk-tŏp'ĭk prĕg'nən-sē), p. 414

episiotomy (e-piz"e-ot'o-me), p. 421

fetal alcohol syndrome (fe'tal al'cuh-hol sin'drŏm), p. 424

premature birth (pre-mah-tyūr'), p. 419

rubella (German measles) (ru-bel'uh), p. 424

spina bifida (spi'nuh bif'ĭ-duh), p. 414

striae gravidarum (strī'e grăv-ĭ-dăr'-um), p. 423

teratogen (ter-ah'to-jen), p. 424

toxoplasmosis (tŏk'sō-plăz-mŏ'sĭs), p. 425

SUMMARY

18.1 Fertilization

During fertilization, a sperm nucleus fuses with the egg nucleus. The resulting zygote begins to develop into a mass of cells, which travels down the uterine tube and embeds itself in the endometrium. Cells surrounding the embryo produce HCG, the hormone whose presence indicates that the female is pregnant.

18.2 Development

A. The extraembryonic membranes, placenta, and umbilical cord allow humans to develop internally within the uterus. These structures protect the embryo and allow it to exchange waste for nutrients and oxygen with the mother's blood.

B. At the end of the embryonic period, all organ systems are established. There is a mature and functioning placenta. The embryo is only about 38 mm (1½ in.) long.

C. Fetal development extends from the third through the ninth months. During the third and fourth months, the skeleton is becoming ossified. The sex of the fetus becomes distinguishable.

D. During the fifth through the ninth months, the fetus continues to grow and to gain weight. Babies born after six or seven months may survive, but full-term babies have a better chance of survival.

E. During pregnancy, the mother's uterus enlarges greatly, resulting in weight gain, standing and walking difficulties, and general discomfort.

18.3 Birth

A. During stage 1 of parturition, the cervix dilates. During stage 2, the child is born. During stage 3, the afterbirth is expelled.

B. During pregnancy, the mother gains weight as the uterus comes to occupy most of the abdominal cavity with resultant annoyances such as incontinence. Many of the complaints of pregnancy, such as constipation, heartburn, darkening of certain skin areas, and diabetes of pregnancy, are due to the presence of the placental hormones.

STUDY QUESTIONS

1. Describe the process of fertilization and the events immediately following it. (pp. 409–410)
2. Name the four extraembryonic membranes, and give a function for each. (pp. 410–411)
3. Label the diagram at right to review the events that occur during pre-embryonic development.
4. What is the basis of the pregnancy test? (p. 414)
5. Specifically, what events normally occur during embryonic development? What events normally occur during fetal development? (pp. 412, 414–417)
6. Describe the structure and function of the umbilical cord. (p. 416)
7. Describe the structure and function of the placenta. (p. 416)
8. What are the three stages of birth? Describe the events of each stage. (pp. 421–422)
9. In general, describe the physical changes in the mother during pregnancy. (p. 423)

2.

ovum nucleus
sperm nucleus

ovum

zona pellucida
corona radiata

single cell = zygote

1.

fimbriae

ovary

uterine tube

5.

inner cell mass

6.

2-cell stage

3.

4-cell stage

8-cell stage

4.

LEARNING OBJECTIVE QUESTIONS

Fill in the blanks.

1. Fertilization occurs when the _____ nucleus fuses with the _____ nucleus.
2. The _____ membranes include the chorion, the _____, the yolk sac, and the allantois.
3. During development, the nutrient needs of the developing embryo (fetus) are served by the _____.
4. The zygote divides as it passes down a uterine tube. This process is called _____.
5. When cells take on a specific structure and function, _____ occurs.
6. Once the blastocyst arrives at the uterus, it begins to _____ itself in the endometrium.
7. During embryonic development, all major _____ form.
8. Fetal development begins at the end of the _____ month.
9. In most deliveries, the _____ appear(s) before the rest of the body.
10. During pregnancy, constipation, darkening of certain areas of the skin, and diabetes are all due to _____ produced by the placenta.

MEDICAL TERMINOLOGY EXERCISE

Consult Appendix B for help in pronouncing and analyzing the meaning of the terms that follow.

1. morphogenesis (mor″fo-jen′ĕ-sis)
2. neonatologist (ne″o-na-tol′o-jist)
3. prenatal (pre-na′tal)
4. hyperemesis gravidarum (hi″per-em′ĕ-sis grav-id-ar′um)
5. dysmenorrhea (dis″men-o-re′uh)
6. pseudocyesis (su″do-si′ĕ-sis)
7. primigravida (pri″mĭ-grav′ ĭ-duh)
8. oligospermia (ol″i-go-sper′me-uh)
9. perineorrhaphy (per″i-ne-or′uh-fe)
10. abruptio placentae (ab-rup′she-o pluh-sen′te)
11. dystocia (dis-to′se-uh)
12. galactostasis (gal″ak-tos′tuh-sis)
13. polyhydramnios (pol″e-hi-dram′ ne-os)
14. amniorrhea (am′ne′-o-re′uh)
15. placenta previa (pla-se′n′tuh pre′ve-uh)

WEBSITE LINK

Visit this text's website at http://www.mhhe.com/longenbaker7 for additional quizzes, interactive learning exercises, and other study tools.

Human Genetics

As you begin to study the topic of human genetics in this chapter, you might be interested to know that genetic chance was, in part, responsible for changing European history. In the early 1900s, the last Russian monarch, Tsar Nicholas Romanov, and his wife, Tsarina Alexandra, had four daughters but were desperate for a male heir for the throne. When their fifth child, crown prince Alexei, was born in 1904, the couple was heartbroken to discover that their son suffered from hemophilia, a sex-linked genetic disorder that causes uncontrolled bleeding in affected boys. When physicians of the time couldn't help their son, the Romanovs turned to Grigory Rasputin, a Russian mystic often called the "mad monk." Historians aren't sure how Rasputin did it, but he was able to alleviate Alexei's suffering. In so doing, Rasputin won the trust of the monarchs, and gained great influence over their decisions. On Rasputin's advice, Tsar Nicholas took control of the Russian army from his generals during World War I, a decision that scholars agree helped to contribute to Russia's defeat. Would Nicholas have been a more effective ruler (and might Europe look different today?) if his only son hadn't suffered from a genetic disorder? No one will ever know. Nicholas abdicated the throne on March 9, 1917, setting the stage for the Russian revolution, and the entire Romanov family was assassinated in July 1918.

Chapter Outline & Learning Objectives

After you have studied this chapter, you should be able to:

19.1 Chromosomal Inheritance

Normally, both males and females have 23 pairs of chromosomes, for a total of 46 chromosomes. The first 22 pairs, or chromosomes 1–44, are called **autosomes** (literally, "self" chromosomes). The final pair, chromosomes 45 and 46, are the **sex chromosomes.** These chromosomes determine the gender of the individual. In humans, males have a **Y chromosome** and an X chromosome. Females have two **X chromosomes.**

Various human disorders result from the inheritance of an abnormal chromosome number, or from inheriting chromosomes that are abnormal in size or shape. Such a disorder may be a **syndrome,** which is a group of symptoms that always occur together. Table 19.1 lists several syndromes that are due to an abnormal chromosome number. It is possible to view an individual's chromosomes by constructing a **karyotype,** a display of the chromosomes arranged by size, shape, and banding pattern. Karyotypes reveal whether an individual has inherited an abnormal number of chromosomes (too many or too few). Karyotyping can also often reveal if a chromosome has an abnormal length (too long or too short) or is otherwise abnormal in shape.

Karyotyping

Any cell in the body except red blood cells, which lack a nucleus, can be a source of chromosomes for karyotyping. In adults, it is easiest to use white blood cells separated from a blood sample for this purpose. The chromosomes of embryos or fetuses can be obtained by either amniocentesis or chorionic villi sampling (CVS). Because each carries a risk of spontaneous abortion (miscarriage), these procedures are not routinely performed for every pregnancy, but only when maternal history and/or prenatal testing indicate a need. For example, a pregnant woman who is over the age of 40 may elect to have CVS or amniocentesis because older women are at increased risk for chromosomal abnormalities in the embryo. Blood testing for pregnant women of any age may indicate the need for follow-up chromosomal studies. A blood test routinely offered to pregnant women measures the levels of four chemicals present in the blood: alpha-fetoprotein, produced by the fetal liver; estriol, an estrogen-like hormone produced by the placenta; human chorionic gonadotropin, the pregnancy hormone, and inhibin-A, a hormone produced by the fetus and placenta. Abnormally high or low blood levels of these chemicals in the blood are linked to certain chromosomal and structural abnormalities in the embryo and fetus. Amniocentesis or CVS may be recommended if abnormal levels are detected. It is important to note that this blood test does not *confirm* any abnormality; it just *suggests* that possibility.

During **amniocentesis,** a sample of amniotic fluid is withdrawn from the uterus of a pregnant woman. Blood tests and the age of the mother are used to determine whether the procedure should be done. Amniocentesis is not usually performed until about the fourteenth to the seventeenth week of pregnancy. A long needle is passed through the abdominal and uterine walls to obtain a small amount of fluid, which also contains fetal cells (Fig. 19.1*a*). Testing the cells and karyotyping the chromosomes may be delayed as long as four weeks so that the cells can be cultured to increase their number. As many as 400 chromosomal and biochemical problems can be detected by testing the cells and the amniotic fluid.

The risk of spontaneous abortion increases by about 0.3% due to amniocentesis, and doctors only use the procedure if it is medically warranted.

Chorionic villi sampling (CVS) is a procedure for obtaining chorionic cells in the region where the placenta will develop. This procedure can be done as early as the fifth week of pregnancy. A long, thin catheter tube is inserted through the vagina into the uterus (Fig. 19.1*b*). Ultrasound, which gives a picture of the uterine contents, is used to place the tube between the uterine lining and the chorionic villi. Remember that the chorion is the most superficial membrane surrounding the embryo, and it is formed from trophoblast cells of the embryo (see Chapter 18, page 412). Thus, the chromosomes in the cells of chorionic villi are the same as those of the embryo, although chorionic villi cells do not contribute to forming the embryo itself. Once the catheter is in place, a tiny sample of chorionic cells is withdrawn by suction. The cells do not have to be cultured, and karyotyping can be done immediately. However, this sampling procedure does not gather any amniotic fluid, so the biochemical tests done on the amniotic fluid following amniocentesis are not possible. Also, CVS carries a greater risk of spontaneous abortion than amniocentesis—0.8% compared to 0.3%. The advantage of CVS is that the results of karyotyping are available at an earlier date.

Table 19.1 Syndromes from Abnormal Chromosome Numbers					
Syndrome	**Sex**	**Disorder**	**Chromosome Number**	**Frequency**	
				SPONTANEOUS ABORTIONS	**LIVE BIRTHS**
Down	M or F	Trisomy 21	47	1/40	1/800
Poly-X	F	XXX (or XXXX)	47 or 48	0	1/1,500
Klinefelter	M	XXY (or XXXY)	47 or 48	1/300	1/800
Jacobs	M	XYY	47	?	1/1,000
Turner	F	XO	45	1/18	1/2,500

a. During amniocentesis, a long needle is used to withdraw amniotic fluid containing fetal cells.

b. During chorionic villi sampling, a suction tube is used to remove cells from the chorion, where the placenta will develop.

c. Cells are microscopically examined and photographed.

d. Normal male karyotype with 46 chromosomes

e. Down syndrome karyotype with an extra chromosome 21

Figure 19.1 Human karyotype preparation. A karyotype is an arrangement of an individual's chromosomes into numbered pairs according to their size, shape, and banding pattern. **a.** Amniocentesis and **b.** chorionic villi sampling provide cells for karyotyping to determine if the unborn child has a chromosomal abnormality. **c.** After cells are treated as described in the text, the karyotype can be constructed. **d.** Karyotype of a normal male. **e.** Karyotype of a male with Down syndrome. A Down syndrome karyotype has three number 21 chromosomes.

Preparing the Karyotype

After a sample of cells has been obtained, the cells are stimulated to divide in a culture medium. A chemical is used to stop mitosis during metaphase when chromosomes are the most highly compacted and condensed. The cells are then spread on a microscope slide and dried. Stains are applied to the slides, and the cells are photographed. Staining produces dark and light cross-bands of varying widths, and these can be used in addition to size and shape to help pair up the chromosomes. Today, a computer may be used to arrange the chromosomes in pairs. It is possible to photograph the nucleus of a cell that is about to divide (the chromosomes are more visible then), so that a picture of the chromosomes is obtained. The picture may be entered into a computer, and the chromosomes electronically arranged by pairs (Fig. 19.1c). The resulting display of chromosomes is the karyotype. Figure 19.1d,e compares a normal karyotype with that of a person who has Down syndrome, the most common autosomal abnormality.

Nondisjunction

An abnormal chromosomal makeup in an individual can be due to nondisjunction. **Nondisjunction** occurs during meiosis I when both members of a homologous pair go into the same daughter cell or during meiosis II when the sister chromatids fail to separate and both daughter chromosomes go into the same gamete (Fig. 19.2). If an ovum with 24 chromosomes is fertilized by a normal sperm cell with 23 chromosomes, the zygote formed has a **trisomy:** One type of chromosome is present in three copies instead of the normal two, and the total chromosome count is 47 instead of the normal 46. Conversely, if an ovum with 22 chromosomes is fertilized with a normal sperm, the zygote formed carries a **monosomy:** One type of chromosome is present in a single copy, and the zygote has only 45 chromosomes. It is important to note that nondisjunction can occur during the formation of gametes in either gender. Thus, although Figure 19.2 shows nondisjunction during the formation of ova, it could also happen during sperm formation as well.

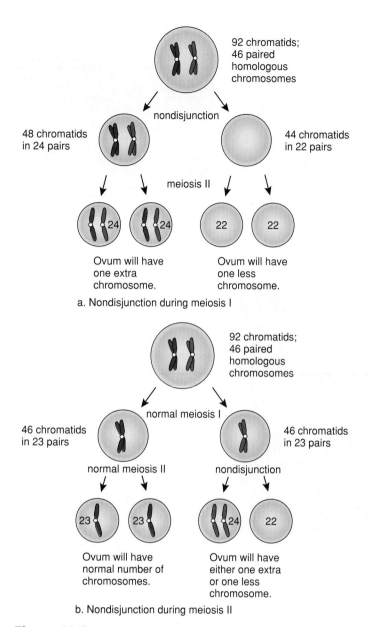

92 chromatids;
46 paired
homologous
chromosomes

nondisjunction

48 chromatids
in 24 pairs

44 chromatids
in 22 pairs

meiosis II

24   24   22   22

Ovum will have
one extra
chromosome.

Ovum will have
one less
chromosome.

a. Nondisjunction during meiosis I

92 chromatids;
46 paired
homologous
chromosomes

normal meiosis I

46 chromatids
in 23 pairs

46 chromatids
in 23 pairs

normal meiosis II   nondisjunction

23   23   24   22

Ovum will have
normal number of
chromosomes.

Ovum will have
either one extra
or one less
chromosome.

b. Nondisjunction during meiosis II

Figure 19.2 Nondisjunction during oogenesis. Nondisjunction can occur **(a)** during meiosis I if homologous chromosomes fail to separate or **(b)** during meiosis II if the sister chromatids fail to separate completely. In either case, abnormal gametes have an extra chromosome or lack a chromosome.

Researchers believe that when the zygote carries either a monosomy or a trisomy, most nondisjunctions produce embryos that cannot survive and develop normally. The result is an early spontaneous abortion (miscarriage), which may occur before a woman even realizes that she is pregnant. Note on Table 19.1 the high rate of miscarriages occurring from monosomy or trisomy.

Down Syndrome

Down syndrome is also called trisomy 21 because the individual usually has three copies of chromosome 21. In most

Figure 19.3 Down syndrome. Down syndrome occurs when the egg or the sperm has an extra chromosome 21 due to nondisjunction in either meiosis I or meiosis II. Characteristics include a wide, rounded face and narrow, slanting eyelids. Developmental delay to varying degrees is usually present.

instances, the egg had two copies instead of one of this chromosome. (In 23% of the cases studied, however, the sperm had the extra chromosome 21.)

Individuals who have Down syndrome share characteristic physical features (Fig. 19.3). Their heads are rounded; they have a short, webbed neck, and eyes that slant upward and have an inner eyelid fold. Facial features are typically small: a small, flattened nose; small ears that may fold over at the top; and a small mouth that often makes the tongue seem large and protruding. The Down syndrome individual is short statured, with small, short hands that have a large, single palmar crease (the so-called simian line). Mild to moderate developmental delays are typical of the syndrome. In rare cases, developmental delays can be severe.

The chance of a woman having a Down syndrome child increases rapidly with age, starting at about age 40. The frequency of Down syndrome is 1 in 800 births for mothers under 40 years of age and 1 in 80 for mothers over 40 years of age. Most Down syndrome babies are born to women younger than age 40, however, because this is the age group having the most babies. Maternal blood testing can screen for abnormal levels of chemicals associated with Down syndrome, but this type of test is not a positive confirmation. Amniocentesis followed by karyotyping is needed to confirm the diagnosis of Down syndrome in the fetus.

It is known that the genes that cause Down syndrome are located on the bottom third of chromosome 21, and extensive investigative work has been directed toward discovering the specific genes responsible for the characteristics of the syndrome. One day it might be possible to control the expression

a.   b.

Figure 19.4 **Syndromes due to an abnormal sex chromosome number. a.** A female with Turner (XO) syndrome has a short thick neck, short stature, and lack of breast development. **b.** A male with Klinefelter (XXY) syndrome has a feminine body shape and some development of the breasts.

of these genes even before birth, so that the symptoms of Down syndrome do not appear.

Begin Thinking Clinically

Individuals with Down syndrome often have a shorter than normal lifespan. What are some of the reasons for this?

Answer and discussion in Appendix C, page 466.

Sex Chromosome Inheritance

As stated, the sex chromosomes in humans are called X and Y. Because women are XX, an ovum always bears an X, but since males are XY, a sperm can bear an X or a Y. Therefore, the gender of the newborn child is determined by the father. If a Y-bearing sperm fertilizes the egg, then the XY combination results in a male. On the other hand, if an X-bearing sperm fertilizes the egg, the XX combination results in a female. All factors being equal, there is a 50% chance of having a girl or a boy.

♀ \ ♂	X	Y
X	XX	XY

Nondisjunction also occurs with regard to the sex chromosomes. Ova or sperm with too many or too few sex chromosomes can occur. Therefore, nondisjunction accounts for the birth of individuals with too few or too many sex chromosomes.

Too Many/Too Few Sex Chromosomes

From birth, an XO individual with **Turner syndrome** has only one sex chromosome, an X; the O signifies the absence of a second sex chromosome. Turner females are short, with a broad chest and a webbed neck. The ovaries, uterine tubes, and uterus are very small and nonfunctional. Turner females do not undergo puberty or menstruate, and their breasts do not develop (Fig. 19.4*a*). They are usually of normal intelligence and can lead fairly normal lives. Some Turner females have even given birth following in vitro fertilization using donor ova.

A male with **Klinefelter syndrome** has two or more X chromosomes in addition to a Y chromosome, and is sterile. The testes and prostate gland are underdeveloped, and the individual has no facial hair. Also, some breast development may occur (Fig. 19.4*b*). Affected individuals have large hands and feet and very long arms and legs. They may have developmental delays, in particular if they inherit more than two X chromosomes.

A **poly-X female** has more than two X chromosomes. Females with three X chromosomes have no distinctive physical appearance, aside from a tendency to be tall and thin. Some exhibit delayed motor and language development, though most are normal. Some may have menstrual difficulties, but many menstruate regularly and are fertile. Their children usually have a normal karyotype.

Females with more than three X chromosomes occur rarely. Unlike XXX females, XXXX females are usually tall and severely developmentally delayed. Various physical abnormalities are seen, but they may menstruate normally.

XYY males with **Jacobs syndrome** can only result from nondisjunction during spermatogenesis. This is because only males have a Y chromosome. Affected males are usually taller than average, suffer from persistent acne, and tend to have speech and reading problems. At one time, it was suggested that these men were likely to be criminally aggressive, but it has since been shown that the incidence of such behavior among them may be no greater than among XY males.

Notice that there are no YO males. This shows that at least one X chromosome is needed for survival. However, XXY individuals are males, not females.

Content CHECK-UP!

1. To describe amniocentesis, you could use the phrase(s):

   a. A long, thin needle is used to penetrate the abdomen and uterus of the pregnant woman, and a sample of amniotic fluid is taken from the amniotic sac surrounding the fetus.

   b. Four hundred or more biochemical and chromosomal tests can be carried out.

   c. The sample can be tested immediately and results are available within days.

   d. Both a and b are correct statements.

   e. All are correct statements.

2. If a sperm cell that has 24 chromosomes fertilizes a normal ovum, the resulting zygote will:

   a. have 47 chromosomes and a trisomy.

   b. have 45 chromosomes and a monosomy.

   c. have 46 chromosomes and be normal.

3. Which is the chromosome designation for Klinefelter syndrome?

   a. 47, XXX   c. 47, XYY

   b. 45, XO   d. 47, XXY

Answers in Appendix C, page 466.

19.2 Genetic Inheritance

Genes, which are sections of chromosomes, control body traits, such as height, eye color, or type of earlobe. Genes can also control the production of specific proteins, such as structural proteins, cell membrane proteins, and enzymes. Recall that as described in Chapter 17, page 377, each cell contains homologous pairs of chromosomes. Alternate forms of a gene having the same position (called the **locus**), on a pair of homologous chromosomes and affecting the same trait are called **alleles.** It is customary to designate an allele by a letter, which represents the specific characteristic it controls. A **dominant allele** is assigned an uppercase (capital) letter,

whereas a **recessive allele** is given the same letter but in lowercase. In humans, for example, unattached (free) earlobes are dominant over attached earlobes, so a suitable key would be *E* for unattached earlobes and *e* for attached earlobes.

Inheritance of Genes on Autosomal Chromosomes

An individual normally has two alleles for an autosomal trait. Just as one member of each pair of chromosomes is inherited from each parent, so too is one of each pair of alleles inherited from each parent. If the two alleles are identical, the pair is termed **homozygous;** if not, the pair is **heterozygous.**

The term **genotype** refers to the genes of the individual. Figure 19.5 shows three possible fertilizations and the resulting genotype of the individual for earlobe attachment. In the first instance, the chromosomes of both the sperm and the egg carry an *E*. Consequently, the zygote and subsequent individual have the alleles *EE*, which can be called a **homozygous dominant** genotype. A person with genotype *EE* obviously has unattached earlobes. The physical appearance of the individual—in this case, unattached earlobes—is called the **phenotype.**

In the second fertilization, the zygote has received two recessive alleles (*ee*), and the genotype is called **homozygous recessive.** An individual with this genotype has the recessive phenotype, which is attached earlobes. In the third fertilization, the resulting individual has the alleles *Ee,* which is a heterozygous genotype. A heterozygote shows the dominant characteristic; therefore, the phenotype of this individual is unattached earlobes.

How many dominant alleles does an individual need to inherit to have a dominant phenotype? These examples show that a dominant allele contributed from only one parent can bring about a particular dominant phenotype. How many recessive alleles does an individual need to inherit to have the recessive phenotype? A recessive allele must be received from both parents to bring about the recessive phenotype.

Sex-Linked Inheritance

The sex chromosomes contain genes just as the autosomal chromosomes do. Some of these genes determine whether the individual is a male or a female. Investigators have now discovered a series of genes on the Y chromosome that determine the development of male genitals, and at least one on the X chromosome that seems to be necessary for the development of female genitals.

Some traits controlled by alleles on the sex chromosomes have nothing to do with the gender of the individual. These traits are said to be **sex-linked.** An allele that is only on the X chromosome is **X-linked** and an allele that is only on the Y chromosome is **Y-linked.** Most sex-linked traits are controlled by alleles on the larger X chromosome. The very small Y chromosome does not have the corresponding allele, simply because it is so short. Thus, the sex chromosomes in males—X and Y—are not truly homologous chromosomes.

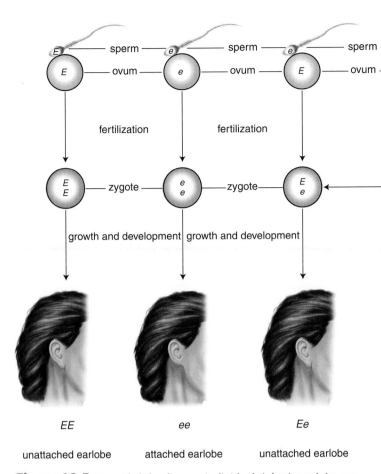

EE
unattached earlobe

ee
attached earlobe

Ee
unattached earlobe

Figure 19.5 Genetic inheritance. Individuals inherit a minimum of two alleles for every characteristic of their anatomy and physiology. The inheritance of a single dominant allele (*E*) causes an individual to have unattached earlobes; two recessive alleles (*ee*) cause an individual to have attached earlobes. Notice that each individual receives one allele from the father (by way of a sperm) and one allele from the mother (by way of an ovum).

As you know, a female has two X chromosomes, and thus her ova will contain either of her two X chromosomes. A male always receives a sex-linked recessive condition from his mother, from whom he inherited his single X chromosome. The Y chromosome from his father does not carry a corresponding allele for the trait. Therefore, males need receive only one recessive allele to have the X-linked disorder. Consider red-green color blindness, for example (Fig. 19.6*a*). This genetic disorder results in an individual who cannot see red or green because the cone cells of the eye are abnormal. It is caused by a recessive gene on the X chromosome. Note that although the mother has one recessive gene on an X chromosome, she has normal vision due to the dominant gene on her other X chromosome. She is referred to as a **carrier** for the trait. A carrier is an individual who is heterozygous for a recessive genetic disorder and therefore has no symptoms. If the carrier's ova are fertilized by sperm from a male with normal vision, any male children she bears will have a 50% chance of inheriting the disorder. Similarly, any female children will have a 50% chance of being a carrier for the disorder.

Would it ever be possible for a female to have red-green color blindness? It is possible, although unlikely. A female must receive two recessive X-linked alleles, one from each parent, before she has an X-linked recessive condition. The inheritance of a dominant allele from either her mother or her father can offset the inheritance of a recessive X-linked allele. A female with red-green color blindness will have inherited a recessive X chromosome from both her carrier mother and her color-blind father (Fig. 19.6*b*).

Punnett Square to Determine Probability

A Punnett square is a "short-cut" method originally devised by a mathematician/geneticist named Reginald Punnett that allows one to predict the *probability* of inheritance of a particular trait. It can be used to answer questions about both autosomal and sex-linked inheritance. Consider this example: An African-American couple is concerned about the possibility of conceiving a child with sickle-cell anemia. As you'll recall from Chapter 11, sickle-cell anemia is a disease characterized by abnormal hemoglobin. Red blood cells containing this hemoglobin are very fragile and easily destroyed (see page 239). Sickle-cell anemia is an autosomal recessive trait, like attached/detached earlobes. Both the husband and wife have sickle-cell trait (i.e., both carry the recessive gene for the trait, though their phenotypes are normal). What is their probability of conceiving a child with the disease?

To solve this problem, first draw a Punnett square. It's just like a tic-tac-toe board:

Then, put the father's genotype in the top boxes. The father is phenotypically normal but carries a recessive gene, so he is heterozygous, Aa. Any symbols can be used to represent the trait, so long as the uppercase/lowercase letters look different (to avoid confusion).

Next, put the mother's genotype in the side boxes. Again, she is heterozygous like the father, since she carries a recessive gene, but is phenotypically normal herself. It makes no difference if the father's symbols are on top, with the mother's on the side, or vice versa—the results will be the same.

Then, bring the father's genes down, as though he was passing them on to his offspring:

	A	a
A	A	a
a	A	a

And finally, bring the mother's genes across, as though she was passing them on to her offspring.

	A	a
A	AA	aA
a	Aa	aa

The four filled boxes give the possibilities for this couple's future children, and allow probability to be determined. The first possibility is 1:4, or 25% probability, of producing a normal child who does not carry the recessive trait, as in the starred box:

	A	a
A	AA*	aA
a	Aa	aa

The second possibility is 2:4, or 50% probability, of producing a child that is phenotypically normal, but carries the recessive gene, just like his/her parents, as in the starred boxes. It's important to note that whether the dominant or recessive trait is written first, the outcome will be the same:

	A	a
A	AA	aA*
a	Aa*	aa

The third possibility is 1:4, or 25% probability, of producing a child that has sickle-cell anemia. Remember that sickle-cell anemia is an autosomal recessive disorder—thus, for the child to be affected, he or she must be homozygous for the recessive gene. That would be the case in the starred box below:

	A	a
A	AA	aA
a	Aa	aa*

Keep in mind that dominant traits are not always "good" or beneficial. Consider the inheritance of achondroplasia, a disorder that results in individuals with very short stature. The torso is normal in size, but the limbs are abnormally short. In this disorder, the normal trait is actually a *recessive* gene. If two heterozygous adults with achondroplasia have children, the Punnett square would be identical to the one above. However, in this case, there would be 75% probability of an affected child and only 25% probability of having a nonaffected child. It's also important to recognize that Punnett squares only allow one to determine the probability of a *single* genetic event. If the parents with sickle-cell trait have a child with sickle-cell anemia, there will still be a 25% probability that any subsequent children will have sickle-cell anemia. As an analogy, consider tossing a coin—with every independent coin toss, there is an equal probability of "heads" or "tails."

The Punnett square method can be used to predict probability in sex-linked inheritance as well. Consider hemophilia, the sex-linked disease inherited by Alexei Romanov, whose history was briefly described in the chapter introduction. Alexei's mother, Alexandra, was the carrier, though she was a phenotypically normal female. Her genotype can be expressed as $X^H X^h$, where X^H is the dominant trait and X^h is the recessive trait. Tsar Nicholas Romanov, Alexei's father, was normal: $X^H Y$. The Punnett square used for sex-linked inheritance is similar to one used for autosomal inheritance; the only difference is the addition of the X and Y chromosomes. The completed square should look like this:

	X^H	X^h
X^H	$X^H X^H$	$X^H X^h$
Y	$X^H Y$	$X^h Y$

The results of this Punnett square show that for every Romanov child conceived, there was a 25% probability of a phenotypically normal noncarrier female ($X^H X^H$), a 25% probability of a phenotypically normal carrier female ($X^H X^h$), a 25% chance of a normal male ($X^H Y$), and a 25% chance of an affected male such as Alexei ($X^h Y$).

As a historical footnote, you might be interested to know that the original carrier female for the hemophilia trait was Queen Victoria of England. Because her descendants were royalty themselves, the hemophilia trait occurred in the royal households of England, Spain, Russia, and Prussia. Eventually, the trait was no longer passed down to Victoria's descendants, so the illness is no longer found among European royalty.

Genetic Counseling

Couples who believe that they may be at risk for having a child with a chromosomal or genetic disorder may choose to seek the advice of a genetic counselor. Genetic counselors are professionals with special training and certification. They help families to understand chromosomal and genetic inheritance and to make informed decisions about pregnancy and childbearing. In order to understand genetic counseling, we will consider the inheritance of cystic fibrosis. **Cystic fibrosis (CF)** is the most common lethal genetic disorder among Caucasians in the United States. About one in 20 Caucasians is a carrier for CF, and about one in 3,000 newborns has the disorder. In children with CF, the mucus in the bronchial tubes and pancreatic ducts is particularly thick and viscous, interfering with the function of the lungs and pancreas. In the past few years, new treatments have raised the life expectancy for CF patients to as much as 35 years of age.

Normal individuals have at least one dominant allele for a plasma membrane channel protein. Figure 19.7 shows the

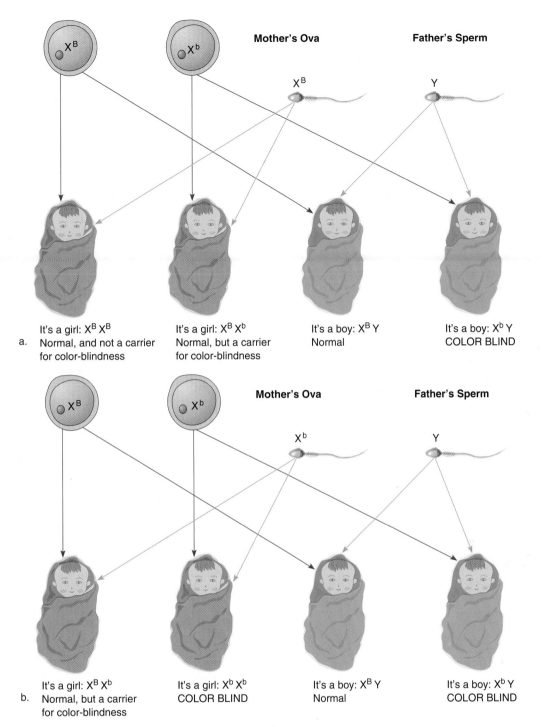

It's a girl: $X^B X^B$
a. Normal, and not a carrier for color-blindness

It's a girl: $X^B X^b$
Normal, but a carrier for color-blindness

It's a boy: $X^B Y$
Normal

It's a boy: $X^b Y$
COLOR BLIND

It's a girl: $X^B X^b$
b. Normal, but a carrier for color-blindness

It's a girl: $X^b X^b$
COLOR BLIND

It's a boy: $X^B Y$
Normal

It's a boy: $X^b Y$
COLOR BLIND

Figure 19.6 Sex-linked inheritance. a. A typical pattern of sex-linked inheritance for the color blindness trait. The mother is a carrier for the trait, and her sons have a 50% chance of inheriting the trait. **b.** A sex-linked trait can be inherited by a daughter only if her father is affected and her mother is a carrier or color blind also.

relationship between inheritance of a normal CF allele on a chromosome (chromosome 7) and development of the channel protein. The illustration emphasizes that alleles are actually a segment of DNA and that alleles cause the production of certain proteins in a cell.

As shown in Figure 19.8a, if a genetic counselor knows the genotype of the potential parents, the counselor can predict the chances that a couple will have a child having a recessive autosomal disorder such as CF. If the parents are both carriers, each offspring has a 25% chance of receiving two recessive alleles and having CF.

If the counselor does not know the inheritance pattern of a disorder, it is sometimes possible to deduce it by studying a pedigree. A **pedigree** is a diagram that depicts the inheritance of a particular trait: Circles are females, and squares are males; shaded-in symbols represent those who have a trait; half-shaded

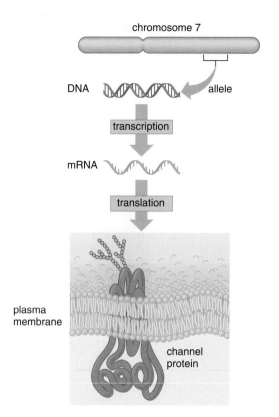

Figure 19.7 **Cystic fibrosis.** A person with cystic fibrosis has an abnormal allele, and the result is an abnormal channel protein in the plasma membrane. Chloride ions cannot exit the cell, resulting in a very thick mucus, particularly in the bronchial tubes and pancreatic duct.

Figure 19.8 **Inheritance pattern for CF, an autosomal recessive disorder. a.** The figures below the parents show four possible combinations of inherited alleles. Therefore, each offspring has a 25% chance of inheriting two recessive alleles and having CF. **b.** A sample pedigree for carrier parents and their children. Note that Sue and Rod are carriers like their parents Joe and Mary, and Tina is affected with CF.

symbols represent carriers; and roman numerals indicate generations. Notice that the pedigree in Figure 19.8b has to be for a recessive disorder because unaffected parents have a child with the disorder. This can only happen when a condition is recessive. If the condition were dominant, one of the parents would be affected!

Prenatal Testing for Genetic Disorders

Commonly inherited genetic disorders are listed in Table 19.2. Several of these disorders do not appear unless multiple abnormal genes are inherited. The inheritance of these conditions, listed in the "Multiple Genes" column in Table 19.2, is complex.

Sometimes parents want to improve their chances of having a child free of a particular genetic disorder that runs in their families. Until recently, the best that could be done was to test for a genetic disorder following amniocentesis or chorionic villi sampling. If the embryo had the disorder, the parents could consider an abortion. Now it is possible to retrieve ova from the ovary to make sure the ovum does not carry an abnormal allele prior to in vitro fertilization or to screen morulas following in vitro fertilization. The Medical Focus reading on page 440 discusses the latter technique. In the future, it might be possible to use gene therapy to cure any genetic defects found in an ovum or embryo.

Content CHECK-UP!

4. The outward, or physical expression of a person's genes for a trait is the:
   a. allele.   b. phenotype.   c. genotype.   d. recessive gene.

5. A couple has just had a child diagnosed with sickle-cell anemia, which is an autosomal recessive trait. Neither parent has the disease. What are the parents' genotypes? (Use the letter *A* to represent the dominant trait, and the letter *a* to represent the recessive trait.)
   a. father *Aa*, mother *aa*   c. father *aa*, mother *Aa*
   b. father *aa*, mother *aa*   d. father *Aa*, mother *Aa*

6. Which of the following statements about sex-linked recessive disorders is correct?
   a. The recessive trait is carried on the X chromosome.
   b. The mother is typically the carrier for the disorder.
   c. If the mother is a carrier and the father is normal, male children have a 50% chance of inheriting the disorder.
   d. Both a and c are correct.
   e. All three statements are correct.

Answers in Appendix C, page 466.

Table 19.2 Inheritance of Some Genetic Disorders

Dominant	Recessive	X-Linked	Multiple Genes
Examples of dominantly inherited disorders include: • Neurofibromatosis—benign tumors in skin or deeper • Achondroplasia—a form of dwarfism • Chronic simple glaucoma (some forms)—a major cause of blindness if untreated • Huntington disease—progressive nervous system degeneration • Familial hypercholesterolemia—high blood cholesterol levels, propensity to heart disease • Polydactyly—extra fingers or toes	Examples of recessive inherited disorders include: • Cystic fibrosis—disorder affecting function of mucous and sweat glands • Galactosemia—inability to metabolize milk sugar • Phenylketonuria—essential liver enzyme deficiency • Sickle-cell disease—blood disorder primarily affecting blacks • Thalassemia—blood disorder primarily affecting persons of Mediterranean ancestry • Tay-Sachs disease—lysosomal storage disease leading to nervous system destruction	Examples of X-linked disorders include: • Agammaglobulinemia—lack of immunity to infections • Color blindness—inability to distinguish certain colors • Hemophilia (some forms)—defect in blood-clotting mechanisms • Muscular dystrophy (some forms)—progressive wasting of muscles • Spinal ataxia (some forms)—spinal cord degeneration	Examples of multifactorial inheritance include: • Cleft lip and/or palate • Clubfoot • Congenital dislocation of the hip • Spina bifida—open spine • Hydrocephalus (with spina bifida)—water on the brain • Pyloric stenosis—narrowed or obstructed opening from stomach into small intestine

Source: Data from the National Foundation/March of Dimes.

19.3 DNA Technology

Previously, you studied the structure of DNA and how it replicates and carries out protein synthesis. **DNA technology** includes our ability to work directly with DNA to determine the relatedness of individuals, to assist forensics in determining whether a person has committed a crime, and to develop new treatments for human illnesses, called gene therapy.

Gene Therapy

Gene therapy is the insertion of genetic material into human cells for the treatment of a disorder. It includes procedures that give a patient healthy genes to make up for faulty genes and also includes the use of genes to treat various other human illnesses, such as cancer and cardiovascular disease. Currently, approximately 1,700 patients are enrolled in over 300 approved gene therapy trials in the United States.

How can a molecule as large as a gene be introduced, intact, into a cell? The most common approach used by scientists is to employ a **vector**—a carrier whose job is to transport the DNA directly into the cell. Viruses are nature's own vectors because they replicate themselves by inserting their own genetic material into a host cell. Scientists studying gene therapy techniques have adapted viruses to function as vectors. The viruses are first modified: Viral genetic material is removed or altered so that it does not cause disease, and human genes are inserted into the virus instead. Next, the virus is allowed to infect human tissue, where it inserts the healthy human genes into tissue cells. Four classes of viruses have been employed in gene therapy research: (1) retroviruses, like the virus that causes AIDS; (2) adenoviruses, which are a form of cold virus;

(3) herpes viruses, which cause cold sores on the mouth; and (4) the most commonly used virus vector, the adeno-associated virus, a very small virus that can have all of its own genetic material removed and still be capable of infecting cells.

Other mechanisms to insert genes into cells are also being investigated. One way is to introduce DNA directly into cells. A second mechanism involves attaching the gene to a liposome. Liposomes are microscopic vesicles that form spontaneously when lipoproteins are put into a solution. Still other scientists attach the DNA to molecules that have specific receptors on the cell membrane, then allowing the cell to incorporate the DNA by phagocytosis.

Among the many gene therapy trials, one is for the treatment of familial hypercholesterolemia, a condition that develops when liver cells lack a receptor for removing cholesterol from the blood. The high levels of blood cholesterol make the patient subject to fatal heart attacks at a young age. In a newly developed procedure, a small portion of the liver is surgically excised and infected with a virus containing a normal gene for the receptor. Several patients have experienced lowered serum cholesterol levels following this procedure.

Cystic fibrosis patients have an abnormal gene for a transmembrane carrier of the chloride ion. Patients often die due to numerous infections of the respiratory tract. In a newly developed procedure, liposomes have been coated with the gene needed to cure cystic fibrosis. The liposomes have then been delivered to the lower respiratory tract.

Genes are also being used to treat medical conditions other than the known genetic disorders. VEGF (vascular endothelial growth factor) can cause the growth of blood vessels. The gene that codes for this growth factor can be

Preimplantation Genetic Studies

We have already seen that in vitro fertilization can lead to therapeutic cloning (see page 413) for the production of various tissues. Now, we will consider the medical implications of the fact that each cell of the pre-embryo is totipotent, meaning that each cell can become a complete embryo because the cells have not yet become specialized. Not surprisingly, then, if a single cell is removed from an eight-celled cleavage pre-embryo, the seven-celled pre-embryo will still go on and develop into a normal newborn (Figure 19A). Researchers are using this knowledge to allow them to carry out preimplantation genetic analysis.

Consider that one partner (or both) of a couple may have one of the genetic disorders listed in Table 19.2. As potential parents, wouldn't the couple want the assurance that the pre-embryo is free of genetic disorders and will develop into a healthy adult? In some instances, it's possible to perform in vitro fertilization and then preimplantation genetic diagnosis (PGD) in order to determine the genotype of the pre-embryo with regard to particular genetic disorders. Then, only a healthy pre-embryo is implanted into the female. Currently, PGD can be used to screen for over 30 chromosomal and genetic abnormalities, including Down syndrome, cystic fibrosis, hemophilia, and sickle-cell anemia. Proponents cite its advantages: By selecting only healthy pre-embryos for implantation, a couple is less likely to suffer a miscarriage that results from chromosomal abnormality. Multiple embryos will not have to be implanted, and multiple pregnancy, with its increased risk of premature birth, can be avoided. Furthermore, the couple need not go through chorionic villus sampling or amniocentesis if pregnancy is achieved, and won't have to face the painful decision to terminate an existing pregnancy because of a genetic or chromosomal disorder. Although the PGD procedure is expensive, proponents argue that lifetime medical care for a child with a genetic or chromosomal disorder costs far more. In the future, it's possible that PGD might be coupled with gene therapy so that any pre-embryo would be suitable for implantation.

Recently, the effectiveness of single-cell PGD has been called into question. Studies have shown that removing the cell can damage the embryo and actually prevent its implantation into the uterus. Further, researchers note that fewer than half of the embryo's chromosomes are currently tested, and other defects may be present on the remaining chromosomes. Newer techniques are being tested on embryos in the blastocyst stage, when 100 cells or more are present. In this later stage, five cells can be removed for testing instead of a single cell, and there is less risk of damaging the embryo. Then, a new screening method called comparative genomic hybridization (CGH) allows all chromosomes in the embryonic cells to be studied.

However, experts in medical ethics continue to question the use of all forms of PGD. Pointing to the fact that each cell in an eight-cell pre-embryo is totipotent, they argue that the single cell used for testing is a potential human being that is destroyed by the testing process. Likewise, defective embryos are potential human beings that will ultimately be destroyed. Further, it's known that many genetic disorders don't appear in the affected

1 cell removed for genetic analysis

7 cells can complete normal development.

Figure 19A **One cell from an eight-celled pre-embryo can be tested for abnormal alleles.** If all tested alleles are normal, the seven-celled pre-embryo can complete normal development.

individual before the third or fourth decade of life. It's feasible that a cure for the disorder could be found during that time period. Finally, medical ethicists stress that although children with genetic or chromosomal abnormalities may have medical issues that must be overcome, their lives can have tremendous meaning and value to their families and to society. Moral philosophers contend that ending a life shortly after conception is an act whose morality must be seriously questioned.

In several instances worldwide, families with a child suffering from a serious blood disorder (leukemia or genetic anemias) have already used PGD to create a "savior baby"—a second child free of the genetic disorder whose umbilical cord blood was then transfused into the ill sibling. This practice creates further concern that PGD could be used to create babies for their "spare parts." Perhaps most disturbing is the notion that PGD could eventually be used to select only for pre-embryos with a couple's desired characteristics: gender, intelligence, physical appearance, or athletic ability. For example, it's possible that couples who already have a daughter could use PGD to select only male embryos for implantation—and any female embryos would subsequently be destroyed. In societies worldwide where male children are valued more than females, unethical practitioners of PGD could find a very lucrative market for their skills.

The debate will undoubtedly continue over the uses of this type of technology, and others like it that will develop in the future. It is important that all of us are well informed and understand the scientific issues so that we can make informed decisions.

Imagine that you're a young man who's been charged with murder. At your trial, a truck driver who gives "eyewitness" testimony states that you were the hitchhiker he picked up near the crime scene—but he's blind in one eye and admits it was very dark that night. A jailhouse snitch tells the judge that you confessed to the murder—and then, rape charges against him are later dropped by the prosecution. A former girlfriend testifies against you—but then recants her testimony two weeks later. A scent-dog handler tells the judge and jury that his dog identified your scent on a bloody T-shirt found near the scene—but the man's credentials and the dog's tracking ability are both later called into question. You maintain that you were 5 miles away from the scene at the time the murder occurred, and others confirm your alibi. However, 5 days later, you're convicted of first-degree murder, sentenced to life in prison, and incarcerated. This was the nightmarish scenario faced by William Dillon in 1981.

Now fast-forward to 2007: With help from the attorneys of the Innocence Project, Dillon's repeated requests for an investigation into his conviction were finally granted by a Florida judge. The Innocence Project—an advocacy group composed of attorneys, law students, and volunteers—uses modern DNA testing in its mission of re-examining past criminal convictions in which evidence is highly suspect. As you remember from Chapter 3, DNA analysis is often a vital tool for criminal identification. However, DNA evidence can also be used to review court findings and potentially overturn a wrongful conviction. In Dillon's case, although nearly all of the evidence from his original murder trial had been lost or destroyed, the blood-stained T-shirt from the crime was fortunately still available. DNA testing was performed on the shirt, and the testing showed that the blood belonged to the victim and to another unidentified man. There was no trace of Dillon's DNA on the shirt. With this new evidence, William Dillon was finally proven innocent, after spending 26 years in prison for a crime that someone else committed. He was cleared of all charges by the prosecution, and was freed from prison on December 10, 2008. The real perpetrator of that 1981 murder has not been found.

William Dillon's case is one of a multitude of decades-old criminal cases that have been reopened through the work of the Innocence Project. To date, the Project's efforts have exonerated more than 240 individuals, including 17 who had been serving time on death row.

injected alone or within a virus into the heart to stimulate branching of coronary blood vessels. Coronary patients report that they have less chest pain and can run longer on a treadmill.

Gene therapy is increasingly used as part of cancer therapy. Genes are being used to make healthy cells more tolerant of, and tumors more vulnerable to, chemotherapy. The gene *p53* brings about the death of cells, and there is much interest in introducing it into cancer cells, and in that way killing them off.

Genomics

Genomics is the molecular analysis of a **genome,** which is all the genetic information in all the chromosomes of an individual. Recall from Chapter 2, Figure 2.14, that the two DNA strands of a double helix are held together by bonding between their bases, designated as A, T, G, and C. Base A is always paired with base T, and base G is always paired with base C. In a worldwide effort known as the Human Genome Project, researchers set out to map the human chromosomes in two ways: (1) by constructing a map that shows the sequence of base pairs along all the human chromosomes, and (2) by constructing a map that shows the sequence of genes along the human chromosomes.

The Base Sequence Map

Researchers completed the first goal in 2003, thanks to an international effort. It took some 15 years for scientists to complete this monumental task. It is now known that human genetic material contains 3.16 billion base pairs, and the order of 99.9% of those base pairs is exactly the same in all people. To date, no one knows how much of the DNA consists of truly functional genes that actually code for proteins. It is estimated that 50% or more is so-called "junk DNA," which does not code for protein at all. The total number of truly functional genes is estimated to be between 30,000 and 50,000—a much lower number than originally projected, and less than 2% of the entire genome.

The Genetic Map

The genetic map tells the location of genes along each chromosome. Sixteen of the 46 human chromosomes have been mapped, including chromosome 22 (the first human chromosome to be completely mapped) and chromosome 4 (the most recently completed). Completing the genetic map should accelerate now that the base sequence map is done. Researchers need only know a short sequence of bases in a gene of interest in order for computers to search the genome for a match. Then, computers can help to pinpoint where this gene is located. However, decades—if not centuries—of research remains to be completed about the genome. Many questions remain unanswered: What does each gene do? How is the gene turned on and off? What is all the junk DNA for? How do genes interact to make each unique individual? What gene, or combination of genes, causes disease? For each question answered, many others will continue to arise.

Many ethical questions also arise regarding how we should use our knowledge of the human genome. Who owns genetic information? How should it be used and accessed? Should humans attempt to control and/or manipulate human genes and, if so, how? Manipulation of human genes is already underway, in laboratories around the world. It's imperative that you be knowledgeable about the human genome so that you can help decide these issues.

Content CHECK-UP!

7. Name four classes of viruses that are being studied as possible gene vectors.

8. An abnormal gene for the carrier that transports chloride across the plasma membrane results in:

   a. familial hypercholesterolemia.

   b. cystic fibrosis.

   c. hemophilia.

   d. sickle-cell anemia.

9. Which percentage of DNA base pairs is the same for all human beings?

   a. 99.9%   b. 75%   c. less than 2%   d. 50%

Answers in Appendix C, page 466.

SELECTED NEW TERMS

Basic Key Terms

allele (uh-lēl'), p. 434

amniocentesis (am"ne-o-sen-te'sis), p. 430

autosome (aw'to-sōm), p. 430

carrier (kăr'e-er), p. 435

chorionic villi sampling (ko-re-on'ik vil'ī sam'pling), p. 430

cystic fibrosis (sĭs'tĭh' fī'bras), p. 437

dominant allele (dom'ī-nant uh-lēl'), p. 434

genes (jēn), p. 434

genome (je'nōm), p. 441

genotype (je'-no-tīp), p. 434

homozygous (ho-mo-zi'-gus), p. 434

heterozygous (het-er-o-zi'gus), p. 434

homozygous dominant (ho-mo-zi'gus dom'ī-nant), p. 434

homozygous recessive (ho-mo-zi'gus re-ses'iv), p. 434

karyotype (kăr'e-o-tīp), p. 430

locus (lōh-kŭs), p. 434

monosomy (mŏ-nōh-sōh-mē), p. 431

nondisjunction (non"dis-junk'shun), p. 431

pedigree (ped-ĭ-gre), p. 437

phenotype (fe'no-tīp), p. 434

recessive allele (re-ses'iv uh-lēl'), p. 434

sex chromosome (seks kro'mo-sōm), p. 430

sex-linked (seks-linkt), p. 434

syndrome (sin'drōm), p. 430

trisomy (trī'sōh-mē), p. 431

X chromosome (x kro'mo-sōm), p. 430

X-linked (x-linkt), p. 434

Y chromosome (y kro'mo-sōm), p. 430

Clinical Key Terms

Down syndrome (down sin'drōm), p. 432

gene therapy (jen thēr'uh-pe), p. 439

Jacobs syndrome (ja'kubs sin'drōm), p. 434

Klinefelter syndrome (klīn'fel-ter sin'drōm), p. 433

poly-X female (pah'le-x fe'māl), p. 433

Turner syndrome (tur'ner sin'drōm), p. 433

XYY male (xyy māl), p. 434

SUMMARY

19.1 Chromosomal Inheritance

A. Normally, an individual inherits 22 pairs of autosomal chromosomes and one pair of sex chromosomes. Females are XX and males are XY.

B. Amniocentesis and chorionic villi sampling are used to provide cell samples for karyotyping fetal chromosomes.

C. Nondisjunction during oogenesis or spermatogenesis explains the inheritance of an abnormal number of chromosomes.

D. The most common autosomal abnormality is Down syndrome, which is due to the inheritance of an extra chromosome 21.

E. Abnormal combinations of sex chromosomes include XO (Turner syndrome), XXX (poly-X), XXY (Klinefelter syndrome), and XYY (Jacobs syndrome).

19.2 Genetic Inheritance

A. Genes control human traits. Uppercase letters designate dominant alleles; lowercase letters designate recessive alleles.

B. The genotype represents the genes of an individual, and the phenotype refers to outward expression of the gene. A homozygous dominant individual inherited two dominant alleles and has the dominant phenotype; a heterozygous individual inherited one dominant and one recessive allele and has the dominant phenotype; a homozygous recessive individual inherited two recessive alleles and has the recessive phenotype. A Punnett square can be used to predict the probability of gene inheritance.

C. The inheritance of X-linked alleles differs in males and females. Males require only one recessive allele to have an X-linked trait; females require two recessive alleles. This means that males are more likely to inherit an X-linked disorder

D. Genetic counselors help couples determine the chances of having children with a genetic disorder, such as cystic fibrosis. They can also determine the pattern of inheritance from studying a family's pedigree.

19.3 DNA Technology
A. Gene therapy, which involves replacing defective genes with healthy genes, is now a reality. Researchers are envisioning various applications aimed at curing human genetic disorders, as well as many other types of illnesses.

B. Genomics is an actively growing field. Because the sequence of base pairs along the human chromosomes has now been determined, new treatments for genetic disorders are expected.

STUDY QUESTIONS

1. What is the normal chromosome inheritance of humans? (p. 430)
2. How is a karyotype prepared? What are the possible sources for cell samples in an adult? In the fetus? (pp. 430–431)
3. What is nondisjunction, and when does nondisjunction occur during meiosis? (pp. 431–432)
4. What are the characteristics of a person with Down syndrome? (pp. 432–433)
5. What are the characteristics of the most common human conditions resulting from inheritance of abnormal numbers of sex chromosomes? (pp. 433–434)
6. Explain autosomal dominant and recessive genetic inheritance. (p. 434)
7. Explain X-linked allele inheritance in humans. (pp. 434–435)
8. Describe how to construct and interpret a Punnett square. (pp. 435–437)
9. What type of information does a genetic counselor give parents who might pass on a genetic disorder? (pp. 437–438)
10. Give examples of dominant, recessive, and X-linked genetic disorders in humans. (p. 439)
11. Describe the function of an allele using the CF allele as an example. (pp. 437–438)
12. What is gene therapy, and what types of genetic disorders have been treated thus far? (pp. 439–441)
13. What is a genome? What is genomics, and what might be the benefits of genomics in the future? (pp. 441–442)

LEARNING OBJECTIVE QUESTIONS

Fill in the blanks.
1. The genes are on the _____.
2. A karyotype shows the individual's _____.
3. The sex chromosomes of a male are labeled _____.
4. A person with Down syndrome has inherited _____ copies of chromosome 21.
5. A person with Klinefelter syndrome has the chromosomes _____.
6. A dominant autosomal genetic disorder only requires the inheritance of _____ (one or two) abnormal gene(s).
7. A male with an X-linked recessive disorder inherited a(n) _____ allele for the disorder.
8. If a person inherits an autosomal genetic disorder and both parents are unaffected by the disorder, the disorder is _____.
9. Replacing defective genes with healthy genes is the goal of _____.
10. The molecular analysis of a genome is called _____.

MEDICAL TERMINOLOGY EXERCISE

Consult Appendix B for help in pronouncing and analyzing the meaning of the terms that follow.

1. neogenesis (ne″o-jen′ĕ-sis)
2. regeneration (re-jen″er′-ashun)
3. fetoscope (fe′to-skōp)
4. polydysplasia (pol″e-dis-pla′ze-uh)
5. congenital (kon-jen′ĭ-tal)

WEBSITE LINK

Visit this text's website at http://www.mhhe.com/longenbaker7 for additional quizzes, interactive learning exercises, and other study tools.

Reference Figures | THE HUMAN ORGANISM

The following series of reference figures show the major organs of the human torso. The first plate illustrates the anterior surface and reveals the superficial muscles on one side. Each subsequent plate exposes deeper organs, including those in the thoracic, abdominal, and pelvic cavities.

Chapters 5–17 of this textbook describe the organ systems of the human organism in detail. As you read them, you may want to refer to these plates to help visualize the locations of organs and the three-dimensional relationships among them

Plate 1 Anterior view of the human torso with the superficial muscles exposed. (m. = muscles; v. = vein)

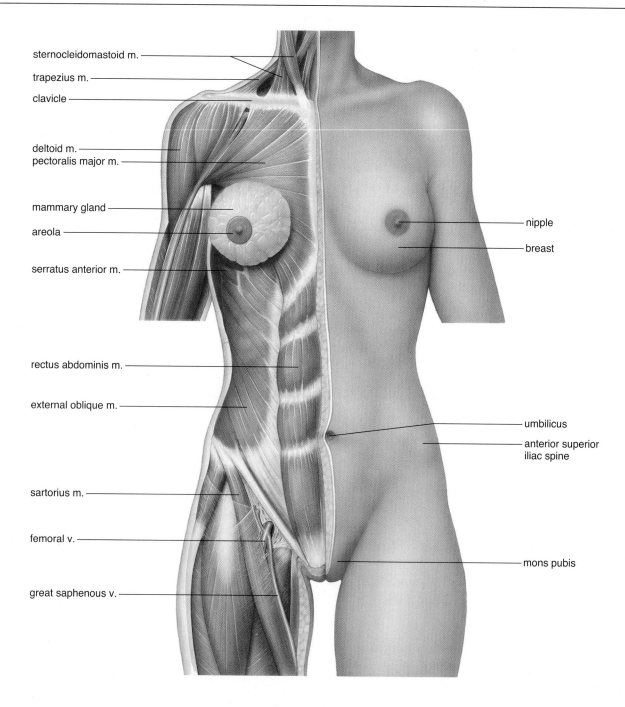

sternocleidomastoid m.
trapezius m.
clavicle
deltoid m.
pectoralis major m.
mammary gland
areola
serratus anterior m.
rectus abdominis m.
external oblique m.
sartorius m.
femoral v.
great saphenous v.

nipple
breast
umbilicus
anterior superior iliac spine
mons pubis

Plate 2 The torso, with the deep muscles exposed. (m. = muscle; n. = nerve; a. = artery; v. = vein)

larynx

sternocleidomastoid m.

clavicle

deltoid m.

pectoralis major m.

latissimus dorsi m.

rectus abdominis m.

external oblique m.

internal oblique m.

transversus abdominis m.

femoral n.

femoral a.

femoral v.

common carotid a.

internal jugular v.

thyroid gland

external intercostal m.

coracobrachialis m.

pectoralis minor m.

long head biceps brachii m.

short head biceps brachii m.

serratus anterior m.

transversus abdominis m.

linea alba

rectus abdominis m. (cut)

tensor fasciae latae m.

sartorius m.

rectus femoris m.

great saphenous v.

Plate 3 The torso, with the anterior abdominal wall removed to expose the abdominal viscera. (a. = artery; v. = vein; m. = muscle; n. = nerve)

common carotid a.

internal jugular v.

external jugular v.

subscapularis m.

teres major m.

latissimus dorsi m.

external intercostal muscles

internal intercostal muscles

liver

falciform ligament

gallbladder

greater omentum

urinary bladder

inguinal canal

spermatic cord

penis

sartorius m.

thyroid cartilage

thyroid gland

trachea

subclavian v.

coracobrachialis m.

sternum

left lung

pericardial sac

diaphragm

stomach

external oblique m.

internal oblique m.

transversus abdominis m.

anterior superior iliac spine

small intestine

femoral n.

femoral a.

femoral v.

Plate 4 The torso, with the anterior thoracic wall removed to expose the thoracic viscera. (a. = artery; m. = muscle; n. = nerve; v. = vein)

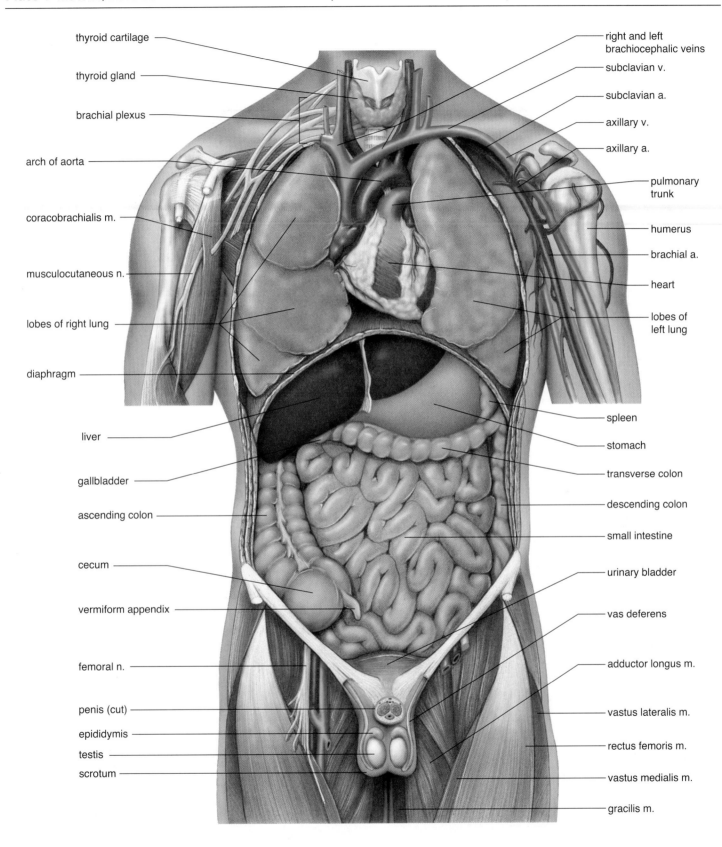

thyroid cartilage

thyroid gland

brachial plexus

arch of aorta

coracobrachialis m.

musculocutaneous n.

lobes of right lung

diaphragm

liver

gallbladder

ascending colon

cecum

vermiform appendix

femoral n.

penis (cut)

epididymis

testis

scrotum

right and left
brachiocephalic veins

subclavian v.

subclavian a.

axillary v.

axillary a.

pulmonary
trunk

humerus

brachial a.

heart

lobes of
left lung

spleen

stomach

transverse colon

descending colon

small intestine

urinary bladder

vas deferens

adductor longus m.

vastus lateralis m.

rectus femoris m.

vastus medialis m.

gracilis m.

Plate 5 The torso as viewed with the thoracic viscera sectioned in a coronal plane; the abdominal viscera as viewed with most of the small intestine removed. (a. = artery; m. = muscle; v. = vein)

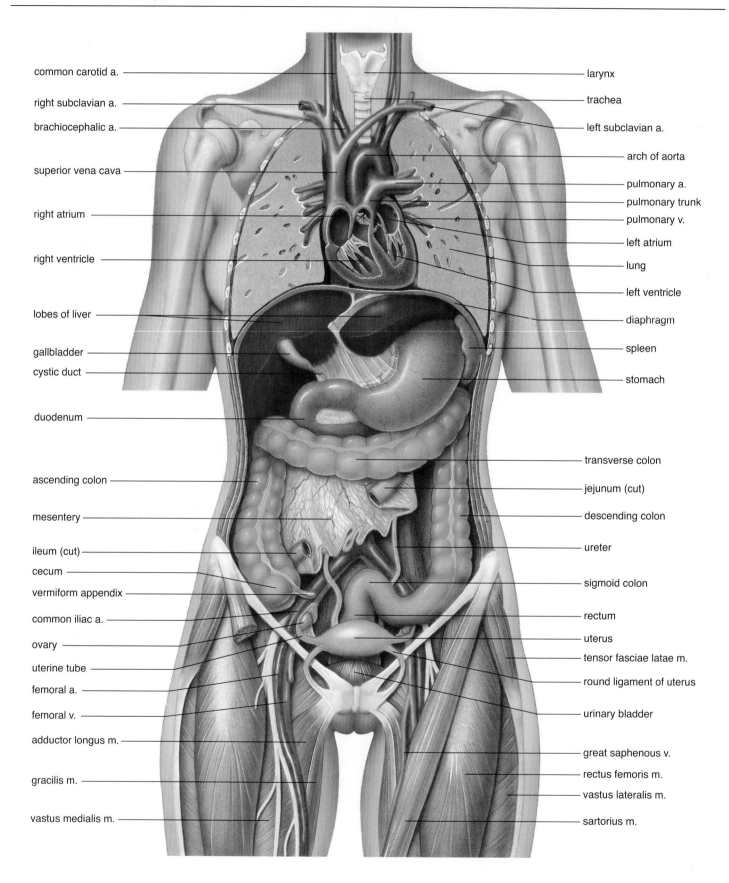

common carotid a.
right subclavian a.
brachiocephalic a.
superior vena cava
right atrium
right ventricle
lobes of liver
gallbladder
cystic duct
duodenum
ascending colon
mesentery
ileum (cut)
cecum
vermiform appendix
common iliac a.
ovary
uterine tube
femoral a.
femoral v.
adductor longus m.
gracilis m.
vastus medialis m.

larynx
trachea
left subclavian a.
arch of aorta
pulmonary a.
pulmonary trunk
pulmonary v.
left atrium
lung
left ventricle
diaphragm
spleen
stomach
transverse colon
jejunum (cut)
descending colon
ureter
sigmoid colon
rectum
uterus
tensor fasciae latae m.
round ligament of uterus
urinary bladder
great saphenous v.
rectus femoris m.
vastus lateralis m.
sartorius m.

Plate 6 The torso as viewed with the heart, liver, stomach, and portions of the small and large intestines removed. (a. = artery; m. = muscle; v. = vein)

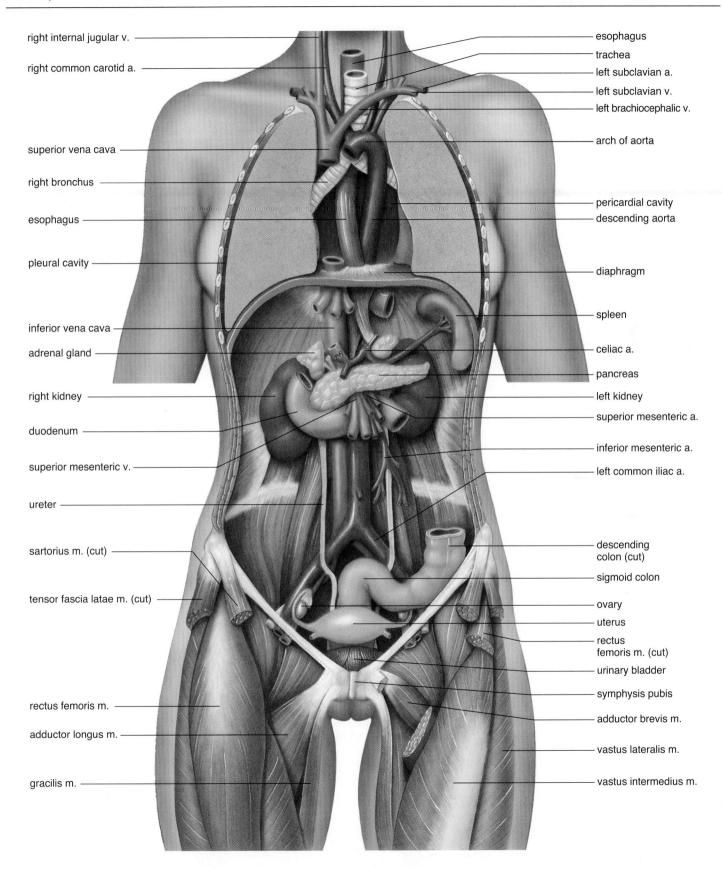

right internal jugular v.

right common carotid a.

superior vena cava

right bronchus

esophagus

pleural cavity

inferior vena cava

adrenal gland

right kidney

duodenum

superior mesenteric v.

ureter

sartorius m. (cut)

tensor fascia latae m. (cut)

rectus femoris m.

adductor longus m.

gracilis m.

esophagus

trachea

left subclavian a.

left subclavian v.

left brachiocephalic v.

arch of aorta

pericardial cavity

descending aorta

diaphragm

spleen

celiac a.

pancreas

left kidney

superior mesenteric a.

inferior mesenteric a.

left common iliac a.

descending colon (cut)

sigmoid colon

ovary

uterus

rectus femoris m. (cut)

urinary bladder

symphysis pubis

adductor brevis m.

vastus lateralis m.

vastus intermedius m.

Plate 7 The torso, with the anterior thoracic and abdominal walls removed, along with the viscera, to expose the posterior walls and body cavities. (a. = artery; m. = muscle)

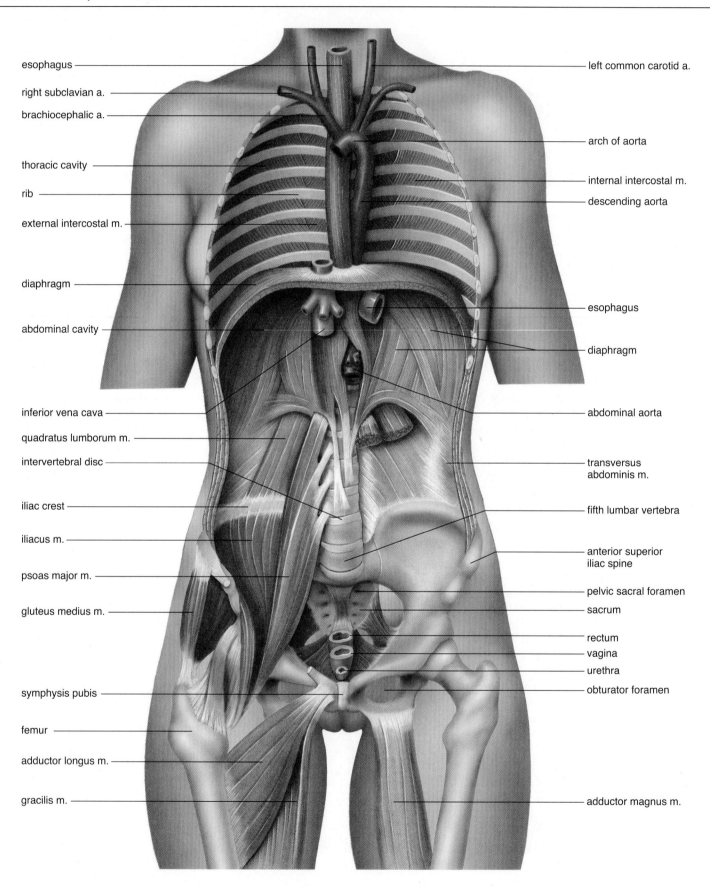

esophagus

right subclavian a.

brachiocephalic a.

thoracic cavity

rib

external intercostal m.

diaphragm

abdominal cavity

inferior vena cava

quadratus lumborum m.

intervertebral disc

iliac crest

iliacus m.

psoas major m.

gluteus medius m.

symphysis pubis

femur

adductor longus m.

gracilis m.

left common carotid a.

arch of aorta

internal intercostal m.

descending aorta

esophagus

diaphragm

abdominal aorta

transversus abdominis m.

fifth lumbar vertebra

anterior superior iliac spine

pelvic sacral foramen

sacrum

rectum

vagina

urethra

obturator foramen

adductor magnus m.

APPENDIX B

Understanding Medical Terminology

Learning Objectives

Upon completion of this section, you should be able to:

1. Discuss the importance of medical terminology and how it can be incorporated into the study of the human body.
2. Differentiate between a root word, a prefix, a suffix, and a compound word.
3. Link word parts to form medical terms.
4. Differentiate between singular and plural endings of medical terms.
5. Pronounce medical words.
6. Separate compound medical words into parts, and analyze their meaning.
7. Recognize root words, prefixes, and suffixes commonly used in medical terminology.

Introduction to Medical Terminology

As students of medical science, we have inherited a wealth of knowledge from scholars of the past. This knowledge was nurtured largely in the atmospheres of European universities, in which Latin and Greek were the languages of lecture and writing. Past scientists sought to define a universal language in which to communicate their findings. Latin and Greek, studied throughout Europe, became the languages of choice for university scholars whose native tongues were English, German, French, Spanish, and so on. Many important early works in medicine were first written in Latin, and their vocabularies remain to this day.

Anatomy and physiology, like many other scientific disciplines, have their own particular language. As the language was developed, names for each new structure to be described were created from existing words. Words were then combined, until they approximated an acceptable description. Medical terminology is simply a catalog of parts that allows us to take apart and reassemble the special language of medicine. The study of medical terminology is easier than it first seems. Relax; you won't need to walk around with a medical dictionary to make sense of medical terminology!

Medical words have three basic parts: a root word, a prefix, and a suffix. The **root word** is the main part of each medical term. Most describe parts of the body, using a Latin or Greek term. Mastering the list of root words in this appendix is an important first step in successfully translating medical terminology. Many of these words will be familiar to you from everyday language: For example, the root word *pyr-* for fire, shows up in the word *pyrotechnics* (fireworks), and the root word for produce, *gen-*, is familiar from the word *generate*.

A **prefix** comes before a root word and alters its meaning. For example, the prefix *hyper-* means over or above. Hyper/kinetic means overactive, hyper/esthesia is overly sensitive, hyper/tension is high blood pressure, and hyper/trophy is overdevelopment.

A **suffix** is attached to the end of a root word and also changes the meaning of the word. For example, the suffix *-itis* means inflammation. Inflammation can occur at almost any part of the body, so *-itis* can be added to root words to make hundreds of words. Dermat/itis is inflammation of the skin, rhin/itis is inflammation of the nose, gastr/itis is inflammation of the stomach, and so on.

Once the root word is known for each part of the anatomy, the prefixes and suffixes can be used to analyze and/or build many medical words. For example, the root word for heart is *cardi*. A few terms in which *cardi* appears are: cardi/algia (pain in the heart), cardio/megaly (enlarged heart), brady/cardia (slow heart), and peri/cardio/centesis (puncture to aspirate fluid from around the heart).

Many medical words have, in addition to a prefix and/or a suffix, more than one root word. These are called **compound words,** and can be analyzed by breaking them into parts. For example, hysterosalpingo-oophorectomy is made up of three root words and a suffix. *Hyster* is the root word for uterus, *salping* is the root word for tube, *oophor* is the root word for ovary, and *-ectomy* is the suffix for cut out. Now we know that hysterosalpingo-oophorectomy means the surgical removal of the uterus, uterine (fallopian) tube, and ovary. (A woman with uterine or ovarian cancer might undergo such a surgery.)

To facilitate pronunciation, word parts need to be linked together. The linkage for word parts is *o*, which may be referred to as a **combining form.** For example, linking the root *cardi* with the suffix *-pathy* would produce a word that would be difficult to pronounce; therefore, an *o* is used to link the root word with the suffix. The complete word is written *cardiopathy* and pronounced kar"de-op'uh-the; the combining form is cardi/o.

When a word is only a root or ends with a root, the word ending depends on whether the word is a noun or an adjective. For example, duodenum (noun) is a part of the small intestine, while duodenal (adjective) is related to the duodenum (for example, duodenal ulcer).

Accurate spelling of each word part is essential:

1. Changing one letter may change the word part. For example, *ileum* is a part of the small intestine, whereas *ilium* is a pelvic bone.
2. Finding a word in the dictionary requires a knowledge of spelling—at least of the beginning of the word. For

example, *pneumonia* and *psychology* have a silent *p, rhinitis* (inflammation of the nose) has a silent *h,* and *eupnea* (easy breathing) has an initial silent *e.*

Plural Endings

In many English words, the plurals are formed by adding *s* or *es,* but in Greek and Latin, the plural may be designated by changing the ending:

Singular Ending	Plural Ending	Examples
-a	-ae	aorta—aortae
-ax	-aces	thorax—thoraces
-en	-ina	lumen—lumina
-ex	-ices	cortex—cortices
-ix	-ices	appendix—appendices
-is	-es	testis—testes
-on	-a	phenomenon—phenomena
-um	-a	medium—media
-ur	-ora	femur—femora
-us	-i	bronchus—bronchi
-x	-ces	calyx—calyces
-y	-ies	anomaly—anomalies
-ma	-mata	adenoma—adenomata

If a word ends in *s* and the vowel in the last syllable is short, the word is singular. If the word ends in *s* and the vowel in the last syllable is long, the word is plural. Any word ending in a consonant is singular (for example, *-urn, -us, -at*).

Pronunciation Key

1. Words are made up of syllables.
2. Syllables are made up of letters—consonants and the vowels *a,e,i,o,u.*
3. Diacritical marks are placed over a vowel to help you understand how to pronounce that syllable. The two diacritical marks you should know are the macron (-) and the breve (˘). The macron designates a long vowel sound, and the breve indicates a short vowel sound. Examples are given below.
4. In this book, vowels standing alone at the end of a syllable with no diacritical mark above are understood to have the "long" sound:

a as in "day"	(matrix = ma′triks)
e as in "be"	(amino = uh-me′no)
i as in "hi"	(ion = i′on)
o as in "no"	(acidosis = as-i-do′sis)
u as in "blue"	(nucleus = nu′kle-us)

5. Vowels embedded within a syllable with no diacritical mark above are understood to have the "short" sound:

a as in "cat"	(lateral = lat′er-al)
e as in "yes"	(pelvic = pel′vik)
i as in "sit"	(distal = dis′tal)
o as in "not"	(abdominal = ab-dom′i-nal)
u as in "cut"	(buffer = buf′er)

6. Vowels in other positions are marked as long (-) or short (˘) with diacritical marks:

membrane = mem′brān
feedback = fēd′bak
peptide = pep′tīd
disease = di-zēz′
cytokinesis = si′to-kĭ-ne′sis
vesicle = ves′ĭ-kl
centriole = sen′tre-ōl
ribosome = ri′bo-sōm

7. Other vowel sounds are expressed as follows:

ah	the sound of **a** as in "father"
aw	the sound of **a** as in "fall"
ar	as in "large" (sarcoma = sar-ko′muh)
ār	as in bare (larynx = lār′inks; paranasal = pār-uh-na′zal)
uh	the sound of **a** as in "about" (anatomy = uh-nat′o-me; negative = neg′uh-tiv)
er	as in "her"
ĕr	as in "very" (therapy = thĕr′uh-pe; glomerulus = glo-mĕr′yu-lus)
ēr	as in peer (delirium = de-lēr′e-um; sarcomere = sar′ko-mĕr)
ow	as in "cow"
oy	as in "boy"
yū	the sound of **u** as in "cute" (mucus = myū′kus; tuberculosis = tu-ber-kyū-lo′sis)

8. The primary accent in a word is indicated by a single accent mark; for example, ostomy (os′to-me).
9. The secondary accent is indicated by a double accent; for example, duodenostomy (du″o-dĕ-nos′to-me).
10. The accent on medical terms is generally on the third from the last syllable.

Practice pronouncing the following words:

1. hematemesis (hem″uh-tem′ĕ-sis), vomiting blood
2. hysterosalpingo-oophorectomy (his″ter-o-sal-ping″go-o″of-or-ek′to-me), surgical excision of the uterus, uterine tube, and ovary
3. phrenohepatic (fren″o-he-pat′ik), pertaining to the diaphragm and liver
4. gastropathy (gas″trop′uh-the), disease of the stomach
5. metatarsus (met″uh-tar′sus), part of the foot between the tarsus (ankle) and toes

Commonly Used Prefixes

Prefix	Meaning	Example
a-, an-, in-	without, negative	a/men/orrhea, without a monthly flow
ab-	from, away from	ab/normal, away from normal
ad-, ac-, as-, at-	to, toward	ad/duct, carry toward
allo-	other	allo/graft, grafted tissue donated from one person to another
amphi-	both sides, surrounding	amphi/arthrosis, union between two bones at a joint
aniso-	unequal	an/iso/cyt/osis, abnormal condition of unequal cells
ante-, pre-	before	anterior, front; pre/natal, before birth
anti-, ant-, ob-	against	anti/pyre/tic, agent used against fever
bi-	two	bi/lateral, two sides
bio-	life	bio/logy, study of life
brachy-	short	brachy/dactyl/ism, short fingers and toes
brady-	slow	brady/cardia, slow heart rate
cata-	down	cata/bolism, breakdown of nutrients
cent-	hundred	centi/meter, 1/100 of a meter
chem-	chemical	chemo/therapy, chemical treatment for cancer
chlor-	green	chlor/opsia, green vision, as in certain cases of poisoning
circum-	around	circum/cis/ion, to cut around
co-, com-, con-	with, together	con/genital, born with
contra-	against	contra/indicated, against indication
cyano-	blue	cyano/sis, blue color to skin
de-	away from	de/hydrate, loss of water
dextr-	right	dextr/o/cardia, heart displaced to the right
dia-	through	dia/rrhea, flow through
diplo-	double, twofold	diplo/pia, double vision
dis-	apart	dis/sect, to cut apart
dys-	bad, difficult	dys/pnea, difficult breathing
e-, ex-	out, out from	ex/cise, to cut out
ect-, exo-, extra-	outside	extra/corporeal, outside the body
en-	in, on	en/capsulated, in a capsule
end-	within	endo/scopy, visualization within
epi-	upon	epi/dermis, upon the skin
eu-	good	eu/phonic, good sound
eury-	broad, wide	eury/gnathous, having a wide jaw
gero-, geronto-	old, elderly	geronto/logy, study of diseases of the elderly
hem-, semi-	half	hemi/gastr/ectomy, surgical removal of half of the stomach
hydro-	water	hydro/cephalus, water on the brain
hyper-	over, above	hyper/kinetic, overactive
hypo-	under, below	hypo/glossal, under the tongue
immun-	free, exempt	immun/ity, exempt from the effects of specific disease-causing agents
in-	not	in/soluble, not able to dissolve
infra-	beneath	infra/mammary, beneath the breast
inter-	between	inter/cellular, between the cells
intra-	within	intra/cranial, within the cranium
iso-	same	iso/coria, pupils of equal size
kil-	thousand	kilo/gram, 1,000 grams
leio-	smooth	leio/myoma, smooth muscle tumor
lepto-	high, thin	lepto/cephaly, having a long, thin head and neck

Commonly Used Prefixes—Continued

Prefix	Meaning	Example
levo-	left	levo/duction of eye, movement of eye to the left
lys-	dissolution, disintegration	lyso/some, organelle that degrades worn cell parts
macr-	large	macro/cyte, large cell
mal-	bad	mal/nutrition, bad nourishment
mes-	middle	mes/entery, middle of intestine
meta-	after, beyond	meta/carpals, beyond the carpals (wrist)
micr-	small	micro/cephal/ic, having a small head
milli-	one-thousandth	milli/liter, 1/1,000 of a liter
multi-	many	multi/para, one who has many children
neo-	new	neo/plasm, new growth
normo-	normal	normo/cyte, normal, healthy red blood cell
olig-	scanty, few	olig/uria, scanty amount of urine
onc-	tumor	onc/ology, study of tumors
opistho-	behind, backward	opistho/tonos, bending backward
pachy-	thick	pachy/derma, thick skin
per-	through	per/cutaneous, through the skin
peri-	around	peri/tonsillar, around the tonsil
photo-	light	photo/phobia, avoidance of light
pleur-	rib, side	pleur/al membranes, serous membranes that enclose the lungs
poly-	much, many	poly/cystic, many cysts
post-	after	post/mortem, after death
pre-	before	pre/natal, before birth
presby-	old	presby/opia, old vision
primi-	first	primi/gravida, first pregnancy
pro-	before	pro/gnosis, prediction of the probable outcome of a disease
re-	back, again	re/generate, produce, develop again
retr-, retro-	behind	retro/sternal, behind the sternum
rhod-	red	rhod/opsin, visual pigment in the eye
semi-	half	semi/comatose, in a partial comatose state; semiconscious
sub-	under	sub/lingual, under the tongue
super-, supra-	above	supra/spinal, above the vertebral column
syn-, sym-	with, together	syn/ergism, working together
tachy-	fast	tachy/phasia, fast speech
un-	not	un/conscious, not conscious

Commonly Used Suffixes

Suffix	Meaning	Example
-algia	pain	dent/algia, pain in the tooth
-ase	suffix for an enzyme	protein/ase, protein-digesting enzyme
-atresia	without an opening	proct/atresia, rectum without an opening
-blast	immature cell form	erythro/blast, immature red blood cell
-cele	hernia	omphalo/cele, umbilical hernia
-centesis	puncture to aspirate fluid	arthro/centesis, puncture to aspirate fluid from a joint
-cept	take, receive	re/cept/or, something that receives again
-cide	kill	bacteri/cidal, able to kill bacteria
-cis	cut	circum/cis/ion, cutting around
-cyte	cell	erythro/cyte, red cell
-denia	pain	cephalo/denia, pain in the head
-desis	fusion	arthro/desis, fusion of a joint
-ectasia	expansion	cor/ectasis, expanding or dilating pupil
-ectomy	cut out, excise	nephr/ectomy, surgically remove kidney
-edema	swelling	cephal/edema, swelling of head
-emesis	vomiting	hyper/emesis, excessive vomiting
-emia	blood	hyper/glyc/emia, elevated blood sugar
-gen	precursor of	terato/gen, agent that causes a malformed fetus
-gnosis	knowledge	dia/gnosis, knowledge through examination (determining cause of disease)
-gram	record	myelo/gram, X ray of the spinal cord
-graphy	making a record	angio/graphy, making a record of vessels
-iasis	condition	chole/lith/iasis, condition of gallstones
-ism	condition, disease	alcohol/ism, addiction to alcohol
-ist	one who	opto/metr/ist, one who measures vision
-itis	inflammation	aden/itis, inflammation of a gland
-lepsy	seizures	narco/lepsy, seizures of numbness, attacks of deep sleep
-lexy	speech, words	dys/lexia, difficulty in reading words
-logist	one who specializes	ophthalmo/logist, one specializing in eyes
-logy	study of	bio/logy, study of life
-lysis, -lytic, -lyze	break down, dissolve	teno/lysis, destruction of tendons
-lyt	dissolvable	electro/lyte, substance that ionizes in water solution
-malacia	abnormal softening	osteo/malacia, abnormal softening of bone
-mania	madness	pyro/mania, irresistible urge to set fires
-megaly	enlargement	spleno/megaly, enlargement of spleen
-meter	measure	thermo/meter, instrument to measure temperature
-oid	resembling	muc/oid, resembling mucus
-oma	tumor	neur/oma, nerve tumor
-opia	vision	ambly/opia, dim vision
-osis	abnormal condition	nephr/osis, abnormal condition of kidney
-osme, -osmia	smell	an/osmia, inability to smell
-ostomy	create an opening	col/ostomy, create an opening in the colon
-otia	ear	macr/otia, large ear
-pathy	disease	encephalo/pathy, disease of the brain
-penia	deficiency, poor	leuko/cyto/penia, deficiency of white cells
-pepsia	digestion	dys/pepsia, bad digestion
-pexy	surgical fixation	nephro/pexy, surgical fixation of kidney

Commonly Used Suffixes—Continued

Suffix	Meaning	Example
-phasia	speak, say	a/phasia, without ability to speak
-philia	love, attraction	chromo/philic, attracted to color
-phobia	abnormal fear	agora/phobia, abnormal fear of crowds
-plasia	formation	hyper/plasia, excessive formation
-plasm	substance	proto/plasm, original substance
-plasty	make, shape	rhino/plasty, to shape the nose
-plegia	paralysis	hemi/plegia, paralysis of one half of body
-pnea	breath	tachy/pnea, fast breathing
-ptosis	prolapse, dropping	hystero/ptosis, prolapse of uterus
-rrhagia	burst forth	metro/rrhagia, hemorrhage from uterus
-rrhaphy	suture, sew	hernio/rrhaphy, suture a hernia
-rrhea	flow, discharge	oto/rrhea, discharge from ear
-rrhexis	rupture	spleno/rrhexis, rupture of the spleen
-scope	instrument for viewing	oto/scope, instrument to look in ears
-scopy	visualization	laryngo/scopy, visualization of larynx
-some, -soma	body	lyso/some, body that lyses or dissolves
-spasm	twitching	blepharo/spasm, twitching of eyelid
-stasis	stop, control	hemo/stasis, control bleeding
-therapy	treatment	hydro/therapy, treatment with water
-tome	instrument to cut	osteo/tome, instrument to cut bone
-tomy	to cut	laparo/tomy, to cut into the abdomen
-tripsy	crushing	nephro/litho/tripsy, crushing stone in kidney
-trophy, -trophic, -trophin	development	hyper/trophy, overdevelopment
-uria	urine	hemat/uria, blood in the urine

Commonly Used Root Words*

Root	Meaning	Example
acro-	extremity, peak	acro/megaly, enlarged extremities; acro/phobia, abnormal fear of heights
aden-	gland	adeno/pathy, disease of a gland
adip-	fat, lipid	adip/ose tissue, fat storage tissue
aer-, aero-	air	aero/phagia, swallowing air
andro-	man, male	andro/gen, agent that causes male development
angi-	vessel	angi/oma, tumor of a vessel
arthr-	joint	arthr/algia, pain in the joint
athero-	soft, pasty material	athero/sclerosis, pasty hardening of an artery
balano-	penis	balano/plasty, surgical repair of penis
blast-	bud, growing thing	neuro/blast, growing nerve cell
blephar-	eyelid	blephar/op/tosis, drooping eyelid
brachi-	arm	brachi/al, pertaining to the arm
bronch-	windpipe	bronch/us, a branch of the trachea (windpipe)
bucco-	cheek	bucco/labial, referring to cheek and lip
calcaneo-	heel	calcaneo/dynia, painful heel
carcin-	cancer	adeno/carcin/oma, cancerous tumor of a gland
cardi-	heart	myo/cardi/tis, inflammation of heart muscle
carp-	wrist	flexor carp/i, muscle to bend wrist
caud-	tail	caud/al, pertaining to tail
celio-	abdomen	celio/tomy, incision of the abdomen
cephal-	head	cephalo/dynia, pain in the head
cervic-	neck, cervix	cervic/itis, inflammation of the neck of uterus
cheil-	lip	cheilo/plasty, shaping the lip
cheir-, chir-	hand	chiro/megaly, large hands
chol-	bile, gall	chole/cyst/ectomy, surgical removal of the gallbladder
chondr-	cartilage	chondro/malacia, softening of cartilage
chrom-	color	poly/chromatic, having many colors
chron-	time	syn/chron/ous, occurring at the same time
cleid-	clavicle	cleido/costal, referring to clavicle and ribs
col-	colon	mega/colon, enlarged colon
colp-	vagina	colp/orrhaphy, suture of vagina
core-	pupil	core/pexy, suturing the iris to change the pupil
cost-	ribs	inter/costal, between the ribs
crani-	skull	crani/otomy, incision into the skull
cry-	cold	cryo/philic, cold loving
crypt-	hidden	crypt/orchid/ism, hidden (undescended) testicle
cutan-, cut-	skin	sub/cutaneous, below the skin
cyan-	blue	acro/cyan/osis, abnormal condition of blueness of extremities
cyst-	bladder	cysto/cele, bladder hernia
cyt-	cell	thrombo/cyte, clotting cell (platelet)
dacry-	tear	dacryo/rrhea, flow of tears
dactyl-	fingers, toes	poly/dactyl/ism, too many fingers and toes
dent-, -odont	teeth	peri/odontal, around the teeth; dent/algia, toothache
derm-, dermat-	skin	intra/dermal, within the skin
desmo-	ligament	desmo/dentium, ligament anchoring tooth
dextr-	right	dextro/cardia, heart displaced to the right
dips-	thirst	poly/dipsia, excessive thirst

Root	Meaning	Example
dors-	back	dors/al, pertaining to the back
duct-	carry	ovi/duct, tube to carry ova (eggs)
encephal-	brain	encephalo/cele, hernia of the brain
enter-	intestine	gastro/enter/itis, inflammation of stomach and intestine
erg-	work	en/ergy, working with
erthyr-	red	erythro/cyto/penia, deficiency of red cells
esthe-	sensation	an/esthe/tic, agent to eliminate sensation
esthen-	weakness	my/esthenia, muscle weakness
facio-	facial, the face	facio/plasty, plastic surgery on the face
febr-	fever	a/febrile, without a fever
flex-	bend	dorsi/flex, bend backward
gastr-	stomach	gastro/scopy, visualization of the stomach
gen-	produce	patho/genic, agent that produces disease
gingiv-	gums	gingiv/ectomy, removal of gums
gloss-	tongue	hypo/glossal, under the tongue
glyc-, glu-	sugar	hypo/glyc/emia, low blood sugar
gnath-	jaw	micro/gnath/ism, small jaw
gonado-	sex organs	gonado/trophin, causes development of sex organs
grav-	heavy, pregnancy	secundi/gravida, second pregnancy
gynec-	female	gyneco/logy, study of female conditions
hem-, hemat-	blood	hemat/emesis, vomiting blood
hepat-	liver	hepato/megaly, enlarged liver
heter-	different	hetero/genous, different origins
hidr-	perspiration	hidro/rrhea, flow of perspiration
hist-	tissue	histo/logy, study of tissue
home-, hom-	same	homeo/stasis, stay same, equilibrium
hydro, hydra-	water	de/hydra/tion, process of losing water
hyster-	uterus	hyster/ectomy, removal of the uterus
iatr-	physician	iatro/genic, produced by the physician
irid-	iris	irid/ectomy, surgical removal of iris
is-	equal	iso/tonic, equal in pressure
kary-	nut, nucleus	mega/karyo/cyte, cell with large nucleus
kerat-	cornea	kerato/plasty, repair of cornea
kin-	move	kinesio/logy, study of movement
labio-	lips	labio/gingival, refers to lips and gums
lacrim-	tear	lacrima/tion, crying
lact-, galact-	milk	lacto/genic, milk-producing
lapar-	abdomen	laparo/rrhaphy, suture of the abdomen
laryng-	larynx	laryngo/scopy, visualization of larynx
later-	side	bi/lateral, two sides
leuk-, leuc-	white	leuko/rrhea, white discharge
lingu-	tongue	sub/lingual, under the tongue
lip-	fat	lip/oma, tumor of fat
lith-	stone	litho/tripsy, crushing a stone
lymphadeno-	lymph node	lymphadeno/pathy, disease affecting lymph nodes
lymphangio-	lymph vessel	lymphangio/graphy, X rays of lymph vessels and nodes
mast-, mammo-	breast	mast/itis, inflammation of the breast; mammo/gram, X ray of breast

Root	Meaning	Example
melan-	black	melan/oma, black tumor
men-	monthly, menstrual	dys/meno/rrhea, difficult monthly flow
mening-	membrane	mening/es, membranes that cover the brain and spinal cord
metr-	uterus	endo/metr/ium, lining of uterus
morph-	shape, form	poly/morphic, pertaining to many shapes
my-	muscle	myo/metr/itis, inflammation of muscle of uterus
myc-	fungus	onycho/myc/osis, fungus condition of the nails
myel-	marrow, spinal cord	myelo/gram, X-ray record of spinal cord
myring-	eardrum	myringo/tomy, opening into eardrum
nas-	nose	naso/pharyng/eal, pertaining to nose and throat
nat-	to be born	pre/nat/al, before birth
necr-	dead	necr/opsy, examining dead bodies; autopsy
nephr-, -ren	kidney	hydro/nephr/osis, abnormal condition of water in the kidney
neur-	nerve	neur/algia, nerve pain
noct-, -nyct	night	noct/uria, voiding at night
nucle-	kernel	nucle/us, dense core (kernel) of an atom
null-	none	nulli/gravida, woman who has had no pregnancies
ocul-	eye	mon/ocul/ar, pertaining to one eye
omphal-	umbilicus	omphalo/rrhea, discharge from the navel
onych-	nail	onycho/crypt/osis, condition of hidden nail (ingrown)
oo-	ova, egg	oo/genesis, producing eggs
oophor-	ovary	oophoro/cyst/ectomy, removal of cyst from ovary
ophthalm-	eye	ex/ophthalmos, condition of protruding eyes
or-	mouth	oro/pharyngeal, pertaining to mouth and throat
orchid-	testis	orchid/ectomy, removal of testis
orexis-	appetite	an/orexis, absence of appetite
orth-	straight	orth/odont/ist, one who straightens teeth
oste-, oss-	bone	osteo/chondr/oma, tumor of bone and cartilage
ot-, aur-	ear	ot/itis, inflammation of the ear; post/auricular, behind the ear
para-	to bear	primi/para, to bear first child.
path-	disease	patho/physio/logy, study of effects of disease on body functioning
pect-	chest	pecto/ralis, chest muscle
ped-	child	ped/iatrician, doctor who specializes in children
peps-	digest	dys/pepsia, bad digestion
phag-	swallow, eat	a/phagia, inability to swallow
phallo-	penis	phallo/dynia, pain in the penis
pharmac-	drug	pharmaco/logy, study of drugs
pharyng-	throat	pharyng/itis, inflammation of the throat
phas-	speak, say	tachy/phasia, speaking rapidly
phleb-	vein	phlebo/thromb/osis, abnormal condition of clot in vein
phon-	voice	a/phonic, absence of voice
phren-	diaphragm	phreno/hepatic, pertaining to the diaphragm and liver
pil-, trich-	hair	tricho/glossia, hairy tongue
pleur-	lining of lung, thorax	pleur/isy, inflammation of the pleura
pneum-	air, breath	pneumo/thorax, air in the chest
pneumon-	lung	pneumon/ectomy, surgical removal of the lung
pod-	foot	pod/iatrist, one who specializes in foot problems
proct-	rectum	procto/scopy, visualization of the rectum
pseud-	false	pseudo/cyesis, false pregnancy

Commonly Used Root Words—Continued

Root	Meaning	Example
psych-	mind	psycho/somatic, pertaining to the mind and the body
pulmon-	lung	cardio/pulmonary, pertaining to heart and lungs
py-	pus	pyo/rrhea, flow of pus
pyel-	kidney	pyelo/nephr/itis, inflammation of the kidney
pyl-	door, orifice	pyl/oric sphincter, ring of muscle that controls food entry into duodenum
pyr-	fire, fever	anti/pyretic, agent used against fever
quadri-	four	quadri/plegia, paralysis of all four extremities
rhin-	nose	rhino/plasty, revision of the nose
salping-	tube	salping/itis, inflammation of the uterine tube
sanguin-	blood	ex/sanguina/tion, process of bleeding out (bleed to death)
sarc-, sarco-	flesh, striated muscle	sarco/lemma, cell membrane of a muscle fiber
scler-	hard	arterio/scler/osis, condition of hardening of arteries
scot-	dark	scot/oma, blind spot in the visual field
sect-	cut	dis/section, cutting apart
sept-	contamination	anti/septic, agent used against contamination
sial-	saliva	sialo/rrhea, excessive salivation
spondyl-	vertebra	spondyl/itis, inflammation of the vertebrae
steato-	fat, lipid	steato/rrhea, fat in the fecal material
sten-	narrow, constricted	pyloric sten/osis, narrowing of pylorus
stomat-	mouth	stomat/itis, inflammation of the mouth
strict-	draw tight	vaso/con/strict/or, agent that compresses vessels
tax-	order, arrange	a/taxic, uncoordinated
ten-	tendon	teno/rrhaphy, suture a tendon
terato-	malformed fetus	terato/gen, agent that can harm a fetus
therm-	heat	hyper/thermia, raising body heat
thorac-	chest	thoraco/centesis, puncture to aspirate fluid from chest
thromb-	clot	thrombo/cyte, clotting cell
tox-	poison	tox/emia, poison in the blood
trache-	windpipe	tracheo/malacia, softening of tracheal cartilages
trachel-	neck	trachel/orrhaphy, suture of cervix (neck of uterus)
traumat-	wound	traumat/ology, study of trauma
tri-	three	tri/geminal, having three beginnings
trop-	turn	ec/tropion, turned out
ur-	urine	ur/emia, urine constituents in the blood
vas-	vessel	vaso/constriction, narrowing of a vessel
veni-	vein	veni/puncture, puncture of a vein
vert-	turn	retro/vert/ed, turned backward
vesic-	bladder	vesico/cele, hernia of the bladder
viscer-	internal organs	e/viscera/tion, process of viscera protruding from abdominal wall
vita-	life	vital, necessary for life

*Words that are the same as anatomical terms used in English (for example, pancreas, tonsil, and so on) have been omitted.

Now you are ready to apply your knowledge! Each time you find a new word in the text, try to analyze its meaning. Break each words into its parts: root, prefix, and suffix. You'll be able to translate the words without using a dictionary. At the end of each chapter, in the Medical Terminology Reinforcement Exercise, you are given an opportunity to reinforce your knowledge. Pronounce the words, and dissect them into parts to arrive at a meaning. You can also begin to build medical words and use them in your everyday conversation. You will be amazed at how rapidly your vocabulary will grow, and how your study of the human body will become easier and more enjoyable.

APPENDIX C

Chapter 1

Content Check-Up!

1. superior; 2. buccal; 3. horizontal; 4. parietal pleura: covers walls of left and right thoracic cavities, visceral pericardium: covers the heart, visceral peritoneum: covers the abdominal organs; 5. chest pain, particularly when a person inhales; 6. endocrine and nervous; 7. cardiovascular and lymphatic; 8. chronic.

Begin Thinking Clinically

Because the child is pointing to the umbilical region, the source of his pain is most likely the small intestine. Two common reasons for intestinal pain are stretching the small intestine (as when the organ overfills with gas) or irritating the intestinal lining (as in bacterial food poisoning).

Learning Objective Questions

1. c; 2. f; 3. g; 4. a; 5. d; 6. c; 7. e; 8. b; 9. a; 10. d; 11. g; 12. f; 13. d; 14. c; 15. e; 16. a; 17. d; 18. b; 19. g; 20. c; 21. d; 22. e; 23. f; 24. a; 25. b; 26. organ; 27. midsagittal or median; 28. homeostasis.

Medical Terminology Exercise

1. above; 2. below; 3. stomach; 4. abdomen; 5. head; 6. thorax; 7. sides; 8. eye; 9. back; 10. study; 11. chest; 12. a; 13. blood; 14. infection or inflammation of the kidney; 15. a.

Chapter 2

Content Check-Up!

1. 20
2. $F=F$; F_2
3. 1×10^{-4} moles per liter, or 0.0001 moles per liter
4. a. urine at pH 5 is more acidic; b. there are 1000 times more free hydrogen ions in pH 5 urine
5. a
6. c
7. b

Begin Thinking Clinically

Because strontium is similar to calcium in structure, it could replace bone calcium and damage bone.

Learning Objective Questions

1. atoms; 2. neutrons; 3. ionic, covalent; 4. hydrogen; 5. hydrogen, lower; 6. glucose, energy; 7. glycerol, fatty acid; 8. amino acids, coil, structure or configuration; 9. enzymes; 10. DNA, nucleotides.

Chapter 3

Content Check-Up!

1. a—2, b—3, c—1
2. e
3. a
4. c
5. b
6. b
7. c
8. b
9. d

Begin Thinking Clinically

When brain tissue is damaged, fluid escapes into the spaces between the nerve cells. This swelling can put pressure on the brain (because it's enclosed by the skull) and cause further brain damage. If a hypertonic solution flows through the blood vessels supplying the brain, osmosis will cause excess fluid to be returned to the bloodstream.

Learning Objective Questions

1. c; 2. e; 3. a; 4. d; 5. b; 6. protein, phospholipid; 7. ribosomes; 8. centrioles; 9. hypotonic; 10. carrier molecule, ATP; 11. endocytosis; 12. 46; 13. cytoplasm, organelles; 14. d; 15. b; 16. a; 17. c.

Chapter 4

Content Check-Up!

1. d
2. d
3. c
4. b
5. c
6. b
7. a—2; b—3; c—1
8. a—3; b—1; c—2

Begin Thinking Clinically 1

Smoking causes constant "wear-and-tear" on the delicate pseudostratified ciliated columnar epithelium of the airways. Over time, this tissue will gradually be replaced by stratified squamous epithelium, which is much sturdier (though unable to move mucus and clear airway pollutants).

Begin Thinking Clinically 2

In young, healthy people, bone will grow much more quickly than cartilage. Because the ligament is dense regular cartilage, the player with the ankle sprain will be sidelined for a longer period of time.

Learning Objective Questions

1. tissues; 2. stratified, cilia, columnar; 3. matrix, fibers; 4. connective; 5. epithelial, loose connective; 6. c; 7. a; 8. b; 9. d; 10. c; 11. a; 12. b.

Chapter 5

Content Check-Up!

1. c
2. b
3. c
4. b
5. a
6. b

Begin Thinking Clinically

In psoriasis, overgrowth of the epidermis causes the skin to have a reddened, crusted appearance. Patches of affected skin are covered with scales, and may itch. Psoriasis is not serious unless it is part of an autoimmune disorder that affects other body tissues (see Chapter 13 for further discussion of autoimmune disorders).

Learning Objective Questions

1. both b and c; 2. c; 3. a; 4. b; 5. a; 6. hair follicles, sebum; 7. temperature; 8. mechanical trauma, pathogen invasion, water gain or loss; 9. D; 10. depth of burn and extent of burn; 11. melanoma, basal cell carcinoma.

Chapter 6

Content Check-Up!

1. b
2. d
3. a
4. d
5. b
6. c
7. b
8. a
9. c
10. b
11. a
12. b

Begin Thinking Clinically

Depending on the severity of the blow, he may have fractured the bones of the anterior orbit, or eye socket. These include the frontal, lacrimal, maxilla, and zygomatic bones. He may also have a fracture of the nasal bone. If he lost consciousness after the blow, he must be carefully monitored for a possible brain injury (see the In Case of Emergency reading in Chapter 8 for a discussion of traumatic brain injury).

Learning Objective Questions

1. c; 2. g; 3. a; 4. d; 5. f; 6. e; 7. d; 8. c; 9. g; 10. a; 11. f; 12. b; 13. e; 14. longer; 15. spongy; 16. sinuses; 17. axial skeleton and/or vertebral column and/or pelvic girdle, axial skeleton and/or rib cage; 18. support and flexibility, support and stability; 19. hand and foot; 20. hinge.

Chapter 7

Content Check-Up!

1. d
2. b
3. c
4. c
5. c
6. c
7. d
8. b
9. b
10. a
11. d
12. c

Begin Thinking Clinically

Compartment syndrome usually occurs in the muscles of the thigh or leg. It can result from severe bruising of the muscle or from bone fracture. Compartment syndrome is treated by elevating the affected limb and using anti-inflammatory drugs to reduce swelling. In severe cases, the fascia that creates the compartment must be cut open, a procedure called fasciotomy.

Learning Objective Questions

1. smooth; 2. endomysium; 3. actin; 4. ATP; 5. muscle tone; 6. prime mover, synergists; 7. biceps brachii; 8. lower spine, humerus; 9. temporalis; 10. deltoid; 11. triceps brachii; 12. gluteus maximus; 13. h; 14. e; 15. c; 16. f; 17. b; 18. a.

Chapter 8

Content Check-Up!

1. d
2. a
3. a
4. c
5. b
6. d
7. d
8. b
9. b

Begin Thinking Clinically

The sensory (receiving) and association (memory) areas for vision are located in the occipital lobe. The person who strikes his or her head in a fall may lose vision or visual memories. Depending on the severity of the injury, the loss may be permanent.

Learning Objective Questions

1. axon; 2. sodium, inside; 3. synaptic cleft; 4. acetylcholinesterase; 5. effectors (muscle cells, glands, organs); 6. interneuron; 7. cerebellum; 8. reticular formation; 9. meninges; 10. cranial nerve, somatic and parasympathetic nervous system functions; 11. brain, spinal cord, nerves, ganglia; 12. learning, memory; 13. sympathetic, parasympathetic; 14. EEG; 15. a. sensory neuron, b. interneuron, c. motor neuron, d. sensory receptor, e. neuron cell body, f. dendrites, g. axon, h. nucleus of Schwann cell, i. node of Ranvier, j. muscle cell.

Chapter 9

Content Check-Up!

1. b
2. a
3. d
4. e
5. a
6. a
7. a
8. d
9. b
10. c
11. b
12. a

Begin Thinking Clinically

An uncinate fit is a form of epileptic seizure. Epilepsy results from excessive neuron activity in a particular brain area. In an uncinate fit, the temporal lobe areas controlling smell may be damaged, and the seizure results.

Learning Objective Questions

1. semicircular canals, utricle, and saccule; 2. chemoreceptors; 3. rods, cones, retina; 4. color, bright; 5. accommodates (becomes thicker); 6. distant objects far away, concave; 7. malleus, incus, stapes; 8. balance, equilibrium; 9. cochlear canal, inner ear; 10. brain.

Chapter 10

Content Check-Up!

1. a
2. c
3. b
4. FSH, LH, and/or TH, and/or ACTH
5. a
6. d
7. a
8. d
9. c
10. e
11. c
12. d
13. d
14. a
15. d
16. d
17. b
18. growth of uterus, vagina; secondary sex characteristics, ovum maturation

Begin Thinking Clinically

A prolactin producing adenoma in a man might cause his mammary glands to swell and produce milk. These adenomas also cause impotence, or inability to achieve erection of the penis.

Learning Objective Questions

1. negative feedback; 2. manufactures, ADH, oxytocin; 3. releasing and inhibiting hormones; 4. anterior; 5. too little thyroid hormone, too little iodine; 6. calcium; 7. cortex; 8. Cushing's syndrome; 9. pancreas, insulin receptors; 10. blood; 11. steroid, protein or amine; 12. medulla, cortex, ACTH (corticotropin).

Chapter 11

Content Check-Up!

1. formed elements and plasma layers; red blood cells, white blood cells, and platelets
2. d
3. by picking up body heat from muscles and transporting it around the body, by maintaining salt-water balance, by buffering acids and bases

4. d
5. d
6. b
7. d
8. c
9. d
10. b
11. c
12. c

Begin Thinking Clinically

Polycythemia can occur whenever the tissues (and the kidney in particular) have a lowered oxygen supply. Thus, heart disease or lung disease can result in polycythemia. Polycythemia also results from unknown causes. Bleeding the person (by removing blood through a vein) cures symptoms.

Learning Objective Questions

1. plasma; 2. oxygen and carbon dioxide, function in defense; 3. oxyhemoglobin; 4. nucleus, 120; 5. neutrophil; 6. antibodies, destroy; 7. fibrin; 8. A, B, neither; 9. Rh^-, Rh^+; 10. c; 11. b; 12. a; 13. d.

Chapter 12

Content Check-Up!

1. d
2. a
3. c
4. b
5. b
6. d
7. c
8. d
9. b
10. c
11. a
12. d
13. a
14. c
15. d

Begin Thinking Clinically

Figure 12.17 shows other ways for blood to return from the foot and leg to the heart. For example, blood could travel from the anterior tibial vein or small saphenous vein to the popliteal vein. However, the clot would cause severe pain, and could break free to form a thromboembolism (see Chapter 11). To prevent blood clotting, drink lots of water, stretch your legs, and walk around as much as possible.

Learning Objective Questions

1. aorta; 2. lungs, pulmonary artery; 3. SA (sinoatrial) node; 4. away from; 5. coronary; 6. subclavian, femoral and common iliac; 7. cerebral arterial circle (Circle of Willis); 8. digestive organs, liver; 9. mean arterial blood pressure; 10. the pumping force of the ventricles, skeletal muscle and respiratory pump and pressure difference between veins and right atrium; 11. systolic, diastolic; 12. cardiac output, resistance; 13. foramen ovale, ductus arteriosus; 14. bicuspid.

Chapter 13

Content Check-Up!

1. d
2. a
3. c
4. d
5. produces thymic hormones, T lymphocytes mature
6. a
7. b
8. a
9. d
10. c
11. c
12. d

Begin Thinking Clinically

If the cancerous tumor has begun to spread throughout the body, a lymph node will often be the first place for cancer cells to lodge. A sentinel lymph node is the first node to receive lymph from a malignant tumor. If this node is free of cancer cells, others are likely to be free as well.

Learning Objective Questions

1. valves; 2. B lymphocytes, T lymphocytes; 3. lymph, blood; 4. neutrophils, eosinophils, monocytes/macrophages; 5. thymus; 6. plasma, memory; 7. antibody; 8. cell; 9. vaccine; 10. histamine, other chemical mediators; 11. passive, active; 12. non-specific; 13. complement.

Chapter 14

Content Check-Up!

1. a
2. trachea, primary bronchi, secondary bronchi, tertiary bronchi, bronchioles, alveoli
3. b
4. external intercostals, diaphragm, accessory muscles
5. b
6. c
7. d
8. b
9. As HCO_3^-, dissolved, attached to Hb
10. d
11. b
12. d

Begin Thinking Clinically

The tissues of the soft palate and tonsils are often enlarged in someone with OSA. The condition is usually treated using a continuous positive airway pressure (CPAP) device, which constantly blows air into the patient's airways to keep enlarged tissues from collapsing and blocking airways.

Learning Objective Questions

1. larynx; 2. alveoli; 3. carbon dioxide; 4. expanded, opened up; 5. diffusion; 6. enters; 7. enters; 8. bicarbonate; 9. smoking; 10. primary bronchi; 11. tidal volume; 12. vital capacity; 13. epiglottis; 14. alveoli, pulmonary capillaries; 15. a. nasal cavity, b. nostril, c. pharynx, d. epiglottis, e. glottis, f. larynx, g. trachea/ cartilage, h. bronchus, i. bronchiole.

Chapter 15

Content Check-Up!

1. c
2. b
3. b
4. d
5. detoxify, produce bile, phagocytize pathogens, gluconeogenesis, urea formation, etc.
6. a
7. b
8. d
9. a

Begin Thinking Clinically

If normal flora bacteria are destroyed by antibiotic abuse, pathogenic microbes can grow in the colon instead. The first symptom caused by these harmful organisms is typically diarrhea. Long-term antibiotic abuse can result in vitamin deficiency.

Learning Objective Questions

1. amylase; 2. epiglottis; 3. esophagus, protein; 4. acidic; 5. portions, rectum, anus; 6. duodenum; 7. glycogen; 8. bile, emulsifies; 9. trypsin, chymotrypsin, and carboxypeptidase, lipase, amylase; 10. lacteals.

Chapter 16

Content Check-Up!

1. a
2. c
3. d
4. c
5. a
6. blood cells and large plasma proteins
7. e
8. c
9. a

Begin Thinking Clinically

Loop diuretics work in the ascending loop of Henle. They inhibit secretion of sodium into the medulla, so sodium is gradually washed away and the sodium concentration gradient disappears. Less water is reabsorbed in the kidneys, and the volume of urine increases as a result. One generic version is furosemide. Aldosterone antagonist diuretics work primarily in the distal convoluted tubule. By interfering with aldosterone, they inhibit the reabsorption of sodium and water. Once again, more urine is formed. Drugs called spironolactones are aldosterone antagonist diuretics.

Learning Objective Questions

1. metabolic wastes; 2. glomerus; 3. urethra; 4. sodium; 5. loop of Henle, distal convoluted tubule, collecting duct; 6. urea; 7. creatinine and/or urochrome; 8. sodium, water, pH; 9. ADH; 10. buffer; 11. gout; 12. ureters; 13. cortex, medulla.

Chapter 17

Content Check-Up!

1. d
2. b
3. d
4. c
5. a
6. a
7. a
8. c
9. b
10. d
11. syphilis, gonorrhea, chlamydia
12. a

Begin Thinking Clinically

An abdominal ectopic pregnancy occurs when an ovum is fertilized outside the uterine tube, and doesn't successfully enter the uterine tube to travel to the uterus. The organs most likely to be able to support a placenta (and allow the fetus to grow) would be the external uterus, urinary bladder, and the intestines. The infant would have to be delivered through the abdominal wall by Caesarian section.

Learning Objective Questions

1. vas deferens; 2. seminal vesicle; 3. blood; 4. testosterone; 5. ovaries; 6. estrogen, progesterone; 7. vesicular follicle of an ovary, uterus; 8. low sperm count, abnormal sperm; 9. seminiferous tubules; 10. sustentacular cells; 11. corpus luteum; 12. ovulations; 13. viruses, bacteria.

free radicals Atoms or molecules with an unpaired electron in their outermost shell; linked to several diseases and play a role in the aging process.

frontal lobe Area of the cerebrum responsible for voluntary movements and higher intellectual processes.

frontal plane Plane or section that divides a structure lengthwise into anterior and posterior portions; pertaining to the region of the forehead.

FSH See *follicle-stimulating hormone.*

functional group In an organic molecule, a particular side chain attached to a larger molecule, which then gives the larger molecule its characteristic properties (i.e., electrical charge or solubility).

fundic region Pouch that is superior to the body of the stomach.

G

gallbladder Saclike organ associated with the liver that stores and concentrates bile.

gamete Sex cell; sperm cells in males and ova in females.

gamma globulin Most common form of plasma protein; a protein fraction of blood rich in antibodies, administered especially for passive immunity against measles, German measles, hepatitis A, or poliomyelitis.

ganglion Collection of neuron cell bodies outside the central nervous system (pl., *ganglia*).

gap junction Junction between cells formed by the joining of two adjacent plasma membranes; it lends strength and allows ions, sugars, and small molecules to pass between cells.

gastric gland Gland within the stomach wall that secretes gastric juice.

gastric inhibitory polypeptide Hormone produced by small intestine cells that inhibits stomach movement and acid secretion; also promotes release of insulin in response to glucose consumption; also referred to as *glucose-dependent insulinotropic peptide.*

gastric juice Thin, watery digestive fluid secreted by the gastric glands in the stomach and containing hydrochloric acid and enzymes, such as pepsin.

gastric phase The time period of gastric juice secretion that is stimulated by gastrin, caused by the presence of food in the stomach and by stomach stretching.

gastric pits Openings into gastric glands.

gastrulation Formation of a gastrula from a blastula; characterized by an invagination of the cell layers to form a caplike structure.

gene Unit of heredity located on a chromosome.

gene therapy Method of replacing a defective gene with a healthy gene.

genital Pertaining to the genitalia (internal and external organs of reproduction).

genital herpes Sexually transmitted disease caused by herpes simplex virus and sometimes accompanied by painful ulcers on the genitals.

genital wart Raised growth on the genitals due to a sexually transmitted disease caused by human papillomavirus.

genome All of the DNA in a cell of an organism.

genotype Combination of genes present within a zygote or within the cells of an individual.

germ cell An ovum or sperm cell, or a cell that gives rise to either ova or sperm.

gestation Period of development, from the start of the last menstrual cycle until birth; in humans, typically 280 days.

GH See *growth hormone.*

ghrelin Hormone manufactured primarily by stomach cells that causes growth hormone secretion, and stimulates fat storage and food intake.

gingivitis Inflammation of the gums.

gland Epithelial cell or group of epithelial cells that are specialized to secrete a substance.

glaucoma Increasing loss of field of vision, caused by blockage of the ducts that drain the aqueous humor, creating pressure buildup and nerve damage.

gliding joint Two bones with nearly flat surfaces joined together.

globulin Type of protein in blood plasma. There are alpha, beta, and gamma globulins.

glomerular capsule Double-walled capsule surrounding the glomerulus of a nephron through which glomerular filtrate passes to the proximal convoluted tubule.

glomerular filtrate Liquid that passes out of the glomerular capillaries in the kidney into the glomerular capsules.

glomerular filtration Process whereby blood pressure forces liquid through the glomerular capillaries in the kidney into the glomerular capsule.

glomerulus Cluster of capillaries surrounded by the glomerular capsule in a kidney nephron.

glottis Slitlike opening between the vocal cords.

glucagon Hormone secreted by the pancreatic islets that causes the release of glucose from glycogen.

glucocorticoid Any one of a group of hormones secreted by the adrenal cortex that influences carbohydrate, fat, and protein metabolism.

glucose Blood sugar that is broken down in cells to acquire energy for ATP production.

glycerol Three-carbon molecule that joins with fatty acids to form fat.

glycogen Polysaccharide that is the principal storage compound for sugar in animals.

glycosuria Presence of glucose in the urine, typically indicative of a kidney disease, diabetes mellitus, or other endocrine disorder.

Golgi apparatus Organelle that consists of concentrically folded membranes and that functions in the packaging and secretion of cellular products.

gonad Organ that produces sex cells: the ovary, which produces eggs, and the testis, which produces sperm.

gonadotropic hormone Type of hormone that regulates the activity of the ovaries and testes; principally, follicle-stimulating hormone and luteinizing hormone.

gonorrhea Sexually transmitted disease caused by the bacterium *Neisseria gonorrhoeae* that causes painful urination and swollen testes in men and is usually symptomless in women, but can cause inflammation of the cervix and uterine tubes.

gout Joint inflammation caused by accumulation of uric acid.

graded potential In a neuron, each of many small signals at a synapse; together their can be summed to bring the axon of the neuron to threshold and cause an action potential.

granular leukocyte White blood cell with prominent granules in the cytoplasm.

Graves disease Autoimmune disease with swollen throat due to an enlarged, hyperactive thyroid gland; patients often have protruding eyes and are underweight, hyperactive, and irritable.

gravitational equilibrium Maintenance of balance when the head and body are motionless.

gray matter Nonmyelinated nerve fibers in the central nervous system.

greater sciatic notch Indentation in the posterior coxal bone through which pass the blood vessels and the large sciatic nerve to the lower leg.

growth Increase in the number of cells and/or the size of these cells.

growth factor Chemical signal that regulates mitosis and differentiation of cells that have receptors for it; important in such processes as fetal development, tissue maintenance and repair, and hematopoiesis; sometimes a contributing factor in cancer.

growth hormone (GH) Hormone released by the anterior lobe of the pituitary gland that promotes the growth of the organism; also known as somatotropin.

gumma Tumor of rubbery consistency that is characteristic of the tertiary stage of syphilis.

gyrus Convoluted elevation or ridge (pl., *gyri*).

H

hair Consists of a cylindrical shaft and a root, which is contained in a flasklike depression (hair follicle) in the dermis and subcutaneous tissue. The base of the root is expanded into the hair bulb, which rests upon and encloses the hair papilla.

hair cell Mechanoreceptor in the inner ear that lies between the basilar membrane and the tectorial membrane and triggers action potentials in fibers of the auditory nerve.

hair follicle Tubelike depression in the skin in which a hair develops.

haploid Having the number of chromosomes of a gamete; half the number characteristic of body cells.

hard palate Anterior portion of the roof of the mouth that contains several bones.

haustra One of the pouches or small sacs into which the large intestine is divided.

hay fever Seasonal variety of allergic reaction to a specific allergen. Characterized by sudden attacks of sneezing, swelling of nasal mucosa, and often asthmatic symptoms.

hCG See *human chorionic gonadotropin.*

head Pertaining to the skeleton, an enlargement on the end of a bone.

heart Muscular organ located in the thoracic cavity that is responsible for maintenance of blood circulation.

heart attack See *myocardial infarction.*

heart block Impairment of conduction of an impulse in heart excitation.

heartburn Burning pain in the chest occurring when part of the stomach contents escapes into the esophagus.

heart murmur Clicking or swishing sounds often due to leaky valves.

heart valve Valve found between the chambers of the heart or between a chamber and a vessel leaving the heart.

heat of vaporization Amount of energy needed to turn water into steam.

helper T cell Secretes lymphokines, which stimulate all kinds of immune cells.

hematocrit Percent of the volume of whole blood that is composed of red blood cells.

hematopoiesis Formation of blood cells in red bone marrow.

heme Iron-containing portion of a hemoglobin molecule.

hemodialysis Mechanical way to remove nitrogenous wastes and to regulate blood pH when the kidneys are unable to perform these functions.

hemoglobin Pigment of red blood cells responsible for oxygen transport.

hemolysis Lysis of red blood cells that releases hemoglobin.

hemolytic anemia Insufficient number of red blood cells caused by an increased rate of red blood cell destruction.

hemolytic disease of the newborn Destruction of a fetus's red blood cells by the mother's immune system, caused by differing Rh factors between mother and fetus.

hemophilia Most common of the severe clotting disorders caused by the absence of a blood clotting factor.

hemorrhage Excessive blood loss from the body.

hemorrhagic anemia Condition in which the blood is deficient in red blood cells, in hemoglobin, or in total volume because of a hemorrhage.

hemorrhagic bleeding Escape of blood from blood vessels.

hemorrhoids Abnormally dilated blood vessels of the rectum.

hemostasis Stoppage of blood flow.

hepatic lobule One of the small masses of tissue of which the liver is made.

hepatic portal system Portal system that begins at the villi of the small intestine and ends at the liver.

hepatic portal vein Vein leading to the liver and formed by the merging blood vessels of the small intestine.

hepatitis Inflammation of the liver; often due to a serious infection by any of a number of viruses.

hernia Protrusion of an organ through an abnormal opening, such as the intestine through the abdominal wall near the scrotum (inguinal hernia) or the stomach through the diaphragm (hiatal hernia).

herniated disk Fibrous ring of cartilage between two vertebrae that has ruptured.

heterozygous Different alleles in a gene pair.

hexose Six-carbon sugar.

hinge joint Type of joint characterized by a convex surface of one bone fitting into a concave surface of another so that movement is confined to one place, such as in the knee or interphalangeal joint.

hippocampus An important part of the limbic system; involved in forming, storing, and processing memory.

hirsutism Excessive body and facial hair in women.

histamine Substance produced by basophil-derived mast cells in connective tissue that causes capillaries to dilate; causes many of the symptoms of allergy.

HLA (human leukocyte-associated) antigen Protein in a plasma membrane that identifies the cell as belonging to a particular in dividual and acts as an antigen in other organisms.

Hodgkin disease Cancer of the lymph glands that is normally localized in the neck region.

homeostasis Constancy of conditions, particularly the environment of body cells: constant temperature, blood pressure, pH, and other body conditions.

homozygous Having two identical genes on paired homologous chromosomes.

homozygous dominant Possessing two identical alleles, such as *AA*, for a particular trait.

homozygous recessive Possessing two identical alleles, such as *aa*, for a particular trait.

hormone Substance secreted by an endocrine gland that is transmitted in the blood or body fluids.

human chorionic gonadotropin (HCG) Hormone produced by the placenta that helps maintain pregnancy and is the basis for the pregnancy test.

human immunodeficiency virus (HIV) Virus responsible for AIDS.

humerus Heavy bone that extends from the scapula to the elbow.

Huntington disease Genetic disease marked by progressive deterioration of the nervous system due to deficiency of a neuro-transmitter.

hyaline cartilage Cartilage composed of very fine collagenous fibers and a matrix of a glassy, white, opaque appearance.

hydrocephalus Enlargement of the brain due to abnormal accumulation of cerebrospinal fluid.

hydrogen bond Weak attraction between a partially positive hydrogen and a partially negative oxygen or nitrogen some distance away; found in proteins and nucleic acids.

hydrolysis reaction Splitting of a bond by the addition of water.

hydrolytic enzyme Enzyme that catalyzes a reaction in which the substrate is broken down by the addition of water.

hydrophilic Type of molecule that interacts with water by dissolving in water and/or forming hydrogen bonds with water molecules.

hydrophobic Type of molecule that does not interact with water because it is nonpolar.

hyperglycemia Excessive glucose in the blood.

hyperopia Inability to see nearby objects.

hyperpnea Deep and labored breathing.

hypertension Elevated blood pressure, particularly the diastolic pressure.

hyperthermia Abnormally high body temperature.

hypertonic solution Solution that has a higher concentration of solute and a lower concentration of water than the cell.

hypertrophy Increase in the size of an organ, usually by an increase in the size of its cells.

hypodermic needle Slender, hollow instrument for introducing material into or removing material from or below the skin.

hypodermis Mainly composed of fat, this loose layer is directly beneath the dermis; subcutaneous.

hypoglycemia Insufficient amount of glucose in the blood.

hypothalamic-inhibiting hormone One of many hormones produced by the hypothalamus that inhibits the secretion of an anterior pituitary hormone.

hypothalamic-releasing hormone One of many hormones produced by the hypothalamus that stimulates the secretion of an anterior pituitary hormone.

hypothalamus Region of the brain; the floor of the third ventricle that helps maintain homeostasis.

hypothermia Abnormally low body temperature.

hypotonic solution Solution that has a lower concentration of solute and a higher concentration of water than the cell.

hypovolemic shock Form of shock caused by excessive blood, tissue fluid, or plasma loss.

hysterectomy Surgical removal of the uterus.

H zone In a sarcomere, the region in the center of an A band where only myosin filaments are found.

I

I band In a sarcomere, the region that is light colored and attached to the Z line; comprised of only thin filaments.

ileum Lower portion of the small intestine.

ilium One of the bones of a coxal bone or hipbone.

immediate allergic response Allergic response that occurs within seconds of contact with an allergen; caused by the attachment of the allergen to IgE antibodies.

immune system All the cells in the body that protect the body against foreign organisms and substances and also against cancerous cells.

immunity Resistance to disease-causing organisms.

immunization Strategy for achieving artificial immunity to the effects of specific disease-causing agents.

immunoglobulin (Ig) Globular plasma proteins that function as antibodies.

immunosuppressive Inactivating the immune system to prevent organ rejection, usually via a drug.

impetigo Contagious skin disease caused by bacteria in which vesicles erupt and crust over.

implantation Attachment and penetration of the embryo to the lining (endometrium) of the uterus.

incontinence Involuntary loss of urine.

incus The middle of three ossicles of the ear; serves with the malleus and the stapes to conduct vibrations from the tympanic membrane to the oval window of the inner ear.

infant respiratory distress syndrome Condition in newborns, especially premature ones, in which the lungs collapse because of a lack of surfactant lining the alveoli.

infarct An area of tissue death resulting from obstruction of blood flow.

inferior Situated below something else; pertaining to the lower surface of a part.

inferior vena cava Large vein that enters the right atrium from below and carries blood from the trunk and lower extremities.

infertility Inability to have as many children as desired.

inflammatory reaction Tissue response to injury that is characterized by dilation of blood vessels and accumulation of fluid in the affected region.

inguinal hernia A hernia in which part of the intestine protrudes into the inguinal canal.

inhibin Hormone secreted by seminiferous tubules that inhibits the release of follicle-stimulating hormone from the anterior pituitary.

inner cell mass An aggregation of cells at one pole of the blastocyte, which is destined to form the embryo proper.

inner ear Portion of the ear, consisting of a vestibule, semicircular canals, and the cochlea, where balance is maintained and sound is transmitted.

inorganic molecule Type of molecule that is not an organic molecule; not derived from a living organism.

insertion End of a muscle that is attached to a movable part.

inspiration The act of breathing in; inhalation.

inspiratory reserve volume Volume of air that can be forcibly inhaled after normal inhalation.

insulin Hormone produced by the pancreas that regulates glucose storage in the liver and glucose uptake by cells.

insulin-dependent diabetes mellitus (IDDM) Type 1 diabetes mellitus characterized by abrupt onset of symptoms, dependence on exogenous insulin, and a tendency to develop ketoacidosis.

insulin shock Severe hypoglycemia caused by excessive insulin; left untreated, can result in convulsions and progressive development of coma.

integration Summing up of excitatory and inhibitory signals by a neuron or by some part of the brain.

integument Pertaining to the skin.

integumentary system Pertaining to the skin and accessory organs.

interatrial septum Wall between the atria of the heart.

intercalated disk Membranous boundary between adjacent cardiac muscle cells.

interferon Protein formed by a cell infected with a virus that can increase the resistance of other cells to the virus.

interleukin Class of immune system chemicals (cytokines) having varied effects on the body.

intermediate filaments Tough protein strands that form cell-to-cell junctions.

internal respiration Exchange of oxygen and carbon dioxide between blood and tissue fluid.

interneuron Neuron found within the central nervous system that takes nerve impulses from one portion of the system to another.

interphase General term for that period of the cell cycle in which cell division is not occurring.

interstitial cell Hormone-secreting cell located between the seminiferous tubules of the testes.

intervertebral disk Layer of cartilage located between adjacent vertebrae.

intestinal phase The time period of gastric juice secretion that is stimulated by the presence of food in the intestine and by intestinal stretching.

intramembranous ossification Bone that forms from membranelike layers of primitive connective tissue.

intrapleural pressure Pressure found in the space between the parietal pleura and the visceral pleura in the thoracic cavity.

intrauterine device Solid object placed in the uterine cavity for purposes of contraception; IUD.

intrauterine insemination (IUI) Process of achieving pregnancy in which donated sperm are deposited in the uterus.

intrinsic factor Protein produced by the normal gastrointestinal mucosa that facilitates absorption of vitamin B_{12}.

inversion Movement of the foot so that the sole is turned inward.

in vitro fertilization (IVF) Process of achieving pregnancy in which eggs retrieved from an ovary are fertilized in a laboratory; viable embryos are then placed into the woman's uterus.

ion A charged atom.

ionic bond Chemical attraction between a positive ion and a negative ion.

ipsilateral Structures located on the same side of the body.

iris Muscular ring that surrounds the pupil and regulates the passage of light through this opening.

iron deficiency anemia Abnormally low amount of red blood cells or hemoglobin, due to a lack of iron in the diet.

ischemic heart disease Insufficient oxygen delivery to the heart, usually caused by partially blocked coronary arteries.

ischial spine Projection of the coxal bone into the pelvic cavity.

ischium Most inferior of the three bones that comprise the pelvic bone.

isotonic solution Solution that contains the same concentration of solutes and water as does the cell.

isotope One of two or more atoms with the same atomic number that differs in the number of neutrons and, therefore, in weight.

IVF See *in vitro fertilization.*

IVI See *intrauterine insemination.*

J

Jacobs syndrome Abnormal condition in a male characterized by a single X chromosome and two Y chromosomes.

jaundice Yellowish tint to the skin caused by an abnormal amount of bilirubin in the blood, indicating liver malfunction.

jejunum Middle portion of the small intestine.

joint Union of two or more bones; an articulation.

jugular Any of four veins that drain blood from the head and neck.

juxtaglomerular apparatus Structure located in the walls of arterioles near the glomerulus that regulates renal blood flow.

K

karyotype Arrangement of all the chromosomes from a nucleus by pairs in a fixed order.

keratin Insoluble protein present in the epidermis and in epidermal derivatives, such as hair and nails.

ketonuria Abnormal presence of acidic molecules called ketones in the urine.

kidney Organ in the urinary system that forms, concentrates, and excretes urine.

Klinefelter syndrome Condition caused by the inheritance of XXY chromosomes.

knee-jerk reflex Automatic, involuntary response initiated by tapping the ligaments just below the patella (kneecap).

kyphosis Increased roundness in the thoracic curvature of the spine; also called "hunchback."

L

labia majora Two large, hairy folds of skin of the female external genitalia.

labia minora Two small folds of skin inside the labia majora and encircling the clitoris.

lacrimal apparatus Structures that provide tears to wash the eye, consisting of the lacrimal gland and the lacrimal sac with its ducts.

lactation Production and secretion of milk by the mammary glands.

lacteal Lymph vessel in a villus of the wall of the small intestine.

lacuna Small pit or hollow cavity, as in bone or cartilage, where a cell or cells are located (pl., *lacunae*).

lambdoidal suture Line of junction between the occipital and parietal bones.

Langerhans cells Specialized epidermal cells that assist the immune system.

lanugo Short, fine hair that is present during the later portion of fetal development.

large intestine Portion of the digestive tract that extends from the small intestine to the anus.

laryngitis Inflammation of the larynx.

laryngopharynx Lower portion of the pharynx near the opening to the larynx.

larynx Structure that contains the vocal cords; also known as the voice box.

lateral Pertaining to the side.

lateral malleolus Rounded protuberance on the lateral surface of the ankle joint.

lens Clear, membranelike structure that is found in the eye behind the iris and that brings objects into focus.

leptin Peptide hormone produced by fat cells; plays a role in body weight regulation by acting on the hypothalamus to suppress appetite and burn fat stored in adipose tissue.

leukemia Form of cancer characterized by uncontrolled production of leukocytes in red bone marrow.

leukocytes Several types of colorless, nucleated blood cells that, among other functions, resist infection; white blood cells.

leukocytosis Abnormally large increase in the number of white blood cells.

leukopenia Abnormally low number of leukocytes in the blood.

LH See *luteinizing hormone.*

ligament Strong connective tissue that joins bone to bone.

limbic system System that involves many different centers of the brain and that is concerned with visceral functioning and emotional responses.

lipase Enzyme secreted by the pancreas that digests or breaks down fats.

lipid Group of organic compounds that are insoluble in water—notably, fats, oils, and steroids.

lipoproteins Any of a large class of proteins composed of protein and lipid.

liver Largest organ in the body, located in the abdominal cavity below the diaphragm; performs many vital functions that maintain homeostasis of blood.

local disease Disease that is confined to a particular area of the body.

locus The position in a chromosome of a particular gene or allele.

loop of Henle Portion of a nephron between the proximal and distal convoluted tubules; functions in water reabsorption.

loose connective tissue Tissue that is composed mainly of fibroblasts separated by collagenous and elastin fibers and that is found beneath epithelium.

lordosis Exaggerated lumbar curvature of the spine; also called "swayback."

lumen Space within a tubular structure such as a blood vessel or intestine.

lung Internal respiratory organ containing moist surfaces for gas exchange.

lung cancer Cancer neoplasm found in the airways or lobes of the lungs.

lunula Pale, half-moon–shaped area at the base of nails.

luteinizing hormone (LH) Hormone produced by the anterior pituitary that stimulates the development of the corpus luteum in females and the production of testosterone in males.

lymph Fluid having the same composition as tissue fluid; carried in lymph vessels.

lymphadenitis Infection of the lymph nodes.

lymphangitis Infection of the lymphatic vessels.

lymphatic organ Organ other than a lymphatic vessel that is part of the lymphatic system; includes lymph nodes, tonsils, spleen, thymus gland, and bone marrow.

lymphatic system Vascular system that takes up excess tissue fluid and transports it to the bloodstream.

lymphatic tissue Reticular connective tissue found in lymphatic organs such as lymph nodes and spleen.

lymphatic vessel Vessel that carries lymph.

lymphedema Edema due to faulty lymphatic drainage.

lymph node Mass of lymphatic tissue located along the course of a lymphatic vessel.

lymphocyte Type of white blood cell characterized by agranular cytoplasm; lymphocytes usually constitute 20–25% of the white cell count.

lymphoma Cancer of lymphatic tissue (reticular connective tissue).

lysis To split or divide.

lysosome Organelle involved in intracellular digestion; contains powerful digestive enzymes.

M

macromolecule Large molecule composed of smaller molecules.

macrophage Enlarged monocyte that ingests foreign material and cellular debris.

macular degeneration Disruption of the macula lutea, a central part of the retina, causing blurred vision.

major histocompatibility complex A group of genes that function especially in determining the tissue compatibility antigen molecules found on cell surfaces; abbreviated *MHC.*

malignant The power to threaten life; cancerous.

malleolus Rounded projection from a bone.

malleus First of three ossicles of the ear; serves with the incus and stapes to conduct vibrations from the tympanic membrane to the oval window of the inner ear.

maltase Enzyme that catalyzes conversion of maltose into glucose.

mammary gland Milk-secreting gland that develops within the breast in pregnancy and lactation; only minimally developed in the breast of a nonpregnant or nonlactating woman.

mast cell Cell to which antibodies, formed in response to allergens, attach, bursting the cell and releasing allergy mediators, which cause symptoms.

mastoiditis Inflammation of the mastoid sinuses of the skull.

matrix Secreted basic material or medium of biological structures, such as the matrix of cartilage or bone.

mechanoreceptor A tactile or stretch receptor that responds to a mechanical stimulus such as touch or stretch.

medial Toward or near the midline.

mediastinum Tissue mass located between the lungs.

medulla oblongata Lowest portion of the brain; concerned with the control of internal organs.

medullary cavity Within the diaphysis of a long bone, cavity occupied by yellow marrow.

megakaryocyte Large bone marrow cell that gives rise to blood platelets.

meiosis Type of cell division in which the daughter cells have 23 chromosomes; occurs during spermatogenesis and oogenesis.

melanin Pigment found in the skin and hair of humans that is responsible for coloration.

melanocyte Melanin-producing cell.

melanocyte-stimulating hormone (MSH) Substance that causes melanocytes to secrete melanin in lower vertebrates.

melanoma Deadly form of skin cancer that begins in the melanocytes, pigment cells present in the epidermis.

melatonin Hormone, secreted by the pineal gland, that is involved in biorhythms.

memory B cell Cells derived from B lymphocytes that remain within the body for some time and account for active immunity.

meninges Protective membranous coverings around the brain and spinal cord (sing., *meninx*).

meningitis Inflammation of the meninges around the brain and spinal cord usually caused by bacterial infection.

meniscus Piece of fibrocartilage that separates the surfaces of bones in the knee (pl., *menisci*).

menopause Termination of the menstrual cycle in older women.

menses (menstruation) Loss of blood and tissue from the uterus.

menstrual cycle Female reproductive cycle characterized by regularly occurring changes in the uterine lining.

Merkel cells Cells that signal the brain that an object has touched the skin; also called *tactile cells.*

mesentery Fold of peritoneal membrane that attaches an abdominal organ to the abdominal wall.

mesoderm The middle of the three primary germ layers of an embryo; source of bone, muscle, connective tissue, and dermis.

messenger RNA (mRNA) Nucleic acid (ribonucleic acid) complementary to genetic DNA; has codons that direct cell protein synthesis at the ribosomes.

metabolism All of the chemical changes that occur within cells.

metacarpal Bone of the hand between the wrist and the finger bones.

metaphase Stage in mitosis when chromosomes align in the center of the cell.

metastasis Mechanism of cancer spread in which cancer cells break off from the initial tumor, enter the blood vessels or lymphatic vessels, and start new tumors elsewhere in the body.

metatarsal bones Bones found in the foot between the ankle and the toes.

microtubule Hollow rod of the protein tubulin in the cytoplasm.

microvillus Cylindrical process that extends from some epithelial cell membranes and increases the membrane surface area (pl., *microvilli*).

micturition Emptying of the bladder; urination.

midbrain Small region of the brain stem located between the forebrain and the hindbrain; contains tracts that conduct impulses to and from the higher parts of the brain.

middle ear Portion of the ear consisting of the tympanic membrane, the oval and round windows, and the ossicles, where sound is amplified.

mineral Inorganic substance; certain minerals must be in the diet for normal metabolic functioning of cells.

mineralocorticoid Hormones the adrenal cortex secretes that influence the concentrations of electrolytes in body fluids.

mitochondrion Organelle in which cellular respiration produces the energy molecule ATP.

mitosis Type of cell division in which two daughter cells receive 46 chromosomes; occurs during growth and repair.

mitral valve Valve in the heart that controls blood flow between the left atrium and the left ventricle; prevents the blood in the ventricle from returning to the atrium (also called *bicuspid valve*).

mixed nerve Nerve that contains both the long dendrites of sensory neurons and the long axons of motor neurons.

molar (tooth) Three posterior pairs of teeth in each human jaw with a rounded or flattened surface adapted for grinding; found behind the incisors and canines.

mole Raised growth on the skin due to an overgrowth of melanocytes.

molecule Smallest quantity of a substance that retains its chemical properties.

monoclonal antibody Antibody of one type that is produced by cells derived from a lymphocyte that has fused with a cancer cell.

monocyte Type of white blood cell that functions as a phagocyte.

mononucleosis Viral disease characterized by an increase in atypical lymphocytes in the blood.

monosaccharide Simple sugar; a carbohydrate that cannot be decomposed by hydrolysis.

monosomy Having one less than the diploid number of chromosomes.

mons pubis The rounded fleshy prominence over the pubic symphysis.

morphogenesis Establishment of shape and structure in an organism.

morula Early stage in development in which the embryo consists of a mass of cells, often spherical.

motor neuron Neuron that takes nerve impulses from the central nervous system to an effector; also known as an efferent neuron.

motor unit Motor neuron and all the muscle fibers it innervates.

mouth Opening through which food enters the body.

MS See *multiple sclerosis*.

mucociliary escalator Protective mechanism involving cilia of the mucous membranes of the respiratory system; transports mucus containing inhaled pollutants to the pharynx where the mucus can be swallowed or expectorated.

mucosa A membrane rich in mucous glands that lines body passages and cavities that communicate directly or indirectly with the exterior, such as the digestive, respiratory, and genitourinary tracts.

mucous membrane Membrane lining a cavity or tube that opens to the outside of the body; also called mucosa.

multiple sclerosis (MS) Disease in which the outer, myelin layer of nerve fiber insulation becomes scarred, interfering with normal conduction of nerve impulses.

multipotent stem cell An unspecialized parent cell that gives rise to differentiated cells.

muscle fiber Muscle cell.

muscle twitch Contraction of a whole muscle in response to a single stimulus.

muscular dystrophy Progressive muscle weakness and atrophy caused by deficient dystrophin protein.

muscularis The smooth muscular layer of the wall of contractile organs, such as the digestive organs and the bladder.

muscular tissue Major type of tissue that is adapted to contract; the three kinds of muscle are cardiac, smooth, and skeletal.

mutation Alteration of the DNA in a tissue cell.

myalgia Pain in a muscle or muscles.

myasthenia gravis Muscle weakness due to an inability to respond to the neurotransmitter acetylcholine.

myelin sheath Fatty plasma membranes of Schwann cells that cover long neuron fibers and give them a white, glistening appearance.

myocardial infarction Damage to the myocardium due to blocked circulation in the coronary arteries; a heart attack.

myocardium Heart (cardiac) muscle consisting of striated muscle cells that interlock.

myofibril Contractile portion of muscle fibers.

myoglobin Pigmented compound in muscle tissue that stores oxygen.

myopia Inability to see distant objects clearly.

myosin Thick myofilament in myofibrils that is made of protein and is capable of breaking down ATP; see also *actin*.

myxedema Condition resulting from a deficiency of thyroid hormone in an adult.

N

nasal cavity Space within the nose.

nasopharynx Portion of the pharynx associated with the nasal cavity.

natural killer cell (NK) Lymphocyte that causes an infected or cancerous cell to burst.

negative feedback Mechanism that is activated by a surplus imbalance and acts to correct it by stopping the process that brought about the surplus.

neoplasm New tissue form that may be harmless, or a precursor to cancer.

nephron Anatomical and functional unit of the kidney; kidney tubule.

nerve Bundle of long nerve fibers that run to and/or from the central nervous system.

nerve deafness Hearing impairment that usually occurs when the cilia on the sensory receptors within the cochlea have worn away.

nerve fiber Thin process of a neuron (i.e., axon, dendrite).

nerve impulse Change in polarity that flows along the membrane of a nerve fiber.

nervous tissue Tissue of the nervous system, consisting significantly of neurons and neuroglia.

neuroglia Nonconducting nerve cells that are intimately associated with neurons and function in a supportive capacity.

neurolemmocyte Type of neuroglial cell that forms a myelin sheath around axons; also called a Schwann cell.

neuromuscular junction Junction between a neuron and a muscle fiber.

neuron Nerve cell that characteristically has three parts: dendrite, cell body, and axon.

neurotransmitter Chemical made at the ends of axons that is responsible for transmission across a synapse.

neutral fat A triglyceride.

neutron Electrically neutral particle in an atomic nucleus.

neutrophil Phagocytic white blood cell that normally constitutes 60–70% of the white blood cell count.

nociceptor Pain receptor.

node of Ranvier Gap in the myelin sheath of a nerve fiber.

nondisjunction Failure of the chromosomes (or chromatids) to separate during meiosis.

noninsulin-dependent diabetes mellitus NIDDM Type 2 diabetes mellitus, usually characterized by gradual onset with minimal or no symptoms of metabolic disturbance and no required exogenous insulin to prevent ketonuria and ketoacidosis.

nonpolar covalent bond Bond created when electrons shared by adjacent atoms are shared equally.

norepinephrine Hormone secreted by the adrenal medulla to help initiate the "fight-or-flight" reaction.

nose Specialized structure on the face that serves as the sense organ of smell and as part of the respiratory system.

nostril One of the external orifices of the nose.

nuclear envelope Membrane surrounding the cell nucleus and separating it from the cytoplasm.

nuclear pore Opening in the nuclear envelope.

nucleic acid Large organic molecule found in the nucleus (DNA and RNA) and in the cytoplasm (RNA).

nucleolus Organelle found inside the nucleus and composed largely of RNA for ribosome formation (pl., *nucleoli*).

nucleotide Building block of a nucleic acid molecule, consisting of a sugar, a nitrogen-containing base, and a phosphate group.

nucleus Large organelle that contains the chromosomes and acts as a cell control center.

O

obesity Excess adipose tissue; exceeding desirable weight by more than 20%.

obstructive sleep apnea Sleep pattern characterized by interruption of breathing; caused by obstruction of the upper airway by weakened or excessive tissues of the pharynx.

occipital condyle One of two processes on the lateral portions of the occipital bone; for articulation with the atlas.

occipital lobe Area of the cerebrum responsible for vision, visual images, and other sensory experiences.

occluded coronary arteries Blocked blood vessels that serve the needs of the heart.

oil Substance, usually of plant origin and liquid at room temperature, formed when a glycerol molecule reacts with three fatty acid molecules.

olfactory cell Cell located high in the nasal cavity that bears receptor sites on cilia for various chemicals and whose stimulation results in smell.

oocyte Developing female gamete.

oogenesis Production of eggs in females by the process of meiosis and maturation.

oophorectomy Surgical removal of one or both ovaries.

opportunistic infection Disease that arises in the presence of a severely impaired immune system.

optic chiasma Structure found on the inferior surface of the brain where optic nerves meet and cross.

optic nerve Nerve composed of the ganglion cell fibers that form the innermost layer of the retina.

organ Structure consisting of a group of tissues that perform a specialized function; a component of an organ system.

organelle Part of a cell that performs a specialized function.

organic molecule Carbon-containing molecule.

organism Individual living thing.

organ of Corti See *spiral organ*.

organ system Group of related organs working together.

organ transplantation Replacement of a diseased or defective organ with a healthy one.

orgasm Physical and emotional climax during sexual intercourse; results in ejaculation in the male.

origin End of a muscle that is attached to a relatively immovable part.

oropharynx Portion of the pharynx in the posterior part of the mouth.

osmosis Movement of water from an area of greater concentration to an area of lesser concentration across the plasma membrane.

osmotic pressure The amount of pressure needed to stop osmosis; the potential pressure of a solution caused by nondiffusible solute particles in the solution.

ossicles Tiny bones in the middle ear; malleus (hammer), incus (anvil), and stapes (stirrup).

ossification Formation of bone.

osteoarthritis Disintegration of the cartilage between bones at a synovial joint.

osteoblast Bone-forming cell.

osteoclast Cell that causes the erosion of bone.

osteon Cylinder-shaped units that comprise compact bone.

osteocyte Mature bone cell.

osteoporosis Weakening of bones due to decreased bone mass.

osteoprogenitor cells Cells found on or near all of the free surfaces of bone, which undergo division and transform into osteoblasts.

otitis media Inflammation of the middle ear.

otolith Granule that lies above, and whose movement stimulates, ciliated cells in the utricle and saccule.

otosclerosis Overgrowth of bone that causes the stapes to adhere to the oval window, resulting in conductive deafness.

ototoxic Damaging to any of the elements of hearing or balance.

outer ear Portion of the ear consisting of the pinna and the auditory canal.

oval window Membrane-covered opening between the stapes and the inner ear.

ovarian cancer Cancer of an ovary.

ovariohysterectomy Surgical removal of the ovaries and uterus.

ovary Female gonad; the organ that produces eggs, estrogen, and progesterone.

ovulation Discharge of a mature egg from the follicle within the ovary.

ovum (pleural, *ova*) Female gamete, contributing most of the cytoplasm of the zygote.

oxygen debt Amount of oxygen required after anaerobic exercise to replace that used during the exercise.

oxygen deficit Amount of oxygen needed to metabolize the lactic acid that accumulates during vigorous exercise.

oxyhemoglobin Hemoglobin bound to oxygen in a loose, reversible way.

oxytocin Hormone released by the posterior pituitary that causes contraction of uterus and milk letdown.

P

pacemaker Small region of neuromuscular tissue that initiates the heartbeat; also called the *SA node*.

pain receptors Receptors that respond to tissue damage or oxygen deprivation.

palatine tonsil Either of two small, almond-shaped masses located on either side of the oropharynx, composed mainly of lymphatic tissue; believed to act as sources of bacteria-killing phagocytes.

pallidotomy Surgical destruction of brain tissue for the treatment of involuntary movements in Parkinson's disease.

pancreas Endocrine organ located near the stomach that secretes digestive enzymes into the duodenum and produces hormones, notably insulin.

pancreatic amylase Enzyme that digests starch to maltose.

pancreatic islets (of Langerhans) Distinctive groups of cells within the pancreas that secrete insulin and glucagon.

papillary muscle Muscle that extends inward from the ventricular walls of the heart and to which the chordae tendineae attach.

Pap smear Sample of cells removed from the tip of the cervix and then stained and examined microscopically.

paranasal sinus One of several air-filled cavities in the maxillary, frontal, sphenoid, and ethmoid bones that is lined with mucous membrane and drains into the nasal cavity.

paraplegia Paralysis of the lower body and legs, due to injury to the spinal cord between vertebrae T1 and L2.

parasympathetic division Portion of the autonomic nervous system that usually promotes those activities associated with a normal state.

parathyroid gland One of four small endocrine glands embedded in the posterior portion of the thyroid gland.

parathyroid hormone (PTH) Hormone secreted by the parathyroid glands that raises the blood calcium level primarily by stimulating reabsorption of bone.

parental cell Cell that divides so as to form daughter cells.

parietal Pertaining to the wall of a cavity.

parietal lobe Area of the cerebrum responsible for sensations involving temperature, touch, pressure, pain, and speech.

parietal pericardium Outer layer of the two layers of the serous pericardium, lining the fibrous pericardium.

parietal peritoneum Lines the abdominal and pelvic walls and the inferior surface of the thoracic diaphragm.

parietal pleura Membrane that lines the inner wall of the thoracic cavity.

parietal serous membrane Serous membrane that covers the inner body wall.

Parkinson's disease Progressive deterioration of the central nervous system due to a deficiency in the neurotransmitter dopamine; also called paralysis agitans.

parturition Processes that lead to and include the birth of a human and the expulsion of the extraembryonic membranes through the terminal portion of the female reproductive tract.

passive immunity Protection against infection acquired by transfer of antibodies to a susceptible individual.

patella Bone of the kneecap.

pathogen Disease causing agents, such as bacteria and viruses.

pathologist Person trained in knowledge of diseases and their symptoms, allowing for diagnosis of disease.

pectoral girdle Portion of the skeleton that provides support and attachment for the upper limbs.

pedigree Chart showing the relationships of relatives and which ones have a particular trait.

pelvic cavity Hollow place within the ring formed by the sacrum and coxal bones.

pelvic girdle Portion of the skeleton to which the lower limbs are attached.

pelvic inflammatory disease (PID) Latent infection of gonorrhea or chlamydia in the vasa deferentia or uterine tubes.

penis Male excretory and copulatory organ.

pentose Five-carbon sugar; deoxyribose is the pentose sugar found in DNA; ribose is a pentose sugar found in RNA.

pepsin Protein-digesting enzyme produced by the stomach.

peptidase Enzyme that catalyzes the breakdown of polypeptides.

peptide bond Bond that joins two amino acids.

peptide hormone Type of hormone that is a protein, a peptide, or derived from an amino acid.

perforin Protein released by cytotoxic T cells that attaches to an antigen.

pericardial cavity Cavity around the heart created by the pericardial sac.

pericardial fluid Fluid found in small amounts in the potential space between the parietal and visceral laminae of the serous pericardium.

pericarditis Inflammation of the pericardium.

pericardium Protective serous membrane that surrounds the heart.

periodontitis Inflammation of the periodontal membrane that lines tooth sockets, causing loss of bone and loosening of teeth.

periosteum Fibrous connective tissue covering the surface of bone.

peripheral Situated away from the center of the body or an organ.

peripheral nervous system (PNS) Nerves and ganglia of the nervous system that lie outside the brain and spinal cord.

peristalsis Rhythmical contraction that moves the contents along in tubular organs, such as the digestive tract.

peritoneum Serous membrane that lines the abdominopelvic cavity and encloses the abdominal viscera.

peritonitis Generalized infection of the lining of the abdominal cavity.

peritubular capillary network Capillary network that surrounds a nephron and functions in reabsorption during urine formation.

pernicious anemia Insufficiency of mature red blood cells, due to poor absorption of vitamin B_{12}.

peroxisomes Small vessicles within a cell which contain enzymes for the break down of fatty acids and hydrogen peroxide.

Peyer patches Lymphatic organs located in small intestine.

phagocytosis Taking in of bacteria and/or debris by engulfing; also called cell eating.

phalanges Bones of the fingers and thumb in the hand and of the toes in the foot (sing., *phalanx*).

pharyngeal tonsil Diffuse lymphatic tissue and follicles in the roof and posterior wall of the nasopharynx.

pharynx Common passageway for both food intake and air movement; the throat.

phenotype Physical manifestation of a trait that results from the action of a particular set of genes.

phlebitis Inflammation of a vein.

phospholipid Lipid that contains two fatty acid molecules and a phosphate group combined with a glycerol molecule.

photoreceptors Cells in the retina of the eye that can respond to light stimuli.

pH scale Measure of the hydrogen ion concentration; any pH below 7 is acidic, and any pH above 7 is basic.

physiology Branch of science dealing with the study of body functions.

pia mater Innermost meningeal layer that is in direct contact with the brain and spinal cord.

pineal gland Small endocrine gland, located in the third ventricle of the brain, that secretes melatonin and is involved in biorhythms.

pinna Outer, funnel-like structure of the ear that picks up sound waves.

pinocytosis Form of endocytosis in which the cell takes in liquids.

pituitary dwarfism Condition in which a person has normal proportions but small stature; caused by inadequate growth hormone.

pituitary gland (hypophysis) Endocrine gland attached to the base of the brain that consists of anterior and posterior lobes.

pivot joint The end of a bone moving within a ring formed by another bone and connective tissue.

placenta Structure formed from the chorion and uterine tissue, through which nutrient and waste exchange occurs for the embryo and later the fetus.

placental membrane Semipermeable membrane that separates the fetal from the maternal blood in the placenta.

plaque Accumulation of soft masses of fatty material, particularly cholesterol, beneath the inner linings of arteries.

plasma Liquid portion of blood.

plasma cell Cell derived from a B lymphocyte that is specialized to mass-produce antibodies.

plasma membrane Membrane that surrounds the cytoplasm of cells and regulates the passage of molecules into and out of the cell.

platelet Cell-like disks formed from fragmentation of megakaryocytes that initiate blood clotting.

platelet plug platelets that stick and cling to each other in order to seal a break in a blood vessel wall.

pleura (pl., *Pleurae*) Serous membrane that covers the lungs and lines the walls of the chest and the diaphragm.

pleurisy Inflammation of the pleura.

pneumonectomy Surgical removal of all or part of a lung.

pneumonia Infection of the lungs that causes alveoli to fill with mucus and pus.

PNS See *peripheral nervous system*.

polar body Small, nonfunctional cell that is a product of meiosis in the female.

polar covalent bond Bond created when electrons shared by adjacent atoms are shared unequally.

polar molecule Combination of atoms in which the electrical charge is not distributed symmetrically.

pole During cell division, the elongated side of a cell.

polycythemia Abnormally high number of red blood cells in the blood.

polydipsia Chronic, excessive intake of water.

polyp Small, abnormal growth on any mucous membrane, such as in the large intestine.

polypeptide A compound formed by the union of many amino acid molecules.

polyphagia Excessive eating.

polyribosome String of ribosomes simultaneously translating regions of the same mRNA strand during protein synthesis.

polysaccharide Carbohydrate composed of many bonded glucose units—for example, glycogen.

polyuria Excessive output of urine.

poly-X female Female who has more than two X chromosomes.

pons Portion of the brain stem above the medulla oblongata and below the midbrain; assists the medulla oblongata in regulating the breathing rate.

portal system A system of veins that begins and ends in capillaries.

portal triad Grouping of the tributaries of the hepatic artery, vein, and bile duct at the angles of the lobules of the liver.

positive feedback Process by which changes cause more changes of a similar type, producing unstable conditions.

posterior Toward the back; opposite of *anterior*.

posterior pituitary (neurohypophysis) Portion of the pituitary gland connected by a stalk to the hypothalamus.

posterior (dorsal)-root ganglion Mass of sensory neuron cell bodies located in the dorsal root of a spinal nerve.

postganglionic fiber In the autonomic nervous system, the axon that leaves, rather than goes to, a ganglion.

power stroke During muscle contraction, the action that pulls thin filaments to the center of the sarcomere.

prefrontal area Association area in the frontal lobe that receives information from other association areas and uses it to reason and plan actions.

preganglionic fiber In the autonomic nervous system, the axon that goes to, rather than leaves, a ganglion.

premature birth Child born before full term and weighing 5 pounds, 8 ounces, or less.

presbycusis Loss of hearing that accompanies old age.

presbyopia A visual condition that occurs with age and is caused by loss of lens elasticity; results in defective accommodation and inability to focus sharply for near vision.

primary germ layers Three layers (endoderm, mesoderm, and ectoderm) of embryonic cells that develop into specific tissues and organs.

primary motor area Area in the frontal lobe where voluntary commands begin; each section controls a part of the body.

primary respiratory center Group of neurons in the medulla oblongata that regulates respiration.

primary somatosensory area Area posterior to the central sulcus where sensory information arrives from the skin and skeletal muscles.

primary spermatocyte Cell dividing into two secondary spermatocytes.

prime mover For muscle action, that muscle or group of muscles that carries out the primary action; for example, the biceps brachii is the prime mover for elbow flexion.

prion Proteinaceous infectious particles; thought to cause neurological disease.

PRL See *prolactin*.

progesterone Female sex hormone secreted by the ovaries that, along with estrogen, promotes the development and maintenance of the primary and secondary female sex characteristics.

prolactin (PRL) Hormone secreted by the anterior pituitary that stimulates milk production in the mammary glands; also known as lactogenic hormone.

promoter Any influence that causes a mutated cancer cell to start growing in an uncontrolled manner.

pronation Rotation of the forearm so that the palm faces backward.

prophase Stage of mitosis when chromosomes become visible.

proprioceptor Sensory receptor that assists the brain in knowing the position of the limbs.

prostaglandins Hormones that have various and powerful effects, often within the cells that produce them.

prostate gland Gland situated around the base of the male urethra; secretes a viscous alkaline fluid that is a major constituent of the ejaculatory fluid.

protease Enzyme that digests protein.

protein Macromolecule composed of amino acids.

prothrombin Plasma protein made by the liver that must be present in blood before clotting can occur.

prothrombin activator Enzyme that catalyzes the transformation of the precursor prothrombin to the active enzyme thrombin.

proton Positively charged particle in an atomic nucleus.

proximal Closer to the midline or origin; opposite of *distal*.

proximal convoluted tubule Highly coiled region of a nephron near the glomerular capsule.

pseudostratified columnar Appearance of layering in some epithelial cells when, actually, each cell touches a baseline and true layers do not exist.

psoriasis Common chronic, inherited skin disease in which red patches are covered with scales; occurs most often on the elbows, knees, scalp, and trunk.

PTH See *parathyroid hormone*.

puberty Stage of development in which the reproductive organs become functional.

pubic symphysis Slightly movable cartilaginous joint between the anterior surfaces of the hip bones.

pubis Anterior-most bone of the three bones that comprise the pelvic bone.

pulmonary artery Blood vessel that takes blood away from the heart to the lungs.

pulmonary capillaries Capillaries found between pulmonary arterioles and pulmonary venules.

pulmonary circuit Path of blood through vessels that take O_2-poor blood to and O_2-rich blood away from the lungs.

pulmonary edema Excessive fluid in the lungs caused by congestive heart failure.

pulmonary embolism Blockage of a pulmonary artery by a blood clot that commonly originates in a vein of the lower legs.

pulmonary fibrosis Accumulation of fibrous connective tissue in the lungs; caused by inhaling irritating particles, such as silica, coal dust, or asbestos.

pulmonary tuberculosis Tuberculosis of the lungs, caused by the tubercle bacillus.

pulmonary vein Blood vessel that takes blood away from the lungs to the heart.

pulse Vibration felt in arterial walls due to expansion of the aorta following ventricular contraction.

pupil Opening in the center of the iris that controls the amount of light entering the eye.

Purkinje fiber Specialized muscle fiber that conducts the cardiac impulse from the AV bundle into the ventricular walls.

pus Thick, yellowish fluid composed of dead phagocytes, dead tissue, and bacteria.

pyelonephritis Inflammation of the kidney; due to bacterial infection.

pyloric region Distal-most portion of the stomach; joins the duodenum of the small intestine.

Q

quadriplegia Paralysis of the entire body and all four limbs, due to injury to the spinal cord between vertebrae C4 and T1.

R

radioactive isotope Atom whose nucleus undergoes degeneration and in the process gives off radiation.

radius Elongated bone located on the thumb side of the lower arm.

receptor potential In sensory receptors, graded potentials that can be summed together to signal other neurons to create an action potential.

recessive allele Hereditary factor that expresses itself only when two copies are present in the genotype.

recruitment Increase in the number of motor units activated as intensity of stimulation increases.

rectum Terminal portion of the intestine.

red blood cell See *erythrocyte*.

red bone marrow Blood cell-forming tissue located in spaces within certain bones.

reduced hemoglobin (HHb) Hemoglobin that is carrying hydrogen ions.

reduction To align the ends of a fractured bone so that proper healing can occur.

referred pain Pain perceived as having come from a site other than that of its actual origin.

reflex Automatic, involuntary response of an organism to a stimulus.

refractory period During an action potential, the time during which a second stimulus won't cause a second action potential.

renal artery Vessel that originates from the aorta and delivers blood to the kidney.

renal corpuscle The part of a nephron that consists of the glomerulus and surrounding Bowman's capsule.

renal cortex Outer, primarily vascular portion of the kidney.

renal medulla Inner portion of the kidney, including the renal pyramids.

renal pelvis Inner cavity of the kidney formed by the expanded ureter and into which the collecting ducts open.

renal tubule The part of a nephron that leads away from a glomerulus; is made up of a proximal convoluted tubule, loop of Henle, and distal convoluted tubule, and empties into a collecting tubule.

renal vein Vessel that takes blood from the kidney to the inferior vena cava.

renin Secretion from the kidney that activates angiotensinogen to angiotensin I.

replacement reactions Chemical reaction involving both decomposition and synthesis, in which two new compounds are formed.

replication Production of an exact copy of a DNA sequence.

repolarization Recovery of a neuron's polarity to the resting potential after the neuron ceases transmitting impulses.

residual volume Volume of air that remains in the lungs after normal exhalation.

respiration Transport and exchange of gases between the atmosphere and the cells via the lungs and blood vessels.

respiratory membrane Alveolar wall plus the capillary wall, across which gas exchange occurs.

resting potential Potential energy of a resting neuron, created by separating unlike charges across the neuron cell membrane.

reticular connective tissue Collagen fibers that form the framework for lymphoid organs.

reticular fibers Thin, highly branched, collagenous fibers that form delicate supporting networks.

reticular formation Nerve cells and fibers situated primarily in the brain stem and functioning upon stimulation, especially in arousal.

retina Innermost layer of the eyeball that contains the rod cells and cones cells.

retinal A form of vitamin A.

retrovirus Any of the family of single-stranded RNA viruses.

rheumatoid arthritis Persistent inflamation of synovial joints, often causing cartilage destruction, bone erosion, and joint deformities.

Rh factor Type of antigen on red blood cells.

rhodopsin Light-sensitive biochemical in the rod cells of the retina; visual purple.

rib cage Bony framework of the thoracic cavity; created by the thoracic vertebrae, ribs, costal cartilages, and sternum.

ribonucleic acid (RNA) Nucleic acid that helps DNA in protein synthesis.

ribosomal RNA (rRNA) RNA (ribonucleic acid) occurring in ribosomes, structures involved in protein synthesis.

ribosome Minute particle, found attached to the endoplasmic reticulum or loose in the cytoplasm, that is the site of protein synthesis.

rickets Defective mineralization of the skeleton, usually due to inadequate vitamin D in the body.

RNA See *ribonucleic acid*.

rod cell Dim-light receptor in the retina of the eye that detects motion but not color.

rotation Movement of a bone around its own longitudinal axis.

rotational equilibrium Maintenance of balance when the head and body are suddenly moved or rotated.

rough ER Endoplasmic reticulum that is studded with ribosomes on the side of the membrane that faces the cytoplasm. See *smooth ER*.

round window Membrane-covered opening between the inner ear and the middle ear.

rubella An acute, infectious disease affecting the respiratory tract in children and nonimmune young adults; characterized by a slight cold, sore throat, and fever, and the appearance of a fine, pink rash.

rugae Deep folds, as in the wall of the stomach.

S

saccule Saclike cavity of the inner ear that contains receptors for gravitational equilibrium.

sacroiliac joint Connection between the coxal bone and the sacrum.

sacrum Bone consisting of five fused vertebrae that form the posterior wall of the pelvic girdle.

saddle joint Two bones joined, having convex and concave surfaces that are complementary.

sagittal plane Plane or section that divides a structure into right and left portions.

sagittal suture Line of junction between the two parietal bones in the cranium.

salivary amylase In humans, enzyme in saliva that digests starch to maltose.

salivary gland Gland associated with the mouth, secretes saliva.

salpingectomy Surgical removal of the uterine tubes.

salt Compound produced by a reaction between an acid and a base.

saltatory conduction Conduction of an action potential down the length of a myelinated axon, by jumping from one node of Ranvier to the next.

SA node See *sinoatrial node*.

sarcoma Cancer that arises in striated muscle, cartilage, or bone.

sarcomere Structural and functional unit of a myofibril.

saturated fatty acid Organic molecule that includes a fatty acid molecule which lacks double bonds between the atoms of its carbon chain.

scapula Large bone in the posterior shoulder area (pl., *scapulae*).

Schwann cell Cell that surrounds a fiber of a peripheral nerve and forms the neurilemmal sheath and myelin.

SCID See *severe combined immunodeficiency disease*.

sclera White, fibrous outer layer of the eyeball.

scoliosis Abnormal lateral (side-to-side) curvature of the vertebral column.

scrotum Pouch of skin that encloses the testes.

sebaceous gland Gland of the skin that secretes sebum.

sebum Oily secretion of the sebaceous glands.

secondary sex characteristic Trait that is sometimes helpful but not absolutely necessary for reproduction and is maintained by the sex hormones in males and females.

secondary spermatocyte One of the two cells into which a primary spermatocyte divides, and which in turn gives rise to spermatids.

secretin Intestinal hormone that stimulates secretion by the pancreas and liver.

selectively permeable Membrane that allows some molecules through but not others.

sella turcica Saddle-shaped area of the sphenoid bone; houses the pituitary gland.

semen Sperm-containing secretion of males; seminal fluid plus sperm.

semicircular canal Tubular structure within the inner ear with ampullae that contain the receptors responsible for the sense of rotational equilibrium.

semilunar valve Valve resembling a half-moon located between the ventricles and their attached vessels.

seminal fluid Sperm-containing secretion of males; also called semen.

seminal vesicle Convoluted, saclike structure attached to vas deferens near the base of the bladder in males; contributes to seminal fluid.

seminiferous tubule Highly coiled duct within the male testes that produces and transports sperm.

sensory neuron Neuron that takes the nerve impulse to the central nervous system; also known as an afferent neuron.

sensory receptor Sensory structure specialized to receive information from the environment and to generate nerve impulses.

serositis Inflammation of a serous membrane.

serous fluid Fluid found in the space between visceral and parietal pleurae.

serous membrane Membrane that covers internal organs and lines cavities lacking an opening to the outside of the body; also called serosa.

serum Light-yellow liquid left after clotting of the blood.

severe combined immunodeficiency disease (SCID) Congenital illness in which both antibody-and cell-mediated immunity are lacking or inadequate.

sex chromosome Chromosome responsible for the development of characteristics associated with maleness or femaleness; an X or Y chromosome.

sex-linked Allele that occurs on the sex chromosomes but may control a trait that has nothing to do with the sex characteristics of an individual.

sexually transmitted infection A bacterial, viral, or parasitic infection resulting from sexual intercourse.

shock A state of profound depression of the vital processes of the body that is characterized by pallor, rapid but weak pulse, rapid and shallow respiration, reduced total blood volume, and low blood pressure; usually is caused by severe allergic reactions, injuries, hemorrhage, burns, or major surgery.

sickle-cell disease Hereditary disease in which red blood cells are narrow and curved so that they are unable to pass through capillaries and are destroyed, causing chronic anemia.

sigmoid colon Portion of the large intestine that is S-shaped and extends from the descending colon to the rectum.

simple columnar epithelium Covering of the internal and external surfaces of the body; composed of a single layer of tall, prismatic cells.

simple cubodial epithelium Covering of the internal and external surfaces of the body; composed of a single layer of cube-shaped cells.

simple diffusion Movement of a substance from high to low concentration, caused by kinetic energy.

simple (endemic) goiter Condition in which an enlarged thyroid produces low levels of thyroxine.

simple squamous epithelium Covering of the internal and external surfaces of the body; composed of a single layer of flattened, platelike cells.

simple tissue Tissue composed of a single layer of cells.

singultus Technical term for hiccup.

sinoatrial (SA) node Small region of neuromuscular tissue that initiates the heartbeat; also called the pacemaker.

sinus Cavity; for example, the sinuses in the human skull.

sinusitis Inflammation of the mucous membrane lining a paranasal sinus.

sister chromatids Two chromatids of a chromosome, held together by a centromere.

skeletal muscle Contractile tissue that comprises the muscles attached to the skeleton; also called striated muscle.

SLE See *systemic lupus erythematosus*.

sliding filament theory Muscles contract when the thin (actin) and thick (myosin) filaments move past each other, shortening the skeletal muscle cells.

small intestine Portion of the digestive tract that extends from the lower opening of the stomach to the large intestine.

smooth ER Synthesizes the phospholipids that occur in membranes, among other functions, depending on the particular cell.

smooth muscle Contractile tissue that comprises the muscles in the walls of internal organs; also called visceral muscle.

soft palate Entirely muscular posterior portion of the roof of the mouth.

solute The substance dissolved in a solution.

solvent Major component of a solution, usually liquid or gas.

somatic system Portion of the peripheral nervous system containing motor neurons that control skeletal muscles.

somatotropin See *growth hormone*.

spasm Sudden, violent, involuntary contraction of a muscle or a group of muscles.

specific heat capacity Amount of energy needed to change an object's temperature by exactly 1°C.

sperm Male gamete having a haploid number of chromosomes and the ability to fertilize an egg, the female gamete.

spermatic cord Cord that suspends the testis within the scrotum; contains the vas deferens and vessels and nerves of the testis.

spermatid Intermediate stage in the formation of sperm cells.

spermatogenesis Sperm production in males by the process of meiosis and maturation.

spermatozoa Developing male gametes.

sphincter Muscle that surrounds a tube and closes or opens the tube by contracting and relaxing.

spina bifida Disorder caused by failure of the neural tube to close.

spinal cord Portion of the central nervous system extending downward from the brain stem through the vertebral canal.

spinal meningitis Inflammation of the meninges of the spinal cord.

spinal nerve Nerve that arises from the spinal cord.

spindle Apparatus composed of microtubules to which the chromosomes are attached during cell division.

spiral organ Structure in the vertebrate inner ear that contains auditory receptors; also called organ of Corti.

spirometer Instrument for measuring the air entering and leaving the lungs.

spleen Large, glandular organ located in the upper left region of the abdomen that stores and purifies blood.

spongy bone Bone found at the ends of long bones; consists of bars and plates separated by irregular spaces.

sprain Joint injury in which some of the fibers of a supporting ligament are ruptured, but the continuity of the ligament remains intact.

squamosal suture Type of suture formed by overlapping of the broad, beveled edges of the participating bones.

squamous cell carcinoma Type of skin cancer that occurs in the epidermis; has warty appearance, bleeds, and/or forms a scab.

squamous tissue Tissue formed from groups of flattened cells.

stapes The last of three ossicles of the ear; serves with the malleus and incus to conduct vibrations from the tympanic membrane to the oval window of the inner ear.

starch Polysaccharide that is common in foods of plant origin.

stem cell A precursor cell.

stent Device used to prop open or reinforce a blood vessel.

sternum Breastbone to which the ribs are ventrally attached.

steroid Lipid-soluble, biologically active molecules having four interlocking rings; examples are cholesterol, progesterone, and testosterone.

steroid hormone Type of hormone that has the same complex of four-carbon rings, but each one has different side chains.

stimulus In nerve physiology, energy needed to cause a neuron to respond.

stomach Saclike, expandable digestive organ located between the esophagus and the small intestine.

strain An overstretching or overexertion of some muscles.

stratified Layered, as in stratified epithelium, which contains several layers of cells.

stratified cuboidal epithelium Tissue created by layers of cube-shaped cells; found in gland ducts.

stratified squamous epithelium Covering of the internal or external surfaces of the body composed of layered, flattened, platelike cells.

stratum basale Deepest layer of the epidermis, where cell division occurs.

stratum corneum Uppermost keratinized layer of the epidermis.

stratum lucidum Layer of epidermis that is found only in thick skin (i.e., the soles of feet).

striae gravidarum Stretch marks of pregnancy; caused by weakening of collagen in the skin.

stroke See *cerebrovascular accident*.

sty Inflammation of a sebaceous gland.

subclavian Either of two arteries branching off the aortic arch and supplying the arms.

subcutaneous injection Introduction of a substance beneath the skin, using a syringe.

subcutaneous tissue Tissue beneath the dermis that tends to contain fat cells.

subdural hematoma Accumulation of blood between the dura mater and the brain.

substrate The reactant in an enzymatic reaction.

summation In muscle contraction, increased contraction until maximal sustained contraction is achieved.

superficial Near the surface.

superior Toward the upper part of a structure or toward the head.

superior vena cava Large vein that enters the right atrium from above and carries blood from the head, thorax, and upper limbs to the heart.

supination Rotation of the forearm so that the palm faces forward when in the anatomical position.

surface tension Force that holds moist membranes together when water molecules attract.

surfactant Agent that reduces the surface tension of water; in the lungs, a surfactant prevents the alveoli from collapsing.

sustentacular cell In the testis, a supporting epithelial cell; also called a *Sertoli cell*.

suture Type of immovable joint articulation found between bones of the skull.

sweat gland Skin gland that secretes a fluid substance for evaporative cooling; also called sudoriferous gland.

sympathetic division Part of the autonomic nervous system whose effects are generally associated with emergency situations.

synapse Region between two nerve cells where the nerve impulse is transmitted from one to the other, usually from axon to dendrite.

synaptic cleft Small gap between the synaptic knob on one neuron and the dendrite on another neuron.

synarthrosis Immovable joint.

syndrome A group of symptoms that characterize a disease condition.

synergist Muscle that assists the action of the prime mover.

synovial fluid Fluid secreted by the synovial membrane.

synovial joint Freely movable joint.

synovial membrane Membrane that forms the inner lining of a capsule of a freely movable joint.

synthesis reaction To build up, as in the combining of two small molecules to form a larger molecule.

syphilis Sexually transmitted disease caused by the bacterium *Treponema pallidum* characterized by a painless chancre on the penis or cervix; if untreated, can lead to cardiac and central nervous system disorders.

systemic circuit Part of the cardiovascular system that serves body parts other than the gas-exchanging surfaces in the lungs.

systemic disease Illness that involves the entire body or several body systems.

systemic lupus erythematosus (SLE) Syndrome involving the connective tissues and various organs, including kidney failure.

systole Contraction of the heart chambers, particularly the left ventricle.

systolic pressure Arterial blood pressure during the systolic phase of the cardiac cycle.

T

tachycardia Abnormally rapid heartbeat.

tactile cells Cells that signal the brain that an object has touched the skin; also called *Merkel cells*.

talus Ankle bone.

tarsal bones Bones of the ankle in humans.

taste bud Organ containing the receptors associated with the sense of taste.

Tay-Sachs Lethal genetic disease in which the newborn has a faulty lysosomal digestive enzyme.

T-cell mediated immune response Immune response caused by lymphocytes that differentiate in the thymus, possess highly specific cell-surface antigen receptors, regulate T- and B-cell maturation and proliferation, and directly lyse antigen-bearing cells.

tectorial membrane Membrane in the spiral organ (organ of Corti) that lies above and makes contact with the receptor cells for hearing.

telophase Stage in mitosis when newly formed cells separate.

template Pattern or guide used to make copies; parental strand of DNA serves as a guide for the production of daughter DNA strands, and DNA also serves as a guide for the production of messenger RNA.

temporal lobe Area of the cerebrum responsible for hearing and smelling and for the interpretation of sensory experience and memory.

tendinitis Inflammation of muscle tendons and their attachments.

tendon Tissue that connects muscle to bone.

teniae coli External longitudinal muscle bands built into the wall of the large intestine.

teratogen Any substance that produces abnormalities during human development.

testis Male gonad; the organ that produces sperm and testosterone (pl., *testes*).

testosterone The most potent of the androgens, the male sex hormones.

tetanic contraction Sustained muscle contraction without relaxation.

tetanus Acute infection caused by the toxin of the tetanus bacterium; results in a rigidly locked jaw.

tetany Severe twitching caused by involuntary contraction of the skeletal muscles due to a lack of calcium.

thalamus Mass of gray matter located at the base of the cerebrum in the wall of the third ventricle; receives sensory information and selectively passes it to the cerebrum.

thermoreceptor Type of sensory receptor that responds to changes in temperature.

thick filaments In the sarcomere, the filaments comprised of only myosin molecules; they appear wider in a photomicrograph.

thin filaments In the sarcomere, the filaments comprised of actin, troponin, and tropomyosin molecules, they appear narrower in a photomicrograph.

thoracic cavity Hollow place within the chest.

thrombin Enzyme derived from prothrombin that converts fibrinogen to fibrin threads during blood clotting.

thrombocyte A blood platelet.

thrombocytopenia Insufficient number of platelets in the blood.

thromboembolism Obstruction of a blood vessel by a thrombus that has dislodged from the site of its formation.

thrombus Blood clot that remains in the blood vessel where it formed.

thymosins Hormones secreted by the thymus.

thymus gland Lobular gland that lies in the neck and chest area and is necessary for the development of immunity.

thyroid gland Endocrine gland, located just below the larynx and in front of the trachea, that secretes thyroid hormones.

thyroid-stimulating hormone (TSH) Hormone that causes the thyroid to produce thyroxine.

thyrotropin Hormone secreted by the anterior pituitary gland that regulates the formation and secretion of thyroid hormone; also called *thyroid-stimulating hormone (TSH)*.

thyroxine (T_4) Hormone produced by the thyroid that speeds the metabolic rate.

tibia Shinbone found in the lower leg.

tic Local and habitual spasm of particular muscles, especially facial muscles.

tidal volume Amount of air that enters the lungs during a normal, quiet inspiration.

tight junction Junction between cells when adjacent plasma membrane proteins join to form an impermeable barrier.

tinnitus A sensation of noise (ringing or roaring) that can only be heard by the one affected; caused by a body condition such as an auditory nerve disorder.

tissue Group of similar cells that performs a specialized function.

tissue thromboplastin Clotting factor, released or derived from tissues, that interacts with platelets, calcium ions, and other clotting factors.

T lymphocyte One of two types of lymphocytes: a killer T cell that interacts directly with antigen-bearing cells and is responsible for cell-mediated immunity, or a helper T cell that stimulates other immune cells.

tone Continuous, partial contraction of muscle.

tonicity Osmolarity of a solution compared to that of a cell. If the solution is isotonic to the cell, there is no net movement of water; if the solution is hypotonic, the cell gains water; and if the solution is hypertonic, the cell loses water.

tonsil Partly encapsulated lymph nodule located in the pharynx.

tonsillectomy Surgical removal of the tonsils.

tonsillitis Inflammation of the tonsils.

toxoplasmosis Disease caused by toxoplasma parasites that invades the tissues and can seriously damage the central nervous system, especially of infants.

trabeculae Branching bony plate that separates irregular spaces within spongy bone.

tracer Substance having an attached radioactive isotope that allows a researcher to track its whereabouts in a biological system.

trachea Windpipe; serves as a passageway for air.

tracheostomy Creation of an artificial airway by incision of the trachea and insertion of a tube.

tract Bundle of neurons forming a transmission pathway through the brain and spinal cord.

transcription Manufacturing RNA from DNA.

transfer RNA (tRNA) Molecule of RNA (ribonucleic acid) that carries an amino acid to a ribosome engaged in the process of protein synthesis.

transient ischemic attack Brief episode of decreased blood flow to the brain; usually characterized by temporary blurring of vision, slurring of speech, numbness, paralysis, or fainting; often predictive of a serious stroke; abbreviated *TIA*; also called *mini-stroke*.

transitional epithelium Tissue that forms the lining of the ureters, urinary bladder, and urethra.

translation Assembly of an amino acid chain according to the sequence of base triplets in a molecule of mRNA.

transverse colon Portion of the large intestine that travels transversely as it extends from the ascending colon to the descending colon.

transverse plane Plane or section that divides a structure horizontally to give a cross section.

transverse tubules In muscle, structures formed from the sarcolemma; they extend into the sarcoplasm and contact but do not touch the sarcoplasmic reticulum.

tricuspid valve Atrioventricular valve between the right atrium and the right ventricle.

triglyceride Lipid composed of three fatty acids combined with a glyercol molecule.

triplet code Three-nucleotide base unit coding for a particular amino acid during protein synthesis.

trisomy State of having an extra chromosome—three instead of the normal two.

trophoblast Outer cells of a blastocyst that help form the placenta and other extra-embryonic membranes.

tropomyosin One of three proteins that comprise the thin filaments in a sarcomere.

troponin One of three proteins that comprise the thin filaments in a sarcomere.

trypsin Protein-digesting enzyme produced by the pancreas.

TSH See *thyroid-stimulating hormone*.

T (transverse) tubule Membranous channel that extends inward from a muscle fiber membrane and passes through the fiber.

tubal ligation Method for preventing pregnancy in which the uterine tubes are cut and sealed.

tubular reabsorption Process that transports substances out of the renal tubule into the interstitial fluid from which the substances diffuse into peritubular capillaries.

tubular secretion Process of substances moving out of the peritubular capillaries into the renal tubule.

tumor Abnormal growth of tissue that serves no useful purpose.

tunica externa The outer layer that makes up a blood vessel; composed of collagenous and elastic fibers.

tunica intima The innermost lining of a blood vessel, consisting of an endothelial layer backed by connective and elastic tissue.

tunica media Middle layer of the walls of arteries and veins; composed of smooth muscle.

Turner syndrome Condition caused by the inheritance of a single X chromosome.

tympanic membrane Membrane located between the external and middle ear; the eardrum.

U

ulcer Open sore in the lining of the stomach; frequently caused by bacterial infection.

ulna Elongated bone within the lower arm.

umbilical Pertaining to the umbilicus.

umbilical artery One of a pair of fetal arteries that pass through the umbilical cord to the placenta to which they carry the deoxygenated blood from the fetus.

umbilical cord Cord through which blood vessels that connect the fetus to the placenta pass.

umbilical vein A vein that passes through the umbilical cord to the fetus and returns the oxygenated and nutrient blood from the placenta to the fetus.

uncinate fit Form of temporal lobe epileptic seizure characterized by hallucinations of taste and odor.

unsaturated fatty acid Organic compound that includes a fatty acid molecule having one or more double bonds between the atoms of its carbon chain.

urea Primary nitrogenous waste of mammals.

uremia High level of urea nitrogen in the blood.

ureters Tubes that take urine from the kidneys to the bladder.

urethra Tube that takes urine from the bladder to the outside of the body.

urethritis Inflammation of the urethra.

uric acid Product of nucleic acid metabolism in the body.

urinalysis Examination of a urine sample to determine its chemical, physical, and microscopic aspects.

urinary bladder Organ where urine is stored before being discharged by way of the urethra.

urticaria Skin eruption characterized by the development of welts as a result of capillary dilation.

uterine tube Tube that extends from the uterus on each side toward an ovary and transports sex cells; also called fallopian tube or oviduct.

uterus Female organ in which the fetus develops.

utricle Saclike cavity of the inner ear that contains receptors for static equilibrium.

uvula A fleshy portion of the soft palate that hangs down above the root of the tongue.

V

vaccine Treated antigens that can promote active immunity when administered.

vagina Female copulatory organ and birth canal.

valve Structure that opens and closes, ensuring one-way flow; common to vessels, such as systemic veins, lymphatic veins, and veins to the heart.

varicose vein Irregular dilation of a superficial vein, seen particularly in the lower legs, due to weakened valves within the veins.

vasa recta In the kidney, hairpin-shaped peritubular capillaries that descend into the renal medulla, reunite as they ascend, and play a role in the concentration of urine.

vascularization Process in which a tumor becomes supplied with blood vessels.

vas deferens Tube connecting the epididymis to the urethra; sperm duct (pl., *vasa deferentia*).

vasectomy Method for preventing pregnancy in which the vasa deferentia are cut and sealed.

vasomotor center Neurons in the brain stem that control the diameter of the arteries.

vein Blood vessel that takes blood to the heart.

venous duct See *ductus venosus*.

ventilation Breathing; the process of moving air into and out of the lungs.

ventral Toward the front or belly surface; the opposite of *dorsal*.

ventricle Cavity in an organ, such as the ventricles of the brain or the ventricles of the heart.

venule Type of blood vessel that takes blood from capillaries to veins.

vermiform appendix Small, tubular appendage that extends outward from the cecum of the large intestine.

vernix caseosa Cheeselike substance covering the skin of the fetus.

vertebra Bone of the vertebral column.

vertebral canal Hollow place within the vertebrae containing the spinal cord.

vertebral column Backbone of vertebrates, composed of individual bones called vertebrae.

vertigo Dizziness and a sense of rotation.

vesicle Small, membranous sac that stores substances within a cell.

vesicular (Graafian) follicle Mature follicle within the ovaries that houses a developing egg.

vestibule Space or cavity at the entrance of a canal, such as the cavity that lies between the semicircular canals and the cochlea.

villi Fingerlike projections that line the small intestine and function in absorption (sing *villus*).

visceral Pertaining to the contents of a body cavity.

visceral serous pericardium The inner layer of the serous pericardium; it is in contact with the heart and the roots of the vessels of the heart.

visceral peritoneum Membrane that covers the surfaces of organs within the abdominal cavity.

visceral pleura Membrane that covers the surfaces of the lungs.

visceral serous membrane Serous membrane that covers the surface of an organ.

visual accommodation Ability of the eye to focus at different distances by changing the curvature of the lens.

visual field Area of vision for each eye.

vital capacity Maximum amount of air a person can exhale after taking the deepest breath possible.

vitamins Organic molecules (usually coenzymes) that must be in the diet and are necessary in trace amounts for normal metabolic functioning of cells.

vitreous body The clear, colorless, transparent jelly that fills the eyeball posterior to the lens; composed of a fiber network filled with vitreous humor.

vitreous humor Watery fluid component of the vitreous body.

vocal cords Folds of tissue within the larynx that produce sounds when they vibrate.

vulva External genitalia of the female that lie near the opening of the vagina.

W

wart Raised growth on the skin due to a viral infection.

water intoxication Disorder caused by excessive water consumption, resulting in edema and electrolyte imbalance.

Wernicke's area Area located in the posterior part of the temporal lobe; plays an important role in the comprehension of language.

white blood cell See *leukocytes*.

white matter Myelinated nerve fibers in the central nervous system.

X

X chromosome Female sex chromosome that carries genes involved in sex determination; see *Y chromosome*.

xenotransplantation Use of animal organs, instead of human organs, in human transplant patients.

X-linked Gene found on the X chromosome that controls traits other than sexual traits.

XYY male Male who has an X chromosome and two Y chromosomes in each nucleus.

Y

Y chromosome Male sex chromosome that carries genes involved in sex determination; see *X chromosome*.

yolk sac Extraembryonic membrane that serves as the first site of red blood cell formation.

Z

Z line Vertical lines found in a sarcomere.

zona pellucida Transparent noncellular outer layer or envelope surrounding an ovum.

zygote Cell formed by the union of the sperm and ovum; the product of fertilization.

CREDITS

Chapter 1
Opener: © Dr. Najeeb Layyous/Photo Researchers, Inc.; 1.4(all): © The McGraw-Hill Companies, Inc./Joe De Grandis, photographer; Medical Focus 1B: © Scott Camazine; Medical Focus 1C, 1D-a: © Mazzlota et al./Photo Researchers, Inc.

Chapter 2
Opener: © Gail Shumway/Getty Inc., 2.2b: © Charles M. Falco/Photo Researchers, Inc.; 2.7: © Jeremy Burgess/SPL/Photo Researchers, Inc.; 2.8: © Don W. Fawcett/Photo Researchers, Inc.; 2.11a: © Warren Toda/epa/Corbis; 2.11b: © Tony Marsh/Reuters/Corbis.

Chapter 3
Opener: © SPL/Photo Researchers, Inc.; 3.3: Courtesy Stephen Wolfe; 3.4: © Barry F. King/Biological Photo Service; 3.5(top): © R. Rodewald/Biological Photo Service; 3.5 (bottom): Courtesy Tim Wakefield, John Brown University; 3.6a: Courtesy Dr. Keith Porter; 3.7a: © Oliver Meckes/Photo Researchers, Inc.; 3.7b: © Manfred Kage/Peter Arnold, Inc.; 3.8a(Basal body): © William L. Dentler/Biological Photo Service; 3.8b(Flagellum): © William L. Dentler/Biological Photo Service; 3.10a-c: © David M. Phillips/Visuals Unlimited; 3.17(all):© Michael Abbey/Photo Researchers, Inc.; Focus on Forensics 3B: Courtesy Dr. Michael Baird, Lifecodes Corporation.

Chapter 4
Opener: © Obstetrics & Gynecology/SPL/Photo Researchers, Inc.; 4.1(right): © Ed Reschke; 4.2(right): © The McGraw-Hill Companies, Inc./Dr. Alvin Telser, photographer; 4.3(right): © Ed Reschke; 4.4(right): © Manfred Kage/Peter Arnold, Inc.; 4.5, 4.6, 4.7, 4.8(right): © Ed Reschke; Medical Focus Box p. 70: Courtesy Forbes Library/Calvin Coolidge Presidential Library & Museum; 4.9(right): © The McGraw-Hill Companies, Inc./Dr. Alvin Telser, photographer; 4.10, 4.11(right): © Ed Reschke; 4.12: © The McGraw-Hill Companies, Inc./Dr. Alvin Telser, photographer; 4.13, 4.14, 4.15, 4.16(right): © Ed Reschke.

Chapter 5
Opener(left): © Digital Vision; opener(right): © BananaStock Ltd.; opener(top) © The McGraw-Hill Companies, Inc./Dr. Alvin Telser, photographer; Medical Focus 5A: Courtesy of Kent Van De Graaff, photo by James M. Clayton; 5.2 b: © The McGraw Hill Companies, Inc./Photo by Dr. Alvin Tesler; 5.5a: © Ken Greer/Visuals Unlimited; 5.5b: © Dr. P. Marazzi/SPL/Photo Researchers, Inc.; 5.5c: © James Stevenson/SPL/Photo Researchers, Inc.; 5.7a: © Neil Borden/Photo Researchers, Inc.; 5.7b: Courtesy Dr. George Bogumill.

Chapter 6
Opener: © SuperStock RF/SuperStock; 6.2: © CABISCO/ Phototake; Focus on Forensics 6Ba: © Ralph T. Hutchings/Visuals Unlimited; Focus on Forensics 6Bb: © Ralph T. Hutchings/Visuals Unlimited.

Chapter 7
Opener(happy): © Fancy Photography/Veer; opener(angry): © Somos Photography/Veer; opener(despair, crying): © Royalty Free Corbis; opener(fright): © image source/Getty RF; 7.1(all): © Ed Reschke; 7.2c: © The McGraw-Hill Companies, Inc./Dennis Strete, photographer; 7.3: Courtesy of Hugh E. Huxley.

Chapter 8
Opener: © Royalty Free Corbis; 8.2(left): © M.B. Bunge/Biological Photo Service; 8.2(right): © Manfred Kage/Peter Arnold,; 8.5: Courtesy Dr. E.R. Lewis, University of California, Berkeley; 8.7a: © Karl E. Deckart/Phototake; 8.7d: Posterior aspect of the cervical segment of the spinal cord (right side) from The Photographic Atlas of the Human Body by Branislav Vidic and Faustino R. Suarez. The C.V. Mosby Co., St. Louis, 1984.; 8.8b: © Colin Chumbley/Science Source/Photo Researchers, Inc.; Medical Focus p. 166: © Reuters/Corbis; 8.11b: © Dr. Richard Kessel & Dr. Randy Kardon/Tissues & Organs/Visuals Unlimited.

Chapter 9
Opener: © Stockbyte/Getty RF; 9.3a: © Coneyl Jay/Photo Researchers, Inc.; 9.3(all): © Omikron/ SPL/Photo Researchers, Inc.; 9.9a: © Lennart Nilsson, from The Incredible Machine; 9.10b: © Biophoto Associates/Photo Researchers, Inc.; 9.13: © P. Motta/ SPL/Photo Researchers, Inc.

Chapter 10
Opener: Photography by Robert Knudsen, White House, In the John F. Kennedy Presidential Library and Museum, Boston; 10.5: © Wellcome Library, London; 10.6: From Clinical Pathological Progression of Acromegaly Conference, "Acromegaly, Diabetes, Hyper Metabolism, Proteniura and Heart Failure," American Journal of Medicine 20 (1956) 133, with permission from Excerpta Medica; 10.7: © Biophoto Associates/SPL/Photo Researchers, Inc.; 10.8: © John Paul Kay/Peter Arnold, Inc.; 10.12a, b: © NMSB/Custom Medical Stock Photo; 10.13a,b: "Atlas of Pediatric Physical Diagnosis," Second Edition by Zitelli & Davis, 1992. Mosby-Wolfe-Europe Limited,London, UK; What's New 10A: Courtesy Robert P. Lanza; 10.15: © James Darell/Stone/Getty.

Chapter 11
Opener: © Frank Conlon/Star Ledger/Corbis; 11.5: © Manfred Kage/Peter Arnold, Inc.; Medical Focus 11B: © Bill Longcore/Photo Researchers, Inc.

Chapter 12
Opener 12-1: 1998: Courtesy NASA/Robert Markowitz/Mark Sowa; Opener 12-2: 1962: Courtesy NASA; 12.3: © Dr. Don W. Fawcett/Visuals Unlimited; Medical Focus 12A: © Biophoto Associates/Photo Researchers, Inc.; 12.6d: © The McGraw-Hill Companies, Inc.; 12.8(left): © Ed Reschke; 12.8(right): © Biophoto Associates/Photo Researchers, Inc.; 12.15: © David Young-Wolff/PhotoEdit.

Chapter 13
Opener: © Photodisc/Getty Images RF; 13.1: Marrow: © R. Valentine/Visuals Unlimited; 13.1(thymus): © Ed Reschke/Peter Arnold; 13.1(lymph node): © Fred E. Hossler/Visuals Unlimited; 13.1(spleen): © Ed Reschke/Peter Arnold; 13.2a-e: © Ed Reschke/Peter Arnold, Inc.; 13.4 b: Courtesy of Dr. Arthur J. Olson, The Scripps Research Institute; 13.7d: © Boehringer Ingelheim International, photo by Lennart Nilsson; Medical Focus p. 297: © Michael Newman/Photo Edit; 13.9: © Estate of Ed Lettau/Peter Arnold, Inc.; What's New 13C: © K. G. Murti/Visuals Unlimited.

Chapter 14
Opener: © Stockbyte/Getty RF; p. 309: © Manfred Kage/Peter Arnold, Inc.; 14.7: © Yoav Levy/Phototake; 14.11a, b: © Martin Rotker/Martin Rotker Photography.

Chapter 15
Opener: © Bettman/Corbis; 15.3b: © Biophoto Associates/Photo Researchers, Inc.; 15.4c: © Ed Reschke/Peter Arnold, Inc.; 15.7(left): © Manfred Kage/Peter Arnold, Inc.; 15.7(right): Photo by Susumu Ito, from Charles Flickinger, "Medical Cellular Biology," W.B. Saunders, 1979; 15.9: © Ed Reschke; 15.13(left): © Tony Freeman/PhotoEdit; 15.13(right): © Donna Day/Stone/Getty Images.

Chapter 16
Opener: © SPL/Photo Researchers, Inc.; 16.3b: © Ralph T. Hutchings/Visuals Unlimited; 16.4(top left): © R.G. Kessel and R. H. Kardon, "Tissues and Organs: A Text Atlas of Scanning Electron Microscopy," 1979.; 16.4(top right, bottom left): © 1966 Academic Press, from A.B. Maunsbach, J. Ultrastruct. Res. 15:242-282; 16.10(left): © Hank Morgan/SPL/Photo Researchers, Inc.

Chapter 17
Opener: © Bruce Grenville Matthews/Getty RF; 17.4b: © Biophoto Associates/Photo Researchers, Inc.; 17.7: © Anatomical Travelogue/Photo Researchers, Inc.; Fig. 17.10: © Ed Reschke/Peter Arnold, Inc.; 17.16a, c, d, f: © The McGraw-Hill Companies, Inc./Bob Coyle, photographer; 17.16b: © The McGraw-Hill Companies, Inc./Vincent Ho, photographer; 17.16e: © Gusto/Photo Researchers, Inc.; Medical Focus 17D (left): © Vol. 161/Corbis RF; Medical Focus 17D(right): © David Raymer/Corbis; Medical Focus 17E: © Argus Fotoarchiv/Peter Arnold, Inc.

Chapter 18
Opener: © Hulton-Deutsch Collection/Corbis; 18.1: © David M. Phillips/Visuals Unlimited; 18.6a: © Lennart Nilsson, A Child Is Born, Dell Publishing; 18.8: © James Stevenson/SPL/Photo Researchers, Inc.; 18.10: © Rune Hellestad/Corbis.

Chapter 19
Opener: © FPG/Getty Images; 19.1(all): © CNRI/ SPL/Photo Researchers, Inc.; 19.3: © Jill Cannefax; 19.4(both): Photograph by Earl Plunkett. Courtesy of G. H. Valentine.

INDEX

Osteoprogenitor cell, 99
Otic region, 4
Otitis media, 319
Otoliths, 201
Otosclerosis, 202, 203
Oval window, 198, 199
Ovarian cancer, 389
Ovariohysterectomy, 388
Ovary (ovaries), 208, 209, 223, 386–387, 448, 449
    cancer of, 389
    cycle of, 390–392
    development of, 418–420
    fetal, 418–420
Ovulation, 386, 387, 390–392, 412
Ovum, 385, 409–410, 412
Oxygen
    transport of, 260–262, 311–312, 316–317, 324
Oxygen debt, 136
Oxyhemoglobin, 317
Oxytocin, 12, 208, 209, 211, 212

P

Pacemaker, cardiac, 256
    ectopic, 256
Pacinian corpuscles, 185
Pain
    back, 111
    referred, 185–186
Pain receptors, 184, 185–186
Palate, 105, 108, 308, 331, 333
Palatine bone, 105, 107, 108
Palatine process, 105, 107
Palatine tonsils, 285, 308, 332
Pallor, 84
Palmar region, 4
Pancreas, 207, 208, 209, 220–222, 330, 342
    in homeostasis, 348
Pancreatic duct, 330
Pancreatic enzymes, 342, 345, 346
Pancreatic islets, 220
    transplantation of, 222
Pancreatic juice, 342
Pandemic, 299
Pantothenic acid, 349
Papillary muscles, 250, 251
Pap smear, 76, 389
Paranasal sinuses, 308–309
    infection of, 103, 319
Paraplegia, 166
Parasympathetic nervous system, 174–176
Parathyroid glands, 207, 208, 209, 216
Parathyroid hormone, 102, 207, 208, 209, 216
Parental cell, 57
Parietal bone, 104, 105, 106, 107
Parietal cell, 335, 336
Parietal lobe, 167
Parietal membrane, 6, 7, 78
Parietal pericardium, 7, 249–250
Parietal peritoneum, 7
Parietal pleura, 6
Parkinson's disease, 168–169, 177, 179
Parotid gland, 330
Parturition, 421, 422, 423
Passive immunity, 297–298
Patch, contraceptive, 396
Patella, 104, 116
Patellar ligament, 148
Patellar reflex, 174
Patellar region, 4
Pathogens, 232
Pathologist, 76
Pectineus, 140, 147
Pectoral girdle, 112, 113

Pectoralis major, 140, 143, 144, 444, 445
Pectoralis minor, 445
Pectoral region, 4
Pedal region, 4
Pedigree, 437–438
Pelvic brim, 115
Pelvic cavity, 6
Pelvic girdle, 115–116
Pelvic inflammatory disease, 400
Pelvic sacral foramen, 450
Pelvis, 115–116
Penis, 379, 382, 383, 446, 447
Pentose, 28
Pepsin, 336, 345, 346
Peptidases, 345, 346
Peptide bond, 32
Peptide hormone, 207
Perforin, 292
Pericardial cavity, 6, 7, 449
Pericardial sac, 446
Pericarditis, 9, 250
Pericardium, 7, 78, 249–250
Perimysium, 129
Perineal region, 4
Periodontal disease, 275, 331
Periosteum, 98, 99, 129
Peripheral, definition of, 4
Peripheral nervous system, 156, 171–174
Peripheral resistance, 263, 264
    blood pressure and, 264
    hormonal regulation of, 264–265
    neural regulation of, 264
Peristalsis, 334, 338, 346, 348
Peritoneum, 7, 78–79, 336
Peritonitis, 9, 339
Peritubular capillary network, 359–360, 361
Pernicious anemia, 240, 245, 336
Peroxisome, 42, 43, 47
Perpendicular plate, 105
Perspiration, 91, 286
Pertussis, immunization against, 296
Pesticides, 398
pH, 25–26, 27, 232, 324
Phagocytosis, 50, 52, 286–288
Phalanges, 104
    of fingers, 114, 115
    of toes, 117
Pharyngeal arch, 417
Pharyngeal tonsil, 285, 308, 309, 333
Pharynx, 307, 308, 309, 330, 332–333
Phenotype, 434
Phenylketonuria, 439
Pheromones, 226
Phlebitis, 262
Phosphate absorption, 366
Phospholipid, 31
Phospholipid bilayer, 43–44
Phosphorus, 350
Photoreceptors, 184, 191, 192, 193
Phrenic nerves, 173, 174
Physiology, 2
Pia mater, 163
Piercing, 93
Pineal gland, 169, 208, 209, 223
Pinna, 198, 201
Pinocytosis, 50, 52
Pituitary dwarfism, 213
Pituitary gland, 208, 209
    anterior, 211–212
    disorders of, 213
    posterior, 211, 212
Pivot joint, 119
Placenta, 273, 410, 416, 417, 421, 422, 423
Placental membrane, 416
Planes, anatomical, 5
Plantar flexion, 120
Plantar region, 4

Plaque, in atherosclerosis, 254
Plasma, 72, 73, 232, 233, 234
Plasma cell, 289
Plasma membrane, 42–44, 65
    diffusion across, 50, 51
    endocytosis by, 50, 52
    exocytosis by, 50, 52
    facilitated diffusion across, 50, 52
    filtration across, 52
    fluid-mosaic model of, 44
    osmosis across, 50–51
Plasma proteins, 232, 233, 234
Platelet, 72, 73, 233, 235, 238–239
Platelet-derived growth factor, 225
Platelet plug, 238–239, 240–241
Pleura, 6–7, 78, 311
Pleural cavity, 6, 7, 449
    infection of, 9
Pleurisy, 9
Pneumococcal vaccine, 296
Pneumonectomy, 323
Pneumonia, 319, 320
Pneumothorax, 311
Podocyte, 360
Poisoning, 26
Polar body, 386
Polar covalent bond, 23–24
Polar molecule, 24
Polio, immunization against, 296
Polycythemia, 236
Polydactyly, 439
Polydipsia, 220
Polypeptide, 32
Polyphagia, 220
Polyps, 341
Polyribosome, 45
Polysaccharides, 29–30
Polyuria, 220
Poly-X syndrome, 430, 434
Pons, 171, 351
Popliteal artery, 268
Popliteal region, 4
Popliteal vein, 270
Portal triads, 343
Positive feedback, 12, 210, 211, 421
Positron emission tomography, 16
Posterior, definition of, 3
Posterior (dorsal) body cavity, 5–6
Posterior superior iliac spine, 115
Potassium, 350
Precapillary sphincter, 260
Prefrontal area, 168
Pregnancy
    bodily changes during, 423
    breast in, 393
    cigarette smoking and, 321
    diabetes mellitus in, 220
    ectopic, 389, 414
    precautions during, 424–425
    premature termination of, 419
    Rh antibody in, 243, 425
Preimplantation genetic diagnosis, 440
Premature birth, 419
Prenatal care, 424–425
Prenatal testing, 439
Prepapillary sphincter, 260
Prepuce, 379, 383
Presbycusis, 203
Pressure
    blood. See Blood pressure
    intrapleural, 313
    osmotic, 50–51, 232, 261
Pressure receptors, 185
Primary germ layers, 414, 415
Primary motor area, 167, 168
Primary respiratory center, 316
Primary somatosensory area, 167, 168
Prime mover, 139
Prions, 34

Proctitis, 402
Progesterone, 208, 209, 223, 390
    in pregnancy, 423
Progesterone injection, for birth control, 394
Progestin injection, for birth control, 396
Prolactin, 208, 209, 212, 213
Promoter, cancer, 76
Pronation, 120
Prophase, of mitosis, 54, 57–58
Proprioceptors, 184
Prostaglandins, 225–226, 380
Prostate cancer, 371
Prostate gland, 359, 379, 380, 382, 383
    enlargement of, 371
Proteins, 32–35
    in active transport, 52
    complement, 288
    denaturation of, 33
    functions of, 32
    on nutrition labels, 29
    plasma, 232, 234
    structure of, 32–33
    synthesis of, 56–57
Proteinuria, 370
Prothrombin, 234, 239, 241
Prothrombin activator, 240
Proton, 21
Proximal, definition of, 3
Proximal convoluted tubule, 360, 361
Pseudostratified ciliated columnar epithe-lium, 67
Pseudostratified columnar epithelium, 64, 66–67
Psoas major muscle, 147, 450
Psoriasis, 84, 88
Puberty, 223, 377, 384–386, 392
Pubic symphysis, 115, 118, 379, 385
Pubis, 115–116
Pulmonary arteriole, 311
Pulmonary artery, 251, 267, 311, 448
    fetal, 273
    right, 252
Pulmonary capillaries, 308, 311, 312
Pulmonary edema, 276, 286
Pulmonary embolism, 262
Pulmonary semilunar valve, 250, 251
Pulmonary trunk, 250, 251, 252, 447, 448
    fetal, 273
Pulmonary vein, 251, 259, 267, 311, 448
    fetal, 273
    right, 252
Pulmonary venule, 311
Pulse, 265
Punnett square, 435–437
Pupil, 189, 190
Purkinje fibers, 253
Pus, 288
P wave, 256
Pyelonephritis, 369
Pyloric canal, 335
Pyloric sphincter, 335
Pyloric stenosis, 439

Q

QRS complex, 256
Quadratus lumborum, 450
Quadriceps femoris, 140, 145, 147, 148
Quadriplegia, 166

R

Radial artery, 268
Radial head, 112, 113
Radial nerve, 173, 174

U

Ulcer, 78
 decubitus, 85, 166
 peptic, 336, 346
Ulna, 104, 113–114
Ulnar artery, 268
Ulnar head, 114
Ulnar nerve, 173, 174
Ulnar notch, 113, 114
Ulnar vein, 270
Ultrasonography, 16
Umbilical artery, 273
Umbilical cord, 410, 414, 416
Umbilical region, 4
Umbilical vein, 273
Umbilicus, 444
Universal donor, 243
Universal recipient, 243
Unsaturated fatty acids, 31
Upper limb
 bones of, 104, 112–115
 muscles of, 139, 144–145
Urea, 344, 357, 373
Uremia, 369
Ureter, 358, 359, 379, 382, 388,
 448, 449
Urethra, 358–359, 379, 380, 382, 385,
 390, 450
Urethritis, 369
Uric acid, 357, 373
Urinalysis, 369, 370
Urinary system, 10, 15, 356–375
 aging effects on, 371
 functions of, 357
 in homeostasis, 13
 organs of, 357–359
 skin effects on, 91
Urination, 359
 aging effects on, 371
Urine, 361–363
 analysis of, 369
 composition of, 362, 363, 368–369
 formation of, 361–363
Urticaria, 88
Uterine tube, 385, 386, 388–389, 448
Uterus, 385, 386, 389, 448, 449
 contractions of, 421
 in menstrual cycle, 390–392
Utricle, 200, 201
Uvula, 308, 331, 332

V

Vaccines, 293, 296–297
 aging effects on, 301
 for birth control, 397
 HIV, 295
Vacuole, 42, 47
Vagina, 385, 386, 389, 450
Vaginal ring, 393, 394
Vagus nerve, 172, 173
Valves
 cardiac, 251–252
 venous, 260, 264
Varicella, immunization against, 296
Varicose veins, 262, 276
Vascular endothelial growth factor, 79
Vascularization, in cancer, 76
Vascular spasm, 238
Vas deferens, 379, 380, 381, 382, 447
Vasectomy, 395
Vasomotor center, 264
Vastus intermedius, 147, 449
Vastus lateralis, 140, 147, 148, 447,
 448, 449
Vastus medialis, 140, 147, 447, 448
Vector, 439
Veins, 262, 269, 270
 aging effects on, 276
 varicose, 262, 276
Vena cava
 inferior, 250, 251, 252, 267, 270, 273,
 449, 450
 superior, 250, 251, 252, 267, 270,
 273, 448, 449
Venous return, 259, 264, 276
Venous sinuses, dural, 162
Ventilation, 312–314
 chemical input to, 316
 control of, 316
 nervous input in, 316
Ventricles
 cardiac, 250, 251, 252, 273, 448
 cerebral, 163, 165
Ventricular diastole, 258
Ventricular systole, 257–258
Venules, 262, 307, 311
Vermiform appendix, 330, 447, 448
Vernix caseosa, 418
Vertebrae, 104, 109, 110
Vertebral artery, 268, 272
Vertebral canal, 6

Vertebral column, 104, 108–109
Vertebral region, 4
Vertigo, 201
Vesicles, 42, 43, 45
Vesicular (Graafian) follicle,
 386, 387
Vestibular nerve, 198, 201
Vestibule
 inner ear, 198, 201
 oral, 330
 vaginal, 390
Vestibulocochlear nerve, 172, 173
Villi
 chorionic, 415, 417, 430
 intestinal, 337
Viral infection
 hepatitis, 296, 297, 298,
 344, 400
 herpes, 400
 HIV, 294–295
Visceral membrane, 6, 7, 78
Visceral nociceptors, 185
Visceral pericardium, 7
Visceral peritoneum, 7
Visceral pleura, 6–7
Visceral sensory system, 171
Visceral serous pericardium, 249
Vision, 188–197. *See also* Eye
 color, 192
 examination of, 194
 focusing for, 191–192
 pathway of, 191, 192,
 196–197
Visual cortex, 196, 197
Visual pathway, 191, 192, 196–197
Vital capacity, 314, 315
 in pregnancy, 423
Vitamin(s), 348–350
 on nutrition labels, 29
 during pregnancy, 424
Vitamin A, 349
Vitamin B$_6$, 349
Vitamin B$_{12}$, 349
Vitamin C, 349
Vitamin D, 91, 102, 349, 357
Vitamin E, 349
Vitamin K, 349
Vitreous humor, 191
Vocal cords, 309
Voice, 309
Vomer bone, 105, 106, 107, 108
Vulva, 385, 386

W

Warts, 89, 400
Water, 24–25
 reabsorption of, 364–366
 skin effects on, 91
Water intoxication, 53
Water-salt balance, 357, 364–366, 373
Weak acid, 26
Weight gain, cardiovascular disease
 and, 274
Wernicke's area, 168
White blood cell(s), 72, 73, 124, 233,
 235, 236, 283, 284
 count of, 238
 differential count of, 240
 mobility of, 236
 in wound healing, 89, 90
White matter, 162, 163–164
Willis, circle of, 271, 272
Withdrawal, for birth control, 394
Wound healing, 89, 90
Wrist, 114
Wryneck, 132

X

X chromosome, 430, 433–435
Xiphoid process, 110, 112
X-ray imaging, 16

Y

Y chromosome, 418–420, 430, 433–435
Yolk sac, 410, 411, 416

Z

Zinc, 350
Z line, 130, 131
Zona pellucida, 409
Zygomatic arch, 106, 107, 108
Zygomatic bone, 106, 107, 108
Zygomatic process, 104, 107
Zygomaticus, 140, 141, 142
Zygote, 379
 cleavage of, 410, 412
 formation of, 409–410